Green Chemistry
in 21st Century and Beyond

Green Chemistry
in 21st Century and Beyond

V.K. Ahluwalia
Prof. (Retd.) Department of Chemistry
University of Delhi, Delhi-110007

Visiting Professor
Dr. B.R. Ambedkar Center for Biomedical Research
University of Delhi, Delhi-110007

★

Sunita Dhingra
Associate Prof. (Retd.) Department of Chemistry
Miranda House
University of Delhi, Delhi-110007

CRC Press
Taylor & Francis Group
Boca Raton London New York

CRC Press is an imprint of the
Taylor & Francis Group, an **informa** business

Manakin
PRESS

First published 2025
by CRC Press
4 Park Square, Milton Park, Abingdon, Oxon, OX14 4RN

and by CRC Press
2385 NW Executive Center Drive, Suite 320, Boca Raton FL 33431

CRC Press is an imprint of Informa UK Limited

British Library Cataloguing-in-Publication Data
A catalogue record for this book is available from the British Library

ISBN: 9781032867809 (hbk)
ISBN: 9781032867816 (pbk)
ISBN: 9781003529163 (ebk)

DOI: 10.4324/9781003529163

Typeset in Times New Roman
by Manakin Press, Delhi

Manakin
PRESS

Contents

Preface

Green Chemistry also called as sustainable chemistry is an area of chemistry and chemical engineering focused on the designing of products and processes that minimize the use and generation of hazardous substances. The emphasis is on reduce, reuse and recycle chemicals, use renewable starting materials, save non renewable energy etc. The guidelines layed down for synthetic chemists helps them to design a bening synthesis.

The book entitled 'Green Chemistry in 21st Century and Beyond' is divided into 20 chapters. Introduction to green chemistry is discussed in Chapter 1. The use of green solvents like water ionic liquids, supercritical solvents like carbon dioxide, polyethylene glycols, fluorous solvents for organic transformations form the subject matter of chapters 2 to 6. Chapter 7 deals with some miscellaneous green solvents. Chapters 9, 10 and 11 deal with organic transformations using microwaves, ultrasound and photochemical activation respectively. All the techniques described in chapters 9 to 11 are helpful in reducing the amount of energy for organic transformations. The use of catalysts (which are also helpful for reducing the amount of energy needed) like phase transfer catalysts, enzymes (biocatalysts) and supported catalysts form the subject matter of chapters 12 to 14. The use of green reagents and renewable starting materials are discussed in chapters 15 and 16 respectively. Electrochemical synthesis, a novel procedure for organic transformations is discussed in chapter 17. Role of green chemistry and green industrial processes for synthesis of pharmaceuticals, polymers and agrochemicals form the subject matter of chapters 18 to 20.

This book will be of immense help to undergraduate, postgraduate and research scholars and for students of engineering colleges. This book will also be of great useful for chemists working in R & D sector and in chemical industries.

The authors take the opportunity of expressing thanks to Mr. Beenu Bhalla (Director, Publishing) of Manakin Press Pvt. Ltd. for all support rendered in the publication of this book.

Any suggestions from the readers are most welcome in order to improve subsequent additions of this book. All such suggestions will be gratefully acknowledged.

Authors

1. INTRODUCTION TO GREEN CHEMISTRY

1

Introduction to Green Chemistry

Scientists have played an unparalleled role in development and providing comforts of life that we enjoy today. Chemists have developed a number of extremely useful products like Pharmaceuticals, diagnostic procedures etc. to protect us from deadly disease and pain, agrochemicals to meet the food requirements of ever increasing population, fuel for vehicles, coolants for our comfort in extreme weather conditions, building materials and other synthetic products like polymers, dyes, cleansing agents, cosmetics etc. Chemists have also played a major role in bridging the gap between demand and supply of essentials from natural resources. Many have spent the entire active life and some have even sacrificed their lives for their passion for chemistry.

Chemicals have been produced from time immemorial. Long back in 7000BC glass was made after purifying alkali and lime stone. The first chemical industry came into existence in first half of nineteenth century. Organic chemical industry came into existence much later. Rayon in 1914, synthetic fertilizers in 1840, petrochemicals from oil, an accidental discovery of a dye by William Henry Perk which he named mauveine are examples of some organic chemicals which were synthesised and form the basis of modern chemical industry. In an attempt to be in the forefront of the race, a very large number of compounds were synthesised, many of which of course were very useful to the mankind and also filled the gap between demand and supply from natural sources. But because of the paucity of information about the toxicity of some chemicals, the choice of starting materials, the process of production of the chemicals and the after use effect of these were not the priority. Till some scientists started raising questions about the adverse effect of some of these chemicals and their indiscriminate use. One such concern was raised by Rachel Carson, an American Marine Biologist in her book-Silent Spring in 1962. This book is credited for sensitizing the scientific community specially the chemists about the adverse effects of some synthetic agrochemicals like DDT on the environment. The book served as a wake up call for the scientists. As a result of this, an initiative was taken up by US government and regulatory bodies like NEPA (National Environmental Policy Act) and EPA (Environmental Protection Agency) were established in US in 1969 and 1970 respectively. A number of legislations followed in US and other countries in the world like Water (Prevention and Control of Pollution) Act (1974), Air (Prevention and Control of Pollution) Act (1981), Environmental Pollution Act and Biodiversity Act (2002). Prior to these development, was considered to be a magic bullet, indiscriminately used to prevent loss of crop by insects till its persistent nature and adverse effects on water soil and life of some animals and man was established. It was subsequently banned for use in many parts of the world. Efforts then followed to clean up the pollution caused by chemical manufacturing units and study the ill effects of chemicals in use. This led to banning the use of a number of chemical products. Until 1970, the main emphasis was on pollution cleanup but very rightly, it shifted to pollution prevention in 1990 (Pollution Prevention Act). Paul Anastas and John C. Warner wrote a book in 1998 entitled "Green Chemistry: Theory and Practice", which gave broad guidelines (12 principles) to synthetic organic chemist to plan a benign organic synthesis. Now very stringent rules and legislations for the production process, trials for a specific use and guidelines for after use effect of chemicals etc. are in place.

Green chemistry also known as sustainable chemistry is the design of chemical products and processes that reduce or eliminate the use or generation of hazardous substances. Green chemistry applies across the life cycle of a chemical substance, including its design, manufacture, use and disposal.

Green Chemistry is a relatively new concept, very important for sustainable future and protection of Environment. It is therefore essential to promote and popularise it. A lot of initiatives have been taken worldwide in this direction. A number of institutes have started academic programmes for introducing green chemistry at grass root level. A comprehensive list of these institutes and the courses offered is given in Appendix 1. Summer training programmes, International workshops, seminars, round table conferences are organised in all parts of the world. Add on courses offered by some institutes are another way of promoting green chemistry at grass root level. A large number of Green Chemistry organisations have been established (Appendix 2). Journals are devoted exclusively to publication of Green Chemistry Research papers (Appendix 3) and a number of prestigious awards announced by American Chemical Society and Royal Society of Chemistry (Appendix 4) strengthen the efforts made in the direction of promotion of Green Chemistry.

1.1 TWELVE PRINCIPLES OF GREEN CHEMISTRY

Green Chemistry is defined as environmentally benign chemical synthesis. The objective of Green Chemistry is basically reduction of use of renewable resources and prevention of pollution which is caused by generation of hazardous substances or by products. The U.S. Presidential green chemistry challenge, March, 1995, defines Green chemistry as "The use of chemistry for source reduction or pollution prevention, more specifically, Green chemistry is the design of chemical products and processes that are more environmentally benign". The objective can be achieved by development of green (benign) synthesis of not only of new products but also development of green synthesis of already existing products/chemicals. The green synthesis can be achieved by following the twelve principles of green chemistry as suggested by Paul T. Anastas and John C. Warner (Green Chemistry, Theory and Practical-Oxford University Press, New York, 1988; J.H. Clark, and D.J. Macquirrie, Handbook of Green Chemistry Technology, Blackwell, Oxford USA). Over the years some additional principles have been added to the original 12.

1. It is better to prevent waste than to treat or clean up waste after it is formed: In any synthesis, the formation of waste or byproduct should be minimum. It is, however, best if the waste or byproducts are not generated. The treatments and disposal of the waste and byproducts involves cost, which is added to the overall cost of production. In case the starting materials and reagents are not fully utilised, the unreacted materials also form part of the waste. Such waste may or may not be hazardous. The best course is to follow the principle 'Prevention is better than cure' and the pollutant formation must be stopped at the source.

2. Synthetic methods should be designed to maximize the incorporation of all materials used in the process into the final product: If in a particular reaction, one mole of a starting material gives one mole of the product the yield is considered to be quantitative (100 percent). However even such a reaction may generate byproducts or waste. Though such a synthesis is 100 percent (as per yield calculated above), it may not be a green synthesis since there may be generation of byproducts. As an example, a Grignard reaction, Wittig reaction and Hoffmann elimination reaction may give 100 percent yield but they do not take into account the byproducts produced.

For a reaction to be considered as Green, there should be 100 percent atom utilization. The extent of incorporation of the starting materials into the product is given by % atom economy, which is calculated as given below (R.A. Sheldon, Chem. Ind. (London), 1992, 903).

$$\%\ \text{Atom economy} = \frac{\text{F.W. of atoms utilized}}{\text{F.W. of the reactants used}} \times 100$$

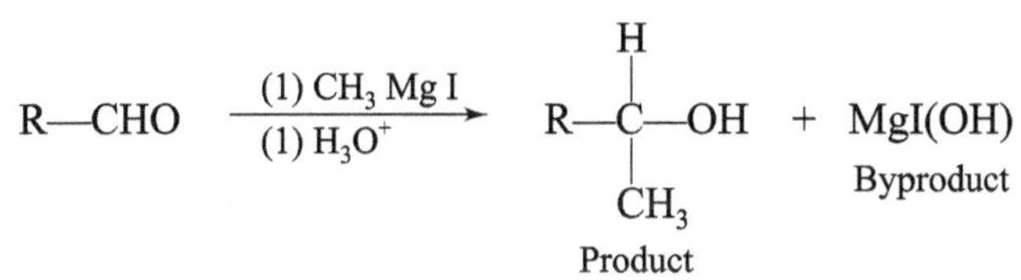

[Grignard reaction]

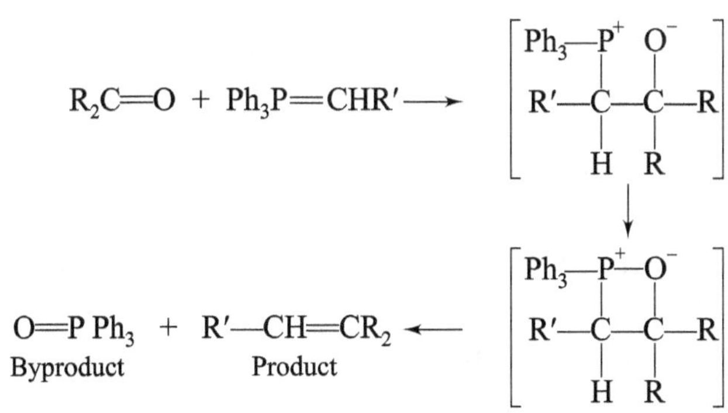

[Wittig reaction]

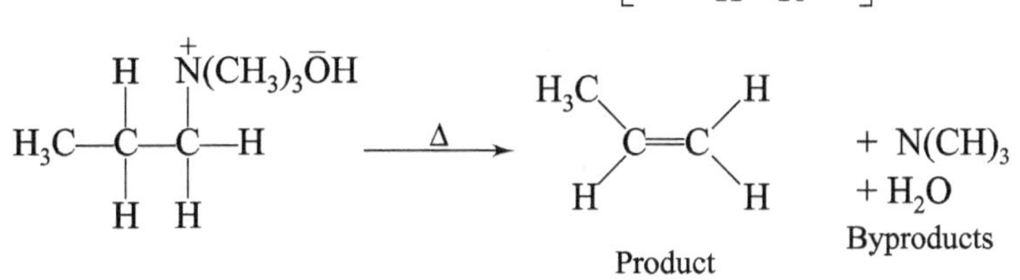

[Hoffmann elimination]

The type of reactions which are commonly used in a synthesis are addition reactions, rearrangement reactions, substitution reactions and elimination reaction. The first two reactions, viz. addition reactions and rearrangement reactions are 100 percent atom economical, since all atoms of the starting materials are incorporated into the final product. These reactions are considered as Green reactions:

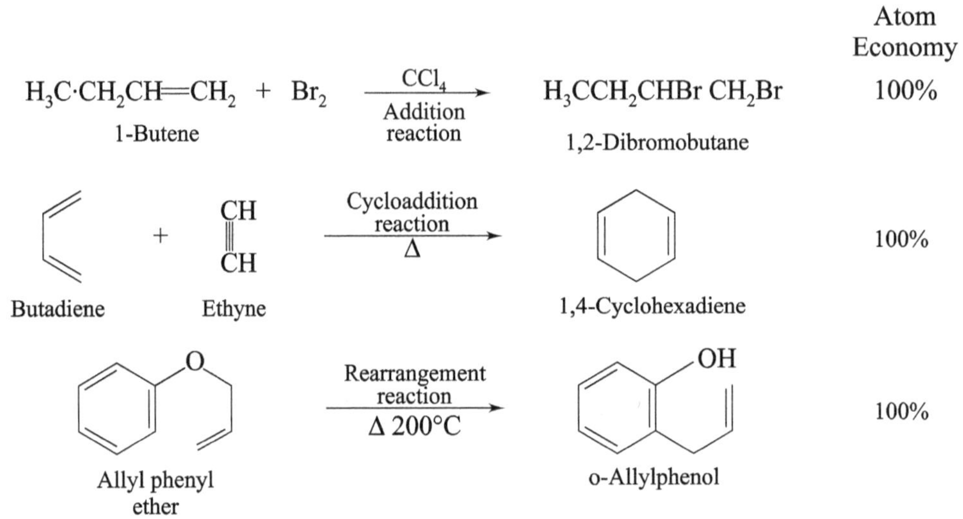

However, substitution reactions and elimination reactions are not 100 percent atom economical.

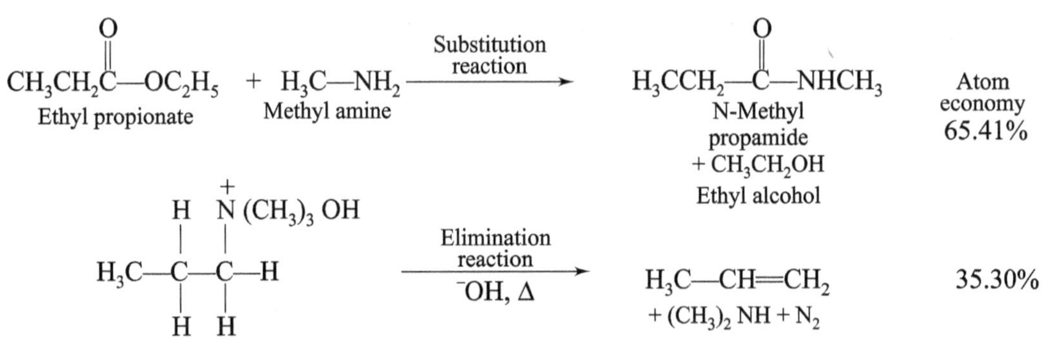

3. Whenever possible, synthetic Methologies should be designed to use and generate substances that possess little or no Toxicity to Human Health and Environment: In the context of Green chemistry, it is important not to use any starting material which is hazardous. Also, it is equally important to ascertain that no hazardous byproducts are generated in any synthetic procedure. Such chemicals are not only toxic to the workers but are also harmful to the environment. The workers using such chemicals should use protective clothing and respirators. The disposal of the hazardous chemicals adds to the overall cost of production. Sometimes more risk is involved if the controls fail as happened in the well known Bhopal Gas Tragedy. In fact, Green Chemistry offers scientific options so that such situations be avoided.

A very interesting and typical example of an extremely useful product but having toxicity to human health and environment is DDT (Dichlorodiphenyl trichloroethane $p\text{-}ClC_6H_4-\underset{\underset{CCl_3}{|}}{C}-C_6H_4Cl(p)$, a common insecticide introduced during World War II and found widespread agricultural use and saved millions of lives by Malaria Control Programme. It was discovered by Paul Muller (a Swedish Chemist) (who prepared it by the reaction of chloral and chlorobenzene) and Won Nobel Prize in chemistry in 1948. Though DDT was a boon at that time, it was banned in USA due to concern about its long term health effects. Widespread use of DDT resulted in pollution of crop lands and a large number of pests became resistant to it. On entering the foodchain, DDT accumulates in the fatty tissue of animals. Many species of hunting birds, particularly those having high concentration of DDT are threatened with possible extinction. The eggs of such birds have thin shells and fragile possibly due to interference with the hormones which control calcium deposition.

DDT is a persistent chemical and once introduced in the environment, it stays for many years. It accumulates in the food chain. Various steps involved are:

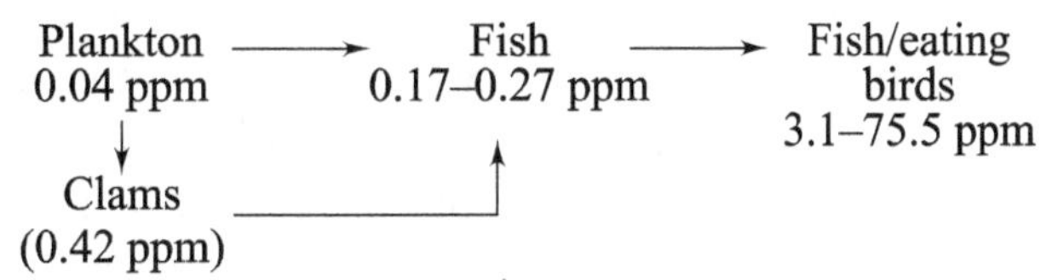

It has also been found that DDT causes liver cancer and also affects the central and peripheral nervous system and also damages reproductive organs. In spite of banning DDT by USA, it is still made in Mexico, China, India and Russia. Attempts are being made for a global treaty to ban the use of DDT.

4. Chemical Products should be designed to preserve efficacy of function while reducing toxicity: The compounds (*e.g.* drugs and other medicines) synthesised should not only be safe to handle but also should not have any toxic effects. An example of a drug having toxic effects is thalidomide. It was used to treat nausea in pregnant women from the late 1950s to 1962.

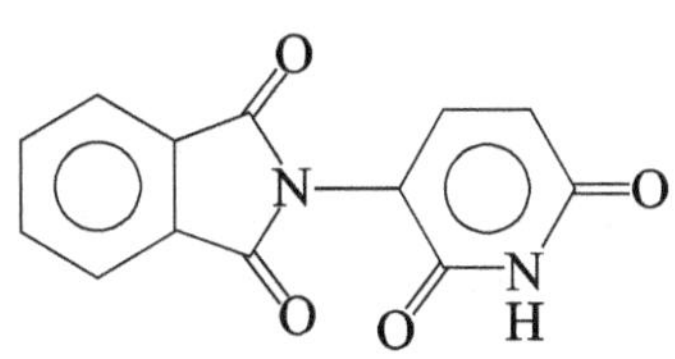

Thalidomide

Subsequent investigations revealed that thalidomide was a mixture of two enantiomers and each of these enantiomers had a different biological activity. Although one of the enantiomer had the desired therapeutic effect, the other isomer was responsible for birth defects.

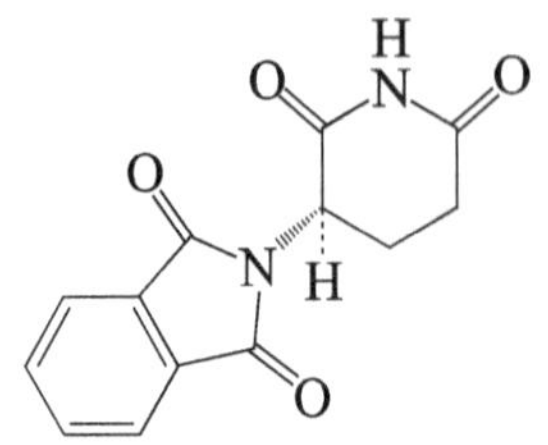

Anti Nausea drug

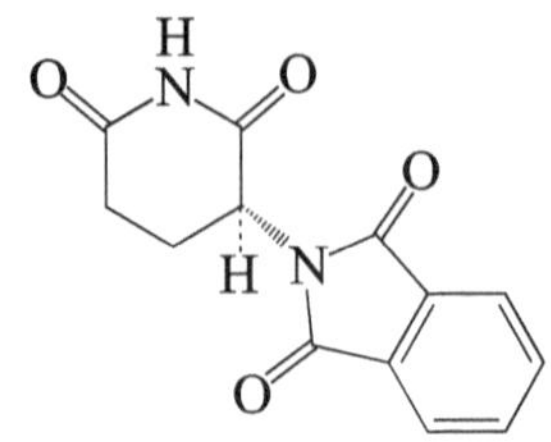

Enantiomer responsible for birth defects

Two enantiomers of thalidomide

Thalidomide was withdrawn after about 8000 children in 46 countries were born with birth defects (Harvard Health Letter, 1998, **23(4)**,4. Finally, the U.S. Food and Drug Administration (FDA) approved the use of the enantiomer (having birth defect) for use for treating painful inflammation of leprosy (R. Hoffmann, Chem. Eng. News. April 29, 1996). It can also inhibit human immunodeficiency syndrome (AIDS).

It is now possible to design safer chemicals due to development and advancement of technology. The molecular structure, in fact, can be appropriately manipulated to have the desired effects.

5. Use of Auxiliary substances (Solvents, separation agents, etc.) should be made unnecessary wherever possible, and innocuous when used: Solvents having excellent solvent properties have been used for organic reactions. Most of these solvents are volatile organic solvents. Examples include methylene chloride, chloroform, perchloroethylene, carbon tetrachloride, benzene and other aromatic hydrocarbons. It has been found that the halogenated solvents and also benzene are human carcinogens. Some of these solvents have also hazardous effects on the environment. As an example, chlorofluorocarbons (CFCs), which were used as cleaning agents, blowing agents and as refrigerants (alternatives to sulphur dioxide and ammonia) are responsible for the depletion of ozone layer, which has serious health effects on human.

Due to the above reasons, it is necessary to use benign solvents called green solvents. Examples include water, super critical carbon dioxide, ionic liquids, polyethylene glycol and its solutions and fluorous solvents. This aspects forms the subject of discussion in a separate section.

Besides using green solvents for organic synthesis, the pathway for a reaction should be such that there is no need for purification or separation. A procedure for achieving this objective is to use polymers supported reagents or substrates. These are discussed subsequently.

6. Energy requirements should be recognised for their Environmental and Economic Impacts and should be Minimised: Energy is required for chemical reactions. This requirement of energy should be kept to a minimum. For example, energy is required to dissolve the reactants in a solvent and for the reaction to go to completion. It should be ascertained that for completion of the reaction, heating be done for minimum time. This can be ascertained by using techniques like TLC. Such a procedure ensures the use of minimum amount of energy for a reaction. The use of suitable catalyst in a reaction is also known to reduce the amount of energy required for a reaction.

The generation of energy from coal also has environmental impacts. It is advantageous to supply energy to a reaction by microwave heating, sonication or photochemical activation. These are discussed subsequenly.

7. A Raw Material or Feed Stock should be Renewable rather than Depleting whenever Technically and Economically Practical: The starting materials for a reaction can be obtained from renewable or non renewable sources. As an example a non renewable source is petroleum, which is a source of petrochemicals, takes millions of years for its formation from animal and vegetable remains. On the other hand, an example of renewable starting materials are agricultural products, which sometimes cannot be obtained in continuous supply due to failure of crops, shortfall of rains etc.

8. Unnecessary Derivatisation (Blocking groups, Protection and deprotection, Temporary Modification of Physical/Chemical Processes) should be avoided wherever possible: In organic synthesis, it is sometimes necessary to protect a particular group so that the reaction can proceed in the desired direction. The reagent to be used for protection of a particular group should be added in stoichiometric amount and after the completion of the reaction, the protecting group is removed in order to get the desired product. Through, in certain cases, protection and deprotection are necessary, but it increases the number of steps in organic synthesis. Additionally, the reagents used for protection and deprotection form wastes. As far as possible the steps of protection and deprotection should be avoided.

Enzymatic transformations, for example are one step process and do not need any protection or deprotection steps.

9. Catalytic Reagents (as selective as possible) are superior to stoichiomatric reagents: In certain reactions, all starting materials are not consumed and the reaction does not go to completion. Such reactions are much less atom economical. The unreacted starting materials form wastes. The catalysts (whenever available) facilitate the transformations without being consumed. Also, the catalysts are selective in their action. As an example, a triple bond reduced to a double bond by palladised barium sulfate. Different types of catalysts are available. Examples include, Phase Transfer catalysts, biocatalysts etc. These are discussed subsequently.

10. Chemical Products should be so designed so that at the end of their function they do not persist in the environment and instead break down into innocuous degradation products: An important concept of Green Chemistry is that the products of a reaction should be biodegradable and not persistant in the environment. Such products (which may be toxic) are taken up by plants and animals; in animals these products accumulate. This is harmful not only for the animal species but also for the humans who consume such plants and animals like fish and meat producing animals. As already stated, DDT was the first pesticide which bioaccumulated in animal species and found its way into the human system causing serious health problems (see also third principle of Green Chemistry). It should be ascertained that the synthetic products should be biodegradable. Introduction of certain groups and some other features in a molecule makes it biodegradable. Groups which are susceptible to hydrolysis, photolysis etc. make the product biodegradable. It is also necessary to ascertain that the products of biodegration should not be toxic to the environment. Though plastics are in general known to be non-biodegradable, but with the advancement of knowledge and technology, it is possible to make biodegradable plastics.

11. Analytical Methodologies need to be further developed to allow real-time, in process monitoring and control prior to the formation of hazardous substances: Methologies, which not only minimise but also prevent the generation of hazardous substances in a chemical reaction are available. It is necessary to instal reliable sensors, monitors and analytical techniques in order to assess that may be present in a process stream. It is thus possible to monitor a chemical process and find if any hazardous products are being generated and also prevent their formation. This is helpful in turn to avoid accidents which may occur. It is appropriate to state that the Bhopal Gas Tragedy is attributed to the failure and malfunction of the inbuilt sensors and monitors.

12. Substances and the form of a substance used in a Chemical Process should be chosen so as to minimise the potential for chemical accidents, including releases, explosions and fires. Necessary and adequate precautions must be taken in order to avoid the occurrence of accidents including releases, explosions and fires in a chemical industry.

A number of explosions as in Bhopal (India), Serveso (Italy) and many of the nuclear accidents are attributed to the failure of taking adequate precautions.

1.2 GENERALIZATION OF THE PRINCIPLES OF GREEN CHEMISTRY

On the basis of the twelve principles of Green Chemistry, (section 1.1) following generalization can be made:

1. The starting materials as far as possible should be from renewable source and volatile organic solvent be not used. Benign green solvents should be used.
2. The procedure should not involve protection and or deprotection.
3. Catalyst should be used wherever possible.
4. Waste generation (byproduct formation) be minimum and product formed be biodegradable.

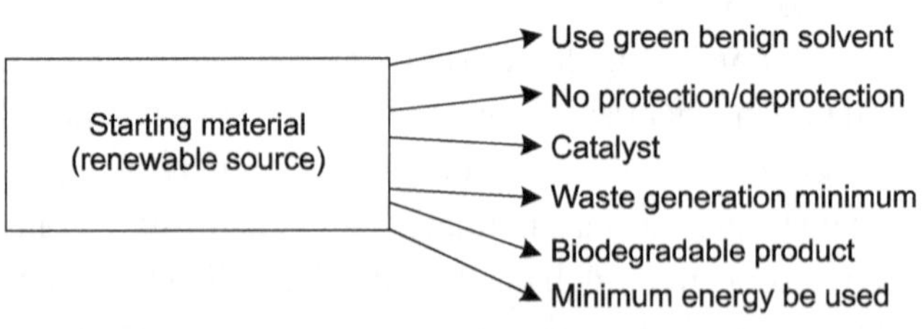

Fig. 1.1: Generalization of Principles of Green Chemistry

1.3 PLANNING A GREEN SYNTHESIS

Following are given the points which should be kept in mind while planning a green synthesis.

1. Atom Economy: In a chemical synthesis, there should be maximum incorporation of the reactants into the final product. An ideal reaction is one which has 100% atom economy i.e. all atoms of the reactants are incorporated into the required product. As has already been started (section 1.1), addition and rearrangement reactions are 100 percent atom economical. On the other hand, substitution and elimination reactions are not.

2. Protection and deprotection of groups be avoided wherever possible.

Protection and deprotection of groups is sometimes necessary particularly to solve a chemoselectivity problems. However, in such cases, the percentage atom utilization becomes much low. If protection and deprotection are necessary steps, the reagents used for these should be used in stoichiometric amounts. As an example, the Grignard reaction of cyclohexanone-2-carboxylate.

The >C=O group is protected first to allow the reaction of the grignard reagent with the less reactive >C=O of the —$COOCH_3$ group.

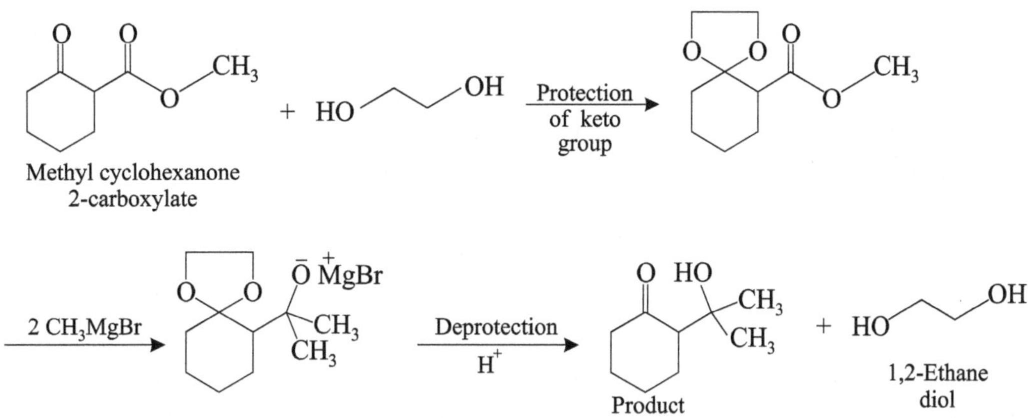

3. Use of environmentally benign solvent: The common solvents used in a synthesis include methylene chloride, chloroform, perchloroethylene, carbon tetrachloride and aromatic hydrocarbons like benzene, toluene etc. These solvents though have excellent solvent properties are volatile and many of these are suspected human carcinogens. Besides, these solvents [particularly the halogenated solvents like methylene chloride, chloroform, perchloroethylene, carbon tetrachloride and chlorofluorocarbons (FCs)] are responsible for environmental pollution particularly destroying the ozone layer in the stratosphere leading to the formation of ozone hole.

In view of the above, it is necessary to carryout the reaction in environmentally benign solvent like water, supercritical carbon dioxide, ionic liquids and polyethylene glycol etc.

In case it is possible, the reaction can be performed in solid state without using a solvent.

4. Energy requirement be minimised: Energy is required for all chemical reactions. However, the use of energy should be kept to a bare minimum. Energy is required for heating the reaction mixture (the time required for completion of the reaction be kept minimum) and for crystallization and purification of the formed product. If possible, the chemical process be designed in such a way that there is no need for purification and or crystallization.

The use of microwaves, sonication and photoactivation are well known procedure for reducing of the energy requirements.

5. Use of catalysts: Catalysts are known to facilitate chemical transformations. Such reactions generally consume less energy and give good yields. An additional advantage is that the catalysts are not consumed in the reaction and can be recycled often.

The catalysts have been used for selective reduction of a triple bond to a double bond.

$$\underset{\text{Propyne}}{H_3CC \equiv CH} + H_2 \xrightarrow{\text{Pd-BaSO}_4} \underset{\text{Propene}}{H_3C - CH = CH_2}$$

A typical example of a catalytic procedure in the synthesis of methyl methacrylate

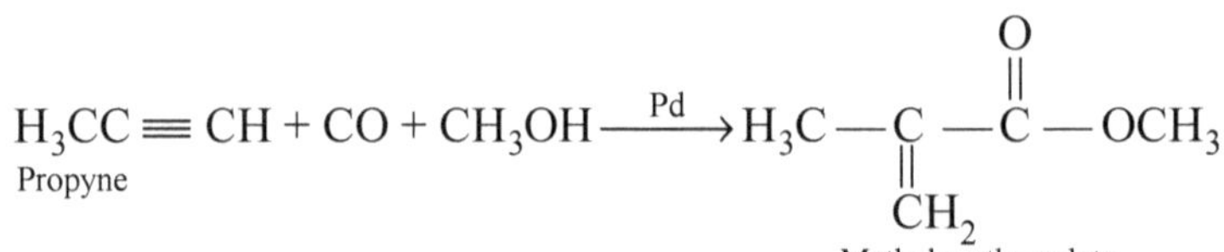

Methyl methacrylate

Biocatalysts as whole cells or purified enzymes are emerging as greener alternatives to chemicals. These are more selective and specific catalyst, most of these catalyst can help reduce the energy requirements of a reaction and allow chemoselectivity and regioselectivity. A number of enzyme catalysed reactions are used for industrial production of useful compounds. Bioengeeneared enzymes and microbes further widen their scope as catalyst.

Another important point that must be considered is the recyclability and reduction in the amount of the catalyst.

Some other well known catalysts that can be used are the phase transfer catalysts and polymer supported catalysts. A discussion on these form the subject matter of subsequent sections.

Besides the points mentioned above, following should also be kept in mind while planning a green syntheses.

- The waste products (or by products) if formed should not be environmentally harmful or toxic to human health.
- The raw materials should be from renewable source rather than depleting resources whenever possible.
- The green aspect *i.e.* reduce, reuse and recycle should be strictly adhered to.

1.4 MEASUREMENT OF GREENNESS OF A REACTION

For the reduction of the environmental impact of an chemical reaction, it is helpful to measure the 'greenness' of a chemical process. A number of metrics have been proposed to make a manufacturer aware of the necessity to change the process of chemical synthesis so as to make them less wasteful (David J.C. Constable, Allan D. Curzons and Virginia L. Cunningham, Green Chemistry, 2002, **4**, 521–527). Some of the metrics which have been proposed include effective mass yield, E-factor, atom economy, mass intensity, carbon efficiency and reaction mass efficiency.

1.4.1 Effective Mass Yield

The metrics, 'effective mass yield' was proposed by Hudlicky *et al*[1]. It is the percentage of the mass of the desired product to the mass of all non-benign materials used in its synthesis.

$$\text{Effective mass yield (\%)} = \frac{\text{Mass of products} \times 100}{\text{Mass of non-benign reagents}}$$

In this metric, it is difficult to define 'non-benign' particularly in complex reagents for which there is limited environmental toxicity information.

1.4.2 Environmental Factor–E-Factor

The second metric, 'E-factor' was proposed by Roger Sheldon[2] and is defined as the total waste (in kg) divided by the mass of the product (in kg)

$$\text{E-factor} = \frac{\text{Total waste (kg)}}{\text{Mass of the product (kg)}}.$$

Assumptions on solvent and other factors can be included in calculations.

1.4.3 Atom Economy

The term 'atom economy' was introduced by Barry Trost[3]. The percentage atom economy was defined by R.A. Sheldon[2] as F.W. of atom utilized divided by F.W. of the reactants and then multiplied by 100 (see third principle of green chemistry)

$$\text{\% Atom Economy} = \frac{\text{F.W. of atoms utilized}}{\text{F.W. of all the reactants used}} \times 100.$$

In the above calculations, the starting materials are those which make up the reactants and so the catalyst and solvents are not considered as starting materials. In fact, the reactants must also include those materials which are incorporated into a reaction intermediate (even though such reactants are not present in the final product). Also, if either of the reactant is used in order to maximise the yield, this is not included in the calculation of atom economy.

In view of the, above, the atom economy should be calculated as follows[4]:

For a reaction which is carried out in following three steps:

$$A + B \rightarrow C$$
$$C + D \rightarrow E$$
$$E + F \rightarrow G$$

The Atom economy

$$= \left(\frac{\text{M.W. of Product G}}{(\text{M.W. of A} + \text{M.W. of B} + \text{M.W. of D} + \text{M.W. of F})} \right) \times 100$$

The metric, atom economy is an indicator of the efficiency of a chemical process. The value 100% indicates that all reactants are converted into products and there is no side products formed as in the case of an addition reaction. However, the calculation does not take into account solvent, catalyst and energy used as already stated.

1.4.4 Mass Intensity

The mass intensity (MI) is defined[5] as

$$\text{MI} = \frac{\text{Total mass used in a process (kg)}}{\text{Mass of product (kg)}}$$

In the above calculations, total mass includes everything that is used in a process (with the exception of water), *i.e.*, reactants, reagents, solvents, catalysts etc.

As seen, the mass intensity (MI) is expressed on a weight/weight basis rather than percentage. By expressing mass intensity as its reciprocal and making it a percentages, the form becomes similar to effective mass yield and atom economy. This metric is called[4] mass productivity

$$\text{Mass productivity} = \frac{1}{\text{MI}} \times 100$$

$$= \frac{\text{Mass of product}}{\text{Total mass used in process}} \times 100$$

The parameter 'mass intensity' indicates the overall efficiency of the process including solvent usage. The ideal value 1 means that no solvents used or recovered/reused and all raw materials are completely converted into the product.

1.4.5 Carbon Efficiency

Carbon efficiency (CE) is defined[4] as the percentage of carbon in the reactants that remain in the final product.

Thus, for a reaction $A + B \rightarrow C$

$$\%\ \text{Carbon efficiency} = \frac{\text{Amount of carbon in product } C \times 100}{\text{Total carbon present in reactants } A + B}$$

For a reaction A + B = Product + Co-product

$$\text{Carbon efficiency} = \frac{\text{No. of moles of product} \times \text{No. of carbon in product} \times 100}{(\text{moles of A} \times \text{carbons in A}) + (\text{moles of B} \times \text{carbons in B})}.$$

1.4.6 Reaction Mass Efficiency

The reaction mass efficiency (RME) is defined as the percentage of the mass of the reactants that remain in the product. Thus, for a reaction $A + B \rightarrow C$, there are two way to calculate RME

$$\text{Reaction mass efficiency} = \left\{\frac{\text{m.w. of product C}}{\text{m.w. of A} + (\text{m.w. of B} \times \text{molar ratio of B/A}}\right\} \times 100$$

or simply

$$\text{RME} = \left\{ \frac{\text{mass of product C}}{\text{mass of A + mass of B}} \right\} \times 100$$

The following example shows how the atom economy, carbon efficiency, reaction mass efficiency, mass intensity and mass productivity are computed

$$\underset{\substack{\text{Benzyl alcohol} \\ \text{10.81g (0.10 mol)} \\ \text{FW 108.1}}}{C_6H_5\,CH_2OH} + \underset{\substack{p\text{-Toluenesulfonyl} \\ \text{chloride} \\ \text{21.9g (0.115 mol)} \\ \text{F.W. 190.65}}}{p\text{-}CH_3C_6H_4SO_2Cl} + \underset{\text{Tolune}}{C_6H_5CH_3} + \underset{\substack{\text{Trimethyl} \\ \text{amine}}}{(CH_3)_3\,N}$$

$$\longrightarrow \underset{\substack{\text{Sulfonate ester} \\ \text{yield 90\% (0.09 mol, 23.6 g)} \\ \text{F.W. 262.24.}}}{C_6H_5CH_2OSO_2C_6H_4.CH_3(p)} + HCl$$

$$\text{Atom economy} = \frac{262.24}{(108.1 + 190.65)} \times 100 = 87.8\%$$

$$\text{Carbon efficiency} = \frac{(0.09 \times 14)}{(0.1 \times 7) + (0.115 \times 7)} \times 100 = 83.7\%$$

$$\text{Reaction mass efficiency} = \frac{23.6}{(10.81 + 21.9)} \times 100 = 70.9\%$$

$$\text{Mass intensity} = \frac{(10.81 + 21.9 + 500 + 15)}{23.6} = 23.2 \text{ g/g} = 23.2 \text{ kg/kg.}$$

$$\text{Mass productivity} = \frac{1}{\text{Mass intensity}} \times 100 = 4.3\%$$

The atom economy is less than 100% since HCl is formed as a byproduct.

The carbon efficiency is also less than 100% and accounts for the excess stoichiometry.

The reaction mass efficiency also takes into account 90% yield and need of a 15% molar excess of *p*-toluenesulfonyl chloride.

Besides what has so far been stated, there are various other parameters for measuring the Greenness of a chemical process. These include:

- Amount of greenhouse gases used or produced per kg of product.
- Amount of ozone-depleting gases produced per kg of product.
- Amount of fresh water used per kg of product.
- Amount of solvent used per kg of the product.
- Amount of solvent lost per kg of the product.
- Amount of water generated per kg of the product.

- Amount of nonbiodegradable material produced per kg. of the product.
- Amount of VOC produced per kg of the product.

[**Source:** Green chemistry and Processes, Mukesh Doble and Anil Kumar Kruthiveto, Elsevier, page 21].

1.5 THE NOBEL PRIZE IN GREEN CHEMISTRY

Yves Chauvin (France), Robert H. Grubbs (USA) and Richard R. Schrock (USA) shared the 2005 Nobel Prize in chemistry for the development of metathesis, which has applications in the polymer industry, the most important aspect being reduction in waste generation via efficient production and manufacturing. This noble prize is expected to be an incentive for the researchers and scientists to wholeheartedly work in the area of green chemistry.

1.6 SUSTAINABLE CHEMICAL INDUSTRY

Industrial sustainability is defined as a continuous process of innovation, improvement and use of benign technologies in order to reduce pollution levels and consumption of resources. It can also be said that industrial sustainability means using technologies and know-how in order to use less material and energy, maximizing use of renewable resources and minimizing generation of pollutants and or harmful wastes during the manufacturing process and producing biodegradable or recyclable products. The aim of sustainable chemical industry is that the process should not degrade the environment and that the products obtained should not be harmful to the human health. All this can be achieved by using the principles of green chemistry and engineering expertise.

In order to achieve sustainable chemical industry, the industrial production must follow the following:

- Use renewable resources.
- Produce Eco-friendly products.
- Increase energy efficiency.
- Reduce waste and reuse if produced.
- Reduce emission of greenhouse gases.
- Make the process safer.

1. Use of Renewable Resources: World wide billions of tons of biomass is produced by photosynthetic process. This natural resource consisting mostly of carbohydrates can be an important source for the production of energy and for the production of chemicals.

Biomass is a renewable raw material and can replace fossil fuel feedstocks. Some of the important uses of biomass include production of sugars (by fermentation), biodiesel (a C –16 – 18 fatty acid methyl ester) (produced from vegetable oil), bioethanol. Oleochemicals are derived from natural fats and oils from animal, marine or vegetable sources find use in detergents and cosmetics, additives for rubber, textiles and plastics. The oleochemical industry uses mainly renewable raw material like palm oil, coconut oil.

Polylactic acid, a synthetic polymer is obtained from corn derived dextrose. Another polymer, polytrimethylene-terephthate (PTT), used for the production of high quality fibre is produced from 1,3-propanediol which is obtained by fermentation of sugars with *Escherichia coli* K12.

L-Threonine, an amino acid used for animal nutrition is also obtained from biomass. Other products obtained from biomass include vitamin C (ascorbic acid), Vitamin B2 (riboflavin), broad spectrum herbicide (by enzymatic oxidation of glycolic acid) and L-Dopa.

Biomass energy is obtained from biomass derived from plant resources, animal waste and waste of various human activities. It is also derived from the byproducts of timber industry, agricultural crops etc.

2. Producing Eco-friendly products: This is an important step in sustainable chemical industry. Some of the pigments like cadmium sulfoselenide, lead molybdate, bismuth vanadate having excellent performance characteristics are toxic to environment. Now non-toxic inorganic pigments based on cerium sulfides (used in automotive coatings, engineering plastics) have been developed. A biodegradable polyester (Ecoflex), a copolymer of 1, 4-butanediol, adipic anul and terephalic acid has good thermoplastic properties and is used as semitransparent packing material (produced by BASF).

3. Using energy efficiently: Development of energy-efficient separation technologies such as membranes, helps in using energy efficiently. Process designs to reduce multiple drying steps in a process also reduces consumption of energy.

Reduction in the use of energy by combining operations in an industries. For example production of methyl acetate has been achieved in a single step by esterification of methyl alcohol with acetic acid and distillation of methyl acetate (Eastern chemical). A number of other procedures have also been described (A. Stankiewics, Chem. Eng. Progress, 2003, **42**, 137.

4. Reduction of waste and their reuse: Reduction of waste is an important step. This can be affect by combining more than one technology. For example, Sumitomo Chemical and Enichem combined two catalytic technologies for caprolactam plant. In place of the classical routes using hydroxylamine sulphate (a toxic reactant), cyclohexane is directly aminoximated by NH_3 and water over a Ti-MFI type zeolite. A catalyst is used (instead of oleum) for the final rearrangement of the intermediate oxime to caprolactam.

The waste (or byproduct) from one industry can be used as a raw material for another industry. An example is the extraction of squalene from olive oil waste streams rather than from Shark livers [process developed by uniquema); http: //www.ici.com/icishe/2000/pages/nes 04. htm#su.]. Another example is the development of a process to recover waste frying oil to produce high grade stearone used in rubber tyres (developed by Uniquema].

5. Reduction of emission of greenhouse gases: Reduction of emission of greenhouse gas is very important for the environment. For example in the extraction industry, supercritical fluids can be used. These are used in decaffination of coffee or drycleaning processes. Union carbide is using co-solvent modified SC CO_2 in place of volatile organic solvents used earlier. This procedure is now being used in automotive and furniture industry (J.M. Barley, Ind. Paint Powders, Nov., 1995).

6. The process should be safe: It should be ascertained that the process is safer and there is no possibility of accidents or evolution of harmful gases and byproducts. The Bhopal gas tragedy in 1984 is a glaring example of unsafe product. A method has been developed by du pond Co. to produce methyl isocyanate on demand so that it need not be stored as in Union Carbide factory at Bhopal

$$CH_3NH_2 + CO \longrightarrow CH_3NHCHO \xrightarrow[>240°C]{O_2} \underset{\text{84-84\% yield}}{CH_3NCO}$$

[J. Haggin, Chem. Eng. News, June 3, 1996, 38; R.F. Service, Science, 1998, **282**, 400.

Avoiding phosphine as raw material, dimethyl carbonate is used (D. Delledonne, F. Rivetti and U. Romano, Appl. Catal, A, 2001, **221**, 241).

CONCLUSION

The only way to prevent environmental pollution by various chemical industries is to follow the principles of green chemistry. It is however necessary for the R and D scientists to ascertain the greenness of a particular plan before it is used for production. Green chemistry is helpful for a Sustainable Chemical Industry. The Nobel Prize of 2005 in Green Chemistry shows the importance of Green Chemistry.

REFERENCES

1. T. Hudlicky, D.A. Frey, L. Koroniak, C.D. Claeboe and L.E. Brammer, Green Chem., 1999, 57.
2. R.A. Sheldon, Chem. Ind. (London), 1992, 905; 1997, 12.
3. B.M. Trost, Science, 1991, **254**, 1471.
4. D.J.C. Constable, A.D. Curzons and V.L. Cunningham, Green Chemistry, 2002, **4**, 521.
5. A.D. Curzons, D.J.C. Constable, D.N. Mortimer and V.L. Cunningham, Creen Chem, 2001, **3,** 1.

2. ORGANIC TRANSFORMATIONS IN WATER

2

Organic Transformations in Water

2.1 INTRODUCTION

The use of solvent is an integral part of a large number of manufacturing processes. Most of the solvents which were used till nineteenth century were volatile organic solvents like, methylene chloride, chloroform, perchloroethylene, carbontetrachloride and aromatic hydrocarbons. All these solvents (known as volatile organic solvents) are known to have disastrous effects in the environment. There has been pressing need to develop alternative solvents for manufacturing process. As already stated, the alternative green solvents include water, supercritical carbon dioxide, ionic liquids, polyethylene glycol and its solutions and florous solvents. In this chapter, the use of water as an alternative solvent for chemical reactions is described.

The advantages of carrying out reactions in using water as a solvent include:

- Being the cheapest solvent, water can be used to make organic transformations economical.
- Unlike organic solvents (VOC), which are inflammable, mutagenic and or carcinogenic and harmful to the environment, water is free from all these disadvantages and is a safe solvent.
- Substances soluble in water can be directly used. Water is particularly useful in carbohydrate and fermentation chemistry.
- In industrial processes, the isolation of the products can be achieved by simple phase separation. It is also easier to control the reaction temperature since water is known to have the largest heat capacities of all substances.
- Water can be readily recycled.
- Organic transformations can be carried out in ordinary liquid water, super critical water (SCW) near critical water (NCW) and high temperature water (HTW). A comparison of the properties of water under different conditions is given below:

	Ordinary water $T < 150°C$, $p < 4$ bar	Near critical water (NCW) $T = 150–350°C$, $p = 4–200$ bar	Super critical water (SCW) $T > 374°C$, $p > 221$ bar	High Temp. water (HCW) $T > 220°C$ $p = 5–190$ bar
Temp. (°C)	25	250	400	2.60
Density (g cm^{-1})	1	0.8	0.17	0.9
Dielectric constant	78.5	27.1	5.9	60-70
pK_w	14	11.2	19.4	40

Note: Data from P. Krammer, H.J. Vogel, Supercritical Fluids, 2000, **16**, 189; P. Krammer *et al.* Chem. Eng. Technol., 1999, **22**, 126.

Transformations in NCW, SCW and HTW form the subject matter of subsequent sections (2.2.2 and 2.2.3). Following is given a brief about organic transformation in water.

2.2 ORGANIC TRANSFORMATIONS IN AQUEOUS MEDIA

A common problem associated with the use of water as solvent is that most of the organic substrates are insoluble in water making the reaction mixture heterogeneous. This can be overcome by using phase transfer catalysts, but this makes the process expensive (for details about use of phase transfer catalysis see chapter 12). Also, the isolation to the products from aqueous media may be difficult. For this, the water has to be evaporated from the reaction mixture, which consumes energy. This problem can be overcome by using microwave heating procedure for reactions in aqueous media (For details about microwave induced reactions in water see chapter 9).

2.2.1 Organic Transformations in Water

Following are given some of the important transformations that have been carried out in aqueous medium.

2.2.1.1 *Pericyclic reactions*

(*a*) Diels-Alder Reaction: An important carbon-carbon bond forming reaction giving cyclic structures was the first reaction to be carried out[1] in aqueous medium (scheme 2.1).

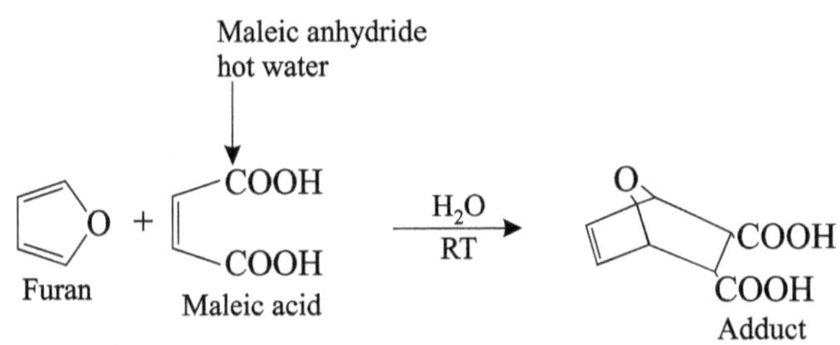

Fig. 2.1: Diels-Alder reaction in water.

The reaction of cyclopentadiene with butenone in water was found[2] to be 700 times faster than in isooctane (Fig. 2.2). The acceleration in reaction was attributed[3] to the 'hydrophobic effect' resulting in bringing together the two nonpolar groups in the transition state.

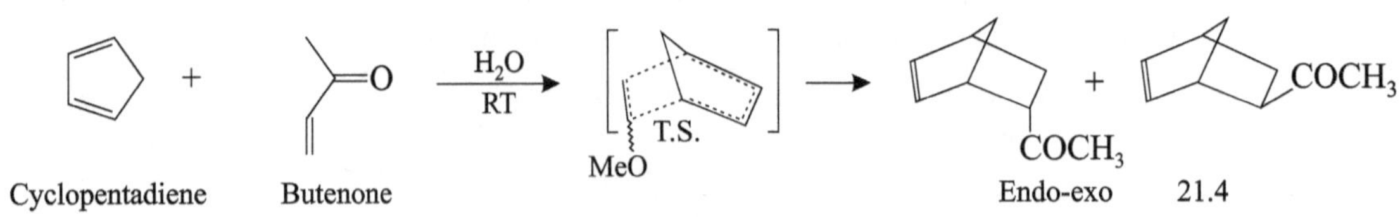

Fig. 2.2: Diel Alder reaction of Cyclopentadiene with butenone.

The rate of aqueous Diels-Alder reaction has been found to increase[4] by the use of β-cyclodextrin (which simultaneously forms an inclusion complex with the diene and dienophile) and also by the use of 4.86 M LiCl aqueous solution as solvent (which salts out nonpolar materials dissolved in water. No change in reactivity was observed with α-cyclodextrin.

The **Retro-Diels-Alder reaction** proceeds much faster in aqueous solution than in organic solvents. An example[4a] is given below [Fig. 2.2(*a*)]

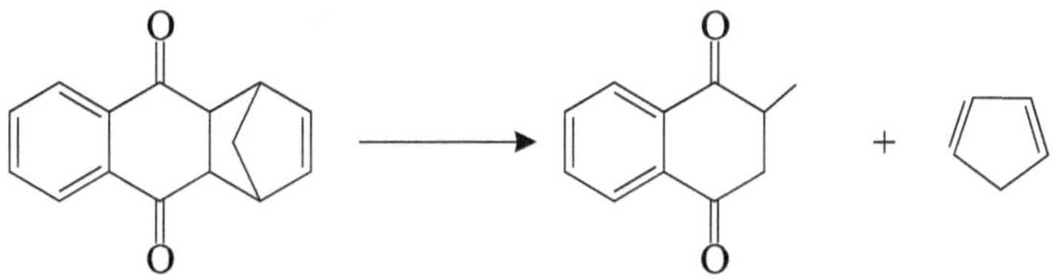

Fig. 2.2(a): Retro-Diels-Alder reaction.

Intramolecular Diels-Alder reaction of a Füran-ene gave two products (Fig. 2.3). Using one equivalent of β-cyclodextrin the yield improved considerably[5].

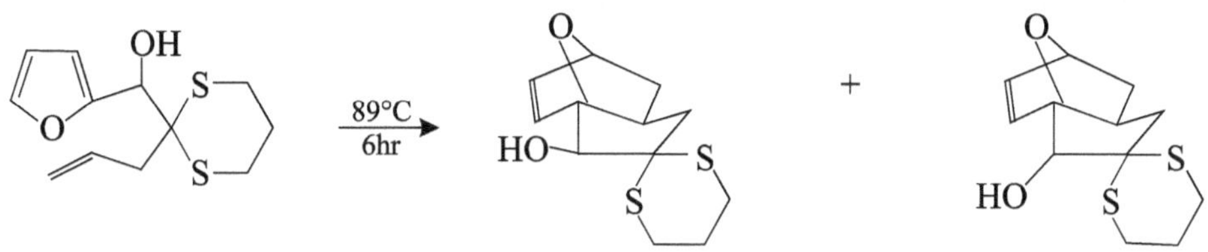

(A) (B)

Water A/B = 1 : 2 (20% yield)
Water + β – CD (1.0) A/B = 1 : 1.5 (91% yield)

Fig. 2.3: Intramolecular Diels-Alder reaction of füran-ene derivative.

Two other examples[6,7] of intramolecular Diels-Alder reaction are given in Fig. 2.4.

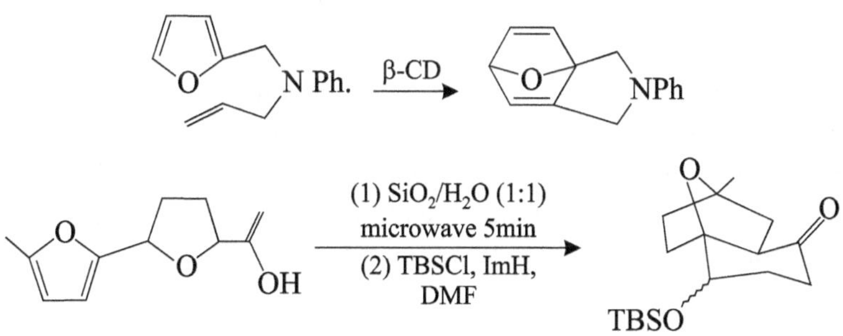

1:1 Mixture

Fig. 2.4: Some examples of intramolecular Diels-Alder reaction.

In Diels-Alder reaction, the use of Lewis acids activates the substrate. The Lewis acids to be used must be air stable and uneffected by water. A typical example[8] of such a Lewis acid is $[Ti(C_p^*)_2(H_2O)_2]^{2+}$. Other examples[9] of air and water stable Lewis acids include scandium triflate, $[Sc(OTf)_3]$ and lanthanide triflate, $[Ln(OTf)_3]$. A typical example[10] is given in Fig. 2.5.

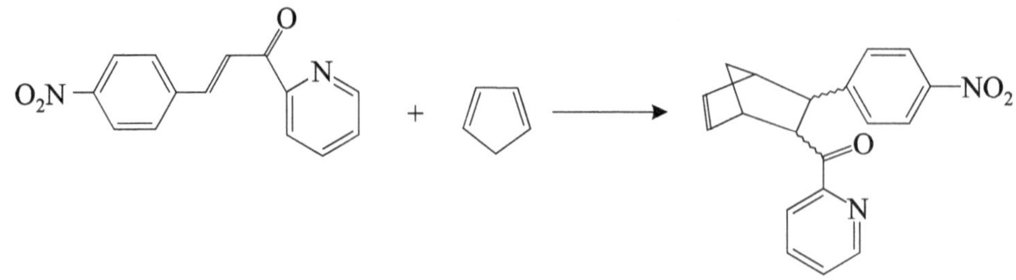

Fig. 2.5: Lewis acid catalysed Diels-Alder reaction.

The above reaction (Fig. 2.5) was 250,000 times faster in presence of aqueous solution containing 0.010 M Cu $(NO_3)_2$. The reaction is also catalysed by bovine serum albumin[11].

A symmetric Diels-Alder reactions have also been conducted[12] in water in presence of various catalysts. A typical example is given[13] in Fig. 2.6.

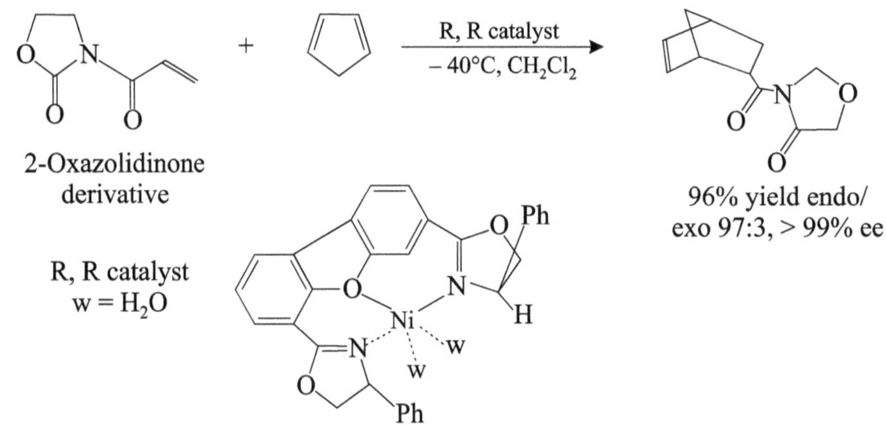

Fig. 2.6: Asymmetric Diels-Alder reaction.

Hetero-Diels-Alder Reaction: These reactions find application in the synthesis of heterocyclic compounds. In these reactions nitrogen or oxygen containing dienophiles are used[14]. The first example of hetero-Diels-Alder reaction with nitrogen containing dienophile in aqueous media was reported[15] in 1985. The method consist in generation of simple iminium salts in situ under Mannich like conditions involving the reaction of amine with formaldehyde in water. The iminium salts on reaction with cyclopentadiene gave on Diel-Alder product (Fig. 2.7).

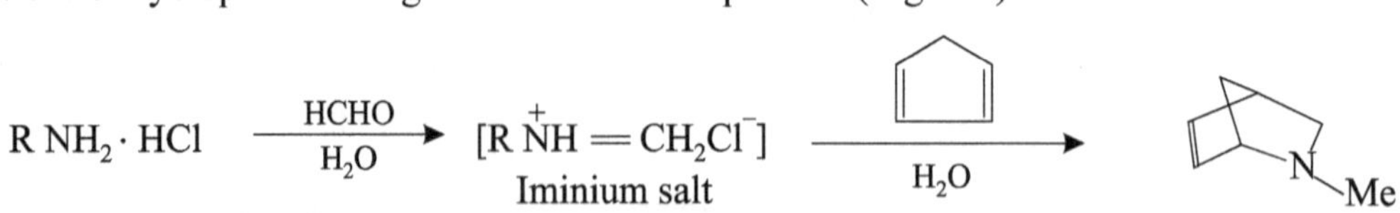

Fig. 2.7: Hetero-Diels-Alder Reaction.

The intramolecular aza-Diels-Alder reaction[16], also occurred in aqueous media giving fused-ring systems having bridgehead nitrogen (Fig. 2.8); such a structure is characteristic of many alkaloids.

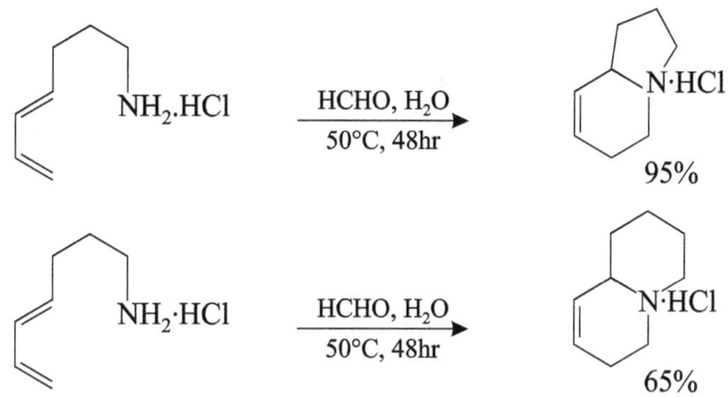

Fig. 2.8: Intramolecular aza Diels-Alder reaction.

Retro-aza-Diels-Alder reaction also occurred readily in water[17]. As an example, 2-azanorborene on treatment with water in presence of acid undergoes acid catalyzed retro-diels alder clevage in water. The formed iminium derivative as in intermediate could be trapped with N-methylmaleimide or is reduced to give primary amine. (Fig. 2.9).

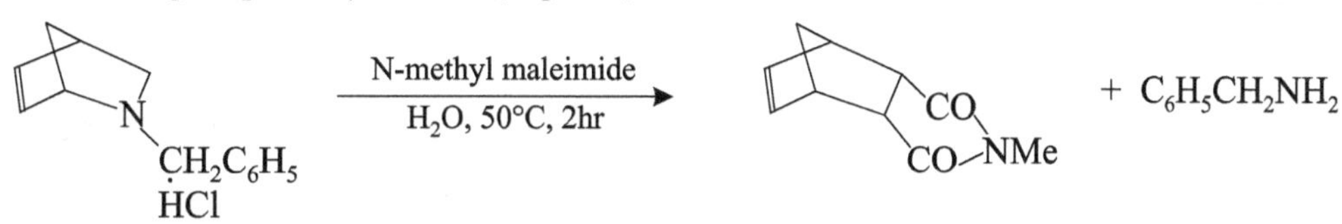

Fig. 2.9: Retro-aza-Diels-Alder reaction.

An analogous aza-Diels-Alder reaction uses (R)- and (S)-amino acid methyl esters and chiral amines as chiral auxiliaries[18]. (Fig. 2.10)

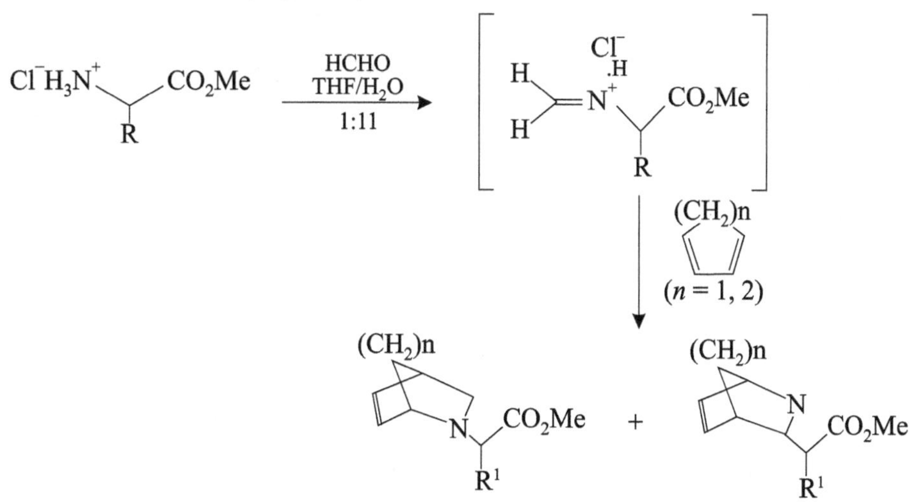

Fig. 2.10: Hetero Diels-Alder reaction using (R)- and (S)-amino acids.

The aza-Diels-Alder reaction has also been used[17] in the asymmetric synthesis of highly functionalized tetracyclic indole derivatives (Fig. 2.11), which are useful in the synthesis of alkaloids[18] of the type yobimbine and reserpine. The reaction (as in the case of Diels-Alder reaction) is catalysed by lanthanide triflates.[19]

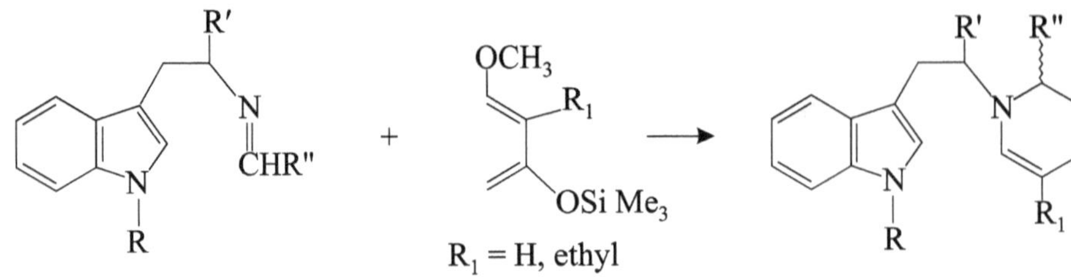

R = H, CHO
R' = CO_2Me, H
R" = Ph, 4-Cl C_6H_4, 4-NO_2 C_6H_5, H

Fig. 2.11: Synthesis of tetracyclic indole derivatives.

Dihydro-4-pyridine derivation could be synthesised[20] in good yields by aqueous Diels-Alder reaction of aldehydes, anilines and Danishefsky's diene in presence of HBF_4 (Fig. 2.12).

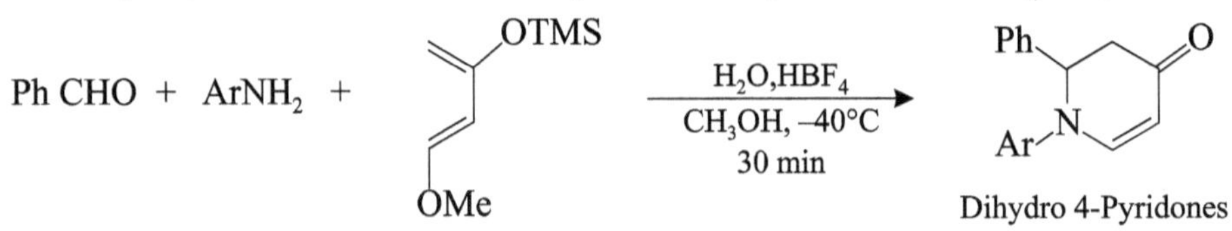

Fig. 2.12: Synthesis of dihydro-4-pyridones.

In hetero-Diels-Alder reaction using an oxygen containing dienophile, glyoxylic acid was reacted with cyclopentadiene or cyclohexadiene in aqueous solution giving α-hydroxy-γ-lactones arising from the cycloadducts by rearrangement[21] (Fig. 2.13)

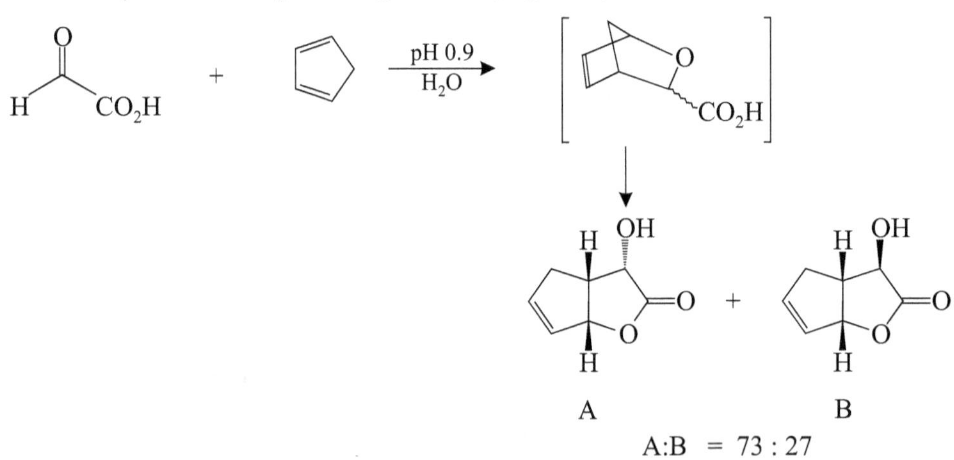

Fig. 2.13: Hetero-Diels-Alder reaction using Oxygen containing dienophile.

The 5+5 fused system generated (Fig. 2.13) has been used for the total synthesis[22] of a number of bioactive molecules like sesbanimides A and B, carbovir and hydroxyacetone moiety of mevinic acid. (Fig. 2.14)

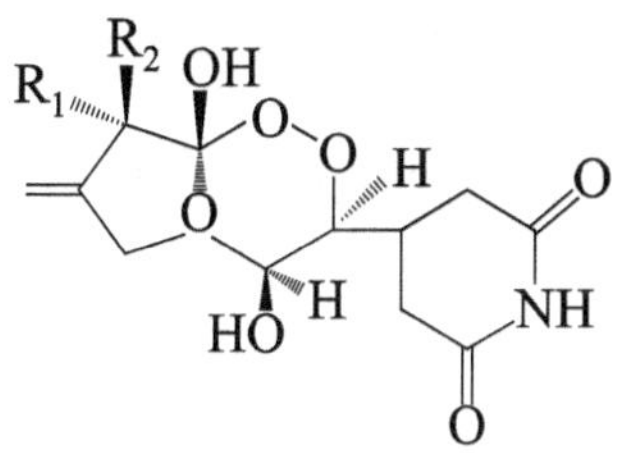

	R_1	R_2
Sesbanimide A	H	Me
Sesbanimide B	Me	H

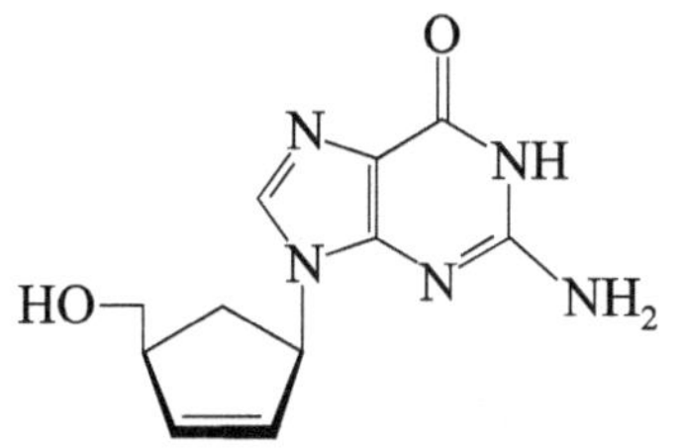

Cabovir

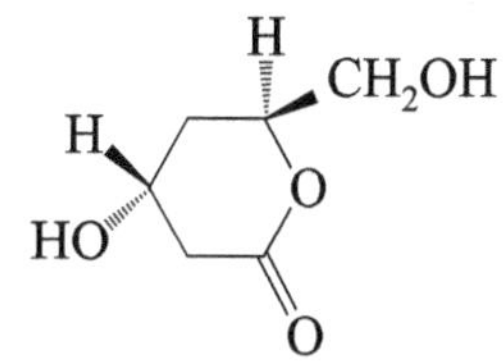

Hydroxylactone moiety

Fig. 2.14: Some bioactive molecules.

Oxo-Diels-Alder reaction can also be carried out with acyclic dienes to give dihydropyran derivatives. (Fig. 2.15)

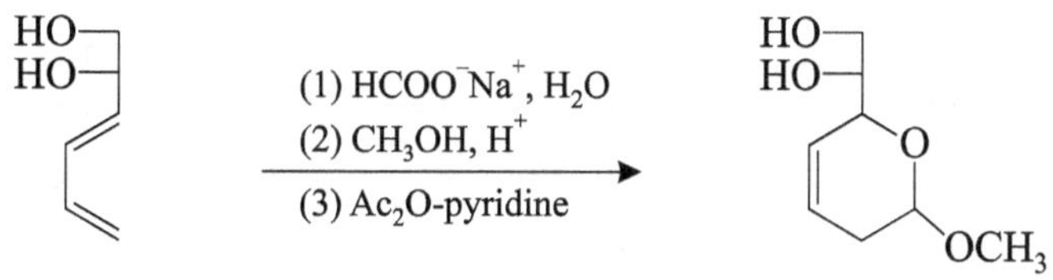

Dihydropyran derive

Fig. 2.15: Synthesis of dihydropyran derivative using Oxo-Diels-Alder reaction.

An application of oxo-Diels-Alder reaction in the synthesis[23] of Sialic acids, 3-deoxy-3-manno-2-octulosonic acid (KDO) (Fig. 2.16)

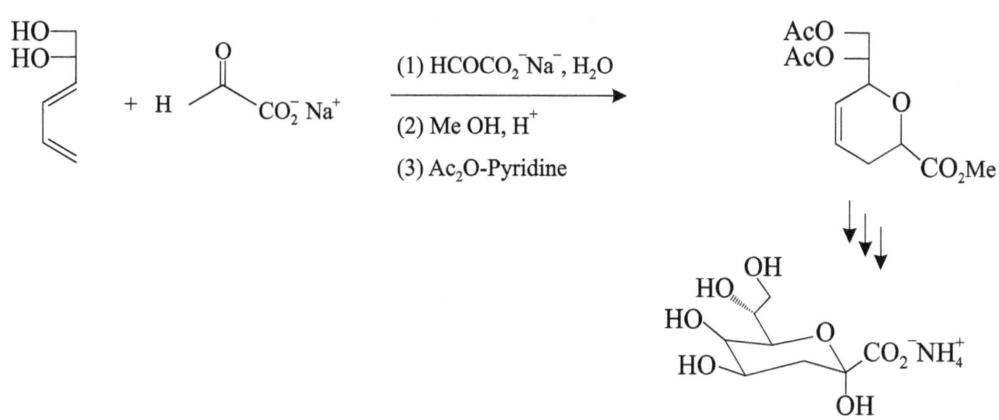

KDO

Fig. 2.16: Synthesis of KDO.

The hetero-Diels-Alder reaction can also be conducted in which both reactive centers are heteroatoms. An intramolecular hetero-Diels-Alder cycloaddition of chiral acylnitroso compounds (generated in situ by periodate oxidation) resulted in trans selectivity in aqueous medium[24] (Fig. 2.17).

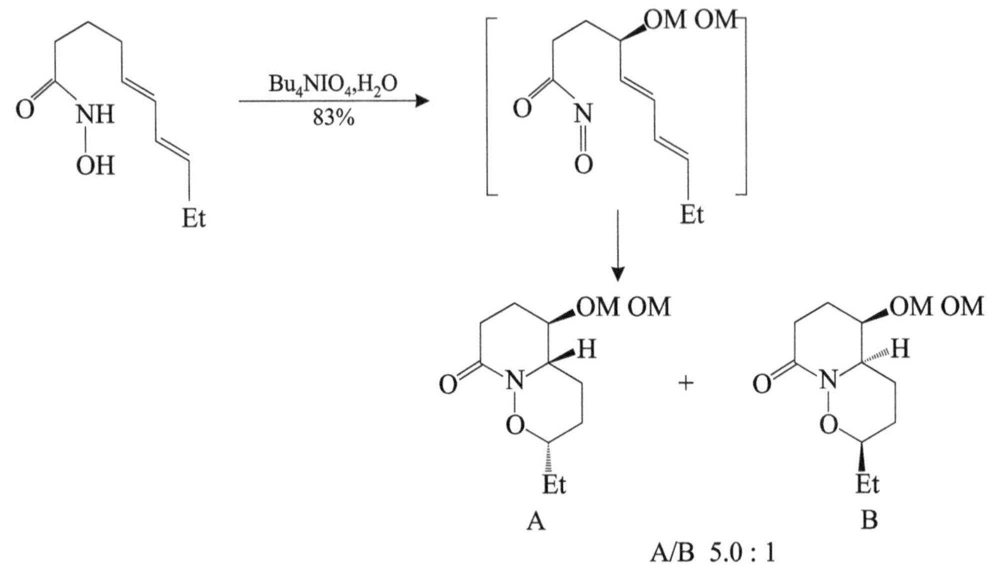

Fig. 2.17: An interesting example of hetero-Diels-Alder reaction.

The reaction has been applied in the total synthesis[25] of (–)-pumiliotoxin C.

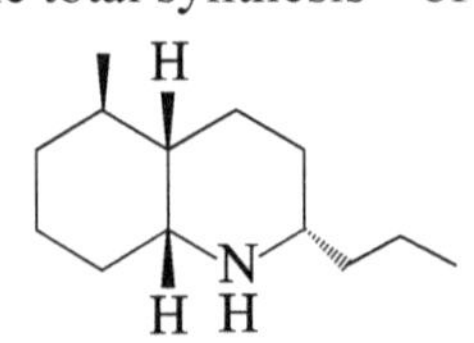

(–) - Pumiliotoxin C

***(b)* Alder-Ene Reaction:** A one pot reactions, it proceeds smoothly in water-THF mixture. It has been used in asymmetric synthesis[26] of pipecolic acid derivatives (Fig. 2.18).

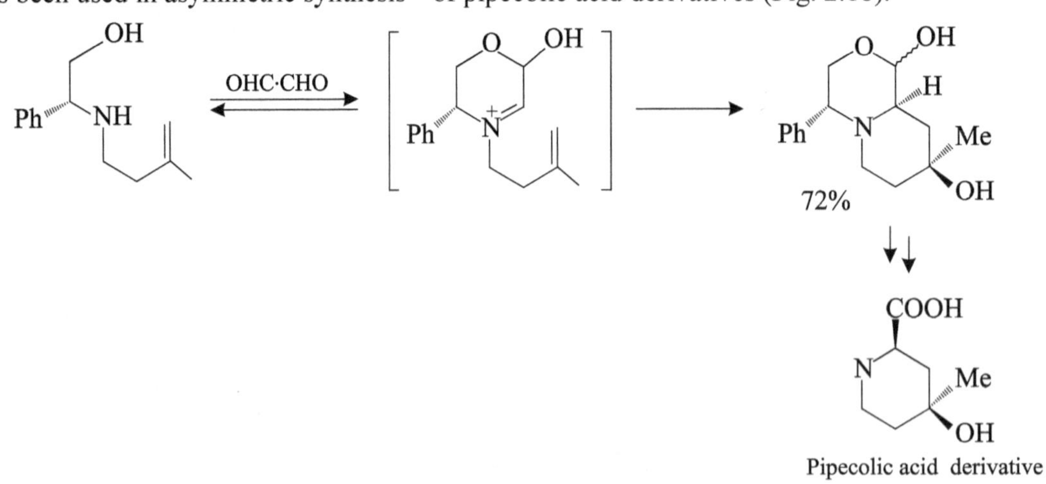

Fig. 2.18: Synthesis of Pipecolic acid Derivatives.

***(c)* 1,3-Dipolar Cycloaddition Reactions:** 1,3-Dipolar cyclization of nitrile oxide with dipolarphiles is used to generate important heterocycles. The reaction is carried out in aqueous-organic biophase system. The nitrile oxide required is generated in situ from oximes or hydrazones. This method is a convenient one-pot procedure for the synthesis of a variety of heterocyclic compounds[27] (Fig. 2.19)

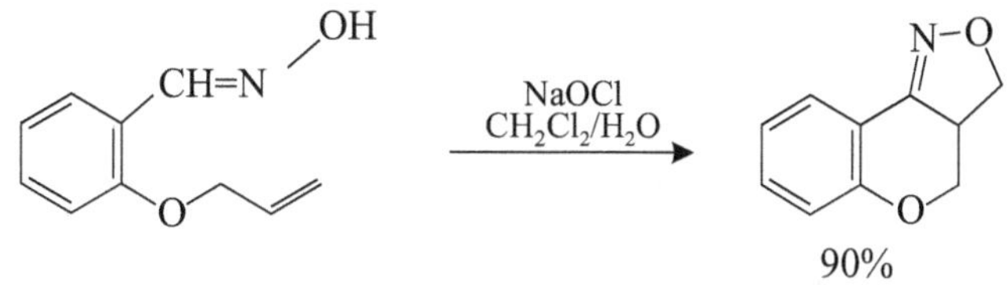

Fig. 2.19: One pot synthesis of heterocycles by 1,3-Dipolar cycloaddition reaction.

1,3-Dipolar cyclization products are also obtained by the reaction of preformed aromatic nitrile N-oxides with disubstituted benzoquinones in aqueous ethanol[28] (Fig. 2.20).

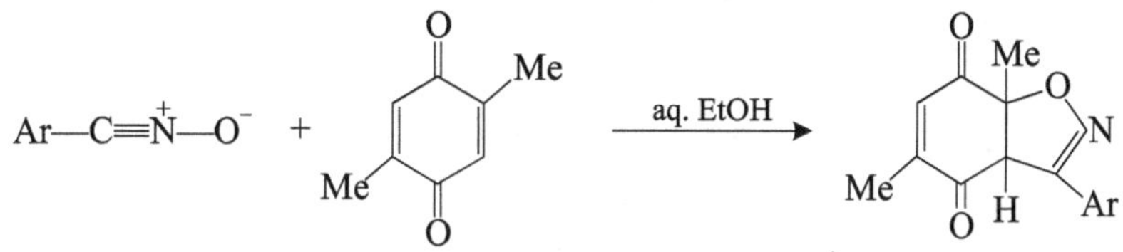

Fig. 2.20: 1,3-Dipolar cycloaddition reaction using aromatic nitrile N-oxides.

The above reaction can be catalysed by Baker's yeast[29] and β-cyclodextrin[30].

An interesting application of 1,3-dipolar cyclization of an azide in water has been used for the synthesis[31] of (+) biotin (Fig. 2.20).

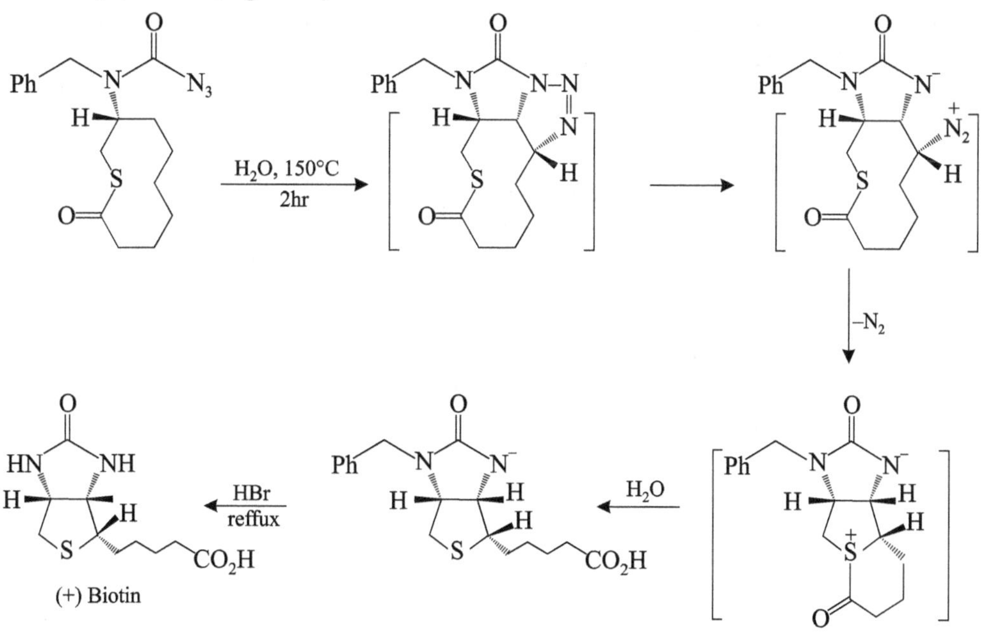

Fig. 2.20(a): Synthesis of (+)-Biotin.

In the above reaction, thermolysis of the azide compound, a mixture of (+)-biotin and its benzylated derivative was directly formed.

***(d)* Sigmatropic Rearrangements:** Claisen rearrangement and cope rearrangement are well known examples of sigmatropic rearrangements.

Claisen rearrangement of chorismic acid was found to be accelerated[32] by the enzyme chorismate (Fig. 2.21).

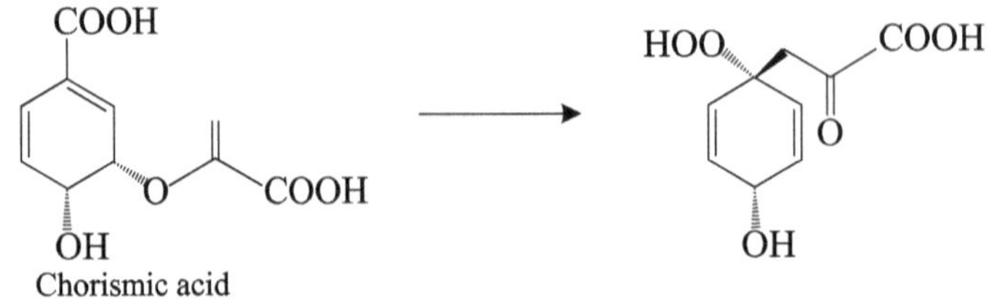

Fig. 2.21: Claisen rearrangement of chorismic acid.

The Claisen rearrangement reactions were found to be considerably accelerated on going from nonpolar to aqueous solvents[33]. In fact, rearrangement of chorismic and related compounds was found to be 100 times faster in water than in methanol[34].

Similarly, [3,3]-sigmatropic rearrangement of an allyl vinyl ether in water gave an aldehyde. The corresponding methyl ester also undervent rearrangement[35] (Fig. 2.22)

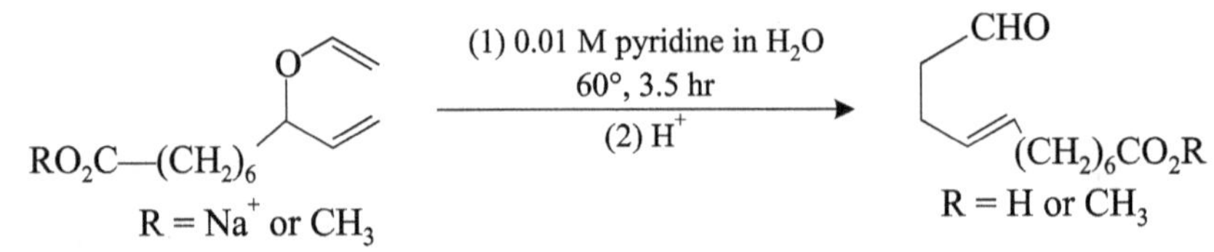

Fig. 2.22: [3,3]-Sigmatropic rearrangement of an allyl vinyl ether or its methyl ester.

The above methodology has considerable potential applications in organic synthesis. Some interesting examples[36] are given in Fig. 2.23.

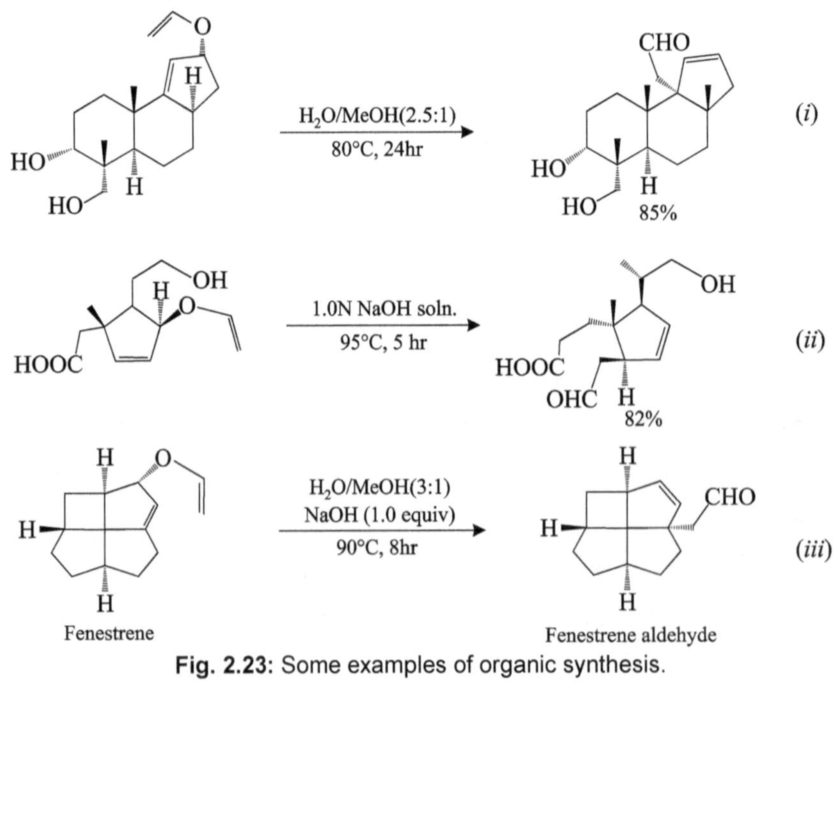

Fig. 2.23: Some examples of organic synthesis.

In the example (*i*), the rearrangement for protected analog had difficulties and resulted in the elimination of aldehyde[37]. In the example (*ii*), the corresponding methyl ester was recovered unchanged on prolonged heating in decalin. Finally in example (*iii*), there is trans ring fusion between the two 5-membered rings and earlier attempts within the carbon framework of fenestrene system were unsuccessful.

Cope rearrangement has also been studied in aqueous medium. An example is the rearrangement of carbaprephenate to carbachorismate, which looses a water molecule to give 3-(2-carboxyallyl)-benzoic acid as the major product[38] (Fig. 2.24)

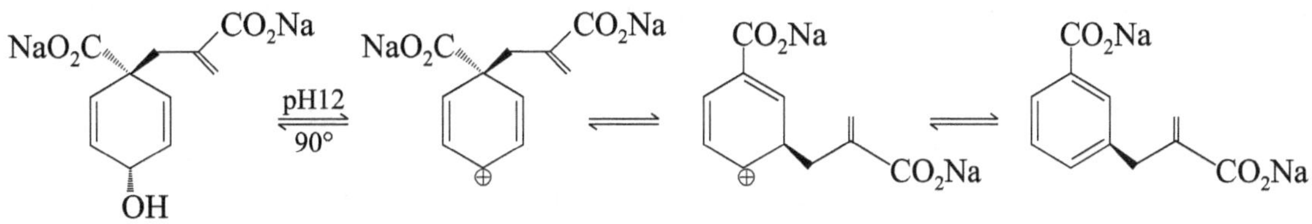

Fig. 2.24: Cope rearrangement in aqueous medium.

An **aza-cope rearrangement** of an eneiminium intermediate in the condensation of a homoallylic alcohol and glyoxal in water gives a new ene-**iminium** compound, which is useful for the synthesis of enantiometrically pure α-amino acids[39].

(*e*) Photochemical Cycloaddition Reactions: Thymine, uracil and their derivatives undergo-photodimerisation[40] in water giving much better yields compared with organic solvents.

Organic compounds having low solubilities in water, such as stilbenes and alkyl cinnamates also photodimerise in water. The same reaction in organic solvents like benzene leads to cis-trans isomerisations[41,42]. Addition of LiCl (increasing hydrophobic effect) increases the yield of dimerization. However, addition to quanidium chloride (decreasing the hydrophobic effect) lowers the yields. The photodimensation of stilbene is shown in Fig. 2.25.

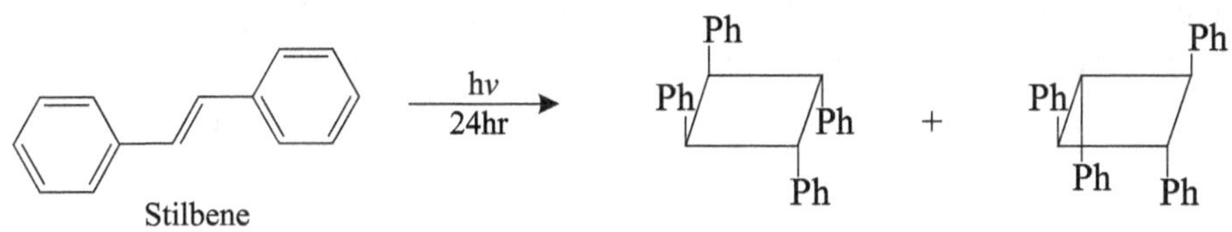

Water	12%	10%
Benzene	0%	0%
Water + LiCl	25%	17%
Water + quinidine chloride	8%	6%

Fig. 2.25: Photodimensation of Stilbene.

Similarly, coumarin photodimensed more efficiently in water than in organic solvents[43] (Fig. 2.26).

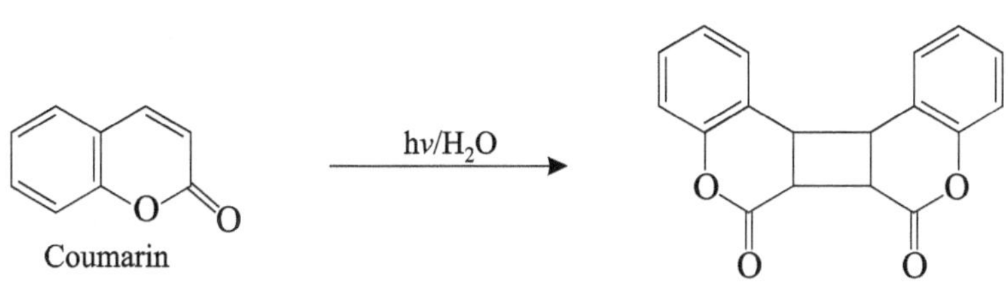

Fig. 2.26: Photodimensation of coumarin in Water.

Photodimensation of anthracene-2-sulfonate ordinarily gives 4 isomers. However, in water only one isomer when β-cyclodixtrin is present.[44]

2.2.1.2 Nucleophilic Addition Reactions

The nucleophilic addition reactions are important reactions for carbon-carbon bond formation. Some of these reactions include aldol condensation, Michael addition, Wittig reaction, Darzen reaction, benzoin condensation and some other miscellaneous reactions. Though these reactions are normally performed in organic solvents, but recently it has been shown that these can be performed in aqueous medium.

***(a)* Aldol Reaction:** It is an useful carbon-carbon bond forming reaction. It involves reaction of two molecules of an aldehyde or ketone in presence of base to form β-hydroxy carbonyl compound, called aldol. Thus, acetaldehyde gives 3-hydroxybutanal (Fig. 2.27).

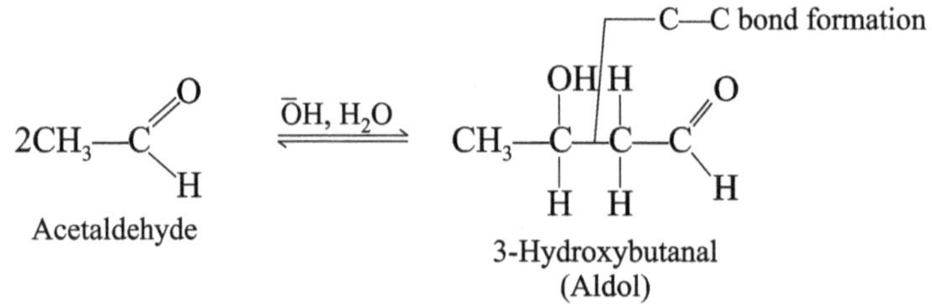

Fig. 2.27: Aldol reaction of Acetaldehyde.

The in above reaction (Fig. 2.27), the presence of surfactants give mainly the dehydration product (*viz.* acryl aldehyde, $CH_3CH{=}CH{-}CHO$). Commonly, aldol reaction is best performed in ethanol, aqueous ethanol and water.[45] For reactive aliphatic aldehydes, aqueous solution of sodium hydroxide-ether two phase system has been found to be the best.[46]

The mechanism of aldol condensation of acetaldehyde is as given in Fig. 2.28.

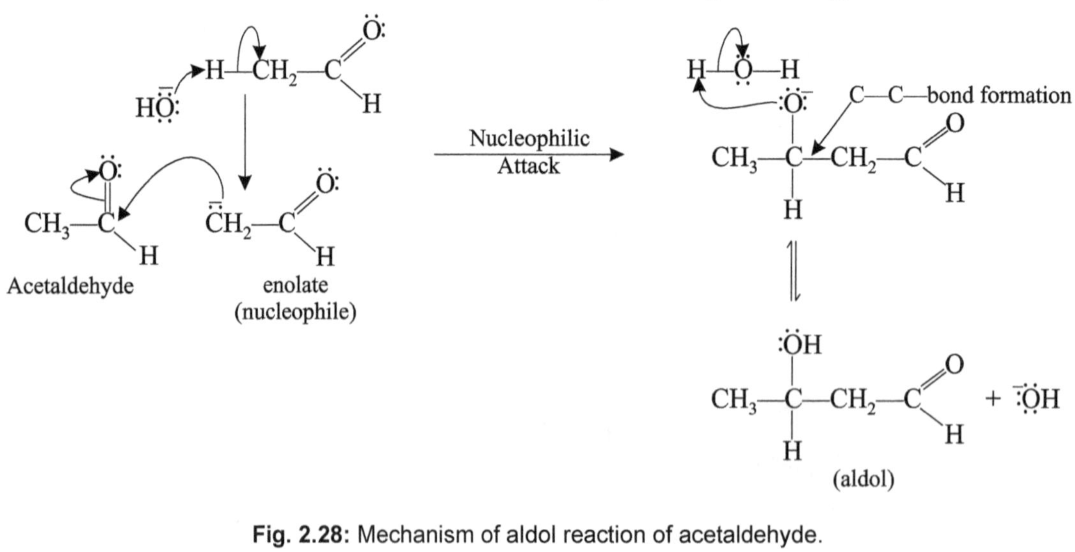

Fig. 2.28: Mechanism of aldol reaction of acetaldehyde.

In aldol reaction, the α-carbon of one carbonyl compound gets bonded to the carbonyl carbon of the other compound.

Lower aliphatic aldehydes containing α-hydrogen undergo aldol reaction with aqueous sodium hydroxide, even in absence of a phase transfer catalyst. However, as the chain length increases, the reaction becomes slower and PTC becomes more effective. It has been reported[46a] that with solid KOH and 3-benzyl-4-methylthiazolium chloride as catalyst, acetoin is obtained from acetaldehyde.

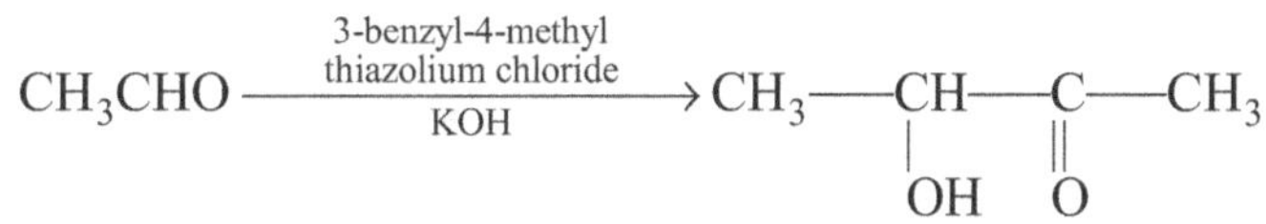

Fig. 2.28(*a*): Catalytic Aldol Condensation.

Acyloin condensation with both aliphatic or aromatic aldehydes are effected easily using N-lauryl thiazolium bromide as catalyst in aqueous phosphate solution[46b].

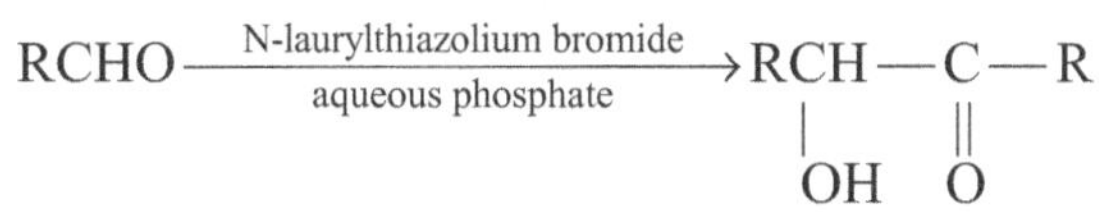

Fig. 2.28(*b*): Catalytic Aldol Condensation.

Aromatic ketones like acetophenone on aldol reaction give alkenes; in this case the intermediate is not isolated, elimination of water is spontaneous (Fig. 2.29).

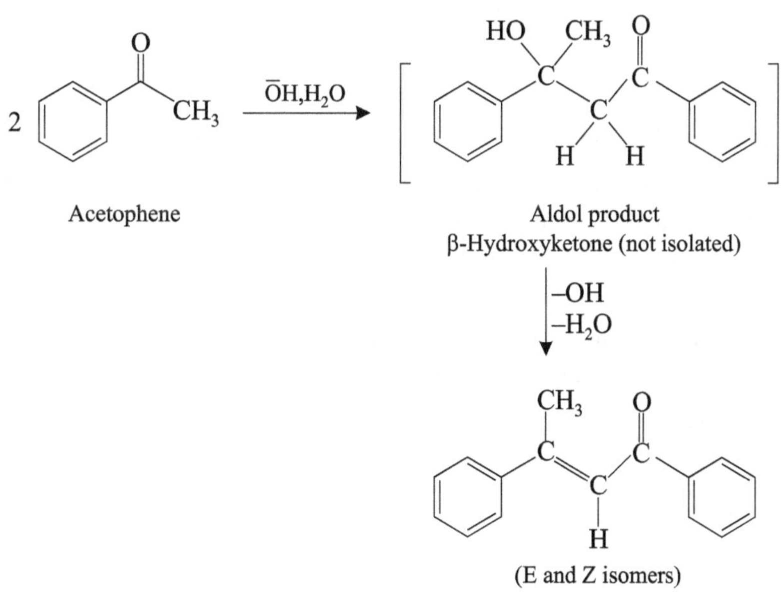

Fig. 2.29: Aldol reaction of Acetophenone.

The mechanism of dehydration of β-hydroxy carbonyl compounds (aldols) take place by E1cB mechanism (Fig. 2.30).

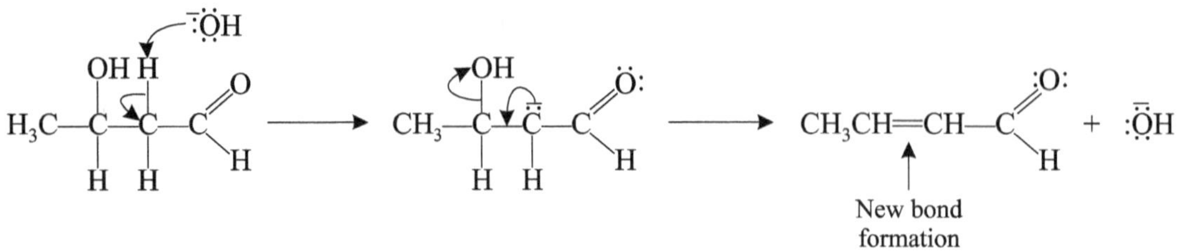

Fig. 2.30: Mechanism of dehydration of β-hydroxy carbonyl compounds (E1cB mechanism)

Problem:

How will you synthesise (*a*)

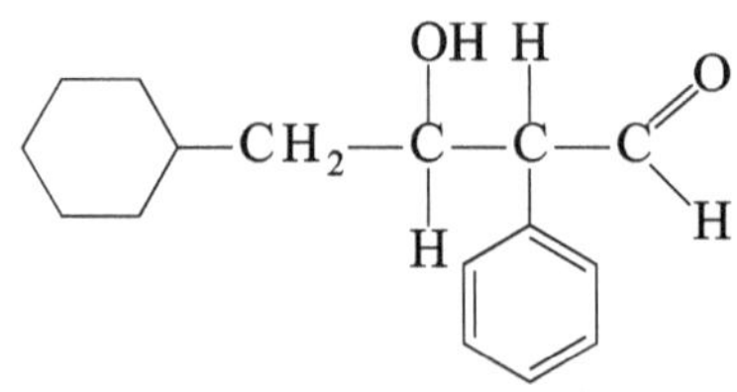

(*b*)

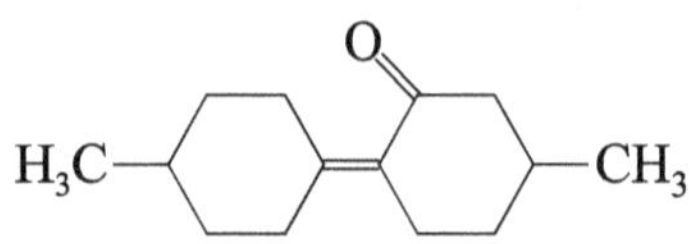

Ans: By Aldol reaction of

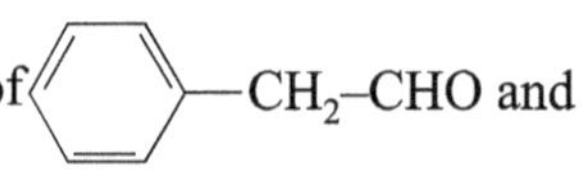

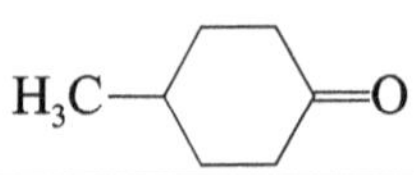
respectively.

Aldol reaction can also be performed in solid state (see chapter 8)

Crossed Aldol Reaction: It is an aldol reaction between two different carbonyl compounds. Aldol reaction between two different aldehyde gives four products and has no synthetic use. Such reactions to give only one product can be performed by directed aldol condensation (See Directed Aldol Reaction).

However, where are carbonyl compound has no α-hydrogen, crossed aldol reaction gives only one product. Some examples of such reactions are given in Fig. 2.31.

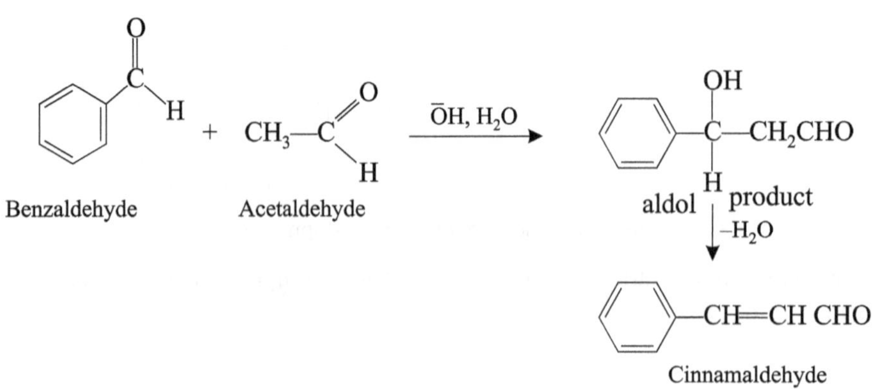

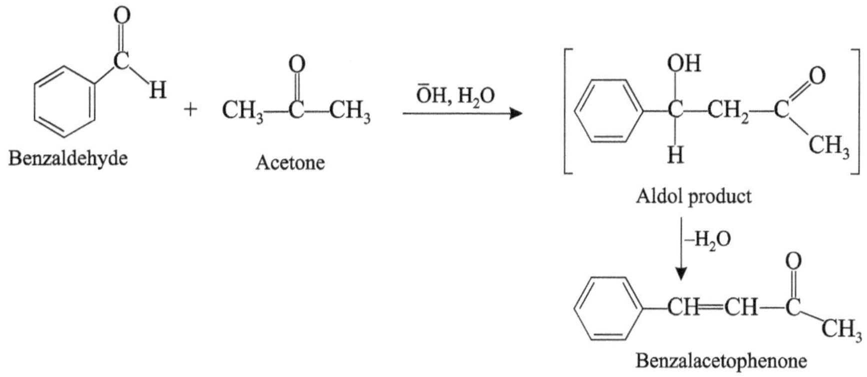

Fig. 2.31: Examples of crossed Aldol Reaction.

Crossed Aldol reaction can also be performed between on aldehyde or a ketone and compounds containing reactive methalene groups (Fig. 2.32).

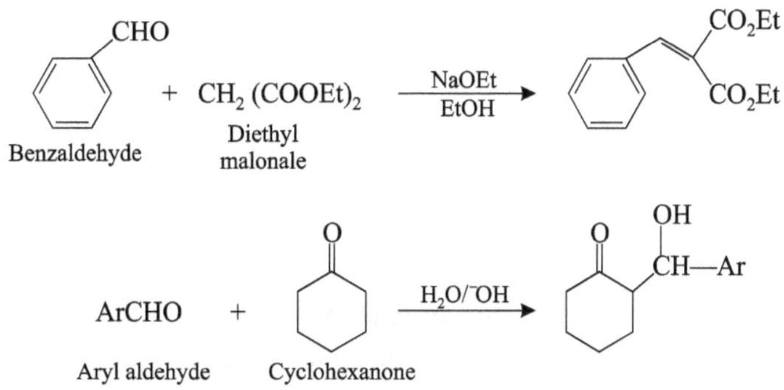

Fig. 2.32: Crossed aldol reaction between aldehyde ketone and compounds containing reactive methyl compounds.

Directed Aldol Reaction: The aldol reaction of two carbonyl compounds can be performed by making enolate of one carbonyl compound with LDA and then adding the second carbonyl compound (the electrophile) to the prepared enolate. Such a reaction gives only one product and is called directed aldol reaction. Thus, the reaction of 2-methyl cyclohexanone and acetaldehyde can be represented as shown in Fig. 2.33.

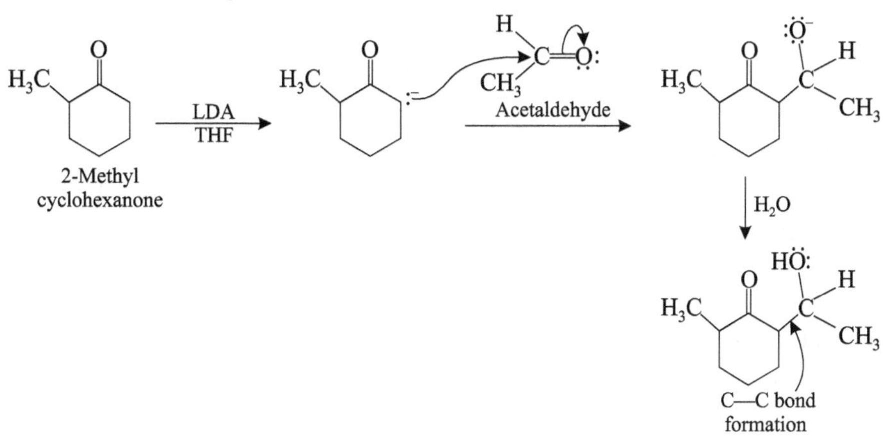

Fig. 2.33: Directed Aldol Reaction.

Intramolecular Aldol Reaction: Five and six membered rings can be made by intramolecular aldol reactions[47] with dicarbonyl compounds. In dicarbonyl compounds one carbonyl group is the nucleophile and the carbonyl carbon of the other carbonyl group is the electrophile. Thus, 2,5-hexane dione (a 1,4-dicarbonyl compound) and 2, 6-heptanedione (*a* 1, 5 dicarbonyl compound) on aldol condensation give the corresponding 5 and 6-membered ring systems (Fig. 2.34).

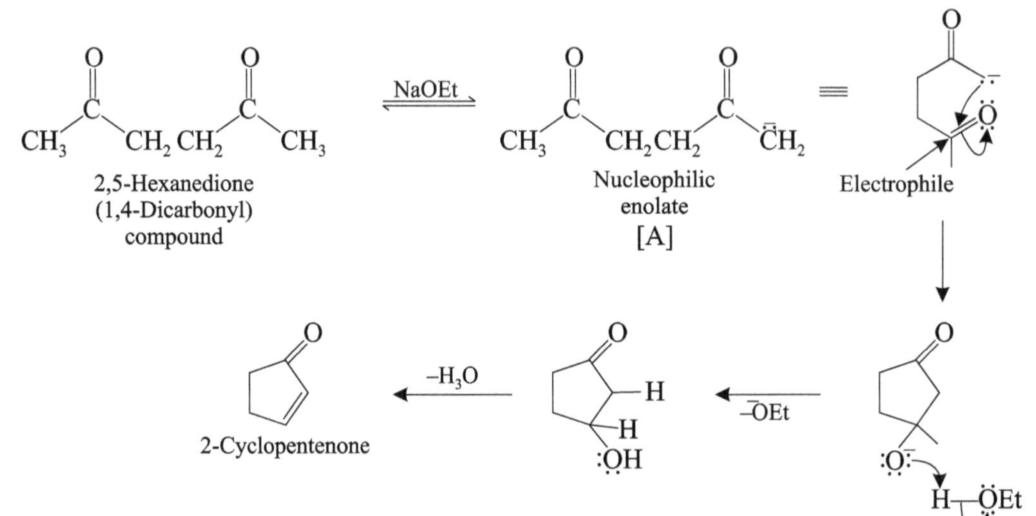

In the above reaction the nucleophilic enolate [A] is formed, though here is possibility of the formation of alternate nucleophilic enolate [B]. This enolate [B] is expected to give three membered ring which is not formed.

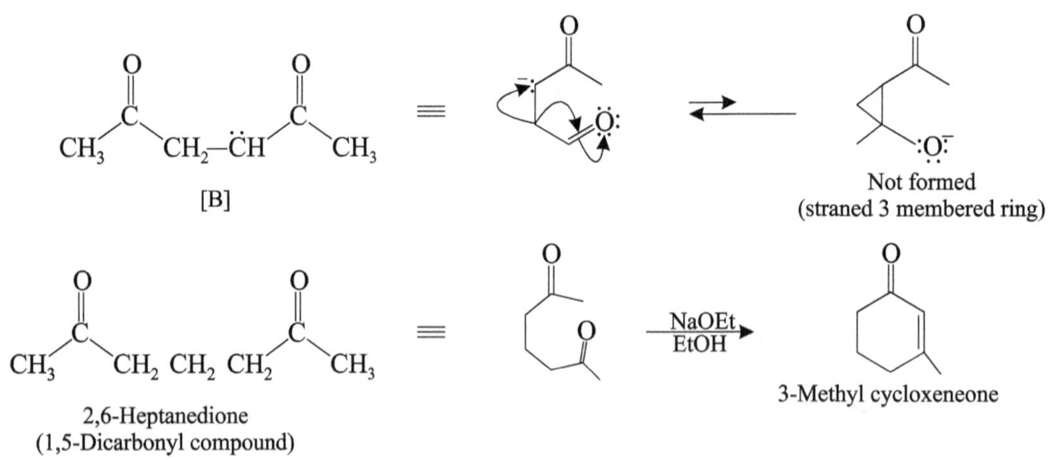

Fig. 2.34: Intramolecular Aldol Reaction.

An interesting example of intramolecular aldol reaction is the synthesis of female sex hormone progesterone (Fig. 2.35).

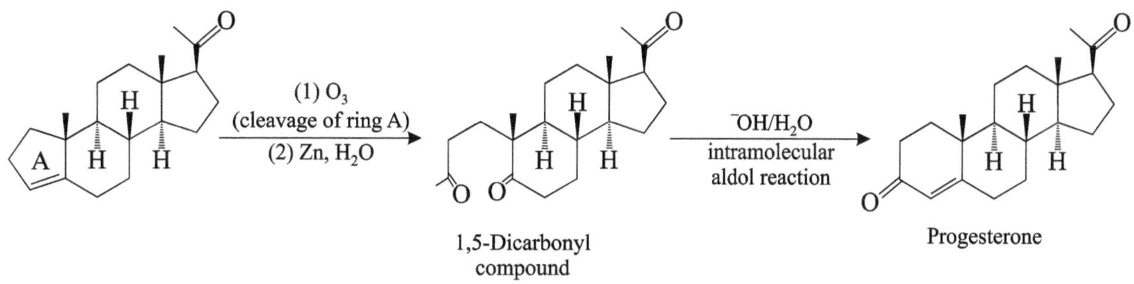

Fig. 2.35: Synthesis of female sex hormone progesterone.

Enzymatic Aldol Reaction: The enzyme aldolases are important catalysts for catalysing carbon-carbon bond formation in nature[48]. A multienzyme system has been used for C–C bond formation.[49]

Mukaiyama Aldol Reaction: It involves cross aldol reaction of silyl enol ethers with carbonyl compounds. The reaction was carried out in aqueous solvents without any acid catalyst[50]. The reaction took several days for completion. The addition of a catalytic amount of lanthanide triflate (*e.g.* ytterbium Triflate) (a stronger lewis acid) improved the yield and rate of the reactions[51]. (Fig. 2.36)

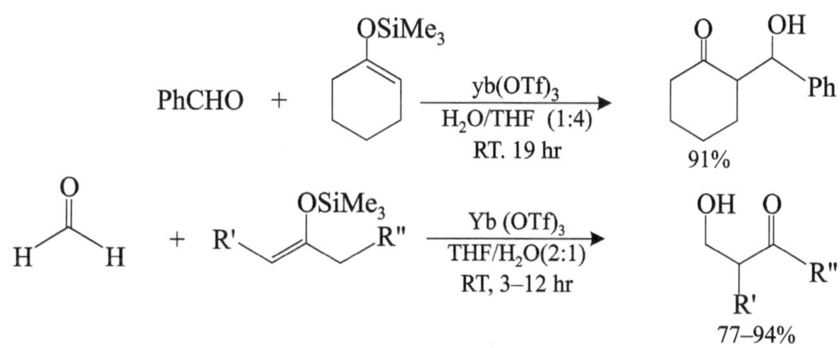

Fig. 2.36: Mukaiyama Aldol Reaction.

Important Synthetic Applications of Aldol Reaction

(*i*) A perfume ingredient with a Jasmine like odour is Flosal. It is synthesised by cross aldol reaction of benzaldehyde with heptanal. (Fig. 2.37)

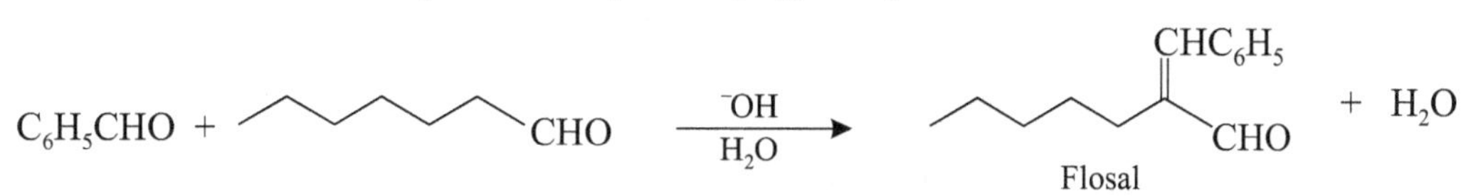

Fig. 2.37: Synthesis of Flosal.

(*ii*) Nabumetone, a pain reliever and anti-inflammatory agent sold under the brand name of Relafen is synthesised by crossed aldol reaction as shown below (Fig. 2.38):

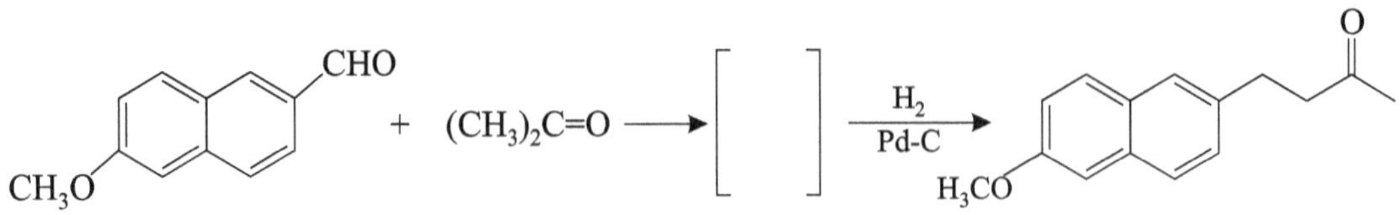

Fig. 2.38: Synthesis of Nabumetone

(*iii*) Periplanone B, the sex pheromone of the female American Cockroach is prepared by directed aldol reaction as shown below. (Fig. 2.38*a*)

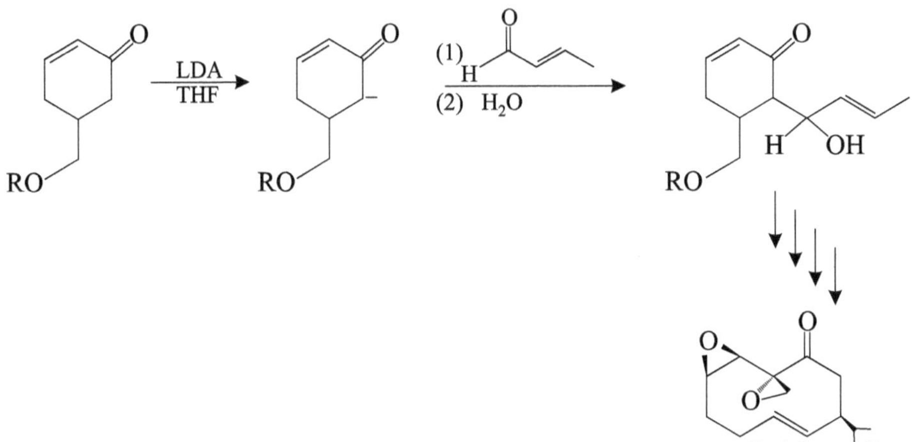

Fig. 2.38a: Synthesis of Periplanone B.

(*iv*) Synthesis of Progesterone (female sex hormone) using intramolecular aldol reaction. (See Fig. 2.35)

(*b*) **Michael Addition:** Like aldol reaction, Michael addition is also a carbon-carbon bond forming reaction involving reaction between a α, β-unsaturated carbonyl compound and a compound with an active methylene group in presence of a base like sodium ethoxide, piperidine etc. (Fig. 2.39)

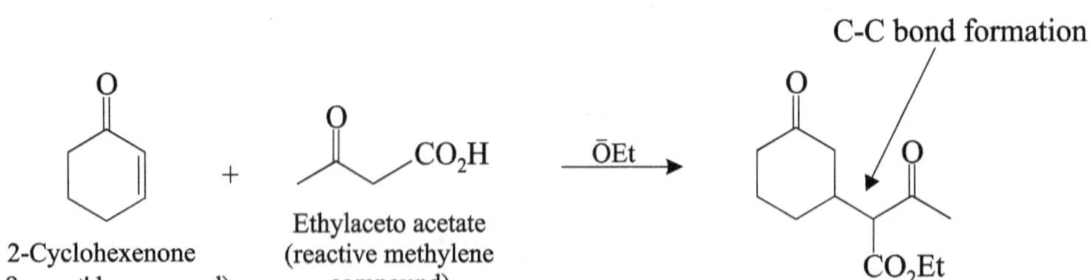

Fig. 2.39: Michael addition.

The reaction is believed to proceed as follows (Fig. 2.40).

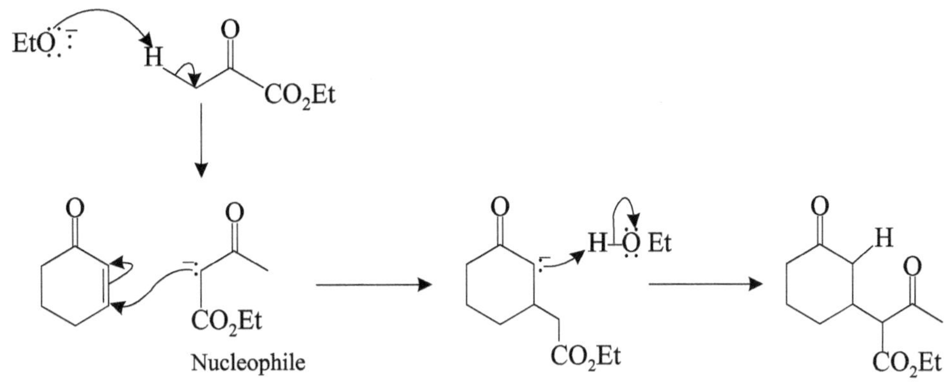

Fig. 2.40: Mechanism of Michael addition

In Michael addition, the nucleophile adds on to the β carbon of α, β-unsaturated ketone.

The first successful Michael addition in water was reported[52-54] in 1970s. Thus, the Michael addition of 2-methylcyclopentane 1,3-dione to vinyl ketone in water gives the corresponding conjugated addition product without using any catalyst. The adduct further cyclises to give a 5,6-fused-ring system (Fig. 2.41).

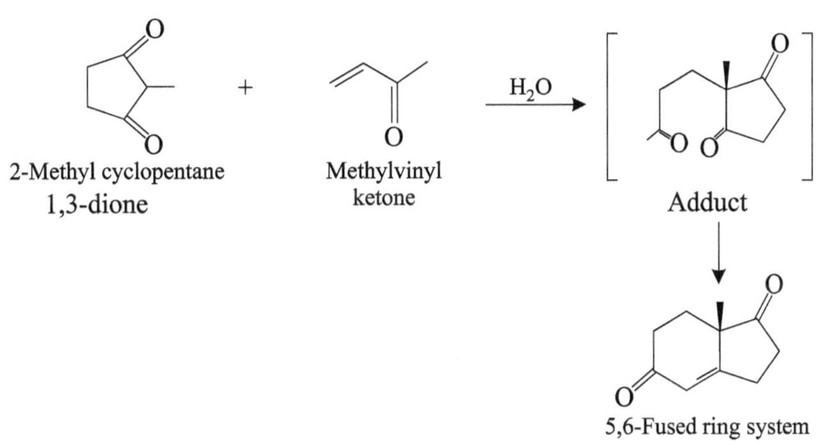

Fig. 2.41: Michael addition in water

As seen, the adduct is a 1,5-dicarbonyl compound and on cyclisation gives a six membered ring by intramolecular aldol reaction (see Fig. 2.34).

The use of water as a solvent in Michael addition gives a pure adduct in much better yield (87.6%) compared to the reaction in methanol/KOH (54% yield).

Similarly, the Michael addition of 2-methyl cyclohexane-1,3-dione to methyl vinyl ketone gave the adduct[55], which finally gave optically pure Wieland-Miescher Ketone (Scheme 2.42)

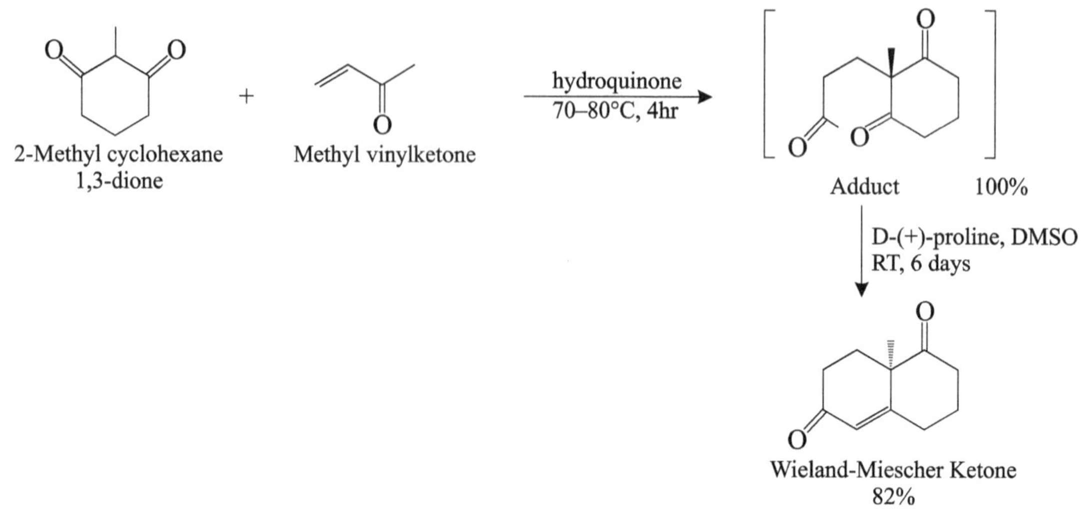

Fig. 2.42: Michael addition of 2-methyl cyclohexane 1,3-dione with methyl vinyl ketone in water.

In place of methyl vinyl ketone used above (Fig. 2.42), acrolein could also be used.[56] The adduct so obtained was used for the total synthesis of 13-α-methyl-14α-hydroxy steroid (Fig. 2.43)

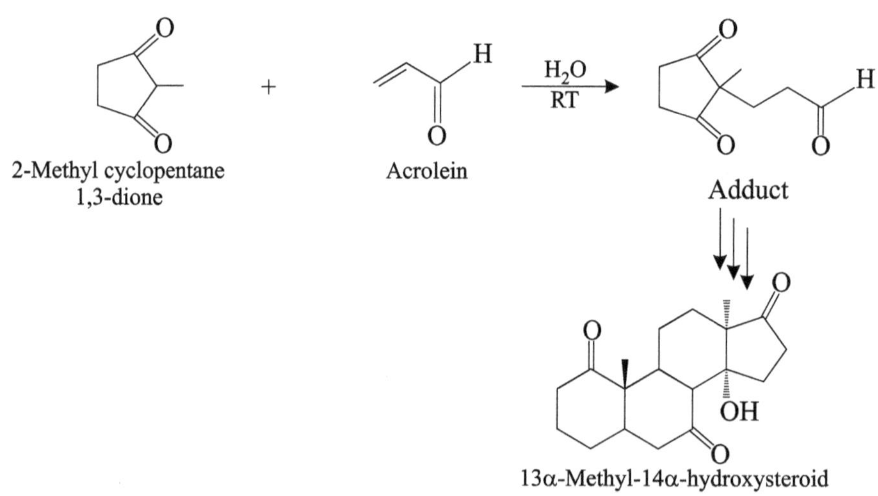

Fig. 2.43: Michael reaction of 2-methylcyclopentane-1,3-dione with acrolein in water.

The rate of the Michael reaction in water could be enhanced by the addition of ytterbium triflate, [yb(OTf)$_3$].[57]

The Michael addition of nitroalkanes to buten-2-one in water gave the adducts in much better yields[58] than in organic solvents (Fig. 2.44). Addition of glucose and saccharose increased further the rate of the reaction.

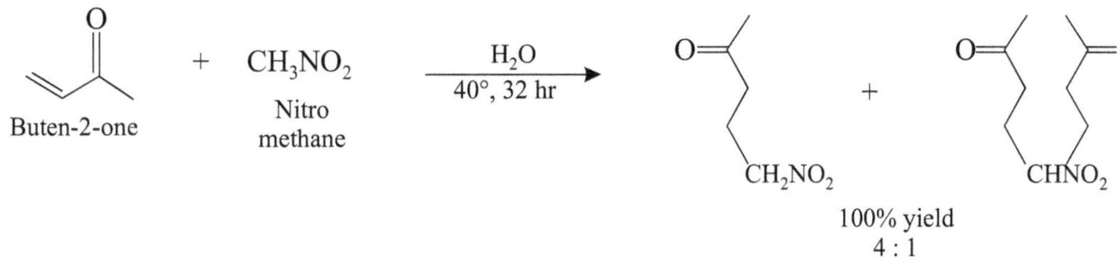

Fig. 2.44: Michael addition of nitromethane to buten-2-one.

The Michael addition reaction between ascorbic acid and acrylic enones in water in presence of inorganic acid, rather than basic catalyst gave the adduct[59]. The Michael addition reaction of benzylidene acetophenone and benzylidene acetone with ethyl aceto acetate and acetylacetone in water in presence of a PTC (trimethyl ammonium bromide) containing aqueous micellar medium gave the adducts.[60] An highly efficient Lewis-acid catalysed, asymmetric Michael addition in water has also been reported.[61]

Synthetic Application: The steroid, a female sex hormone has been prepared from an intermediate obtained by Michael addition as shown below:

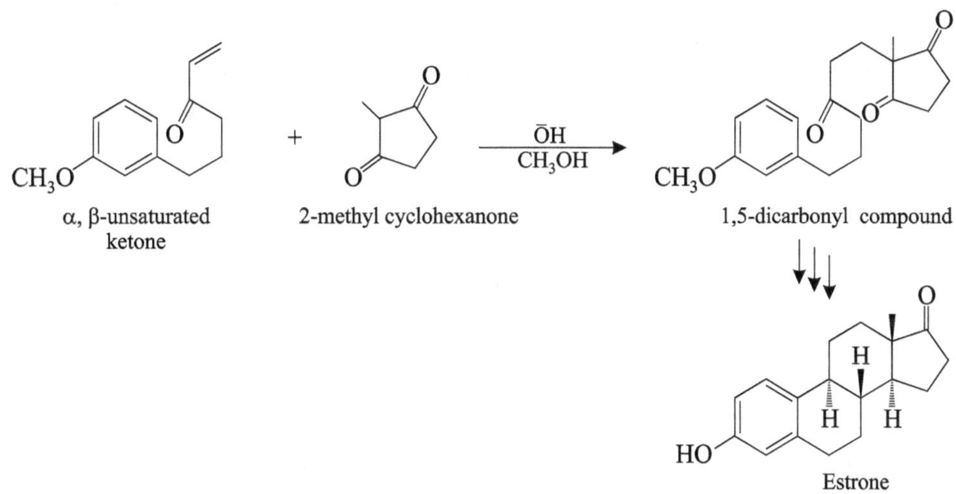

***(c)* Wittig Reaction:** The well known Wittig reaction[62] is useful for the preparation of olefins from alkylidene phospharanes (ylids) and carbonyl compounds (Fig. 2.45).

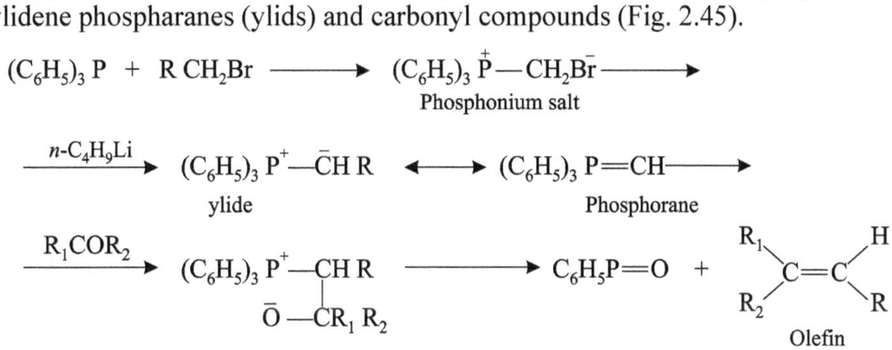

Fig. 2.45: Wittig Reaction

A modification of the above reaction is **Wittig-Horner reaction** or **Wittig-Wadsworth-Emmons reaction** uses stabilized ylides. Thus, the reaction of ethyl bromoacetate with triphenylphosphite gives the phosphate ester, which on reaction with cyclohexanone in presence of base (NaH) gives α, β-unsaturated ester ethyl cyclohexylidene carboxylate in 70% yield. [Fig. 2.46]

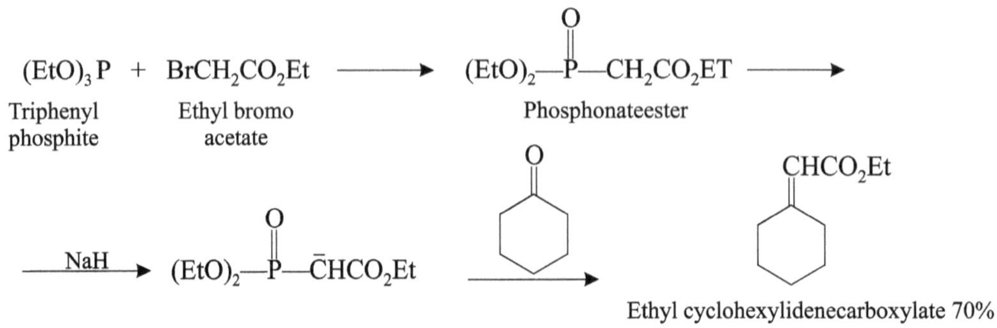

Fig. 2.46: Wittig-Horner reaction.

The above reaction can at times performed in organic/water biphase system[63]. Sometimes, a phase transfer catalyst can also be used. Recently, the use of water alone as solvent has been investigated[64]. It has been found that the reaction proceeds very well with a much weaker base like K_2CO_3 or $KHCO_3$. No PTC is required. Even compounds having acid and base-sensitive functional group can also be used directly. As an example, β-dimethyl hydrazone acetaldehyde can be olefinated easily[65] (Fig. 2.47).

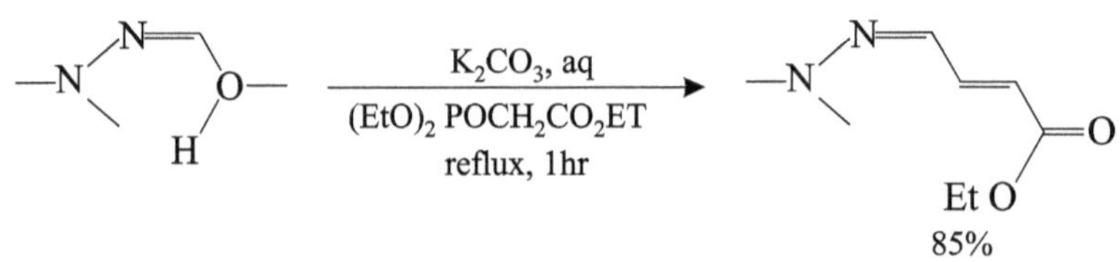

Fig. 2.47: Wittig-Horner reaction in H_2O/K_2CO_3.

A number of water soluble phosphonium salts have been synthesised and their Wittig reactions studied in water[66] (Fig. 2.48)

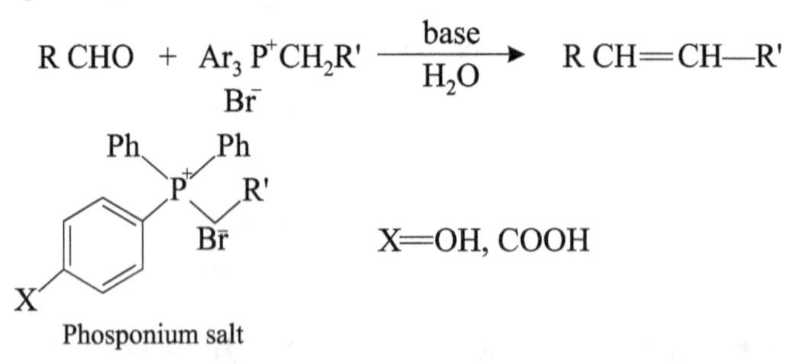

Fig. 2.48: Wittig reaction in water using water soluble phosphonium salts.

1. Synthetic applications of Wittig Reaction: Since in Wittig reaction, two carbon-carbon bonds are formed (in a single reaction), it has been used to synthesise many natural products. A typical example is synthesis of β-carotene, (the orange pigment found in carrots), a Vitamin A precursor as shown below:

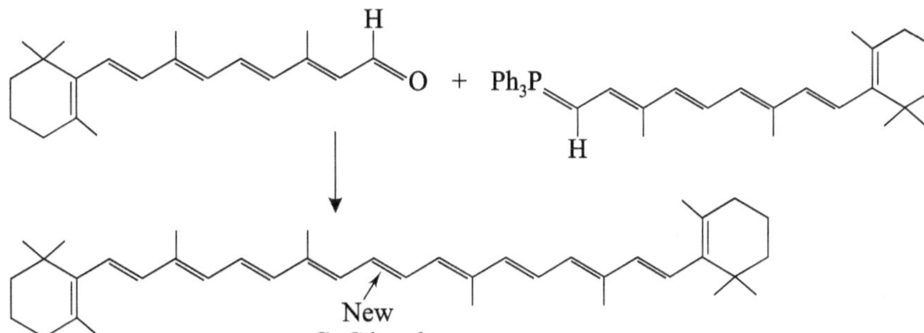

β-carotene

2. Synthesis of Trans Stilbene: It is prepared[67] by the reaction of benzyl chloride with triphenyl phosphene followed by reaction of the formed Wittig reagent with benzaldehyde in aqueous sodium hydroxide in presence of a PTC.

$$C_6H_5CH_2Cl + (C_6H_5)_3\,P \xrightarrow{^{-}OEt} (C_6H_5)_2\,P^{+} - CH_2C_6H_5Cl^{-}$$

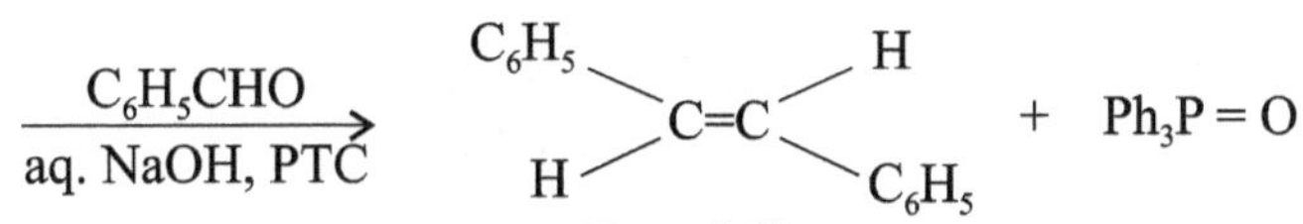

Trans Stilbene

(*d*) Darzen Reaction: Also called glycidic ester condensation it consists in the reaction of aldehydes or ketones with α-haloesters in presence of base to give α, β-epoxyesters (glycidic esters)[68] (Fig. 2.49)

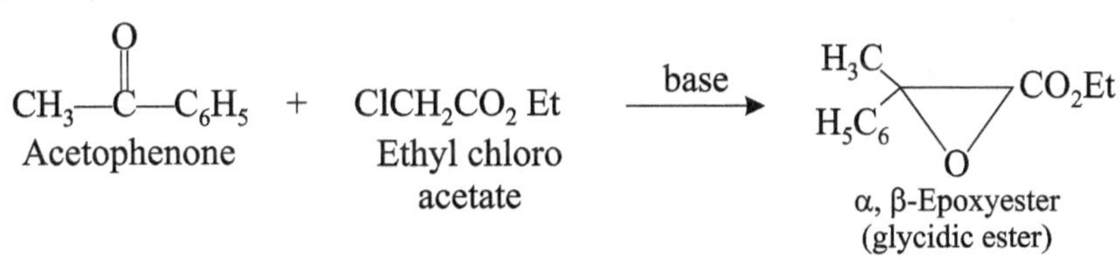

Fig. 2.49: Darzen Reaction.

In order to avoid S_{N2} displacement of the chloride, a hindered base like potassium tertiary butoxide is used in Darzen reaction. The mechanism of the reaction is given in Fig. 2.50.

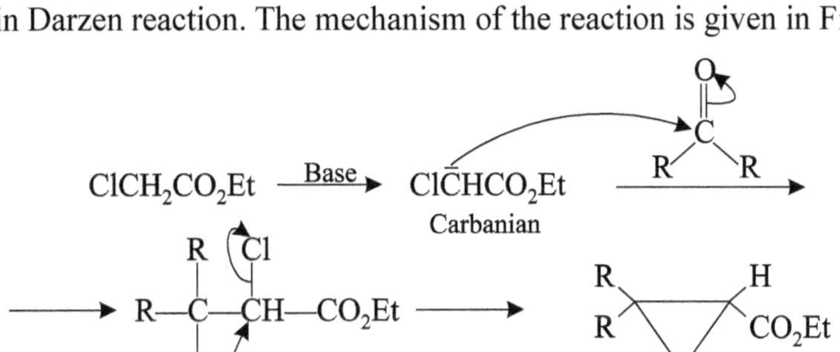

Fig. 2.50: Mechanism of Darzen reaction.

The use of phase transfer catalyst has made Darzen reaction simple and the yields are good[69] (Fig. 2.51)

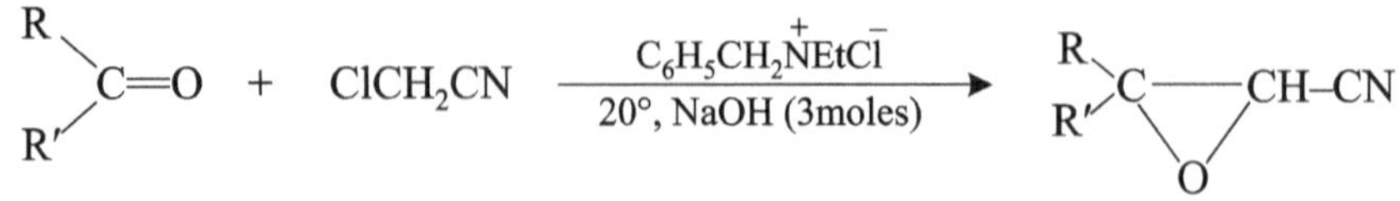

Fig. 2.51: PTC Catalysed Darzen reaction.

With unsymmetrical ketones and aldehydes, both possible stereoisomers are formed. With more acidic ketones (like phenyl acetone, the ketone carbanion is formed leading to alkylation of ketone [Fig. 2.51(*a*)].

$$C_6H_5CH_2COCH_3 + ClCH_2CN + NaOH \xrightarrow{Q^+X^-} \underset{\displaystyle CH_2CN}{C_6H_5\overset{}{C}HCOCH_3}$$

Fig. 2.51(*a*): Darzen reaction with acidic ketones.

In place of PTC, crown ethers can also be used[70] (Fig. 2.52).

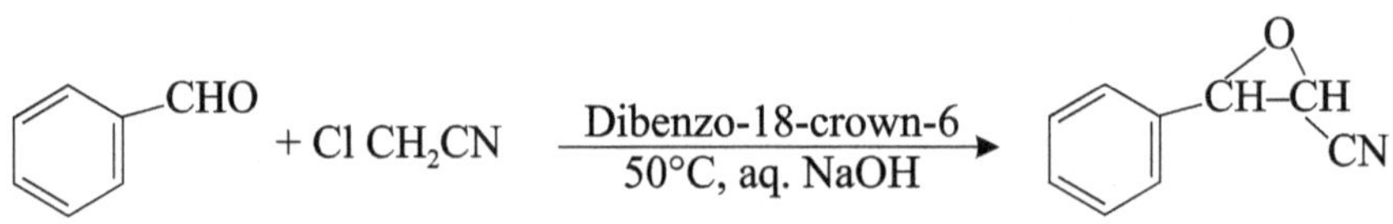

Fig. 2.52: Darzen reaction in presence of crown ether.

Intramolecular Darzen Reaction: The reaction of α-thiolcarbonyl compounds with 2-chloroacrylonitrile gives 2-cyano-2,3-epoxy-tetrahydrothiophenes via the addition of thiol followed by an intramolecular Darzen Reaction[71] (Fig. 2.53).

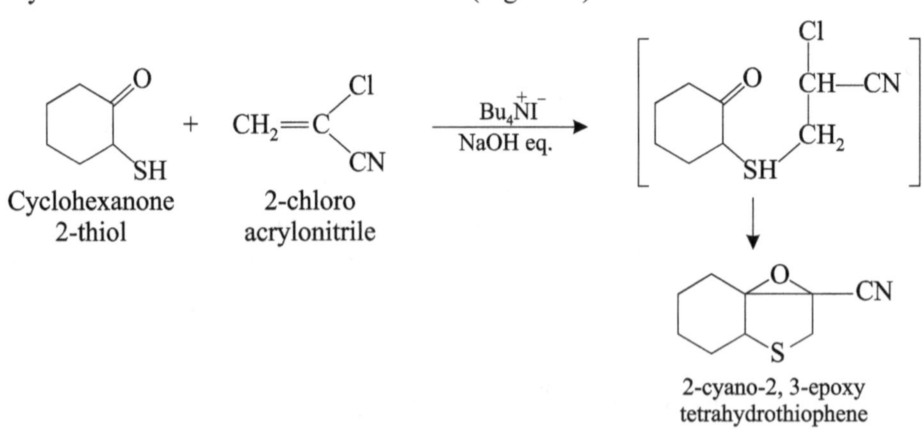

Fig. 2.53: Intramolecular Darzen Reaction.

Synthetic Applications of Darzen Reaction

(*i*) The glicidic ester obtained in the Darzen reaction are useful intermediates for the synthesis of aldehydes or ketones. An example is given below:

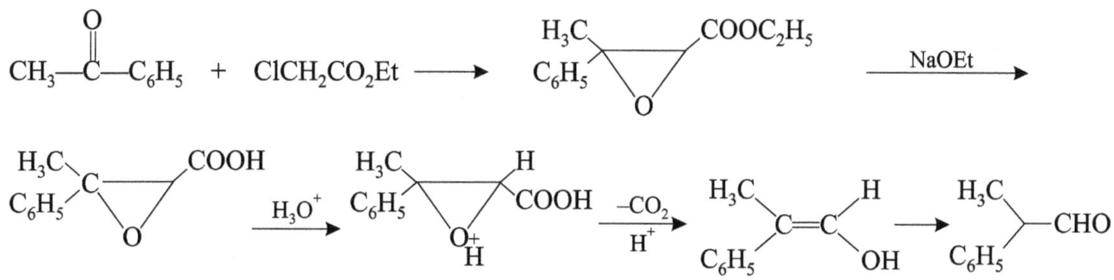

As seen the overall process involves the addition of one more carbon atom, as CHO to a carbonyl group of the starting ketone.

(*ii*) Synthesis of 1-oxaspiro-[2,5]-octane-2-carbonitrile[69].

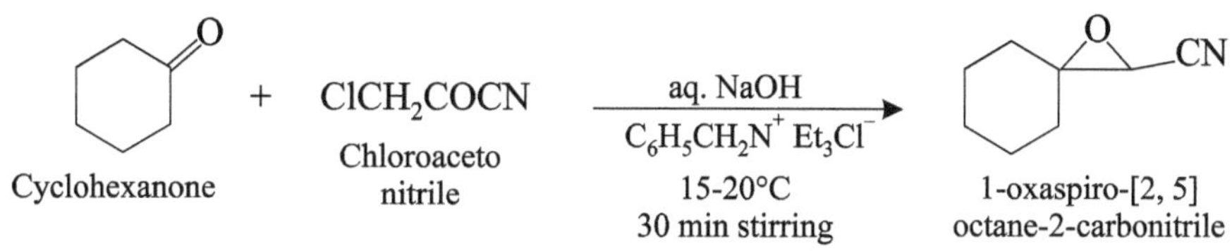

(*e*) Benzoin Condensation: It involves self condensation[72] of aromatic aldehydes (having no α-hydrogen) in presence of cyanide ion to give benzoin (Fig. 2.54).

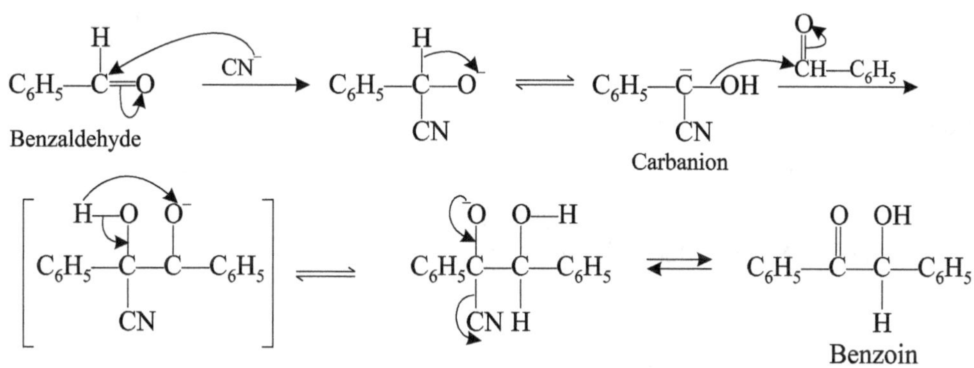

Fig. 2.54: Benzoin Condensation

Catalytic Benzoin Condensation: Aromatic aldehydes undergo benzoin condensation with aqueous sodium cyanide in presence of quaternary ammonium salts[73]. (Fig. 2.55)

$$\mathrm{ArCHO} \xrightarrow[\text{NaCN}]{Q^+X^-} \mathrm{Ar-\underset{\underset{OH}{|}}{CH}-\underset{\underset{O}{||}}{C}-Ar}$$

Fig. 2.55: Catalytic benzoin condensation.

Similarly aliphatic or aromatic aldehydes undergo acyloin condensation using N-laurylthiazolium bromide with aqueous phosphate solution[74] (Fig. 2.56)

$$\mathrm{RCHO} \xrightarrow[\text{aqueous phosphate}]{\text{N-Laurylthiazolium bromide}} \mathrm{R-\underset{\underset{OH}{|}}{CH}-\underset{\underset{O}{||}}{C}-R}$$

16 – 95%

Fig. 2.56: Acyloin Condensation.

In the above reaction (Fig. 2.56), aromatic aldehydes reacted much faster (few minutes) compared to aliphatic aldehydes (5-10 hr.)

Benzoin condensation in aqueous media using inorganic salts (*e.g.*, LiCl) is about 200 times faster than in ethanol without any salt[75]. The condensation is also accelerated by the addition of γ-cyclodextrin whereas addition of β-cyclodextrin inhibited the condensation.

In biological systems, benzoin condensation is catalysed by thiamine[76] (Fig. 2.57)

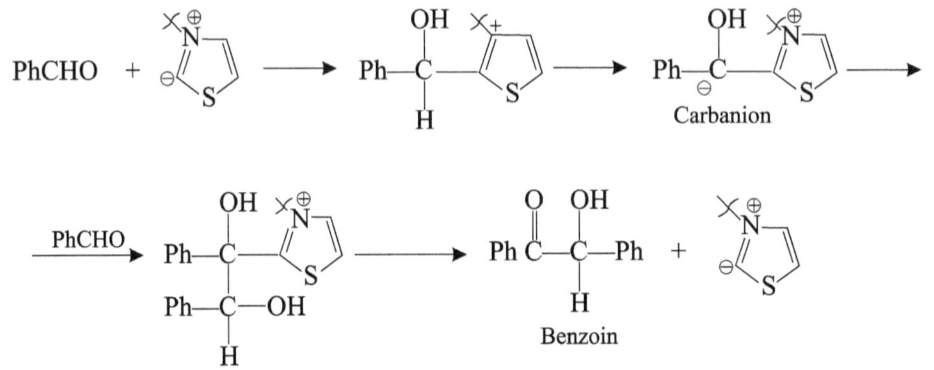

Fig. 2.57: Benzoin Condensation Catalysed by Thiamine

As seen, in the above condensation, a carbanion is generated (Stablized by thiazolium group) but in benzoin condensation catalysed by cyanide, the condensation proceeds through a carbanion stabilised by cyano group (see Fig. 2.54). In fact Breslow has prepared a number of γ-cyclodextrin thiazolium salts, which mimic the action of thiamine and catalyse benzoin condensation efficiently.[77]

(*f*) Miscellaneous Nucleophilic Additions

(*i*) Mannich Reaction: It involves condensation at a compound containing atleast one active hydrogen (like ketones, nitroalkanes, β-ketoesters, β-cyanoacids) with formaldehyde and primary or secondary amine or ammonia (in the form of hydrochloride salts) to give β-amino carbonyl compounds known as Mannich base.[78] (Fig. 2.58)

$$\underset{\text{Acetophenone}}{C_6H_5\overset{\overset{\displaystyle O}{\|}}{C}\,CH_3} + \underset{\text{Formaldehide}}{H-\overset{\overset{\displaystyle O}{\|}}{C}-H} + \underset{\substack{\text{Dimethylamine}\\ \text{hydrochloride}}}{(CH_3)_2\overset{+}{N}H_2\,Cl^-} \longrightarrow \underset{\substack{\beta\text{-aminocarbonyl compound}\\ \text{Mannich base}}}{C_6H_5-\overset{\overset{\displaystyle O}{\|}}{C}-CH_2CH_2\,\overset{+}{N}H(CH_3)_2\,Cl^-} + H_2O$$

Fig. 2.58: Mannich Reaction.

In Mannich reaction, other activated hydrogen compounds like indole, furan, pyrrole, phenols etc. can also be used. The Mannich base obtained is of great synthetic use. For example, on heating Mannich base, amine hydrochloride is elimated to give α, β-unsaturated ketones. Also, the dimethylamino group in Mannich base can be replaced by cyanide on heating with KCN (Fig. 2.59).

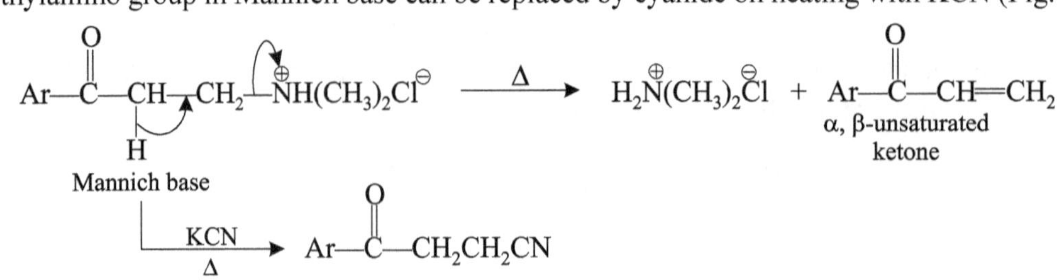

Fig. 2.59: Some reactions of Mannich Base.

The mechanism of Mannich reaction is as given in Fig. 2.60.

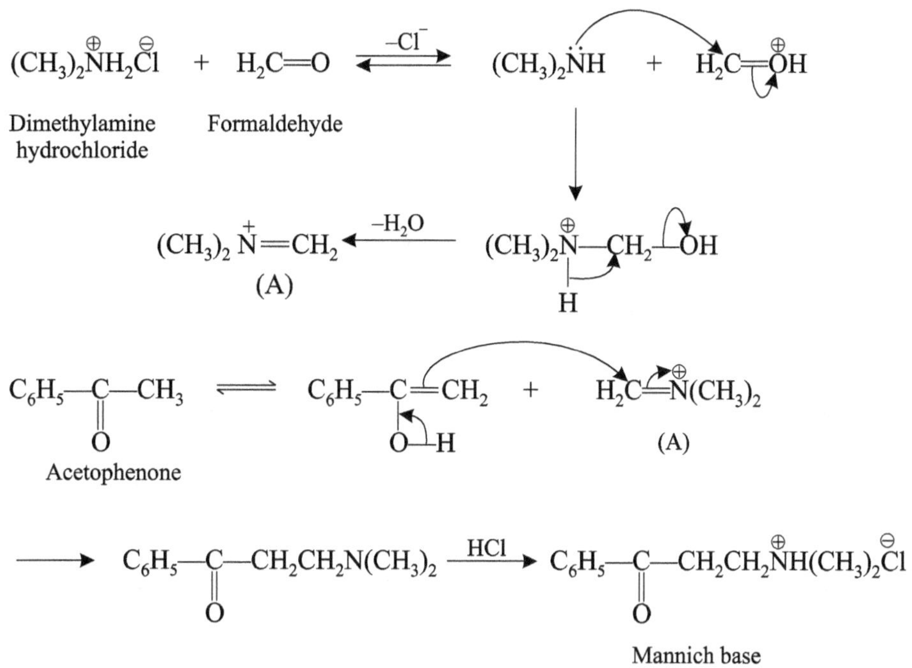

Fig. 2.60: Mechanism of Mannich Reaction.

Mannich Reaction in Water

The original Mannich reaction, needs drastic conditions and the yields are low. An aqueous-MW assisted Mannich reaction[79] involving reaction of acetophenone, secondary amines (in the form of their hydrochlorides) and trioxy methylene (a source of formaldehyde) gives much better yield of the products and in short time (Fig. 2.61).

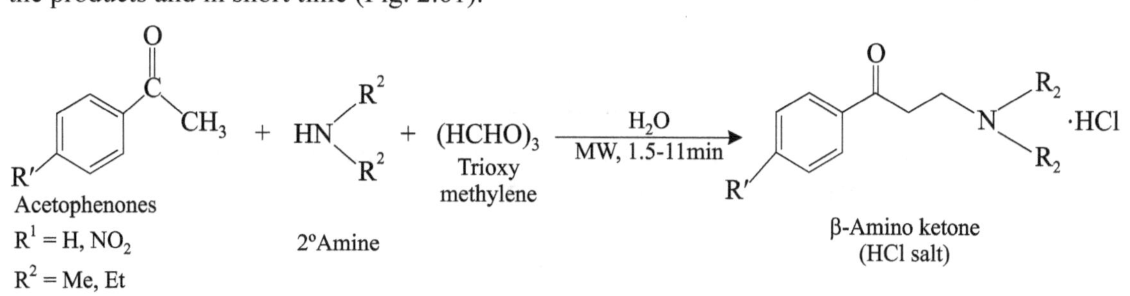

Fig. 2.61: Aqueous Mannich Reaction.

The above reaction (Fig. 2.61) can be performed in much shorter times (20-50 sec) by using combined microwave and ultrasound conditions[80].

A **novel type of Mannich reaction[81]** involves the reaction of an aldehyde, a primary or secondary amine and a terminal alkyne in presence of cuprous iodide gives the Michael type product (Fig. 2.62).

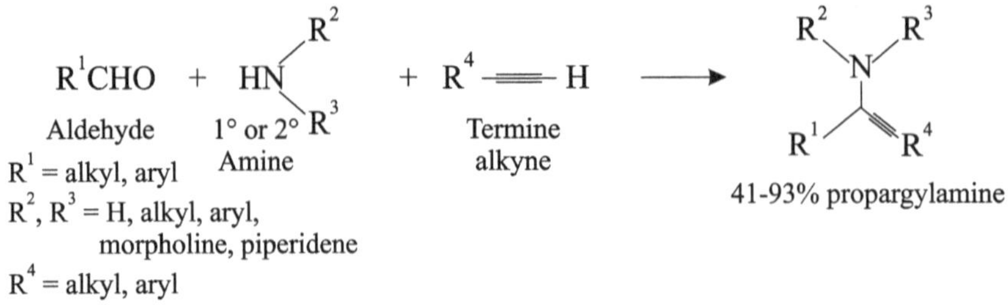

Fig. 2.62: A novel type of Mannich reaction.

Using (S) proline methyl ester as chiral amine in the above procedure (Fig. 2.62), an asymmetric synthesis of propargyl amine has been developed with high distereoselectivity (95 : 5 for $R^1 = R^4 = Ph$).

Mannich Type Reactions: In a Mannich type reaction, β-amino ketones are synthesised by the reaction of imines with enolates (especially trimethyl silyl ethers)[82] (Fig. 2.63)

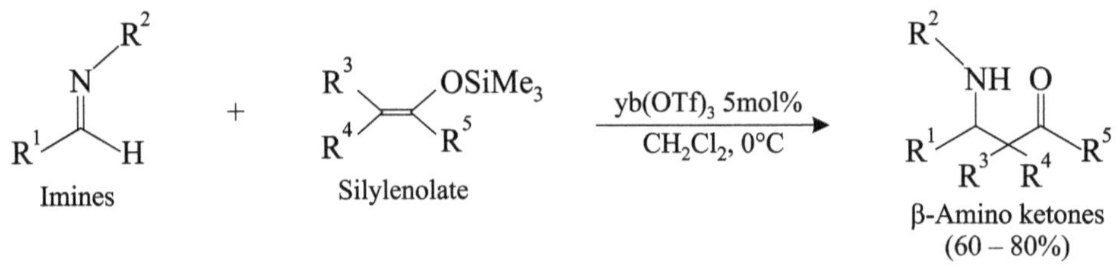

Fig. 2.63: Synthesis of β-amino ketones.

Mannich type product is also obtained[83], by the reaction of aldehyde with an amine and a vinyl ether in THF-H_2O (9 : 1) in presence of Ytterbium triflate (Fig. 2.64)

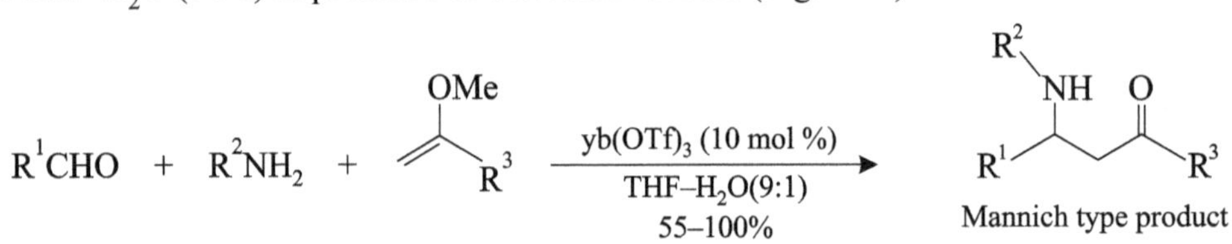

Fig. 2.64: Mannich Type reaction in THF-H_2O.

Mannic-Type reaction was also reported in the reaction of indole with benzaldehyde using lanthanium triflate as catalyst in EtOH-H_2O (Fig. 2.65).

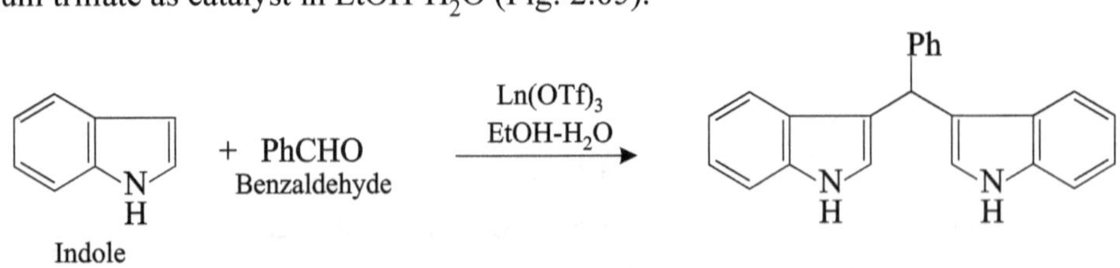

Fig. 2.65: Reaction indole with benzaldehyde.

In the above reaction (Fig. 2.65) use of organic solvent like chloroform gave oxidised byproduct.[84]

The reaction of diazonium cations (generated by the decomposition of diazonium salts) with iminium salts (formed in situ from aromatic amines and aldehydes) in presence of $TiCl_3$ in aqueous media gave secondary amines[85] (Fig. 2.66).

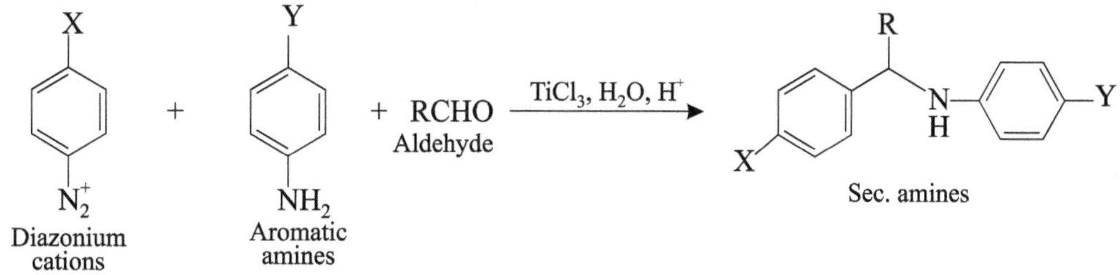

Fig. 2.66: Synthesis of 2° amines.

***(ii)* Prins Reaction:** A useful C—C bond forming reaction[86] consisting in the reaction of olefins with aldehydes in water in presence of acid catalyst to give tetrahydropyran derivatives (Fig. 2.67).

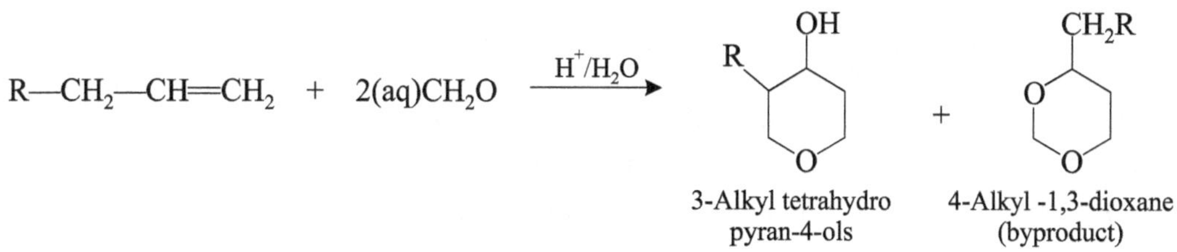

Fig. 2.67: Prins Reaction in Water

***(iii)* Reformatsky Reaction:** It involves[87] the reaction of an α-halo ester (usually a bromoester with an aldehyde or a ketone in presence of zinc in an inert solvent to produce β-hydroxy ester (Fig. 2.68)

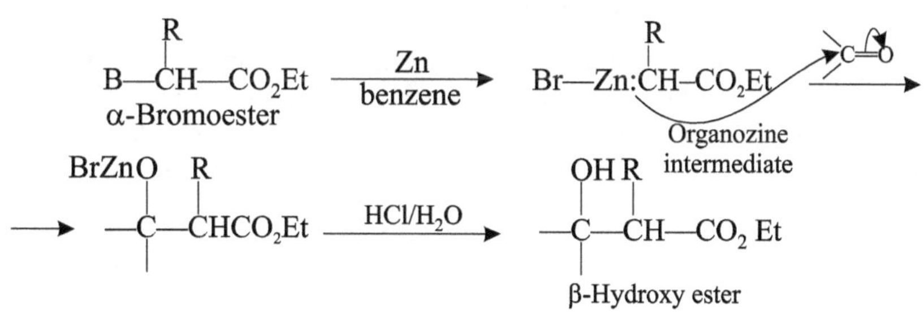

Fig. 2.68: Reformatsky Reaction.

In the above reaction, the organo zinc reagent being less reactive than organo magnesium reagent (grignard reagent), it does not add to the ester group.

A Reformatsky type reaction occurred by the reaction of α-bromoester in water mediated by zinc.[88] (Fig. 2.69)

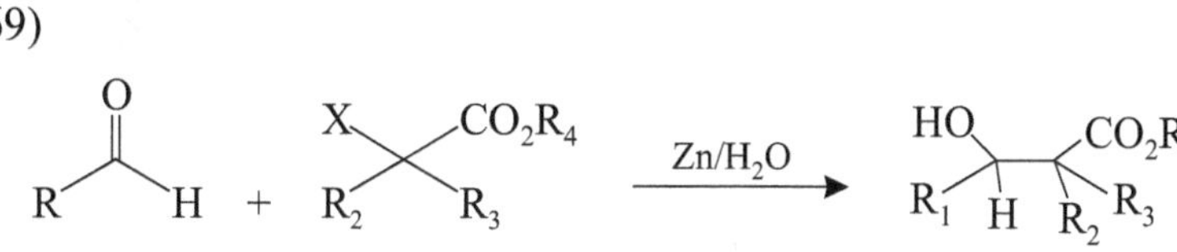

Fig. 2.69: Reformatsky reaction in Water.

Excellent yields are obtained in Reformatsky reaction by using sonication[89] (Fig. 2.70).

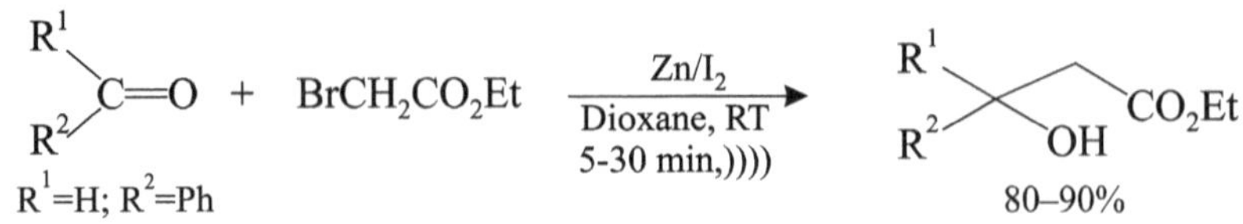

Fig. 2.70: Sonical Reformatsky Reaction.

The Reformatsky reaction with nitriles gives imines, which are hydrolysed to ketones in good yields[90] (Fig. 2.71).

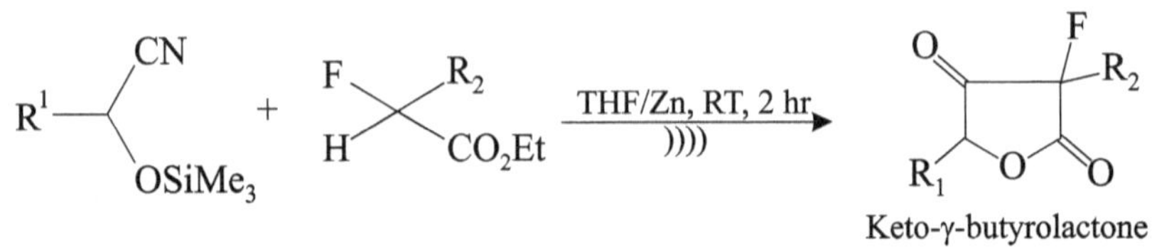

Fig. 2.71: Synthesis of keto γ-butyrolatone by Reformatsky reaction.

The Reformatsky reaction of bromoacetates with aromatic aldehydes is greatly enhanced by adding catalytic amount of benzoyl peroxide or peracides[91]. The indium catalysed aqueous Reformatsky reaction has been used for the synthesis of α, α-difluoro-β-hydroxyketones[92].

The Reformatsky reaction has been used for the synthesis of citral, citric acid and Vitamin A_1.

***(iv)* Pinacol Coupling:** It involves the coupling of carbonyl compounds to give 1, 2-diols (Fig. 2.72).

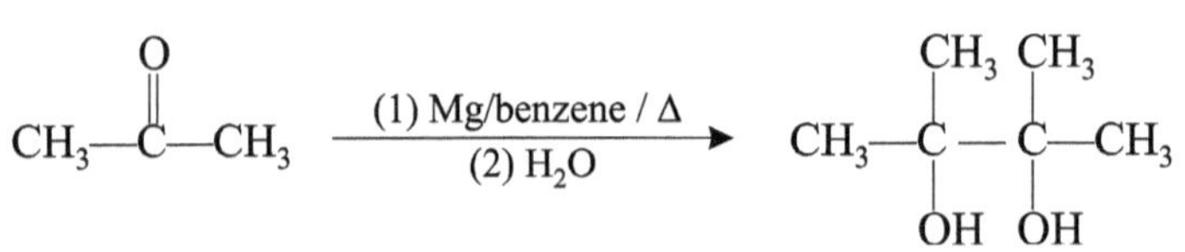

Fig. 2.72: Pinacol coupling.

This reaction can be carried out in aqueous media using unsaturated aldehydes[94] in presence of Zn-Cu couple. Subsequently, chromium, vanadium[95] and some ammoniacal based $TiCl_3$[96] based reducing agents were used. It has been found that aromatic ketones and aldehydes homocoupled by $TiCl_3$ in aqueous solution under alkaline conditions[97]. However, under acidic conditions aliphatic

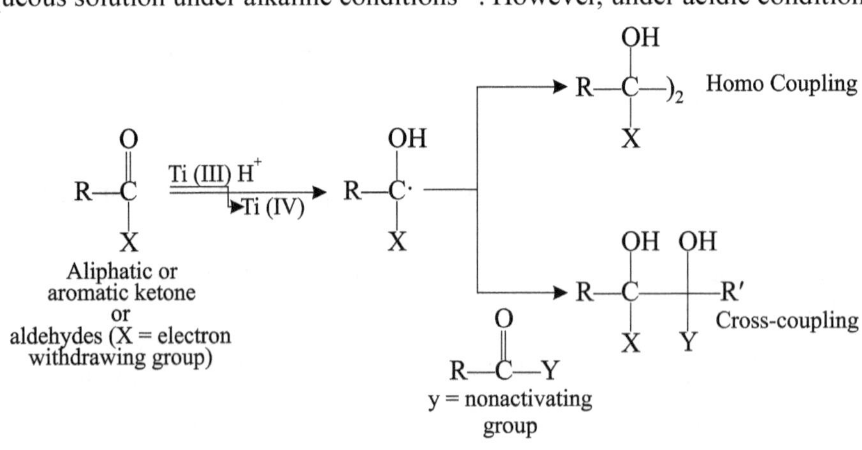

Fig. 2.73: Mechanism of Piracol Coupling.

or aromatic ketones and aldehydes containing strongly electron withdrawing groups (like CN, CHO, COMe, COOH, pyridyl) gave homocoupling products[98,99]. But cross coupling products[100] were obtained by using non-activated carbonyl compounds. A radical mechanism has been suggested (Fig. 2.73).

α, β-unsaturated carbonyl compounds cross coupled with acetone is presence of Zn-Cu couple and ultrasonic irradiation in aqueous acetone suspension[101] (Fig. 2.74a)

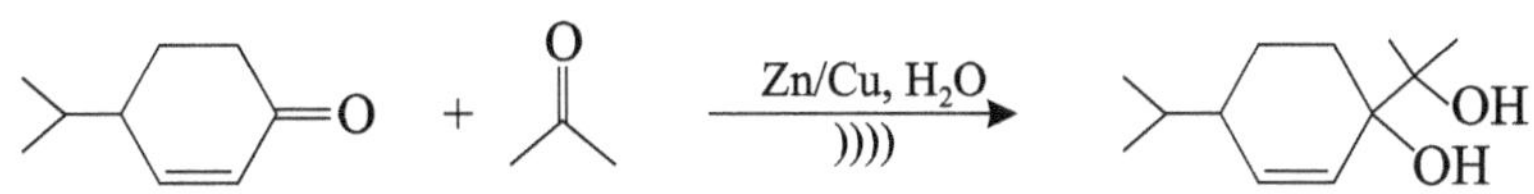

Fig. 2.74a: Pinacol coupling using ultrasonic irradiation.

Stereo selective pinacol coupling has been reported in presence of a cyclopentadiene titanium complex[102].

Aldimines could be reductively coupled to vicinal diamines in aqueous ethanol in presence of indium[103] (Fig. 2.74b).

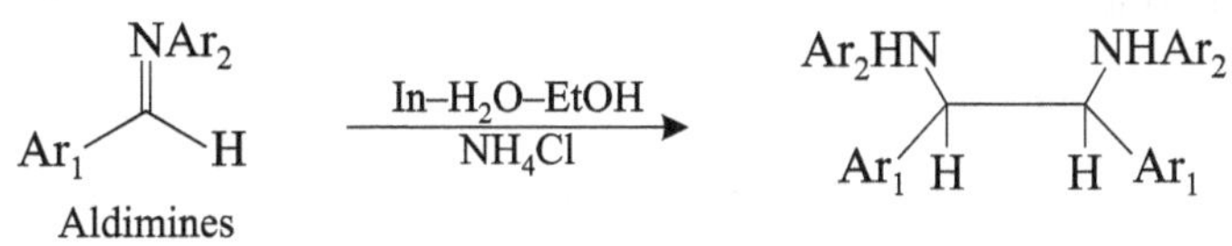

Fig. 2.74b: Coupling of aldimines.

(*v*) Barbier-Grignard-Type Reaction

A very important reaction involving the use of reagents used in this reaction as nucleophiles results in carbon-carbon bond formation. When the carbonyl compound is the electrophile, the reaction is referred to as the Barbier-Grignard-type reaction (Fig. 2.75). The intermediate organometallic reagent can be generated in situ (Barbier) or stepwise (Grignard).

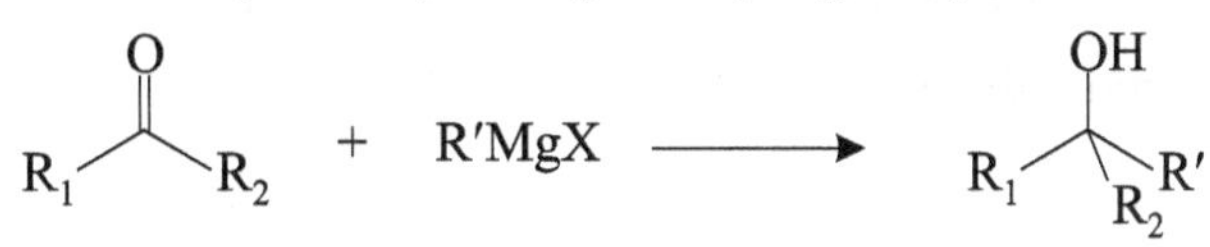

Fig. 2.75: Barbier-Grignard Type Reaction.

The disadvantage associated with Barbier reaction of carbonyl compounds, only reactive alkyl halides (for the generation of Grignard reagent) can be used. This difficulty has been overcome by carrying out the reaction by sonication[104]. In this procedure even imperfectly dried alkyl halides give excellent yield. Also, the reactive halides, *e.g.*, allyl or benzyl halides which generally give Wurtz coupling can be used (Fig. 2.75a).

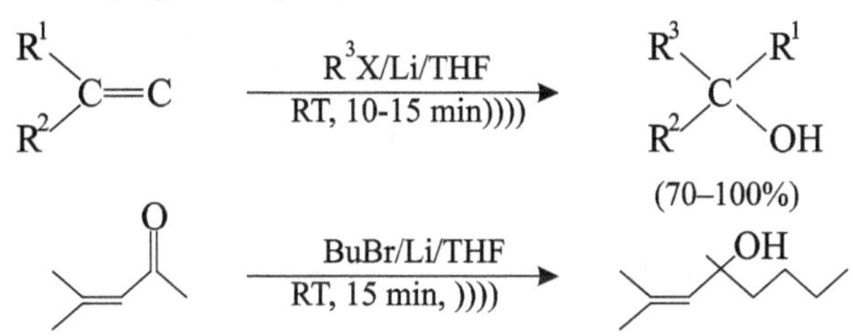

Fig. 2.75a: Sonochemical Barbier Reaction.

It has been found[105] that Barbier type reaction can be performed in water. Thus, allylation of carbonyl compounds with diallyltin bromide in ether by the addition of water to the reaction mixture gave the product in 75% yield (Fig. 2.76).

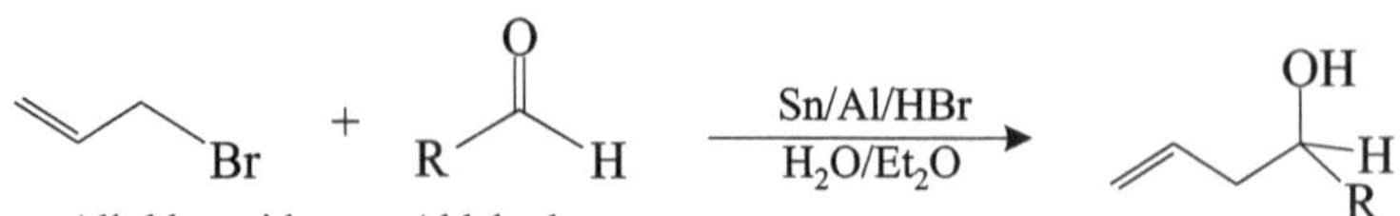

Allyl bromide Aldehyde

Fig. 2.76: Barbier reaction using Sn/Al/HBr in Water/Ether.

The allylation of aldehydes and ketones can also be effected in aqueous media using zinc as the metal and THF as cosolvent under sonication[106] (Fig. 2.77).

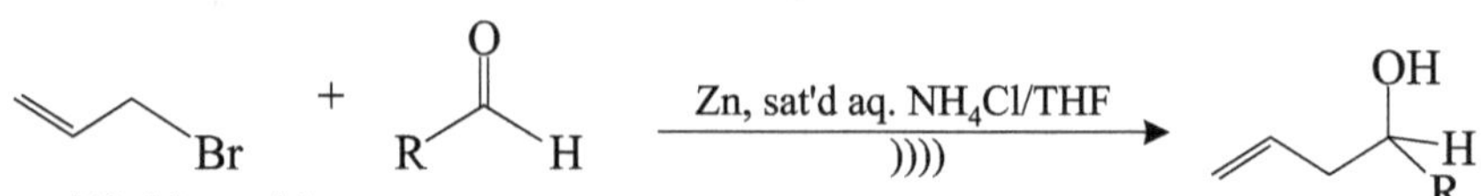

Allyl bromide

Fig. 2.77: Sonochemical Barbier reaction using Zinc

Barber-Grignard-type reactions in water could be performed using indium[107] in water or Bismuth[108] in water.

Applications of Barbier-Grignard-type Reactions

(*i*) Synthesis of Cyclopentanone[109]

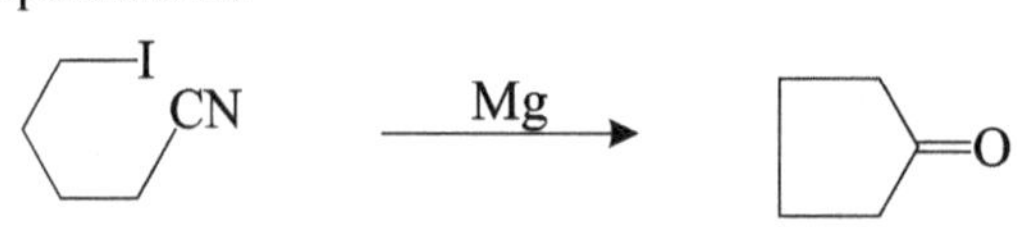

5-Cyanoiodobutane Cyclopentanone 79%

(*ii*) Synthesis of a intermediate required for the synthesis of a sesquiterpene[110]

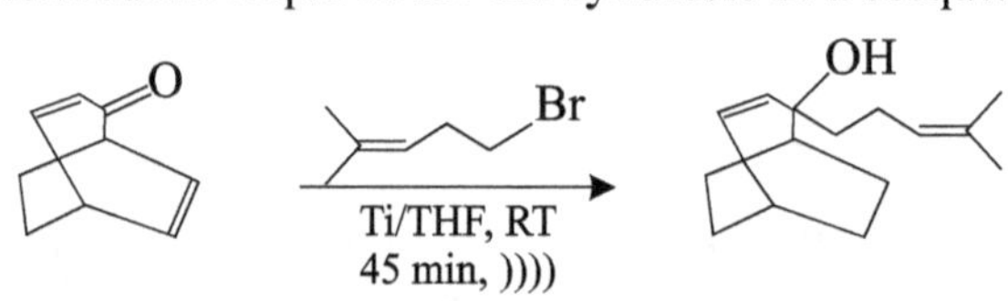

(*iii*) Synthesis[111] of aldehydes from alkyl/aryl halides

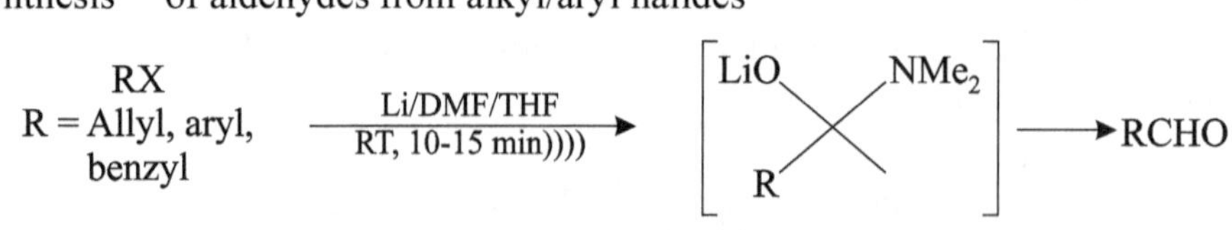

(***vi***) **Wurtz Coupling:** It involves the coupling of alkyl halides with sodium in dry ether to give[112] coupled products (Fig. 2.78).

$$\underset{n\text{-Propyl bromide}}{CH_3CH_2CH_2Br} \xrightarrow[\text{ether}]{\text{Na}} \underset{n\text{-Butane}}{CH_3(CH_2)_4CH_3}$$

Fig. 2.78: Wurtz coupling.

Wurtz type coupling can also take place in water in presence of zinc[113] (Fig. 2.79)

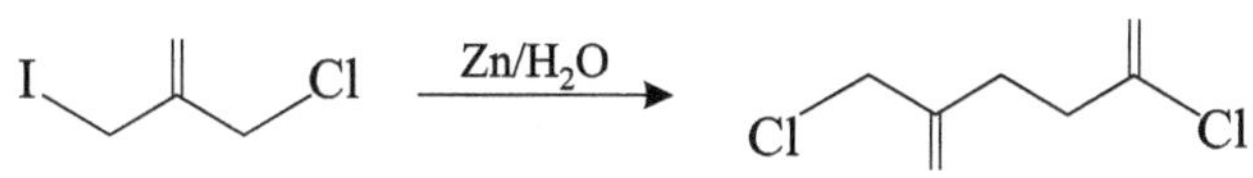

Fig. 2.79: Wurtz type coupling in water.

Much better yields are obtained if the reaction is carried out under sonication.

(*vii*) **Ullmann Coupling:** It involves the coupling of aryl halides in presence of copper to give diaryls[114] (Fig. 2.80).

$$2C_6H_5I + Cu \xrightarrow[\Delta]{C_6H_5NO_2} C_6H_5 - C_6H_5 + CuI_2$$

Fig. 2.80: Ullmann coupling.

Much better yields are obtained in case aryl bromides and chlorides are activated by a substituent like NO_2 group (Fig. 2.81).

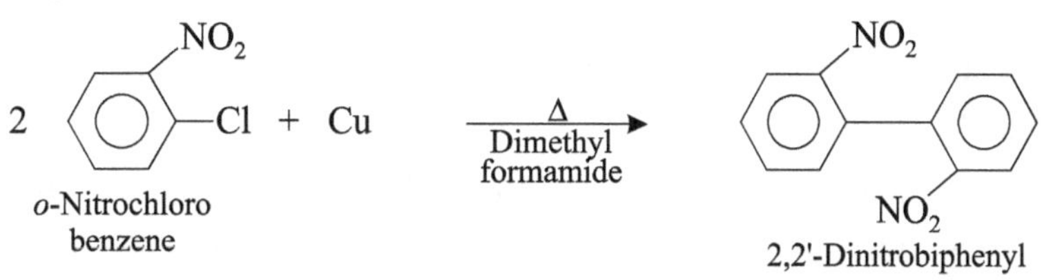

Fig. 2.81: Ullmann coupling of nitro chlorobenzene

The reaction is much faster giving much better yields in presence of sonication[115] (Fig. 2.82)

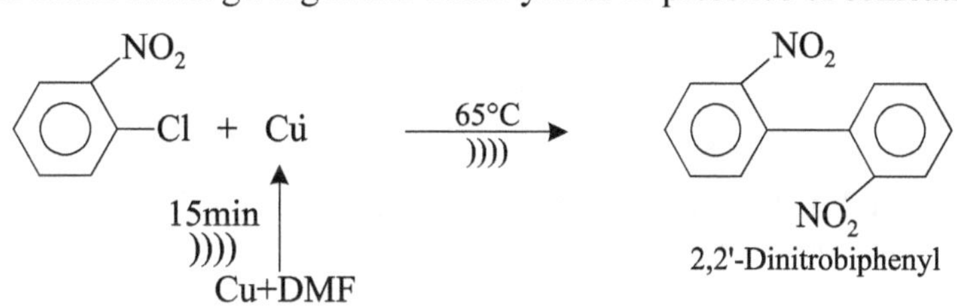

Fig. 2.82: Sonochemical Ullmann's Reaction.

Ullmann Type Coupling of aryl sulfinic acids takes place in aqueous solvents in presence of Pd (II) to give biaryls[116] (Fig. 2.83).

$$2ArSO_2Na + Na_2PdCl_4 \xrightarrow{H_2O} Ar - Ar + 2SO_2 + Pd + 4NaCl$$

Fig. 2.83: Ullmann Type Coupling of arylsulfinic acid in water.

The condensation of aryl halides with phenol and anilines could take place effectively in presence of copper in water[117] (Fig. 2.84).

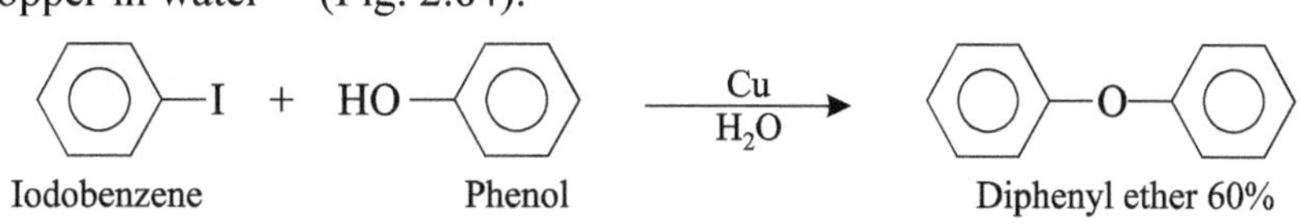

Fig. 2.84: Synthesis of diphenyl ether.

The aryl halides homocoupled in presence of hydrogen gas and stoichiometric amounts of palladium[118] (Fig. 2.85).

$$2\text{ArX}\xrightarrow[\text{H}_2\text{O, BuOH}]{\text{H}_2\text{, Cat. PdCl}_2\text{, K}_2\text{CO}_3}\text{Ar}-\text{Ar}$$
$$\text{X} = \text{I, Br; Ar} = \text{Ph, p-MePh, etc.}$$

Fig. 2.85: Homocoupling of aryl halides in H_2O in presence of Pd

A facile coupling of aryl halide via a palladium catalysed reducting coupling using Zn in air and aqueous acetone at room temperature using Pd/C gave coupled products[119]. The use of 18-crown-6 gave much better yields[120] (Fig. 2.86).

$$2\text{ArX}\xrightarrow[\substack{\text{Zn, H}_2\text{O/18 crown-6}\\ \text{air atmosphere RT}}]{\text{Pd/C}}\text{Ar}-\text{Ar}$$

Fig. 2.86: A facile coupling aryl halides in water.

2.2.1.3 Some Organic Reactions involving Carbon-carbon bond Formation using Pd Catalyst

A number of interesting reactions have been used for carbon-carbon bond formation. These reaction include Heck reaction, Suzuki reaction, still reaction and Trost-Tsuji Reaction.

(*i*) Heck Reaction: It involves[121] the coupling of an alkene with a halide or triflate in presence of Pd (O) catalyst to form a new alkane (Fig. 2.87)

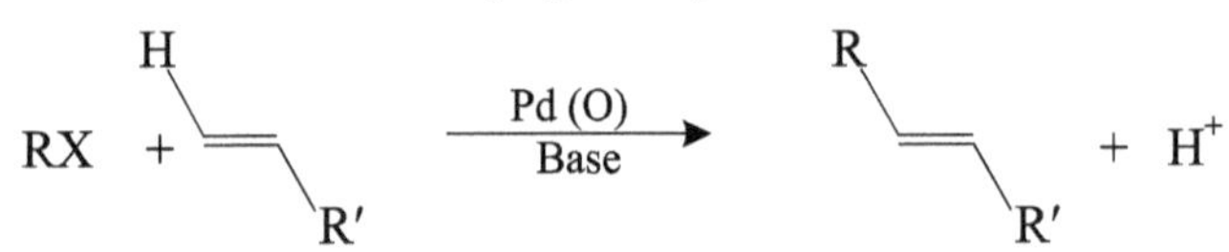

R = Aryl, Vinyl or alkyl group without β-hydrogen on a sp^3 carbon atom
X = halide or triflate (OSO_2CF_3)

Fig. 2.87: Heck Reaction.

In the original Heck Reaction, a mild base like Et_3N or anions like $\overline{O}H$, $\overline{O}COCH_3$, CO_3^{2-} etc. and the reaction is carried out in anhydrous polar solvents like DMF, MeCN etc.

Both inter- and intramolecular Heck reactions have been performed in aqueous media. Thus, the palladium-catalyst reactions of aryl halides with acrylic acid or acrylonitrile gave the corresponding coupled product in high yields in presence of a base in water[122] (Fig. 2.88).

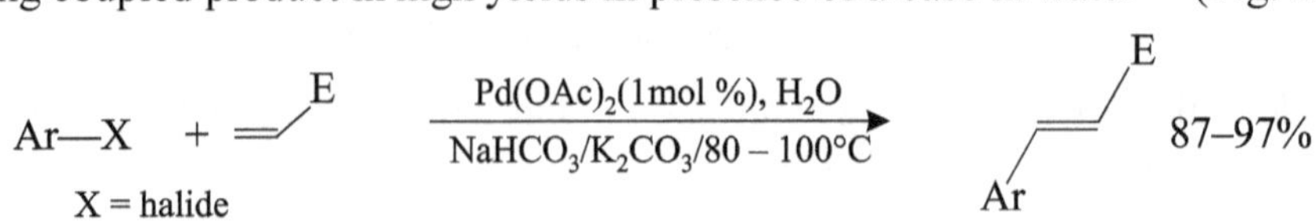

X = halide
E = COOH or CN

Alkenes

Fig. 2.88: Heck Reaction in Water.

This reaction provides a simple method for the synthesis of substituted cinnamic acids and cinnamonitriles.

The Heck reaction can also be carried out at room temperature[123] if diaryliodium salts are used as the arylating agent. It was found that at room temperatures, only one aryl group of iodinium salt is transferred to the product. However, at 100°C, both aryl groups of iodinium salt are transferred (Fig. 2.89).

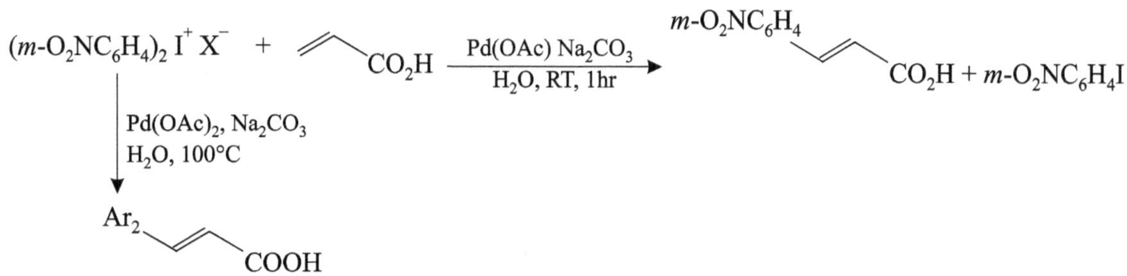

Fig. 2.89: Heck reaction using diaryliodium salts.

Heck reactions (as in Fig. 2.88) have also been carried out using a water-soluble phosphate ligand[124], Poly-supported carbene ligands[125] and dendrimer-encapsulated nanopartical catalysts[126] for Heck coupling in water.

The aqueous Heck reaction can also be applied to the formation of C–C bond in solid-phase synthesis[127] (Fig. 2.90).

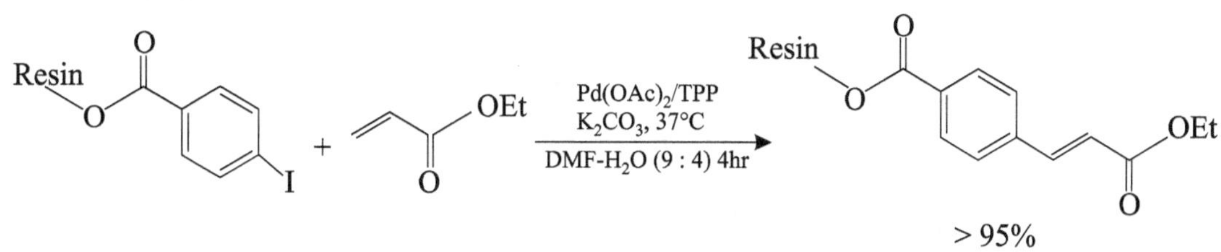

Fig. 2.90: Solid Phase Synthesis using Heck Reaction.

The Heck reaction in water using Microwave heating gives the coupled products[128] (Fig. 2.91).

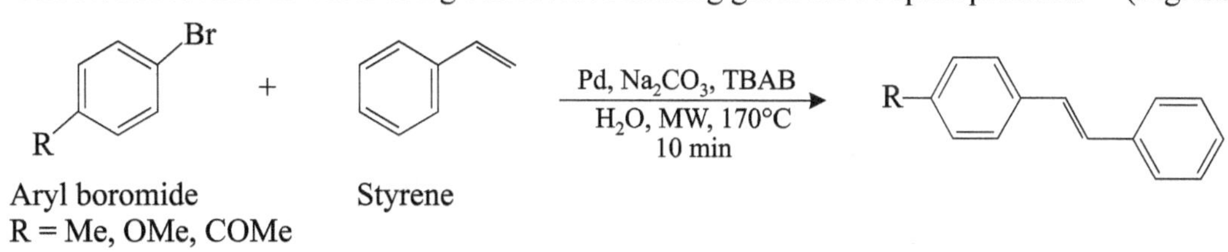

Fig. 2.90: Heck reaction in water using microwave heating.

Heck reaction can also be conducted in supercritical carbon dioxide, ionic liquids, polyethylene glycol and using Fluorous phase Technique. All these are discussed subsequently.

***(ii)* Suzuki Reaction:** It involves[129,130] palladium catalysed cross-coupling of alkenyl and aryl halides with organoborane derivatives and a base. It is a convenient method for the synthesis of biaryals (Fig. 2.91).

$$RX + R'B \xrightarrow{[Pd]} R - R'$$
$$R = 1-\text{alkenyl or aryl; } R' = \text{aryl or alkyl}$$
$$X = \text{Br or I}$$

Fig. 2.91: Suzuki Reaction.

The Suzuki reaction was initially carried out in a mixture of benzene and aqueous Na_2CO_3. The reaction, however, proceeds rapidly in aqueous DME[131]. Thus, the reaction of alkenylborane with vinyl bromide in refluxing THF in presence of Pd $(PPh)_4$ and aqueous NaOH solution yielded dienes[132] (Fig. 2.92).

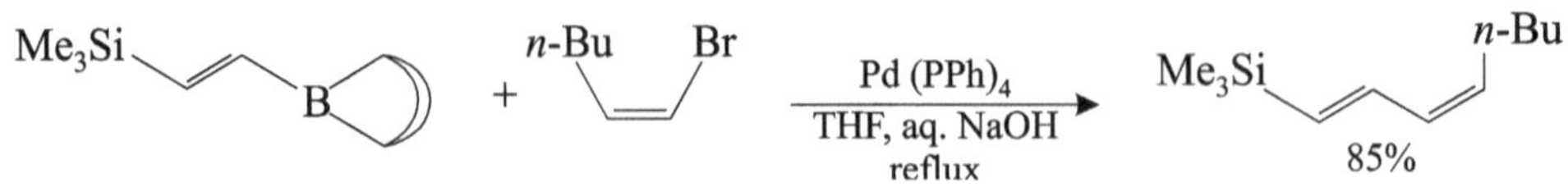

Fig. 2.92: Synthesis of dienes using Suzuki reaction in Aq NaOH.

In the above reaction, the use of TlOH in place of NaOH increased the rate of the reaction. In Kishi's palytoxin synthesis, cross coupling between alkenylboronic acid and iodoalkene was stereoselectively accomplished at room temperature[133]. Using modified method (*i.e.*, using TlOH in place of NaOH) the synthesis of Kijanimicin[134], 12R-HETE[135] [(12R)-hydroxyeicosatetraenoic acid] and rutarmycin B[136] was carried out.

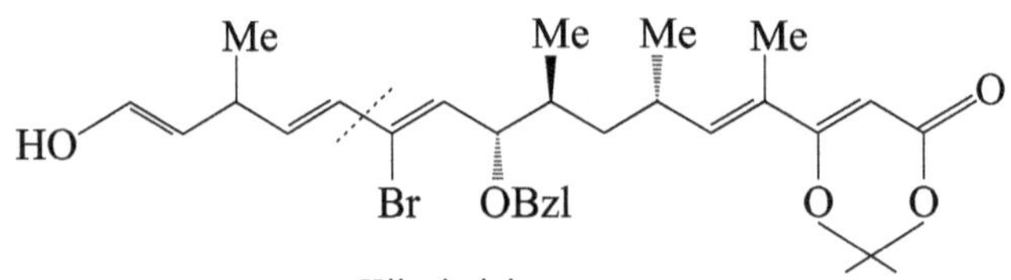

Kijanimicin precursor

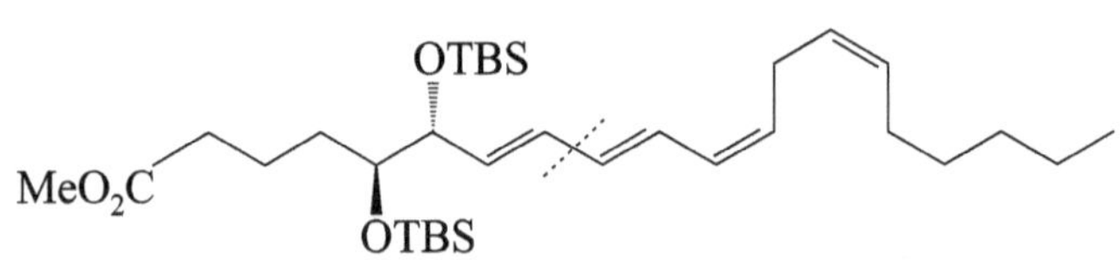

(12R)–HETE

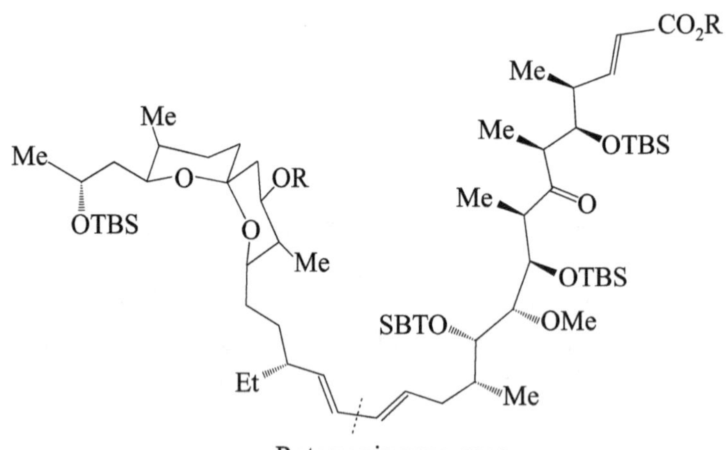

Rutamycin precursor

Using Pd/C-mediated Suzuki-coupling approach[137], biphenyl carboxylic acid derivatives could be synthesised. The Suzuki reaction in aqueous medium has been used in synthesising a combinational library of biaryls[138].

Also using water-soluble palladium (O) catalyst, Pd $[PPh_2\ (m\text{-}C_6H_4SO_3M)_4]$ ($M = Na^+, K^+$), a number of aryl bromides and iodides reacted with aryl and vinyl boronic acids, terminal alkynes and dialkyl phosphites to give cross-coupling products in good yields in water[139].

Terminal alkynes also coupled with allyl bromides using copper (I) and phase-transfer catalyst.[133,134]. In fact use of copper salt as cocatalyst improved the yield significantly[140].

Suzuki reaction has also been used for the synthesis of benzofuran derivatives[141] by the coupling of 2-iodoaniline or 2-iodophenol with terminal alkynes (Fig. 2.93).

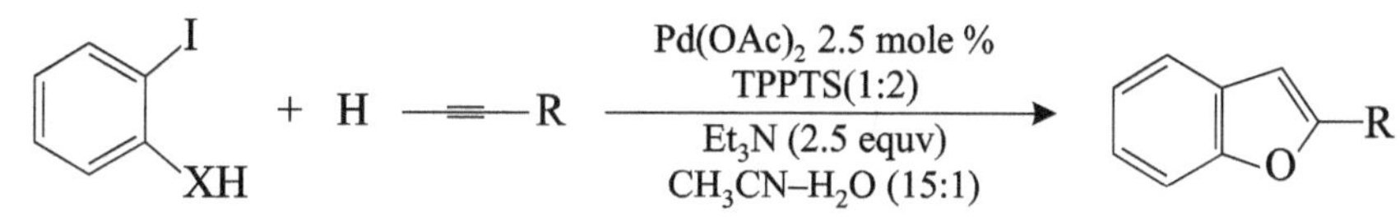

X = NH, NR, O

Fig. 2.93: Synthesis of benzofurans/indoles.

Aryl halides could couple with acetylene in a mixture of acetonitrile and water[142] (Fig. 2.94).

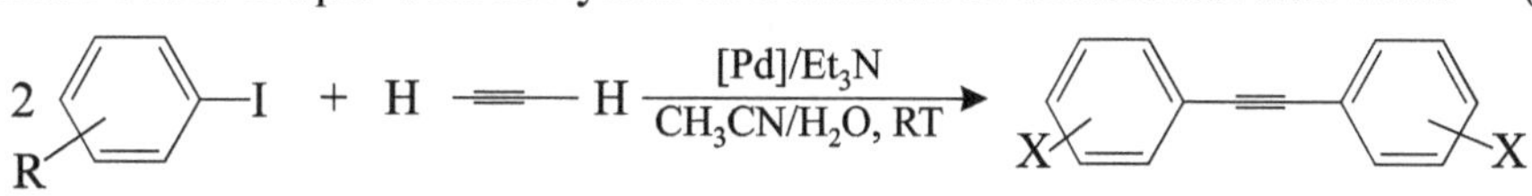

Fig. 2.94: Suzuki coupling of arylhalides with acetylene.

Suzuki type reactions in water are facilitated by using microwave heating[143]. It is found that palladium nano particles are efficient catalyst for Suzuki reaction in aqueous medium[144].

The aryl halides homocoupled in presence of hydrogen to give biaryls in aqueous/organic microemulsion[145] (Fig. 2.95).

$$2\text{Ar X} \xrightarrow[\text{H}_2\text{O}-\text{BuOH}]{\text{H}_2,\text{cat.PdCl}_2,\text{K}_2\text{CO}_3} \text{Ar}-\text{Ar}$$

$$X = \text{I}, \text{Br}$$

$$\text{Ar, Ph}, p\text{-MeC}_6\text{H}_4,\ p\text{-MeOC}_6\text{H}_4, p\text{-ClC}_6\text{H}_4$$

Fig. 2.95: Homocoupling of aryl halides.

Aryl halides also coupled with 2° amines in alkaline aqueous ethanol emulsion in presetnce of Pd $(OAc)_2$ and a PTC (CTAB)[146] (Fig. 2.96).

$$\text{Ar}_2\text{NH} + y - \text{C}_6\text{H}_4 - \text{I} \xrightarrow[\text{H}_2\text{O–BuOH, CTMAB}]{\text{Pd (OAc)}_2,\text{K}_2\text{CO}_3,\text{CuI}} \text{Ar}_2\text{NH} - \text{C}_6\text{H}_4 - y$$

$$\text{Ar}_2\text{NH} - \text{Ph}_2\text{NH}$$

$$y = \text{H}, p\text{-Cl C}_6\text{H}_4, p\text{-MeC}_6\text{H}_4, p\text{-MeO C}_6\text{H}_4$$

Fig. 2.96: Coupling of 2° amines with aryl halides.

Suzuki Reaction can also be performed in ionic liquids and polyethylene glycol. This is discussed subsequently.

***(iii)* Still Reaction:** It involves coupling of alkenyl and aryl halides with orgastannanes in presence of palladium catalyst[147] (Fig. 2.97).

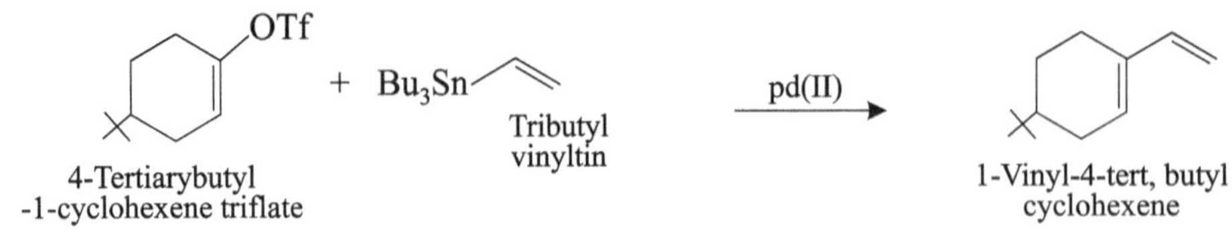

Fig. 2.97: Still Reaction

It was known[148] that addition of water in stille reaction increase yields and also affected the regio and stereochemistry. Subsequently still coupling reaction was investigated in water. It was found that coupling in aqueous alcohol increased the yield of the coupled product, which hydrolysed in situ[149] (Fig. 2.98).

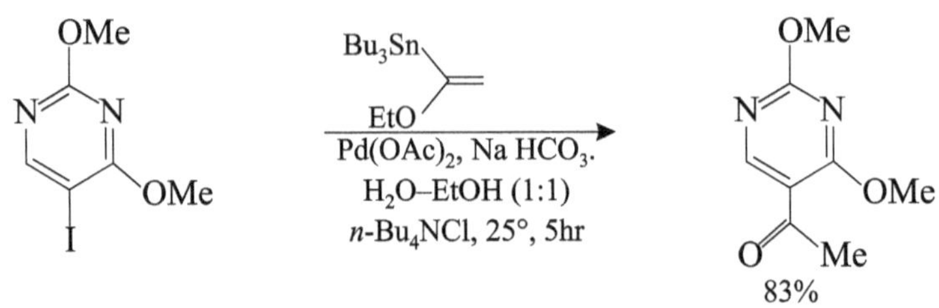

Fig. 2.98: Still coupling in water.

Though still coupling can proceed without using a phosphine ligand, addition of a water-soluble ligand greatly improved[150] the yield. Thus, water-soluble aryl and vinyl halides coupled with alkyl, aryl and vinyltrichloro stannane derivatives (Fig. 2.99)

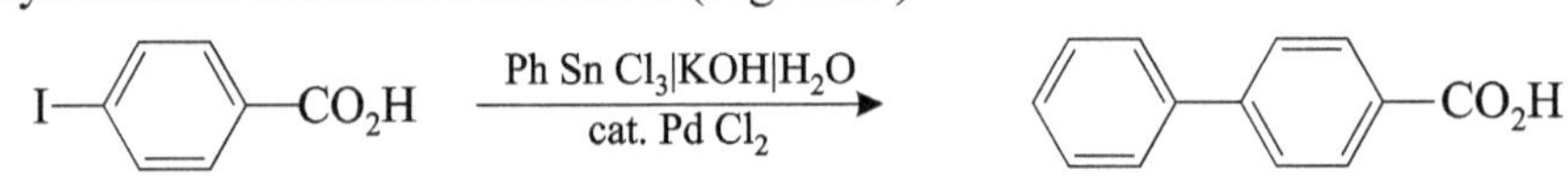

no ligand..81%

with Ph P(m-C_6H_4–SO_3Na)$_2$...........95%

Fig. 2.99: Still reacting using water soluble with Vinyl-trichloro stannane derivative.

Arenediazonium chlorides and hydrogen sulfates were found to react with tetramethyl tin in aqueous acetonitrile in presence of palladium acetate (catalytic amount) to give excellent yields of substituted toluenes[151] (Fig. 2.100)

$$\underset{X = Cl \text{ or } HSO_4}{ArN_2X} + Me_4Sn \xrightarrow[20°C,\ 2\ hr.]{Pd(OAc)_2,\ H_2O–MeCN\ (1:1)} \underset{30\text{-}88\%}{ArMe}$$

Fig. 2.100: Still reaction of aryl dazonium salts with tetramethyl tin.

An efficient still cross-coupling reaction using different aryl halides in neat water has been developed[152]. Using palladium-phosphinous acid catalyst [(t-Bu)$_2$ P (OH)]$_2$ $PdCl_2$, biaryls could be synthesised from aryl chlorides and bromides in good yields.

Still type coupling has been reported[153] in water under an atmosphere of air.

Still coupling reaction has also been conducted in SC–CO_2, ionic liquids and by using Fluorous Phase Technique. All these are discussed subsequently.

***(iv)* Trost-Tsuji Reaction:** It involves the coupling reaction involving π-allyl palladium intermediates. These reactions have been investigated in aqueous media[154]. An example of coupling of allyl acetates with α-nitroacetate in water in presence of Et_3N (Fig. 2.101).

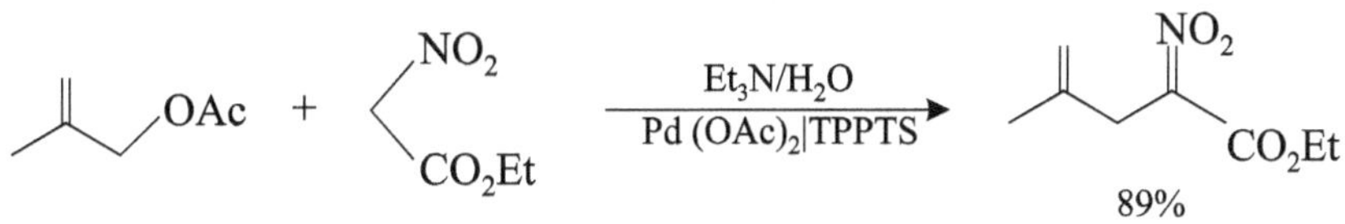

Fig. 2.101: Trost-Tsuji reaction in Water.

In the above reaction, other compounds having an active hydrogen can also be used in place of α-nitroacetates. Even heteroatom nucleophiles can be used. Thus the reaction of an amine with allyl acetate gives the N-allylation product[155]. Azides and toluenesulfinates also reacted to give corresponding allyl azide and allylsulfonate in quantitative yields[156].

Water-mediated transition-metal free Tsuji-Trost-type reactions have been reported[157]. As an example tretatment of 1-acetoxy-1, 3-diphenyl propene by C—, O—, S— and N-nucleophiles in basic alkaline aqueous media yielded the corresponding substitution products in the absence of a transition metal catalyst (Fig. 2.102).

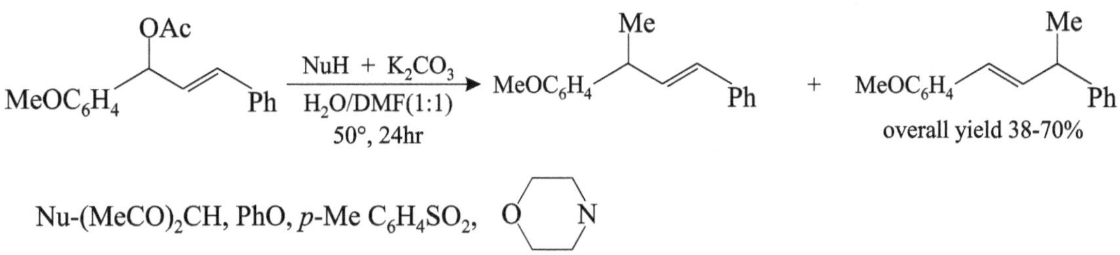

Fig. 2.102: Tsuji-Trost Type reaction using C, O, S and N-nucleophiles

2.2.1.4 Formation of C—C Bonds from sp C—H bonds, sp^2C—H bonds and sp^3-C—H bonds

[A] Formation of C–C Bonds from *sp* C–H bonds

The hydrogen atom of terminal alkyne (*sp* C–H) are more acetic than those of ethene (sp^2 C–H) or ethane (sp^3 C–H).

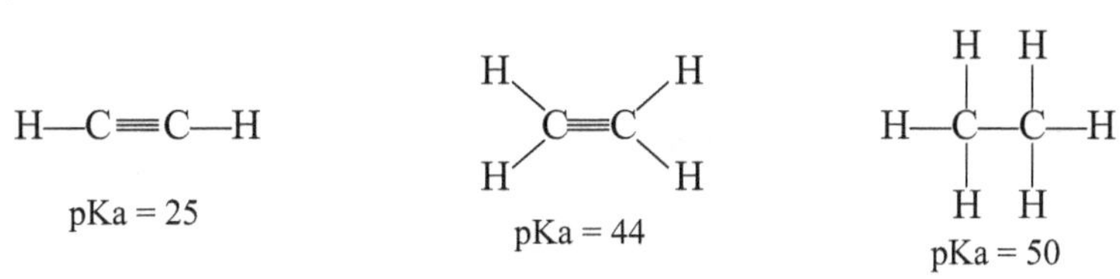

It is well known that in the sp^3 C—H bond, the H can be removed by treatment with a suitable base to give alkynide, which in turn are useful intermediates for the synthesis of other alkynes. A typical example is given below (Fig. 2.103).

$$R\text{–}C{\equiv}C\text{–}H + NaNH_2 \xrightarrow{\text{liq. } NH_3} R\text{–}C{\equiv}C\!:^{-}Na^{+} + NH_3$$

$$\underset{\text{Sod. alkynide}}{R\text{–}C{\equiv}C\!: - Na^{+}} + R'CH_2\text{–}Br \xrightarrow[S_N2]{} \underset{\text{Alkyl substituted alkyne}}{R\text{–}C{\equiv}C\text{–}CH_2R'} + NaBr$$

Fig. 2.103: Acidic character of *sp*–CH bond

The above reaction can be recognised as a case of nucleophilic substitution (S_N2).

Following are given some of the examples of *sp* C–H bonds using terminal alkynes as an example.

***(i)* Dimerisation:** The coupling of terminal alkynes is catalysed by copper and is an example of classical reaction known as **Strauss coupling**. However, under the classical conditions a mixture of regioisomers is obtained[158].

Dimerization of alkynes is possible by using a rhodium-catalysed polymerisation[159]. It was subsequently found that using a catalytic amount of Pd $(OAc)_2$ and triphenyl phosphine as catalyst in dichloro-ethane gave good yield of coupled product of terminal alkynes[160].

Disubstituted alkynes could be obtained by Cu(II) mediated coupling of alkenyldialkyl or trialkylboranes with alkynyl copper compounds (generated in situ in presence a solvent and a small amount of H_2O)[161] (Fig. 2.104).

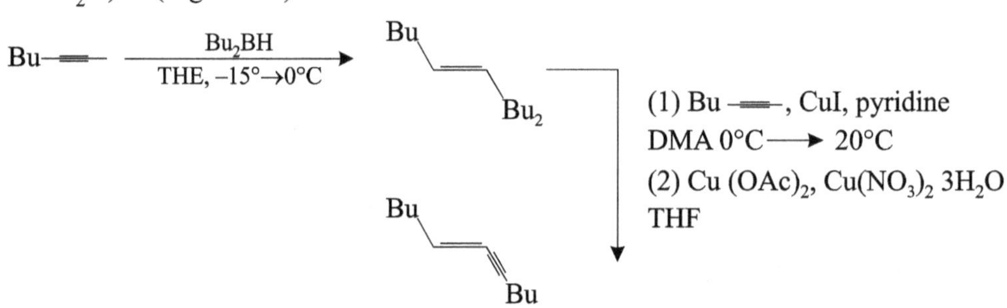

Fig. 2.104: Coupling of alkynes with alkenes.

***(ii)* Coupling of terminal alkynes with organic halides:** Terminal alkynes could be coupled with aryl and vinyl halides in presence of Pd/Cu to give unsymmetrical alkynes. The reaction is known as **Sonogashira reaction**[162] (Fig. 2.105).

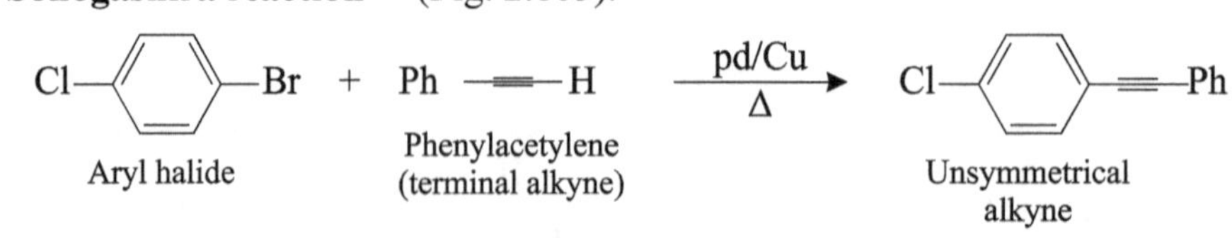

Fig. 2.105: Sonogashira Reaction.

An sonogashira-type coupling reaction proceeds in water as the sole solvent[163], without the need for copper (I) or any transition-metal phosphine complex, which overcomes the problem of toxicity and air-sensitivity of transition-metal complexes and the use of costly phosphine ligands (Fig. 2.106).

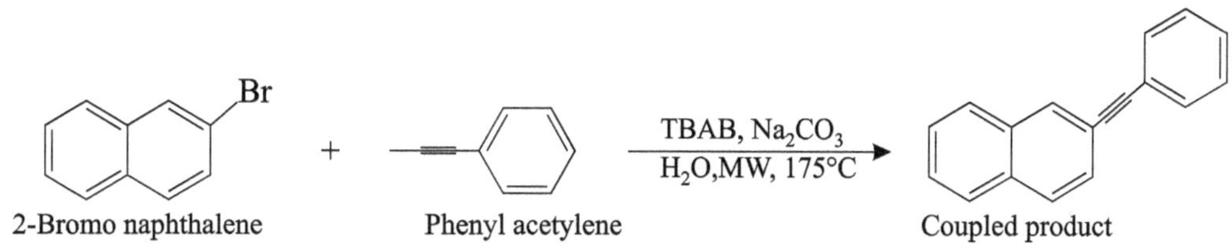

Fig. 2.106: Sonogashira-type coupling reaction in water.

Sonogashira coupling of aryl bromides and iodides with phenylacetylene could also be achieved by using polymer complex (A) a monomeric complex (B) as catalyst, pyrrolidine as base and TBAB as additive[164] under microwave heating (Fig. 2.107). Use of catalyst (B) gave better yields.

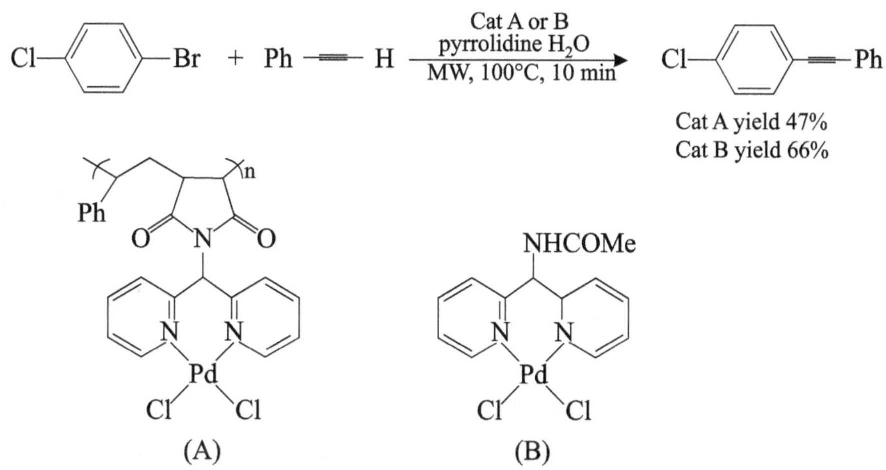

Fig. 2.107: Catalytic Sonogashira coupling.

Sonogashira coupling of aryl and heteroaryl bromides and iodides with phenyl acetylene occurred in presence of heterogeneous catalysts[165] (C) under MW heating to give coupled product (Fig. 2.108).

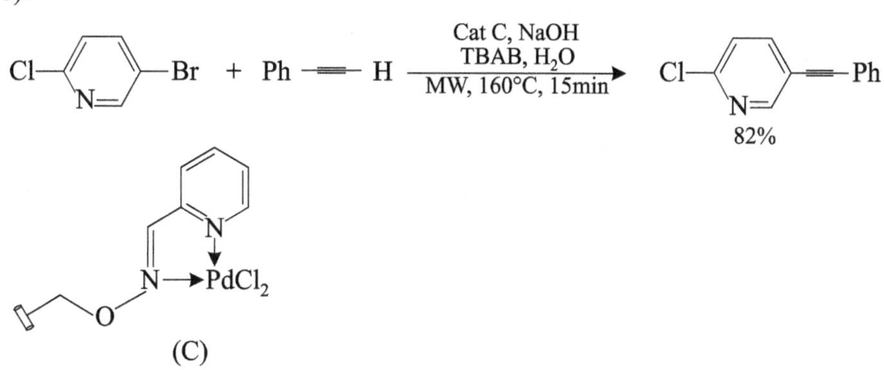

Fig. 2.108: Sonogashira coupling of heteroaryl bromides/iodides.

Sonogashira coupling under transition-metal free conditions[166] involving low Pd concentrations as contaminations[167] and using NaOH as base, Polyethylene glycol (PEG) as PTC and water as solvent under MW heating gave the coupled products in good yield (Fig. 2.109).

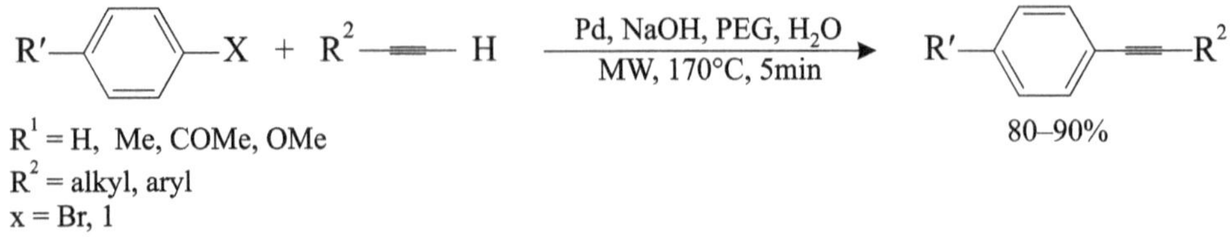

Fig. 2.109: Sonogashira coupling using PGE as PTC

Synthesis of indoles or benzofuran derivatives has been achieved by the two step coupling of 2-iodoaniline or 2-iodophenol with terminal alkynes is good yield[168] (Fig. 2.110).

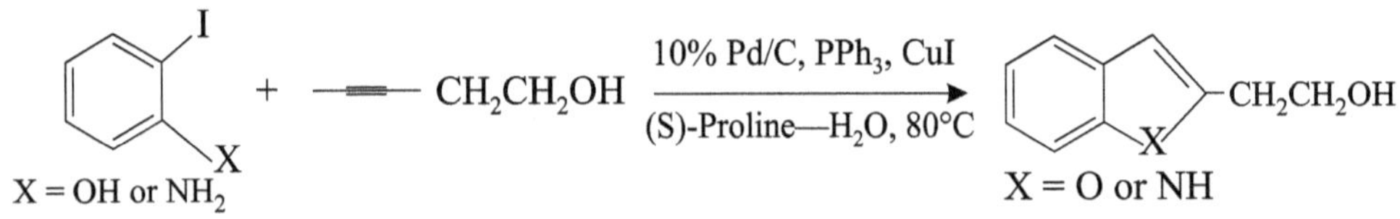

Fig. 2.110: Synthesis of benzofurans and indoles.

The coupling of acetylene with aryl halides in acetonitrile-water gives the coupled products[169] (Fig. 2.111)

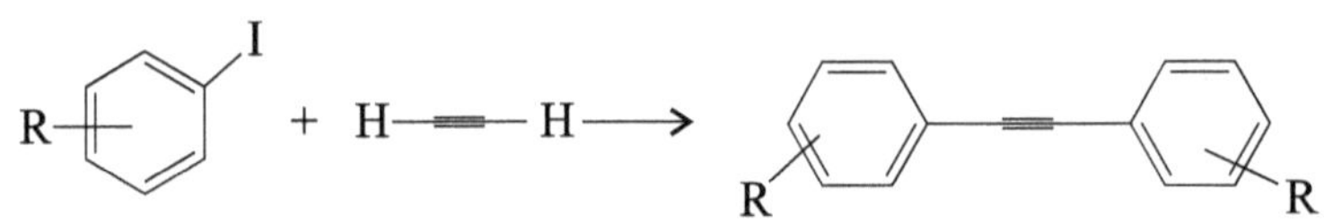

Fig. 2.111: Coupling of aryl halides with acetylene

A microwave assisted Sonogashira-type coupling reaction in water without the use of copper (I) or a transition metal phosphene complex between aryl and hetero-aryl halides and terminal acetylene gives the coupled products is good yields[170] (Fig. 2.112).

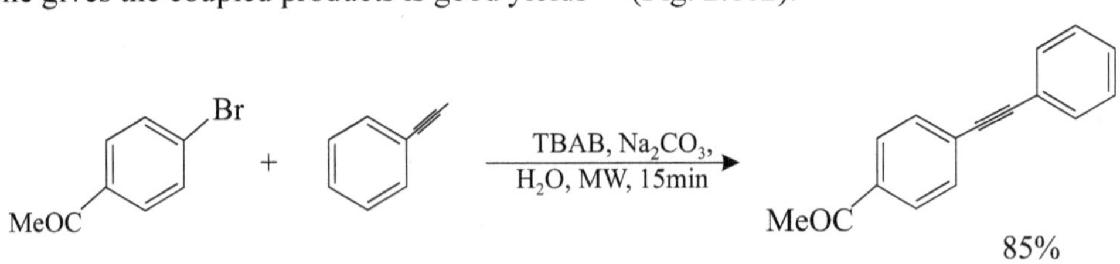

Fig. 2.112: Microwave assisted Sonoashira-type reaction.

Sonogashira coupling of acid chlorides with terminal alkyne catalysed by $PdCl_2$ $(PPh_3)_2$/CuI in presence of catalytic amount of sodium lauryl sulfate as the surfactant and K_2CO_3 in water gives 98% yield of coupled product[171] (Fig. 2.113)

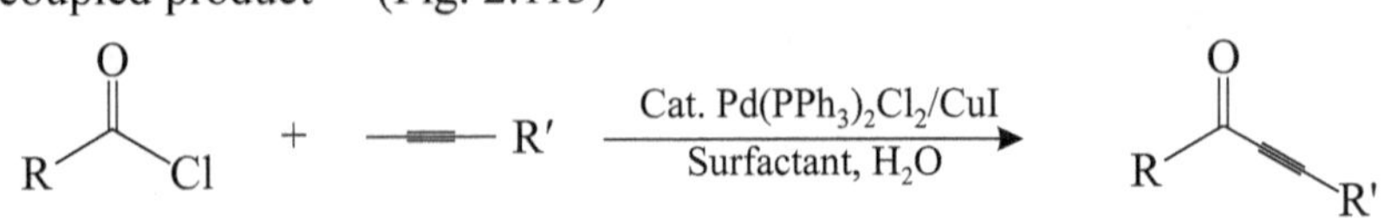

Fig. 2.113: Sonogashira coupling of terminal alkyne with acid chlorides.

***(iii)* Coupling of terminal akynes with carbonyl compounds:** Coupling of terminal alkynes with aldehydes in water in presence of a ruthenium-indium bicatalyst system gives coupled products[172] (Fig. 2.114).

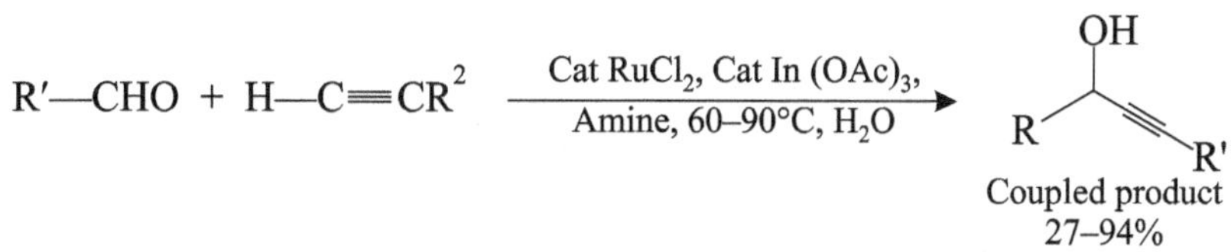

Fig. 2.114: Coupling of terminal alkynes with carbonyl compounds.

***(iv)* Reaction of terminal alkynes with imines:** Terminal alkynes on reaction with imines and their derivative is a convenient route for the synthesis of propargylamines. An efficient route for the synthesis of propargyl amines consist is reaction of an aldehyde, alkyne and amine in water or without solvent has been reported[173]. The reaction takes place in presence of catalytic amount of Cu(I) in aqueous media to give the adducts. Use of $RuCl_3$ as cocatalyst improved the yield[174] (Fig. 2.115).

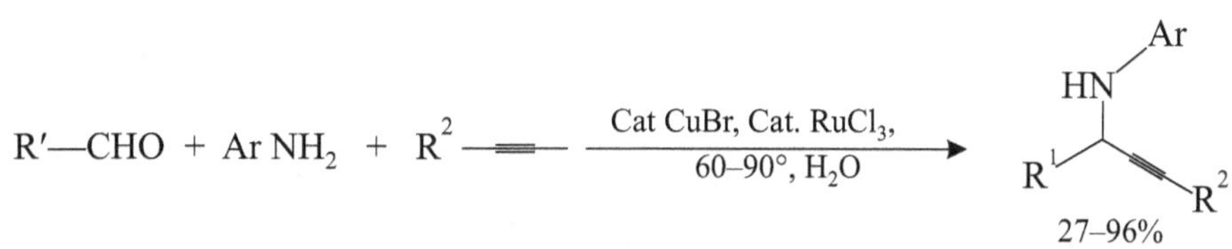

Fig. 2.115: Reaction of terminal akyne, aldehyde and amine in water.

The above reaction become efficient in presence of gold catalyst[174] (Fig. 2.116). In this case no cocatalyst or activator is required.

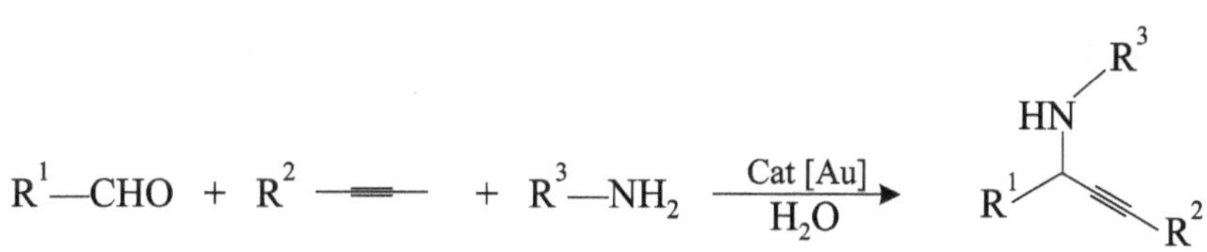

Fig. 2.116: Synthesis of propargyl amines using gold as catalyst.

In the above reaction (Fig. 2.116), it was found that AgCl, AgBr and AgI also showed good catalytic activity[175].

Use of an asymmetric aldehyde in the above reaction (Fig. 2.116) in presence of the tiridentate *bis* (oxazolinyl) pyridines (pybox) with Cu(OFT) yielded a product in excellent yield and enantioselectivity (99.6ee)[176].

The reaction of aldehyde, alkyne and amine could also be accelerated in presence of CuI in H_2O using microwave irradiation[177]. Using (*s*)-proline methyl ester as chiral source a highly diasteroselective method was developed for the synthesis of chiral propargyl amines.

***(v)* Addition of terminal alkynes to unsaturated carbonyl compounds:** Terminal alkynes were found to react with activated alkylidene Meldrums acid in presence of alkynyl copper reagents (generated in situ from terminal alkynes and catalytic $Cu(OAc)_2$ in presence of sodium ascorbate to give the adducts in water[178] (Fig. 2.117)

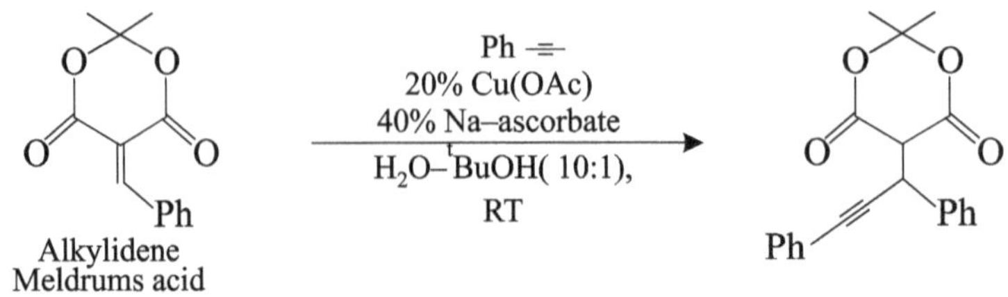

Fig. 2.117: Addition of terminal alkynes to alkylidene Meldrum's acid

The addition of terminal alkynes to activated alkynes in water catalysed by copper-palladium gave the coupled product[179]; in this case there is no homocoupling of terminal alkynes (Fig. 2.118)

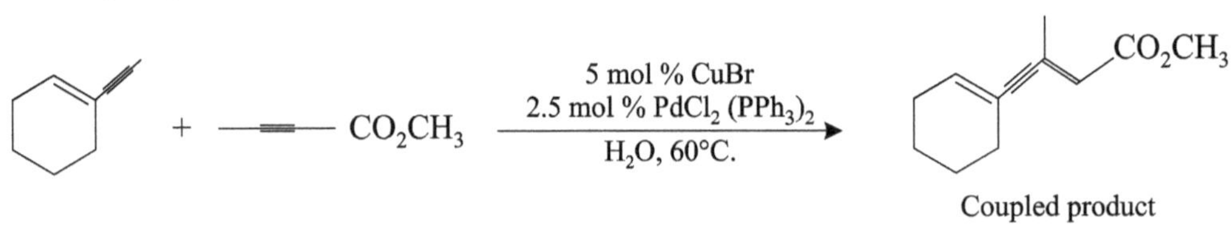

Fig. 2.118: Coupling of terminal alkynes to activated alkynes.

Terminal alkynes add on to α, β-unsaturated carbonyl compounds in water or acetone in air in presence of [Pd] gave the coupled product[179] (Fig. 2.119).

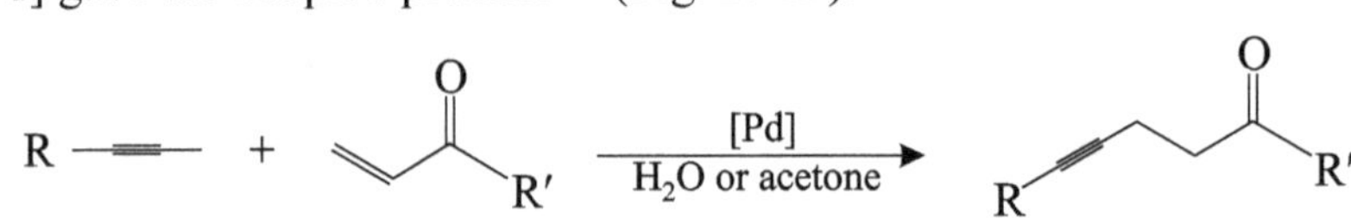

Fig. 2.119: Coupling of terminal alkynes with α, β-unsaturated carbonyl compounds.

(*vi*) Addition of water to terminal alkynes: It is well known that water adds on to terminal alkynes in presence of $HgSO_4/H_2SO_4$ to give ketones (Fig. 2.120).

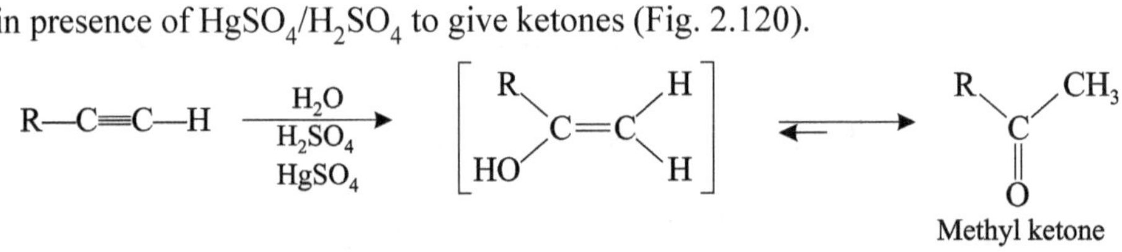

Fig. 2.120: Addition of H_2O to terminal alkyne in presence of $HgSO_4/H_2SO_4$.

The above conversion (Fig. 2.120) can be conveniently carried out in water-methanol in presence of gold catalyst[180] (Fig. 2.121).

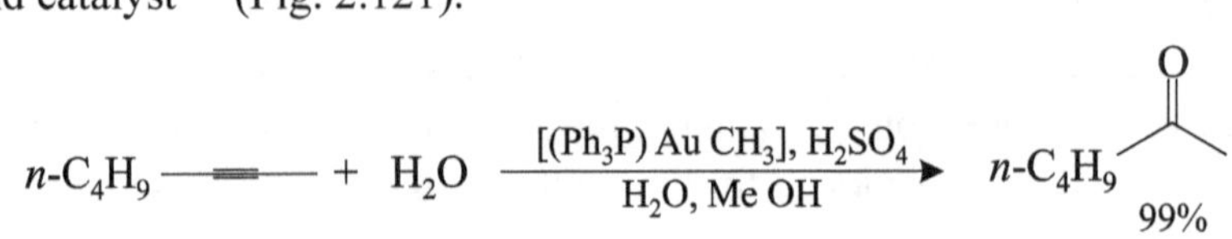

Fig. 2.121: Hydration of terminal alkyne in presence of gold catalysts.

(*vii*) Reaction of terminal alkynes with carbon monoxide-water (carbonylation of alkynes): The reaction of terminal alkynes with carbon monoxide-water in presence of catalytic amount of palladium iodide and potassium iodide in water-dioxane at 80° and 10 atm. of carbon monoxide gives 3-alkyl-or-3-aryl-substituted furan-2 (5H) ones. (Fig. 2.122)

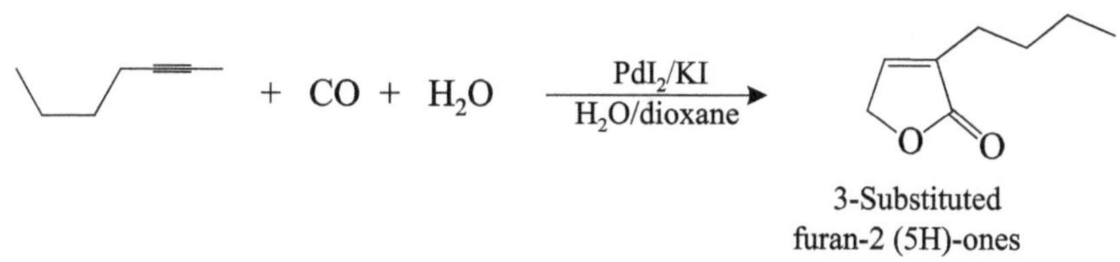

Fig. 2.122: Synthesis of 3-substituted furan-2 (5H) ones.

[B] Formation C–C bond from sp^2 C–H bonds

The hydrogen atoms of alkenes (sp^2 C–H) are much less acidic than those of ethyne (*sp* C–H) and are comparatively more defficult to remove. Some transition metals like palladium, ruthenium and rhodium etc. have been used to activate C–H bonds on arenes[181]. Some reactions of alkenes involving sp^2 C–H bonds are given below.

(*i*) Prins Reaction: It involves the reaction of olefins with aldehydes to give tetrahydropyran derivative (see Fig. 2.67). A direct formation of tetrahydropyranol derivatives in water has been reported using a cerium catalysed cyclization in aqueous ionic liquids[182] (Fig. 2.123).

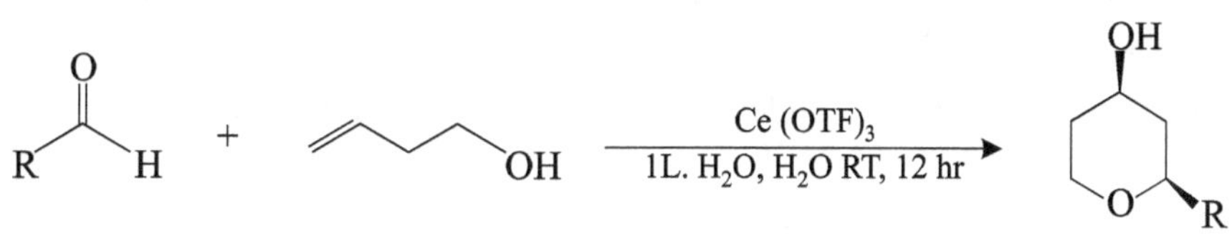

Fig. 2.123: Synthesis of tetrahydropyranol derivatives.

An **intramolecular prins type reaction** involved indium trichloride cyclisation of appropriately substituted alkene in aq. THF gave bicyclic compounds[183] (Fig. 2.124).

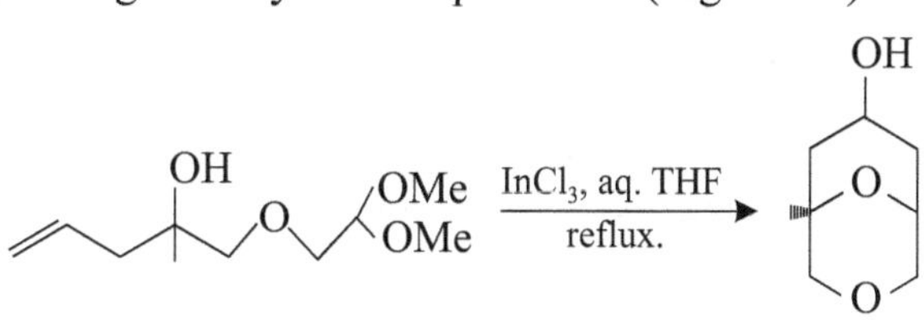

Fig. 2.124: Intramolecular prins type reaction.

(*ii*) Polymerization of Alkenes: Free radical polymerization can be carried out on a industrial scale in aqueous conditions[184].

(*iii*) Radical Additions: The radical addition of 2-iodoalkanamide to alkenols in water in presence of a water-soluble radical initator gives γ-lactones[184] (Fig. 2.125).

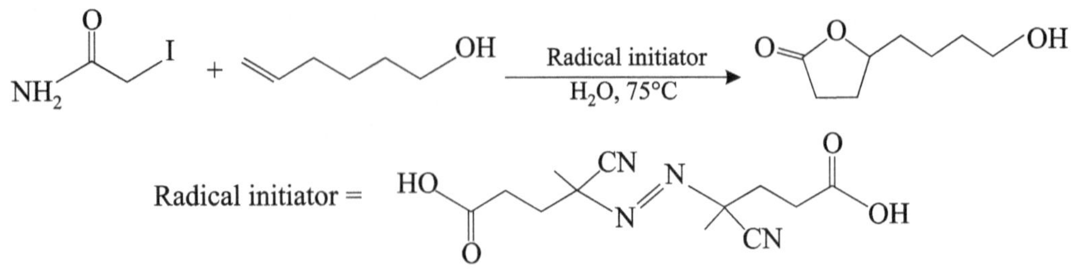

Fig. 2.125: Radical additions.

***(iv)* Oxidative dimerization of benzene:** Benzene on oxidative dimerization using a combination of Pd (OAc), a co-catalyst in presence of O_2 in AcOH (H_2O) give the dimerized product, biphenyl[185] (Fig. 2.126)

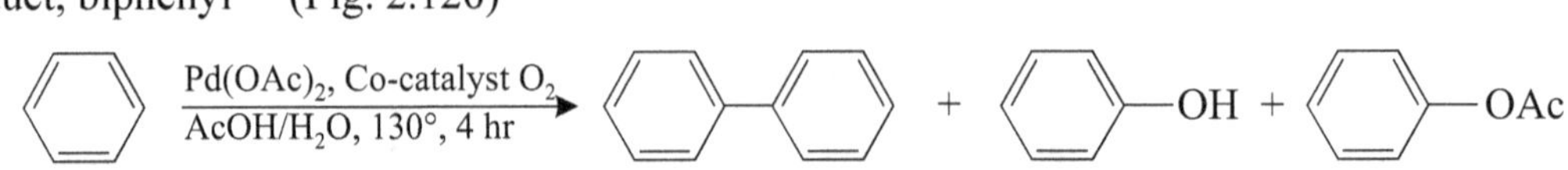

Fig. 2.126: Oxidative dimerization of benzene.

***(v)* Heck Reaction:** It involves coupling of an alkene with a halide or triflate in presence of Pd(O) catalyst to form a new alkene. For details see section 2.2.1.3 (Fig. 2.87, 2.88).

***(vi)* Electrophilic substitutions of aromatic compounds:** Electrophilic substitutions of aromatic hydrogens proceed under aqueous conditions. A common example is the known **Friedal-crafts type reactions**. Indole derivatives on reaction with equimolar amounts of 3 % aqueous CH_2O and 33% aqueous Me_2NH at 75°C in 95% ethanol gave Mannich type products[186].

The reaction of indole with benzaldehyde in presence of a lanthanide catalyst gave the product is ethanol-water[187] (Fig. 2.127).

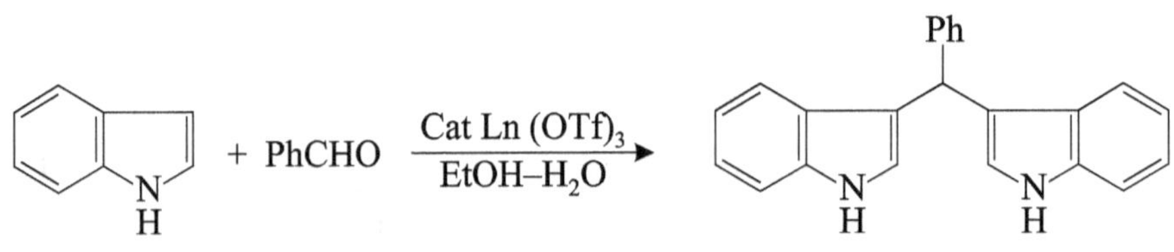

Fig. 2.127: Reaction of Indole with benzaldehyde.

Friedal-crafts reaction of N-methyl indole and N-methyl pyrrole with $OHCCO_2Et$ in aqueous solution gave substituted indoles and pyrroles without using any metal catalyst[188]. An example is given in Fig. 2.128.

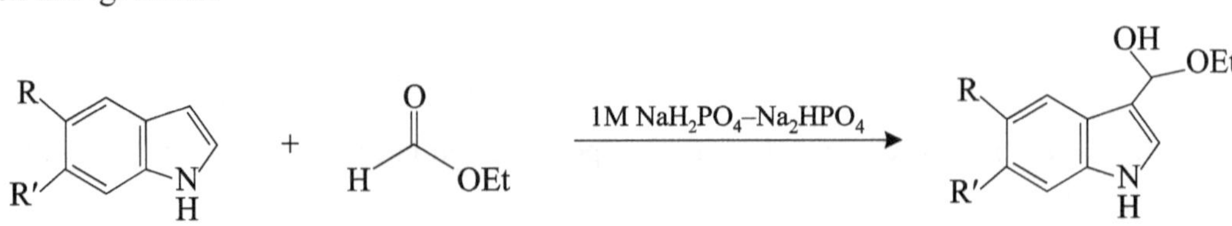

Fig. 2.128: Friedel-crafts reaction of indoles with ethyl formate.

Aromatic compounds on Friedal-crafts reaction with methyl trifluoropyruvate in water gave the corresponding α-hydroxy ester[189] (Fig. 2.129).

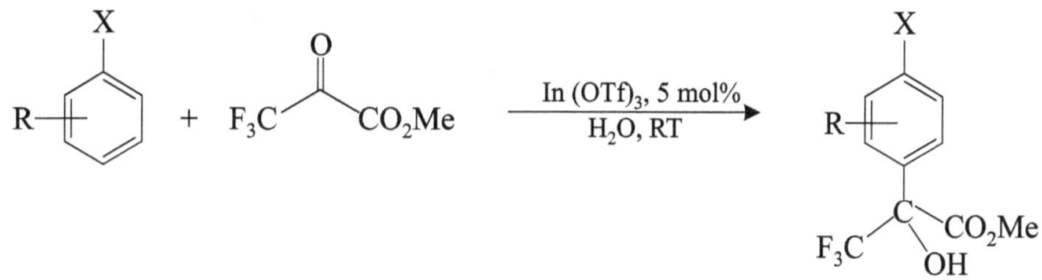

Fig. 2.129: Friedal-crafts reaction of aromatic compounds with methyl trifluoropyruvate.

[C] Formation of C–C bond from sp^3 C–H bonds

The sp^3 C–H bonds are considered to be least reactive compared to sp^2 C–H and *sp* C–H bonds. In fact, in conventional organic chemistry sp^3 C–H bonds (alkanes) are considered to be nonreactive. However, the alkanes react only under drastic conditions. In fact, some examples have been reported during the past 20 years or so in which the sp^3 C–H bonds take part in reactions under milder conditions. Following are given some of the examples of this type.

***(i)* Reaction of methane with carbon Monoxide:** Methane couples with CO to generate acetic acid in water in presence of catalyst[190] (Fig. 2.130).

$$CH_4 + CO \xrightarrow[\text{Water, pH 7.3}]{\text{air, Na VO}_3\text{, 80°C}} CH_3COOH + MeOH + HCHO \cdot$$

Fig. 2.130: Reaction of methane with CO.

***(ii)* Carbene Insertion into sp^3 C–H bond:** Many reactive carbenes insert into a C–H bond either by one step process or a two step process (Fig. 2.131).

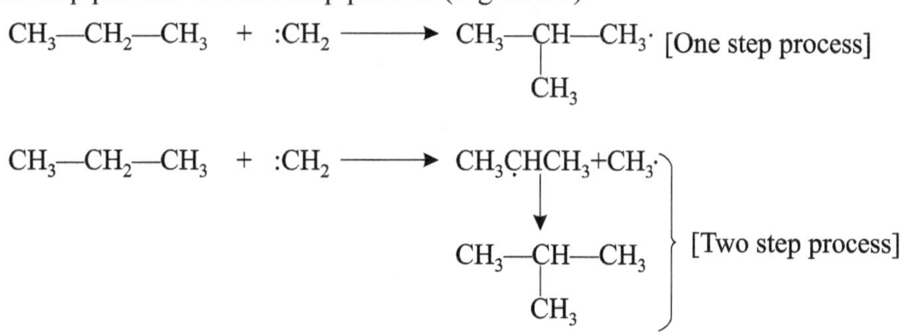

Fig. 2.131: Insertion of carbene into sp^3 C–H bond.

Carbethoxy carbene (generated in situ from ethyl diazoacetate) insert at tertiary C–H bonds about 3 times as fast as C–H bonds in simple alkene[190].

Intramolecular insersion reactions usually occur with C–H bond that is closest to the carbene and good yields are obtained. These reactions provide routes to highly strained structures that are difficult to obtain by other procedures. Rhodium carboxylates have been found to be effective catalysts for intramolecular C–H insersion reactions of α-diazoketones and esters[191].

Dichlorobenzene (generated by the PTC technique)[192] undergoes insersion reactions into C–H bonds. Some examples are given below[193,194] (Fig. 2.132).

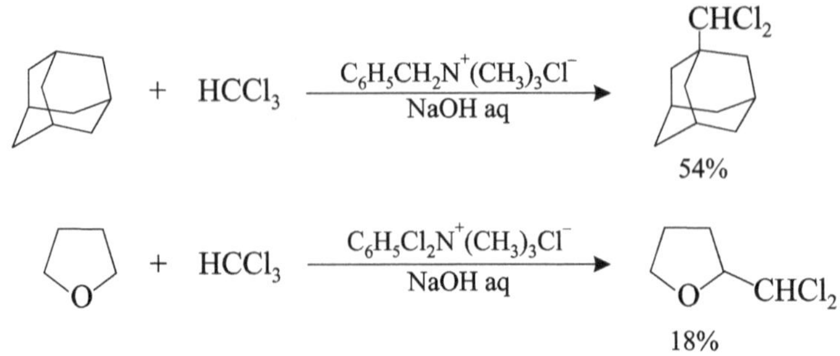

Fig. 2.132: Some examples of insersion of dichloro carbene into C–H bonds.

2.2.1.5 Miscellaneous organic reactions in water

In the foregoing sections a number of reactions in water have been described. Besides these following are given some other organic reactions in water.

***(i)* Acetoacetic Ester Synthesis:** The reactive methylene group in acetoacetic ester can be alkylated with alkyl halide in presence of aqueous KOH, K_2CO_3 and or PTC (tetrabutyl ammonium chloride) and irradiation with microwaves[195] (Fig. 2.133).

$$CH_3COCH_2CO_2Et \xrightarrow[\text{MW 3 mm}]{\text{RX, KOH, } K_2CO_3\text{, PTC}} CH_3CO\underset{}{\overset{\displaystyle R \atop |}{C}}H\ CO_2Et$$

$$R = CH_3CH_2CH_2,\ PhCH_2,\ m\text{-}CH_3O\ C_6H_4\ CH_3,\ p\ Cl\ C_6H_4\ CH_2CH(CH_3)_2$$

Fig. 2.133: Alkylation of ethyl acetoacetate.

The formed substituted ethyl acetoacetate on hydrolysis with dilute alkali or HCl followed by decarboxylation gives substituted carboxylic acids or ketones.

***(ii)* Acyloin Condensation:** The acyloin condensation can be carried out under green conditions[196] using coenzyme thiamine (Fig. 2.134).

$$CH_3CHO \xrightarrow[H_2O]{\text{Thiamine}} CH_3CO\ \underset{\displaystyle | \atop \displaystyle OH}{C}HCH_3$$

Acyloin

Fig. 2.134: Acyloin condensation using thiamine

The acyloin condensation of aliphatic or aromatic aldehyde can also be performed by stirring with a PTC in water[197].

***(iii)* Alkylations:** As already stated compounds containing reactive methylene group can be alkylated in a MW owen with alkyl halides in presence of KOH, K_2CO_3 and a PTC (see Fig. 2.133). Some other examples include alkylation of phenyl sulfonyl acetate[198] [Fig. 2.135, equation (*i*)],

Nitriles having α-hydrogens[199] [Fig. 2.135, equations (*ii*), (*iii*) and (*iv*)], aldehydes containing α-hydrogen[199] [Fig. 2.135, equation (*v*)] and ketones[200] [Fig. 2.135, equation (*v*)].

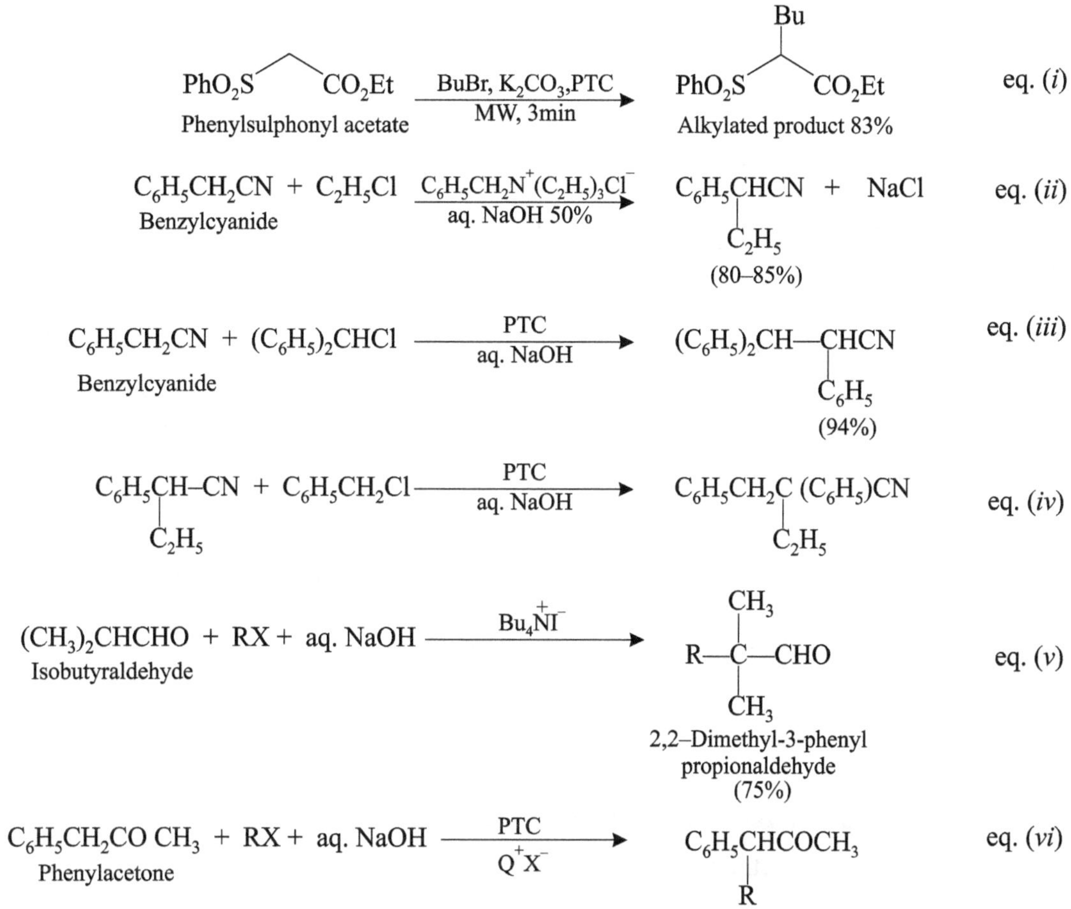

Fig. 2.135: Alkylations of phenylsulphonyl acetate, nitriles, aldehydes and ketones.

***(iv)* Baker-Venkataraman Rearrangement:** It involves[201] the synthesis of diketones (one step process) involving the reaction of *o*-hydroxyacetophenones with benzoyl chloride in presence of a PTC and aq. K_2CO_3 (Fig. 2.136).

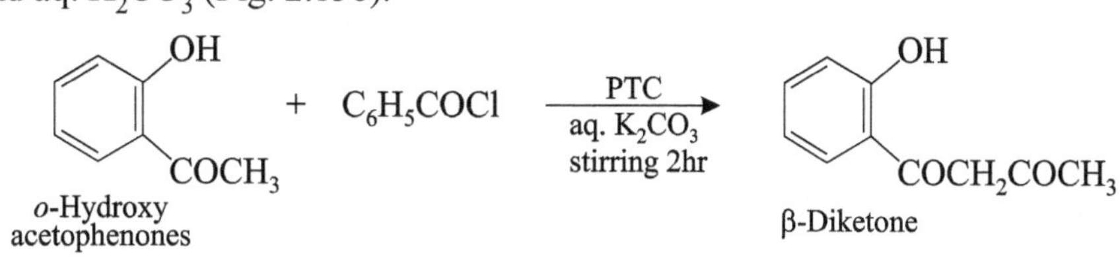

Fig. 2.136: Synthesis β-Diketones.

***(v)* Claisen Rearrangement:** The claisen rearrangement of allyl phenyl ether in aqueous media using microwave irradiation gave the rearranged product in 10 mim[202] (Fig. 2.137).

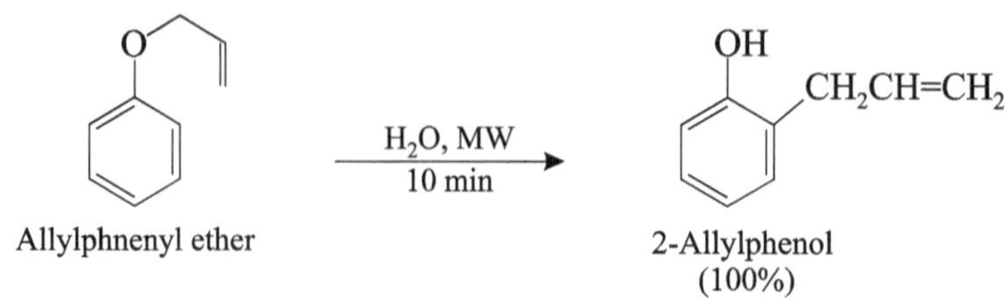

Fig. 2.137: Claisen Rearrangement in water

***(vi)* Darzen Reaction:** The Darzen Reaction occurs in alkaline solution in presence of PTC (benzyltriethyl ammonium chloride)[203] (Fig. 2.138).

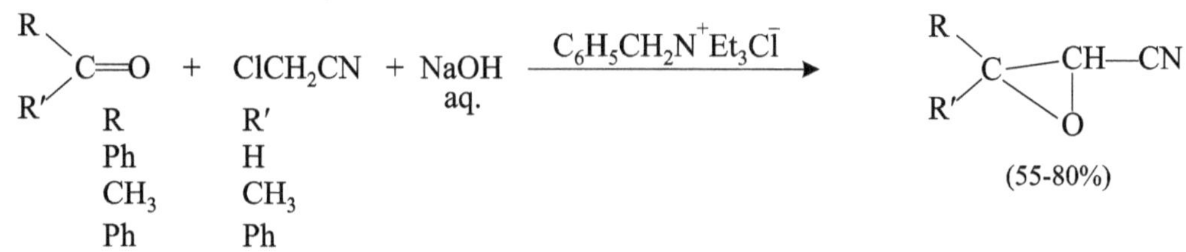

Fig. 2.138: Darzen reaction in water.

With aldehydes and unsymmetrical ketones both possible stereoisomers are obtained. However, with more acidic ketones, such as phenyl acetone, the ketone carbanion is formed rather than from the nitrile (which is normally the case) leading to alkylation of the ketone (Fig. 2.139).

$$\underset{\text{Phenyl acetone}}{C_6H_5CH_2COCH_3} + \underset{\substack{\text{Chloro}\\ \text{acetonitrile}}}{ClCH_2CN} + \underset{\text{aq}}{NaOH} \xrightarrow{\text{PTC}} C_6H_5 - \underset{\substack{|\\ CH_2CN}}{CHCOCH_3}$$

Fig. 2.139: Darzen reaction of phenyl acetone with chloroacetonitrile in water.

In place of PTC, crown ethers can also be used[204].

***(vii)* Photochemical Cycloaddition Reactions:** Photodimensation of thymine, uracil and their derivatives increased considerably in water[205] in comparison to other organic solvents (Fig. 2.140).

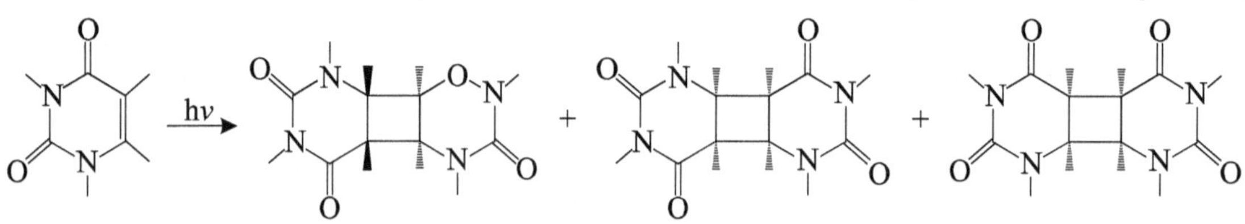

H_2O	27.8%	63.1%	9.1%	θ=0.013
Acetonitrile	24.9%	68.2%	6.7%	θ=0.004
Methanol	31.4%	68.6%	—	θ=0.004

Fig. 2.140: Photochemical cycloaddition in water.

A number of other photochemical cycloadditions like that of stilbene, coumarin and anthracene-2-sulphonate have also been reported in water[206].

(*viii*) Pinacol-Pinacolone Rearrangement: Pinacol-Pinacolone rearrangement can be performed in water[207,208] by heating in a microwave oven at 270°. At this temperature water acts as an acid catalyst.

(*ix*) Reimer-Tiemann Reaction: Phenols on reaction with chloroform in presence of aqueous sodium or potassium hydroxide gives *o*-hydroxy aldehydes as the major product[209] (Fig. 2.141).

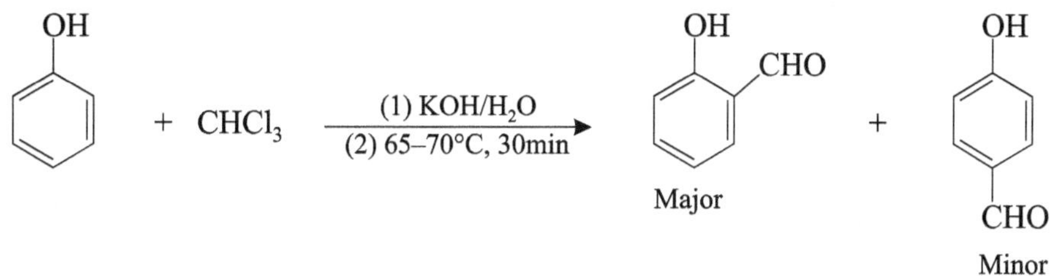

Fig. 2.141: Reimer-Tiemann Reaction.

Much better yields are obtained if the reaction is carried out in presence of a PTC (Tetrabutyl ammonium bromide)[210].

(*x*) Claisen-Schmidt Reaction: The reaction of aromatic aldehydes (without α-hydrogen) with aliphatic aldehydes or ketones (having α-hydrogen) in presence of 10% aqueous NaOH give α, β-unsaturated aldehydes or ketones[211] (Fig. 2.142)

$$C_6H_5CHO + CH_3CHO \xrightarrow{\text{10\% aq. NaOH}} \underset{\text{Cinnamaldehyde}}{C_6H_5CH = CHCHO}$$

$$C_6H_5CHO + CH_3COCH_3 \xrightarrow[\text{2-3 days 7\%}]{\text{10\% aq. NaOH}} \underset{\text{Benzyldene acetone (Benzalacetone)}}{C_6H_5CH = CHCOCH_3}$$

$$C_6H_5CHO + CH_3COC_6H_5 \xrightarrow[\text{<30°, 85\%}]{\text{10\% NaOH}} \underset{\text{Benzylidene acetophenone (Benzalacetophenone)}}{C_6H_5 - CH = CHCOC_6H_5}$$

Fig. 2.142: Claisen-Schmidt Reaction.

Much better yields are obtained if the above reaction is carried out in presence of a PTC (such as cetyl trimethyl ammonium compounds) in presence of dilute aq. alkali[212] (Fig. 2.143)

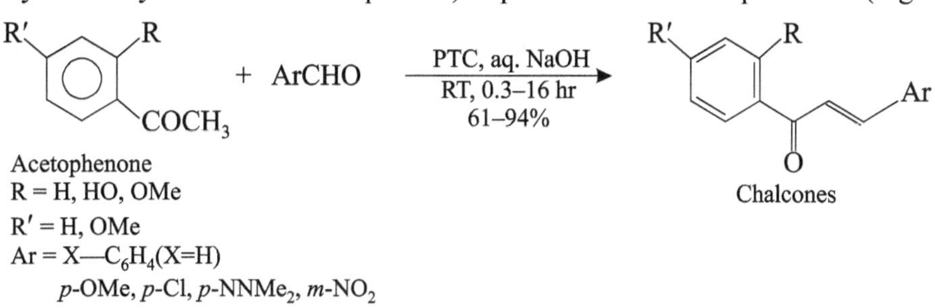

Fig. 2.143: Claisen Smith reaction in presence of PTC.

(*xi*) Knoevenagel Condensation: The knoevengel reaction between aldehydes and acronitrile has been performed in water. Thus, the reaction of salicyaldehyde with malononitrile in heterogeneous alkaline medium at room temperature gave *o*-hydroxy benzylidene malononitrile[213] which on heating with acid gave 3-cyanocomarins in good yield. (Fig. 2.144)

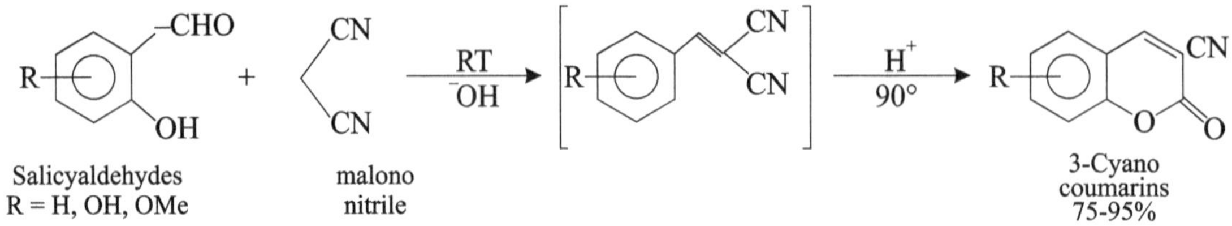

Fig. 2.144: Knolvenagel condensation.

The reaction of substituted malonunitriles, R–CH_2CN (R = CN, CO_2Et, NO_2, 2-Py) requires the presence of PTC (CTABr).

(*xii*) Meyer-Schuster Rearrangement: The acid catalysed rearrangement of acetylenic alcohols gives α, β-unsaturated carbonyl derivatives[214] (Fig. 2.145)

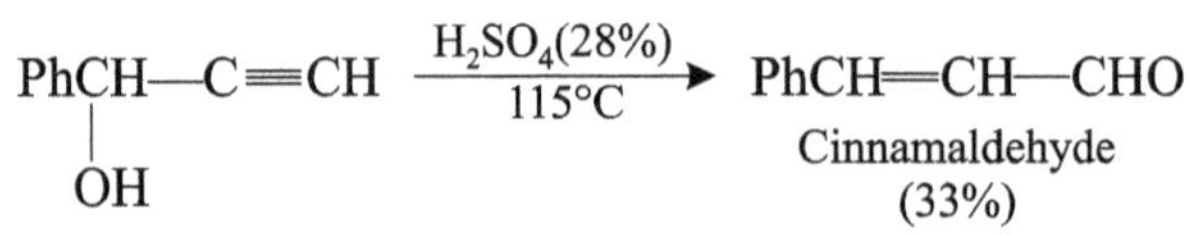

Fig. 2.145: Meyer-Schuster Rearrangement.

2.2.1.6 Formation of C–O Bonds

Next to C–C bond formation, C–O bond formation is important for organic synthesis. Following are given some of the methods used for C–O bond formation.

(*i*) Baeyer-Villiger Oxidation: Ketones on oxidation with H_2O_2 or with peracids (RCO_3H) give esters[215]. This reaction has been satisfactorily carried out in aqueous heterogeneous medium with *m*-chloroperbenzoic acid[216] (*m*CPBA) (Fig. 2.146)

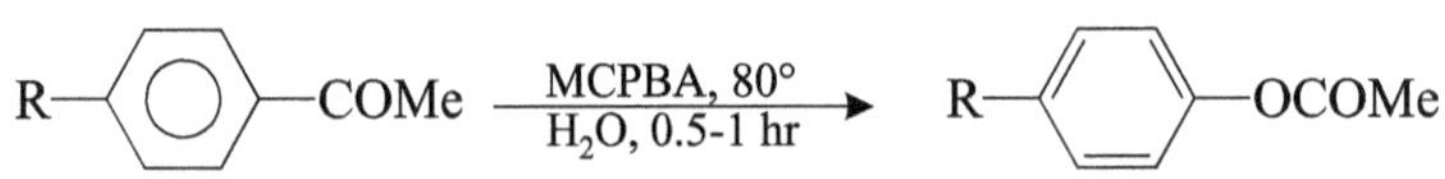

Fig. 2.146: Bayer-Villiger oxidation in water.

(*ii*) Benzil-Benzilic acid Rearrangement: α-Diketones (benzils) undergo base catalysed reaction to give benzilic acid[217]. (Fig. 2.147)

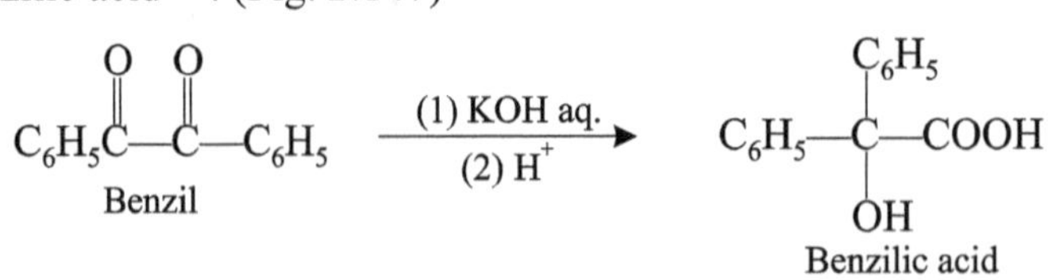

Fig. 2.147: Benzil-Benzilic acid rearrangement .

(*iii*) Cannizzaro Reaction: Aldehydes without α-hydrogen on treatment with conc. aqueous alkali undergo self oxidation and reduction to give an alcohol and the salt of the corresponding carboxylic acid[218]. (Fig. 2.148)

$$\underset{\text{Benzaldehyde}}{C_6H_5CHO + C_6H_5CHO} \xrightarrow{KOH} \underset{\text{Benzyl alcohol}}{C_6H_5CH_2OH} + \underset{\text{Pot. benzoate}}{C_6H_5CO\bar{O}\bar{K}}$$

Fig. 2.148: Cannizzaro Reaction

An interesting example is the conversion of formaldehyde into methanol (Fig. 2.149).

$$\underset{\text{Formaldehyde}}{HCHO} + \underset{\text{aq.}}{NaOH} \xrightarrow{\Delta} \underset{\text{Methyl alcohol}}{CH_3OH} + \underset{\text{Methyl formate}}{HCOONa}$$

Fig. 2.149: Conversion of formaldehyde into methyl alcohol.

***(iv)* Dakins Reaction:** The oxidation of hydroxy aldehydes with H_2O_2/aq. NaOH gives dihydroxy compound (–CHO or –$COCH_3$ is converted into OH)[219] (Fig. 2.150)

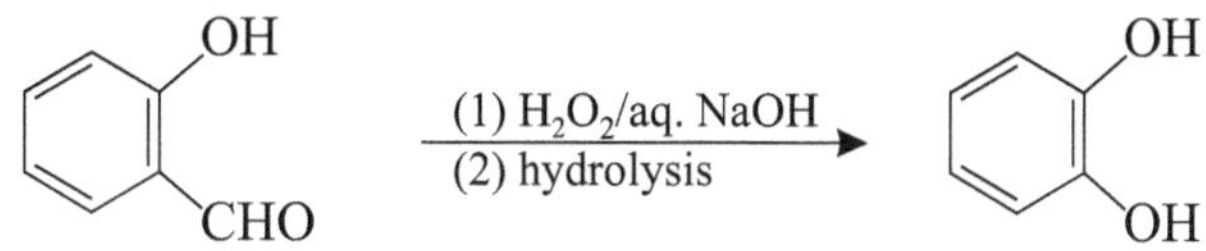

Saliylaldehyde Catechol

Fig. 2.150: Dakins oxidation.

***(v)* Elb Persulphate Oxidation:** Phenols on oxidation with potassium persulphate[220] in aqueous alkaline solution give hydroxy phenols. The –OH group enters para position with respect to the original OH group (Fig. 2.151)

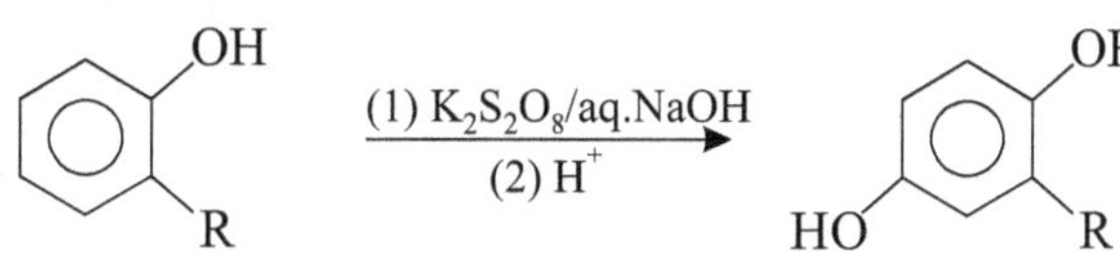

R = H, Cl, CHO, NO_2

Fig. 2.151: Elb Persulphate Oxidation.

***(vi)* Transition-Metal-Catalysed Oxidations:** It was believed that hydroxylation of linear alkanes or methane to give terminal alcohol is very difficult. The first example of electrophilic activation of alkanes in aqueous solution was carried out by Shilvov *et al* (1970s) using platinum complexes in aqueous solutions[220,221] (Fig. 2.152)

$$R - H + PtCl_6^{2-} + \text{cat Pt } Cl_4^{2-} \xrightarrow[100-120^\circ]{H_2O} R - OH$$

Fig. 2.152: Hydroxylation of linear alkanes in water.

Though the yields in the above hydroxylation is low, but the system is highly attractive due to its unique selectivity[221] under mild conditions. It has been found that linear alkenes are functionalized at the terminal position of stronger C–H bonds rather than the weaker nonterminal secondary C–H bonds.

The system, known as Shilov system has been used for the oxidation of methane and ethane to give methanol and ethyene glycol respectively[222]. For these oxidations K_2PtCl_6/K_2PtCl_4 system[223] was used.

Using a combination of metallic palladium and copper chloride in presence of molecular oxygen and carbon monoxide as coreductant methane could be oxidised to methanol[224] (Fig. 2.153).

$$CH_4 \xrightarrow[\substack{CO,\ O_2 \\ H_2O,\ CF_3COOH(1:3) \\ 80\text{–}85\%\ ^\circ C,\ 20\ hr}]{Cu(II)Pd(O)} CH_3OH$$

Fig. 2.153: Oxidation of methane to methanol.

Some other transition-metal-catalysed oxidations has also been reported[225].

Besides the metal complexes used for activation of C–H bonds, other methods have also been used. These include techniques like photochemical oxidations, free radical reactions and carbene and superacid reactions[226]. Such techniques have also been used at industrial scales[227].

Oxyfunctionalization of C–H bonds can also occur with dioxiranes[228] in water. Thus, alkylketones and alkylketoesters on treatment with dioxiranes (generated in situ by oxone in H_2O/MeCN (1:1.5) gave[229] the corresponding δ-hydroxy ketone, which cyclised to give a hemiketal (which prevents further oxidation at δ-position (Fig. 2.154).

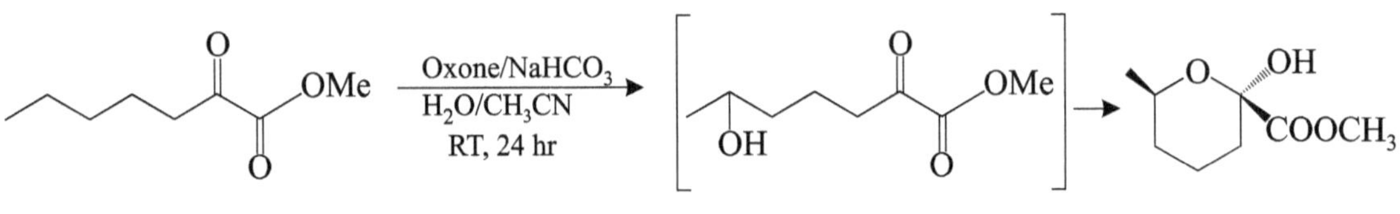

Fig. 2.154: Use of dioxiranes for oxyfunctionalization of C–H bonds.

(*vii*) Enzymatic oxidation: A large number of enzymatic oxidations have been carried out in aqueous medium for the formation of C–O bonds. The aspect has been discussed subsequently (see Chapter 13.

2.1.1.7 Formation of C-Halogen Bonds

(*i*) Aqueous Phase Reactions: It is well known that C-halogen bonds are formed via free radical halogenations[226] of alkanes (Fig. 2.155).

$$\underset{\text{excess}}{CH_4} + Cl_2 \xrightarrow[\text{or heat}]{\text{light}} CH_3Cl + HCl$$

$$\underset{\text{excess}}{CH_4} + Br_2 \xrightarrow{\text{light}} CH_3Br + HBr$$

Fig. 2.155: Free radical halogenation of alkanes to form C-halogen bond.

It has, however, been found that water is an excellent solvent for radical reactions due to the presence of strong O–H bond due to which abstraction of hydrogen from the solvent is most unlikely. Thus, halogenation (chlorination, bromination and iodination) of various alkanes could be carried out by reaction of alkanes with CCl_4, CBr_4 or CHI_3 in aqueous medium in presence of a phase transfer catalyst, a quaternary ammonium salt[230] (which prevents over functionalization of the products). This reaction is highly regionselective in the attachment of halogen in the tertiary C–H bonds. As an example in adamantane, the halogenation takes place in the bridge head position even in presence of methyl group[231] (Fig. 2.156).

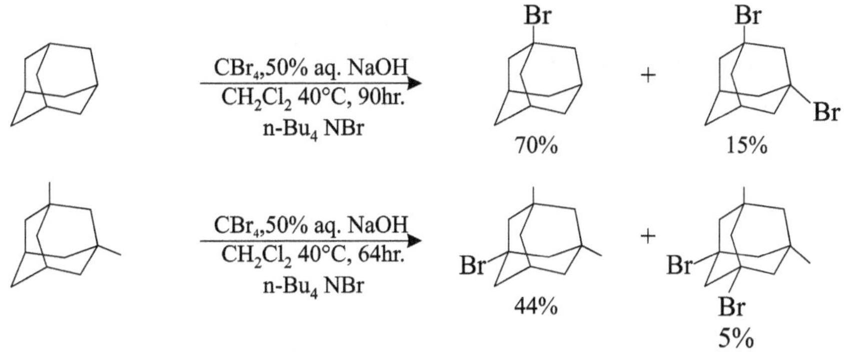

Fig. 2.156: Bromination of adamantanes.

It is appropriate to state that dichloro carbene (generated by PTC technique) ungoes insersion reactions into C–H bonds particularly in case of adamantane (see Fig. 2.132).

Free radical halogenation also takes place in the photochemical bromination of a number of alkanes and alkyl arenes with bromine in water[232]. It is believed that the use of water provides a diluting medium in which free radicals are uniformly distributed, thus reducing the overbromination of alkanes.

Bromination of alkylarenes has also been reported[233] with a mixture of $NaBrO_3/NaHSO_3$ in aqueous media under mild conditions.

***(ii)* Formation of C-halogen bonds from other functional groups:** Besides the procedures (stated above) for C-halogen bond formation, these can also be formed by functional group transformations, some of which are given below.

***(a)* From alcohols:** Alcohols on reaction with dichlorocarbene generated in situ by PTC procedure give good yields the corresponding chlorides[234] (Fig. 2.157)

$$R-OH+CHCl_3+\underset{\text{aq.}}{NaOH}\xrightarrow{C_6H_5CH_2N^+Et_3Cl^-}R-Cl+NaCl+H_2O$$

Fig. 2.157: Conversion of alcohol into the corresponding chlorides.

***(b)* Hunsidiecker Reaction:** The reaction of silver salt of carboxylic acids with Br_2/CCl_4 give alkyl/aryl bromides[235] (Fig. 2.158).

$$\underset{\text{Carboxylic acid}}{R-COOH}\xrightarrow{AgNO_3}\underset{\text{Silver salt}}{RCOOAg}\xrightarrow[CCl_4]{Br_2}R-Br+AgBr+CO_2$$

Fig. 2.158: Hunsidiecker reaction

The above transformation can also be performed by the reaction of carboxylic acids with Br_2 by photolysis in presence of mercuric oxide (Fig. 2.159).

$$\underset{p\text{-Chloro benzoic acid}}{ClC_6H_4COOH}+Br_2\xrightarrow{HgO,hv}\underset{\substack{p\text{-Bromochlorobenzene}\\80\%}}{ClC_6H_4Br}$$

Fig. 2.159: Photolytic bromination of carboxylic acids.

The above reaction is known as **Hunsidiecker-Borodin-crystol-Forth reaction**[236].

***(c)* Hell-volhard-Zelinsky Reaction:** Carboxylic acids having CH_2 group in the α-position on treatment with Br_2/P give the corresponding α-bromo compounds[237]. (Fig. 2.160)

$$C_6H_5-CH_2COOH \xrightarrow{Br_2/P} C_6H_5\underset{\displaystyle |\atop Br}{CH}-COOH$$

Fig. 2.160: HCl-Volhard-Zelinsky Reaction.

***(d)* Sandmeyer Reaction:** Diazotisation of arylamines with sodium nitrite and dilute HCl followed by treatment of the formed diazonium salt with cuprous halides gives[238] C-halogen bonds (Fig. 2.161).

$$ArNH_2 \xrightarrow[0°]{NaNO_2/\text{dil HCl}} Ar\ N_2^+ \xrightarrow[X=I,\ Br,Cl]{CuX} Ar-X$$

Fig. 2.161: Sandmeyer's reaction.

***(e)* From alkanes:** Unsymmetrical alkanes on treatment with HBr gives bromocompounds as per Markovnikov rule. However, in presence of peroxide, anti-Markovnikov addition takes place (Fig. 2.162).

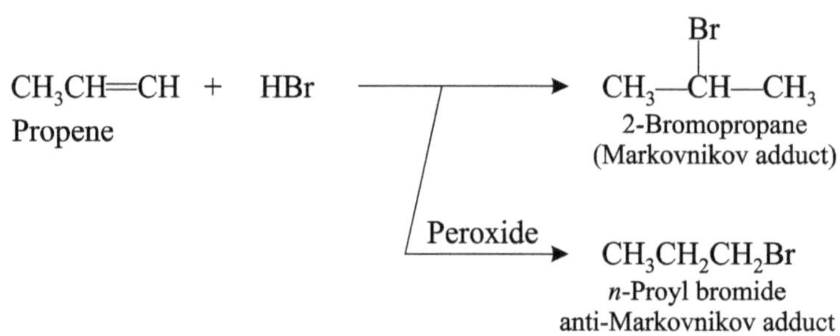

Fig. 2.162: Preparation of bromo compounds from unsymmetrical alkenes.

2.2.1.8 Formation of C–N bonds

The carbon-nitrogen can bond by single, double or triple bonds, all of which are discussed in this section. Following are given some of the reactions used for C–N bond formations. As will be seen, in most of the reaction along with C–N bonds, C–C bonds are also formed.

***(i)* Aza-Diels-Alder Reaction:** The reaction of dienes and iminimum salts (nitrogen containing dienophiles) generated in situ under Mannich like conditions[239] (Fig. 2.163)

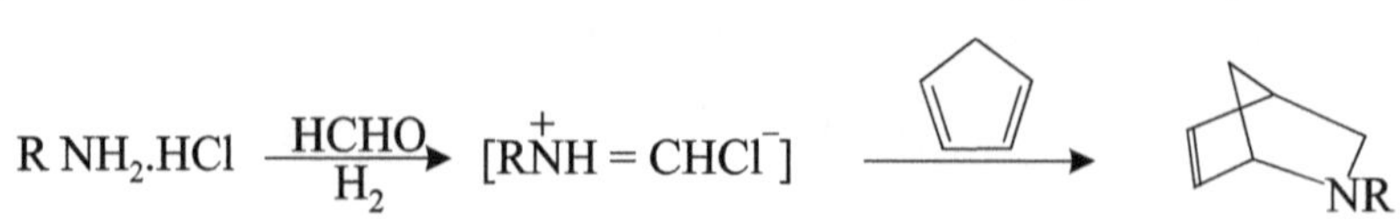

Fig. 2.163: Aza-Diels-Alder-Reaction.

***(ii)* Beckmann Rearrangement:** Normally in Beckmann reaction, the ketoximes are heated with acidic reagents like PCl_5, HCOOH, $SOCl_2$ etc. to give anilides[240]. However a green procedure[241] is carried out in solid state microwave assisted reaction, involves the oxime is mixed with montmorillonite K-10 clay in dry medium and mixture heated in a microwave for 7-10 min to give corresponding anilide in 91% yield (Fig. 2.164).

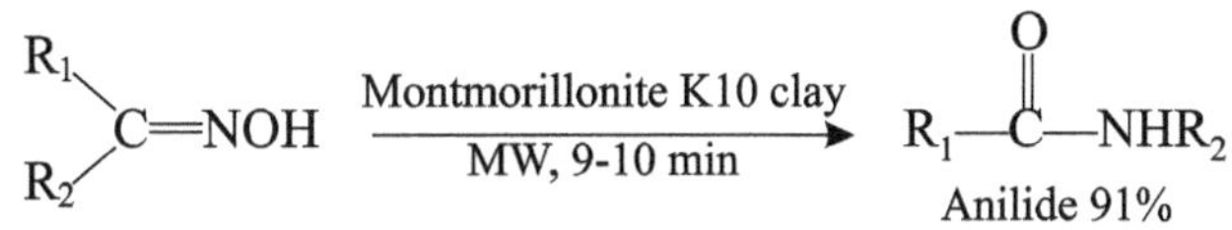

Fig. 2.164: A Green Beckmann Reaction.

***(iii)* Beginelli Reaction:** Acid catalysed condensation of an aldehyde, α, β-ketoester and urea gives tetrahydropyrimidones[242] (Fig. 2.165)

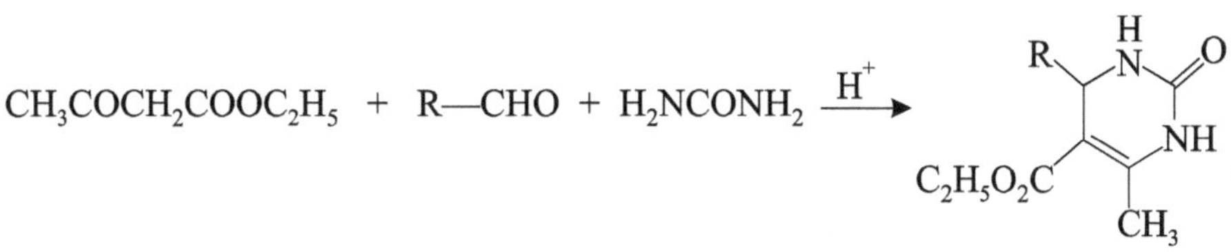

Fig. 2.165: Beginelli Reaction.

A green Beginelli reaction consist in mixing the reagents in presence of catalytic amount of clay and heating in a microwave oven[243] (60–95% yield). Alternatively, ionic liquids such as [b_{min}] [BF_4] and [b_{min}] [PF_6] have also been used as catalyst for the Beginelli reaction[244].

***(iv)* Curtius Rearrangement:** Thermal rearrangement of acyl azides (prepared from acyl chlorides by reaction with sodium azide or hydrazine) give isocyanates, which on hydrolysis give amines[245] (Fig. 2.166).

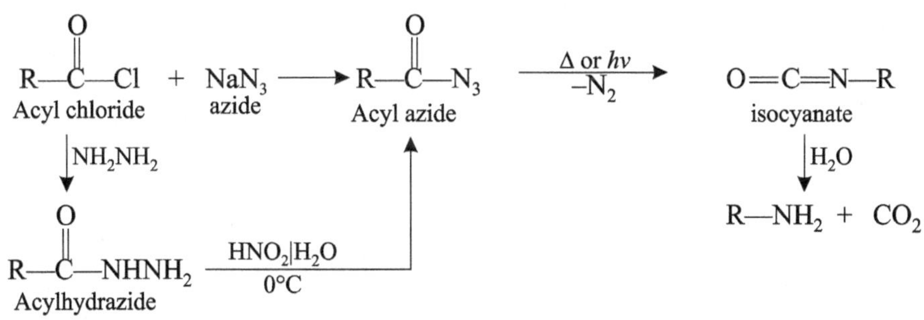

Fig. 2.166: Curtius rearrangement.

Tertiary alkyl azides on curtius rearrangement gives imines[246] (Fig. 2.167)

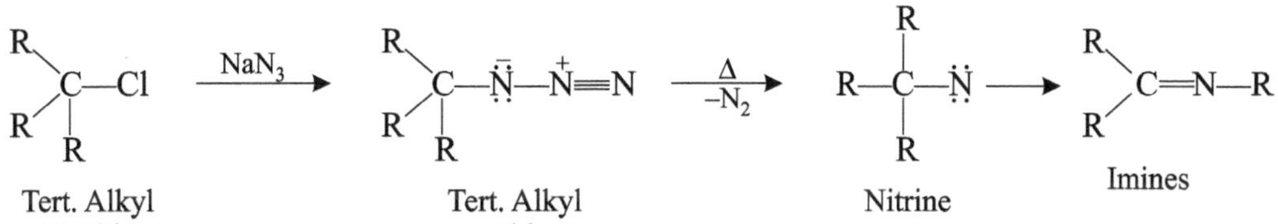

Fig. 2.167: Curtius arrangement of tertiary alkyl azides

(*v*) **Doebner-Miller Synthesis:** It involves the reaction of primary amines with aldehydes in presence of H_2SO_4 to give quinoline derivatives[247] (Fig. 2.168)

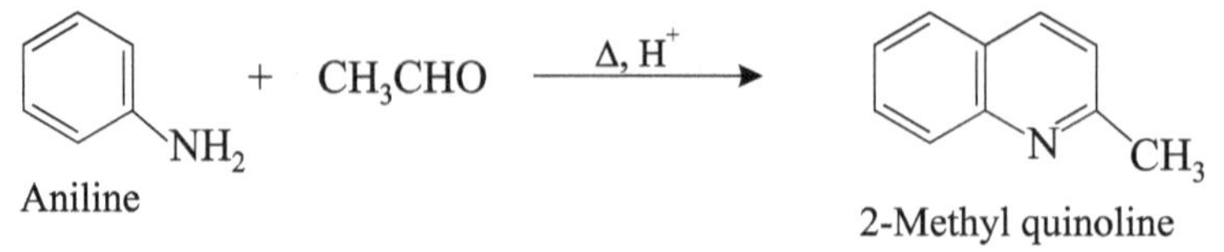

Fig. 2.168: Doebner-Miller Synthesis.

(*vi*) **Fischer-Indole Synthesis:** It consists in heating aryl hydrozones of aldehydes and ketones in presence of $ZnCl_2$ or polyphosphoric acid (PPA) or BF_3 to give indoles[248] (Fig. 2.169).

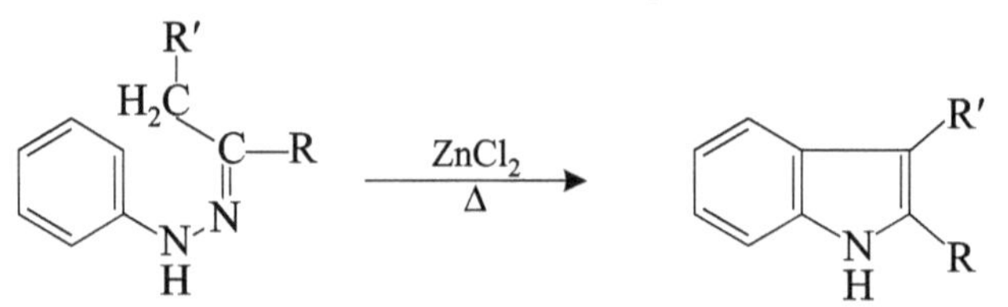

Fig. 2.169: Fischer-Indole Synthesis.

The reaction could be conducted in solid state by heating cyclohexanone and phenyl hydrazine in presence of montomorillonite KSF clay using microwave irradiation to give carbazole[249]. Alternatively, the Fischer Indole reaction can also be performed in water by heating at 220° a mixture of phenyl hydrazine and ethyl methyl ketone in microwave oven to give 2,3-dimethyl indole[250] (Fig. 2.170).

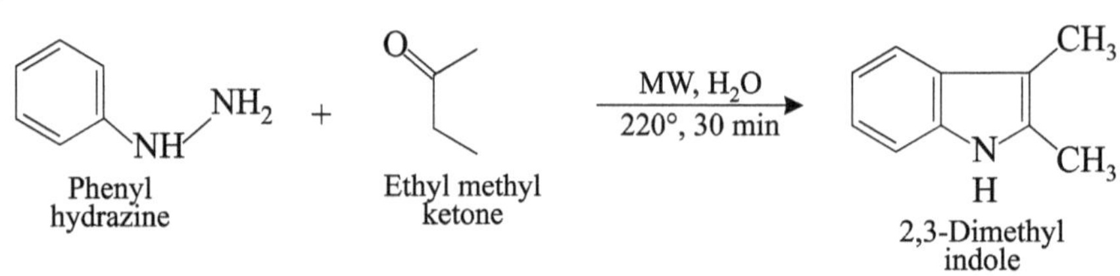

Fig. 2.170: Fischer-Indole reaction in water

(*vii*) **Friedlander Synthesis:** It involves[251] base catalysed reaction of 2-aminobenzaldehydes with ketones to give quinoline derivatives (Fig. 2.171).

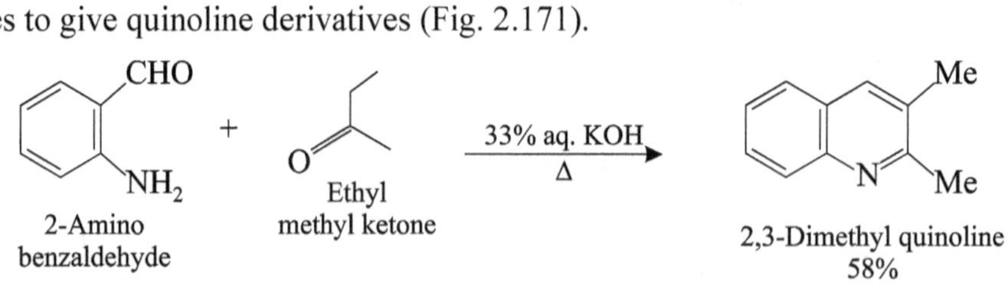

Fig. 2.171: Friedlander Synthesis.

A Green Friedlander synthesis involves reaction of 2-amino aldehydes or ketones with carbonyl compounds containing α-methylene group in presence of KSF clay and heating in a MW oven to give quinoline derivatives[252] (Fig. 2.172).

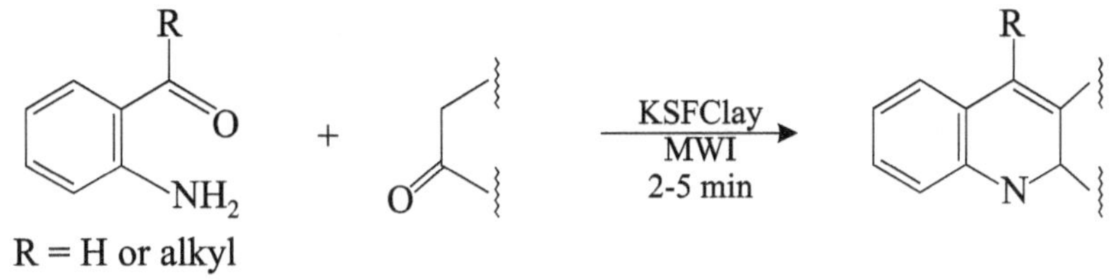

Fig. 2.172: A Green Friendlander synthesis.

(*viii*) **Graebe-Ullmann Synthesis:** Diazotisation of 2-aminodiphenyl amine gives benzotriazole, which on decomposition gives carbazoles[253] (Fig. 2.173)

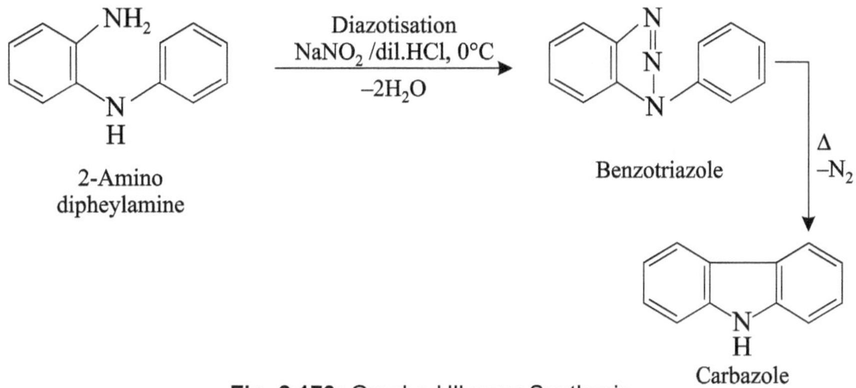

Fig. 2.173: Graebe-Ullmann Synthesis.

A Green Graebe-Ullmann synthesis involves heating a mixture of benzotriazole and chloropyridine in microwave oven in presence of pyrophosphoric acid to give β-carboline[254] (Fig. 2.174)

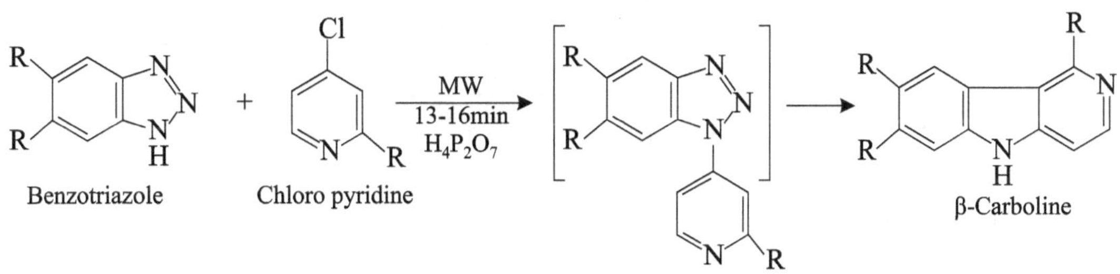

Fig. 2.174: A Green Graebe-Ullmann Synthesis.

(*ix*) **Hantzsch Pyridine Synthesis:** Reaction of ethyl acetoacetate with an aldehyde in presence of ammonia followed by oxidation of the resultant dihydropridine derivative gives[255] pyridine derivative (Fig. 2.175).

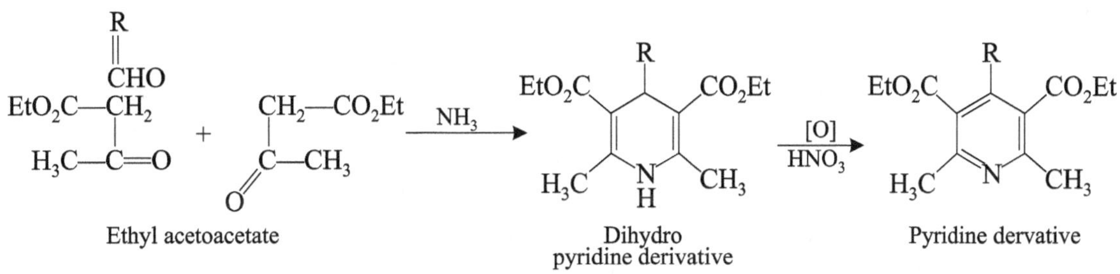

Fig. 2.175: Hantzsch Pyridine Synthesis.

A Green Hantzch pyridine synthesis consist in heating a mixture of 1,3-dicarbonyl compounds and aryl aldehydes and ammonium nitrate (as a source of ammonia) in a MW oven to give pyridine derivatives[256] (Fig. 2.176).

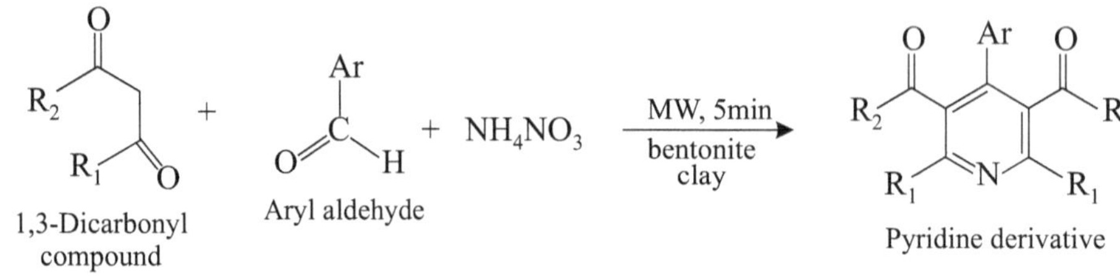

Fig. 2.176: A Green Hantz Pyridine Synthesis

(*x*) Hentzsch Pyrrole Synthesis: The reaction of β-ketoesters with α-chloroacetone in presence of ammonia gives pyrrole derivatives[257] (Fig. 2.177)

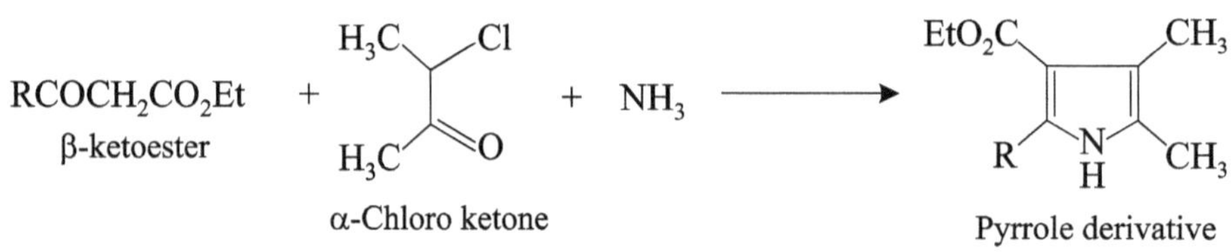

Fig. 2.177: Hentzsch Pyrrole Synthesis.

(*xi*) Hetero Diels-Alder Reaction: Also known as **aza-Diels-Alder Reaction,** it involves reaction of 2-azabutadienes with dienophills[258] (Fig. 2.178).

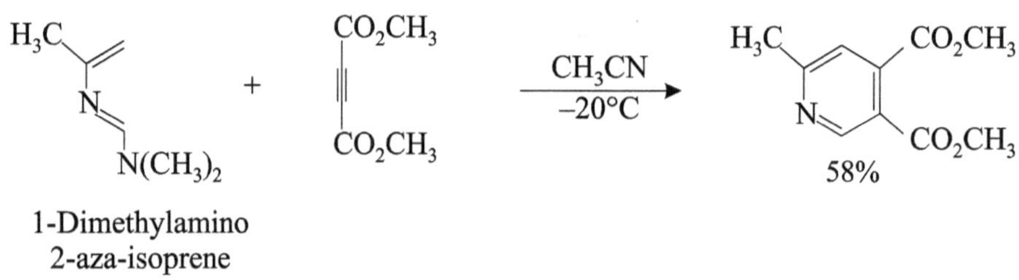

Fig. 2.178: Hetero-Diels-Alder-Reaction.

Hetero-Diels-Alder-Reaction can also be carried out in water. Thus, the reaction of iminium salts (generated in situ under Mannich-like conditions) on reaction with dienes give aza-Diels-Alder Products[259]. (Fig. 2.179)

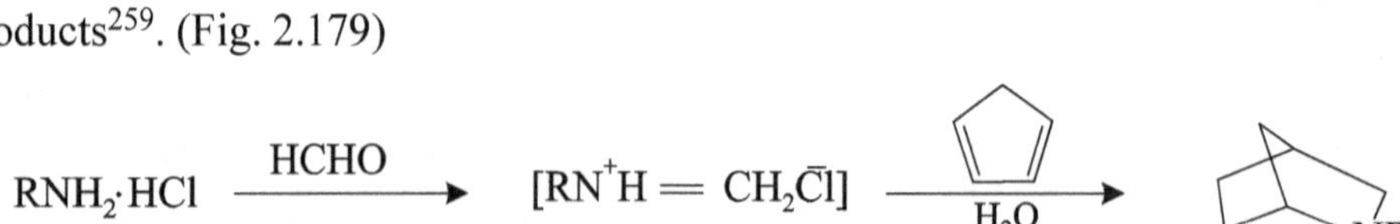

Fig. 2.179: Green hetero-Diels-Alder-Reaction

The **intramolecular aza-Diels-Alder Reaction** also occurs in water[260] to give fused ring systems with bridged head nitrogen (Fig. 2.180).

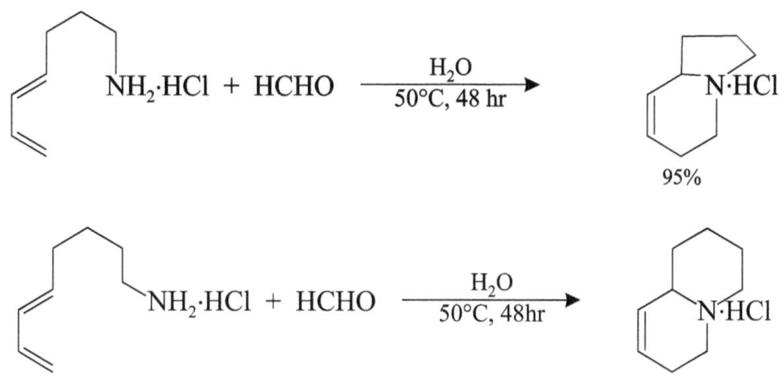

Fig. 2.180: Intramolecular aza-diels-Alder-Reaction in Water.

(*xii*) **Knoevenagel Condensation:** The reaction of salicyaldehyde with malononitrile in heterogeneous aqueous alkaline medium at room temperature gives *o*-hydroxy benzylidiene malonitrile, which on heating with acid gives 3-cyanocoumarins[261] (Fig. 2.181).

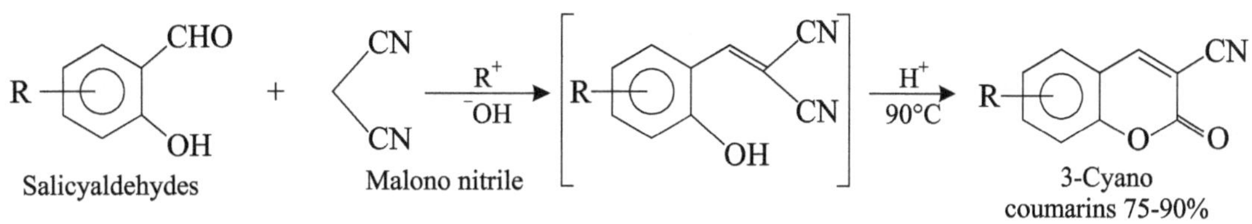

Fig. 2.181: Knoevenagel Condensation in Water.

(*xiii*) **Knorr Pyrazole Synthesis:** The reaction of 1,3-dicarbonyl compounds (β-diketones) with substituted hydrazines give two structurally isomeric pyrazoles[262]. However use of symmetrical β-diketones with hydrazines gives only one pyrazole (Fig. 2.182).

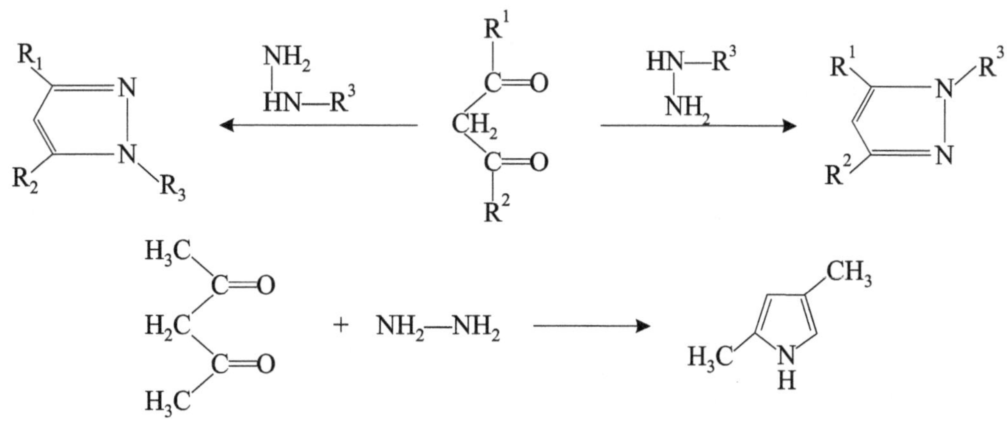

Fig. 2.182: Knorr Pyrazole Synthesis.

(*xiv*) **Knorr Quinoline Synthesis:** The reaction of β-ketoesters with aryl amines followed by cyclisation of the formed anilides with conc H_2SO_4 gives α-hydroxy quinolines[263] (Fig. 2.183).

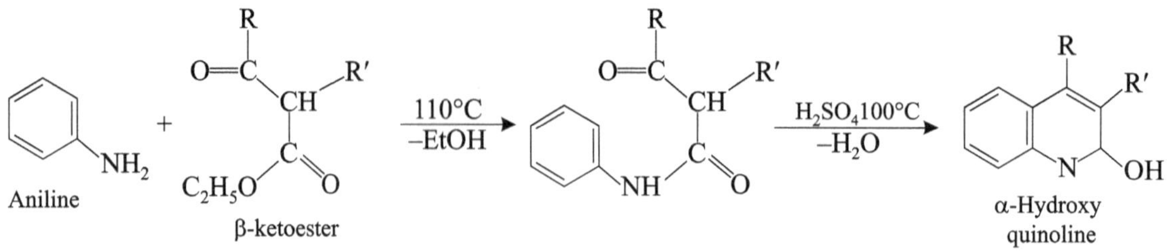

Fig. 2.183: Knorr Quinoline Synthesis.

(*xv*) Skarup Reaction: Synthesis of quinoline by heating aniline with glycerol in presence of conc. H_2SO_4 and nitrobenzene[264] (Fig. 2.185)

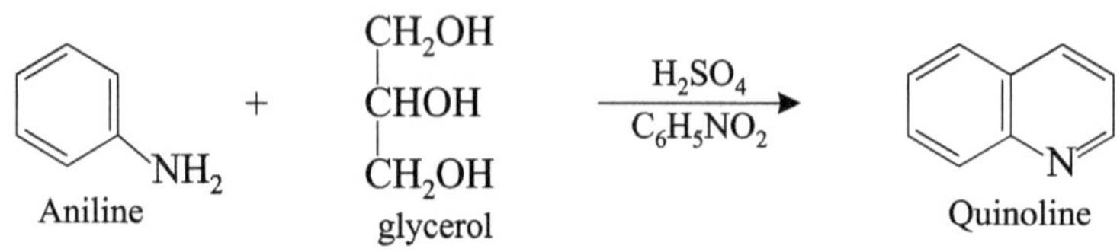

Fig. 2.184: Skarup Reaction.

(*xvi*) Thorpe Reaction: Base catalysed condensation of nitriles yield imines, which tautomerize to enamines[265].

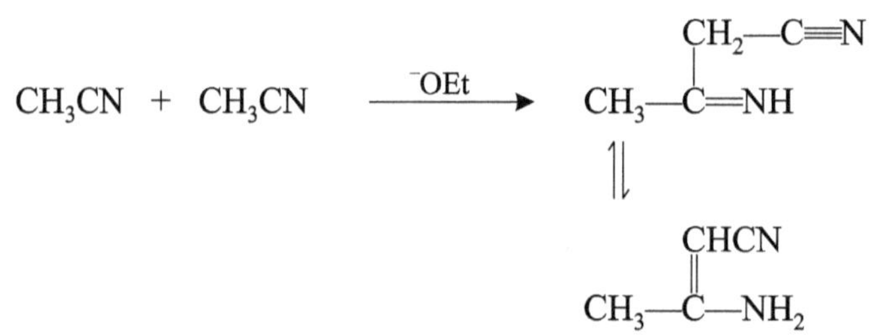

Fig. 2.185: Thorpe Reaction.

A intramolecular version is called Thorpe-Ziegler reaction[266] (Fig. 2.186).

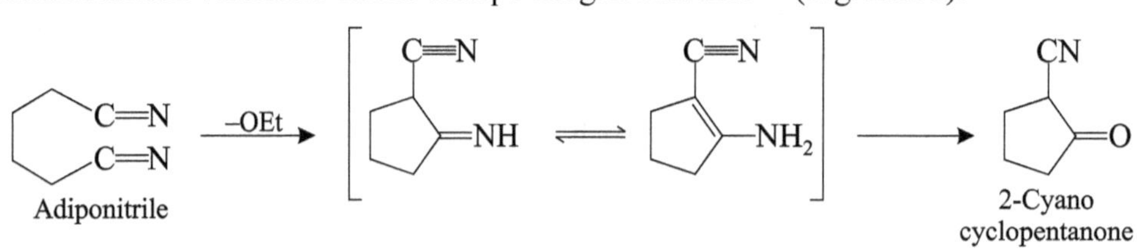

Fig. 2.186: Thorpe-Ziegler Reaction.

(*xvii*) Ullmann Reaction: The reaction of acetanilide with bromobenzene in presence of Copper bronze gives diphenyl amine (Fig. 2.187).

$$\underset{\text{Acetanilide}}{C_6H_5-NHCOCH_3} + \underset{\text{Bromo benzene}}{C_6H_5Br} \xrightarrow[\text{reflux}]{\text{Cu}} \underset{\text{Diphenylamine}}{C_6H_5NHC_6H_5} + CH_3COOH + HBr$$

Fig. 2.187: Ullmann Reaction.

***(xviii)* Miscellaneous methods:** It is well known that cyanides or nitriles are the only compounds having C≡N group. These are prepared from alkyl halides[267], aldehydes[268], alkynes, grignard reagents[269], amines and dichlorobenzene reaction with amides etc. (Fig. 2.188).

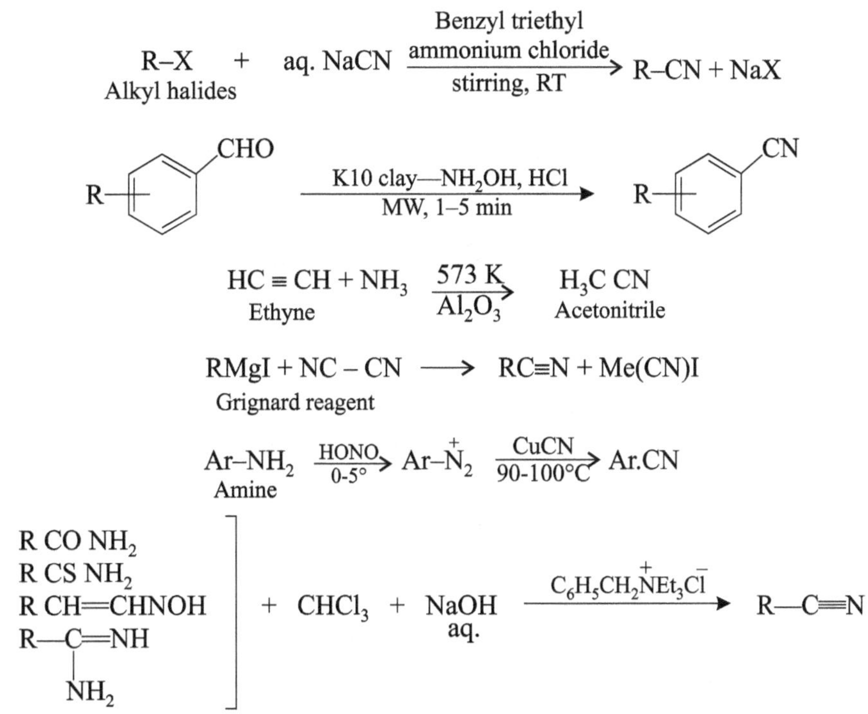

Fig. 2.188: Miscellaneous methods for the preparation of nitriles.

2.2.1.9 Formation of Carbon-Sulphur Bonds

***(i)* Hinsberg Thiophene Synthesis:** The reaction of α-diketones with diethyl thioacetate in presence of base gives thiophene derivatives[270] (Fig. 2.189)

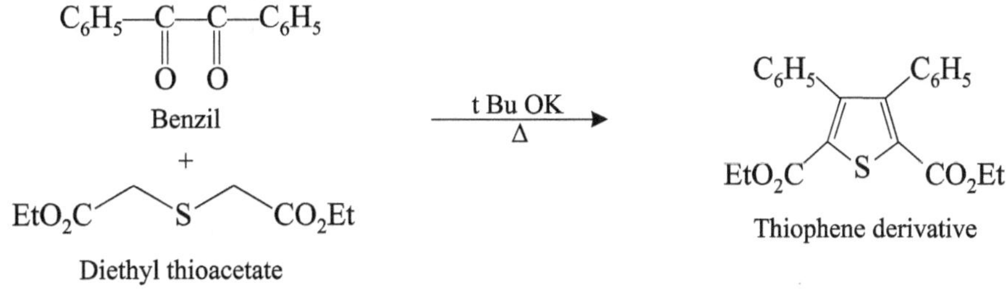

Fig. 2.189: Hinsberg Thiophene Synthesis.

***(ii)* From Epoxides:** The reaction of epoxides with thiophenol in presence of catalytic amount of aq. NaOH on heating in MW oven gives the corresponding β-hydroxy sulphides[271] (Fig. 2.190).

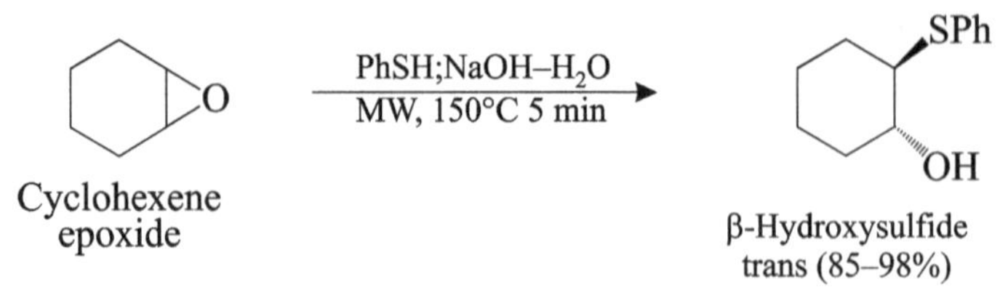

Fig. 2.190: Carbon-sulphur bond formation from epoxides.

The oxygen of epoxide can be replaced by sulphur by reaction with potassium thiocyanate in $[b_{min}]$ $[PF_6] - H_2O$ (2 : 1) at room temperature[272] (Fig. 2.191).

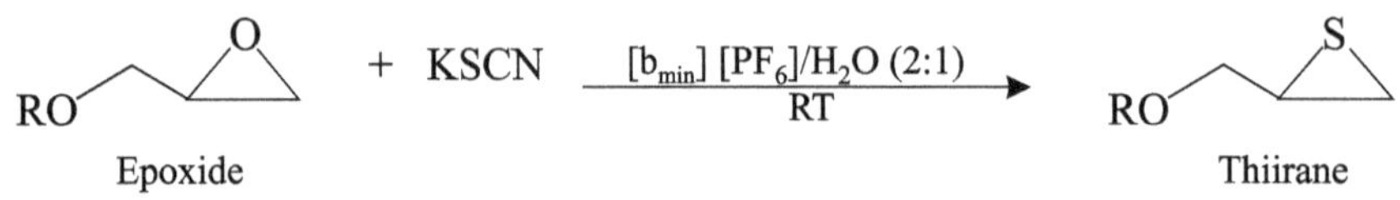

Fig. 2.191: Conversion of epoxide into thiirane.

***(iii)* Thiocyanation of Alkyl halides:** Alkyl halides on treatment with ionic liquid $[b_{min}]$ [SCN] give the corresponding thiocyanates at room temperature[273] (Fig. 2.192)

$$C_6H_5–CH_2Cl + [b_{min}]\,[SCN] \xrightarrow{RT} C_5H_5SCN + [b_{min}]\,[Cl]$$

1.2 equiv. 80-95%

KCNS (1.5 eg)

Fig. 2.192: Thiocyanation of alkyl halides.

***(iv)* Thionation Reaction:** An ecofriendiely process involving formation of thioketones, thioamides, thioethers and thioflavonoids by heating the corresponding oxygen containing compounds with **Lawesson's reagent** (0.5 equiv.)[274]. (Fig. 2.193)

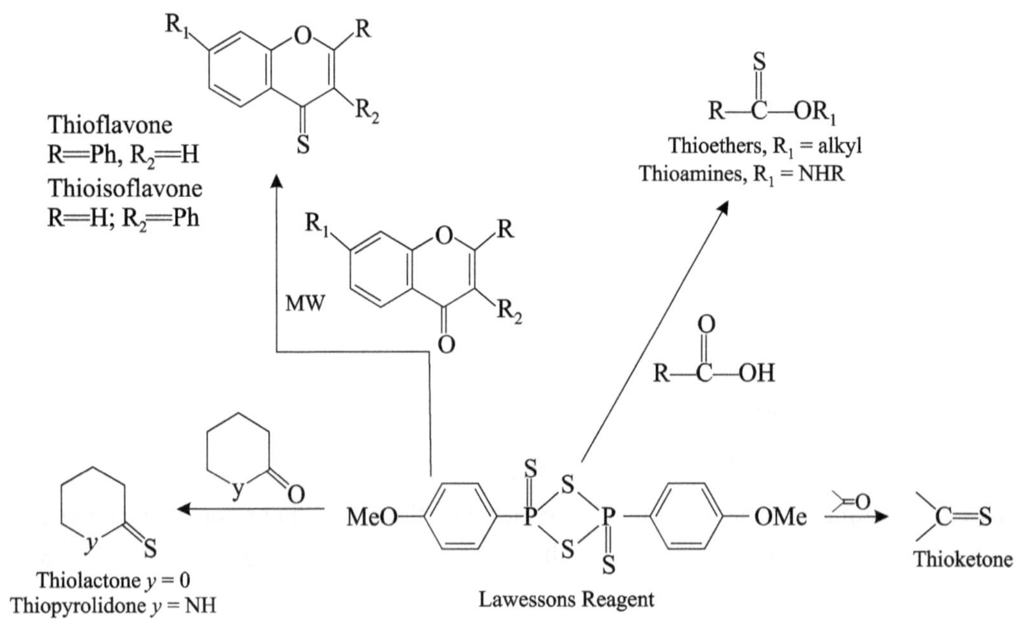

Fig. 2.193: Thionation Reaction

(v) From Thiiranes: Thiiranes as obtained by the reaction of epoxides with $[b_{min}]$ $[PF_2]/H_2O$ (2 : 1) (Fig. 2.191) has been used as synthon for the synthesis of a variety of products (Fig. 2.194).

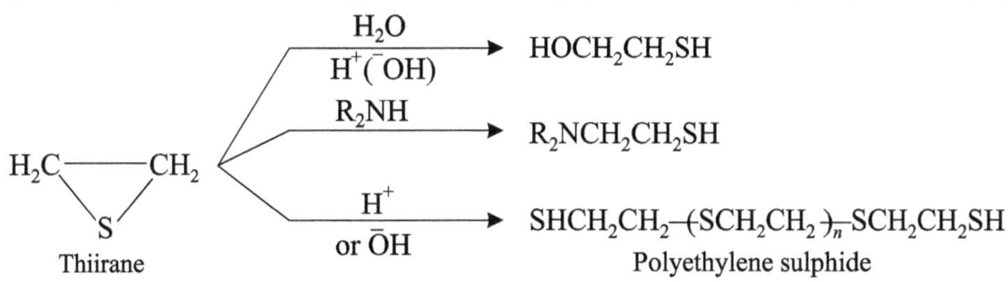

Fig. 2.194: Use of thiiranes for the synthesis of a variety of products.

2.2.1.10 Cross Dehydrogenative Coupling (CDC) Formation of C–C Bonds

We have so for discussed the formation of C–C bonds using Pd catalyst (section 2.2.1.3) and also from *sp* C–H bonds, sp^2 C–H bonds and sp^3 CH-bonds (section 2.2.1.4). Here in we discuss CDC methodology for formation of C–C bonds directly using two C–H bonds[275] (Fig. 2.195). This method is based on multi C–H activations for the synthesis of target molecules[276]. Some of the reactions using CDC methodology are discussed in this section.

$$C-H+H-C\xrightarrow{\text{Cat. M}}C-C$$

Fig. 2.195: Cross coupling method for C–C bond formation.

As an example, coupling reactions[277] of aromatic compounds and 1,4-benzoquinone has been carried out in water in presence of In $(OTf)_3$ (Fig. 2.196).

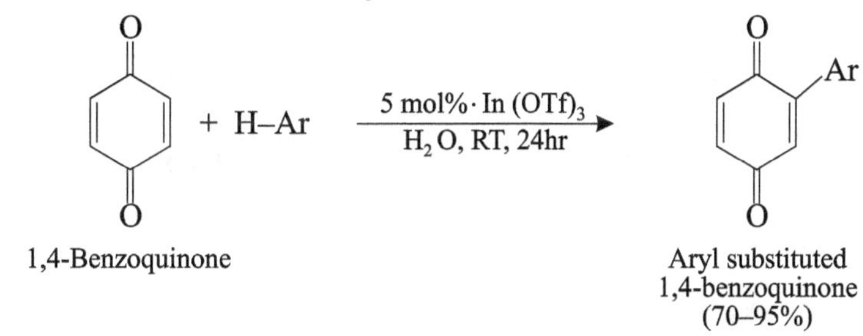

Fig. 2.196: Coupling of 1,4-benzoquinone with aromatic compounds in water.

The CDC reaction of benzoquinone with indoles give 3-indolylquinones[278] in water without any catalyst (Fig. 2.197).

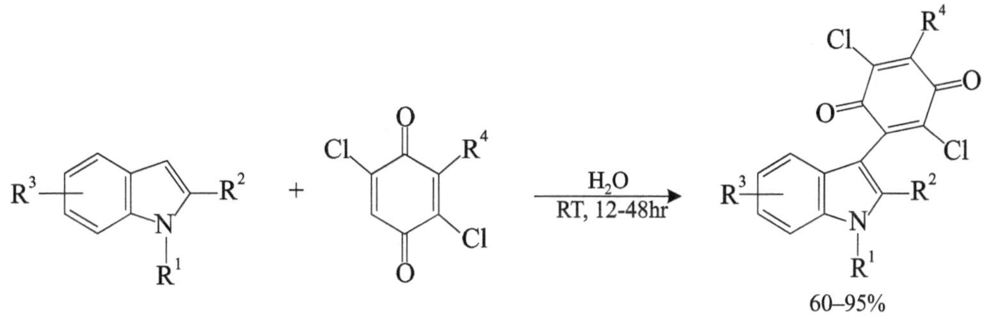

Fig. 2.197: CDC reaction of benzoquinones with indoles.

2.2.1.11 Homocoupling Reactions

The homocoupling reaction are well known reactions for chemical synthesis and are fully established compared to CDC Reactions. As an example coupling of 2-naphthols and substituted phenols is promoted by a supported Ru catalyst using O_2 as oxidant in water[279, 280] (Fig. 2.198).

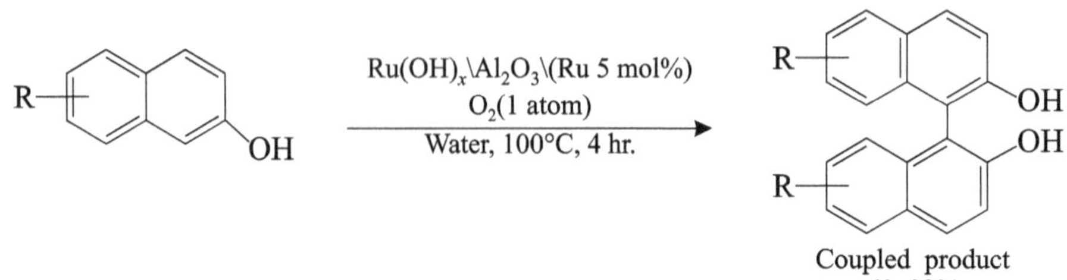

Fig. 2.198: Homocoupling of 2-naphthols in water.

The above reaction (Fig. 2.198) is believed to involve radical coupling. The homocoupling of 2-naphthols and substituted phenols occurred successfully in water in presence of $FeCl_3$[280].

Homocoupling is also useful for obtaining benzidine derivatives[281] (Fig. 2.199).

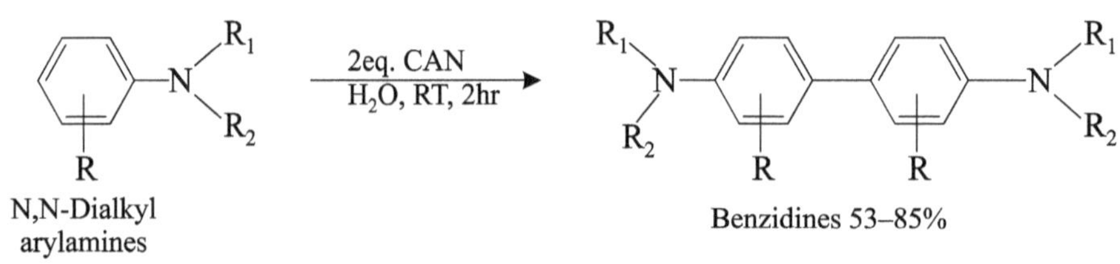

Fig. 2.199: Synthesis of benzidine derivatives.

2.2.2 Organic Transformation in Super Critical Water (SCW)

2.2.2.1 Introduction

Water at high temperature and high pressure behaves very differently[282, 283] compared to ordinary water. Such a difference in behaviour has not been observed with any other solvent. As an illustration, the electrolytic conductance of aqueous solutions increase with increase in pressure compared to decrease in electrical conductivity with increase in pressure in case of other solvents. The unusual behaviour of water is attributed to its peculiar associative properties[284].

In case of ordinary water, its thermal expansion becomes less with increase in temperature. Also, with increase in pressure, the vapour density of water increases. It is found that the density of water at room temperature 1.0 g/cm^3 and it becomes 0.7 g/cm^3 at 306°C. The densities of the two phases becomes identical at the critical point and they become a single phase. It is in this phase that the resultant liquid is called **supercritical fluid**. The density of water at this critical point is about 0.3 g/cm^3. The phase diagram of water near the critical region is given in Fig. 2.200.

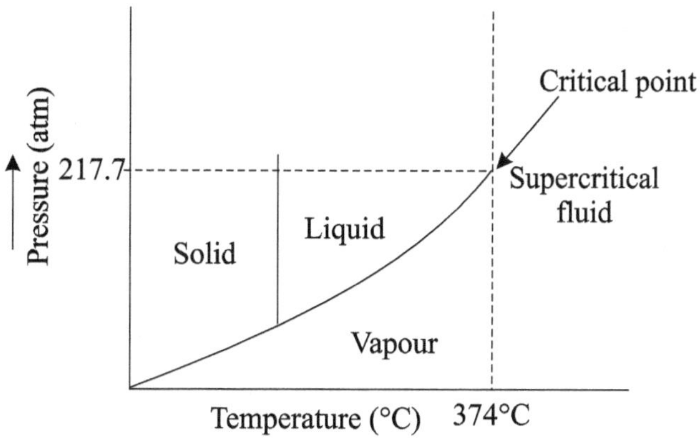

Fig. 2.200: Phase diagram of Water near the Critical Point.

The properties of water near its critical point are quite different than those of liquid water at ordinary temperature. Some of such properties include:

- The dielectric constant at near the critical point is much lower. In fact, it varies from 5 to 20 on variation of applied pressure.
- The electrolytic conductance of water rises sharply from ambient to critical point and becomes independent of the pressure. This is attributed to decrease in the viscosity of water as the temperature increases.
- Near the critical point of water, the number and persistance of hydrogen bonds both become less. Due to this water near its critical point behaves as most of the organic solvents and so the organic compounds dissolve in super critical water.
- As water approaches its critical point, its dissociation constant (K_w) becomes three times that of ambient liquid water. At this point, the concentration of H^+ and OH^- increase considerably and so acid and base catalysed reactions can be conveniently performed. This aspect is discussed subsequently.
- Gases like oxygen, carbondioxide, methane and other alkanes are soluble in super critical water and so combustion can occur in the phase. This leads to the application of super critical water for the treatment of organic hazardous wastes. Thus, non-polar wastes like polychlorinated biphenyles (PCBs) become miscible in super critical water and in presence of an oxidizer react to produce carbon dioxide, water, chloride salts and other small molecules. In fact, oxidation in supercritical water in short time can transform about 99.9% of hazardous organic materials into forms which are environmentally acceptable.
- Another form of water is **Near Critical Water** (NCW). In a microwave oven pure water reaches an autogenic pressure of 50 bar at 250°C. This condition is referred to near critical water region between 200 and 300°C (G.W. Sokel, S.A. DiBase and B.A. Lipiska, Tetrahedron Lett., 1976, 3495). NCW in generated is stainless steel pressure vessels or microwave-irradiated Teflon or glass reactors.

2.2.2.2 Organic Transformations in Super Critical Water

Super Critical Water (SCW)] or water near its critical point (Near critical water NCW) being an environmentally benign solvent, a reactant and a catalyst is known to serve as a medium for a number of organic transformation[285-287]. Some of the organic transformations are given below.

***(a)* Conversion of Cyclohexanol to Cyclohexanone and benzene:** The reaction of cyclohexanol with PtO_2 in SCW gave benzene as the major produced[288] while in presence of NH_4OH, cyclohexanone was the major product (Fig. 2.201).

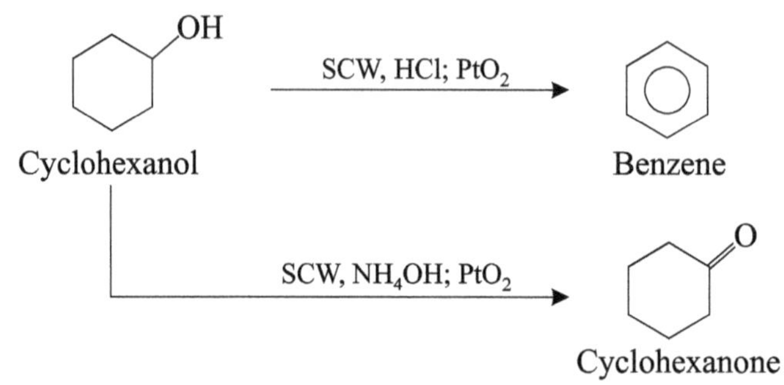

Fig. 2.201: Conversion of cyclohexanol to benzene or cyclohexanone.

In this above transformation PtO_2 was converted into the metallic platinum, which was the active species.

***(b)* Diels-Alder Reaction:** Diels-Alder reactions have been conducted in SCW[289]. A number of different diene/dienophile combination can be used.

(c) Heck Coupling Reactions: The coupling of an alkene with a aryl halide in presence of Pd(O)/base give the coupled product. The reaction is known as Heck coupling reaction.[290] (Fig. 2.202).

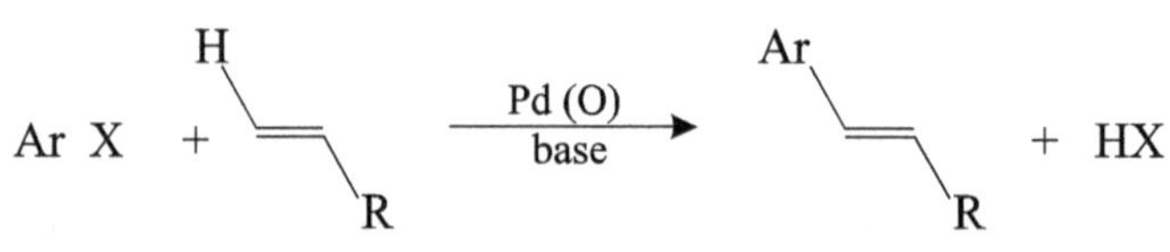

Fig. 2.202: Heck coupling reaction.

The above coupling reaction can be conveniently performed in SCW. Under these conditions, in place of alkenes, compounds likes alcohols, halides and carboxylic acids can be used. These compounds in SCW get converted into unsaturated compounds, which thus generated in situ undergo Heck coupling reaction[291]. As an example use of 2-phenylethanol gives styrene, which couple with an arene to give the Heck product. Thus a large number of starting compounds could be used in Heck coupling in SCW.

***(d)* Claisen Rearrangement:** Normally Claisen rearrangement of allylphenyl ether at 200°C gives 2-allylphenol in about 84%. The reaction can also be performed in H_2O by heating at 240°C. However, if allyl phenyl ether is heated in near critical water (NCW) at 250°C for about 6 min. the product obtained is 2-methyldihydrobenzofuran (Fig. 2.203).

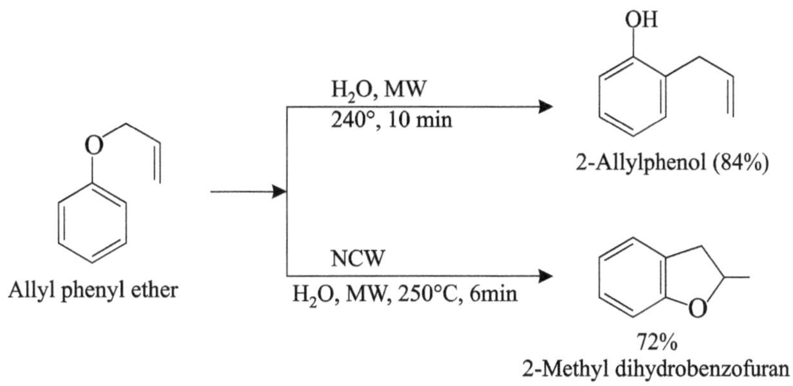

Fig. 2.203: Claisen rearrangement in NCW.

(*e*) **Carbon-Carbon bond cleavage** is also found to occur in near-critical and supercritical water. As an example, 2,5-dimethylfuran on heating at 250°C gives 2, 5-hexanedione in quantitative yield in presence of acid catalyst[292].

(*f*) **Rearrangement of Cyclohexane to methyl cyclopentene:** Cyclohexane was found[288] to rearrange to methyl cyclopentene in SCW in presence of a mineral acid or acidic metallic salts. This ring contraction is attributed to acid catalysts and rearrangement of a carbocation.

(*g*) **Dehydration of alcohols:** The conversion of tert. butyl alcohol to isobutylene in near critical water at about 250° was found[293,294] to be rapid in presence of H_2SO_4 or NaOH. Other alcohols like cyclohexanol[288], 2-methyl cyclohexanol[292] and 2-phenylethanol[291] also undergo dehydration though is less yield.

(*h*) **Hydration of alkenes:** Alkenes undergo hydration in SCW. As an example, cyclohexene in SCW gave Cyclohexanol in presence of PtO_2. Even alkynes gave the corresponding ketones[287].

(*i*) **Decarboxylation of carboxylic acids:** Carboxylic acid like cinnamic and indole-2-carboxylic acid under go facile decarboxylation[287] in SCW. An interesting example is conversion of citric acid to methacylic acid and crotonic acids[295] (Fig. 2.204).

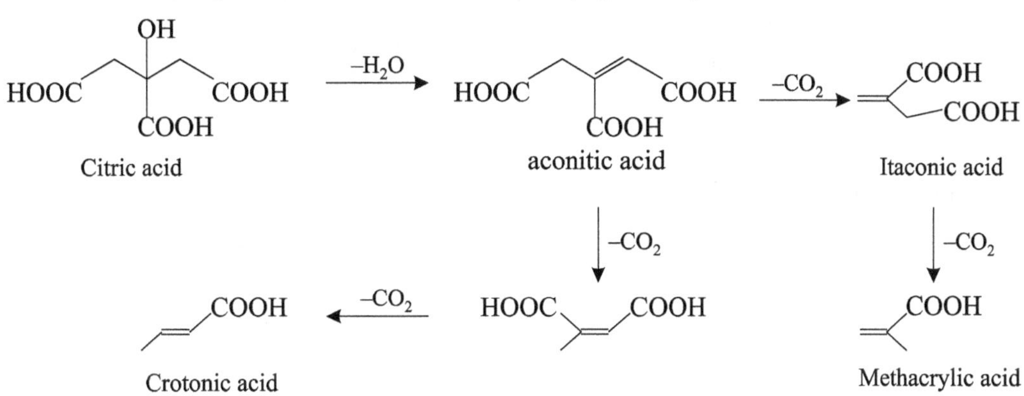

Fig. 2.204: Conversion of citric acid into methacrylic acid and crotonic acid.

(*j*) **Hydrolysis:** Esters and nitriles underso hydrolysis[286,287] in SCW (Fig. 2.205).

$$\text{RCOOEt} \xrightarrow{\text{SCW}} \text{RCOOH} + \text{EtOH}$$
$$\text{RC} \equiv \text{N} \xrightarrow{\text{SCN}} [\text{RCONH}_2] \longrightarrow \text{RCOOH}$$

Fig. 2.205: Hydrolysis of esters and nitriles

(*k*) **Synthesis of Isocoumarins:** Medicinally important isocoumarins are synthesised[296] by the reaction of activated methylene compounds with 2-iodobenzoic acid in NCW in presence of KOH and Cu^+. (Fig. 2.206)

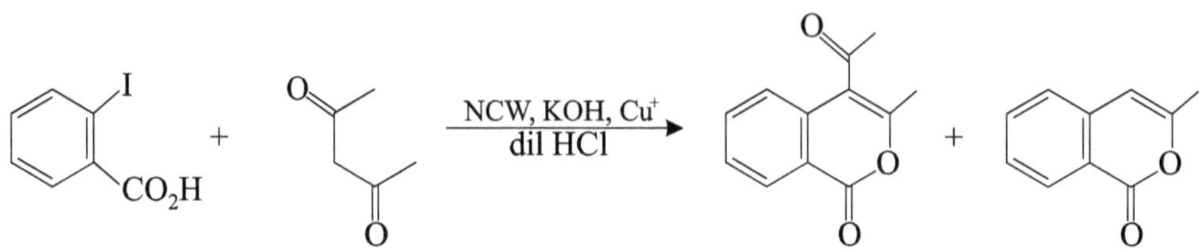

Fig. 2.206: Synthesis of isocoumarins in NCW medium.

(*l*) Dehydration of cyclohexanol to cyclohexene (Fig. 2.207).

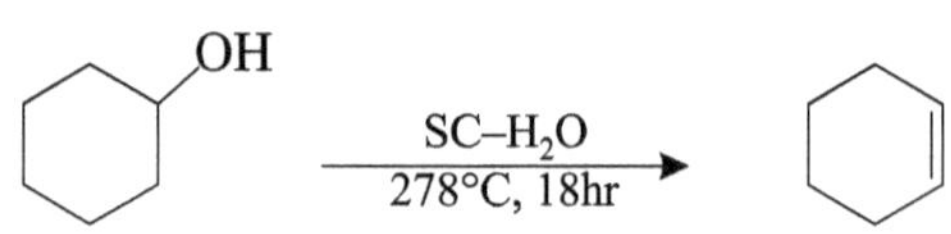

Cyclohexanol Cyclohexene

Fig. 2.207: Dehydration of cyclohexanol to cyclohexane

(*m*) Pinacol-Pinacolone rearrangement (Fig. 2.208)

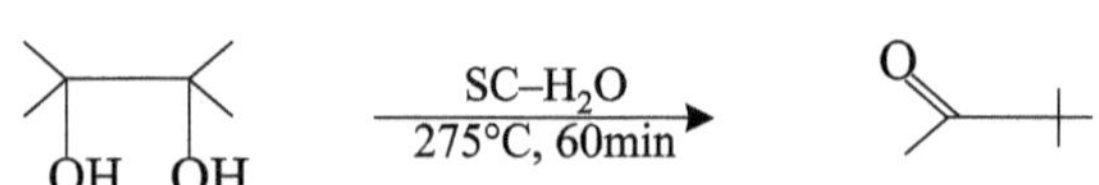

Pinacol Pinacolone (100%)

Fig. 2.208: Pinacol-Pinacolone rearrangement

(*n*) Ring opening of 2, 5-dimethyl furan (Fig. 2.209).

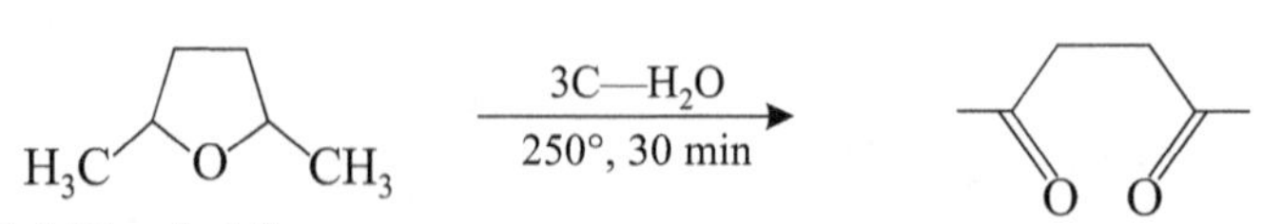

2,5-Dimethyl furan 2,4-Hexadiane (100%)

Fig. 2.209: Ring opering of 2,5-dimethylfuran.

(*o*) Hydrolysis of acetals and esters (Fig. 2.210).

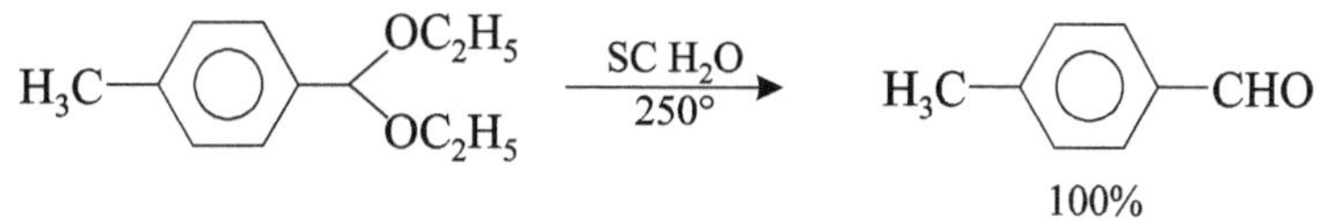

Fig. 2.210: Hydrolysis of acetals and esters.

(*p*) Conversion of Glucose into erythrose (Fig. 2.211)[296a]

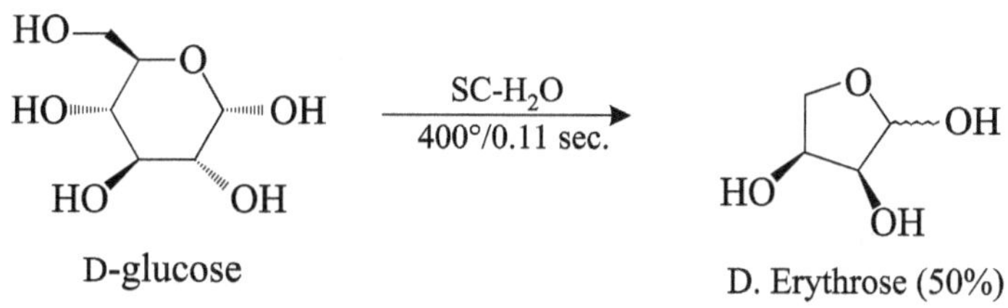

Fig. 2.211: Conversion of D-glucose into D-erythrose

2.2.3 Organic Transformation in High Temperature Water

2.2.3.1 Introduction

High-temperature water (HTW) in general, is defined[297] as liquid water above 200°C (compare with SCW, section 2.2.2.1). The properties of HTW are quite different from those of ambient liquid water. As example, the dielectric constant of HTW is lower compared to ordinary water and has fewer and weaker hydrogen bonds and higher isothermal compressibility. Organic compounds (particularly smaller compounds) are more soluble in HTW (but completely miscible in SCW). As in the case of SCW, most of the gases are soluble in HTW. In fact, in HTW, the solubility of gases initially decreases as the temperature is increased but after some time the solubility increases with rise of temperature. Also the ionic product (Kw) for high-temperature liquid water is about three times higher that for ambient water. The unique properties of HTW support ionic, polar non-ionic and free radical reactions.

Following are given some of the transformations which are carried out in HTW.

(*a*) Hydrolysis: Compounds containing a saturated carbon atoms attached to a heteroatom-containing functional groups are susceptible to hydrolysis in HTW. Following are given the expected products[297] for the hydrolysis of ethers, esters, amides, amines, nitroalkanes and aryl halides (Fig. 2.212).

$$\underset{\text{Ethers}}{R-O-R'+H_2O} \xrightarrow{\text{HTW}} ROH + R'OH$$

$$\underset{\text{Esters}}{R-COOR'+H_2O} \xrightarrow{\text{HTW}} RCOOH + R'OH$$

$$\underset{1^{\circ}\text{ A min es}}{R-NH_2+H_2O} \xrightarrow{\text{HTW}} ROH + NH_3$$

$$\underset{2^{\circ}\text{ Amines}}{R-NH-R'+H_2O} \xrightarrow{\text{HTW}} ROH + R'NH_2$$

$$\underset{3^{\circ}\text{ Amines}}{RR'NR''+2H_2O} \xrightarrow{\text{HTW}} ROH + R'OH + R''NH_2$$

$$\underset{1^{\circ}\text{ nitro alkanes}}{RCNO_2+H_2O} \xrightarrow{\text{HTW}} \underset{\text{aldehydes}}{RCHO + HNO_3}$$

$$\underset{2^{\circ}\text{ Nitroalkanes}}{RCNO_2R'+H_2O} \xrightarrow{\text{HTW}} \underset{\text{ketones}}{RCR'O + HNO_3}$$

$$\underset{1^{\circ}\text{ gem-dihalides}}{RCX_2H+H_2O} \xrightarrow{\text{HTW}} \underset{\text{aldehydes}}{RCHO + 2HX}$$

$$\underset{2^{\circ}\text{ gem-dihalides}}{RCX_2R'+H_2O} \xrightarrow{\text{HTW}} \underset{\text{ketones}}{RCR'O + 2HX}$$

Fig. 2.212: Hydrolysis of various compounds in HTW.

***(b)* Synthesis of 2-phenyl benzimidazole:** Medicinally important, 2-phenyl benzimidazole can be synthesised[298] from 1, 2-phenylenediamine and benzoic acid in HTW (Fig. 2.213).

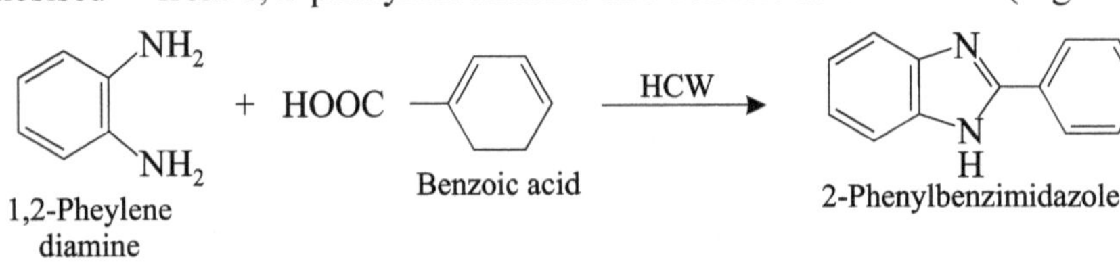

Fig. 2.213: Synthesis of 2-phenylbenzemodular in HTW.

***(c)* Aldol Condensation:** Aldol condensation reactions have been accomplished in HTW. Thus, 2,5-hexanedione which is unreactive in pure water undergoes intramolecular aldol condensation in HTW in presence of small amount of NaOH to give 3-methylcyclopent-2-enone in 81% yield.[299]

***(d)* Rearrangement Reactions:** It has been found[299] that pinacols undergoes quantitative rearrangement to the corresponding ketones in pure water at 275°C (HTW). Also, Claisen, Rupe and Meyer-Schuster rearrangements could also be performed[287] in pure water (HTW) at elevated temperatures.

***(e)* Fischer-Indole Synthesis:** The reaction of phenylhydrazine and ethyl methyl ketone in HTW (220°C) gave 2,3-dimethyl indole[300] (Fig. 2.214).

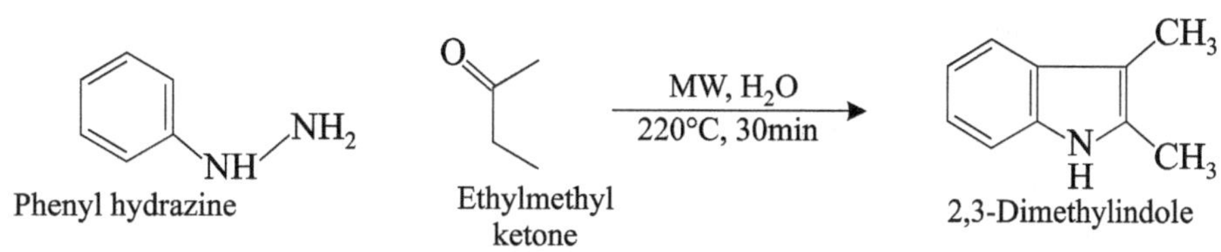

Fig. 2.214: Synthesis of 2,3-dimethylindole in HTW.

***(f)* Synthesis of p-isopentyl phenol:** It is prepared[301] by the decomposition of bisphenol A by performing the reaction at 350° for 20 min (Fig. 2.215).

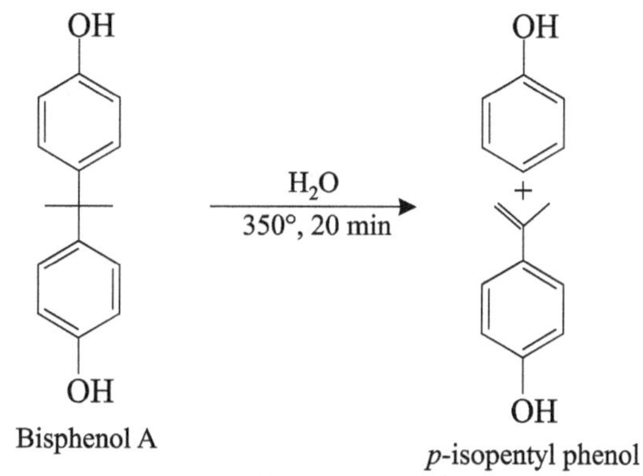

Fig. 2.215: Synthesis of *p*-isopentyl phenol.

The formed *p*-isopentyl phenol could be separated by simply cooling the reaction mixture at room temperature.

A large number of organic transformations have been carried out in water using microwave heating. These transformations could be considered as being carried out using high temperature water, since microwave heating produces temperature in the range 100-250°C. A discussion on the use of microwave assisted transformations in water forms the subject matter of a subsequent section (see Chapter 9, Section 9.2).

Most of the organic transformations using biocatalysts are performed using water as solvent (For details see Chapter 13). Also organic transformations using phase transfer catalysts uses water along with an organic solvent (For details see Chapter 12).

BIBLIOGRAPHY

1. Organic Reactions in aqueous media C.-J.Li and T.-K. Chan, Willy, New York, 1997.
2. Organic Synthesis in water. P.A. Grieco, Blackil, London, 1998.
3. Organic Reactions in aqueous media with a focus on C–C bond formations. C.–J. Li, Chem. Rev., 2005, **105**, 3095-3166 and the references cited there in.
4. Reactions of C–H bonds in water, C.I. Herrerias, X. Yao, Z. Li and C.–J. Li, Chem. Rev., 2007, **107**, 2546-2562 and the references cited therein.
5. Microwave-Assisted Synthesis in Water as Solvent. D. Dallinger and C.O. Kappe, Chem. Rev., 2007, **107**, 2563-2591 and the references cited therein.

6. Organic Reactions in Supercritical Water, P.A. Savage, Chem. Rev., 1999, **99**, 603-621 and the references cited therein.
7. Role of water for Chemical Reactions in High-Temperature Water., N. Akiya and P.E. Savage, Chem. Rev., 2002, **102**, 2725-2750 and the references cited therein.
8. Water as a reaction medium for clean chemical processes, W. Wei, C.C.K. Keh, C.-J.Li and R.S. Varma, Chean Techn Environ Policy, 2005, **7**, 62-69 [http://dx.doi.org/10.1007/s10098-003-0242-7]
9. Alternative solvents for Green Chemistry, F.M. Kerton. RSC Publishing.

Conclusion

Water is the greenest alternative amongst environmentally benign solvent. It has been used as solvent in pericyclic reactions, nucleophilic addition reactions, carbon-carbon bond formations, carbon-carbon bond formations from *sp* C–H bonds, sp^2 C–H bonds and sp^3C–H bonds, formation of C–halogen, C–N and C–S bonds. Water in the form of supercritical water (SCW) and high temperature water has also been used for a number of organic transformations. Though at present water has not been used as industrial solvent, but it is hoped that in times to come the use of water as industrial solvent will be possible. This aspect is especially important since water is the ideal green solvent.

REFERENCES

1. R. Breslow, Acc. Chem. Res., 1991, **24**, 159; P.A. Grieco, Adrichim. Act, 1991, **24, 59.**
2. D.C. Rideout and R. Breslow, J. Am. Chem. Soc., 1980, **102**, 7816.
3. A Ben-Naim, Hydrophobic Interactions, Plenum Press, New York, 1980; C. Tanford. The Hydrophobic Effect, **2nd** ed., Wiley, New York, 1980.
4. P.H. Von Hippl and T. Schleich, Acc. Chem. Res. 1969, **2**, 257.

4(a). J.W. Wijnen and J.B.F.N. Engberts, J. Org.Chem., 1997, **62,** 2039.

5. D.D. Sternbach and D.M. Rossana, J. Am. Chem. Soc., 1982, **104,** 5853.
6. W.B. Wang and E.J. Roskamp, Tetrahedron Lett; 1992, **33**, 7631.
7. B.A. Keay, J. Chem. Soc., Chem. Commun., 1987, 420.
8. T.K. Hollis, N.P. Robinson and B. Bosnich, J. Am. Chem. Soc., 1992, **114**, 5464.
9. S. Kobayashi, I. Hachiya, M. Araki and H. Ishitani, Tetrahedron Lett., 1993, **34**, 3755.
10. S. Otto and J.B.F.N. Engberts, Tetrahedron Lett., 1995, **36**, 2645.
11. S. Colunna, A. Manfredi and R. Annunziata, Tetrahedron Lett., 1988, **29**, 3347.
12. Chao-Jun Li, Chem. Rev., 2005, **105**, 3095.
13. S. Kanemasa, Y. Oderatoshi, H. Yamamoto, J. Tanaka, E. Wada and D.P. Curran, J. Org. Chem., 1997, **62**, 6454.
14. D. Boger and S. Weiveb, Hetero-Diels-Alder Methology in Organic Syntehsis, Academic Press, Orlando, 1987.
15. P.A. Grieco and S.D. Larsen, J. Am. Chem. Soc., 1985, **107**, 1768.
16. W. Oppolzer, Angew. Chem. Int. Ed. Engl., 1972, **11**, 1031.

17. P.A. Grieco, D.T. Parker, W.F. Fobace and R. Ruckle, J. Am. Chem. Soc., 1987, **109**, 5859.
18. H. Waldmann, Angew, Chem. 1988, **100**, 307; Ingew. Chem. Int. Ed. Engl., 1988, **27**, 274; H. Waldmann, Liebigs Ann. Chem., 1989, 231; H. Waldmann and M. Braun, Liebigs Ann., 1991, 1045; R. Lock and H. Waldmann, Tetrahedron Lett., 1996, **37**, 2753.
19. L. Yu, D. Chen and P.G. Wang, Tetrahedron Lett., 1996, **37,** 2169.
20. T. Akiyama, J. Takaya and H. Kagoshima, Tetrahedron Lett., 1999, **40**, 7831.
21. A. Lubineau, J. Auge and N. Lubin, Tetrahedron Lett, 1991, **32**, 7529; P. Grieco, K.J. Henery, J.J. Nunes and J.E. Matt, Jr., J. Chem. Soc. Chem. Commun, 1992, 368.
22. R. McCague, H.F. Olivo and S.M. Roberts, Tetrahedron Lett., 1993, **34**, 313.
23. A. Lubineau, J. Auge and N. Lubin, Tetrahedron, 1993, **49**, 4639; A. Lubineau and Y.J. Queneau, Carbohydrate Chem., 1995, **14** 1295.
24. M. Naruse, S. Aoyogi and C. Kibayashi, Tetrahedron Lett, 1994, **35**, 595.
25. M. Naruse, S. Aoyogi and S. Kibayaski, Tetrahedron Lett, 1994, **35**, 9123; M. Naruse, S. Aoyogi and C.J. Kibayashi, Chem. Soc. Perkin Trans. I, 1996-1113.
26. C. Agami, F, Couty, M. Poursoulis and J. Vaissermann, Tetrahedron, 1992, **48**, 431.
27. G.A. Lee, Synthesis, 1982, 508.
28. Y. Inoue, K. Araki and S. Shiraishi, Bull. Chem. Soc. Jpn., 1991, **64**, 3079.
29. K.R. Rao, N. Bhanumati, T.N. Srinivasan and P.B. Sattur, Tetrahedron Lett., 1990 **31,** 899.
30. K.R. Rao, N. Bhanumati and P.B. Sattur, Tetrahedron Lett., 1990, **31**, 3201.
31. A. Lubineau, G. Bouchain and Y.J. Queneau, J. Chem. Soc. Perkin Trans. I. 1955, 2433.
32. B. Ganem, Angew. Chem. Int. Ed. Engl., 1996, **35**, 936
33. A.K. Ponaras, J. Org. Chem., 1983, **48**, 3866; R.M. Coates, B.D. Rogers, S.J. Hubbis, D.R. Peck and D.P. Curran, J. Am. Chem. Soc., 1987, **109**, 1160; J.J. Gajewski, J. Jurayj, D.R. Kimbrough, M.E. Gande, B. Ganem and B.K. Carpenter J. Am. Chem. Soc., 1987, **109**, 1170.
34. S.D. Copley and J.J. Knowles, J. Am. Chem. Soc., 1987, **109,** 5008.
35. E.B. Brandes, P. Grieco and J.J. Gajewski, J. Org. Chem., 1989, **54**, 515
36. P. Grieco, E.B. Brandes, S. McCann and J.D. Clark J. Org. Chem., 1989, **54**, 5849.
37. J.E. McMurry, A. Andrus, G.M. Ksander, J.H. Musser and M.A. Johnson, Tetrahedron, 1981, **37**, 319 (Suppl. 1).
38. A. Aemissegger, B. Jaun and D. Hilvert, J. Org. Chem., 2002, **67**, 6725.
39. C. Agami, F. Couty and M. Poursoulis, Synlett, 1992, 847.
40. R. Ramamurthy, Tetrahedron, 1986, **42**, 5753.
41. M.S. Soyamala and V. Ramamurty, J. Org. Chem., 1986, **51**, 3712.
42. Y. Ito, T. Kajita, K. Kunimoto and T. Matsuura, J. Org. Chem., 1989, **54**, 587.
43. K. Muthuramu and V. Ramamurthy J. Org. Chem., 1982, **47**, 3976.
44. T. Tamaki, Chem. Lett., 1984, 53; T. Tamaki and T.J. Kokuba, Inclusion Phenom, 1984, **2**, 815.
45. A.T. Nielsen and W.J. Houlihan, Org. React, 1468, **1**, 16.
46. V. Grignard and M. Dubien, Ann. Chem. (Paris) 1924, 2(10), 282.

46(a). A.F. Babicheva, A.I. Vovk and A.A. Yasnikov, Ukr. Kim. Zn., 1976, **42**, 259.

46(b). W. Tagaki and H. Hara, J. Chem. Soc., Chem. Commun., 1973, 891.

47. K. Bowden and A.J. Brownhill, J. Chem. Soc., Perkin 2, 1997, 997.

48. T.D. Machajewski and C.-H. Wong, Angew. Chem. Int. Ed. 2009, **39**, 1352; W.-D. Fessener and V. Helaine, Curr. opin. Biotechnol, 2001, **12**, 574; M. Breuer and B. Hauer, Curr. opin. Biotechnol, 2003, **14**, 570.

49. L. Sanchez-Moerno, J.F. Garcia-Garcia, A. Basticla and E. Garcia-Juncada, Chem. Common, 2004, 1634.

50. A. Lubineau, J. Org. Chem., 1986, **51**, 2142; A Lubineau and E. Meyer, Tetrahedron, 1988, **44**, 6065.

51. S. Kobayashi and J. Hachiya, J. Org. Chem., 1994, **59**, 3590; Synlett, 1994, 689.

52. A. Michael, J. Prak. Chem., 1887, **35(2),** 349; E.D. Bergmann, D. Ginsburg and R. Pappo, Org. Reactions, 1959, **10**, 179.

53. Z. G. Hajos and D.R. Parrish, J. Org. Chem., 1974, **39**, 1612.

54. U. Elder, G. Sauver and R. Wiechert, Angew. Chem. Int. Ed. Engl, 1971, **10**, 496.

55. N. Harada, T. Sugioka, H. Uda and T. Kuriki Synthesis, 1990, 53.

56. J.F. Lavellee and P. Deslongchamps, Tetrahedron Lett., 1988, **29**, 6033.

57. J.E. Keller and B.L. Feringa, Tetrahedron Lett., 1996, **37**, 1870.

58. A. Lubineau and J. Auge, Tetrahedron Lett., 1992, **33**, 8073.

59. K. Sussangkarn, G. Fodor, I. Karle and C. George, Tetrahedron, 1988, **44**, 7047.

60. C.D. Mudaliar, K.R. Navalkar and S.H. Mashraqui, Org. Prep. Proced. Int., 1997, **29**, 584.

61. S. Kobayashi, K. Kakumoto, H. Mori and M. Kei, Isr. J. Chem., 2002, 247.

62. G. Wittig and U. Schollkopf, Ber., 1954, **87**, 1318; G. Wittig and W. Hang, Ber., 1955, **88**, 1654; U. Schollkopf, Angew. Chem. 1959, 71, 20, S. Trippet, Quart. Rev., 1963, **17**, 406.

63. C. Piechucki, Synthesis, 1976, 187; M. Mikolajczyk, S. Crzejszczak, S. Midura and A. Zatovski, Synthesis 1976, 396.

64. M. Rambaud; A. de Vecchio and J. Villieras, J. Synth. Commun, 1984, **14**, 833.

65. M. Schimtt, J.J. Bourguignon and C.G. Wermuth, Tetrahedron Lett., 1990, **31**, 2145.

66. M.G. Russell and S. Warren, J. Chem Soc. Perkin I, 2000, **4**, 505; M.G. Russell and S. Warren, Tetrahedron Lett., 1998, **39**, 7995.

67. John C. Warner, Paul T. Anastas and Jean-Pierre, Anselme, J. Chem. Education, 1985, **62**, 346.

68. G. Darzens, Compt. Rend., 1904, **139**, 1214; 1905, **141,** 766; 1906, **142**, 214; M. Ballister, Chem. Rev. 1955, **55**, 283.

69. A. Jonczyk, M. Fedorynski and M. Makosa, Tetrahedron Lett., 1972, 2395.

70. M. Makosza and M. Ludwikow, Anglew.Chem. Int. Ed. Engl., 1974, **13**, 655.

71. J.M. Mclntosh and H. Khalili, J. Org. Chem. 1977, **42**, 2123.

72. A.J. Lapworth, J. Chem Soc. 1903, **83**, 955; 1904, 85, 1206; J. Solodar, Tet. Lett., 1971, 287.

73. J. Solodar, Tetrahedron Lett., 1971, 287.

74. W. Tagaki and H. Hara, J. Chem. Soc. Chem. Commun., 1973, 891.

75. E.T. Kool and R. Breslow, J. Am. Chem. Soc., 1988, **110**, 1596.

76. R.D. Breslow, J.Am. Chem. Soc., 1958, **80**, 3719.

77. R.D. Breslow and E. Kool, Tetrahedron Lett., 1988, **29**, 1635.
78. C. Mannich and W. Kroschl, Arch. Pharm. 1912, 250, 648; F.F. Blicke, Org. Reactions, 1942, **1**, 303.
79. Y. Peng, R. Dou, G. Song and J. Jiang, Synlett, 2005, 2245.
80. P. Peng and G. Song, Green Chemistry, 2001, **3**, 302.
81. L. Shi, Y.Q.Tu, M. Wang, F.M. Zhang and C.A. Fan, Org. Lett., 2004, **6**, 1001.
82. I. Ojima, S.-I. Inaba and K. Yoshoda, Tetrahedron Lett., 1977, 3643.
83. S. Kobayashi and H. Ishitani, J. Chem. Soc. Chem. Commun., 1995, 1571.
84. D. Chen., L. Yu and P.G. Wang, Tetrahedron Lett., 1996, **37**, 4467.
85. A. Cleria and O. Porta, Tetrahedron Lett., 1990, **31**, 2069.
86. R.W. Stevens, N. Iwasawa and T. Mukaiyama Chem. Lett., 1982, 1459; A. Lubineau, J. Org. Chem., 1986, 51, 2143; A. Lubineau and E. Meyer, Tetrahedron, 1988, **44**, 6065.
87. S. Reformatsky, Ber., 1887, **20**, 1210; J. Russ Chem. Soc., 1890, **22**, 44; R.L. Shriner, Organic Reactions, 141, **1**, 1.
88. T.H. Chan, C.J. Li, Z.Y. Wei, J. Chem. Soc. Chem Commun., 1990, 505.
89. B. Han and P.J. Boudjouk, J. Org. Chem., 1982, **47**, 5030.
90. T. Kitazume, Synthesis 1986, 855.
91. X.H. Yi, Y. Meng and C.J. Li, Tetrahedron Lett., 1997, **38**, 4731.
92. W.J. Chung, S. Higashiya and J.T. Welch, J. Fluorine Chem., 2001, **112**, 343.
93. B.E. Kahn and R.D. Ricke, Chem. Rev., 1988, **88**, 733; J.M. Pons and M. Santelli, Tetrahedron, 1988, **44**, 4295.
94. Grinder, Ann. Chem. Phys., 1892, **26**, 369.
95. J.B. Conant and H.B. Cutter, J.Am. Chem. Soc., 1926, **48**, 1016.
96. P. Karrer, Y. Yen and I. Reichstein, Hel. Chem. Acta, 1993, **13**, 1308.
97. A. Clerici and O. Porta, Tetrahedron Lett., 1982, **23**, 3517.
98. A. Clerici and O. Porta, J. Org. Chem., 1982, **47**, 2852.
99. A. Clerici, O. Porta and M. Riva, Tetrahedron Lett, 1981, **22**, 1043 and references cited therein.
100. A. Clerici, O.Porta, J. Org. Chem., 1983, **48**, 1690, A. Clerici and O. Porta, Tetrahedron, 1983, **39**, 1239; A. Clerici, O. Porta and P. Zago, Tetrahedron, 1986, **42,** 561; A. Clerici and O. Porta, J. Org. Chem., 1989, **54**, 3872.
101. P. Delair and J.L. Luchi, J. Chem. Soc., Chem. Commun., 1989, 398.
102. M.C. Barden and J.J. Schwartz., J. Am. Chem. Soc. 1996, **118**, 5484.
103. N. Kalyanam and V.G. Rao, Tetrahedron Lett., 1993, **34**, 1647.
104. J.L. Luche and J.C. Damiano, J. Am. Chem. Soc., 1980, **102**, 7926.
105. J. Nokami, S. Wakabayashi and R. Okawara, Chem. Lett., 1984, 869.
106. C. Petrier and J.L. Luche, J. Org. Chem., 1985, **50**, 910; C. Einhorn, J.L. Luche, J. Organomet. Chem., 1987, **322**, 177.
107. C.J.Li and T.H. Chan, Tetrahedron Lett., 1991, **32**, 7017.
108. M. Wada, H Ohki and K.Y. Abik, Bull Chem. Sec., Jpn., 1990, **63**, 1738; J. Chem. Soc., Chem. Commun., 1987, 708.

109. C. Bloomberg, Synthesis, 1977, 18.

110. T. Uyechara, K. Yamada and T. Kato, Bull. Chem. Soc. Japan, 1985, **58**, 211.

111. C. Petrier, A.L. Gemal and J.L. Luche, Tetrahedron Lett., 1982, **23**, 3361; J. Einhorn and J.L. Luchi, Tetrahedron Lett., 1986, **27**, 1791.

112. A. Wurtz, Ann. Chem. Phys., 1855, **(3)44**, 275, Ann., 1855, **96**, 364; R.E. Buntrock, Chem. Rev., 1968, **68**, 209.

113. C.J. Li and T.H. Chan, Organometallics, 1991, **10**, 2548.

114. F. Ullmann, Ann., 1904, **332,** 38; F. Ullmann and P. Sponagel, Ber., 1905, **38**, 2211; P.E. Fanta, Chem. Rev., 1946, **38**, 139.

115. J.Lindley, T.J. Manson and J.P. Lorimer, Ultrasonics, 1987, **25**, 45.

116. K. Garres, J. Org. Chem., 1970, **35**, 3273.

117. S. Mukhopadhyay, A.V. Joshi, L. Peleg and Y. Sassan, Org. Process Res. Dev., 2003, **7**, 44.

118. D.V. Davydov and I.P. Beletskaya, Russ. Chem. Bull., 1995, **44**, 1139.

119. S. Venkataraman and C.–J. Li, Org. Lett., 1991, **1**, 1133.

120. S. Venkataraman and C.–J. Li, Tetrahedron Lett., 2000, **41**, 4831.

121. R.F. Heck, Acc. Chem. Res., 1979, **12**, 146; Organic Reactions, 1982, 27, 345; H.A. Diech. J. Organometallic Chem., 1975, **93**, 259.

122. N.A. Bumagin, N.P. Andryuchova and I.P. Beletskaya, Izu. Akad. Nauk SSSR, 1988, **6**, 1949; N.A. Bumagin, P. More and I.P. Beletskaya, J. Organomet. Chem., 1989, **371,** 397; N.A. Bumagin, V.V. Bykou, L.I. Sukhomlinova, T. Tolstaya and I.P. Beletskaya, J. Organomet. Chem., 1995, **486,** 259.

123. Organic Synthesis in water, Ed. Paul A. Grieco, Blackie Academic and Professionals, London (1998), pp.181-188 and references cited therein.

124. J.P. Genet, E. Blart and M. Savignac, Synlett., 1992, 715; A. Hessler, O. Stelzer, H. Dibavski, K. Worm and F.P. Schmidtchen, J. Org. Chem., 1997, **62**, 236.

125. D. Schoenfelder, K. Fischer, M. Schmidt, O. Nuyken and R. Weberskirch, Macromolecules, 2005, **38**, 254.

126. R.W.J. Scott, O.M. Wilson nd R.M. Crooks, J. Phys. Chem. B. 2005, **109**, 692.

127. M. Hiroshige, J.R. Hauske and P. Zhou, Tetrahedron Lett., 1995, **36**, 4567.

128. R.K. Arvela and N.E. Leadbeater, J. Org. Chem., 2005, **70**, 1786.

129. N.E. Leadbealer, Chem. Commun., 2005, 2881.

130. N. Miyaura and A. Suzuki, Chem. Rev., 1995, **95**, 2457.

131. B.I. Alo, A. Kandil, P.A. Patil, M.J. Sharp, M.A. Siddique and V. Sbieckus, J. Org. Chem., 1987, **52**, 3932.

132. J.C. Colberg, A. Rane, J. Vaquer and J.A. Soderquist, J. Am. Chem. Soc., 1993, **115**, 6065.

133. R.W. Armstrong, J.M. Beau, S.H. Cheon, W.J. Christ, H. Fujioka, W.H. Harm, L.D. Hawkin, H. Jin, S.H. Kang, Y. Kishi, M.J. Martinelli, W.W. McWhorter, M. Mizuno, M. Nakata, A.E. Stutz, F.X. Talamas, M. Taniguchi, J.A. Tino, K. Ureda, J.I. Uenishi, J. White and M. Yonaga. J. Am. Chem. Soc., 1989, **111**, 7525.

134. W.R. Roush and B.B. Broun, J. Org. Chem., 1993, **58**, 2162.

135. K.C. Nicolaou, J.Y. Ramphal, J.M. Palazon and R.A. Spanevello, Angew. Chem. Int. Ed. Engl., 1989, 27, 587.

136. D.A. Evans, H.P. Ng and D.L. Rieger, J. Am. Chem. Soc., 1993, **115**, 11446.

137. D.S. Ennis, J. McManus, W. Wood-Kaczmar, J. Richardson, G.E. Smith and A. Carstairs, Org. Process Res. Dev., 1993, **3**, 248.

138. Y. Uozumi and Y. Nakai, Org. Lett., 2002, **4**, 2997.

139. A.L. Casalnuovo and J.C. Calabrese, J. Am. Chem., Soc., 1990, **112**, 4324.

140. K. Sonogashira, Y. Tohda and N. Hagira, Tetrahedron Lett., 1975, 4467.

141. C. Amatore, E. Blart, J.P. Genet, A. Jutand, S. Lemaire-Andoire and M. Savignce J. Org. Chem., 1995, **60**, 6829.

142. J.C. Li, in Environmentally Benign Organic Synthesis, P.T. Anastas and T.C. Williansons, eds., Oxford University Press.

143. N.E. Leadbeater and M. Marco, Org. Lett., 2002, **4**, 2973; N.E. Leadbeater and M. Marco, J. Org. Chem., 2003, **68,** 888; L Bai, J.-X. Wang and Y. Zhang, Green Chem., 2003, **5**, 615.

144. Y. Li, X.M. Hong, D.M. Collard and M.A. El-Sayed, Org. Lett., 2002, **2**, 2385; F. Lu and D. Astruc, Tetrahedron Lett., 2004, 45, 9443; F. Lu, J. Ruez and D. Astrac, Tetrahedron Lett., 2004, **45**, 9943; A.G. Kuznetsov, D.N. Korolev and N.A. Bumagin. Russ. Chem. Bull., 2003, **52**, 1882.

145. D.V. Davydov and I.P. Beletskaya, Russ. Chem. Bull., 1995, **44**, 1139.

146. D.V. Davydov and I.P. Beletskaya, Russ. Chem. Bull 1995, **44**, 1141.

147. J.F. Stille, Angew. Chem. Int. Ed. Engl, 1986, **25**, 508; J. Org. Chem., 1990, **55,** 3019; J. Am. Chem. Soc., 1984, **106,** 4630.

148. D.R. Tueting, A.M. Echavarren and J.K. Stille, Tetrahedron, 1989, **45**, 979.

149. H.C. Zhang and G.D. Davis, Jr. Organometallics 1993, **12**, 1499.

150. R. Rai, K.B. Aubrecht and D.B. Collum, Tetrahedron Lett., 1995, **36,** 3111.

151. N.A. Bumagin, L.I. Sukhomlinova, T.P. Tolstaya and I.P. Beleskya, Russ. J. Org. Chem., 1994, **30**, 1605.

152. C. Wolf and R. Lerebours, J. Org. Chem., 2003, **68,** 7551.

153. S. Venkataraman and C.J. Li, Adv. Syn. Catal, 2002, **344**, 399.

154. M. Safi and D. Sinou, Tetrahedron Lett., 1991, **32**, 2025. J. Genet, E. Blart and S. Savignac, Synlett., 1992, 715; C. Chervin, J. Le Bras, F. Henin and J. Muzart, Tetrahedron Lett., 2003, **44**, 8099; H. Kinoshita, H. Shinokubo and K. Oshima, Org. Lett., 2004, **6**, 4085.

155. S. Sigismondi, D. Sinou, M. Perez, M. Moreno-Manas, R. Plexats and M. Villarroya, Tetrahedron Lett., 1994, **35**, 7085.

156. E.Blart, J.P. Genet, M. Safi, M. Savignac and D. Sinou, Tetrahedron, 1994, **50,** 505.

157. C. Chevrin, J. Le Brass, F. Henin and J. Muzart, Tetrahedron Lett., 2003, **44**, 8099.

158. F. Strauss, Annalen., 1905, **342**, 201.

159. W. Baidossi, N. Goren and J. Blum, J. Mol. Catal., 1993, **85**, 153.

160. B.M. Trost, C. Chan and G. Ruhter, J.Am. Chem Soc., 1987, **109,** 3486.

161. Y. Masuda, M. Murata, K. Sato and S. Watanabe, Chem. Commun., 1988, **7**, 807.

162. R. Chinchilla and C. Najera, Chem. Rev., 2007, **107**, 874, K. Sonoyashira, In Handbook of Organometallic chemistry for organic synthesis, E. Negishi Ed., Wiley, Hoboken, NJ, 2002, p. 493.

163. P. Appukkutan, W. Dehaen and E.V. dev Eycken, Eur. J. Org. Chem., 2003, 4713.

164. J. Gil-Molto, S. Karlström and C. Najera, Tetrahedron Lett., 2005, **61**, 12168.

165. K.M. Dawood, W. Solodenko and A. Kirshning ARKIVOC, 2007, 104; K.M. Dawood and A. Kirshning, Tetrahedron, 2005, **61**, 12121.

166. N.E. Leadbeater, M. Marco and B.J. Tominace, Org. Lett., 2003, **5**, 3919.

167. R.K. Arrela, N.E. Leadbeater, M.S. Sangli, V.A. Williams, P. Granadus and R.D. Singer, J. Org. Chem., 2005, **70,** 161.

168. M. Pal., V. Subramanian and K.R. Yeleswarapu, Tetrahedron Lett., 2003, **44**, 8221; Y. Uozmi and Y. Kobayashi, Heterocycles, 2003, **59**, 71; B. Liang M. Dai, J. Chen and Z. Yang, J. Org. Chem., 2005, 70, 391.

169. C.J. Li, D.L. Chen and C.W. Costello, Org. Res. Process Dev., 1997, **1**, 315.

170. I.P. Beletskaya, G.V. Latyshev, A.V. Tsvetkov and N.V. Lukashev, Tetrahedron Lett., 2003, **44**, 5011.

171. L. Chen and C.-J. Li, Org. Lett., 2004, **6**, 3151.

172. C.M. Wei and C.J. Li, Green Chem., 2004, **4,** 39.

173. C.J. Li and C.M. Wei. Chem. Commun., 2002, 268.

174. C.M. Wei and C.J. Li, J. Am. Chem. Soc., 2003, **125,** 9584.

175. C.M. Wei, Z.G. Li and C.J.Li, Org. Lett., 2003, **5**, 4473.

176. C.M. Wei and C.J. Li., J. Am. Chem. Soc., 2002, 124, 5638; C.M. Wei, J.T. Mague and C.J. Li, Proc. Natl. Acad. Sci; USA, 2004, **101**, 5949.

177. L. Shi, Y.-Q. Tu, M. Wang, F.M. Zhang and C.A. Fan, Org. Lett., 2004, **6**, 1001.

178. T.F. Knoepfel and E.M. Carreira, J. Am. Chem. Soc., 2003, **125**, 6054.

179. L. Chen and C.J. Li, Tetrahedron Lett., 2004, **45**, 2771.

180. E. Mizushima, K. Sato, T. Hayashi and M. Tanaka, Angew. Chem. Int. Ed., 2002, **41,** 4563.

181. C. Jia, T. Kitamura and Y. Fujiwara, Acc. Chem. Res., 2001, **34**, 633.

182. C.C.K. Keh, V.V. Namboodiri, R.S. Varma and C.J.Lee, Tetrahedron Lett., 2002, **43**, 4993.

183. Y.S. Cho, H.Y. Kim, J.H. Cha, A.N. Pae, H.Y. Koh, J.H. Choi and M.H. Chang, Org. Lett., 2002, **4,** 2025.

184. M. Takebayashi and T. Shingaki, Bull. Chem. Soe. Jpn: 1953, **26**, 137; F.M. Bauers, Angew Chem. Int. Ed., 2002, **41**, 544; D. Grande, S. Baskaran and E.L. Chaikof, Macromolecules, 2001, **34**, 1640.

185. H. Yorimitsu, K. Wakabayaski, H. Shinokubo and K. Oshima, Tetrahedron Lett., 1999, **40**, 519; H. Yorimitsa, K. Warkabayashi, H. Shinokubo and K. Oshmitsa, Bull. Chem. Soc. Jpn, 2001, **74**, 1963.

186. J. Thesing and P. Binger, Chem. Ber., 1957, **90,** 1419.

187. D. Chen, L. Yu and P.G. Wang, Tetrahedron Lett., 1996, 37, 4467.

188. W. Zhuang and K.A. Jorgensen, Chem. Commun., 2002, 1336.

189. R. Ding, H.B. Zhang, Y.J. Chen, L. Liu, D. Wang and C.J. Li, Synlett., 2004, 555.

190. W.V.E. Doering and L.H. Knox, J. Am. Chem. Soc., 1961, **83**, 1989.

191. D.F. Taber and E.H. Petty, J. Org. Chem., 1982, **47**, 4808, D.F. Taber and R.E. Ruckle, Jr., J. Am. Chem. Soc., 1986, **108**, 7686.

192. A.P. Kreshkov, E.N. Sugushkima and B.A. Krozdov, J. Appl. Chem. USSR (Engl. Transl.), 1965, **38**, 2357.

193. T. Greibrookk, Acta. Chem. Scand., 1973, **27**, 3207.
194. L.V. Chau and M. Schlosser, Synthesis, 1973, 112.
195. R. Runhua, W. Yuliang and Y. Yazhong, Synth. Commun, 1994, **24,** 111.
196. V.K. Ahuwalia, Green Chemistry, Environmentally Benign Reactions, Ane Books, India, 2006, P. 20 and references cited there in.
197. W. Tagaki and H. Hara, J. Chem. Soc. Chem. Commun, 1973, 891.
198. W. Yaliang and J. Yaozhong, Synth. Commun. 1992, **22**, 2287.
199. M. Makosza, Tetrahedron, 1968, **24**, 175.
200. A. Jonck, M. Fedorynski and M. Makosa, Tetrahedron Lett., 1973, 1273.
201. V.K. Ahluwalia et al, unpublished results.
202. R.J. Giguere, A.M. Namen, O. Lopez, A. Arepally, D.E. Ramos, G. Majetich and J. Defaw, Tetrahedron Lett., 1987, **28**, 6553.
203. A. Jonczky, M. Fedorynski and M. Makosaza, Tetrahedron Lett., 1972, 2395.
204. G.W. Gokel, S.A. DiBiase and B.A. Lipiska, Tetrahedron Lett., 1976, 3495.
205. R. Ramamurty, Tetrahedron, 1986, **42**, 5753.
206. For details see Organic reactions in aqueous media, C.-J.Li. and T.-H. Chan, John Wiley, New York, Page 41-43.
207. B. Kuhlman, E.M. Amett and M. Siskin, J. Org. Chem., 1994, **59**, 5377.
208. J.M. Kremsner and C.O. Kappe, Eur. J. Org. Chem., 2005, 3672.
209. K. Reimer and F. Tiemann, Ber., 1876, **9**, 824, 1285; H. Wynberg, Chem. Rev., 1960, **60**, 169.
210. S.L. Regan and A. Singh, J. Org. Chem., 1982, **47**, 1587.
211. L. Claisen and A. Claparedl, Ber., 1881, **14,** 2460; J.G. Schmidt, Ber., 1881, **14**, 1459.
212. K.R. Nivaklar, C.D. Mudaliar and S.H. Mashraqui, J. Chem. Res. (S), 1992, 98.
213. K. Takai, C.H. Hearthcock, J. Org. Chem., 1985, **50,** 3247, A.E. Vougioakas and H.B. Kazan, Tetrahedron Lett, 1987, **28**, 5513; K. Mikami, M. Terada and N. Nakai J. Chem. Soc. Chem. Commun, 1935, 343 and the references cited there in.
214. K.M. Meyer and K. Schuster, Chem. Ber, 1922, 55, 819.
215. A.V. Baeyer and V. Villager, Ber, 1899, 32, 3625.
216. F. Fringulli, R. Germani and G. Savalli, Gazz. Chem. Ital., 1989, 119, 249.
217. S. Selman and J.F. Eastham, Quart. Rev., 1960, **14**, 221.
218. S. Cannizzaro, Ann., 1853, **88**, 129; K List and H. Limprechl, Ann., 1854, **90**, 180.
219. H.D. Dakin, OS, 1941, **1**, 149; J.E. Letter, Chem. Rev, 1949, **45**, 385; H.D. Dakin, J.Am. Chem. Soc., 1909, 42, 477; H.D. Dakin, Org. Synth, Coll Vol. I, 1941.
220. N.F. Goldshleger, V.V. Eskova, A.E. Shilov and A.Z. Shteinman, A.A. Zh, Fiz. Khim, 1972, 46, 1353 (Engl. Trans. 1972, **12**, 895).
221. V.V. Eskova, A.E. Shilov and A.A. Shteinman, A.A. Kinet, Katal., 1972, **13,** 534.
222. A Sen, M.A. Senvenuto, M. Lin, A.C. Hutson and N. Basickes, J. Am. Chem. Soc. 1994, **116**, 998.
223. L.-C. Kao and A. Sen, J. Chem Soc. Chem. Commun., 1991, 1242.
224. M. Lin, T. Hogan and A Sen, J. Am Chem. Soc., 1997, **119**, 6048.

225. C.I. Herrerias, X. Yao, Z.-Li and C.-J. Li, Chem. Rev., 2007, **107**, 2546-2562 and the references cited there in.

226. A.E. Shilov, G.B. Shulpin in activation and catalytic reactions of saturated hydrocarbons in presence of metal complexes, Kluwer Academic Publishers, Dordrecht, 2000, P. 21; A.A. Fokin and P.R. Scheiner, Adv Synth. Catal., 2003, **245**, 1035 and references cited therein; P.P. Schreiner and A.A. Fokin, Chem. Record 2004, **3**, 247 and references cited therein.

227. L. Buchholz and C. Münnica, In Handbook of C–H Transformations, G. Dyker, Ed., Wiley-VCH, Weinheim, 2005 p. 11 and references cited therein.

228. W. Adam, Peroxide Chemistry. Mechanistic and Preparative aspect of oxygen transfer, Wiley-VCH, Weinheim 2000; W. Adam, C.R. Saha-Mollar and P.A. Ganeshpure, Chem. Rev., 2001, **101**, 3499; Y Shi, Acc. Chem. Res., 2004, **37**, 488; R. Curci, L.D' Accolti Accolti and C. Fusco, Acc. Chem. Res., 2006, **39**, 1.

229. D. Yang, M.-K. Wong, X.-C. Wang and Y.-C. Tang, J.Am. Chem. Soc., 1998, **120**, 6611.

230. P.R. Schreiner, O. Lauenstein, I.V. Kolomitsyn, S. Nadi and A.A. Fokin, Angew. Chem. Int. Ed., 1998, 37, 1895; P.R. Schreiner, O. Lauenstein, E.D. Butova and A.A. Fokin, Angew. Chem. Int. Ed., 1999, **38**, 2786; O. Laurenstein, A.A. Fokin and P.R. Schreiner, Org. Lett., 2000, **2**, 2201; A.A. Fokm, O. Lauenstein, P.A. Gunchenko and P.R. Schreiner, J.Am. Chem. Soc., 2001, **123**, 1842; P.R. Schreiner, O. Lauenstein, E.D. Butova, P.A. Gunchinko, I.V. Kolomitsen, A. Wittkopp, G. Feder and A.A. Fokin, Chem. Eur. J., 2001, 7, 4997; P.R. Schreiner, A.A. Fokin, O. Lauenstein, Y. Okamoto, T. Waketa, C. Rinderspacher, G.H. Robinson, J.K. Vohs and C.F. Campana, J. Am. Chem. Soc., 2002, **124**, 13348; A.A. Fokin, R.P. Schreiner, R. Berger, G.H. Robinson, P. Wei and C.F. Campana, J. Am. Chem. Soc., 2006, **128**, 5332.

231. Cited in reference 225.

232. H. Shaw, H.D. Perlmutter and C. Gu, J. Org. Chem., 1997, **62**, 236.

233. D. Kikuchi, S. Sakorguchi and Y. Ishii, J. Org. Chem., 1998, **63**, 6023.

234. I. Tabushi, Z. Yoshida and N. Takahashi, J. Am. Chem. Soc, 1971, **93**, 1820.

235. C.H. Hunsdiecker, Chem. Ber., 1942, **75**, 291.

236. A.J. Meyers, J. Org. Chem., 1979, **44**, 3405; A. Borodin, Ann., 1942, **75**, 291.

237. C. Hell, Ber., 1881, **14**, 891; J. Volhard, Ann., 1887, **242**, 141; Z. Zelinsky, Ber., 1887, **20** 2026; N.O.V. Sonntag, Chem. Rev., 1953, **52**, 237.

238. T. Sandmeyer, Ber., 1984, **17**, 1633, 2650; L. Gattermann, Ber, 1890, **23**, 1218.

239. P.A. Grieco and S.D. Larsen, J. Am Chem. Soc., 1905, **107**, 1768; D. Boger and S. Weireb, Hetero-Diels-Alder Methodology in Organic Synthesis, Academic Press, Orlendo, 1987.

240. E. Beckmann, Chem. Ber., 1886, **19**, 988; R.T. Conley, J. Org. Chem., 1963, **28,** 210.

241. I. Almena A. Diaz-Ortiz, E. Diaz-Barra and L. Loupy, Chem. Lett., 1996, **333**; S. Caddick, Tetrahedron, 1995, 10400; A.I. Bosch., P. de La Cruez, E. Diez-Barra and F. Lanya, Synlett., 1995, 1259.

242. P. Bignilli, Bev. 1891, **24**, 1317, 2962; 1893, **26**, 447; Zaugg, Martin, Organic Reactions, 1965, **14**, 88.

243. C.O. Kappe, D. Kumar and R.S. Varma, Synthesis, 1999, 1799, M. Kidwai, S. Saxena, R. Mohan and R. Venkataraman, J. Chem. Soc. Perkin Trans I, 2002, 1845.

244. J. Peng and Y. Deng, Tetrahedron Lett., 2001, **42**, 5917.

245. T. Curtius, J. Prakt. Chem, 1894, **[2]50**, 275; J.H. Saunders and R.J. Sloccombl, Chem. Rev., 1948, **43**, 205.

246. W. Lwowski, S. Linke and G.T. Tisue, J. Am. Chem. Soc., 1967, **89**, 6308.

247. A.A. Mortan, The Chemistry of Heterocyclic compounds, McGraw Hill, Inc. New York, Y.K. 1946, 263; Cambell and Schaffner. J. Am. Chem. Soc., 1945, **67**, 86.

248. E. Fischer and F. Jourdan, Ber, 1883, **16**, 2241; E. Fisher and O. Hess, Ber., 1804, **17**, 559

249. D. Villemin, B. Labiad and Y. Ouhilal, Chem. and Ind., 1989, 607.

250. C.R. Strauss and R.W. Trainer, Aust. J. Chem., 1998, **51**, 703.

251. P. Friendlander, Chem. Ber., 1882, **15**, 2572; P. Friendlander and C.P. Gohring, Ber., 1883, **16**, 1833; Manske, Chem. Rev. 1942, **30**, 124.

252. G. Sabitha, R.S. Babu, B.V.S. Reddy and J.S. Yadav, Synth. Commun., 1999, **29(24)**, 4403.

253. G. Graebe and F. Ullmann, Ann., 1896, **291**, 16; F. Ullmann, Ann., 1904, **332**, 82.

254. A. Mohna, J.I. Vaguero, J.J. Garefa and J. Alvarez-Builla, Tetrahedron Lett, 1993, **34**, 2673.

255. A. Hantzsch, Ann., 1882, **215**, 1, 72, Ber. 1885, 18, 1744; 1886, **19**, 289.

256. I.C. Cottrill, A.Y. Usyatinsky, J.M. Arnold, D.S. Clark, J.S. Domick, P.C. Michels and Y.L. Khmelnisky, Tetrahedron Lett., 1998, **39**, 1117.

257. A. Hentzsch, Ber., 1890, **23**, 1474.

258. D.L. Boger, Tetrahedron, 1983, **39**, 2869.

259. P.A. Greco and S.D. Larsen, J. Am. Chem. Soc. 1985, **107**, 1768.

260. W. Oppolzer, Angew Chem. Int. Ed. Engl., 1912, **11**, 1031.

261. K. Takai and C.H. Heathcook, J. Org. Chem., 1985, **50**, 3247.

262. L. Knorr, Ber., 1883, **16**, 2587.

263. L. Knorr. Ann., 1886, **236**, 69; 1888, **245,** 357, F.W. Bergston, Chem. Rev., 1944, **35** 157; 1948, **40**, 67.

264. Z.H. Sikraup, Ber., 1880, **13**, 2086; R.H. Manske, M. Kukla, Org. Reactions, 1953, **7**, 80; Bergstrom, Chem. Rev., 1944, **35**, 152.

265. H. Barton, F.G.P. Remtry and Y.F. Thorpe, J. Chem. Soc., 1904, **85**, 1726; K. Ziegler, E. Eberle and M. Ohlinger, Ber., 1933 **504**, 94; J.P. Schaefer and J.J. Bloomfield, Org. Reactions, 1967, 151.

266. J.J. Bloomfield and P.V. Fennessey. Tet. Lett., 1964, 2273.

267. C.M. Starks, J. Am. Chem. Soc., 1971, **93**, 195; N. Sugimoto, T. Fujita, N. Shigematsu and A. Ayada, Chem. Pharm. Bull., 1962, 10, 427 (Jap. Patent, 1961/1963).

268. R.S. Varma, K.P. Naicker, Molecules on line, 1982, **2**, 94; R.S. Varma, K.P. Nariker D. Kumar, R. Dahiya and P.J. Liesen, J. Microwave Power Engineering Energy, 1999, 34, 113.

269. A.P. Kreshkov, E.N. Suguskima and B.A. Krozdov, J. Apple Chem. USSR (Engl. Transl.), 1965, 38, 2357.

270. O. Hinsberg. Chem. Ber., 1910, **42**, 901; N. Wynberg, J. Org. Chem., 1964, **29**, 1919; N. Wynberg, J. Am. Chem. Soc., 1965, **87**, 1739.

271. V. Pirotic and S. Colonna, Green Chem. 2005, **7**, 43.

272. J.S. Yadav, B.V.S. Reddy, C.R. Reddy and K. Rajaskhar, J. Org. Chem., 2003, **68**, 2525.

273. A Kamal and G. Chauhan, Tetrahedron Lett., 2005, **46**, 1489.
274. R.S. Varma and D. Kumar, Org. Lett., 1999, **1**, 697.
275. Z. Li and C.-J. Li, J. Am. Chem. Soc., 2006, **128**, 56 and references cited therein.
276. Y. Luo, Z. Li and C.-J. Li, Org. Lett., 2005, 7, 2675.
277. H.-B. Zhang, L.Liu, Y.-J. Chen, D. Wang and C.-J. Li, Adv. Synth. Catal., 2006, 348, 229.
278. H.-B. Zhang, L. Liu, Y.-J. Chen. D. Wang and C.-J.Li., J. Org. Chem., 2006, 869.
279. M. Matsushita, K. Kamata, K. Yamaguchi and N. Mizuno, J. Am Chem Soc., 2005, 127, 6632.
280. P.J. Wallis, K.J. Booth, A.F. Patti, and J.L. Scott, Green Chem., 2006, **8**, **333**.
281. Y. Jiang, C. Xi and X. Yang, Synlett., 2005, 1381.
282. C.Xi, Y. Jiang and X. Yang, Tetrahedron Lett., 2005, 46, 3909.
283. S.B. Brummer and A.B. Gancy, in water and Aqueous solutions, Structure, Thermodynamics and Transport Processes. R.A. Horne ed., Wiley-Interscience, 1972; E.U. Franck, C & EN, Dec. 23, 1991, p. 26.
284. J.L. Kavanau, Water and Solute Water Interactions, Holden-Day, San Francisco, 1964.
285. E.J. Parson, CHEMITCH, 1966, **26(6),** 30.
286. A.R. Ktrisky, S.M. Allin and M. Siskin, Acc. Chem. Res., 1966, **29(8)** 399.
287. J. An, L. Bagnell, T. Cablewski, C.R. Strauss and R.W. Trainer J. Org. Chem., 1997, **62(8),** 2505.
288. R. Crittendon and E.J. Parson, J. Organometallics, 1994, **13(7),** 2587.
289. M.B. Korzenski and J.W. Kolis, Tetrahedron Lett., 1997, **38 (32),** 5611.
290. T. Jeffery, Tetrahedron Lett., 1994, **35**, 3051.
291. J. Diminnie, S. Metts and E.J. Parson, Organometallics 1995, **141(8),** 4023.
292. B. Kuhlmann, E.M. Arnett and M. Siskin, J. Org. Chem., 1994, **59(11)**, 3098.
293. X.Xu and M.J. Antal, Jr., AICHE J., 1994, **40(9)**, 1524.
294. X.Xu and M.J. Antal, Jr., Ind. Eng. Chem. Res., 1997, **36(1)**, 23.
295. M. Carlson, C. Habenicht, L.C. Kam, M.J. Antal, Jr., N. Bian, R.J. Cunningham and M. Jones, Jr., Ind. Eng Chem. Res., 1994, **33(8)**, 1989.
296. T.A. Bryson, J.J. Stewart, J.M. Gibsen, P.S. Thomas and J.K. Berch, Green Chemistry 2003, **5**, 174.
296a. B.M. Kabyemela, T. Adschiri, R.M. Malaluan, K. Arai, H. Ohzei, Ind. Eng. Chem. Res., 1997, **36**, 5063.
297. A. Naoko and E.P. Savage, Chem. Rev., 2002, **102**, 2725.
298. Cited in Water as a reaction medium for Clean Chemical Processes, W.W. Charlene, C.K. Keh, C.-J. Li and R.S. Varma, Clean Techn. Environ. Policy, 2005, **7**, 62-69 (http://dx.doi.org)/10.1007/s10098-003-0247-7); L.M. Dud, E. Venardon, E. Garcia-Verdugo, P. Licence, A.J. Blake, C. Wilson and M.C. Poliakoff, Green Chem., 2003, **5**, 187.
299. B. Kuhlmann, E.M. Arnett and M. Siskin, J. Org. Chem., 1994, **59(18),** 5377.
300. C.R. Strauss and R.W. Trainer, Aust. J. Chem., 1998, **51**, 703.
301. S.E. Hunter, C.A. Felczak and P.E. Savage, Green Chem., 2005, **7,** 586.

3. TRANSFORMATIONS IN IONIC LIQUIDS

3

Transformations in Ionic Liquids

3.1 INTRODUCTION

Ionic liquids, as we know, are made up of atleast two components (the cation and the anion) which can be varied. The solvent can be designed with a particular end use in mind or to have a particular set of properties. So these are commonly known as 'designer solvents'[1]. These have emerged as replacement of volatile organic compounds that have been so far used as industrial solvents and have serious environmental issues. The ionic liquids are considered to be green solvent which is expected to be used in future. A number of ionic liquids are commercially available. These are referred to as 'room temperature ionic liquids'[2] since these exist in the liquid state at or around ambient temperature. These room temperature ionic liquids possess many properties[3] which make them attractive media for a number of organic reactions. These have virtually no vapour pressure, possess good thermal stability and are able to dissolve a wide range of organic, inorganic or organometallic compounds. Gases like H_2, CO, O_2 or CO_2 are soluble in ionic liquids.

The ionic liquids commonly compose of following organic cations.

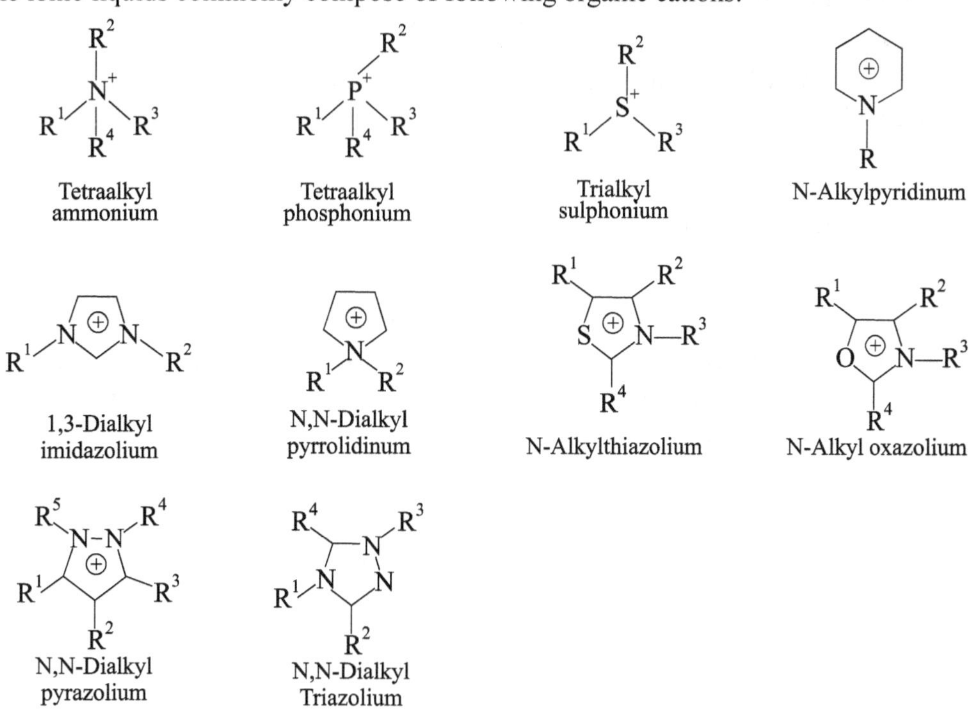

The common anions are BF_4^-, PF_6^-, SbF_6^-, NO_3^-, $CF_3SO_3^-$, $(CF_3SO_2)_2N^-$, $ArSO_3^-$, CF_3COO^-, $CH_3CO_2^-$, $ZnCl_3^-$, $CuCl_2^-$, $SnCl_3^-$, $(C_2F_5SO_2)_2N^-$, $(FSO_2)_2N^-$, $C(CF_3SO_2)_3^-$, $CH_3SO_3^-$.

The ionic liquids being polar and ionic in nature, couple with microwave irradiation efficiently and so chemical reactions can be conducted efficiently[4].

1,3-Dialkylimidazolium halides could be prepared conveniently using microwave heating reducing reaction time from several hrs. to minutes and avoiding use of excess of alkyl halides and organic solvents[5–8]. This procedure has been extended for the preparation of other ionic acids bearing tetrafluoroborate anions that involves N, N′-di alkylimidazolium chloride and ammonium tetrafluoro borate to MW irradiation[7].

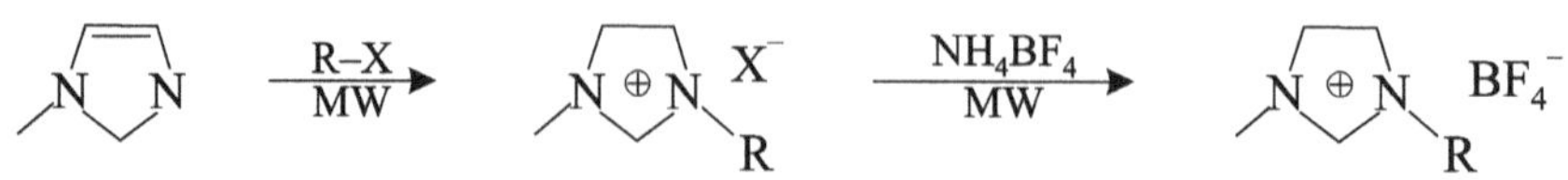

(Microwave assisted preparation of room temperature ionic liquids)

An improvement contrasting the MW approach involves preparation of ionic liquids at room temperature under ultrasonic irradiation conditions[9].

See Also Chapter 9, Section 9.5.13.

3.2 SYNTHETIC APPLICATIONS

Some of the synthetic applications of ionic liquids are discussed below.

3.2.1 Baylis-Hillman Reaction

It is a useful C–C bond forming reaction and has been shown to involve reaction[10] between an α-position of an activated alkene and an aldehyde in presence of a base diazobicyclo [2.2.2] octane (DABCO) (Fig. 3.1).

X-electron withdrawing group

Fig. 3.1: Baylis-Hillman Reaction.

It was found[11] that Baylis-Hillman reaction between benzaldehyde and methyl acrylate in ionic liquid, [b_{min}] [PF_6] was 33 times faster than the reaction in CH_3CN, though only moderate yields of the desired product was obtained (Fig. 3.2).

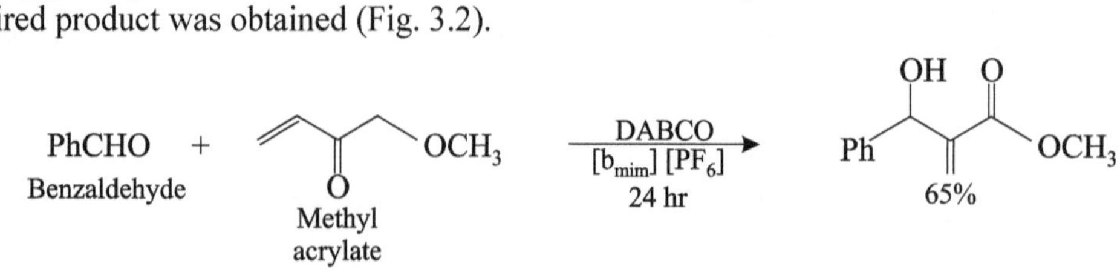

Fig. 3.2: Baylis-Hillman reaction in ionic liquid [b_{min}] [PF_6].

The low yields in the above reaction was because under basic conditions, the aldehyde was being consumed in a side reaction with the imidazolium cation[12]. The C(2) hydrogen of the imidazolium cation (being acidic) was responsible for the side reaction (Fig. 3.3).

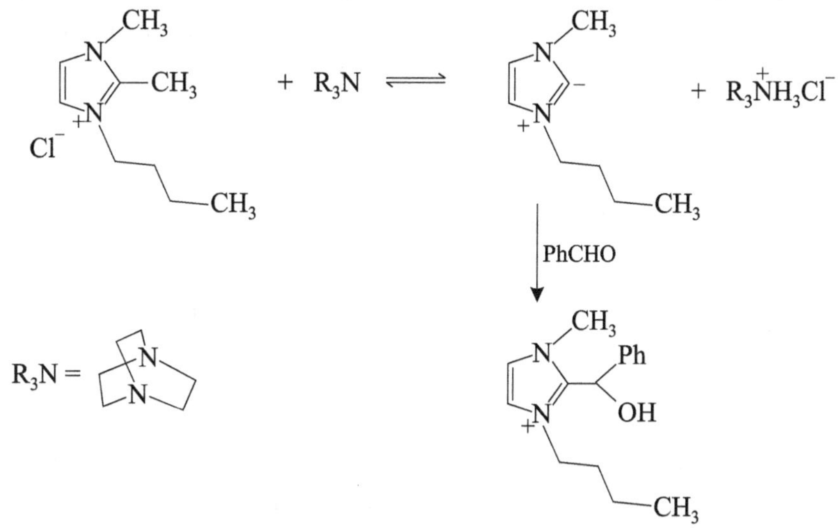

Fig. 3.3: The acidic nature of C(2) hydrogen of the imidazolium cation.

In view of the above, caution must be exercised when iodic liquid from one reaction is used for another reaction. The problem associated with the use of imidazolium based ionic liquid could be solved by using ionic liquid substituted at 2-position. It was found[13] that Baylis-Hillman reaction between a number of aldehyde and methyl acrylate proceeded smoothly in the ionic liquid $[b_{m\ \mathrm{min}}]$ $[PF_6]$, in contrast to the results obtained with the ionic liquid, $[b_{\mathrm{min}}]$ $[PF_6]$.

It has been found that imidazolium based ionic liquids containing a phenyl group at C(2) position (cation = $[\mathrm{m\ Ph_{min}}]$) are stable to a variety of strongly basic conditions[14] and can be profitably used for reactions like the Baylis-Hillman Reaction.

3.2.2 Knoevenagel Condensation

The knoevenagel condensation of benzaldehyde with malononitrile in presence of KOH in the ionic liquid $[b_{\mathrm{min}}]$ $[PF_6]$ gave the styrene product is low yields[15] (Fig. 3.4).

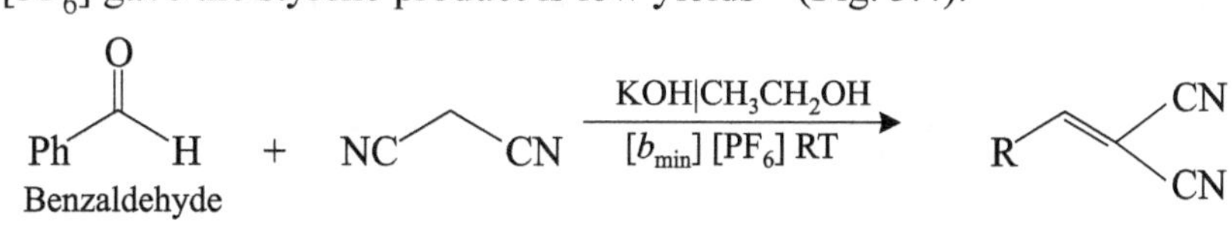

Fig. 3.4: Knoevenagel Condensation in ionic liquid.

The low yield was due to the fact that the product formed was soluble in the ionic liquid. The yields of the styrene product improved as the concentration of the substrate increased and also the ionic liquid was reused. The ionic liquid could be re-used up to five times with addition of additional base.

3.2.3 Claisen-Schmidt Condensation

Claisen-Schmidt condensation of benzaldehyde and acetophenone in presence of NaOH and ethyl alcohol in the ionic liquid [b_{min}] [PF_6] gave the condensation product in low yields as in the case of Knoevenagel condensation. (Fig. 3.5). In this case ethyl benzoate was obtained as a byproduct and the base was depleted after two cycles of reuse of ionic liquid.

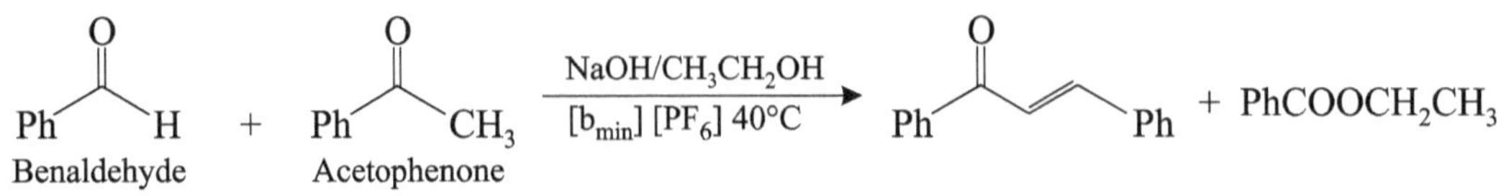

Fig. 3.5: Claisen-Schmidt Condensation in ionic liquid.

It was found that in the reaction of ionic liquid with base in the absence of substrate—the base was consumed by the reaction with imidazolium cation, particularly at higher temperature. This was not a problem in Knoevenagel condensation which was carried out at room temperature. On the basis of controlled experiments, it was shown that the source of ethyl group was sodium ethoxide and not ethanol. It was concluded on the basis of results obtained that the imidazolium ionic liquids are suitable under basic conditions only in few cases and care must be taken when reaction was carried out in these ionic liquids with bases particularity at higher temperatures.

3.2.4 Horner-Wadsworth-Emmons Reaction

The Horner-Wadsworth-Emmons reaction between the aldehyde and phosphonoacetates in the ionic liquid [e_{min}] [BF_4] or [e_{min}] [PF_6] also gave low yields of either of the products[16] (Fig. 3.6).

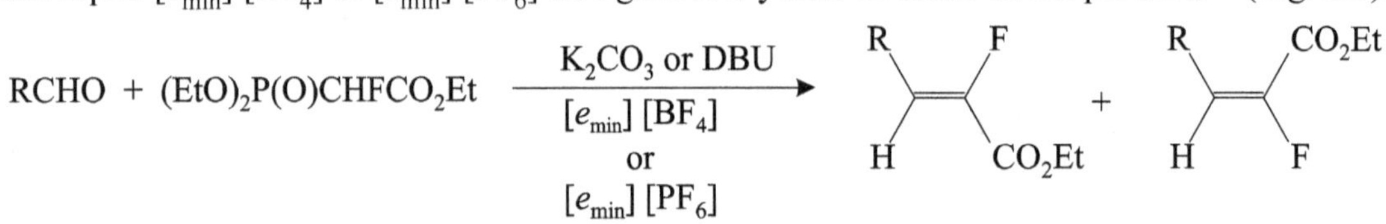

Fig. 3.6: Horner-Wadsworth-Emmons reaction in ionic liquids.

Much better yields of the products were obtained when the above reaction was carried out in the ionic liquids 8-ethyl-1,8-diazabicyclo [5,4,0]-7-undecene trifluoromethyl sulfonate (A) and 8-methyl-1, 8-diazabicyclo [5, 4, 0]-7-undecene trifluoromethyl sulfonate (B).

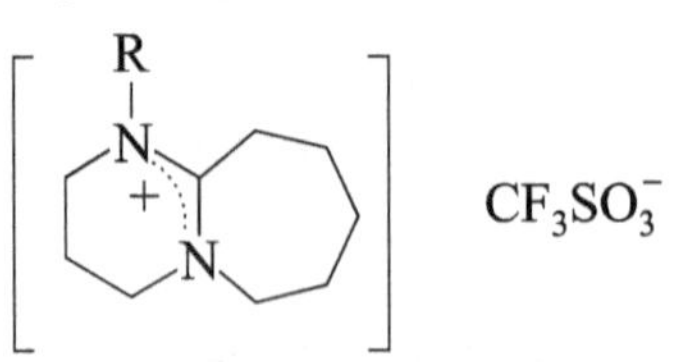

(A) R = Et
(B) R = Me

3.2.5 Heck Reaction

A carbon-carbon bond forming reaction, it involves palladium-catalysed coupling of an aryl or alkenyl halide with an olefin (Fig. 3.7).

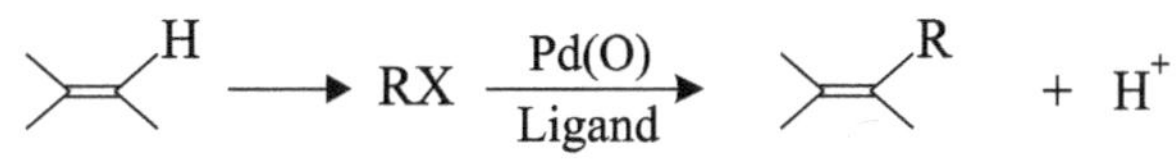

Fig. 3.7: Heck reaction.

The Heck reaction uses mild conditions and can also be performed with less reactive halides (bromides and chlorides). In order to stabilise the active palladium species a phosphine ligand is required. The main drawback of the Heck reaction is that the palladium catalyst cannot be recovered for reuse. The recycling of the catalyst can be achieved by using ionic liquids[17].

The Heck reaction of iodobenzene with ethyl acrylate in the ionic liquid N-hexylpyridinium [C_6Py] or N, N′dialkylimideazo based ionic liquid gave the corresponding ethyl cinnamate (Fig. 3.8); higher yields were obtained in the former ionic liquid. The addition of a phosphine ligand to the palladium species in the reaction in the pyridinium based ionic liquid decreased the yield and higher temperatures were necessary to force the reaction to go to completion.

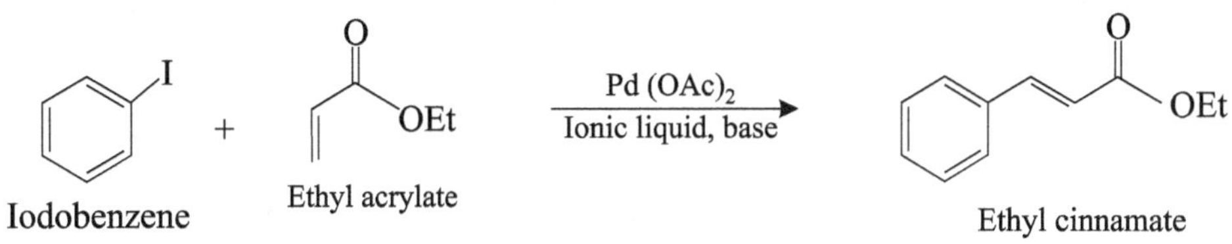

Fig. 3.8: Heck reaction in ionic liquid.

The low yields in imidazolium ionic liquids is believed to be due to the formation of carbene which reacted with palladium to form a mixture of palladium-carbene complex.

Heck reaction was also carried out by the reaction of aryl halides and acrylates and also with styrene in presence of $Pb(OAc)_2$ in the ionic liquids [b_{min}] [BF_4] and [b_{min}] [Br] (Fig. 3.9)[18].

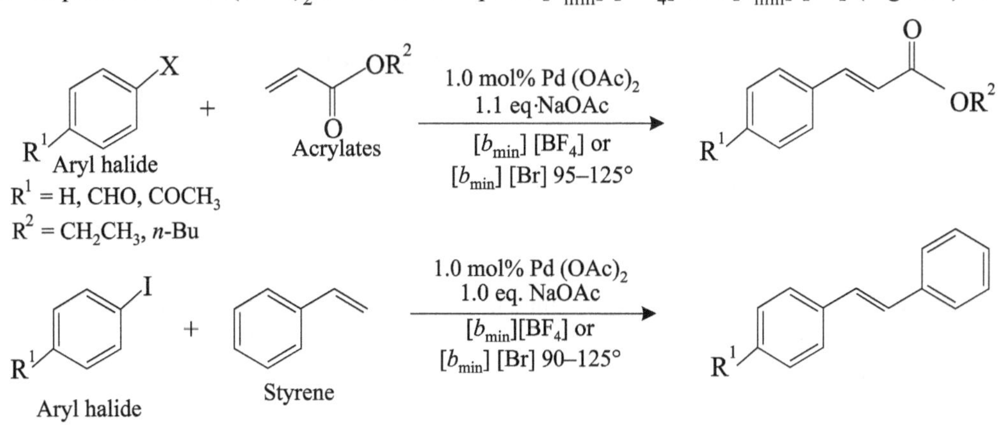

Fig. 3.9: Heck reaction of aryl halides with acrylates and styrene in ionic liquids.

Much better conversion and selectivity was observed in reactions carried out in [b_{min}] [Br]. In both ionic liquids, a homogeneous yellow solution was formed followed by precipitation of platinum black. The low yields in case of [b_{min}] [BF_4] was attributed to the formation of carbene intermediate which complexed with palladium to give a mixture of palladium-carbene complex.

It was believed that in the Heck reaction, the active catalyst is palladium nanoparticles generated in situ from palladium-carbene species.[18] It has been shown that solutions of ammonium stabilized Pd clusters are useful catalysts in the Heck reaction[19]. Thus, the Heck reaction of iodobenzene and *n*-butyl acrylate in presence of $[N_{8,8,8,8}]$ [Br] stabilized 3 nm Pd clusters gave *n*-butyl cinnamate (Fig. 3.10).

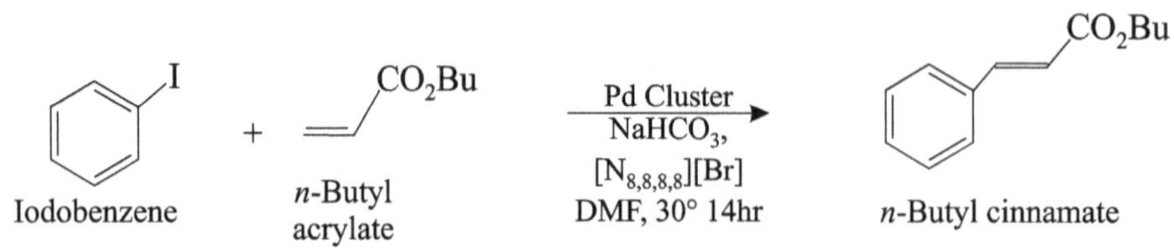

Fig. 3.10: Heck reaction of iodobenzene with n-butyl acrylate in presence of Pd cluster.

Intramolecular Heck reactions have been reported in ionic liquid $[C_{4min}]$ $[BF_4]$ using $PdCl_2$ as catalyst[19a]. Thus, Heck reaction of ortho-iodobenzyl allyl ethers gave substituted benzofurans at 60° [Fig. 3.10(*a*)]

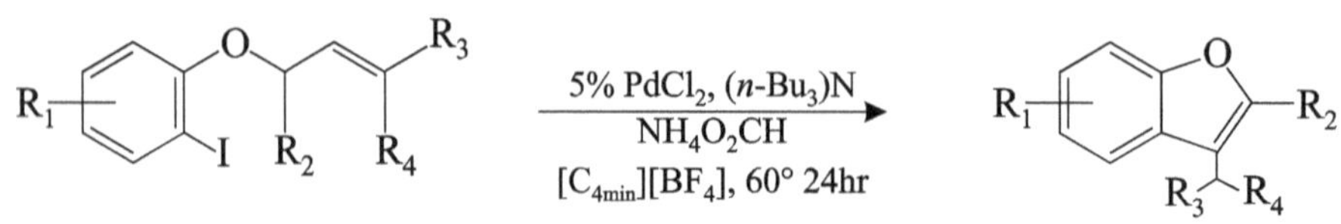

Fig. 3.10(*a*): Intramolecular Heck reaction.

Diazonium salts have also been used for the Heck reaction [Fig. 3.10(*b*)]. It has been shown that $Pd(OAc)_2$ may be used for the reaction of olefins with substituted tetrafluoroborate arene diazonium salts in $[C_{4min}]$ $[PF_8]$ with good yields using methyl acrylate or methyl acetonitrile at room temperature at 50°[19b] [(Fig. 3.10(*b*)].

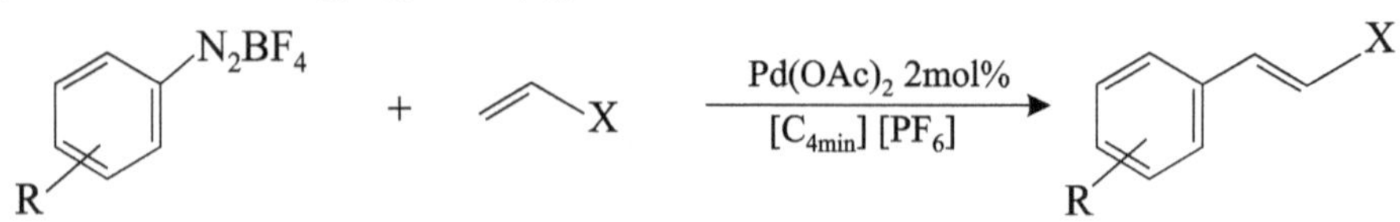

Fig. 3.10(*b*): Heck reaction with diazonium salts.

3.2.6 Suzuki Coupling

A C–C bond forming reaction, it is a palladium catalysed coupling reaction between aryl halides or aryl triflates and aryl boronic acids to give biaryls[20,21]. The Suzuki coupling reaction in ionic liquids give good yields[22-24]. (Fig. 3.11)

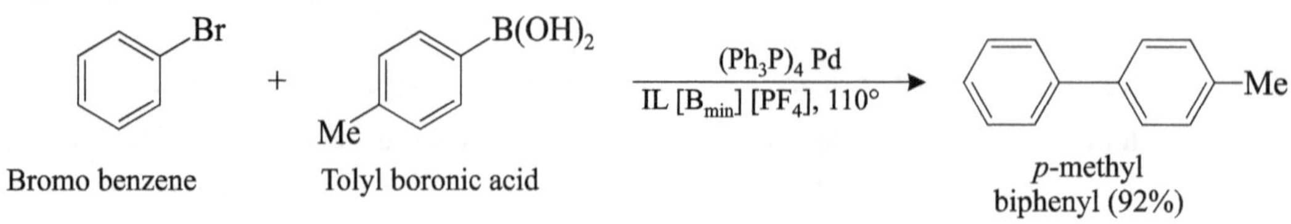

Fig. 3.11: Suzuki coupling in ionic liquid.

It was shown that 1,3-bis-(2, 4, 6-Trimethylphenyl) imidazol-2-ylidene(A) catalysed the Suzuki coupling reaction[25] between 4-chlorotolune and phenyl boronic acid in dioxane at 80°. As the imidazol-2-ylidene carbenes are not very stable to air and moisture, the carbene ligand is generated in situ from the salt (A) using CS_2CO_3 as the base [Fig. 3.12].

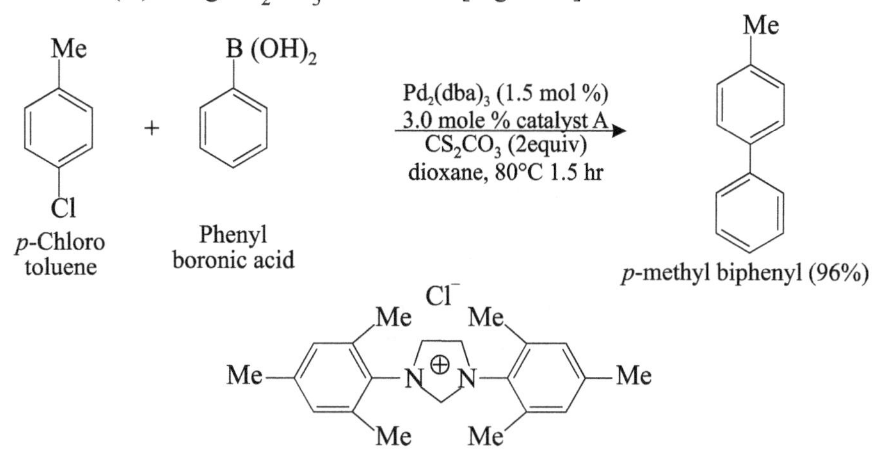

(A)

Fig. 3.12: Catalytic Suzuki coupling reaction.

On the basis of results obtained, it was suggested that in Suzuki coupling there is involvement of N-heterocyclic carbene.

The Suzuki coupling reaction of halo benzenes with phenyl boronic acid has also been carried out under milder conditions using methanol as co-solvent using ultrasound (Fig. 3.13)[26]. The co-solvent is used to solubilize phenyl boronic acid.

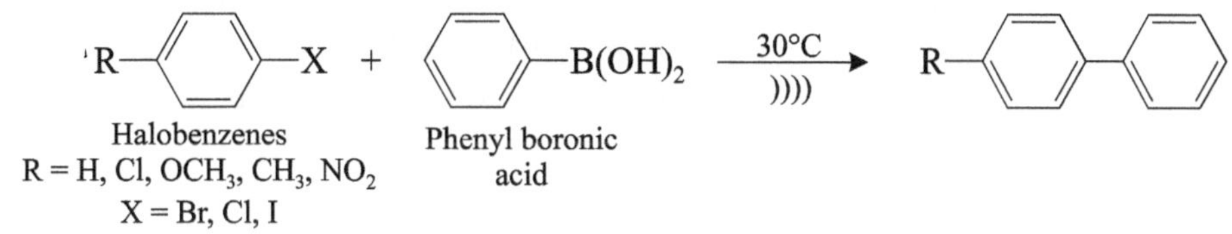

Fig. 3.13: Ultrasound assisted Suzuki coupling reaction

In the above reaction, the inactive Pd black was invariably formed thus preventing recycling of the catalyst. In a modified process the Pd-biscarbene complex (B) was used as a catalyst using only methanol under sonochemical conditions.

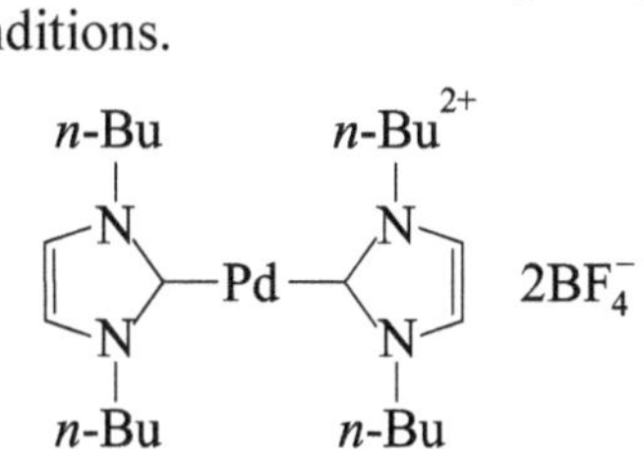

(B) Pd-biscarbene complex

3.2.7 Stille Coupling Reaction

The reaction is one of the most widely used step in the synthesis of a variety of materials including polyarenes and diaryl and aromatic carbonyl compounds.

The still coupling reaction is performed in ionic liquid [C_{4min}] [BF_4] using bis (benzonitrile) palladium (II) chloride, triphenylarsine and copper (I) iodide as catalyst system (Fig. 3.14)[27]. Using this solvent for the reaction of α-iodoenone and vinyltributyl tin gave results quite comparable by using NMP as solvent. Better results were obtained by using phenylbutyl tin.

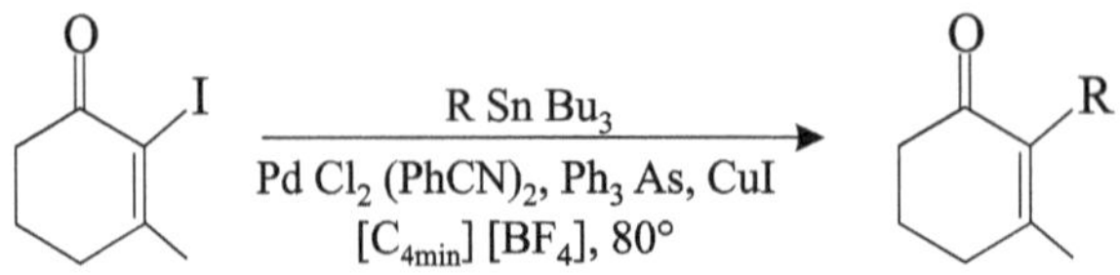

Fig. 3.14: Still coupling reaction between α-iodoenone and vinyltributyl tin.

3.2.8 Negishi Cross-Coupling Reaction

It is a palladium-catalysed cross coupling reaction. Thus the cross-coupling of organozinc reagents has been achieved in 1-butyl-2, 3-dimethylimidazolium tetrafluoroborate ([bd_{min}] [BF_4]) ionic liquid using a novel phosphine (C), which is prepared by the reaction of PPh_2Cl with [b_{min}] [PF_6][28]. A number of substrates were used. Best results were obtained for aryl iodides. The ionic liquid containing the catalyst could be recycled thrice, but the yield of the product decreased in each cycle.

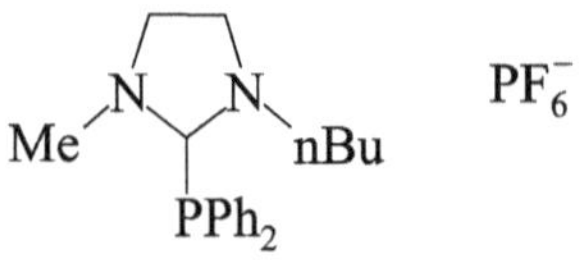

(C)

3.2.9 Trost-Tsuji Coupling Reaction

A C–C bond forming reaction, it involves nucleophillic allylic substitution catalysed by Pd(O) complexes in ionic liquids (mono-and biphasic system) using Pd $(OAc)_2$ – PPh_3/K_2CO_3 in [b_{min}] [BF_4][29] and $PdCl_2$–TPPTS (TPPTS = triphenyl phosphine trisulphone, sodium salt) in[b_{min}] [Cl]/ cyclohexene[30] respectively (Fig. 3.15).

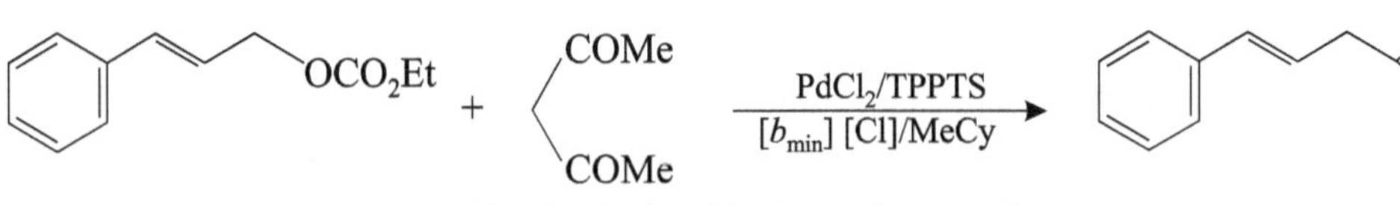

Fig. 3.15: Trost Tsuji coupling reaction.

3.2.10 Sakuai Reaction

A C–C bond forming reaction, it involves the reaction of α, β-unsaturated ketones with allyl trimethylsilane in the presence of $InCl_3$ in ionic liquid [C_{4min}] [PF_6] or [C_{min}] [BF_4] as solvent (Fig. 3.16)[31]

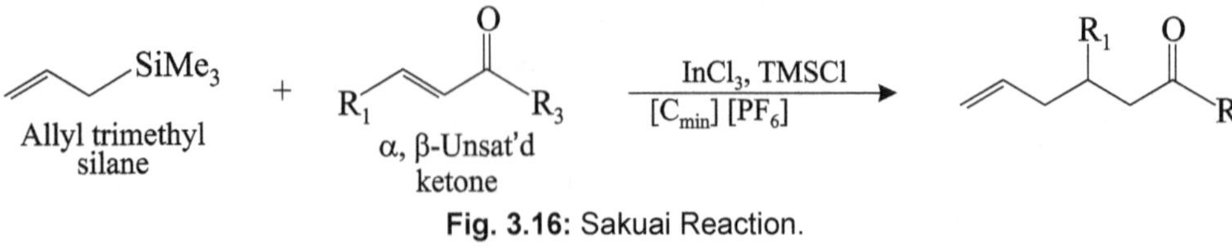

Fig. 3.16: Sakuai Reaction.

3.2.11 Henry Reaction

It involves the reaction of nitroalkanes having α-hydrogen undergoing aldol type reaction with aliphatic and aromatic carbonyl compounds to give β-hydroxy nitro compounds. This reaction can be accelerated with chloroaluminate ionic liquids[32]. The 1,1,3,3-tetramethyl guanidine (trifluoroacetate and lactate) based IL has been reported to be recyclable catalyst for Henry reaction to produce 2-nitro alcohols (Fig. 3.17).[33]

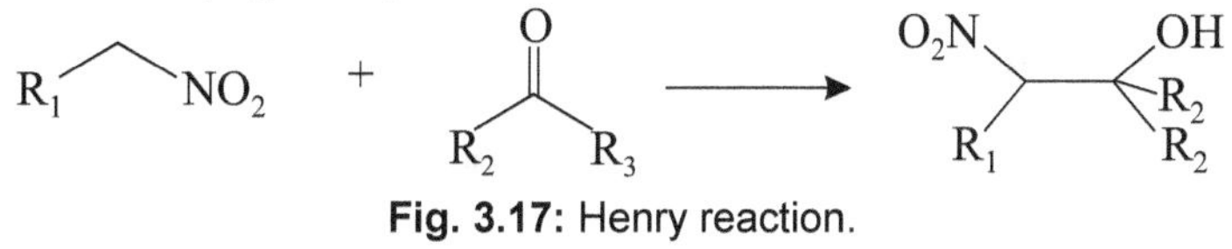

Fig. 3.17: Henry reaction.

3.2.12 Stetter Reaction

It involves the reaction of aldehydes and olefins to give 1, 4-dicarbonyls. It is now possible to carry out the reaction in ionic liquid using Et_3N as catalyst (Fig. 3.18)[34].

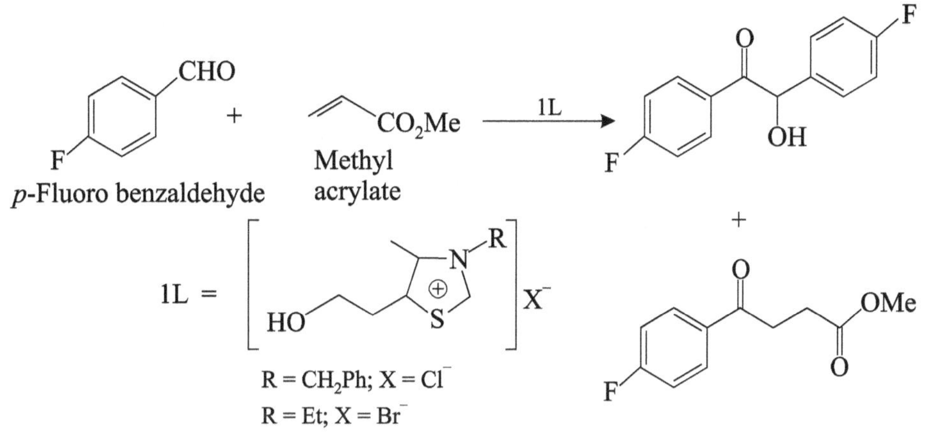

Fig. 3.18: Stetter reaction.

3.2.13 Sonogashira Reaction

It is a C—C bond forming reaction involving palladium and copper catalysed coupling of terminal alkynes with vinyl and aryl halides to give unsymmetrical alkynes (Fig. 3.19)[35].

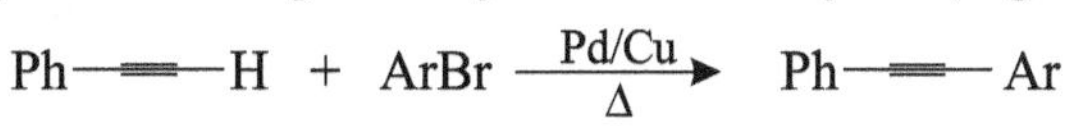

Fig. 3.19: Sonogashira reaction.

The reaction can also be performed in ionic liquids[36]. Thus, aryl halides couple with aryl or cyclohexyl acetylenes using Pd(O) nanoparticles as catalyst generated in situ by sonication of $PdCl_2$ is the ionic liquid [b_{bim}] [BF_4] (Fig. 3.20)

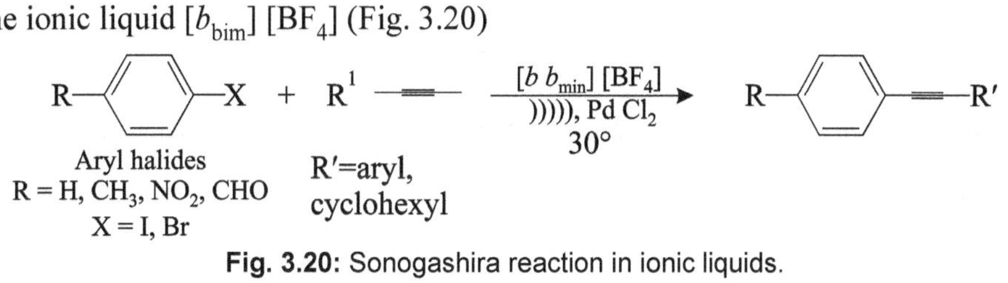

Fig. 3.20: Sonogashira reaction in ionic liquids.

3.3 ACIDIC AND BASIC IONIC LIQUIDS

A large number of ionic liquids have been designed with a specific reactivity in mind. These ionic liquids are known as **task specific ionic liquids** (TSILs) and serve the role of a catalyst and also a reaction medium. These are of two types, *viz.* bronsted-acidic ionic liquids and bronsted-basic ionic liquids.

3.3.1 Bronsted-acidic Ionic Liquids

Following are given some of the important reactions which have been performed in bronsted-acidic ionic liquids.

3.3.1.1 Friedel-Crafts Alkylation

It is a key step in the manufacture of linear alkyl benzenes, which are widely used as surfactants, detergents, emulsifiers, dry-cleaning agents, lubricants and greases. The alkyl benzenes are obtained by the reaction of aromatics with acetylchloride, formalin and olefins using chloroaluminate ionic liquids $[e_{min}]$ $[Al_2C_7]$, $[e_{min}]$ $[AlCl_4]$, [bmin] $[PF_6]$. The Friedel Crafts alkylation and acylation are of considerable significance and are one of the most important industrial applications for the synthesis using ionic liquids[37]. Modification of the chloroaluminate ionic liquids with HCl provides a reaction medium resulting in approximately 100% yield (Fig. 3.21)[38].

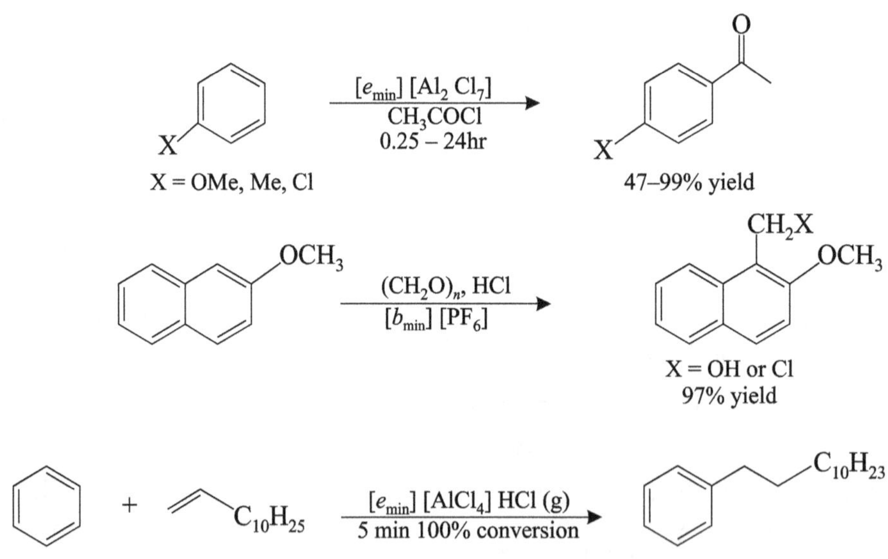

Fig. 3.21: Fridel Crafts reaction in ionic liquids.

3.3.1.2 Esterification, ether formation and pinacol-pinacolone rearrangement

The ionic liquid (D)[39] has been used to effect a number of transformations such as esterification, ether formation and pinacol-pinacolone rearrangements (Fig. 3.22).

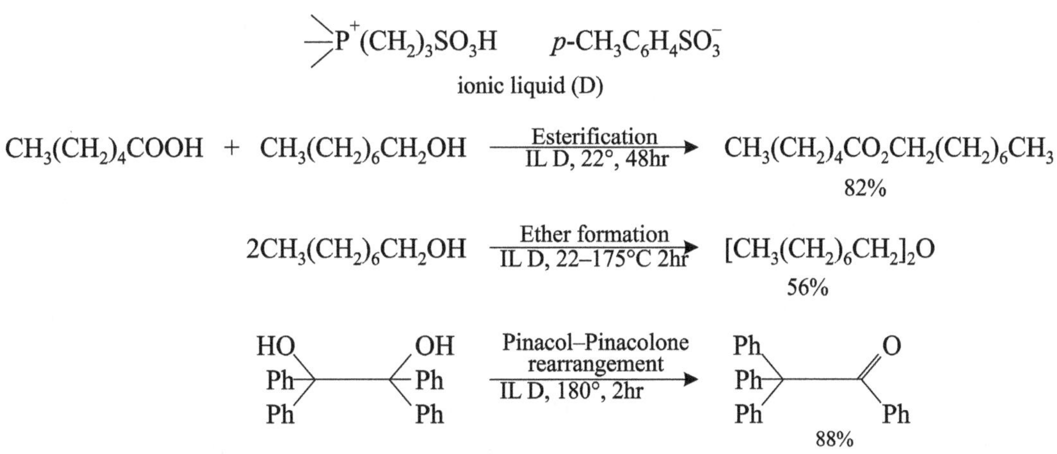

Fig. 3.22: Esterification, ether formation and Pinacol-pinacolone rearrangement using ILD

3.3.1.3 Pechmann Condensation

The Pechmann condensation of a number of phenols with methyl acetoacetate in presence of ionic liquid (which acts as catalyst and solvent) gave the corresponding 4-methyl cumarins (Fig. 3.23)[40].

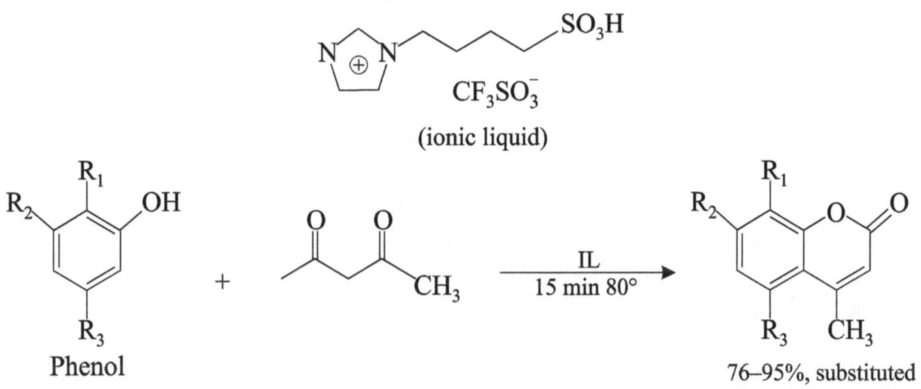

R_1	R_2	R_3
H	OH	H
H	OH	OH
H	OMe	H
OMe	OH	H
OH	OH	H
Me	H	H

Fig. 3.23: Pechmann condensation in ionic liquid.

3.3.1.4 Synthesis of Caprolactam

Caprolactam has traditionally been prepared on an industrial scale by the **Beckmann rearrangement** of cyclohexanone oxime with corrosive oleum, which is subsequently neutralised with NH_4OH. This gives large amounts of $(NH_4)_2SO_4$ as byproduct.

It has been found that the Beckmann rearrangement of cyclohexanone oxime in presence of caprolactum based ionic liquid coupled with a dynamic exchange between the caprolactam product and the ionic liquid enabled facile product isolation (Fig. 3.24)[41].

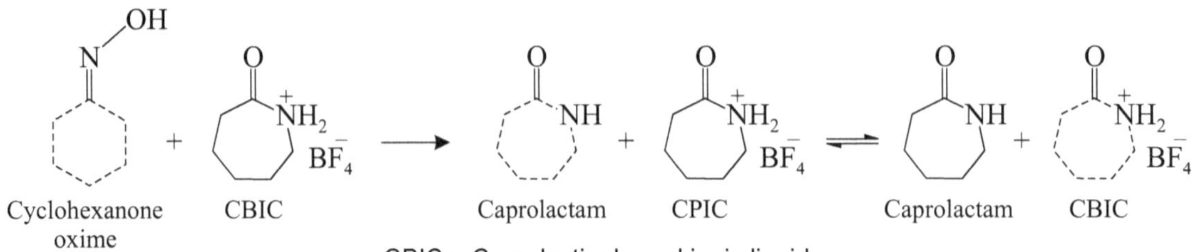

CBIC = Caprolactim based ionic liquid

Fig. 3.24: Synthesis of Caprolactam.

In the above synthesis of caprolactum (Fig. 3.24) much better yields (80%) are obtained when the product was isolated by chromatography.

3.3.1.5 Synthesis of 3,4-dihydropyridimine--2-(1H) Ones

The reaction of aromatic or aliphatic aldehydes with ethyl acetoacetate and urea in Bronsted-acidic ionic liquid, $[H_{min}]$ $[BF_4]$ at room temperature under sonochemical condition gave 3,4-dihydropyridimine-2-(1H) ones (Fig. 3.25).[42]

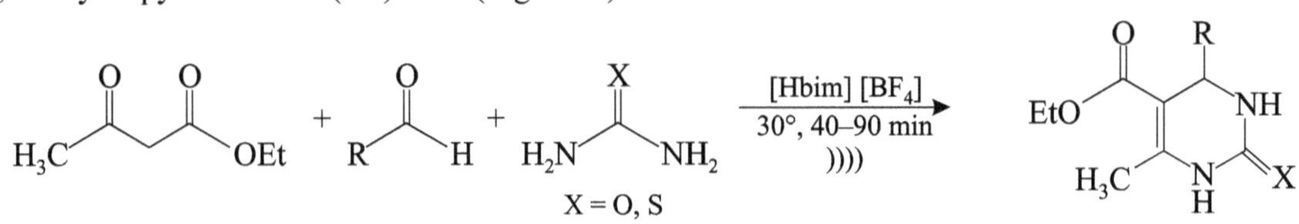

Fig. 3.25: Synthesis of 3, 4-dihydropyridimine-2-(1H) ones.

3.3.1.6 Synthesis of β-enamines

These were synthesised by the reaction of 1,3-diketones with amines in Bronsted-acidic ionic liquid $[H_{min}]$ $[BF_4]$ (Fig. 3.26)[43].

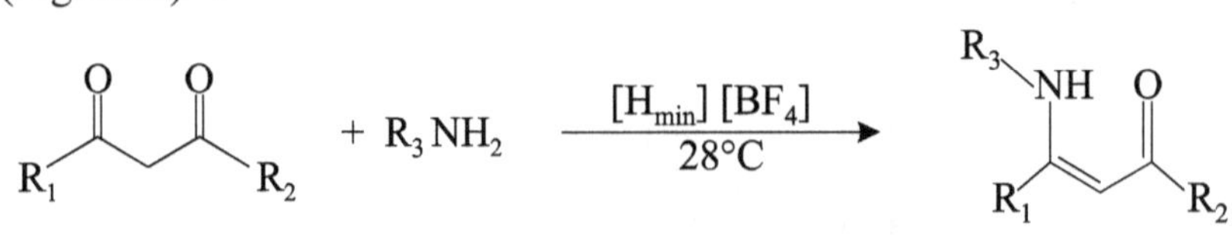

Fig. 3.26: Synthesis of β-enaminones

3.3.2 Bronsted-basic Ionic Liquid

3.3.2.1 Michael addition

Bronsted-basic ionic liquid, $[b_{min}]$ [OH] has been used as a catalyst and reaction medium for Michael addition of active methylene compounds to conjugated alkenes (Fig. 3.27)[44]. Different types of active methylene compounds with a number of α, β-unsat'd ketones, carboxylic esters and nitriles underwent Michael additions to give the corresponding adducts.

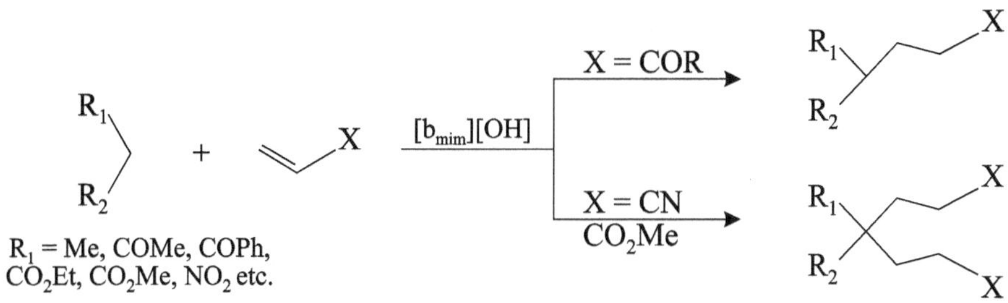

Fig. 3.27: Michael addition.

3.3.2.2 Markovnikov addition of N-heterocycles to vinyl ethers

A number of N-heterocycles underwent Markovnikov addition to vinyl ethers using bronsted-basic ionic liquid, [b_{mim}] [OH] (Fig. 3.28)[45]

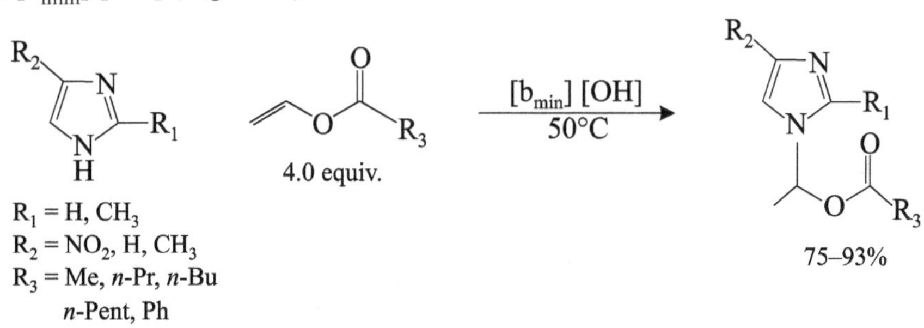

Fig. 3.28: Markovnikov addition of N-heterocycles to vinyl ethers.

3.4 MISCELLANEOUS APPLICATIONS

3.4.1 Conversion of Epoxides to Halohydrins

Normally the epoxides are converted into halohydrins by using HX or hypohalite-water, but result in the formation of byproducts. However, a better procedure involves the use of ionic liquids like [Ac_{mim}] [X] (Fig. 3.29).[46]

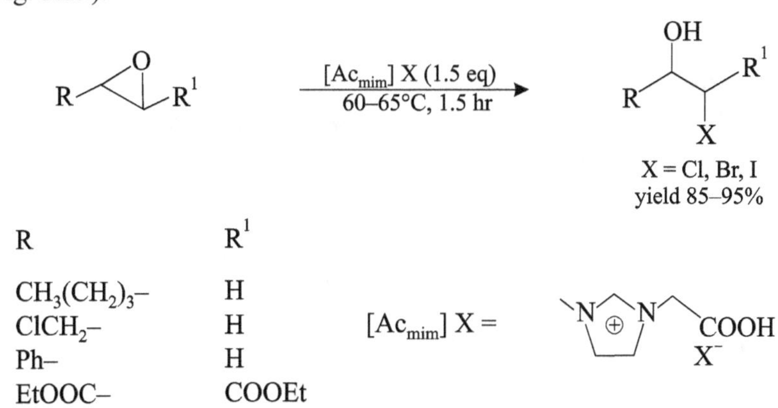

Fig. 3.29: Conversion of epoxides into halohydrins

An exception to the above procedure is that trans-stilbene oxide does not give the corresponding halodydrin. Instead, it was found that the epoxide underwent rearrangement to give deoxybenzoin (Fig. 3.30)[47]

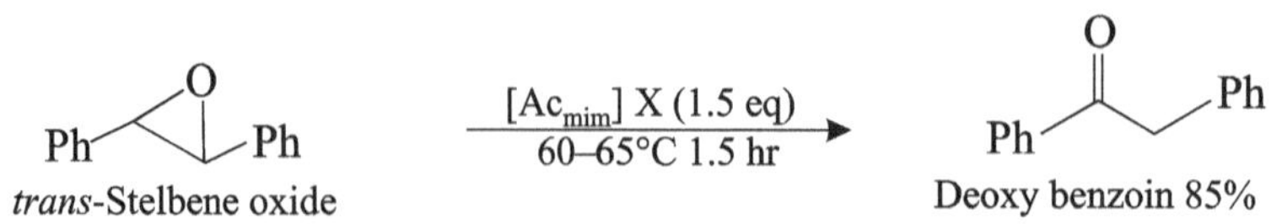

Fig. 3.30: Conversion of trans-stilbene oxide into deoxybenzoin

3.4.2 Conversion of Alkyl Halide into Alkyl Thiocyanate

The reaction of alkyl halide with ionic liquid [b_{mim}] [SCN] gave the corresponding thiocyanate at room temperature (Fig. 3.31)[48]

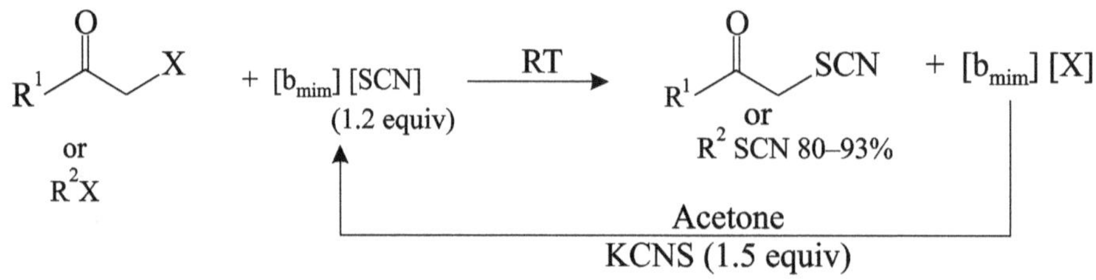

$R^1 = C_6H_5$, 4-BrC_6H_4, 4-Cl C_6H_4, 4-Me C_6H_4, 2-thienyl

$R^2 = C_6H_5CH_2$, HOC_2H_4, HOOC C_3H_6; X = Cl, Br

Fig. 3.31: Conversion of alkyl halides into thiocyanate.

3.4.3 Synthesis of Cyclic Carbonates

The coupling reaction of carbondioxide with epoxide in presence of tetrahaloindate (III)-based ionic liquids (Fig. 3.32)[49] gave cyclic carbonates.

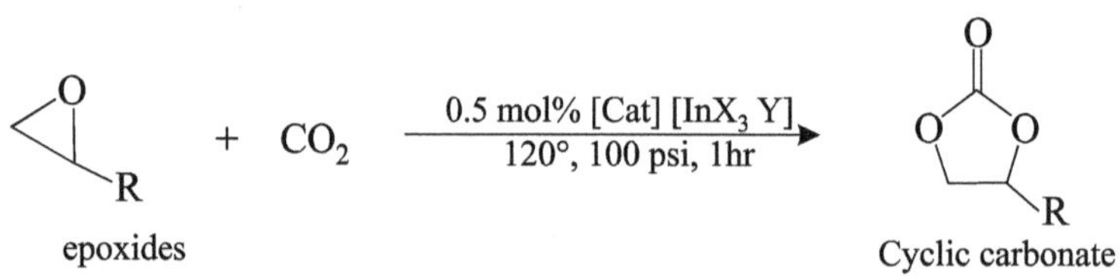

Cat = imidazolium, phosphonirum, ammonium, pyridinum
X = Cl, BrI; Y = Cl, Br

Fig. 3.32: Synthesis of Cyclic Carbonates.

For the above reaction, the thermally stable ionic liquids were synthesised by the microwave promoted reaction of Indium Chloride, In X_3, with a number of ionic liquids.

3.4.4 Rosenmund-Von Braun Reaction

Ionic liquids based on 1-butyl-3-methylimidazolium halides have been used as reusable reaction medium for the Rosenmund-Von Braun reaction of aryl halides with CuCN or NaCN and Cu(I) salts as catalyst (Fig. 3.33)[50]

$$\text{ArX} \xrightarrow[\text{ionic liquid, }\Delta]{\text{CuCN or NaCN, Cu catalyst}} \text{Ar CN}$$

Fig. 3.33: Rosenmund-Von-Braun reaction in ionic liquid.

3.4.5 Synthesis of Poly (Aryl Ether Ketones) (PAEKS)

The poly (aryl ether ketones) (PAEKs) are special engineering thermoplastics, which are known to exhibit special characteristics including excellent mechanical, thermooxidative and chemical resistance properties.[51]

The PAEKs have been synthesised by the reaction of 4, 4′-difluorobenzophenone (DFB) with bisphenol A *via* S_N AR mechanism in presence of anhydrous K_2CO_3 using ILs as reaction media. (Fig. 3.34)[52]

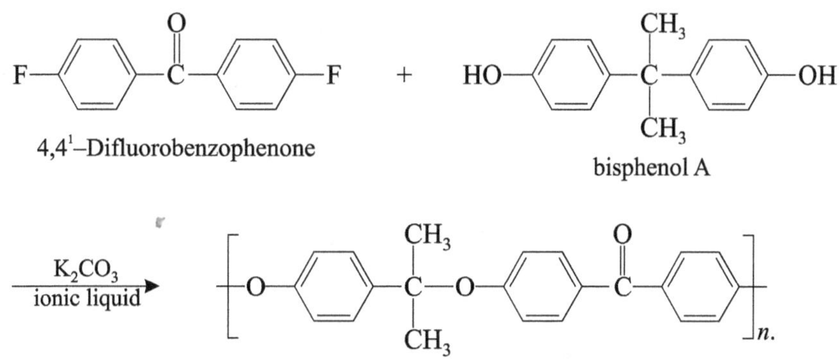

PAEKs

1L = 1, 3-dialkylimidazolium based

Fig. 3.34: Synthesis of PAEKs in ionic liquids.

This is an outstanding application of ionic liquids in the synthesis of polymers.

3.4.6 Oxidation Reactions

(*i*) **Oxidation of alcohols to aldehydes and ketones:** Alcohols on oxidation using tetrapropyl ammonium perruthenate (TPAP) and N-methyl morpholine-N-oxide (NMO) in ionic liquid, $[b_{mim}]$ $[BF_4]$ give the corresponding aldehydes[53]. It is found that benzylic alcohols give good yields. However, aliphatic alcohols require greater reaction times and give poor yields. The reaction products could be removed easily from the reaction mixture with ether extraction. The catalyst and the ionic liquids can be recyclised after extraction with ether. A typical example of oxidation is given below (Fig. 3.35)[53]

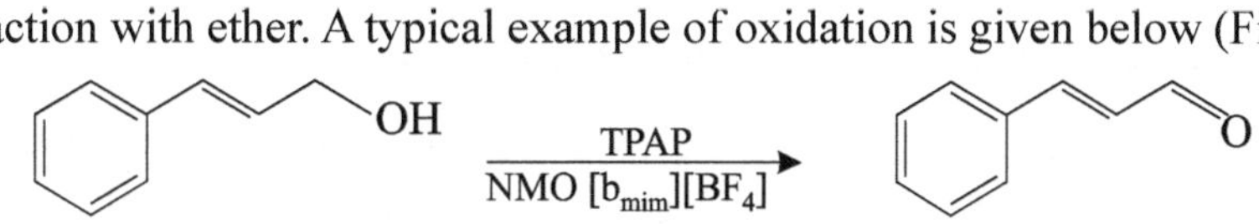

Fig. 3.35: Oxidation of alcohols to aldehydes.

(*ii*) **Oxidation of aldehydes to carboxylic acids:** A number of aldehydes on oxidation using nickel catalysts and oxygen (1 atms) in ionic liquid give the corresponding carboxylic acids (Fig. 3.36).

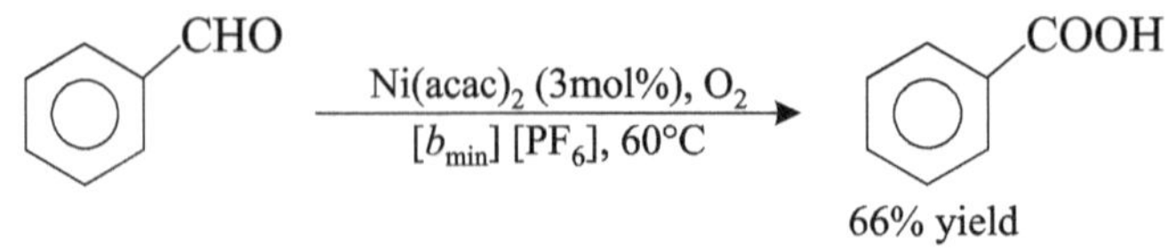

Fig. 3.36: Oxidation of aldehydes to carboxylic acids in ILs.

(*iii*) **Epoxidation of alkenes:** Alkenes could be epoxidated with methyltrioxorhenium and urea—H_2O_2 in ionic liquid (Fig. 3.37)[54]

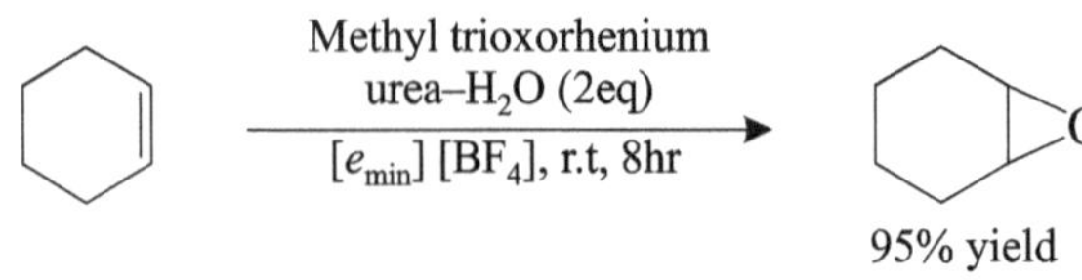

Fig. 3.37: Epoxidation of alkenes.

Asymmetric epoxidations in ionic liquid using Jacobson's chiral (Salen) Mn (III) gives the corresponding epoxide (Fig. 3.38)[55]

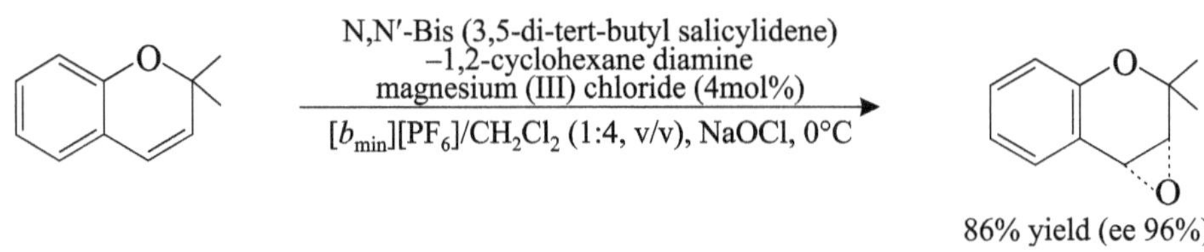

Fig. 3.38: Asymmetric epoxidation in IL.

3.4.7 Diels-Alder Reaction

The Diels-Alder reaction of cyclopentadiene with methyl acrylate in a number of ionic liquids give endo adducts[56]. Synthetically important Diels-Alder reaction has also been conducted in Lewis acidic ionic liquids derived from choline chloride–MCl_2 (1 : 2) (M = Zn or Sn) (Fig. 3.39)[56].

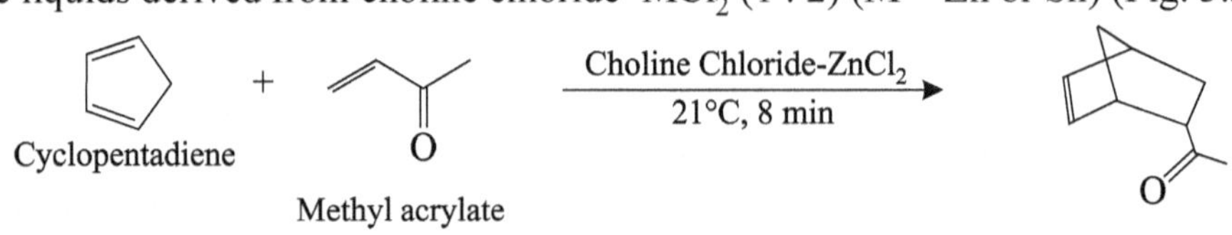

Fig. 3.39: Diels-Alder reaction in choline-derived ionic liquids.

The ionic liquids after separation of the product can be reused without any drop in activity.

3.4.8 Conversion of Oxiranes (epoxides) into Thiiranes

The reaction of oxiranes with potassium thiocyanate in the ionic liquid [b_{mim}] [PF_6]-H_2O (2 : 1) at room temperature gave the corresponding thiiranes in good yield (Fig. 3.40)[57]

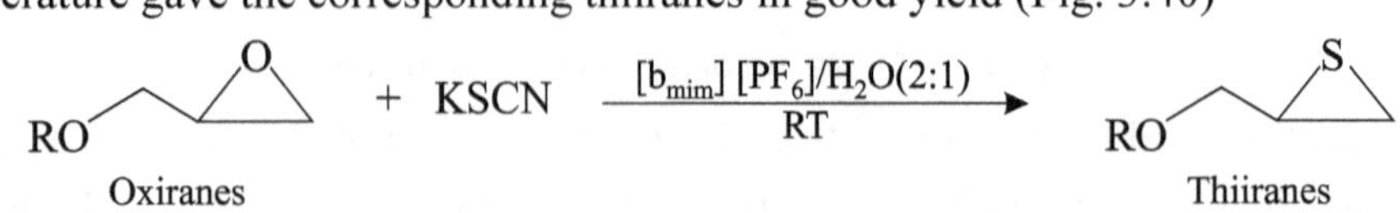

Fig. 3.40: Conversion of oxiranes with Thiiranes.

3.4.9 Beginelli Reaction

Ionic liquids such as $[B_{mim}]$ $[BF_4]$ and $[b_{mim}]$ $[PF_6]$ catalyse the Beginelli reaction under solvent-free conditions (Fig. 3.41)[58]

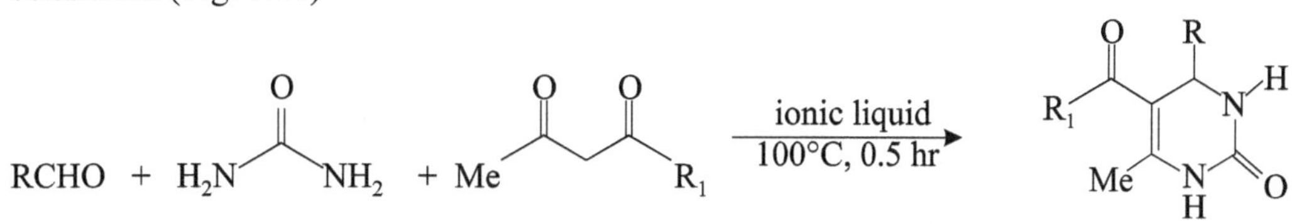

Fig. 3.41: Beginelli Reaction catalysed by ionic liquid.

See also section 10.2.26.2.

3.4.10 Synthesis of 3-Acetyl-5 [(z) Arylmethylidene] 1,3-Thiazolidine-2,4-diones

These exhibit various biological and pharmaceutical activities besides being useful as intermediates in organic synthesis[59,60]. A one pot synthesis of the title compounds involves the use of ionic liquid as solvent (Fig. 3.42)[61].

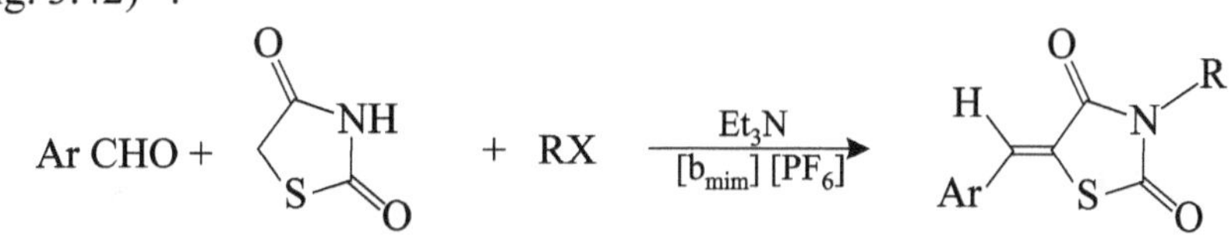

Fig. 3.42: One pot synthesis of 3-acetyl-5[(z)-aryl methylidene] 1,3-thiazolidene-2,4-diones

3.4.11 Synthesis of Symmetric Urea Derivatives

N,N-Disubstituted ureas, a group of important chemicals are manufactured by the reaction of amines with phosgene. An effective process for the direct synthesis and separation of symmetric ureas involves the reaction of amines with CO_2 in ionic liquids[62]. The recyclable catalytic system used is an ionic liquid $[b_{min}]$ $[BF_4]$, $[b_{min}]$ $[PF_6]$ or $[b_{min}]$ $[Cl^-]$ and a base CeOH.

3.4.12 Synthesis of Aryl Amines

These are structural components in biologically active products and are prepared by nucleophilic substitution of activated aryl halides with secondary amines in ionic liquids $[b_{mim}]$ $[BF_4]$ and $[b_{mim}]$ $[PF_6]$ at room temperature (Fig. 3.43)[63].

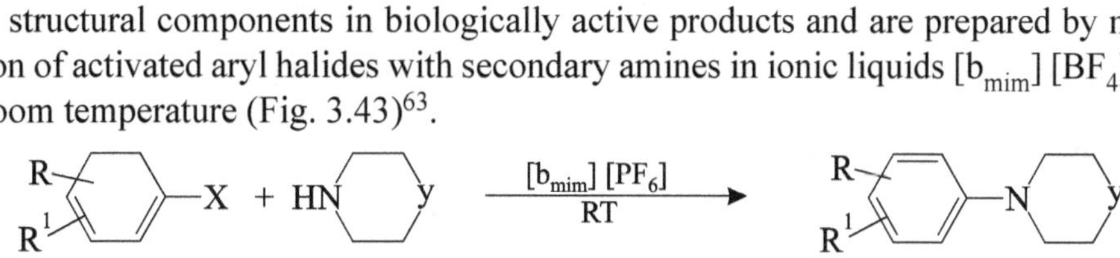

Fig. 3.43: Synthesis of aryl amine in ionic liquids.

3.4.13 Synthesis of α-aminonitriles

These are routenly synthesised by the well known **Strecker reaction**. However, a one-pot three component coupling of aldehydes, amines and trimethylsilylcyanide using ionic liquids as novel promoters is a better route[64].

3.4.14 Synthesis of Homolytic Amines

These are prepared from imines (derived in situ from aldehydes and amines) by nucleophilic addition with allyl tributyl stannane in [b_{min}] [BF_4] in high yield[65]. This method is particularly useful for acid sensitive aldehydes.

3.4.1.5 Conjugate Addition of Thiols to α, β-unsat'd Ketones in [b_{min}] [PF_6]/H_2O

α, β-Unsaturated ketones undergo rapid addition with thiols in [b_{min}] [PF_6]/H_2O (2 : 1) in the absence of any acid catalyst to give the corresponding **Michael adducts** is high yields with excellent 1,4-selectivity (Fig. 3.44)[66]

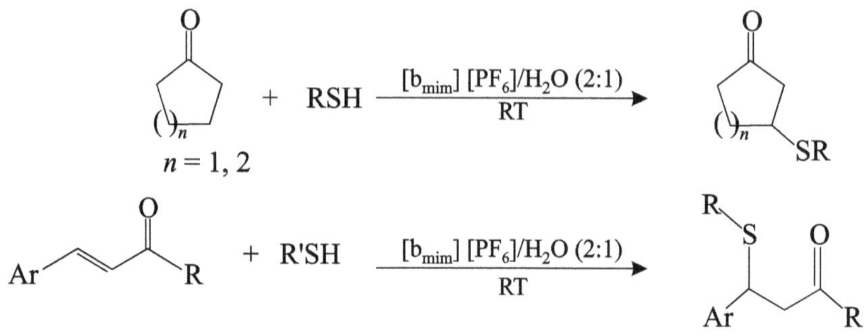

Fig. 3.44: Conjugate addn. of thiols to α, β-unsat'd ketones in ionic liquids.

3.4.16 Nucleophilic Displacement Reactions

These reactions are generally carried out using PTC. However, ionic liquids can also be used for such reactions. Thus, nucleophilic fluorination of 2-(3-methanesulphonyloxypropyl)-naphthalene with metal fluorides (*e.g.* KF) in ionic liquid [b_{min}] [BF_4] gives the corresponding substitution product (Fig. 3.45)[67]

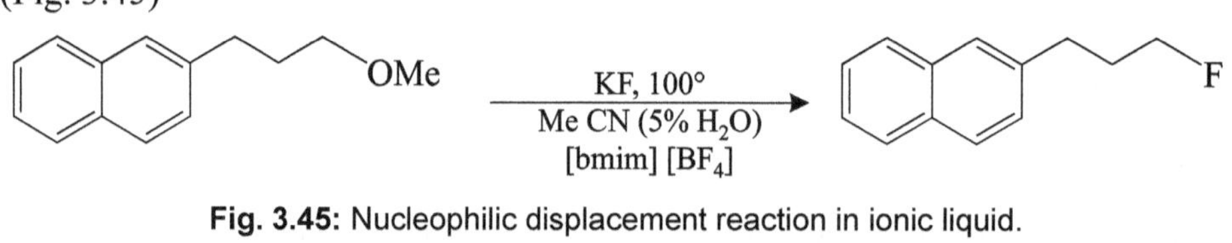

Fig. 3.45: Nucleophilic displacement reaction in ionic liquid.

The above reaction does not take place in the absence of ionic liquid.

3.4.17 Bromination of Alkynes

Arylalkynes on bromination with bromine in the ionic liquid [b_{min}] [Br] gave the anti-addition product (Fig. 3.46) following a second order rate law. However, in the ionic liquid [b_{min}] [PF_6] a mixture of syn and anti products are obtained and the reaction followed a second or third order rate law, depending on the structure of the alkyne and the concentration of bromine[68].

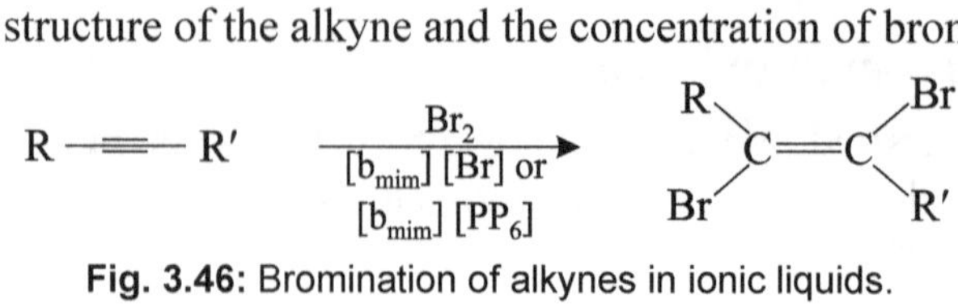

Fig. 3.46: Bromination of alkynes in ionic liquids.

3.5 BIOTRANSFORMATIONS

The most important application of ionic liquids has been in the field of biotransformations[69]. It has been found that most of the enzymes tolerate aqueous ionic liquid mixtures as the reaction medium. The ionic liquids are tolerated to higher concentrations than water miscible solvents. The activities of enzymes in ionic liquids are comparable or higher than those observed in conventional solvents. In fact, the use of ionic liquids present a promising green alternative to organic solvents for enzyme catalysed reactions. Following are given some of the important biotransformations in ionic liquids.

3.5.1 Synthesis of Z-Aspartame

A precursor to artificial sweetner (aspartame), Z-aspartame has been synthesised by the reaction of carbobenzoxy-L-asparate and L-phenylalanine methyl ester in ionic liquid. Z is a protecting group which is subsequently removed. The enzyme used is thermolysis that exhibited excellent stability in the ionic liquid (Fig. 3.47)[70]

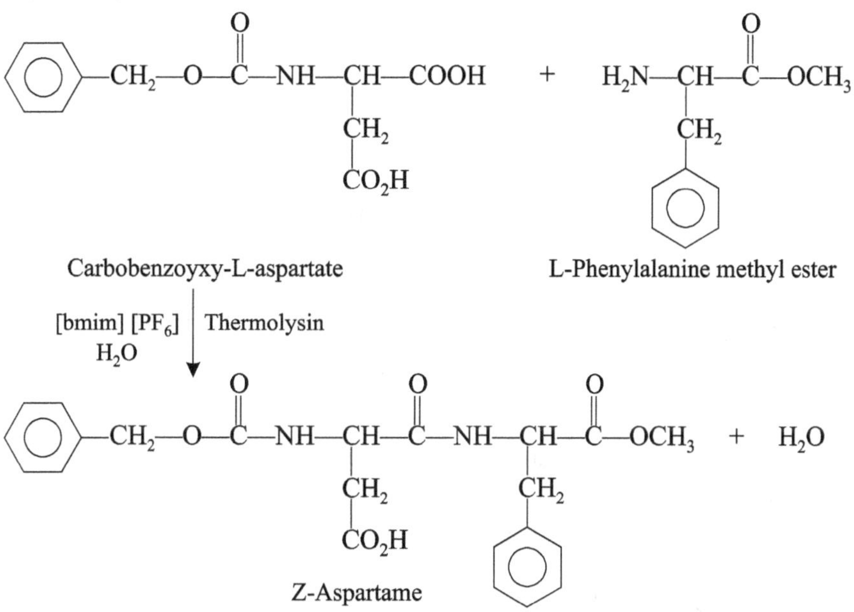

Fig. 3.47: Synthesis of Z-Aspartame

3.5.2 Conversion of 1,3-dicyanobenzene into 3-cyanobenzamide and 3-cyano-benzoic acid

1,3-Dicyanobenzene on treatment with nitrile hydratase from Rhodocococus312 give 3-cyanobenzamide and 3-cyanobenzoic acid. The conversion was carried out in biphasic water. $[b_{mim}]$ $[PF_6]$ system[71].

3.5.3 Transesterifications

These could be carried out using candida antarctica lipase B (CaLB) either as free enzyme (SP 525) or in an immobilised form (Novozym-435) in the ionic liquid $[b_{min}]$ $[BF_4]$ or $[b_{min}]$ $[PF_6]$[72] [Fig. 4.48).

$$R^1CO_2Et + R^2OH \xrightarrow[\text{or } [b_{min}][BF_4],\ 40°C]{CaLB[b_{min}][PE_6]} R^1CO_2R^2 + EtOH$$

Fig. 3.48: Transesterification in ionic liquid.

Transesterification of N-acetyl-L-phenyl alanine ethyl ester with 1-propanol in the ionic liquid $[b_{mim}]$ $[PF_6]$ and 1-octyl-3-methylimidazolium hexathiophosphate $[O_{mim}]$ $[PF_6]$ is carried out with the enzyme α-chymotrypsin (Fig. 4.49)[73]

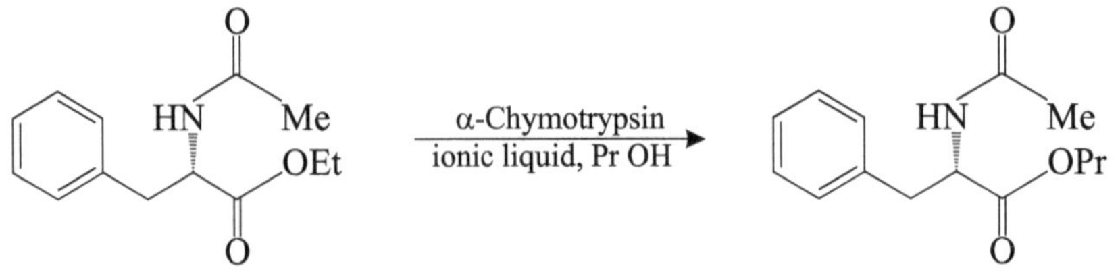

N-Acetyl-L-phenyl alanine ethyl ester

Fig. 3.49: Transesterification of N-acetyl-L-phenylalanine ethyl ester with PrOH in ionic liquid.

The transesterification of ethyl butomate with butan-1-ol in ionic liquid $[b_{mim}]$ $[BF_4]$ or $[b_{mim}]$ $[PF_6]$ gave excellent yield butylbutanoate with supported Cal B[72] (Fig. 3.50).

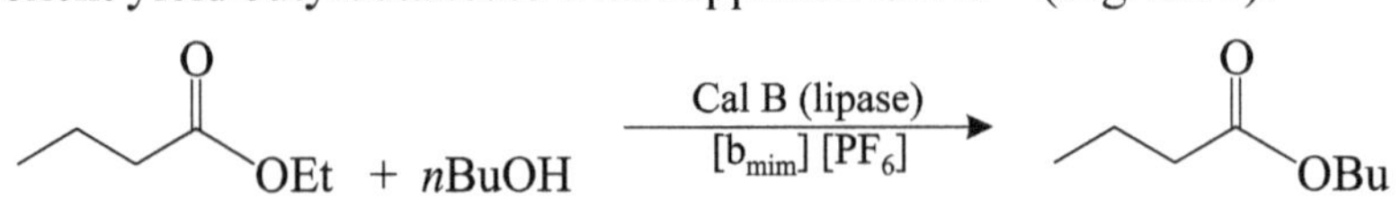

Fig. 3.49: Transesterification of ethylbutanate in ionic liquid with Cal B.

Transesterification of 2-hydroxymethyl-1,4-benzodioxane using Vinyl acetate (Fig. 3.50)[74]

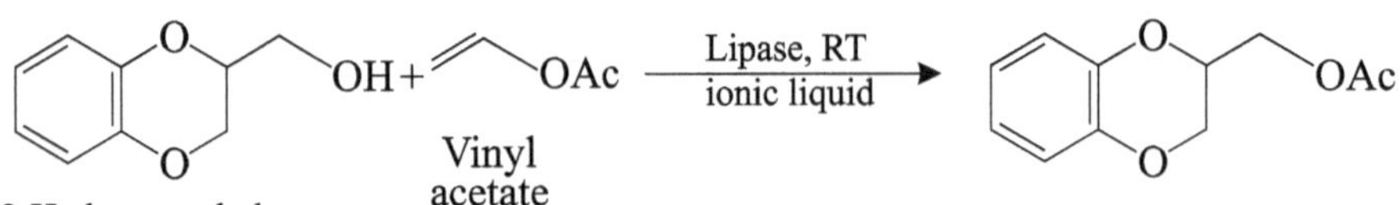

2-Hydroxymethyl-1,4-benzodioxane

Fig. 3.50: Transesterification of 2-hydroxymethyl-1, 4-benzodioxane with Vinyl Acetate.

3.5.4 Ammoniolysis of Carboxylic Acids

The ammoniolysis of carboxylic acids could be carried out with ammonia in presence of NOVOZYM-435 at 40° in the ionic liquid $[b_{mim}]$ $[BF_4]$ (Fig. 3.51).[72]

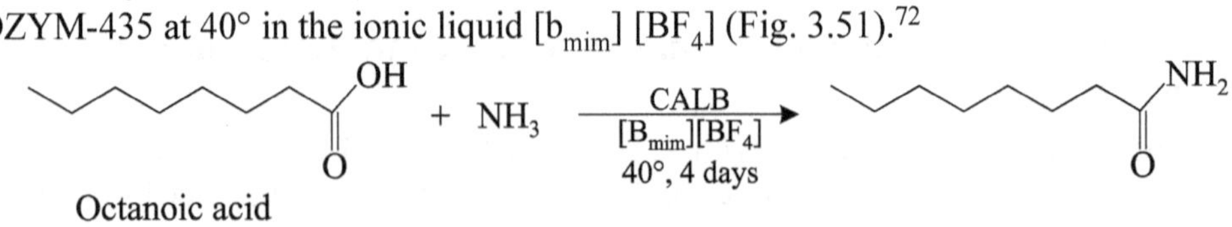

Octanoic acid

Fig. 3.51: Ammoniolysis of octonic acid.

3.5.5 Synthesis of Geranyl Acetate

The natural fragrance geranyl acetate is prepared from geraniol by esterification in presence of a lipase, immobilised CaLB (NOVOZYM 435) in $[B_{mim}]$ $[PF_6]$ (Fig. 3.52)[75].

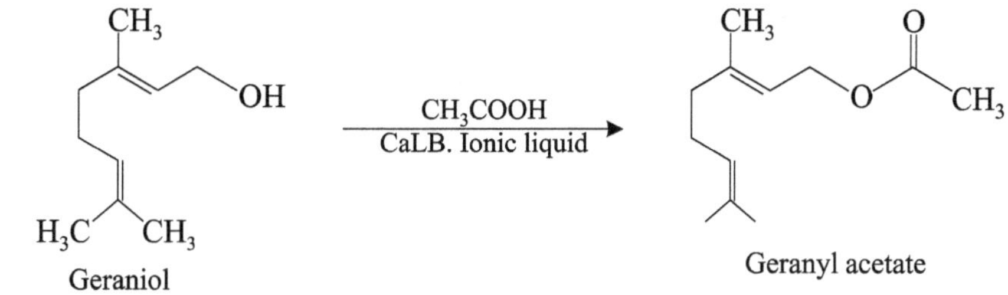

Fig. 3.52: Synthesis of geranyl acetate.

3.5.6 Transesterification of Glucose

Glucose on transesterification with vinyl acetate in the ionic liquid [e_{mim}] [BF_4] is completely selective in presence of CaLB to give monoacetyl glucose (Fig. 3.53)[76]

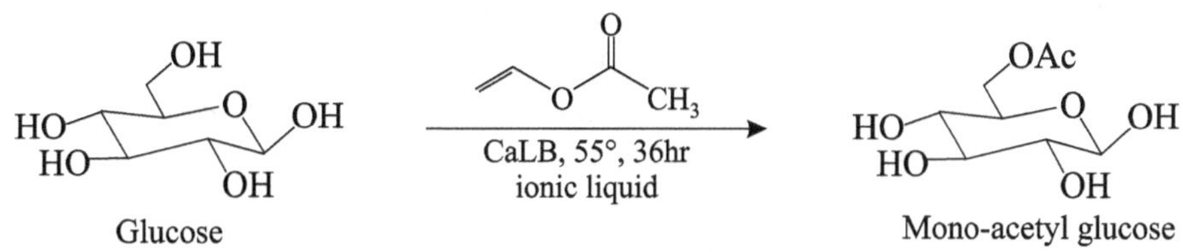

Fig. 3.53: Transesterification of Glucose

3.5.7 Transesterification of L-Ascorbic Acid

Ascorbic acid can be esterified with oleic acid in presence of CaLB in [b_{mim}] [BF_4] (Fig. 3.54)[77]

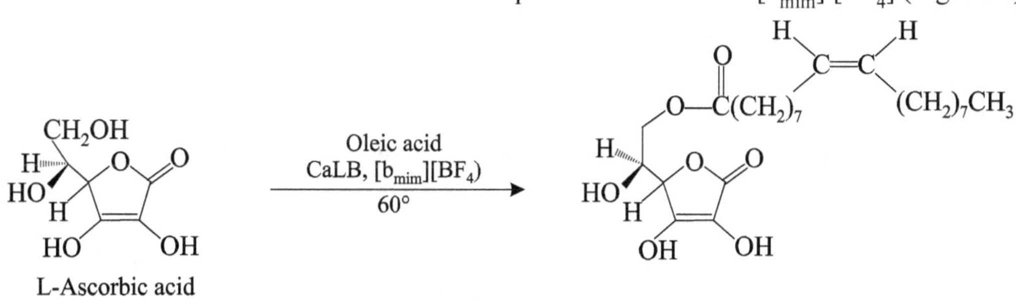

Fig. 3.54: Transestirification of L-ascorbic acid.

3.5.8 Enantioselective hydrolysis of a Prochiral Malonic Ester

Enantioselective hydrolysis of a prochiral malonic ester in presence of Pig liver esterase (PLE) in presence of a trace amount of ionic liquid gave the corresponding carboxylic acid. The yield is improved considerably by addition of 10% of a cosolvent like isopropyl alcohol (Fig. 3.55)[78]

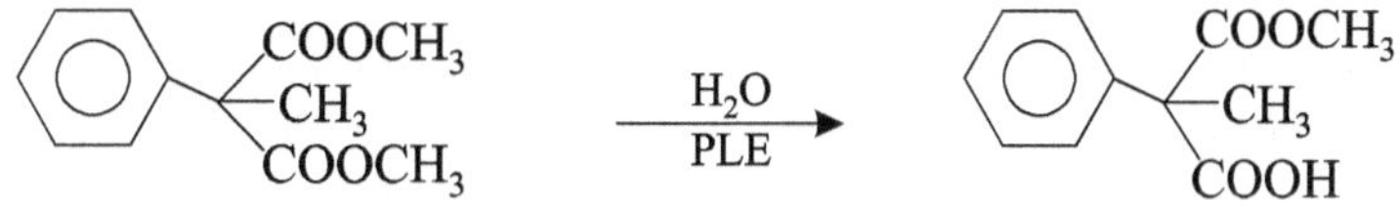

Fig. 3.55: Enantioselective hydrolysis of a prochiral malonic ester.

3.5.9 Enantioselective Hydrolysis of Naproxen Methyl Ester

Enantioselective hydrolysis of Naproxen methyl ester in presence of candida Lipase (CrL) in the ionic liquid containing (B_{mim}) and [H_{mim}] cations and [PF_6], [BF_4] anions gave (S)-Naproxen which is the pharmaceutical enantiomer (Fig. 3.56)[79].

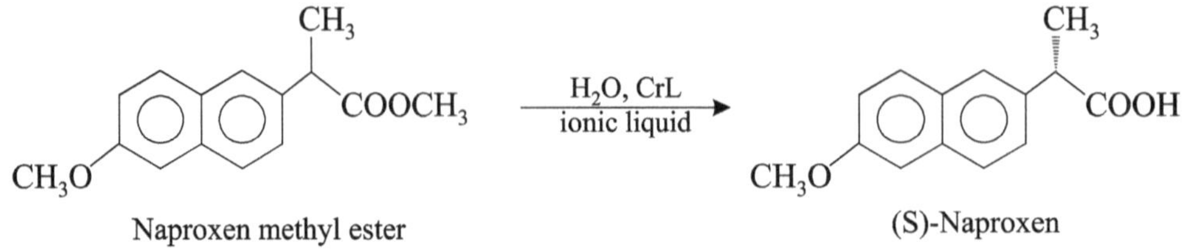

Fig. 3.56: Synthesis of (S)-Naproxen

3.5.10 Enantioselective Esterification of Ibuprofen

Enantioselective esterification of Ibuprofen with butyl alcohol in presence of CrL gives the (S)-ester when the reaction was conducted in isooctane. Better selectivity was however obtained in ionic liquid [B_{mim}] [PF_6] (Fig. 3.57)[80]

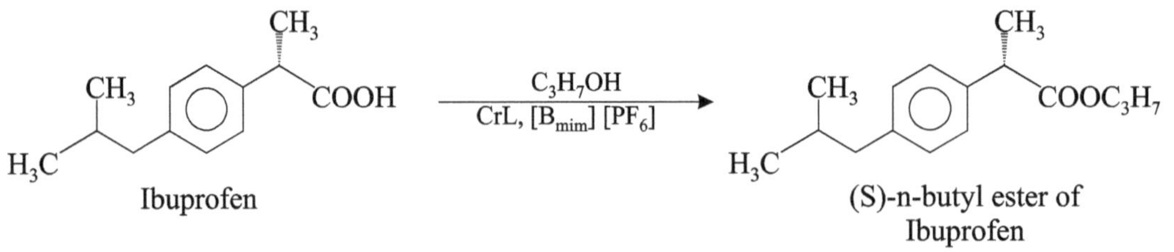

Fig. 3.57: Enantioselective esterification of Ibuprofen.

Conclusion

Ionic liquids have a good potential for pharmaceutical industries. The main problem which has to be addressed is removal of last traces of ionic liquid solvent which remains with the products. The chemists in industry are not able to adopt the use of ionic liquids only because these are considered to be green. They look for economic and practical benefits. The self catalysing nature of acidic ionic liquids provide green potential since it can eliminate the need of environmentally damaging Lewis acids. The biotransformations has special advantage since the enzymes are stable in ionic liquids and can profitably used in ionic liquids.

REFERENCES

1. M.J. Earle, K.R. Seddon, Pure Appl. Chem., 2000, **72(7),** 1391.
2. T. Welton, Chem. Rev., 1999, **99**, 2071.
3. R. Sheldon, Chem. Commun., 2001, 2399.
4. N.E. Leadbeater and H.M. Treniun, J. Org. Chem., 2002, **67**, 3145; E. Van der Eycken, P. Appukkuttan. W. De Borggraeve, W. Dehaln, D. Dallinger and C.O. Kappe, J. Org. Chem., 2002, **67**, 7904.
5. R.S. Varma and V.N. Nambodiri, Chem. Commun., 2001, 643.
6. R.S. Varma and V.N. Namboodiri, Pure Applied Chem. 2001, **73**, 1307.
7. V.N. Namboodiri and R.S. Varma, Tetrahedron Lett., 2002, **43**, 5381.

8. V.N. Namboodiri and R.S. Varma, Chem. Commun. 2002, 342.
9. V.N. Namboodiri and R.S. Varma, Org. Lett., 2002, **4**, 3161.
10. D. Basavaiah, P.D. Rao and R.S. Hyma, Tetrahedron, 1996, **52**, 8001.
11. J.N. Rosa, C.A.M. Afonso and A.G. Santos, Tetrahedron, 2001, **57**, 4189.
12. V.K. Aggarwal, I. Emme and A. Mereu, Chem. Commun, 2002, 1612.
13. J.-C. Hsu, Y.-H. Yen and Y.-H. Chu, Tetrahedron Lett., 2004, **45**, 4673.
14. V. Jurc'ik and R. Wilhelm, Green Chem., 2002, **7**, 844.
15. P. Formentin, H. Garcia and A. Leyva, Molecular catal. A: 2004, **214**, 137.
16. T. Kitazume and G. Tanaka, J. Fluorine Chem., 2000, **106**, 211.
17. A.J. Carmichael, M.J. Earle, J.D., Holberg, P.B. McCormac and K.R. Seddon, Org. Lett., 1999, **1**, 997.
18. L. Xu, W. Chen and J. Xiao, Organometallics, 2000, **19**, 1123.
19. M.T. Reetz, R. Breinbauer and K. Wanninger, Tetrahedron Lett., 1996, **37**, 4499.
19a. X.Xie, B. Chen., J.Lu, J. Han., X. She and X. Pan, Tetrahedron Lett, 2004, **45**, 6235.
19b. G.W. Kabalka, G. Dong and B. Venkataiah, Tetradron Lett., 2004, **45**, 6243.
20. N.Miyaura and A. Suzuki, Chem. Rev., 1995, **95**, 2457.
21. A. Suzuki, J. Organometallic Chem., 1999, **576**, 147.
22. C.J. Mathews, P.J. Smith and T. Welton, Chem. Commun, 2000, 1249.
23. C.J. Mathews, P.J. Smith, T. Welton A.J.P. White and D.J. Williams, Organometallics, 2001, **20**, 3849.
24. F. McLachlan, C.J. Mathews, P.J. Smith and T. Welton, Organometallics, 2003, **22**, 5350.
25. C. Zhang, J. Huang, M.L. Trudell and S.P.J. Nolan, J. Org. Chem., 1999, **64**, 3804.
26. R. Rajagopal, D.V. Jarikote and K.V. Srinivasan, Chem. Commun, 2002, 616.
27. S.T. Handy and X. Zhang, Org. Lett., 2001, **3**, 233.
28. J. Sirtieix, M. Ossberger, B. Betzemeier and P. Knochel, Synlett, 2000, 1613.
29. W. Chen, L. Xu, C. Chatterton and J. Xiao, Chem. Commun. 1999, 1247.
30. C.de Bellefon, E. Pollet and P. Grenouillet, J. Mol. Catal. A: Chem., 1999, **145**, 121.
31. J. Hawarth, P. James and J. Dai, J. Mol Catal. A 2004, **214**, 143.
32. A. Kumar and S.S. Pawar, J. Mol. Catal. A, 2005, **235**, 244.
33. T. Jiang, H. Gao, B. Han., G. Zhao, Y. Chang, W. Wiu, L.Gao and G. Yang, Tetrahedron Lett., 2004, **45**, 2699.
34. S. Anjaiah, S. Chandrasekhar and R. Gree, Adv. Synth. Catal, 2004, 346, 1329.
35. A. Chinchilla and C. Najera, Chem. Rev., 2007, **107**, 874.
36. A.R. Gholap, K. Venkatesan, R. Pasricha, T. Daniel, R. Lahoti and K.V. Srinivasan, J. Org. Chem., 2005, **70**, 4869.
37. R. Sheldon, Chem. Commun., 2001, 2399-2407; P. Formentin and H. Garcia, Catal. Lett., 2002, **78**, 115-118.
38. K.Qiao and Y. Deng, J. Molecular Catal. A, Chemical, 2001, **171**, 81-84.
39. A.C. Cole, J.L. Jensen, I. Ntai, K.L.T. Tran, K.J. Weaver, D.C. Forbes and J.H. Davis, Jr., J.Am. Chem. Soc., 2002, **124**, 5962.
40. Y. Gu, J. Zhang, Z. Duan and Y. Deng, Adv. Synth. Catal., 2005, **347**, 512.
41. S. Guo, Z. Du, S. Zhang, D.Li and Y. Deng Green Chemistry, 2006, **8**, 296.
42. A.R. Gholap, K. Venkatesan, T. Daniel, R.J. Lahoti and K.V. Srinivasan, Green Chem., 2004, **6**, 147.
43. A.R. Gholap, N.S. Chakor, T. Daniel, R.J. Lahoti and K.V. Srinivasan, J. Mol. Catal.A: Chem. 2006, **245**, 37.

44. Z. Duan, Y. Gu and Y. Deng, Synth. Commun, 2005, **35**, 1939.
45. J.-M. Xu, K.-B. Liu, W.-B. Wu, C. Qian, Q.Wu and L.-F. Lin, J. Org. Chem., 2006, **71**, 3991.
46. B.C. Ranu and S. Banerjee, J. Org. Chem., 2005, **70**, 4517.
47. H.O. House, J. Am. Chem. Soc., 1955, **77**, 3070.
48. A. Kamal and G. Chouhan, Tetrahedron Lett., 2005, **46**, 1489.
49. Y.J. Kim andR.S. Varma, J. Org. Chem., 2005, **70**, 7882.
50. J.X. Wu, B.Beck and R.X. Ren, Tetrahedron Lett., 2002, **43**, 387.
51. M. Chang and X.Zhang, Chin. Sci. Bull., 1993, **38**, 208.
52. J. Wang and Z. P. Liu, Chin. Sci. Bull., 2013, **58**, 1262.
53. V. Farmer and T. Welton, Green Chem., 2002, **4**, 97.
54. G.S. Owens and M.M. Abu-Omar, Chem. Commun., 2000, 1165.
55. O. Bortolini, V. Conte, C. Chiappe, G. Fantin, M. Fogaygnolo and S. Maietti, Green Chem., 2002, **4**, 94.
56. A. Aggarwal, N.L. Lancaster, A.R. Sethi and T. Welten, Green Chem., 2002, **4**, 517.
57. L. Gallion and F. Bedioue Chem. Commun, 2001, 1458.
58. J. Peng and Y. Deng, Tetrahedron Lett., 2001, **42**, 5917.
59. E.B. Grant, D. Guiadein, E.Z. Baum and B.D. Foleno, Bioorg. Med. Chem. Lett, 2000, **10**, 2179.
60. R.G. Giles, N.J. Lewis, J.K. Quick, M.J. Sasse, M.W.J. Urquhart and L. Youssef, Tetrahedron 2000, **56**, 4531.
61. D.-H. Yang, Z.-C. Chen, S.-Y. Chen and Q.-G. Zheng, Synthesis 2003, **12**, 1891.
62. F. Shi, Y. Deng, T. SiMa, J. Peng, Y. Gu, B.Qiao, Angew. Chem. Int. Ed., 2003, **42**, 3257.
63. H. Kotski, T. Sakai and T. Shinohara, Synlett., 2000, 116.
64. J.S. Yadav, B.V.S. Reddy, B. Eshwaraiah, M. Srinivas and P. Vishnumurthy, New J. Chem., 2003, **27**, 462.
65. J.S. Yadav, B.V.S. Reddy and A.K. Raja, Synthesis, 2003, **6**, 6883.
66. J.S. Yadav, B.V.S. Reddy and G. Baishya, J. Org. Chem., 2003, **68**, 7098.
67. D.W. Kim, C.E. Soong and C.Y.J. Chi, Am. Chem. Soc., 2002, **124**, 10278; M.R.C. Gerstenberger and H. Mass, Angew Chem. Int. Ed., 1981, **20**, 647.
68. C. Chippl, V. Conte, Pioraccini, Eur. J. Org. Chem., 2002, 2831.
69. U. Kragl, M. Eckstein and N. Kaftzik, Curr. Opin. Biotechnol, 2002, **13**, 565.
70. A. Kumar, N. Jain and S.M.S. Chauhan, Synthetic Communication, 2004, **34**, 2835.
71. H.S. Schöfer, N. Kaftzik, P. Wasserscheid and U.Kragl, Chem. Common, 2001, 425.
72. R.M. Lau, F. Van Rantwijk, K.R. Seddon and R.A. Sheldon, Org. Lett, 2002, **2**, 4189.
73. J.A. Laszlo and D.L. Compton, Biotechnol. Bioeng, 2000, **75**, 181.
74. S.J. Nava, J.R. Harjani and M.M. Salunkhe, Tetrahedron Lett., 2002, **43**, 2979.
75. D. Bdrahona, P.H. Pfromm and M.C. Rezac, Biotechnol. Bioeng., 2006, **93**, 318.
76. S. Park and R.J. Kazlauskas, J. Org. Chem. 2001, **66**, 8395.
77. S. Park, K. Vikund, K. Hult and R.J. Kuzlauskas, Green Chem., 2003, **5**, 715.
78. S. Wallert, K. Drauz, I. Grayson, 11, Gröger, P. Dominquez de Maria and C. Bolm, Green Chem., 2005, **7**, 602.
79. J.-Y. Xin, Y.-J. Zhao, G.-L. Zhao, Y. Zheng, X.-S.Ma, C.-G. Xia and S.-B. Li, Biocat. Biotrasformation, 2005, **23**, 353.
80. H. Yu, J. Wu and C.B. Ching, Chirality, 2005, **17**, 16.

4. TRANSFORMATIONS IN SUPER CRITICAL CARBON DIOXIDE

4

Transformations in Super Critical Carbon Dioxide

4.1 INTRODUCTION

Carbon dioxide is known to be non-toxic, inert, nonflammable, abundant and is much cheaper. As a solvent, it is used as supercritical carbon dioxide (*i.e.*, above its critical temperature of 31.1°C and critical pressure of 7.38 MPa). In this region the density of SC-CO_2 and its solvent properties can be varied by changing the pressure and temperature[1]. SC-CO_2 is a green solvent and has been used since early times for the extraction of flavour from hops, caffeine from coffee, fats and cholesterol from foods[2], ginseng (from which it does not extract the pesticide residue)[3] and ginger[4] etc. Addition of a small percentage of ethanol to SC-CO_2 has been used for the extraction of oil from cotton seed, oil from sunflower seeds and phospholipids from canola oil.

SC-CO_2 has been used for the transformations of organic substrates. A special advantage associated with SC-CO_2 is that after the completion of the reaction, the isolation of the final product is much easier by either reducing the pressure or by removal of CO_2.

The replacement of traditional solvents (like carbon tetrachloride, methylene chloride etc.) with the environmentally benign carbon dioxide is useful in pharmaceutical processings[5]. Drugs could be microencapsulated[6] 'by the rapid expansion of SC-CO_2 containing the drug. Solutions of drugs such as progesterone in SC-CO_2 on releasing through a micronozzle produce 4.5 mm particles[7], which are finer than the 7.5-mm particles produced from Jet milling.

4.2 REACTIONS IN Sc-CO_2

Supercritical carbondioxide has be used as solvent for a number of reactions[8], some of which are discussed below:

4.2.1 Hydrogenation in Supercritical Carbon Dioxide

Hydrogen is completely miscible with Sc-CO_2. This has been utilised for one-port synthesis of formic, alkyl formates and formamides[9], Dimethylformamide[10] and methyl formate[11].

4.2.2 Hydrogenation of Acetophenone

Acetophenone could be hydrogenated in Sc-CO_2 using a polyseloxane supported palladium catalyst[12] using 2 equivalents of hydrogen give the reduced product ($-\overset{\overset{\displaystyle O}{\|}}{C}-Me \longrightarrow -\overset{\overset{\displaystyle OH}{|}}{C}-Me$).

At higher temperature (300°C) and hydrogen ratio 6:1, fully reduced product is obtained in >95% yield. At intermediate temperature (200°C) ethyl benzene and completely reduced alcohol are obtained[13] (Fig. 4.1)

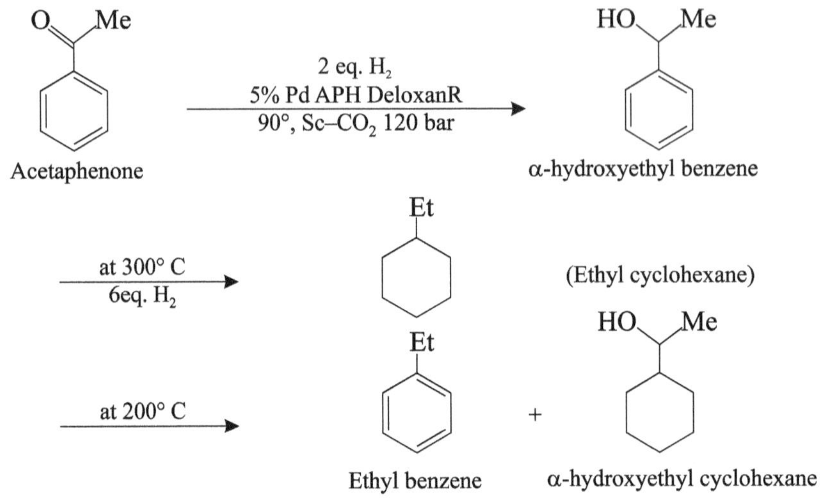

Fig. 4.1: Products obtained by hydrogenation of acetophenone.

4.2.3 Semihydrogenation of Dehydroisophytol

Dehydroisophytol, a propargylic alcohol on semihydrogenation gave the corresponding alkenol, isophylol by using the glassy alloy, Pd_{81} Si_{19} supported on Pd/SiO_2 catalyst[14]. The alloy Pd_{81} Si_{19}, exhibited more than 50 times higher turnover frequency compared to the conventional silica-supported catalyst under similar conditions. The combined use of glassy palladium-silicon alloy along with Sc-CO_2 is promising for this type of **Lindlar reaction** (Fig. 4.2).

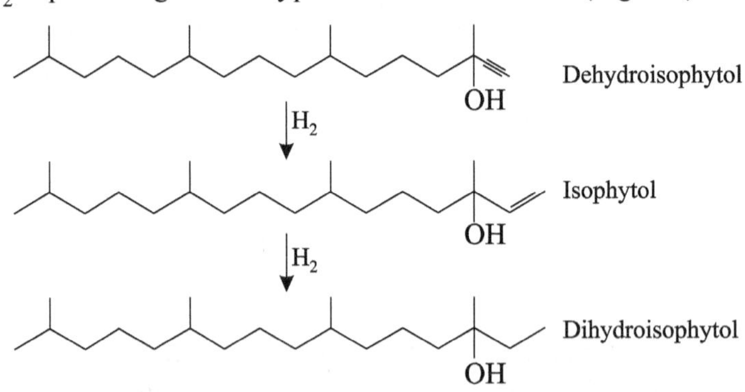

Fig. 4.2: Hydrogenation of dehydroisophytol

4.2.4 Hydrogenation of Maleic Anhydride

Hydrogenation of maleic anhydride with hydrogen in presence of Pd/Al_2O_3 using Sc-CO_2 as solvent gives α-butyrolactone (Fig. 4.3)[15], which is one of the most valuable alternative to the environmentally harmful chlorinated solvents used in polymer and paint industries.

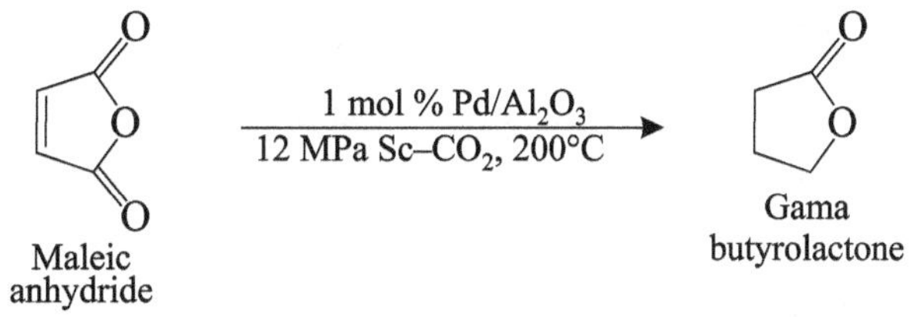

Fig. 4.3: Hydrogenation of maleic anhydride in Sc-CO_2

4.2.5 Asymmetric Hydrogenation of Ethyl Pyruvate

The asymmetric hydrogenation of ethyl pyruvate gives (R)-ethyl lactate (Fig. 4.4)[16] using heterogeneous catalyst system. The enantioselectivity in this case is derived from a chiral cinchonidine modifier absorbed on the surface of supported platinum. However, there was a strong catalytic deactivation in the enantioselective hydrogenation in Sc-CO_2 due to CO poisoning of the Pt, which was formed by the reduction of CO_2 to CO on the surface of the catalyst. The problem was solved by using Sc-ethane.

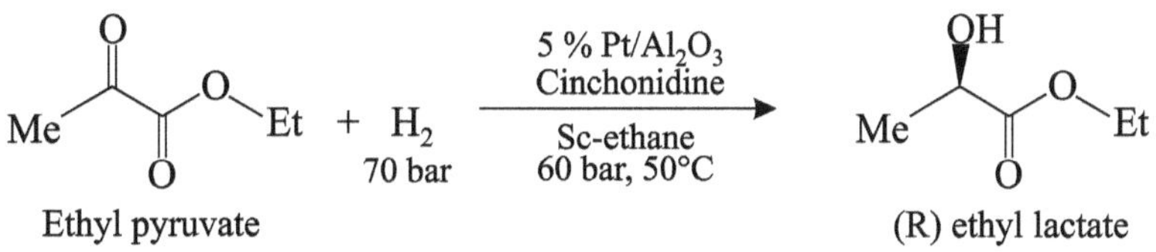

Fig. 4.4: Asymmetric hydrogenation of ethylpyruvate to ethyl lactate

4.2.6 Enantioselective Hydrogenation of Prochiral α-enamides

The enantioselective hydrogenation of prochiral α-enamides in Sc-CO_2 in presence of a cationic rhodium species (A), which incorporates the chiral bidentate DuPHOS ligand (B) and tetrakis-3,5-bis (trifluoromethyl) phenyl borate (BARF) counter ion (C) gives high enantioselectives with 0.2 mol% catalyst loadings. The results are comparable to the reduction in conventional solvents (hexane-methanol) (Fig. 4.5)[17].

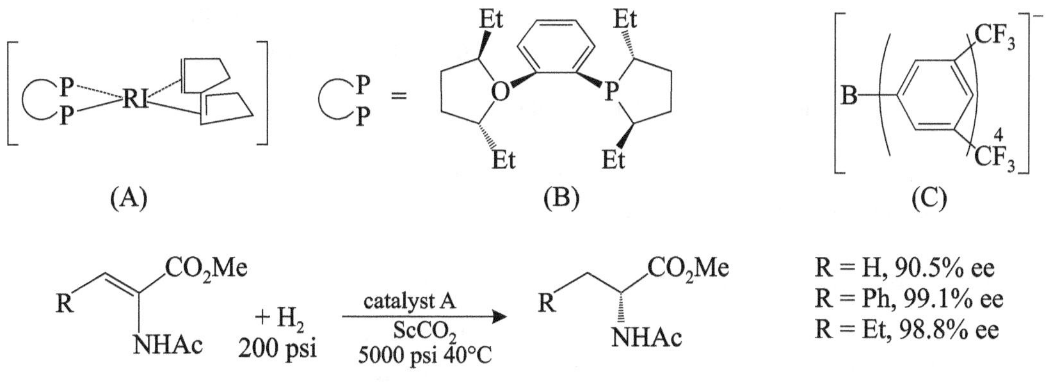

Fig. 4.5: Asymmetric Hydrogenation α-enamides in $ScCO_2$.

4.2.7 Asymmetric Hydrogenation of β, β-disubstituted enamide

The asymmetric hydrogenation of β, β-disubstituted enamide in Sc-CO_2 in presence of catalyst (A) gave the valine derivative (Fig. 4.6)[16] in 85% ee.

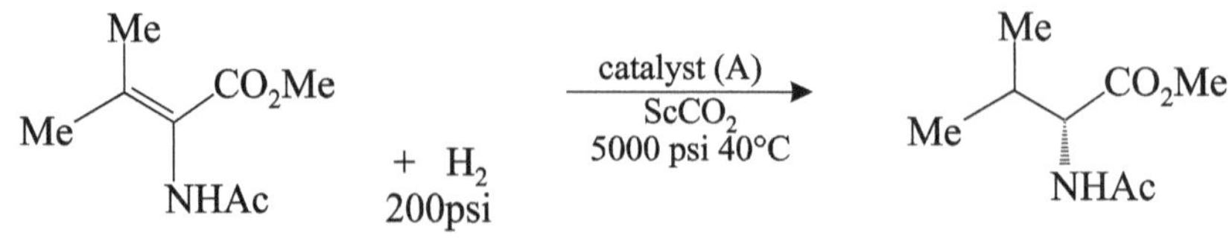

Fig. 4.6: Asymmetric hydrogenation of β,β-disubstituted enamides in $ScCO_2$

4.2.8 Hydrogenation of Tiglic Acid

Hydrogenation of tiglic acid in Sc-CO_2 in presence of catalyst (D) gave (S)-2-methyl butanoic acid (Fig. 4.7)[18]

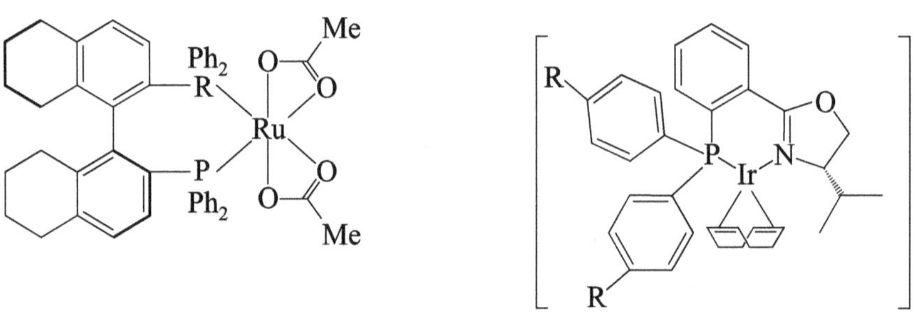

(D) (E)

R = $CH_2CH_2C_6F_{13}$

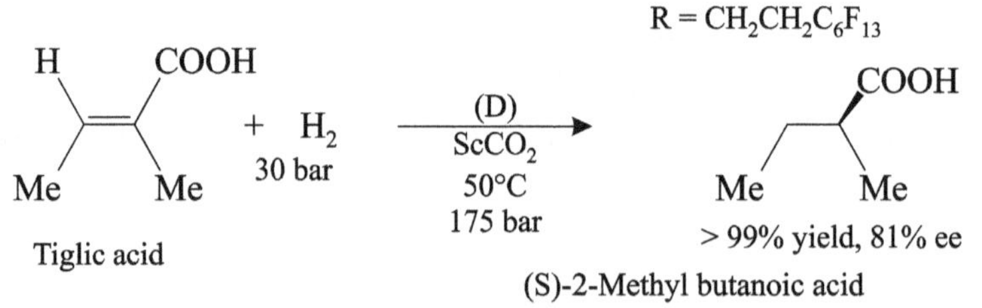

Fig. 4.7: Asymmetric hydrogenation of Tiglic Acid in $ScCO_2$

In the above hydrogenation (Fig. 4.7), introduction of perfluoroalcohol CF_3 (CF_2) CH_2OH (E) caused increase in conversion (100%, at 5 bar H_2 pressure) and enantioselectivity (89% ee)

4.2.9 Enantioselective Hydrogenation of Imines

Enantioselective hydrogenation of imines utilizes the cationic iridium (I) complexes with chiral phosphinodihydrooxazole, which was modified with perfluoroalkyl group for increased CO_2-philicity[19]. The product obtained was (R)-N-phenyl-1-phenylethylamine (Fig. 4.8)

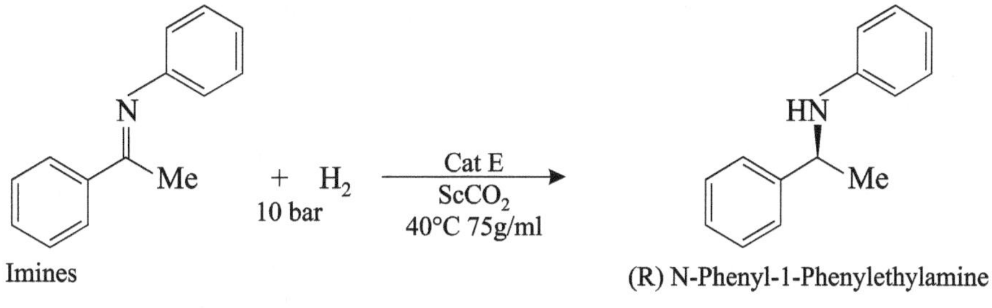

Fig. 4.8: Asymmetric hydrogenation of imines in $ScCO_2$

4.3 HYDROFORMYLATION IN SUPERCRITICAL CARBON DIOXIDE

Hydrofarmylations are normally carried out using hydrogen and carbonmonoxide in presence of catalysts.

4.3.1 Hydroformylation of Styrene

Asymmetric hydroformylation of styrene in presence of $Rh(CO)_2$ (acac) and (R, S)-BINAPHOS (F) as ligand in Sc-CO_2 gave the branched product (Fig. 4.9)[20] as the major product alongwith achiral regioisomar.

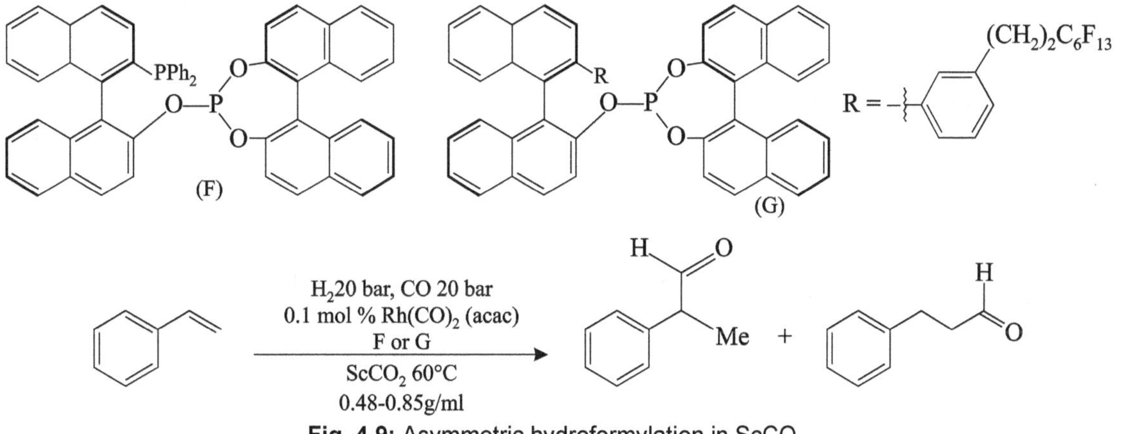

Fig. 4.9: Asymmetric hydroformylation in $ScCO_2$.

4.3.2 Hydroformylation of Acrylates

Though acrylates are least reactive olefins towards hydroformylation, the reaction becomes rapid and stereoselective in Sc-CO_2 in presence of [Rh (acac) $(CO_2)_2$] and the fluoroalkylated phosphine ligand P $(p\text{-}C_6H_4C_6H_{13})_3$ (Fig. 4.10)[21]. The use of heterogeneous immobilised rhodium-diphosphine complex (H) gives much better yields[22].

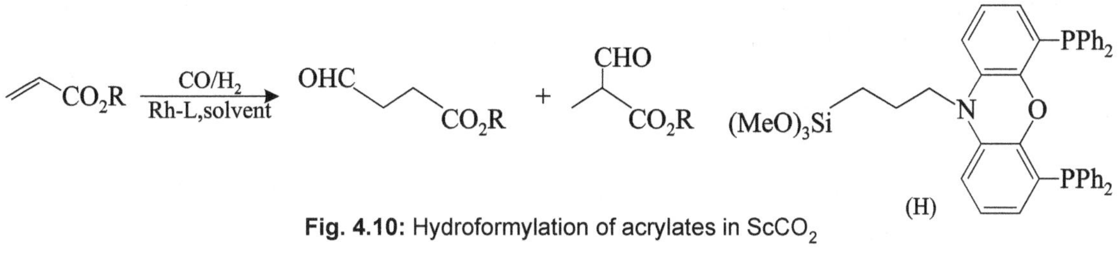

Fig. 4.10: Hydroformylation of acrylates in $ScCO_2$

4.4 OXIDATIONS IN SUPERCRITICAL CARBONDIOXIDE

The high solubility of oxygen in Sc-CO_2 is useful for oxidations as in the case of hydrogenations and hydroformylations.

The partial oxidation of alcohols to aldehydes in Sc-CO_2 over supported metal catalysts is an environmentally friendly process. This medium is especially useful for the oxidation of weakly polar, water insoluble alcohols[23,24]. The oxidation of alcohols to carbonyl compounds proceeds with high selectivity at enhanced rate[25]. These oxidations are performed is a continuous-bed reactor over promoted noble metal catalyst (Pd-Pt.-Bi/C ratio 4 : 1 : 5). Using this method benzyl alcohol gives the corresponding aldehyde with 99% selectivity.

A method for using oxygen is the transition metal catalysed **epoxidations** is by using an aldehyde as a sacrificial oxygen transfer agent. Such expoxidations occur smoothly in SC-CO_2[26]. Thus compounds containing internal double bonds (*e.g.*, *cis*-cyclooctene) could be epoxised without addition of metal catalyst with complete selectivity to give the epoxide (epoxycyclooctane). In this case 2-propionaldehyde was used as sacrificial oxygen transfer agent (Fig. 4.11).

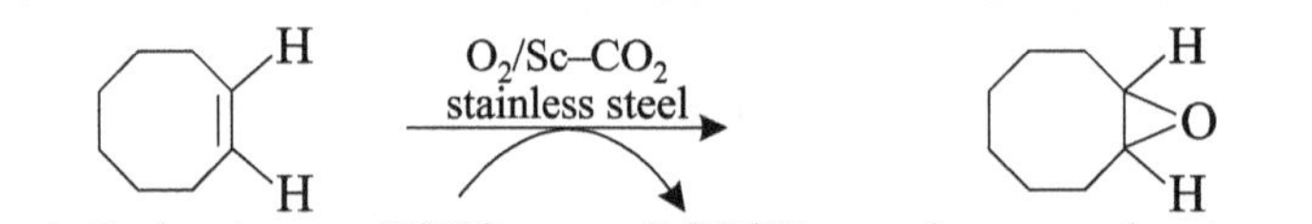

Fig. 4.11: Epoxidation of alkenes using sacrifical aldehyde in Sc-CO_2

In a similar way, cyclohexene and trans-3-hexene could be quantitatively epoxided (> 99% selectivity) (Fig. 4.12).

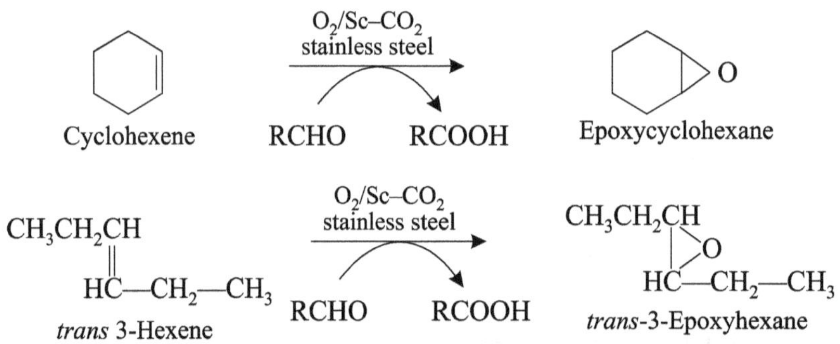

Fig. 4.12: Epoxidation of cyclohexene and trans 3-hexene.

The above epoxidations are catalysed by stainless steel reactor which facilitates the initial formation of acylperoxy radical, allowing the epoxidation to occur via a non-catalyst radical pathway.

Alkenes can also be epoxidised with *tert*-butyl hydroperoxide (TBHP) in presence of catalytic amount of Mo $(CO)_6$[27]. It is found that using aqueous TBHP solution (70 wt%), trans diols were obtained. On the other hand, using anhydrous decane solutions of TBHP gave the epoxide. *Cis*-allene were found to react much faster than the trans alkene.

Allylic alcohols could be epoxidised with aqueous TBHP is Sc-CO_2 using a Salen catalyst (X)[28]. Good distereoselectivity was obtained in the epoxidation of alkenes having secondary alcoholic group (Fig. 4.13).

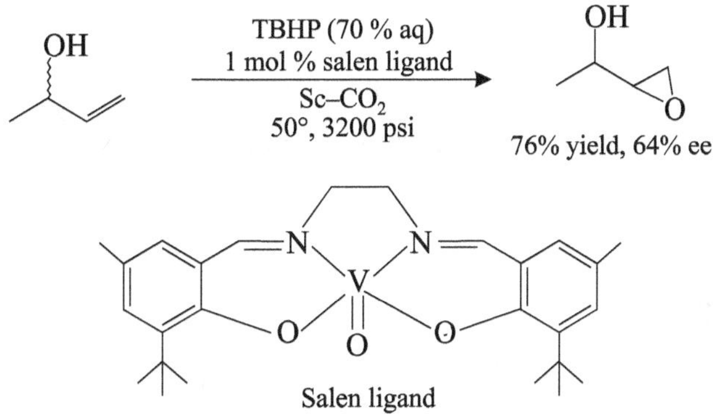

Fig. 4.13: Diastereoselective epoxidation in Sc-CO_2

The expoxidation of allylic alcohols with anhydrous TBHP in presence of catalytic oxovanadium (V) tri (isopropoxide) was possible in Sc-CO_2 at 25°C. The reaction rate was three times faster than in hexane solution. **Sharpless asymmetric epoxidation** of allylic alcohols was found to be effective by using titanium (IV) tetra (isopropoxide) and a chiral di-isopropyl-L-tartarate (DIPT) ligand at 0° in Sc-CO_2 (Fig. 4.14)[29]

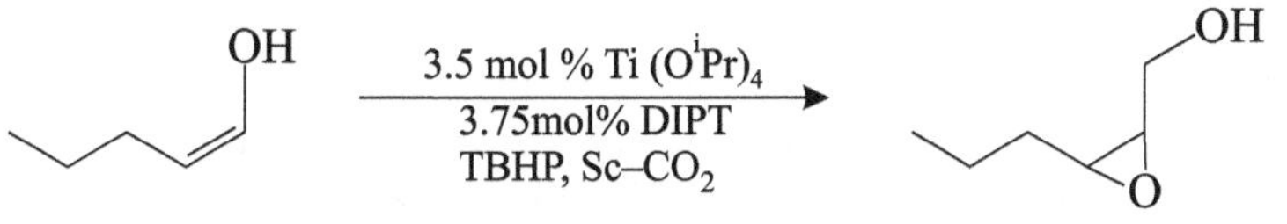

Fig. 4.14: Sharpless asymmetric epoxidation in Sc-CO_2

Selective oxidation of sulphides to sulfoxides can be achieved by using TBHP as oxidant and Amberst ™15 as heterogeneous acid catalyst[30]. Using this method diastereoselective sulfoxidation of cystein derivative in Sc-CO_2 could be affected to give anti-diasteomer as the sole product[31] (Fig. 4.15).

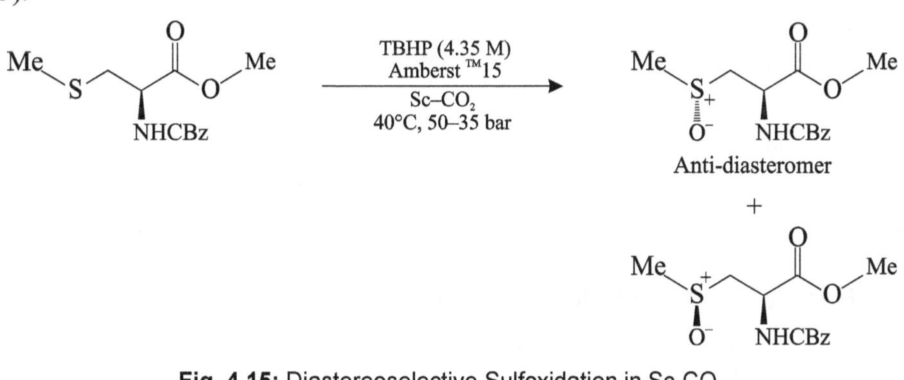

Fig. 4.15: Diastereoselective Sulfoxidation in Sc-CO_2

An interesting reagent for epoxidations is more reactive peroxy acid, which is obtained in situ from H_2O_2 and CO_2.[32] Thus, the epoxidation of cyclohexene (Fig. 4.16) and 3-cyclohexene-1-carboxylate sodium salt in a biphasic system composed of Sc-CO_2/olefin phase and an aqueous H_2O_2 phase yielded the corresponding epoxides and diols. The yield of the epoxide could be improved by addition of dimethyl formamide (DMF), which increases the aqueous solubility of the olefin.

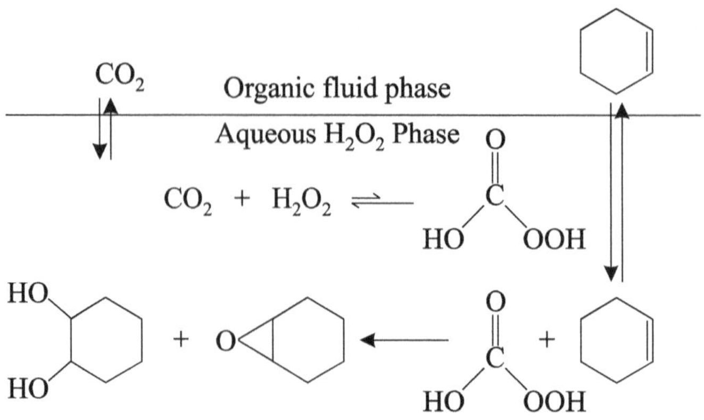

Fig. 4.16: Epoxidation of cyclohexene in aqueous H_2O_2/Organic fluid phase.

4.5 RADICAL REACTIONS IN SUPERCRITICAL CARBON DIOXIDE

The supercritical fluids are of low density and low viscosity and so are useful for radical processes. Some of the radical reactions are discussed below.

4.5.1 Free Radical Halogenation

These reactions are normally carried out in CCl_4. However, Sc-CO_2 is an attractive alternative. Thus the reaction of bromine with toluene and ethyl benzene give the corresponding benzylic bromides (Fig. 4.17)

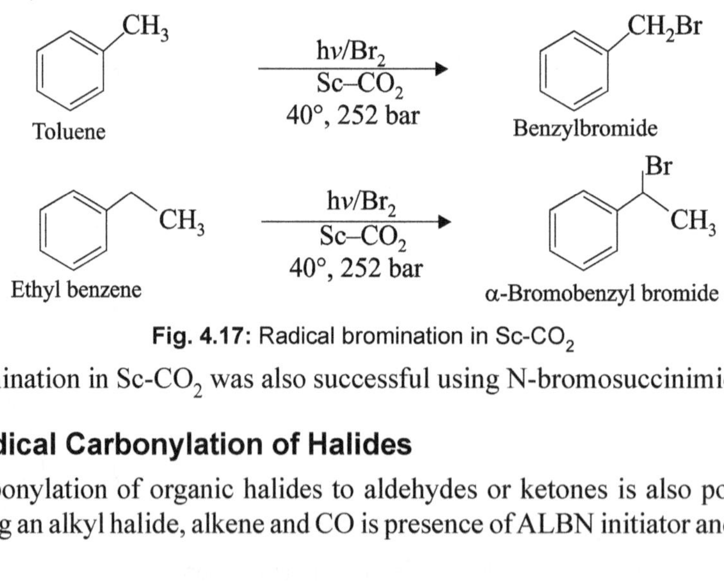

Fig. 4.17: Radical bromination in Sc-CO_2

Ziegler bromination in Sc-CO_2 was also successful using N-bromosuccinimide (NBS)[33].

4.5.2 Free Radical Carbonylation of Halides

Free radical carbonylation of organic halides to aldehydes or ketones is also possible in Sc-CO_2 (Fig. 4.18)[34] using an alkyl halide, alkene and CO is presence of ALBN initiator and Silane mediated catalyst.

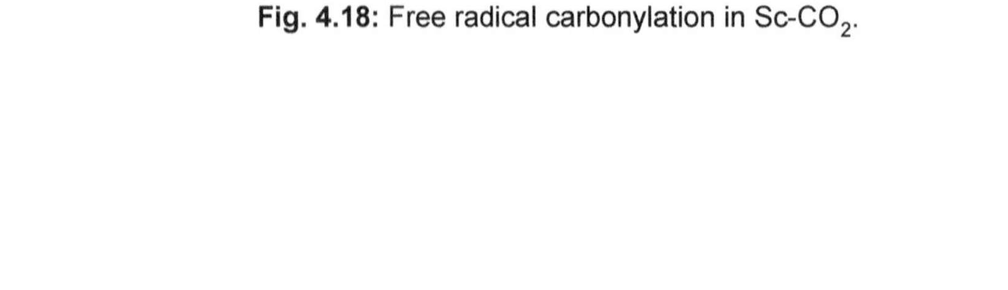

Fig. 4.18: Free radical carbonylation in Sc-CO_2.

4.5.3 Reduction of Bromoadamantane in Sc-CO_2

Bromoadamantane could be reduced by ALBN is presence of a catalyst, tributyltin hydride (Bu_3SnH) (A) or *tris* (perfluoro hexylethyl) tin hydride ($(CF_3CF_2CF_2CF_2CF_2CF_2CF_2CF_2)_3$ SnH (B) in Sc–CO_2 gave the corresponding reduced product (Fig. 4.19)[35]

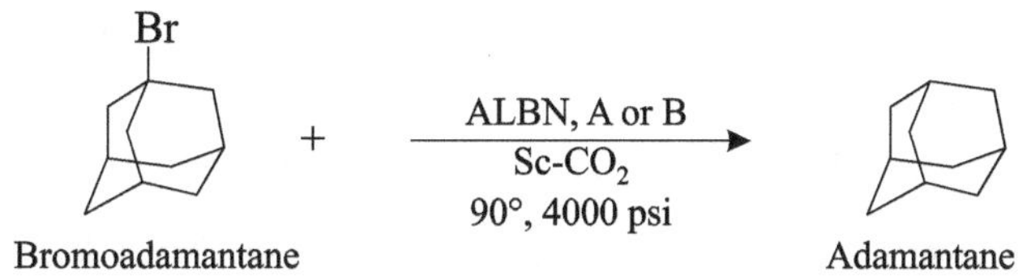

Fig. 4.19: Reduction of bromoadamantane in Sc-CO_2.

4.5.4 Radical Cyclisations in Sc-CO_2

Radical cyclisaltion of 1,1-diphenyl-6-bromo-1-hexene with (B) gave the 5-exo product. In a similar way reduction of aryl iodide with (B) also gave the cyclised product in quantitative yield. In both the cases no reaction was observed with tributyltin hydride (A) (Fig. 4.20).

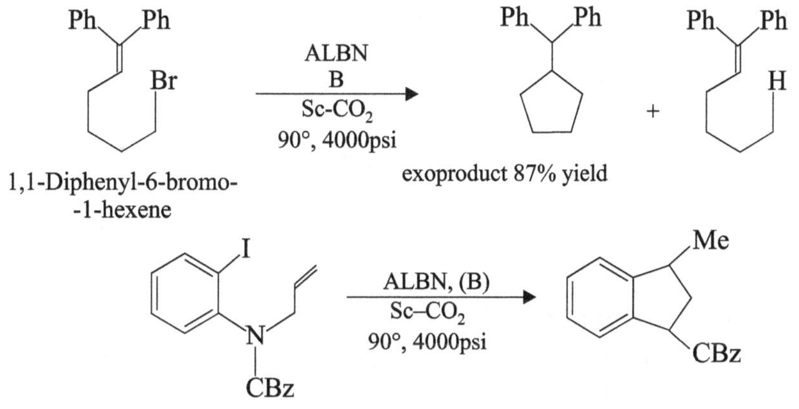

Fig. 4.20: Radical cyclisations in Sc-CO_2

4.6 DIELS-ALDER REACTION IN Sc-CO_2

A well known carbon-carbon bond formation reaction[36].

4.6.1 Diel-Alder Reaction of Isoprene and Methyl Acrylate

In this case a mixture of para and meta isomers are obtained (Fig. 4.21)[36]

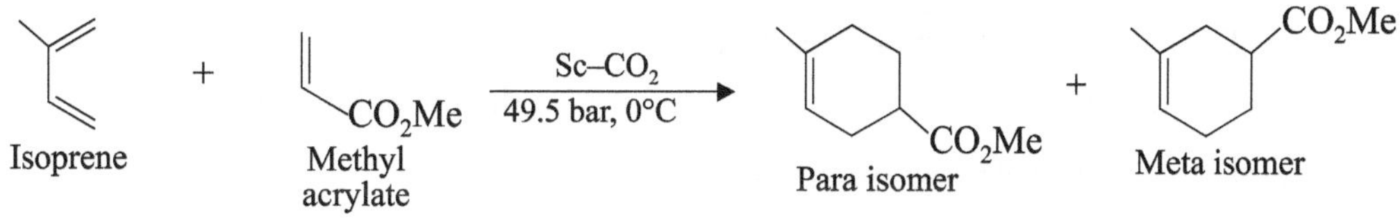

Fig. 4.21: Diels-Alder reaction of isoprene and methylacrylate.

4.6.2 Diels-Alder Reaction of 2-*t*-butyl-1,3-butadiene and Methyl Acrylate

In this case also a mixture of para and meta isomers are produced (Fig. 4.22)[36]

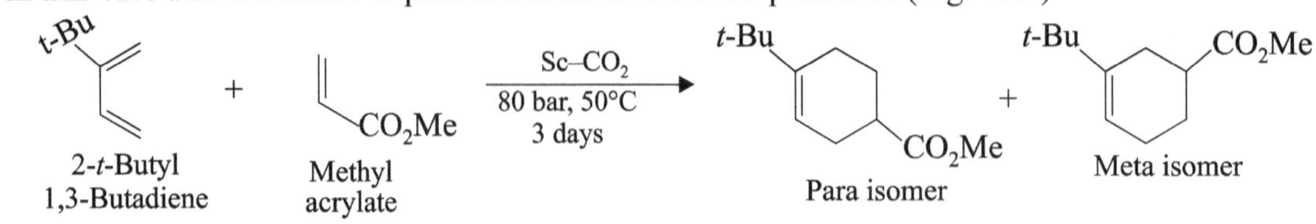

Fig. 4.22: Diels-Alder reaction of 2-t-butyl-1,3-butadiene and methylacrylate

4.6.3. Diels-Alder Reaction of Cyclopentadiene and Ethyl Acrylate Derivatives

A mixture of endo and exo isomers are produced (Fig. 4.23)[37].

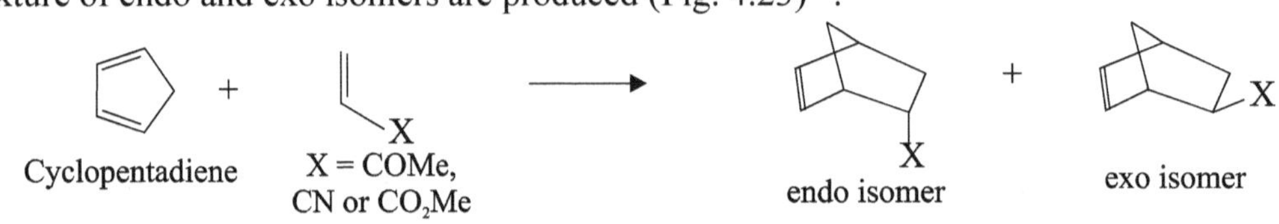

Fig. 4.23: Diels-Alder reaction of cyclopentadiene and ethyl acrylate derivative.

In Diels-Alder reaction, interesting results were obtained in presence of Lewi's acid catalyst like scandium trifluoromethane sulfonate. In this case optimum selectivity was observed around critical point of the reaction mixture (Fig. 4.24)[38].

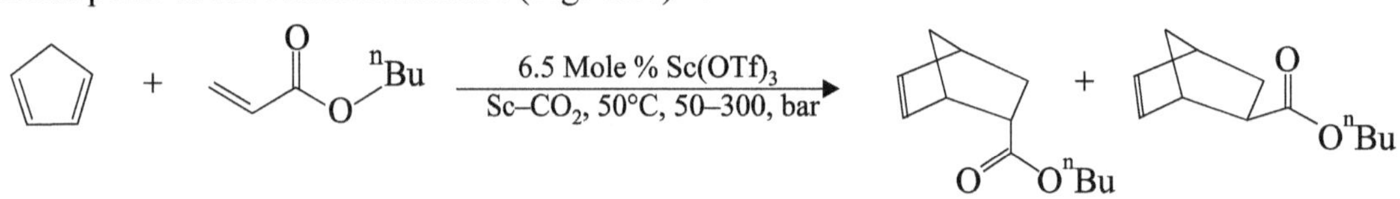

Fig. 4.24: Diels-Alder reaction is Sc-CO_2 in presence of $Sc(OTf)_3$ catalyst.

4.6.4 Aza-Diels-Alder Reaction

An aza-Diels-Alder reaction of Danishefsky's diene with an imine gave the aza-Diels-Alder adduct in presence of scandium perfluoroalkanesulfonate, Sc $(OSO_2C_8H_{17})_3$. (Fig. 4.25)[39]. In this case silica gel has been reported to be an efficient catalyst.[40]

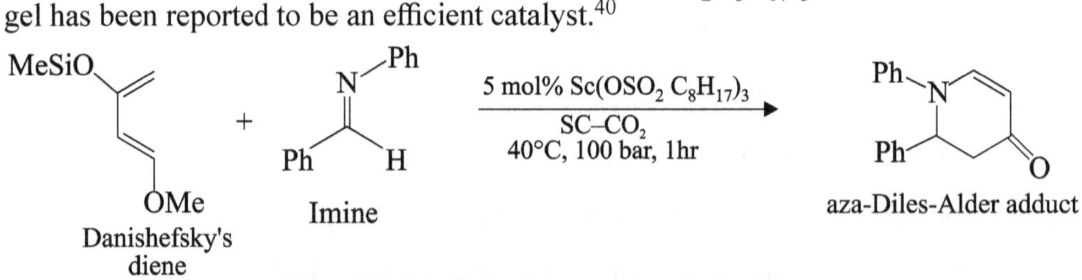

Fig. 4.25: Aza-Diels-Alder reaction in Sc-CO_2.

4.7 ACID CATALYSED REACTIONS

Following are given some of the acid catalysed reaction conducted in Sc-CO_2.

4.7.1 Friedel Crafts Alkylation of Aromatics

Friedel crafts alkylation of aromatics has been performed using a fixed bed heterogeneous polysiloxane supported catalyst in super critical media[41]. Thus mesitylene could be alkylated in

Sc-propene (Tc = 91.9°C, Pc = 46.0 bar). In this case Sc-propene acts as solvent and alkylating agent (Fig. 4.26).

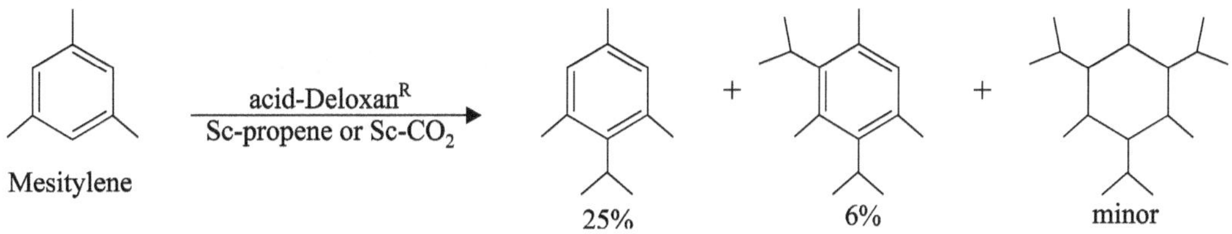

Fig. 4.26: Friedel-crafts alkylation of mesitylene in SCFs

In the above reaction (Fig. 4.26) using a molar ratio of mesitylene to propan-2ol of 2 : 1, a pressure of 20 bar. catalyst temperature 250° and a flow rate of 0.60 g/min. the monoalkylated product was obtained as the only product. See also section 4.7.4 (Fig. 4.29).

4.7.2 Synthesis of Tetrahydrofuran

The dehydration of 1,4-butandiol using the conditions as in Friedel-crafts alkylation, THF was obtained in quantitative yield (Fig. 4.27)[42].

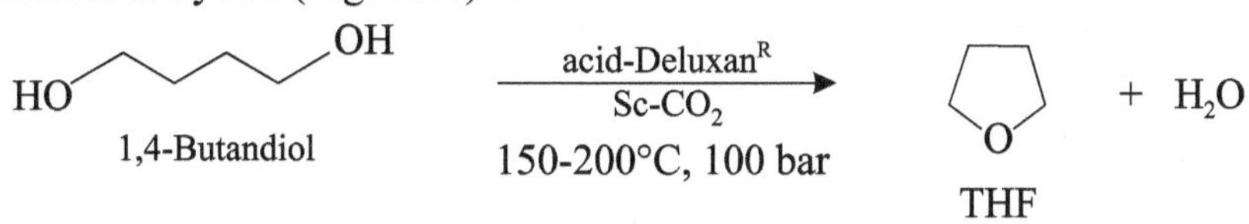

Fig. 4.27: Synthesis of THF

4.7.3 Mannich Reaction and Aldol Reaction

These reactions are performed by using a variety of Lewis acid catalysts in SC-CO_2 in presence of poly (ethylene glycol) PEG derivatives by the reaction of silyl enolates with aldehydes and imines (Fig. 4.28)[39, 43].

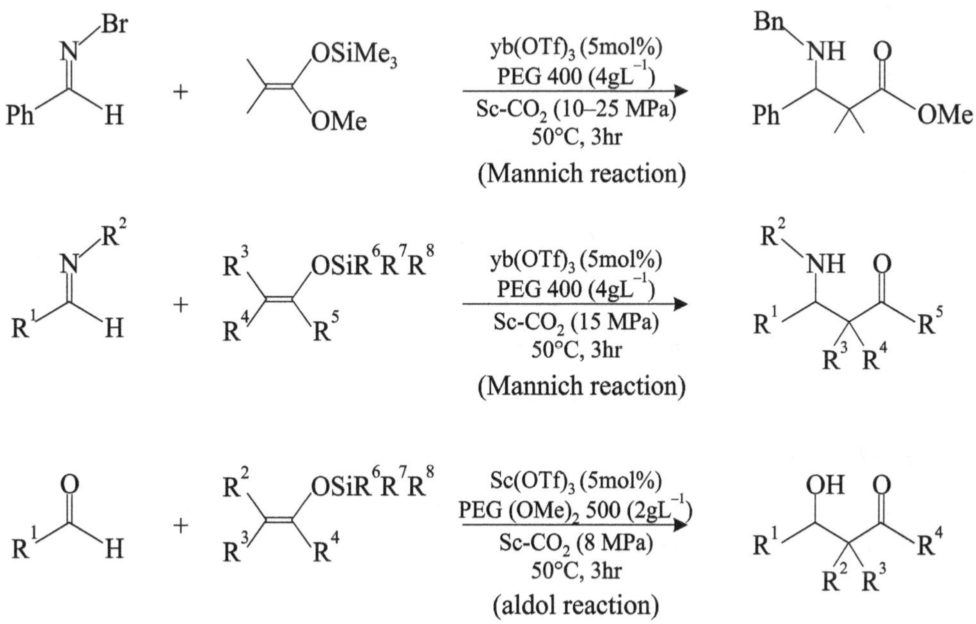

Fig. 4.28: Mannich and aldol reactions in Sc-CO_2 in presence of PEG derivatives.

In the above reactions (Fig. 4.28), PEGs act as surfactants forming colloidal dispersions in Sc-CO_2, which accelerate the reactions.

4.7.4 Fridel-crafts Alkylation of Indole Derivatives in Sc-CO_2

Fridel-crafts alkylation of indole derivatives in Sc-CO_2 in presence of 1-dodecyloxy-4-heptadeca-fluorooctyl benzene (A) gave the alkylated product (Fig. 4.29).[44]

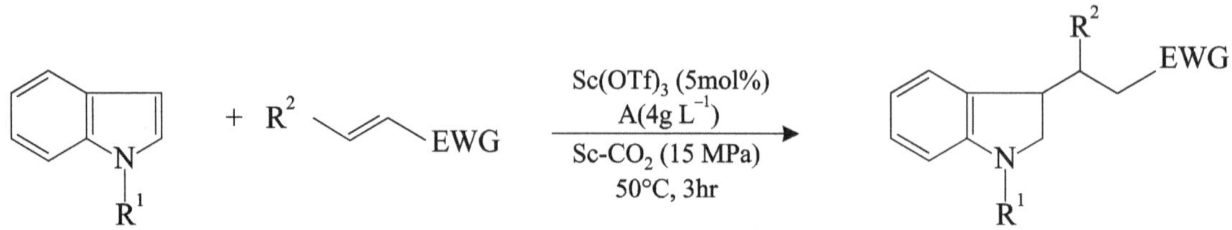

R^2 = H, Ph

EWG = COMe, COPh, NO_2

Fig. 4.29: Fridel-crafts alkylation of indole derivatives in Sc-CO_2.

4.8 COUPLING REACTIONS

The palladium catalysed coupling reactions, have been performed in Sc-CO_2. The main problem that the solubility of phosphine ligands (*e.g.* PPh_3) normally used along with palladium catalyst like $Pd(OAc)_2$ is low in Sc-CO_2. However, use of a fluorinated phosphine ligand (X) gave of palladium ligand with much enhanced solubility[45]. (Fig. 4.30).

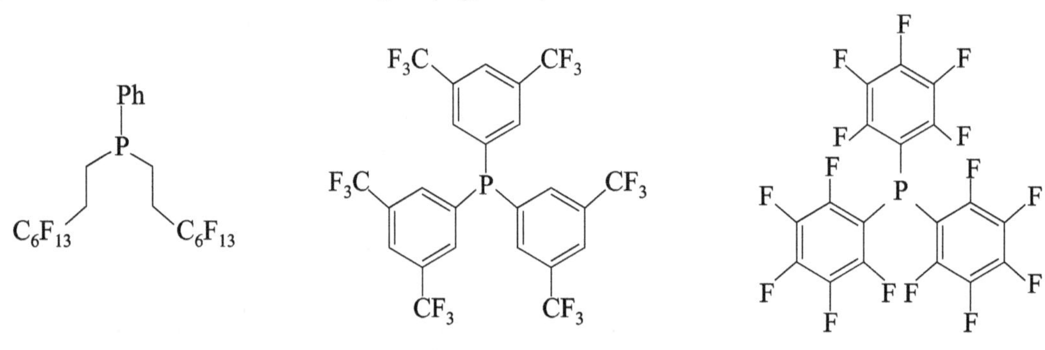

(X) (Y) (Z)

Fig. 4.30: Fluorinated phosphine ligands.

Following are given some of the coupling reactions.

4.8.1 Heck Reaction

The Heck reaction between iodobenzene and methyl acrylate in presence of $Pd(OAc)_2$ and the ligand (X) gave 92% yield of methyl cinnamate (Fig. 4.31).

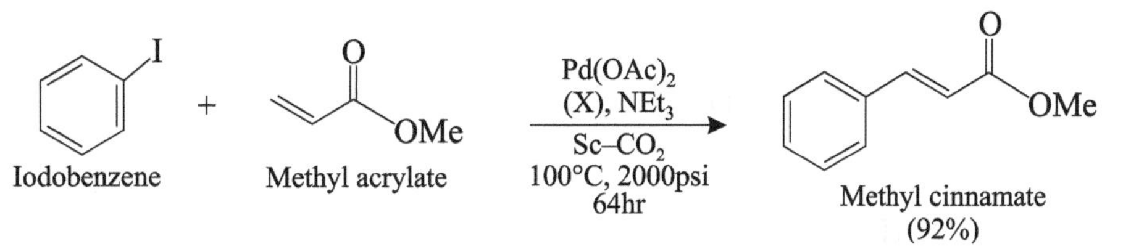

Fig. 4.31: Heck reaction using Pd-mediated coupling using a fluorous ligand.

Heck coupling reaction has also been reported[46] in Sc-CO_2 using the ligands (Y) and (Z).

A number of palladium catalysed coupling reactions in Sc-CO_2 were developed using commercially available ligands (*e.g.*, PBu_3, PCy_3). These are of special interest if the procedures are to be adopted widely[47]. These catalytic systems are useful for performing Heck coupling reactions. Thus, intramolecular Heck reactions showed reduced double bond isomerisation when carried out in Sc-CO_2 compared to conventional procedures (Fig. 4.32).

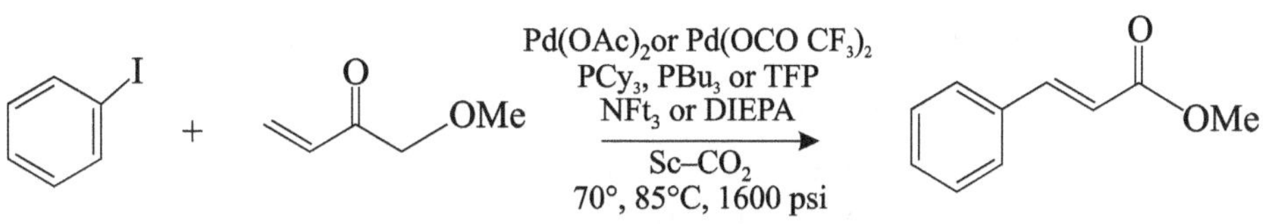

Fig. 4.32: Heck reaction using fluorinated palladium sources in $ScCO_2$

The Heck reaction has also been carried out using water-soluble catalyst in Sc-CO_2/water biphasic system[48]. Thus, the coupling of iodobenzene with butyl acrylate in Sc-CO_2 using $Pd(OAc)_2$ and triphenylphosphine trisulfonate sodium salt (TPPTSS) in presence of polar co-solvent like water or ethylene glycol gave much better conversion (Fig. 4.33).

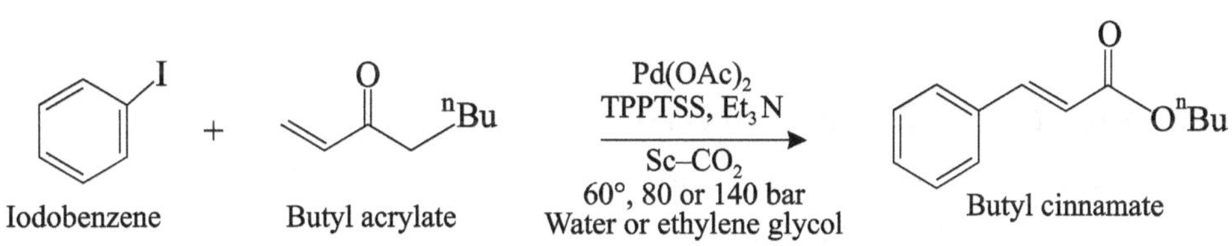

Fig. 4.33: Heck reaction in water/Sc-CO_2 biphasic system.

Enhanced selectivity has been observed in the **Mizoroki-Heck arylation reaction** of aryl halides with ethylene in Sc-CO_2-liquid biphasic system (Fig. 4.34)[49]. The product obtained is exclusively the styrene derivative.

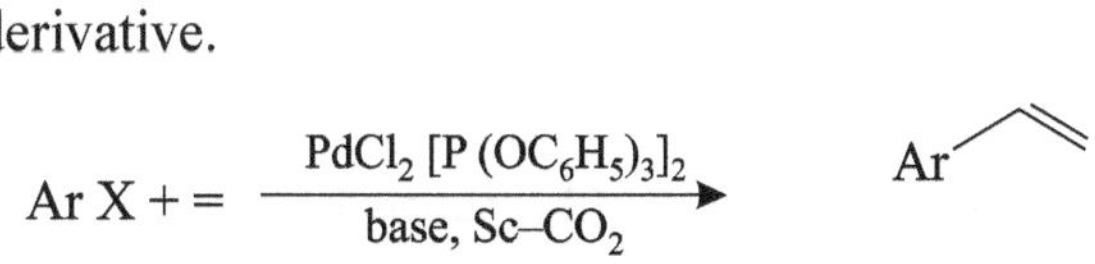

Fig. 4.34: Mizoroki-Heck arylation in Sc-CO_2.

4.8.2 Still Coupling

Still coupling mediated by fluorous tagged phosphines (PTP-I and PTP-II) has been achieved in Sc-CO_2 using $(n\text{Bu})_4$ NCl as catalyst (Fig. 4.35)[50].

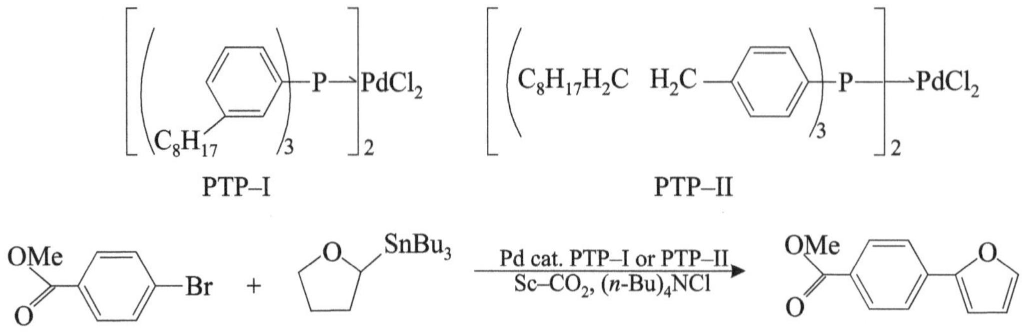

Fig. 4.35: Still coupling mediated by fluorous tagged phosphines.

Still coupling and also **Suzuki coupling** can also be carried out in water/SC-CO_2 biphasic system as in the case of Heck reaction (Fig. 4.33).

4.8.3 Synthesis of Biaryls

Homocoupling of iodobenzenes in Sc-CO_2 in presence of Pd-catalysis give biaryls (Fig. 4.36)[51].

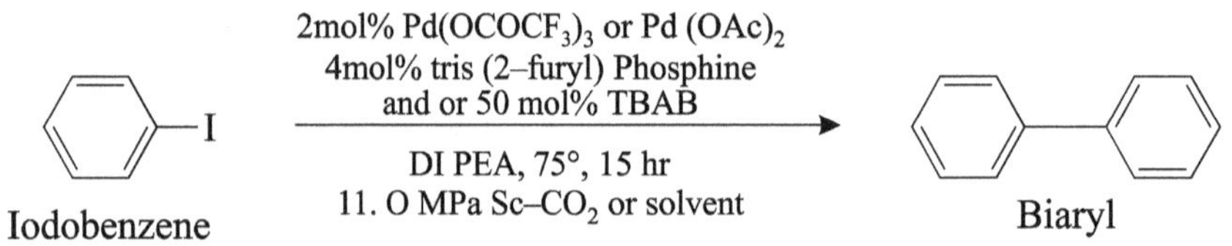

Fig. 4.36: Pd catalyst biaryl formation in Sc-CO_2.

The above reaction (Fig. 4.36) is favoured by the presence of electron donating groups (*e.g.*, MeO) in Sc-CO_2.

4.8.4 Carbonylation of 2-iodobenzyl Alcohol

The palladium catalysed carbonylation of 2-Iodobenzyl alcohol using the palladium complex $PdCl_2[P(OEt)_3]_2$ in Sc-CO_2 and CO in presence of Et_3N gave the phthalide (Fig. 4.37)[52,53].

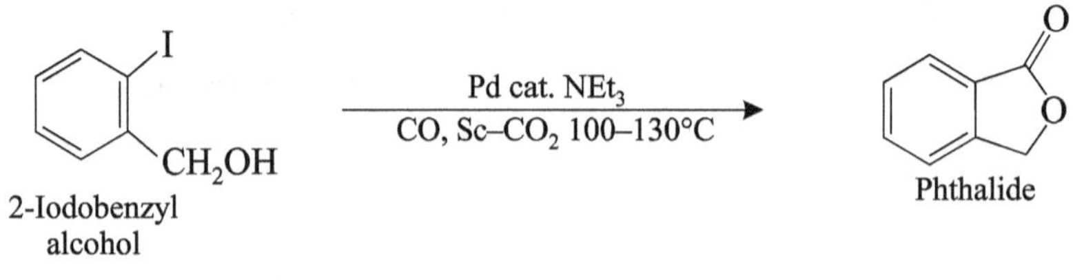

Fig. 4.37: Carbonylation of 2-iodobenzylalcohol in Sc-CO_2.

4.8.5 Carbonylation of *n*-butylamine

Carbonylation of *n*-butylamine in presence of $PdCl_2$ and $CuCl_2$ in MeOH is $ScCO_2$ gave 1,1′-oxalyldibutylamine. However, carboxylation in presence of oxygen gave methyl-N-*n*-butyl carbamate (Fig. 4.38)[54].

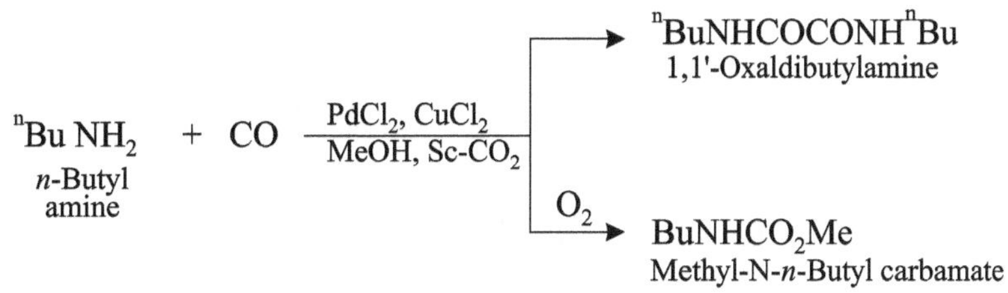

Fig. 4.38: Carbonylation of *n*-butylamine.

4.8.6 Wacker Reaction

Oxidation of higher terminal alkenes with $PdCl_2/CuCl_2$ using SC-CO_2 and methanol as cosolvent gave methyl ketones (Fig. 4.39)[55].

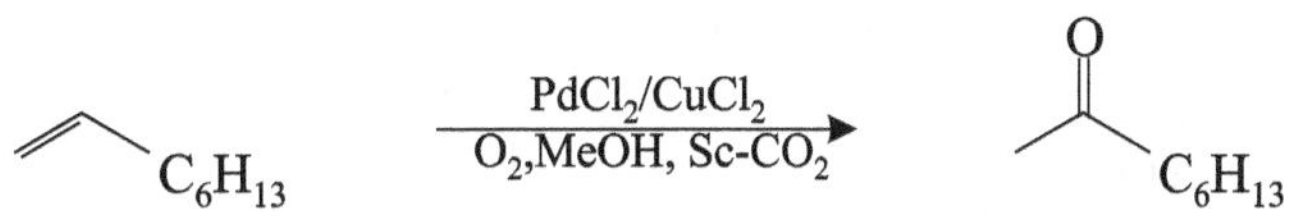

Fig. 4.39: Wacker reaction of Oct-1-ene in Sc-CO_2.

4.9 MISCELLANEOUS REACTIONS IN SC-CO_2

4.9.1 Synthesis of 2-pyrones

These are obtained by the coupling of two molecules of alkynes with CO_2 in presence of Nickel catalyst. Thus 3-hexyne on reaction with CO_2 in presence of a catalyst generated from $Ni(Cod)_2$ and 1,4-diphenyl phosphinobutane, $[Ph_2P(CH_2)_4PPh_2]$ gives tetraethyl-2-pyrone (Fig. 4.40)[56,57]

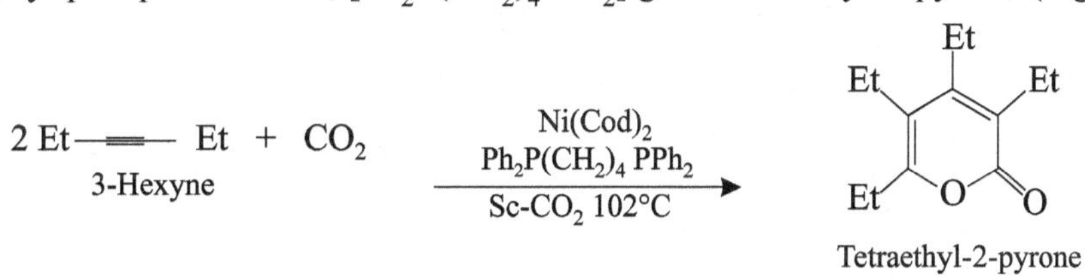

Fig. 4.40: Synthesis of pyrone in Sc-CO_2

4.9.1 Pauson-Khand Reaction

The reaction of an alkyne with an alkene and carbonmonoxide in Sc-CO_2 in presence of $Co_2(CO)_8$ gives cyclopentanones (Fig. 4.41)[58].

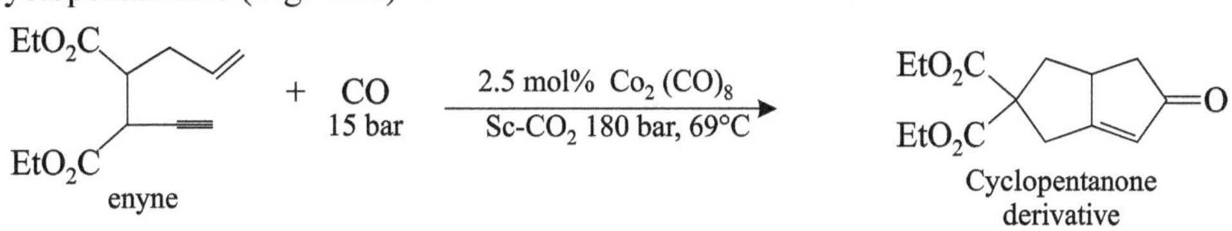

Fig. 4.41: Passon-Khand reaction in Sc-CO_2.

An intramolecular Pauson-Khand reaction is also reported (Fig. 4.42).

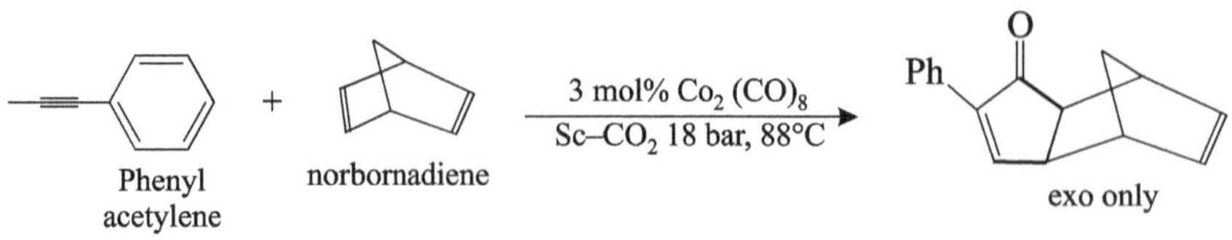

Fig. 4.42: Intramolecular Pauson-Khand reaction in Sc-CO_2

4.9.2 Asymmetric Cyclopropanation

Cyclopropanation reaction between styrene and methyl phenyldiazoacetate in presence dirhodium species in Sc-CHF_3 gave diastereomeric products (Fig. 4.43)[59]

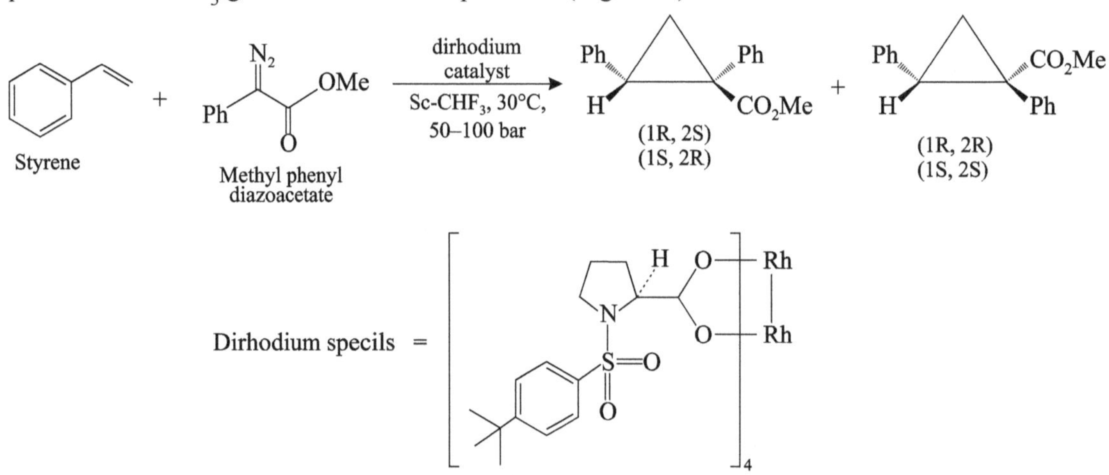

Fig. 4.43: Asymmetric cycloproponation in SCF.

4.9.3 Enantioselective Hydrovinylation of Styrene

Enantioselective hydrovinylation of styrenes in Sc-CO_2 in presence of a catalytic system n^3-allyl-Ni (a dimeric azaphospholene ligand and tetrakis-3,5-*bis* (trifluoromethyl) phenyl borate (BARF) counter ion gave quantitative conversion with high regioselectivity (> 70%) and excellent enantioselectivity (83.2-91.6%) (Fig. 4.44)[60].

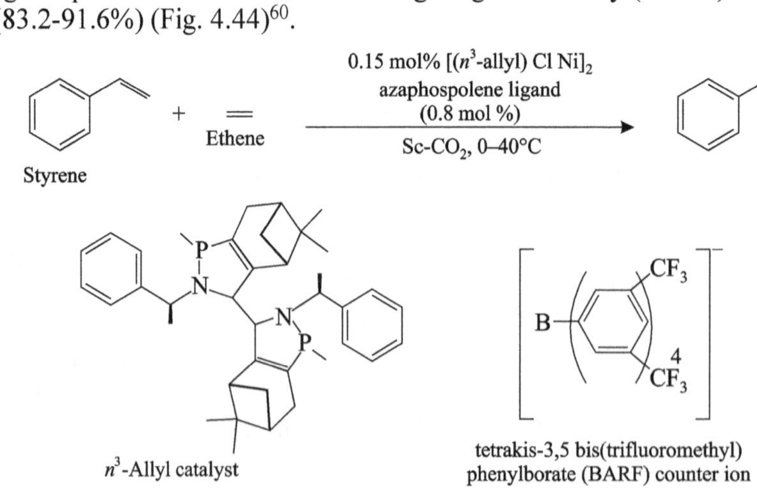

Fig. 4.44: Enantioselective vinylation of styrene in Sc-CO_2.

4.9.4 Hydroboration of Styrene

Hydroboration of styrene with catecholborane (Hbat) in presence of rhodium catalyst precursor [(hfacac) Rh (cyclooctene)$_2$] (A) and phosphorus ligand [$Cy_2P(CH_2CH_2C_6F_{13})$] (B) gave quantitative conversion in Sc-CO_2 (Fig. 4.45)[61].

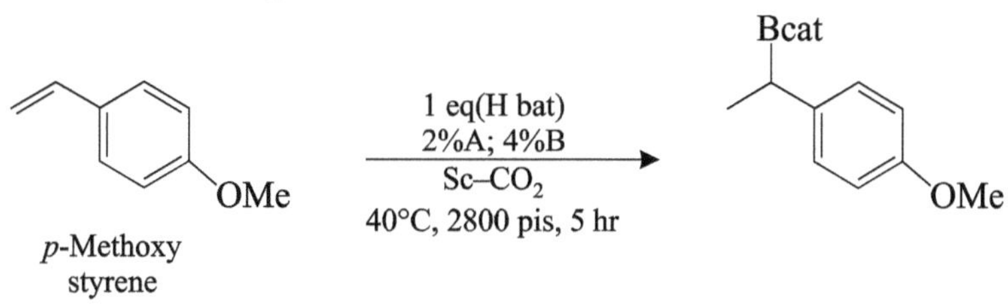

Fig. 4.45: Hydroboration of styrene derivative in Sc-CO_2 .

4.9.5 Synthesis of Carbamates

The reaction of primary, secondary and aromatic amines with butyl chloride in presence of ammonium salt and potassium carbonate in Sc-CO_2 as solvent gave the corresponding carbamates (Fig. 4.46)[62].

$$\underset{\substack{\text{R, R}'\text{=H, Me, Et,}\\ \text{Bu, Bn, Ph}}}{RR'NH + BuCl + K_2CO_3} \xrightarrow[\substack{\text{Sc-CO}_2\\ 100°\text{C, 80 bar,}\\ \text{2hr.}}]{10\%\ Bu_4NBr} \underset{\substack{72\text{-}90\%\\ \text{Carbamates}}}{RR'NCOOBu}$$

Fig. 4.46: Synthesis of Carbamates in Sc-CO_2.

In the above synthesis, use of potassium phosphate in place of potassium carbonate showed that the carbonyl source was CO_2 and not the carbonate. The above procedure is an one pot, alternative to the use of phosgene in the synthesis of urethane.

4.9.6 Synthesis of Vinyl Carbamate

These are synthesised by the reaction of secondary amines with terminal alkynes in Sc-CO_2 in presence of Ru catalyst [$RuCl_2(C_5H_5N)_4$] (Fig. 4.47)[63].

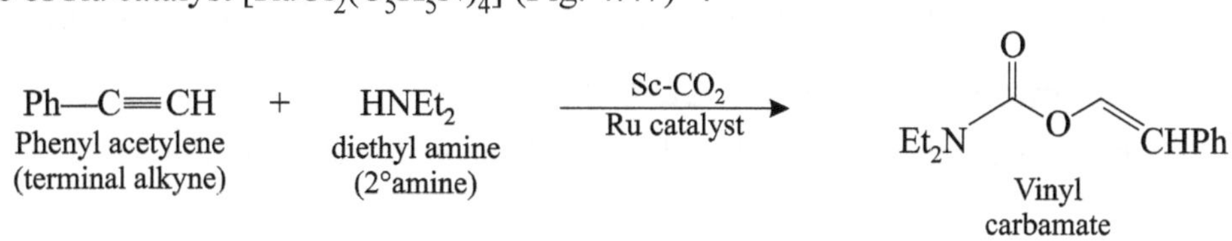

Fig. 4.47: Synthesis of Vinyl Carbamates in Sc-CO_2.

4.9.7 Baylis-Hillman reaction

This reaction has been carried out in Sc-CO_2 and gives better conversion and reaction rates compared to solution phase reaction[64]. It is interesting to note that carrying out the reaction in presence of an alcohol, the major produce is an ether which results from a three component coupling reaction, which occurs only in presence of Sc-CO_2 (Fig. 4.48).

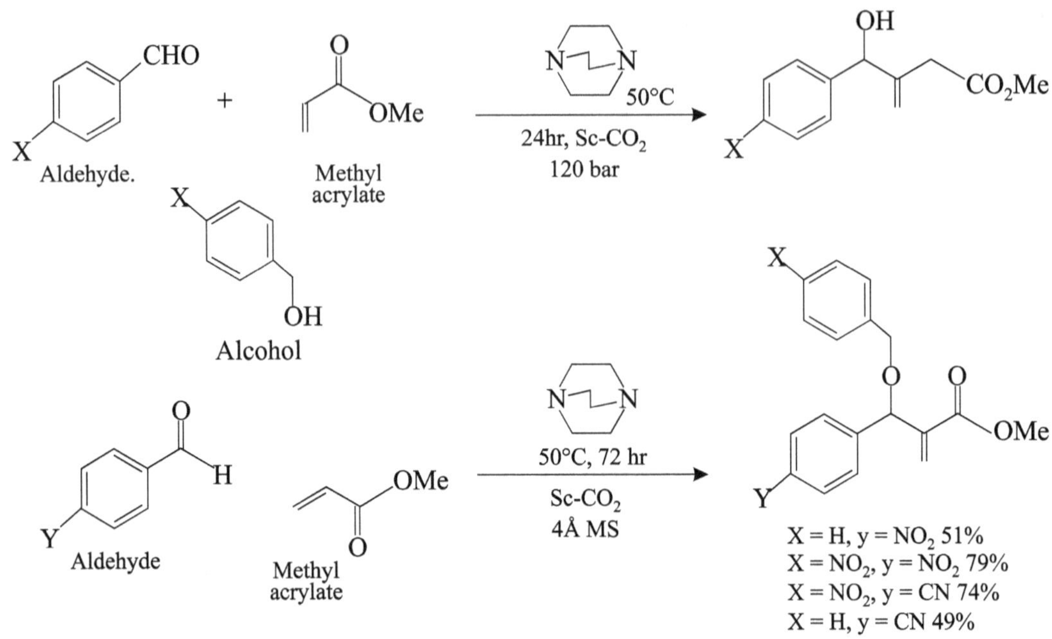

Fig. 4.48: Baylis-Hillman reaction in Sc-CO_2.

4.9.8 Henry Reaction

The reaction of *p*-nitrobenzaldehyde with nitropropane in Sc-CO_2 in presence of NEt_3 give the coupled product.

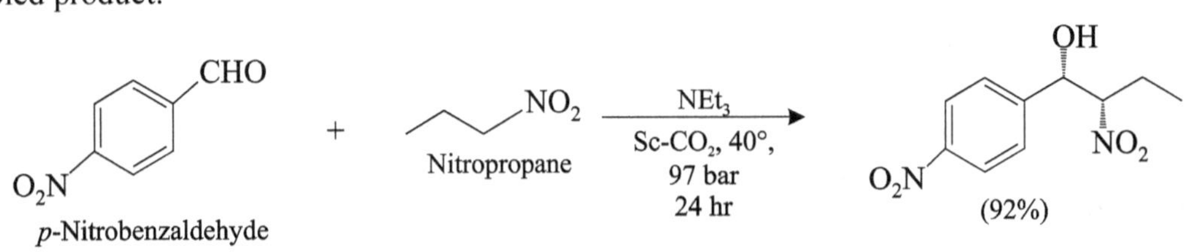

Fig. 4.49: Henry Reaction.

(Cited in Alternative solvents for Green Chemistry F.M. Kerton, RSC Publication)

4.9.9 Photochemical Reactions in SC-CO_2

The photoinduced reaction of an aldehyde to *p*-benzoquinone give 2-acyl hydroquinones (Fig. 4.50)[64].

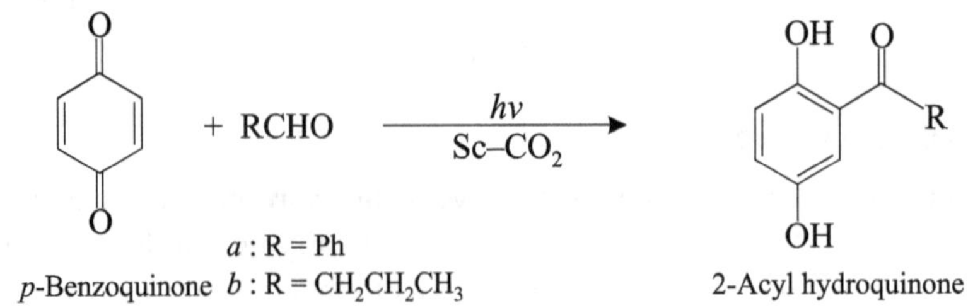

Fig. 4.50: Photochemical reaction in Sc-CO_2.

The above process provides a viable environmentally benign alternative to the well known Friedel-Crafts acylation (Fig. 4.51).

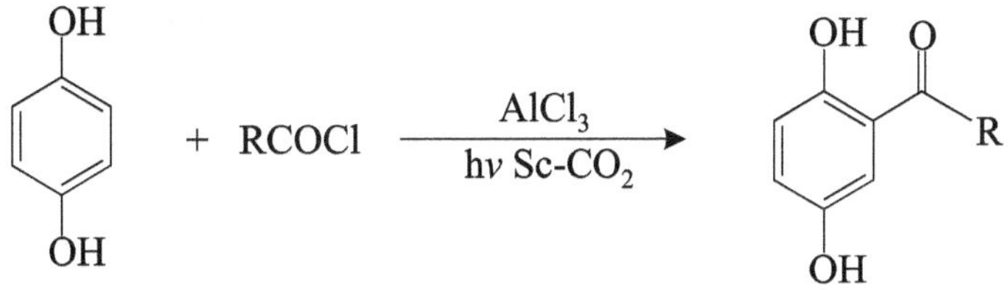

Fig. 4.51: Photochemical reaction leading to Friedel-Crafts acylation in $Sc\text{-}CO_2$.

The results obtained demonstrate that $Sc\text{-}CO_2$ is an excellent replacement for photoinduced addition of aldehydes to α, β-unsaturated carbonyl compounds. The problem of insolubility of organic solvent could be overcome by using an alcohol co-solvent (particularly *t*-BuOH) in order to increase the polarity of the solvent. As an example the photochemical reaction of aldehyde with α,β-unsat'd ketone gives[65] the product as shown in Fig. 4.52.

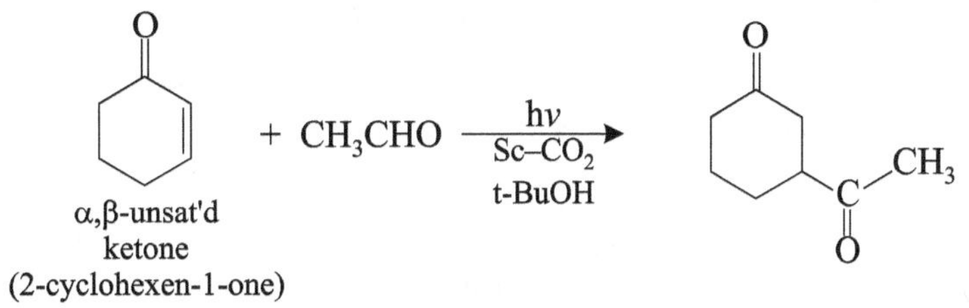

Fig. 4.52: Photochemical Reaction of acetaldehyde with 2-cyclohexen-1-one in $Sc\text{-}CO_2$.

4.10 ENZYMATIC TRANSFORMATION IN SC-CO_2

A number of examples have been reported[66,67] using enzymes as catalysts in $Sc\text{-}CO_2$ and performing hydrolysis, oxidations, esterifications and transesterifications. Thus, as an example the enzyme alkaline phosphate (EC 3.1.3.1) has been found to be active in $Sc\text{-}CO_2$ for the hydrolysis of disodium *p*-nitrophenylphosphate to give *p*-nitrophenol. The reaction was conducted at 35°C and 100 bar with a 0.1 volume percent water concentration.[68]

Following are given some of the enzymatic transformations in $Sc\text{-}CO_2$.

4.10.1 Kinetic Resolution of Racemic 3-(4-methoxyphenyl) Glycidic Acid Methyl Ester

The kinetic resolution of racemic 3-(4-methoxyphenyl) glycidic acid methyl ester by immobilised Mucor Miehei Lipase (Lypozyme 1M 20) in $Sc\text{-}CO_2$ could be achieved in 53% conversion with 87% *i.e.* (Fig. 4.53).[69]

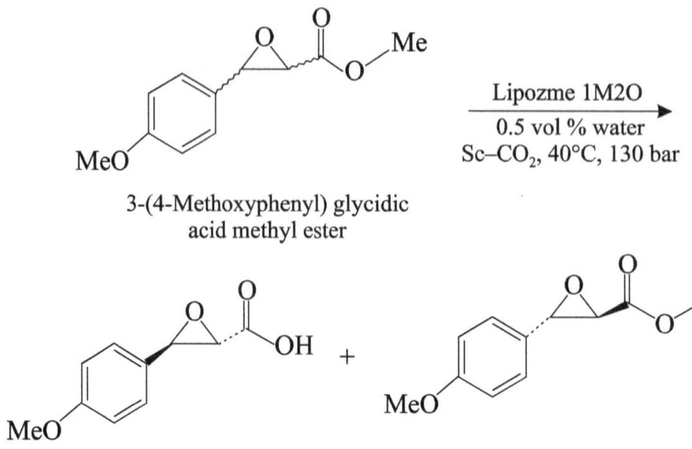

Fig. 4.53: Kinetic resolution of esters in Sc-CO_2

4.10.2 Transesterification of N-acetyl-1-phenylalanine Chloroethyl Ester

The transesterification of N-acetyl-1-phenylalanine chloroethyl ester with ethyl alcohol catalysed by *Subitilisn carlsberg* gave quantitative conversion in Sc–CO_2 (Fig. 4.54)[70]. The enzyme could be recovered and reused with virtually, no loss of activity after 3 cycles.

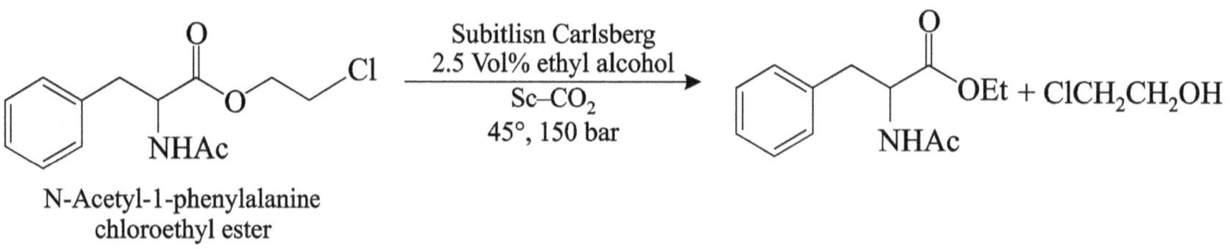

Fig. 4.54: Enzymatic transesterification in Sc-CO_2.

4.10.3 Trans Acetylations of l-O-p-nitrophenyl-β-D-glactopyranoside

Transacetalisation of I-O-p-nitrophenyl-β-D-glactopyranoside with 5-phenylpentan-1-ol in Sc-CO_2 at 40°/150 bar in presence of lipid coated enzyme (LCE) prepared for Bacillus circulans gave the product in 72% yield after 3 hrs. (Fig. 4.55)[71].

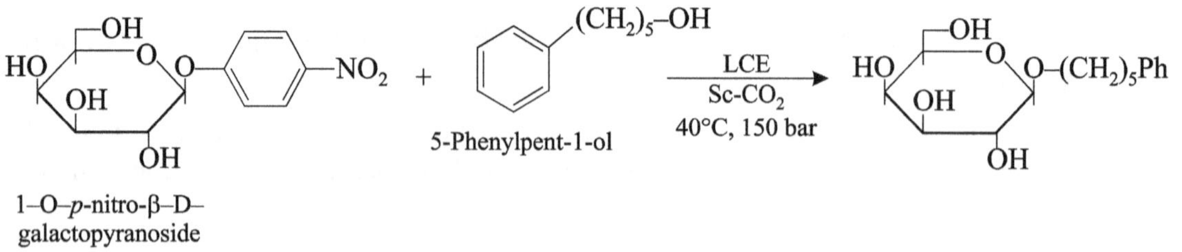

Fig. 4.55: Enzymatic trans acetalisation in Sc-CO_2.

4.10.4 Esterification of Lauric Acid

Esterification of lauric acid with glycenride in Sc-CO_2 in presence of LCE prepared from Rhizopus delemar gave di and triglycerides. (Fig. 4.56)[72].

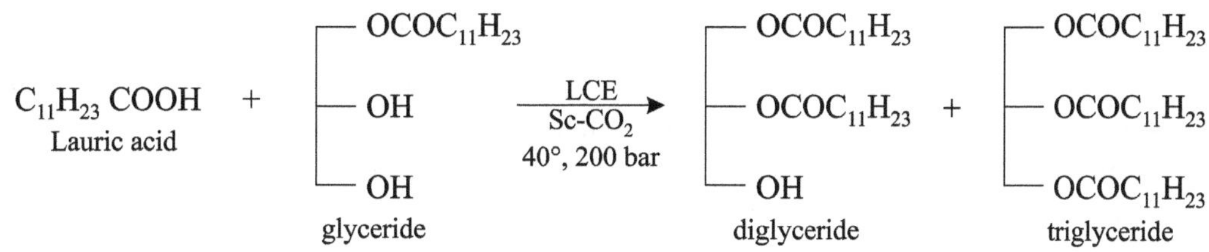

Fig. 4.56: Enzymatic Esterification of Lauric acid in Sc-CO_2

4.10.5 Esterification of Citronellol in SC-CO_2

The esterification of terpene alcohol citronellol with *n*-valeric acid using the enzyme cadida cylindracea lipase (CCL) in Sc-CO_2 give the ester (Fig. 4.57)[73].

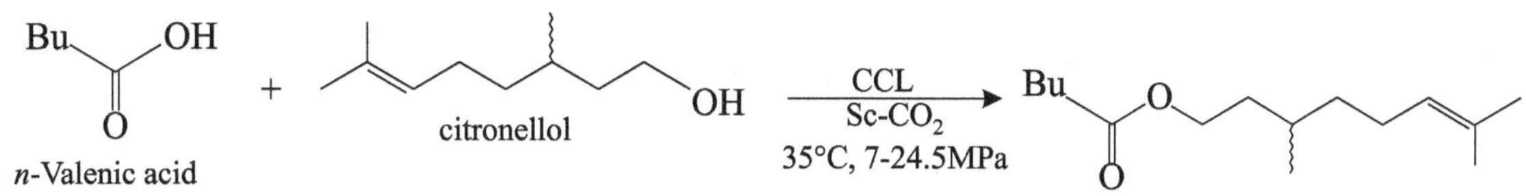

Fig. 4.57: Enzymatic esterification of citronellol in Sc-CO_2.

The esterification of oleic acid with citronellol in Sc-CO_2 in presence of the enzyme CCL gave 3,7-dimethyl-6-octenyl ester (Fig. 4.58).

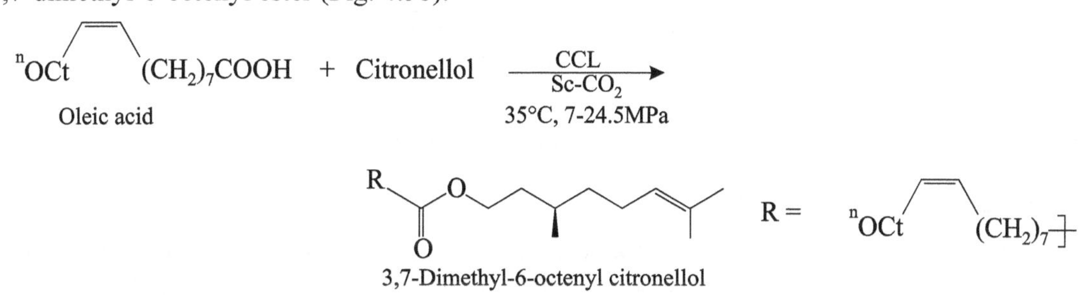

Fig. 4.58: Enantioselective enzymatic esterification of oleic acid with citronellol in Sc-CO_2.

4.10.6 Kinetic Resolution 1-phenylethanol with Vinyl Acetate

The kinetic resolution of 1-phenylethanol with vinyl acetate in SC-CO_2 in presence of enzyme Novozym 435 (EC3.1.1.3) from candidia antarcitia B gave (R)-1-phenlethylacetate is > 99% ee with 55% conversion (Fig. 4.59)[74].

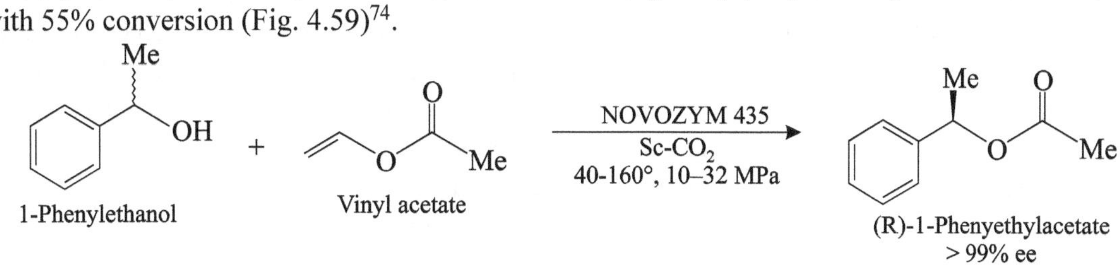

Fig. 4.59: Kinetic resolution of 1-phenylethanol in Sc-CO_2.

A similar resolution of 1-(*p*-chlorophenyl)-2,2,2-trifluoroethanol under similar condition has also been reported[75].

4.10.7 Enantioselective Acetylation of Alcohols in SC-CO_2

The enantioselective acetylation of alcohols (*e.g.*, trans-3-penten-2-ol) with acetic anhydride using an immobilised lipase (Pseudomonas *sp.* from Amano P) in Sc-CO_2 has been reported (Scheme 4.60)[76]. There was 50% conversion after 250 min. at 40°C compared to reaction in toluene in which only <20% conversion was achieved. Also, the enentioselectivity was higher in SC-CO_2.

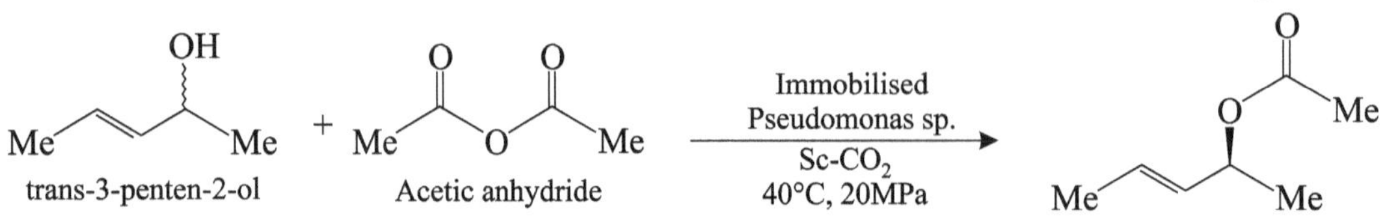

Fig. 4.60: Enantioselective acetylation of alcohols in Sc-CO_2

4.10.8 Synthesis of Pyrrole-2-carboxylic Acid

Pyrrole could be converted into pyrrole-2-carboxylic acid in Sc-CO_2 using cells of Bacillus megaterium PYR 2910 which catalyse the fxatia of CO_2.[77] The reaction could be conducted by adding CO_2 (10 MPa) to the reactant at 40°C to give 54% yield after 1 hr. The maximum yield was at pressures below the critical pressure of CO_2 (4-7 MPa) (Fig. 4.61).

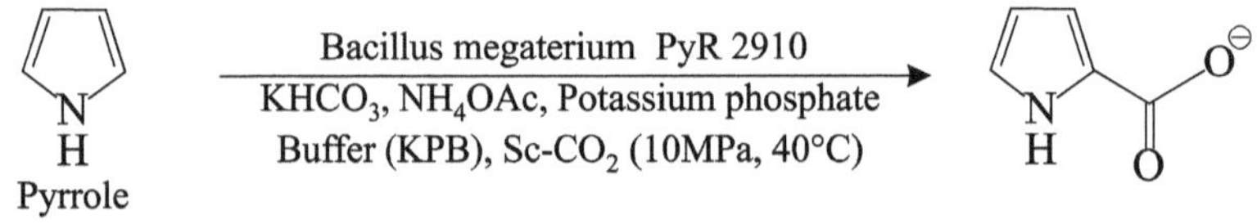

Fig. 4.61: Enzymatic synthesis of pyrrole-2-carboxylic acid in Sc-CO_2.

4.10.9 Polymerisations in Sc-CO_2

Sc-CO_2 has been used[78] successfully as a solvent in different types of polymerisations. These include free radical solution polymerisation, emulsion polymerisation, suspension polymerisation, heterogeneous polymerisation, chain polymerisation etc. For details see chapter 19.

4.10.10 Formation of Silica Nanoparticles

Silica nanoparticles are well known for their abresive, optical, electrical and thermal properties. Besides, these are also used as catalysts or catalyst supports, insulating materials, binding agents for fibrous or granular materials and polishing agents[79]. A number of methods have been used for the preparation of nano particles. A novel method for their preparation involves using[80] Sc-CO_2, which acts as a solvent and also as a reactant. A water-in-oil microemulsion of an aqueous solution of sodium silicate in isooctane or *n*-heptane is injected into Sc-CO_2 using a micronozzule; this results in the formation of small droplets. The Sc-CO_2 rapidly extracts the solvent from the droplets and reacts with exposed surfactant-supported aqueous sodium silicate reverse miscelles forming silica nanoparticles.

$$2NaOH.SiO_2 + CO_2 \xrightarrow{\text{(in } n\text{-heptane)}} Na_2CO_3 + 2SiO_2 + H_2O$$

The precipitated silica particles are washed with water and sodium carbonate free particles are obtained by ultrafiltration. This method gives silica nanoparticles having sizes in the range 20-800 nm. In fact, silica nanoparticles of adjustable size can be produced by using supercritical or gaseous CO_2.

Conclusion

Supercritical carbon dioxide is emerging as a versitle medium for carrying out different types of chemical reactions including enzymatic transformation and polymerisations. There is absolutely no doubt that Sc-CO_2 offer environmental benefits. Besides, the isolation of the product by using Sc-CO_2 is very simple and convenient.

REFERENCES

1. J.F. Brennecke, Chem. Ind., 1996, 831; M. Poliakoff and S. Howdle, Chem. Ber., 1995, **310**, 118; R. Noyori, ed. Chem. Rev., 1999, **99**, 353.
2. No Maness, D. Chrz, T. Pierce and G.H. Brusewiz. J.Am. Oil Chem. Soc., 1995, **72**, 665.
3. S. Samdani, Chem. Eng., 1995, **102(8)**, 23.
4. B.C. Roy, M.Goto and T. Hirose, Ind. Eng. Chem. Res., 1996, **35**, 607.
5. B. Subramaniam, R.A. Rajeinski and K. Snavely, J. Pharmaceutical Society, 1997, **86(8)**, 885.
6. J.-H. Kim, T.E. Paxton and D.L. Tomusko, Biotechnology Prog., 1996, **12**, 650.
7. D. Alessi, A. Cortesi, F. Kikic, N.R. Foster, S.J. Macnaughton and I. Columbo, Ind. Eng. Chem. Res., 1996, **35**, 4718.
8. A. Baiker, Chem. Rev., 1999, **99**, 453; J.A. Darr, M. Poliakoff, Chem. Rev., 1999, **99**, 495.
9. P.G. Jessop, Y. Hsiao, T. Ikariya and R. Noyori, J.Am. Chem. Soc., 1996, **118**, 344.
10. P.G. Jessop, Y. Hsiao, T. Ikariya and R. Noyori, J.Am. Chem. Soc., 1994, **116**, 8851.
11. P.G. Jessop, Y. Hsiao, T. Ikariya and R. Noyori, J. Chem. Soc. Chem. Commun, 1995, 707.
12. M.G. Hitzler and M. Poliakoff, J. Chem. Soc. Chem. Cummun., 1997, 1667.
13. M.G. Hitzler, F.R. Smail, S.K. Ross and M. Poliakoff, Org. Proc. Res. and Dev., 1998, **2**, 137.
14. R. Tschan, R. Wandeler, M. Schneider, M. Burgener, M. Schubert and A. Baiker, Applied catalyst A: General, 2002, 223, 173.
15. U.R. Pillai and E. Sahle-Demessi, Chem. Commun, 2002, 422.
16. B. Minder, T. Mallat, K.H. Pickel, K. Steiner and A. Baiker, Catal. Lett., 1995, **34**, 1.
17. M.J. Burk, S. Feng, M.F. Gross and W. Tumas, J.Am. Chem. Soc., 1995, **117**, 8277.
18. J. Xiao, S.C.A. Nefkens, P.G. Jessop, T. Ikariya and R. Noyori, Tetrahedron Lett., 1996, **37**, 2813.
19. S. Kainz, A. Brinkmann, W. Leitner and A. Pfalz, J. Am. Chem. Soc., 1999, **121**, 6421.
20. G. Francio and W. Leitner. J. Chem. Soc. Chem. Commun 1999, 1663.
21. Y. Hu, W. Chen, A.M.B. Osuna, J. Xiao, A.M. Stuart and E. Hope, Chem. Commun, 2001, 725.
22. N. Meenan, M. Poliakoff, A.J. Sandel, J.N.H. Reck and P.C. Kamer, Chem. Commun, 2000, 1497.
23. A. Baiker, Chem. Rev., 1999, **99**, 473.
24. P.G. Jessop and W. Leitner, Chemical Synthesis using supercritical fluids, Wiley-VCH, Weinheim, 1999.

25. G. Jenzer, D. Sueur, D. Mallat and A. Baikar, Chem Commun, 2000, 2247.
26. F. Loeker and W. Leitner, Chem. Eur. J., 2000, **6**, 2011.
27. G.R. Haas and J.W. Kolis, Organometallics, 1998, **17**, 4454.
28. G.R. Haas and J.W. Kolis, Tetrahedron Lett., 1998, **39**, 5923.
29. D.R. Pesiri, D.K. Morita, W. Glaze and W. Tumas, J. Chem. Soc. Chem. Commun., 1998, 1015.
30. R.S. Chain, in Ph.D. Thesis Univ. of Leeds, Leeds, UK, 2000.
31. R.S. Oakes, A.A. Clifford, K.D. Bortle, M. Thornton-Pett and C.M. Rayner, J. Chem. Soc. Chem. Commun, 1999, 247.
32. S.A. Nolen, J. Lu, J.S. Brown, P. Pollet, B.C. Eason, K.N. Griffith, R. Glaser, D. Bush, D.R. Lamb, C.L. Liotta, C.A. Eckert, G.F. Thiele and K.A. Bartels, Ind. Eng. Chem. Res., 2002, **41**, 316.
33. J.M. Tanko, J.F. Blackert, Science 1994, **263,** 203.
34. I. Ryu and N. Sorada, Angew. Chem. Int. Ed. Engl, 1996, **35**, 1051.
35. S. Hadida, M.S. Super, E.J. Beckman and D.P. Curran, J.Am. Chem. Soc., 1997, **119**, 7406.
36. A.R. Renslo, R.D. Weinstein, J.W. Tester and R.L. Danherisel, J. Org. Chem., 1997, **62**, 4530.
37. R.D. Weinstein, A.R. Renslo, R.L. Danheiser and J.W. Tester, J. Phy. Chem. B. 1999, 103(15), **2878.**
38. R.S. Oakes, T.J. Happenstall, N. Shezad, A.A. Clifford and C.M. Rayner, Chem. Commun, 1999, 1459.
39. J. Matsuo, T. Tsuchiya, K. Odashima and S. Kobayashi, Chem. Lett., 2000, 178.
40. R.D. Weinstein, A.R. Renslo, R.L. Danheiser and J.W. Tester, J. Phys. Chem. B., 1999, **103**, 2878.
41. M. Hitzler, F.R. Smail, S.K. Ross and M. Poliakoff, J. Chem. Soc. Chem. Commun, 1998, 359.
42. W.K. Gray, F.R. Smail, M.G. Hitzler, S.K. Ross and M. Poliakoff, J. Am. Chem. Soc., 1999, **121,** 10711.
43. I. Komoto and S. Kobayashi, Chem. Commun, 2001, 1842.
44. I Komoto and C. Kobayashi, Org. Letters, 2002, **4**, 1115.
45. M.A. Carroll and A.B. Holmes, J. Chem. Soc. Chem. Commun., 1998, 1395.
46. D.K. Morita, D.R. Pesiri, S.A. David, W.H. Glaze and W. Tumas, J. Chem. Soc. Chem. Commun., 1998, 1397.
47. N. Shezad, R.S. Oakes, A.A. Clifford and C.M. Ruyner, Tetrahedron Lett., 1999, **40**, 2221.
48. B.M. Bhanage, Y. Ikushima, M. Shirai and M. Arai, Tetrahedron Lett., 1999, **40**, 6427.
49. Y. Kayaki, Y. Noguchi and T. Ikariya, Chem. Commun, 2000, 2245.
50. S. Schneider and W. Bannwarth, Angew. Chem. Int. Edn., 2009, **39**, 4142.
51. N. Shezad, A. Clifford and C.M. Rayner, Green Chem, 2002, **4**, 64.
52. T. Ikariya, Y. Kayaki, Y. Kishimdo and Y.Noguchi, Progress in nuclear Energy, 2000, **37**, 429.
53. Y. Kayaki, Y. Noguchi, S. Iwasa, T. Ikariya and R. Noyori, Chem. Soc. Chem. Commun., 1999, 1235.
54. J. Li, H. Jiang and M. Chen., Green Chem., 2001, **3**, 137.
55. H. Jiang, L. Jia and J. Li, Green Chem., 2002, **2**, 161.
56. M.T. Reetz, W. Komen and T. Strack, Chimia, 1993, **47**, 493.
57. P.Y. Jessop and W. Leitner, Chemical Synthesis using Supercritical Fluids, Willey VCH, Weinheim, 1999.
58. H. Jeong, S.H. Hwang, Y.W. Lee and J.S. Lim, J. Am. Chem. Soc., 1997, **119**, 10549.

59. D.C. Wynne, M.M. Olmstead and P.G. Jessop. J. Chem. Soc., 2000, **122**, 7638.
60. A. Wegner and W. Leithner, J. Chem. Soc. Chem. Commun, 1999, 1583.
61. C.A.G. Carter, R.T. Baker, S.P. Nolam and W. Tumas, Chem. Commun., 2000, 347.
62. M. Yoshida, N. Hara and S. Okuyama, Chem. Commun., 2000, 151.
63. M. Rohr, C. Geyer, R. Wandeler, M. Schneider, A. Murphy and A.Baiker, Green Chem., 2001, **3**, 123.
64. G.A. Kraus and M. Kirihara, J. Org. Chem., 1992, **57**, 3256.
65. R. Pacut, M.L. Grimm, G.A. Krauss and M. Tanko, Tetrahedron Lett., 2001, **42**, 1415.
66. A.J. Mesiano, E.J. Beckman and A.J. Russell, Chem. Rev., 1999, **99**, 623.
67. P.G. Jessop and W. Leitner, Chemical Synthesis using supercritical Fluids-Wiley-VCH, Weinheim, 1999.
68. T.W. Randolph, H.W. Blanch, J.M. Papusitz and C.R. Wicke, Biotech. Lett., 1985, **7**, 325.
69. M. Rantakyla, M. Alkio and O.Aaltonen. Biotech. Lett, 1996, **18**, 1089.
70. P. Pasta, G. Mazzola, G. Carrea and S. Riva, Biotech. Lett., 1989, **11**, 643.
71. T. Mori and Y. Okahata, J. Chem. Soc. Chem. Commun., 1998, 2215.
72. T. Mori, A. Kobayashi and Y. Ohahata, Chem. Lett., 1998, 921.
73. G. Ikushima, N. Saito, K. Hataketa and O. Sato, Chem. Eng. Sci, 1996, **51**, 2817.
74. A. Overmeyer, S. Senrader-lippelt, V. Kasche and G. Brunner, Bootech Lett., 1999, **21**, 65.
75. T. Matsuda, R. Kanamaru, K. Watanabe, T. Harada and K. Nakamura, Tetrahedron Lett., 2001, **42**, 8319.
76. E. Cerina, C. Palocci, F. Gaparrni, D. Misiti and N. Fagnano, J. Mol. Cat., 1994, **89**, L11.
77. T.Motsuda, Y. Ohasti, T. Harada, R. Yanagihara, T. Nagasawa and K. Nakamura, Chem. Commun., 2001, 2194.
78. E.J. Beckman, J. Supercritical Fluids, **2003, 28(1),** 1 and the references cited there in.
79. P. Pasta, G. Mazzola, G. Carrea and S. Riva, Biotech. Lett., 1989, **11,** 643.
80. P. Chattopadhyay and R.B. Gupta, Ind. Eng. Chem. Res., 2003, **42**, 465.

5. TRANSFORMATIONS IN POLYETHYLENE GLYCOL AND ITS SOLUTIONS

5

Transformations in Polyethylene Glycol and its Solutions

5.1 INTRODUCTION

Polyethylene glycol (PEG, HO—$CH_2CH_2O)_n$–H, is available in a variety of molecular weights ranging from 200 to tens of thousands. At ambient temperature PEG–H_2O solution and hygroscopic polymer is a colourless liquid of molecular weight less than 600 and a waxy white solid of molecular weight more than 800[1]. The numerical designation of PEGs indicates the number of average molecular weight (*e.g.*, PEG-2000). Liquid PEG is completely miciable with water and solid PEG is highly soluble in water.

Polyethyline glycol is known to have a number of benign characteristics[2]. It has been approved by the FDA for internal consumptions[3, 4] PEG is weakly, if at all, immunogenic and this has led to the development of PEG-protein conjugates as drugs[5]. The aqueous solutions of PEG being biocompatible are used in tissue culture and for the preservations[2] of organs.

The low molecular weight PEGs, unlike VOCs are nonvolatile. Its vapour density is greater than 1 relative to air as per available MSDS data[6]. This is consistent with the industry standard for the selection of alternative solvents to VOCs[7]. Also PEG has low flammability and is biodegrable.

PEG is found to be stable to acid, base, high temperature, O_2, H_2O_2 high oxidation systems and $NaBH_4$ reduction systems. PEG can also be recovered from aqueous solutions or extraction with a suitable solvents or by direct distillation of water or solvent[9].

It is believed that PEG may prove to be a green reaction medium of the future[8].

5.2 USE OF PEG IN ORGANIC TRANSFORMATIONS

PEG has been used successfully in a number of organic transformations, some of which are given below.

5.2.1 Substitution Reactions

(*i*) Alkyl halides of the type RCH_2Br react with nucleophiles like CH_3COO^-, I^-, CN^- etc. to give[10] the corresponding substitution products (Fig. 5.1)

$$RCH_2Br + X^- \xrightarrow{\text{PEG.400}} RCH_2X$$

$$R = C_6H_5,\ C_3H_7 \text{ etc.},\ X^- = CH_3COO^-,\ I^-,\ CN^-$$

Fig. 5.1: Reaction of alkyl hallides with nucleophiles

(*ii*) Potassium thiocyanate in PEG-400 has been used as a nucleophilic reagent to substitute[11] alkyl halide of the type R_1–CHX – R_2 (Fig. 5.2).

$$R_1 - CHX - R_2 + CH_3COS^- \xrightarrow{\text{PEG-400}} R_1 - \underset{\text{SCOR}}{\underset{|}{CH}} - R_2$$

$$R_1 = C_6H_5, C_7H_{15}, C_9\,H_{19}, C_{10}\,H_{21}$$
$$R_2 = H, CH_3$$
$$X = Cl, Br, I, CH_2\,C_6\,H_4\,SO_2, CH_2\,SO_2$$

Fig. 5.2: Reaction of secondary alkyl halides with thiocyanate

(*iii*) Tertiary butyl chloride on treatment with water in PEG-300 undergoes hydrolyses to give[12] *tert* butyl alcohol. (Fig. 5.3)

$$(CH_3)_3\,C\,Cl + H_2O \xrightarrow{\text{PEG-300}} (CH_3)_3C - OH$$

Fig. 5.3: Hydrolysis of *tert.* butyl chloride.

(*iv*) 1-Chloro-2-methyl propene on reaction with water in PEG–425 gives[13] a mixture of alcohols (Fig. 5.4)

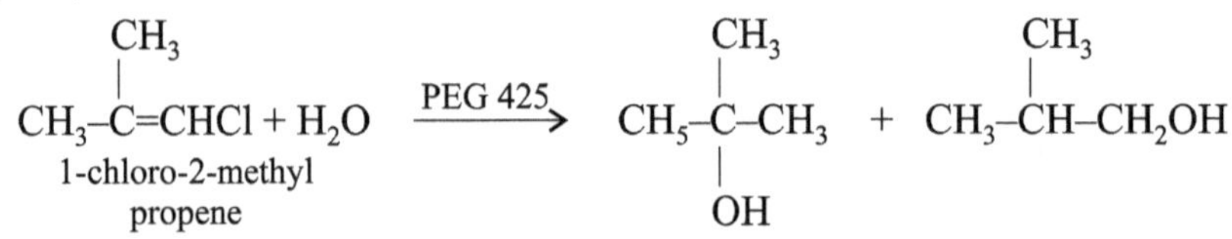

Fig. 5.4: The reaction of 1-chloro-2-methyl propene with water.

(*v*) The Diels-Alder reaction of 2,3-dimethyl -1, 3-butadiene with acrolein gave[14] the adduct in good yield. (Fig. 5.5)

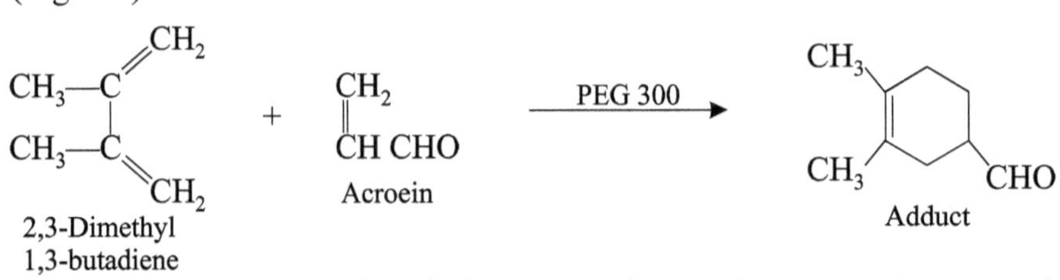

Fig. 5.5: Diet-Alder reaction.

In the above reaction there was 14 fold increase in yield compared to the reaction in methanol.

(***vi***) The aryl amines on diazotisation followed by Sandmeyer reaction gave [15] halogeno arenes and cyanoarenes (Fig. 5.6). The reaction could be carried out even in 10 mM dilute substrate

$$ArNH_2 \xrightarrow[\substack{\text{(2) Sandmeyers} \\ \text{Reaction} \\ \text{X or CN}^- \\ \text{PEG}}]{\text{(1) Diazotisation}} ArX \text{ or } ArCN$$

Ar = C_6H_4, *p*-OMe C_6H_4, o–OMe–C_6H_4–, *m*-OMeC_6H_4

Fig. 5.6: Sandmeyers reaction.

***(vii)* Heck Coupling reaction:** The reaction of aryl bromides with alkenes in molten liquid PEG-2000 at 80°C gave the coupled product in high yield[16] (Fig. 5.7). The reaction rates and regio and streo selectivities in PEG are comparable to that in the conventional organic solvents like DMF, DMSO, CH_3 CN and the ionic liquid, 1-butyl-3-methylimidazolium tetrafluoroborate. The PEG and the catalysed Pd $(OAc)_2$ can be recyclised by simple ether extraction of the product

$$RC_6H_4Br + CH_2 = CHX \xrightarrow[Pd(OAc)_2]{PEG-2000} R - C_6H_4 - CH = CHX$$

$$R = Cl, OH; X = C_6H_5, CO_2Et$$

Fig. 5.7: Heck coupling reaction.

The Heck coupling reaction forms the subject matter of a subsequent section (see section 5.6).

5.2.2 Oxidations

A number of oxidations can be conducted in PEG as solvent, some of which are given below.

***(i)* Oxidation of Benzyl bromide to aldehyde:** Potassium dichromate, being soluble in PEG-400, oxidises bromide to benzaldehyde (Fig. 5.8)[17]

$$\underset{\text{Benzyl bromide}}{C_6H_5CH_2Br} \xrightarrow[\text{Oxidation}]{K_2Cr_2O_7\,/\,PEG-400} \underset{\text{Benzaldehyde}}{C_6H_5CHO}$$

Fig. 5.8: Oxidation of benzyl bromide to benzaldehyde is PEG-400.

The above oxidation is similar to the reaction of $Na_2Cr_2O_7$ in hexamethylphosphoramide (HMPA) and crown ether.

***(ii)* Oxidation of alcohols to aldehydes:** The aerobic oxidation of alcohols is catalysed by a polyoxometalate. H_5PV_2 Mo_{10} O_{40} in PEG-200 or PEG-400 to give the corresponding aldehyde (Fig. 5.9)[18]

$$\begin{array}{l} R - CH_2OH \\ R = R'C_6H_4 \end{array} \xrightarrow[\text{PEG 200 or PEG-400}]{\substack{\text{aerobic oxidation} \\ \text{catalysed by } H_5PV_2MO_{10}O_{40}}} R - CHO$$

Fig. 5.9: Aerobic oxidation of alcohols to aldehyde.

In a similar way sulfides (R—S—R) could be oxidised to sulfoxides $(R–SO_2–R)$[18].

***(iv)* Oxidation of Terminal Alkenes:** Aerobic oxidation of terminal alkene is catalysed by $H_5PV_2Mo_{10}O_{40}$ and palladium to give ketones (Fig. 5.10)[18].

$$\underset{\text{Propene}}{CH_3CH = CH_2} \xrightarrow[\text{PEG-200}]{\substack{\text{aerobic oxidation catalysed} \\ \text{by } H_5\ PV_2\ Mo_{10}O_{40}}} \underset{\text{acetone}}{CH_3COCH_3}$$

Fig. 5.10: Aerobic oxidation to terminal alkenes

(v) Dihydroxylation of alkenes: The dihydroxylation of alkenes could be achieved using PEG-400 as solvent and OsO_4 as catalyst[19] (Fig. 5.11).

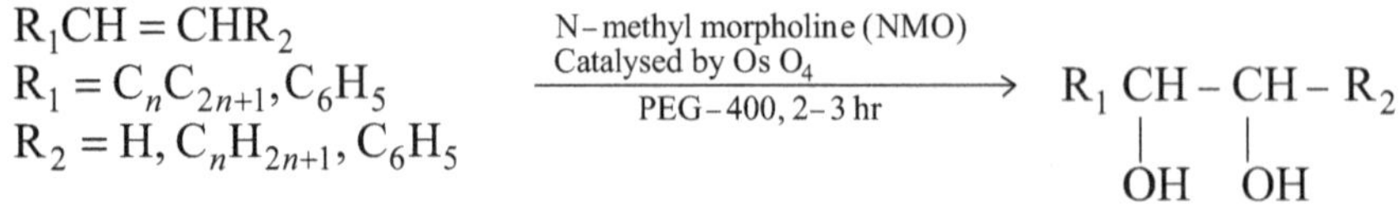

Fig. 5.11: Dihydroxylation of alkenes

In the above reaction the solvent (PEG-400) and the catalyst could be reused by simple extraction of the product using ether.

The products of dihydroxylation have a promient role in pharmaceuticals and fine chemicals[20]. In the earlier methods of dihydroxylation the catalyst invariably remained in the product.

However, in the present method (Fig. 5.11) no catalyst remained in the product. Additionaly the reaction is rapid and high yielding at very low concentration of $Os\,O_4$ (0.5 mol %).

Dihydroxylation of styrene[19] and α-methyl styrene[19] using $Os\,O_4$ (0.5 mol %) in PEG-400 using NMO (1.3 eq.) as the reoxidant gave the diols within 2 hr. in 94 and 97 percent yields (Fig. 5.12).

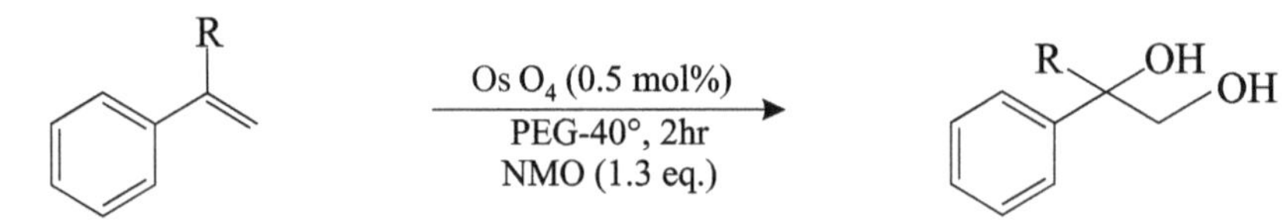

Styrene R = H
α-Methyl
Styrene R = Me

Diol R = H, 94%
Diol R = Me, 97%

Fig. 5.12: Dihydroxylation of styrene and α-methyl styrene

A number of other olefins were also dihydroxylated using the above procedure. These olefines include trans stilbene, trans ethyl cinnamate, cyclohexene and 4-methyl pentene.

The asymmetric dihydroxylation of olefins was carried out as per Sharpless procedure by using $(DHQD)_2$ PHAL (2 mol %) and $Os\,O_4$ (0.5 mol %). The results obtained are given below (Fig. 5.13)[19].

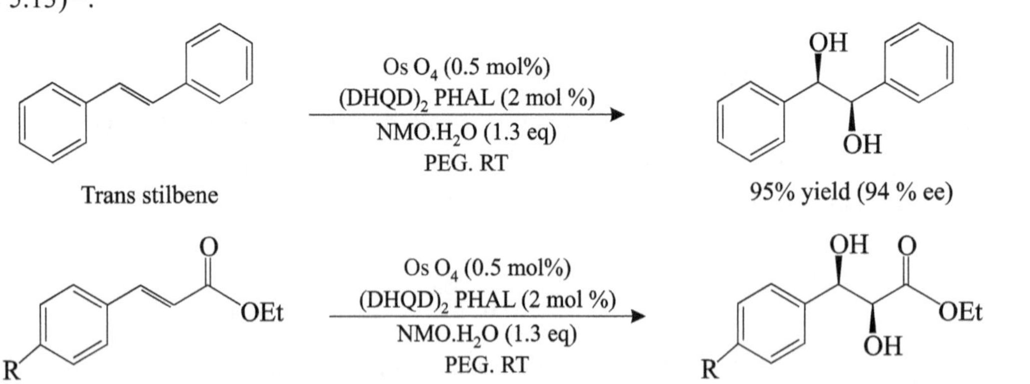

Ethyl cinnimate R = H
p-methoxyethyl cinnamate R = OMe

R = H yield 92% (91% ee)
R = OMe yield 94% (96 % ee)

Fig. 5.13: Asymmetric dihydroxylation of alkenes

5.2.3 Reductions

PEG has been successfully used as solvent for a number of reductions, some of which are given below.

***(i)* Reduction of Carbonyl compounds:** In PEG-400, carbonyl compounds could be reduced by $NaBH_4$ to the corresponding alcohols[21]. The reaction is much faster than that in THF (Fig. 5.14).

$$CH_3CO\,C_6H_{13} + NaBH_4 \xrightarrow{\text{PEG-400}} CH_3CH\,OH\,C_6H_{13}.$$

Fig. 5.14: Reduction of carbonyl compounds by $NaBH_4$

***(ii)* Reduction of esters:** The reduction of alkyl and aryl esters to the corresponding alcohols by $NaBH_4$ is considerably enhanced in PEG-400, whereas these compounds are considered to be inert in other organic solvents[22] (Fig. 5.15).

$$\begin{array}{l} R-COOR'+NaBH_4 \xrightarrow{\text{PEG-400}} RCH_2OH \\ R=\text{alkyl, aryl} \\ R'=CH_3, C_2\,H_5 \end{array}$$

Fig. 5.15: Reduction of esters by $NaBH_4$ in PEG

***(iii)* Reduction of secondary alkyl halides:** The reduction of secondary alkyl halides of the type R_1–CHX–R_2 by $NaBH_4$, in PEG-400 is quite comparable to that in polar protic solvents like DMSO, HMPA, DMF etc. (Fig. 5.16)[23].

$$\begin{array}{l} R_1CHX-R_2+NaBH_4 \xrightarrow{\text{PEG-400}} R_1-CH_2-R_2 \\ R_1=\text{alkyl, aryl} \\ x=Cl, Br, I \\ R_2=H, CH_3, C_4H_9 \end{array}$$

Fig. 5.16: Reduction of secondary alkyl halides by $NaBH_4$ in PEG

***(iv)* Reduction of acyl chlorides:** The reduction of acyl chlorides with $NaBH_4$ in PEG can be considered as a substitute for the inert solvent dioxane[24] (Fig. 5.17).

$$\begin{array}{l} R-COCl+NaBH_4 \xrightarrow{\text{PEG-400}} R-CHO \\ R=C_6H_5,\ p-BrC_6H_4 \end{array}$$

Fig. 5.17: Reduction of acyl chloride by $NaBH_4$ in PEG.

***(v)* Reduction of Triple bonds to double bond:** Lindlar's catalyst, Pd-$CaCO_3$ poisoned with PbO reduces alkynes to *cis*-defin in PEG-400[25]. The PEG and the catalyst could be used a number of times without loss of activity.

***(vi)* Reduction of Styrene:** The reduction of styrene using waxy solid PEG-900 at 155 bar pressure in Sc-CO_2 at 40° using the catalyst RhCl $(PPh_3)_3$ gave ethyl benzene. The ethyl benzene could be extracted into Sc-CO_2 and the catalyst containing PEG phase could be reused[26] (Fig. 5.18).

$$\underset{\text{Styrene}}{C_6H_5CH=CH_2+H_2} \xrightarrow[\text{RhCl }(PPh_3)_3]{\text{PEG-900}} \underset{\text{Ethylbenzene}}{C_6H_5CH_2CH_3}$$

Fig. 5.18: Reduction of styrene.

5.3 PEG AS PHASE-TRANSFER CATALYSTS (PTC)

Phase-transfer catalysts are known to transport an aqueous reagent into an organic phase in an activated phase and so the reaction can proceed[27]. In fact, PEG has the ability to serve as a PTC as the polyethylene oxide chain can form complexes with metal cations.[28] The PEG-metal cation complexes, in order to maintain electroneutrality bring an equivalent anion into the organic phase and thus make the anion available for the reaction with the reactants. A number of factors affect phase catalytic activity, such as PEG molecular weight, chain end effects, and the nature of the associated cations and anions.

PEGs are used in many commercial process in order to replace expensive and environmentally-harmful PTCs[28-32]. Amongst the commonly used PTCs, linear, PEGs are much cheaper than the analogous macrocyclic crown ethers and cryptands[33]. PEGs are in general more stable at high temperatures up to 150-200°C and also show higher stability to acidic and basic conditions than the quaternary onium salts[27].

Following are given some to the examples in which PEGs have been used as PTC.

5.3.1 Williamsons Ether Synthesis

Williamsons ether synthesis is an important nucleophilic substitution reaction (SN_2) involving synthesis of ethers using alkyl halides and an alkoxide in alcoholic solution (Fig. 5.19).

$$R-OH+R'X \xrightarrow{\text{PEG}} R-O-R'$$

Fig. 5.19: Williamsons ether synthesis

As an example, the reaction of decan-1-ol with *n*-butyl halide in presence of PEG gives the corresponding ether in 84% yields (Fig. 5.20)[34].

$$\begin{array}{l} C_{10}H_{21}OH + C_4H_9X \xrightarrow{\text{PEG-300-2000}} C_{10}H_{21}-O-C_4H_9 \\ X = Cl, Br, I \end{array}$$

Fig. 5.20: Synthesis of ether using PEG.

A number of other examples of Williamsons ether synthesis have been recorded[8].

5.3.2 Substitution Reactions using PEGs as PTC

As PTC, the PEGs are used in nucleophilic substitution reactions. The common anionic nucleophilic reagents include hydroxides, halides, sulfides, cyanides, cyanamides, carboxylates etc. As an example, the synthesis of 1,4-phenylene dioxydiacetic acid and diaryl 1,4-phenylene dioxydiacetate could be achieved using PEG-400 as PTC good yields (Fig. 5.21)[35].

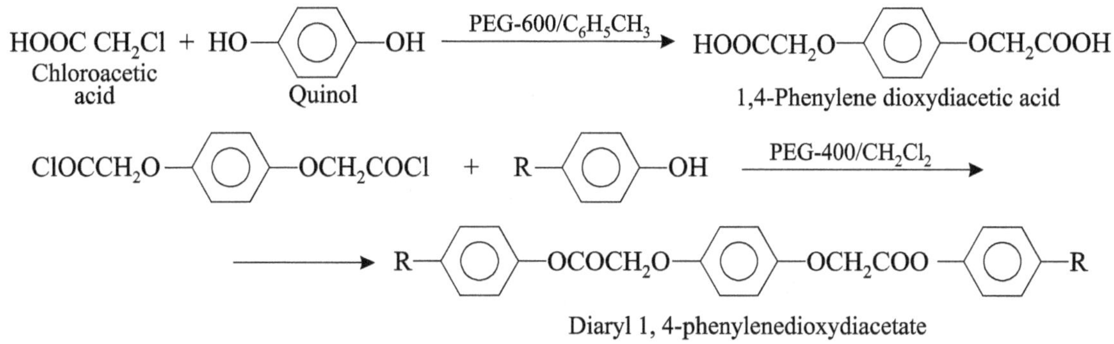

Fig. 5.21: A typical substitution reaction using PEG.

PEG as PTC has also been used in Sc–CO_2 solvent to convert benzyl chloride into phenylacetonitrile by treatment with KCN (Fig. 5.22). In this case, the yield is lower than in a similar reaction using tetraheptylammonium cyanide. However, the use of PEG as PTC is a green reaction[36].

$$\underset{\text{Benzylchoride}}{C_6H_5-CH_2Cl+KCN} \xrightarrow{\text{PEG/Sc–CO}_2} \underset{\text{Phenylacetonitrile}}{C_0H_5CH_2CN}$$

Fig. 5.22: Conversion of benzyl chloride into Phenylacetonitrile using PEG as PTC

Use of KCNS in the above reaction gives the corresponding sulphocyanide[36].

Mono and di-halobenzenes on treatment with alkoxide in PEG and KOH give[37] monoalkyl benzene (Fig. 5.23).

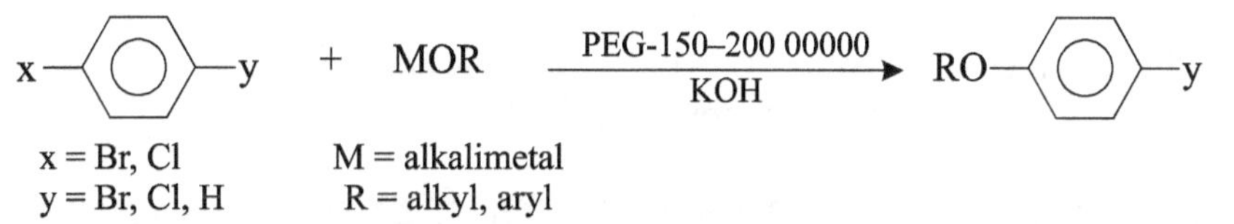

Fig. 5.23: Reaction of mono- and dihalobenzene with alkoxide using PEG as PTC.

The reaction of *o*-chlorobenzoyl chloride with ammonium thiocyanate using PEG-400 as PTC gave the corresponding thiocyanate (Fig. 5.24)[38].

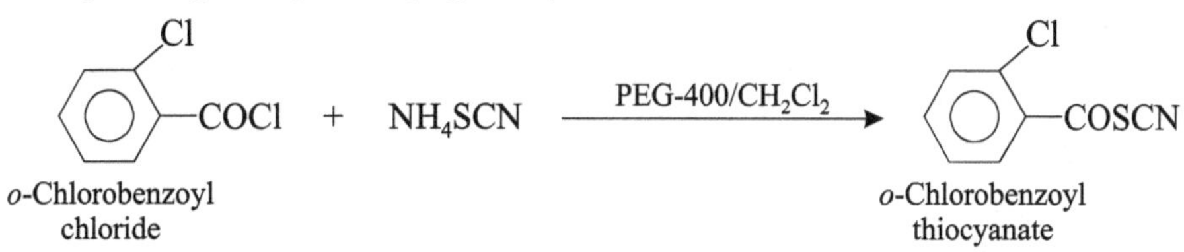

Fig. 5.24: Conversion of acid chloride into the corresponding thiocyanate using PEG as PTC

5.3.3 Oxidations using PEG as PTC

PEG has been used as PTC in a number of oxidations. As an example, the cobalt catalysed carbonylation of benzyl halides using PEG-400 is much cheaper than using quatenary ammonium salts, or crown ethers (Fig. 5.25)[39].

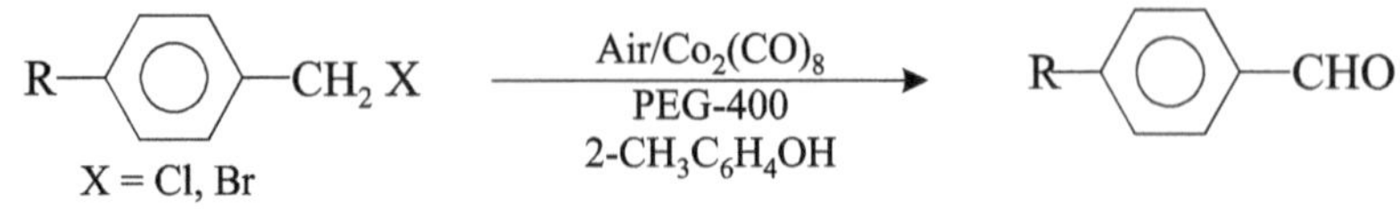

X = Cl, Br

Fig. 5.25: Cobalt catalysed carbonylation of benzyl halides.

The base catalysed oxidation of alkylnitro benzene using PEG as PTC in presence of ultrasound gave[40] *p*-nitrobenzoic acid (Fig. 5.26).

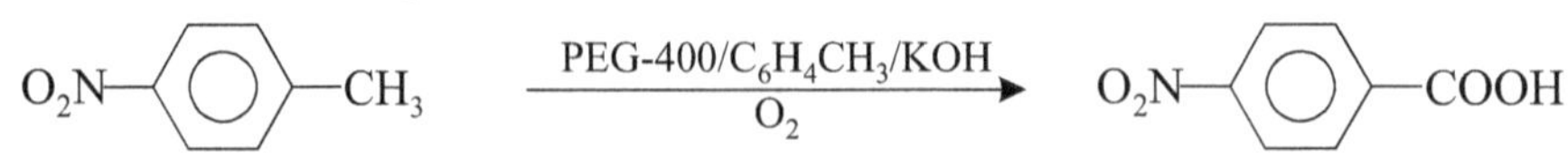

p-Nitrotoluene *p*-Nitrobenzoic acid

Fig. 5.26: Oxidation of p-nitrotoluene to p-nitrobenzoic acid.

Iodoalkanes and iodoarenes on oxidation with oxygen in PEG-400 and cheaper $CoCl_2$ in KCN–BF_3–Et_2O – $FeCl_2$ gave[41] carboxylic acids. The reaction conditions in this oxidations are much milder than that using platinum catalyst (Fig. 5.27)[41].

$$\underset{R=CH_3C_6H_4CH_3(CH_2)_n}{RI} \xrightarrow[\text{PEG-400/KCN/KOH}]{\text{air/CoCl}_2} RCOOH$$

Fig. 5.27: Oxidation of iodoalkanes or iodoarenes to carboxylic acid.

5.3.4 Reductions using PEG as PTC

PEG with $NaBH_4$ and PEG-$NaBH_4$ complexes are used for the reduction of[42-44] both aldehydes and ketones. Free PEG catalyses reduction of ketones by $NaBH_4$[44], whereas PEG-Na BH_4 derivative can reduce aldehydes selectively in presence of ketones[42].

1-Bromoalkyne can be reduced in PEG-400 which has been shown to be superior to the same reaction using the PTC benzyltriethylammonium salts.[45].

5.4 L-PRODINE CATALYSED ASYMMETRIC ALDOL REACTIONS

Aldol reaction is an effective C–C bond forming reaction. However, asymmetric version of aldol reaction has been useful for enantioselective C–C bond formation. It has been found that L-Proline catalysed asymmetric aldol reaction of acetone with aromatic and aliphatic aldehyde in PEG gives[46] asymmetric aldol products (Fig. 5.28).

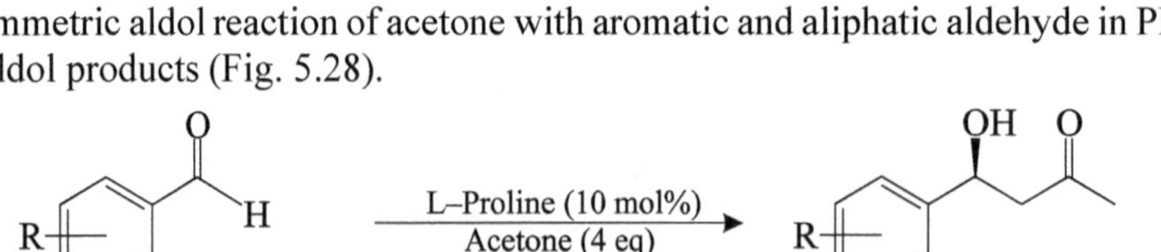

Aldehydes
R = H; 2-NO_2, 3-NO_2, 4-NO_2
4-Br, 2-Cl, 5-NO_2

Asymmetric aldol product
yield 80-90%, ee 60-70%

Fig. 5.28: Asymmetric aldol reactions of aromatic aldehydes.

The aliphatic aldehydes, isobutyraldehyde and cyclohexane carboxaldehyde have also been subjected to L-Proline catalysed aldol reaction to give the products is in 90 and 65% yield respectively and ee in 84% and 71% respectively (Fig. 5.29)[46].

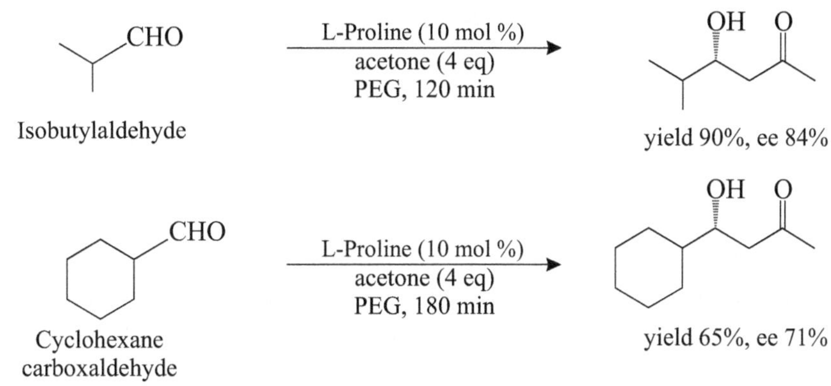

Fig. 5.29: Asymmetric aldol reaction of aliphatic aldehydes.

The above asymmetric aldol reactions are green reactions, since PEG (a biologically compatible product) is used as a solvent, which is much cheaper compared to ionic liquids and L-Proline is used in small concentration. Also, the solvent and the catalyst can be recycled.

5.5 L-PROLINECATALYSEDASYMMETRICTRANSFERALDOLREACTION

Transfer aldol reactions involve diacetone alcohol as a source of ketone (acetone). In these reactions Aldol-Tischeno product (A) is obtained[47] as byproduct in the usual Lewis acid catalysed reaction.

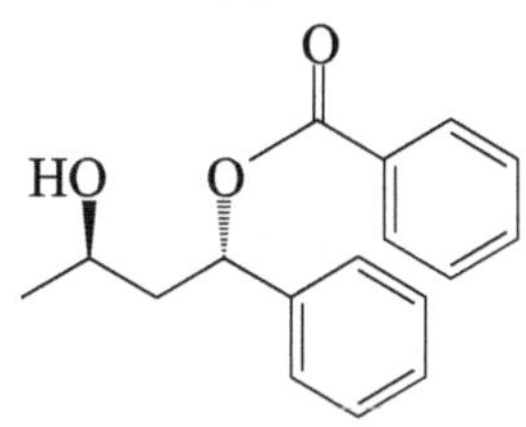

Aldol-Tischeno product (A).

However, transfer aldol reaction between diacetone alcohol and various aldehydes in presence of L-Proline gave the aldol product without the formation Tischeno-Aldol product A. A typical L-proline catalysed asymmetric transfer Aldol Reaction[46] is given in Fig. 5.30.

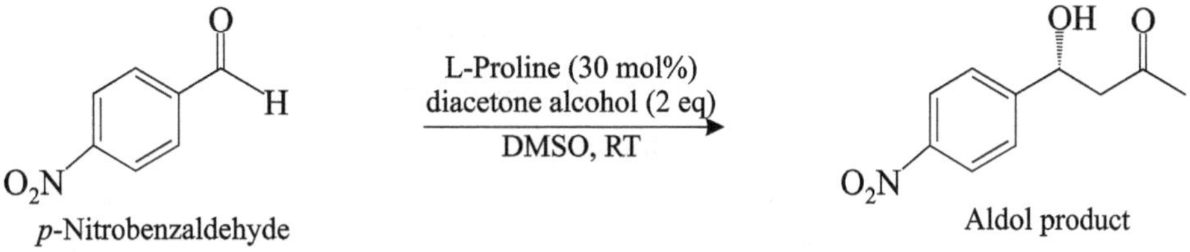

Fig. 5.30: L-Proline catalysed asymmetric transfer Aldol Reaction.

Similarly, other aldehydes like 3-nitro benzaldehyde, 2-chloro-5-nitrobenzaldehyde and 2-chlorobenzaldehyde gave the corresponding aldol products up to 88% yield and ee's upto 71%.

The aliphatic aldehydes, isobutyraldehyde and cyclohexane carboxaldehyde gave the aldol products in 80 and 96% yield respectively with ee 84% and 86% respectively. A plausible reaction mechanism involving a cyclic transition state (re-facial attack of aldehyde to proline) has been proposed.[46].

5.6 REGIOSELECTIVE HECK REACTION

PEG having molecular weight 2000 or lower have been used[48], as efficient reaction medium for Pd-catalysed C–C bond forming Heck reaction. The transformations are rapid, high yielding and easily recycled efficiently. It is found[48] that the stereo and regioselectivities are different than those with conventional solvents including ionic liquids. As an example, the reaction of bromobenzene with ethyl acrylate (1:1), Pd $(OAc)_2$ and TEA in PEG 2000 at 80° for 8 hr gave ethyl cinnamate in 90% yield and purity (Fig. 5.31).

However, reaction with electron rich vinyl butyl ether gave clear formation butyl styryl ether. This reaction when conducted in conventional solvents (DMF, DMSO, CH_3CN) gave[49] mixture of products.

The Heck reaction of bromobenzene with styrene gave transtilbene in 93% yield (Fig. 5.31).

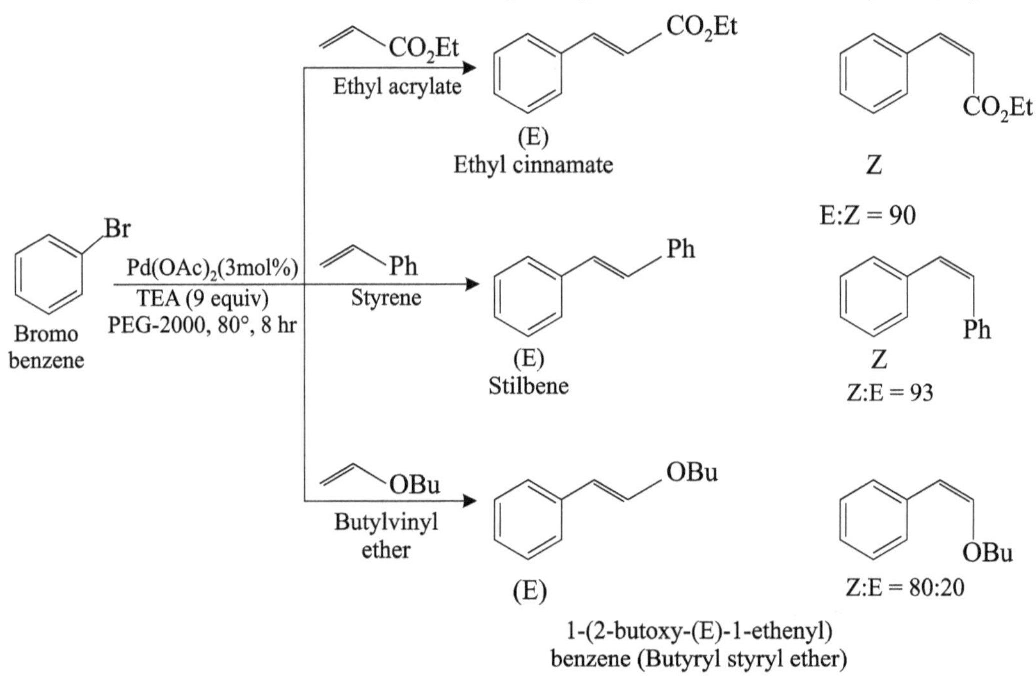

Fig. 5.31: Heck reaction of bromobenzene in PEG-2000

The Heck reaction of bromoanisol with the three olefins (*viz.* ethyl acrylate, styrene and butyl vinyl ether) gave excellent regio and steroselectives of 4-methoxyethyl cinnamate (E : Z = 91), 4-methoxy stilbene (E : Z = 85) and 1-(2-butoxy–(E)-1-ethenyl 4-methoxybenzene (E : Z = 82) respectively. In the case of napthyl bromide the products obtained had selectivities similar to that in bromobenzene. It was believed that PEG acted not only as an efficient solvent medium but also as a PTC for the formation C–C bond.

5.7 BAYLIS-HILLMAN REACTION

The Baylis-Hillman Reaction[50], is a C–C bond forming reaction between an aldehyde and an electron deficient olefin. In this reaction a new C–C bond between the aldehyde and α-carbon of the olfin is formed. The products obtained are used as synthons for other synthesis. The reaction is one of the reactions in which atom economy[51] is 100%. The draw back in the usual reaction is low yield, requirements of high concentration of catalyst (DABCO) and longer reaction times. Also, the reaction is inert to enones, α, β-substituted aldehydes and hindered aldehydes. A number of modifications[52] have been used including use of new catalysts, solvents like Sc-CO_2, ionic liquids, polar solvents, ultra sound and high pressure etc.

It has been found[52] that PEG-400 is a recyclable solvent for the rapid Baylis-Hillman reaction between unreactive aldehydes and activated olefins in presence of DABCO (Fig. 5.32).

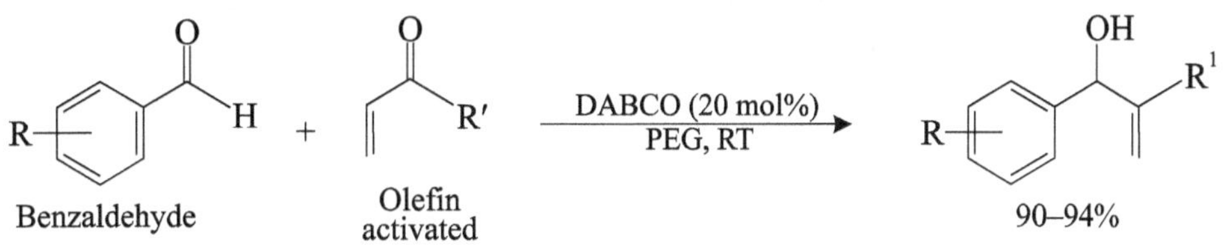

R	R^1
H	OEt
H	H
$4-NO_2$	OEt
4 – F	OEt
4 – OMe	OEt

Fig. 5.32: Baylis-Hillman reactions of benzaldehydes with activated olefins in presence of DABCO and PEG

In a similar way all the aldehydes (as mentioned in Fig. 5.32) reacted with acrylonitrile to give the Baylis-Hillman products (Fig. 5.33).

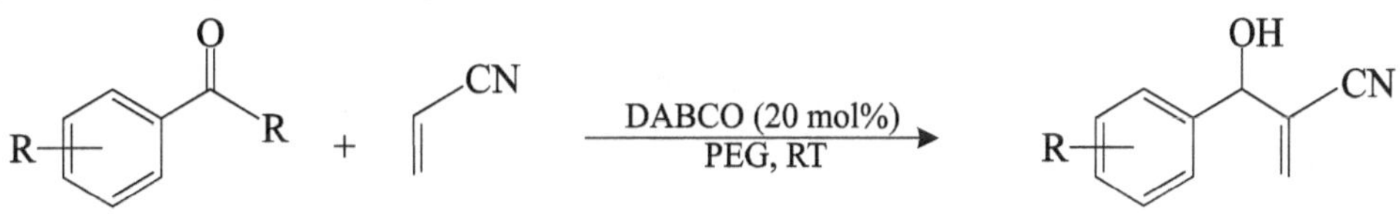

Fig. 5.33: Baylis-Hillman reaction of banzaldehydes with acrylonitrite in presence of DABCO and PEG

Other aldehyde like 2-chloro-5-nitro benzaldehyde, 2-furaldehyde, 2-thiophene carboxaldehyde, 3-phenyl propanal, isobutyraldehyde, hexanal and trans-Cinnamaldehyde also underwent Baylis-Hillman reaction in PEG in presence of DABCO[52].

It was found that PEG could be recycled and no additional catalyst was required for subsequent runs.

5.8 SUZUKI CROSS COUPLING REACTION IN PEG

The Suzuki cross coupling reaction is a Pd catalyzed C–C coupling reaction of organoboranes with aromatic aldehydes to give biaryls[53]. The biaryls have been used as pharmacophores in a number of different biologically active molecules[54].

The method[55] involved consist of the reaction of substituted aromatic bromides or iodides with aromatic boronic acid in PEG-400 in presence of potassium fluoride as a base. The reaction is conducted in a microwave oven for 50 sec. (Fig. 5.34).

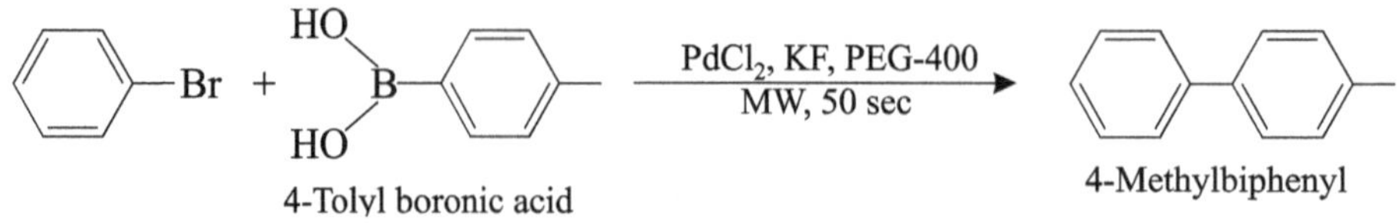

Fig. 5.34: Suzuking coupling reaction in PEG

In the above reaction, the formation of small amount of symmetrical biaryl due to self coupling of boronic acid[56] takes place.

The aliphatic aldehyde, 1-butyl iodide did not undergo reaction under the above conditions. However, the reactions could be conducted by heating in an oil bath at 100°C for 15 min.

It is found that the use of microwave oven is very convenient and clean. Using this procedure a variety of aromatic halides (Fig. 5.35), *p*-tolyl boronic acids (Fig. 5.36) could be used.

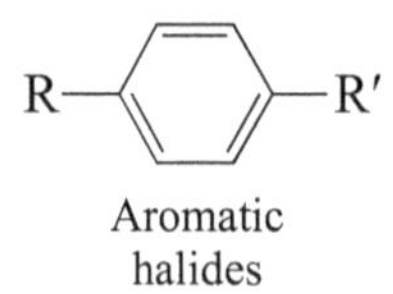

R	R
Br	Me
Br	CHO
Br	OMe
Br	COMe
I	Me
I	OMe
I	NO_2
I	F

Fig. 5.35: Aromatic aldehydes used for Suzuki coupling.

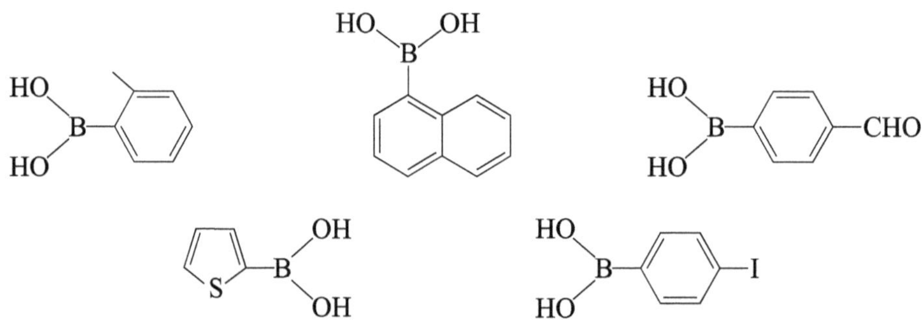

Fig. 5.36: Different boronic acids used in Suzuki coupling.

5.9 SYNTHESIS OF AZO COMPOUNDS USING PEG

Azo compounds are utilised as dyes, analytical reagents and also as materials for non-linear optics and for storage of optic information in laser dishes[57]. These have played a significant role in machanistic and synthetic organic chemistry[58]. The azo compopents have been found to possess excellent memory and photoelectric properties.[59].

A number of methods[60] are used for the synthesis a azo compounds. In most of the methods there is evolution of large amounts nitrogen oxides, which cause air pollution. A green route[61] for the synthesis of azo compounds has been developed. It involves the absorption of NO_2 in PEG-400 to form an adduct, PEG. NO_2. The absorption efficiency of NO_2 as determined by Saltzan method[62] is found to be about 97%. The concentration of NO_2 in PEG was found to be 10.53 mmol; this was determined by reaction the adduct (PEG. NO_2) with excess aqueous NaOH and then back titrating with HCl[63]. The adduct PEG. NO_2 has been proved to be efficient, clean and moderate oxidant (Fig. 5.37).

$$\text{PEG} + NO_2 \longrightarrow \text{PEG.}\ NO_2$$

Fig. 5.37: Formation of PEG. NO_2

Using the adduct PEG.NO_2, hydrazo derivatives are converted[61] to the corresponding azo compounds (Fig. 5.38).

$$\underset{\text{Hydroazo deriv.}}{R'NHNHR^2} + \text{PEG.}NO_2 \xrightarrow{20-30\ \text{min}} \underset{\text{Azocompound } 75-80\%}{R'N=NR^2} + N_2$$

$$R' = R^2 = p-MeC_6H_4$$

$$R' = m\text{-Me}\ C_6H_4;\ R' = \text{p-ClC}_6H_4$$

Fig. 5.38: Conversion of hydrazo derivatives into azo compounds by PEG. NO_2.

5.10 OXIDATION OF CYCLOHEXENE TO ADIPIC ACID WITH HYDROGEN PEROXIDE IN PRESENCE OF Na_2WO_4 USING PEG/$NaHSO_4$

Adipic acid is an essential intermediate for the manufacture of nylon-66. Most of the industrial production of adipic acid is by the oxidation of cyclohexene and or cyclohexane with HNO_3. These processes involve production of oxides of nitrogen which is a source of environmental pollution. Subsequently environmentally benign procedures were developed based on catalytic oxidation of cyclohexene with conc. aq. H_2O_2.

For elegant, environmentally benign synthesis of adipic acid from cyclohexene in polyethylene based aqueous biphasic system using sodium tungstate and H_2O_2 was developed[64] (Fig. 5.39)

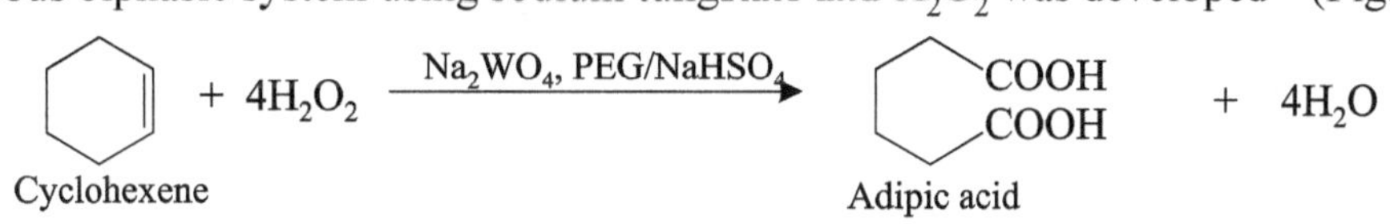

Fig. 5.39: Environmentally benign synthesis of adipic acid.

5.11 ENZYMATIC TRANSFORMATIONS IN PEG

A number of important enzyme catalysed biomass hydrolysis and biosynthesis including processes applicable to cellulose, antibiotics, starch etc. have been conducted in PEG aqueous two phase system. Following are given some of the important transformation.

(*i*) The biotransformation of cellulose to glucose, the enzymes for hydrolysis and enzyme recycling constitute the major portion of the cost. For the biotransformation of cellulose to glucose the enzyme Endo-β-glucanase and Exo-β-glucanase are used in PEG 4000 and 200000 and Dextrin (Fig. 5.40)[65-70].

$$\underset{(C_6H_{10}O_5)_n}{\text{Cellulose}} + \underset{\text{Exo-}\beta\text{-glucanase}}{\text{Endo-}\beta\text{-glucanse}} \xrightarrow[\text{Dextrin (MW.1,4,11,50,200} \times 10^4)]{\text{PEG}-4000,200000} \underset{(C_6H_{12}O_6)}{\text{Glu cos e}}$$

Fig. 5.40: Biotransformation of cellulose into glucose.

(*ii*) The biotransformation of starch or native starch to glucose can be accomplished[71-74] by using the enzyme α-amylase, glucoamylase and amyloglucosidase in PEG-6000, 20000 and Dextrin (M.W = 5.7×10^4) (Fig. 5.41).

$$\text{Starh or native starch} + \alpha\text{-Amylase, glucoamylase, amyloglicosidase} \xrightarrow[\text{Dextrin (MW 5-7}\times 10^4)]{\text{PEG6000,20000}} \text{Maltose + glucose}$$

Fig. 5.41: Biotransformation of starch or native starch into maltose and glucose

(*iii*) Cephalexin is synthesised[75–77] in ABS using penicillin G acylase as catalyst and 7-amino-deacetoxicephalosoporanic acid (7-ADCA) and phenyl glycine methyl ester (PGME) as substrate in 60% yield (Fig. 5.42)

$$\text{7-Amino-deactoxicephalosp organic acid (7-ADCA)} + \text{Phenol glycine methylester (PGM)} \xrightarrow[\text{PEG-400, MgSO}_4]{\text{Penicillin g-acylase}} \text{Cephalixin}$$

Fig. 5.42: Synthesis of Cephalexin

5.12 SOME MISCELLANEOUS REACTIONS IN PEG

5.12.1 Diels-Alder Reaction

The reaction of 2, 3-dimethyl-1, 3-butadine with nitrosobenzene in PEG-300 or PEG-425 gave 3.3 fold increase in the rate compared with dichloromethane and 2.5 fold increase compared with ethanol[78] (Fig. 5.43).

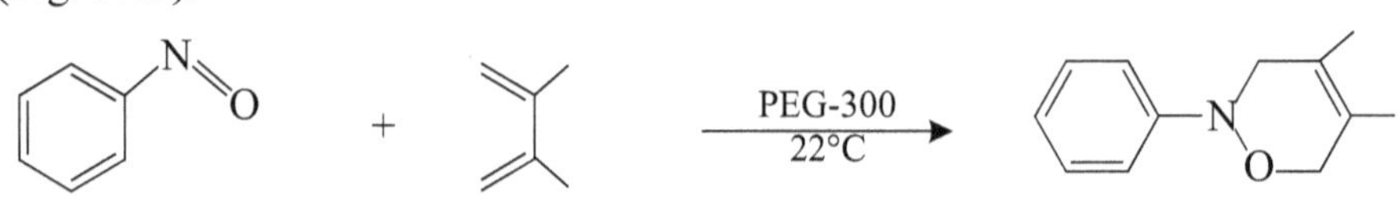

Nitroso benene

2,3-Dimethyl--1,3-butadiene

Fig. 5.43: Diels-Alder Reaction

5.12.2 Michael Addition Reaction

Michal addition proceeds efficiently in PEG in the absence of added catalyst[79] (Fig. 5.44).

Fig. 5.44: Michael addition of an amine to a conjugated alkene

It is however, found that for other Michael addition reactions in PEG, added catalyst are used. As an example 0.5 mol% $RuCl_3$ in PEG was used in **aza-Michael addition** of α, β unsaturated carbonyls with aliphatic and aromatic amines and also with thiols and carbamates[80].

Michael additions in PEG has also been used for the synthesis of N3-functionalized dihydropyrimidines using K_2CO_3 as catalyst[81].

5.12.3 Wacker Oxidation

Wacker oxidation of styrene in presence of CuCl/$PdCl_2$ in PEG/Sc-CO_2 gives ketones. However, in absence of $CuCl_2$ benzaldehyde is obtained (Fig. 5.45). The catalyst could be reused[82].

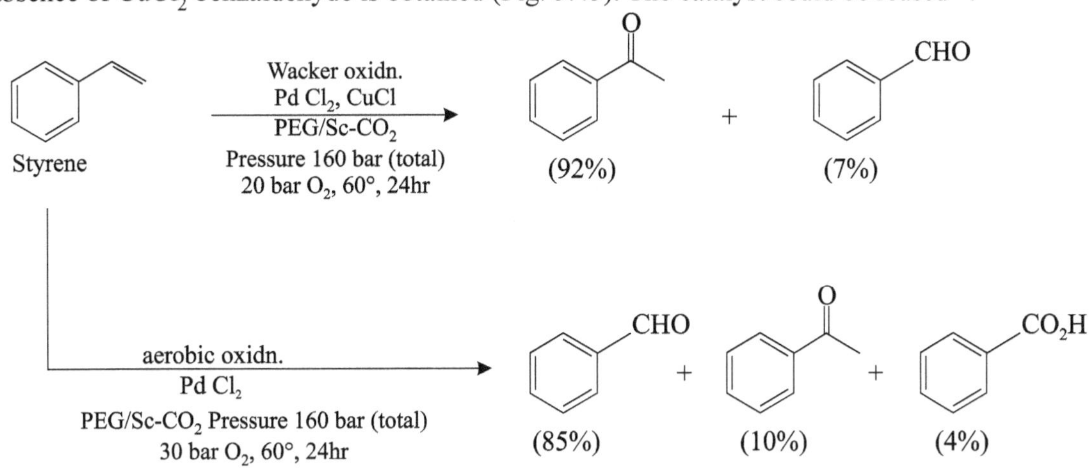

Fig. 5.45: Wacker oxidin. and aerobic oxidation of styrene.

Conclusion

Polyethylene glycol and its aqueous solutions has benign characteristics, is approved by FDA for internal consumption, is non volatile and biodegradable. It is considered to be a green solvent for future use. It has been used for a number is organic transformations. Besides it is also used as PTC is oxidations, reductions etc. PEG have also been used in L-Proline catalysed asymmetric aldol reaction, asymmetic dihydroxylations L-Prolines catalysed asymmetric trans aldol reaction, regioselective Heck reaction, Baylis-Hillman reaction, Suzuki cross compling reaction. PEG. NO_2 complex has been used for the conversion of –NH–NH– group to –N=N–group. It is also used for the manufacture of adipic acid and for a number of enzymatic transformations.

REFERENCES

1. F.E. Bailey, Jr., and J.V. Koleske, Poly (Ethylene Oxide), Academic Press, New York, 1976.
2. P.Å. Alberston, Partition of cell Particles and Macromolecules, Wiley, New York, 3rd Edn., 1986.
3. J.M. Harris in Polyetylene Glycol Chemistry, Biotechnical and Biomechanical applications, ed. J.M. Harris, Plenum Press, New York and London, 1992, P.7.
4. D.A. Herold, K. Keil and D.E. Bruns, Biochem Pharmacol., 1989, 38, 73.
5. M.G. Peglation, Cancer Treat. Rev., 2002, **28**, 13; D. Bhadra, S. Bhadra, P. Jain and N.K. Jain Pharamazil, 2002, **57**, 5.
6. Carbowax PEG-2000 MSDS of canadian cancer for occupational health and safety (CCOHS), Record Number 2426725, Oct., 2002, http://www.Safety. Vanderbilt edu/pdf/hes_msdv/carbowax PEG 200. Pdf.
7. P.T. Anastas, in clean solvents, Alternative Media for Chemical Reactions and Processing, ed. M.A. Abraham and L.Moens, ACS Symposium series 819. American chemical Society Washington, DC, 2002, P.1.

8. Cited in Ji Chen, S.K. Spear, J.G. Huddeleston and R. Rogers, Green Chem, 2005, 7, 64 and the references cited there in.
9. E. Santaniello, A Manzocchi and P. Sozzani, Tetrahedron Lett., 1979, **20**, 4581.
10. E. Santaniello in crown ethers and Phase transfer catalysts in polymer science, ed. L.J. Mathias and C.E. Carraher Jr, Plenum Press, New York, 1984, P. 397.
11. P. Ferravoschi, A.Foecchi, P. Griser Di., E. Santoniello and S. Trave, Synth. commun., 1987, **17**, 1569.
12. N.F. Leininger, R. Clontz, J.L. Gainer and D.J. Kirwan, in clean solvents Alternative media for Chemical Reactions and Processing, ed. M.A. Abraham and L. Moens. ACS Symposium series 819, American Chemical Society, Washington, D.C. 2002, P. 208.
13. L.F. Leininger, J.L. Gainer and D.J. Kirwan AIChE J, 2004, **50**, 511.
15. N. Suzuki, T. Azuma, Y. Kaneko, Y. Izawa, H. Tomioka and T. Nomoto, J. Chem. Soc., Perkin Trans., 1987, 1, 645.
16. S. Chandrasekhar, Ch. Narsihmulu, S.S. Sultana and N.R. Reddy, Org. Lett., 2002, 4399.
17. E. Santaniello, A. Manzoochi and Sozzani, Tetrahedron Lett., 1979, **20**, 4581.
18. A. Haimov and R. Neumann, Chem. consumur, 2002, 876.
19. S. Chandra Shekher, Ch. Narsihmulu, S.S. Sultana and N.R. Reddy, Chem. Commun, 2003, 1716.
20. E.J. Corey and J.P. Dittami, J. Am. Chem. Soc., 1985, **107**, 256; X. Lu, Z.Xu and G. Yang, Org. Process Res, Dev., 2000, 4, 575; L. Ahrgren and L. Sutin, Org. Process Res. Dev., 1997, **1**, 425; Z. –M. Wang and B. Sharpless, J. Org. Chem., 1994, **59.** 8302.
21. E. Santaniello, A. Manzocchi and P. Sozzani, Tetrahedron Lett., 1979, **20,** 4581.
22. E. Suaniello, P. Ferraboschi and P. Sozzani, J. Org. Chem., 1981, **46**, 4584.
23. E. Santaniello, A Fiecchi, A. Manzocchi and P. Ferrabosai. J. Org. Chem., 1983, **48**, 3074.
24. E. Santaniello, A. Fiecchi, A. Manzocchi and P. Ferraboschi, J. Org. Chem., 1983, **48**, 3074.
25. S. Chandrasekhar, Ch. Narsihmulu, G. Chandra Shekhar and T. Shyamsunder. Tetrahedron Lett, 2004, **45**, 2421.
26. D.J. Heldebrant and P.G. Jessop, J. Am. Chem. Soc., 2003, **125**, 5600.
27. S.D. Naik and L.K. Doraiswamy, AIChE J., 1998, **44**, 612.
28. G.E. Totten, N.A. Clinton and P.L. Matlock, J. Macromol, Sci., Rev. Macromol. Chem. Phys. 1998, **C38**, 77.
29. G.E. Totten and N.A. Clinton, J. Macromol Sci., Rev. Macromol. Chem. Phys., 1988, **C28**, 293.
30. C.M. Starks, C.L. Liotta and M. Halpen, Phase transfer catalyst. Fundamentals, Applications and Industrial Perspectives, Chapman and Hall, New York 1994, P. 158.
31. Phase-Transfer catalysts: mechanism and synthesis ed. M.E. Halpen, ACS sympossium series 659, American chemical society, Washington, DC, 1997.
32. J.M. Harris, N.H. Hundley, T.G. Shannon and E.C. Struck, in crown ethers and Phase-Transfer catalysis in polymer science ed. L.J. Mathias and C.E. Carraher, Jr., Plenum Press, New York, 1984, P.371.
33. C.B. Dartt and M.E. Davis, Ind. Eng. Chem. Res., 1994, **33**, 2887.
34. B. Abribat, Y. Le Bigot and A. Gaset. Synth. Commun, 1994, **24**, 2091.
35. T.B. Wei, J.C. Chen, X.C. Wang, Y.M. Zhang and L.L. Wang, Synth. Commun., 1996, **26**, 1447.
36. K. Chandler, C.W. Culp, D.R. Lamb, C.L. Liotta and C.A. Eckert, Ind. Eng., Chem. Res., 1998, **37**, 3252.

37. R. Neumann and Y, Sassan, Tetrahedron, 1983, **39**, 3437.
38. Y.M. Zhang, T.B. Wei and L.L. Wang. Synth. Commun., 1997, **27**, 751.
39. C. Zucchi, G. Palyi, V. Galamb, E. Sampar-Szerencsis, L. Marko, P. Li and H. Alber, Organometallics, 1996, **15**, 3222.
40. R. Neumann and Y. Sasson, Chem Commun. 1985, 616.
41. J.T. Lee and H. Alper, Organometallics, 1990, **9**, 3064.
42. J.R. Blanton, Synth. Commun, 1997, **27**, 2093.
43. J.R. Blanton, Rect. Funct. Polym., 1997, 33, 61.
44. B.G. Zupancic and M. Kokalj, Syn. Commun, 1982, **12**, 881.
45. P. Li and H. Alpert, J. Org. Chem., 1986, **51**, 4354.
46. S. Chandrashekhar, N.R. Reddy. S.S. Sultana and K.V. Reddy, Tetrahedron, 2006, **62**, 338.
47. L. Simpura and V. Nevalainen, Angew. Chem. Int. Ed., 2000, **39**, 3422; Tetrahedron Lett., 2001, **42**, 3905.
48. S. Chandrashekhar; Ch. Narsihmulu, S.S. Sultana and N.R. Reddy, Org. Letters, 2002, 25, 4399 and the refrences cited there in.
49. W. Cabri, I. Candiani, A Bedeschi and S. Penco, J. Org. Chem., 1992, **57**, 1481.
50. A. B. Baylis, M.E.D. Hillman, German Patent 2, 155, 113 (1972); CA, 1972, **77**, 34174 q; J.E. Drewis, G.H.P. Roos, Tetrahedron, 1988, **44**, 4653; D. Basavaiah. P.D. Rao R.S. Hyma, Tetrahedron, 1996, 52 8001; P. Langer, Angew chem Int. Ed., 2000, **39**, 3049.
51. B.M. Trost, Science, 1991, **254**, 1471.
52. Cited in S. Chandrasekhar, Ch. Narsihmulu, B. Saritha and S.S. Sultana, Tetrahedron Lett., 2004, **45**, 5865 and the references cited therein.
53. N. Miyaura and A. Suzuki, Chem. Rev.,1995, **95**, 2457; T. Ishiyama, H. Kizaki, T. Hayashi, A. Suzuki and N. Miyaura, J. Org Chem., 1998, 63, 4720; A.R. Martin and Y. Yang, Acta Chem. Scand., 1993, **47**, 221.
54. Y.F. Hallock, K.P. Manfredi, J.W. Blunt, J.H. Cardellina, M. Schaffer. K.P. Gulden, G. Bringmann, A.Y. Lee, J. Clardy, G. Francois and M.R. Boyd. J. Org. Chem., 1994, **59**, 6349.
55. V. V. Namboodiri and R.S. Varma, Green Chemistry, 2001, **3**, 146.
56. M. Moreno-Manas, R. Perez and R. Pleixats, J. Org. Chem. 1996, **61**, 2346.
57. H. W. Russ and H. Tappe, Eur. Pat. Appl., EP, 629, 667 (1994); T. Ikeda and 0. Tsutumi, Science 1995, **268**, 1873; X.J. Peng and J.Z. Yang, Chim. Image Sci. Practise 1998, 4, 5.
58. R.D. Little and M. Y. Venegas, J. Org. Chen., 1978, **43**, 2921; A.B. Hashim, A. J. Elie and C. Noel, Tetrachedran Lett., 1996, **37**, 2951.
59. Z.F. Liu, K. Hashimoto and A. Fujishima, Nature, 1990, **347**, 658.
60. R.D. Little and M. G. Venegas, Org. Synth. 1983, **61**, 17; M.A. Brimble, G.H. Heathcock and G. N. Nobin, Tetrahedron Asymmetry, 1996, **7**. 2007; F.S, Loew and C.W. Weiss, J. Heterocycl. Chem., 1976. **13**, 826.
61. R.-Z. Qiao, Y. Zhang, X-P, Hui, P. – F. Xu, Z. – Y. Zhang, X. –Y. Wang and Y. – L. Wang, Green Chem., 2001, **3**. 186.
62. R.W. Boubel, Foundation of Air Pollution, Academi Press, Inc., New York, **2nd** ed, 1992, P. 765.
63. Q. Z. Zhang, R.Z. Qiao and S.S. Zhang, Hua Xue Tong Bao (China), 1994, **4**, 55.

64. J. Chen, S.K. Spear, J.G. Huddleston, J.D. Holbreg, R. P. Swatloski and R.D. Rogers, Ind. Eng. Chem. Res, 2004, **43**, 5358.
65. F. Tjerneld, I. Persson, P.Å. Albortsson and B. Hahn-Hägerdal, Biotechnol. Bioeng. 1985, **27**. 1036.
66. F. Tjerneld, I. Persson, P. Å-Albertson and B. Han-Hägerdal, Biotechol. Bioeng., 1985, **27**, 1044.
67. I. Persson, H Stalbrandl, F. Tjerneld and B. Hahn-Hagerdel, Appl. Biochem. Biotechnol. 1992, **27**, 27.
68. C.F. Mandenius, B. Nilsson, I. Persson and F. Tjerneld, Biotchnol. Bioeng., 1988, **31**, 203.
69. F. Tjerneld, I. Persson, P.Å. Albertsson and B. Haln-Hägerdel, Biotechnol. Bioeng. Symp, 1986. **15**, 419.
70. F. Tjerneld, I. Persson and J.M. Lee, Biotechnol. Bieng, 1991, **37**, 876.
71. A.Kondo, T. Urabe and J. Higahitani. J. Ferment Bioeng. 1994, 77, 700.
72. M. Larsson, V. Arasaratnam and B. Mattiasson, Biotechnol. Bioeng, 1989, **33**, 758.
73. M.Li, J. W. Kim and T.L. Peeples, Biochem. Eng., 2002, **11**, 25.
74. M.Li, J.W. Kim and T.L. Peeples, J. Biotechnol. 2002, **93**, 15.
75. D.Z, Wei, J.H. Zhu and X. J. Cao, Biochem. Eng. J., 2002, **11**, 95.
76. L.C. Liao, C.S. HO and W.T. Wu, Process Biochem, 1999, **34**, 417.
77. E. Anderson, B. Mattiasson and B. Hähn- Hägendal, Enzyme Microb Technol, 1984, **6**, 301.
78. N.F. Leininger, R. Clontz, J.L. Gainer and D.J. Kirwan, Chem. Eng. Commun., 2003, **190**, 431.
79. R. Kumar, P. Chaudhary, S. Nimesh and R. Chandra, Green Chem., 2006, **8**, 356.
80. H. X. Zhang, Y.H. Zhang, L.F. Liu, H.L. Xu and Y.Y Wang, Synthesis, 2005, 2129.
81. X.C. Wang, Z. J.Quan and Z. Zhang, Tetrahedron, 2007, **63**, 8227.
82. J.Q. Wany, F. Cai, E. Wang and L.N. He, Green Chem., 2007, **9**, 882.

6. TRANSFORMATIONS IN FLUOROUS SOLVENTS

6

Transformations in Fluorous Solvents

6.1 INTRODUCTION

Fluorous solvents are mostly perfluorohydrocarbons like perfluorohexane, perfluoroheptane, perfluoromethyl cyclohexane and perflorodecalin. Perfluorotributylamine is also used as a fluorous solvent. These fluorous solvents are useful in synthetic organic chemistry[1]. It was only in 1994, that the concept of a fluorous biphasic system (FBS) and fluorous biphasic catalyst (FBC) was developed[2]. The FBC is based on catalysts or reagents equipped with perfluorinated carbon chains and are soluble in fluorous solvents as FC-72 (a mixture of perfluorohexanes). The fluorous solvents are virtually insoluble in common organic solvents at ambient temperature and form biphasic mixture in presence of organic solvents like toluene or dichloromethylene. The miscibity increases with increase of temperature.[3] The organic substrate are dissolved in organic phase and the perfluoro-tagged catalyst or reagents in the fluorous solvent added. At higher temperature the biphasic heterogeneous mixture becomes monophasic permitting the reaction to take place. After cooling, the two phases (organic phase and fluorous phase) are reformed. The formed product is recovered by decantation of the upper organic phase. The catalyst remains in the lower fluorous phase and can be reused. The FBS technique (as stated above) is comparable with the solid phase organic chemistry (SPOC, Fig. 6.1), which is widely used pharmaceutical industry.

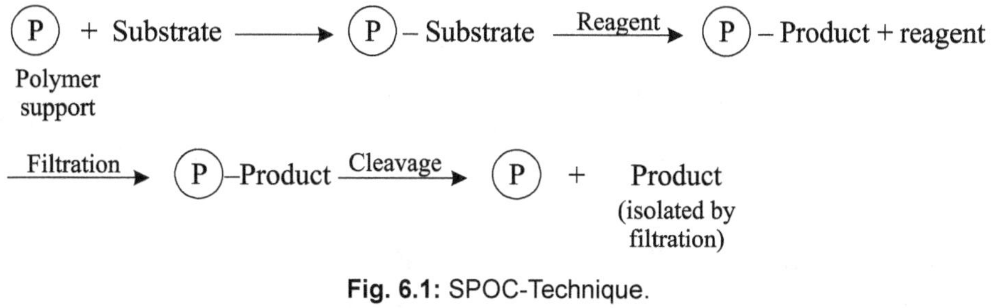

Fig. 6.1: SPOC-Technique.

Both SPOC-technique and FBS-technique offer similar opportunities for the separation of product/catalyst. However, the FBS-technique (Fig. 6.2) seems to offer advantage over the SPOC-technique.

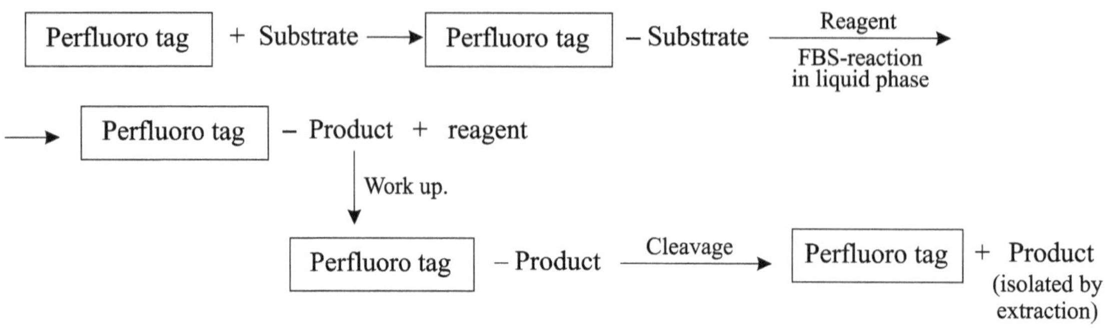

Fig. 6.2: FBS-Technique.

6.2 CHARACTERISTICS OF FLUOROUS LIQUIDS

The fluorous liquids are attractive alternatives to conventional organic solvents. This is due to some unique properties which they exhibit. They are particularly inert towards chemical treatments, are non-polar and have high thermal stability and so reactions even under drastic conditions can be conducted if necessary[4]. The boiling points of fluorous alkanes are in general lower than the corresponding alkanes; this is due to decreased van der Waals interactions. Their density is higher than water and most of the organic solvents. The commercially available perfluorous alkanes also contain isomers and small amounts of their homologues. Physical properties of commonly available perfluorinated solvents are given in Table 6.1.

Table 6.1: Physical Properties of Some of the Commonly available fluorinated Solvents

Perfluorinated Solvent	M.P. (°C)	B.P (°C)	Density (g/cm^2)
Perfluorohexane (C_6F_{14})	– 87	57	1.68
Perfluoroheptane (C_7H_{16})	–78	82	1.73
Perfluoromethyl Cyclohexane (C_7F_{14})	– 45	76	1.79
Perfluorodecalin ($C_{10}H_{18}$)	– 10	142	1.95
Perfluorotributyl amine ($C_{12}F_{27}N$)	–50	173	1.88

Data taken from Reference 5 and 6.

6.3 FLUOROUS LABELLING OF ORGANIC MOLECULES

Fluorous labelling of organic molecules (*viz*., catalyst, reagent or the product) is needed in order to realize the earlier mentioned fluorous biphasic system (FBS). The solubility of a compound in perfluorous solvents can be incrased by covalently attaching perfluoroalkyl chains to the molecule. During planning of a reaction procedure involving fluorous chemistry, it is necessary to remove the covalently bound fluorous tag from the desired product without effecting the organic moity chemically. This is referred to as phase-switching (Fig. 6.3). For such a phase switching, mild and efficient protocols are required.

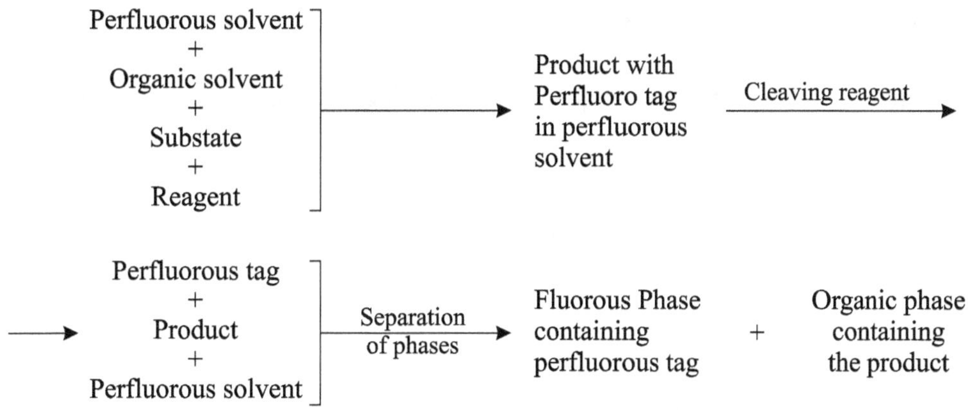

Fig. 6.3: Phase switching using fluorous tag.

A method commonly used is Fluorous Reversed Phase Silica Gel aided work up procedure (PRPSG)[7,8]. In this procedure, interactions between the perfluoroalkyl chains on the surface of the gel and the fluorous ponytail at the labelled molecule lead to retardation–the principle –which is applied for the purification of a single compound or isolation of various molecules with fluorous tag of different length by HPLC (fluorous mixture synthsis, Fig. 6.4)[9–13].

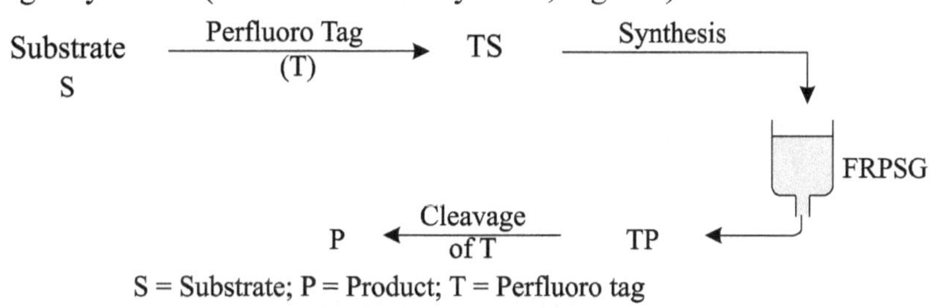

Fig. 6.4: Synthesis using FRPSY.

The above procedure was used to isolate perfluoro-alkyl compounds, products of radical allylations of perfluoroalkyl iodides (Fig. 6.5).

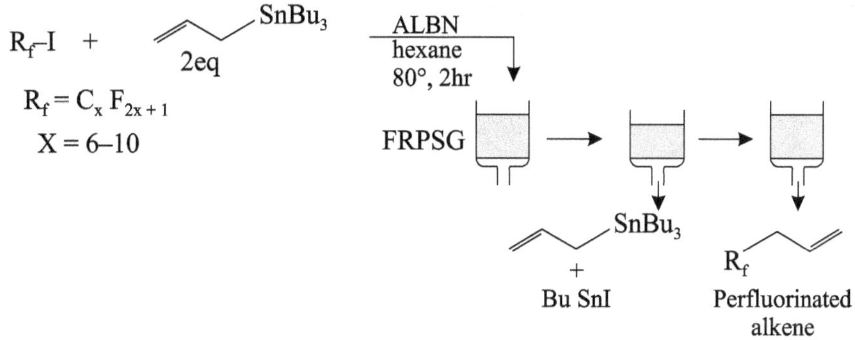

Fig. 6.5: Synthesis of perfluorinated alkene including workup by filtration through FRPSG

6.4 PERFLUORINATED CATALYSTS

A number of perfluorinated catalysts involving transitions metals have been developed for use in organic synthesis involving FPS technique. Some of such catalyst are given in Fig. 6.6.

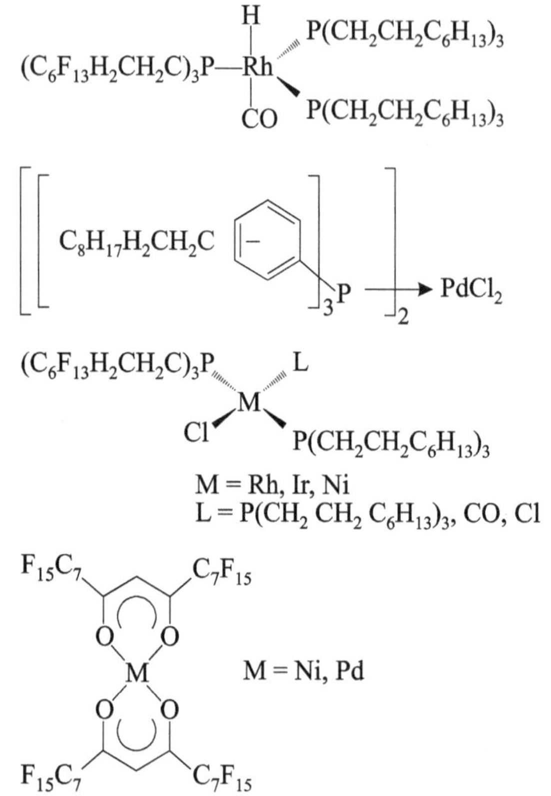

(A)

(B)

(C)

(D)

Fig. 6.6: Perfluorinated catalysts

6.5 SOME APPLICATIONS OF FLUOROUS PHASE TECHNIQUE

(*i*) **Hydroformylation of Olefins** using the fluorous biphasic catalyst (FBC) in triphasic system [perfluoro (methyl cyclohexane) (PFMC)/toluene, CO, H_2], which became biphasic at elevated temperaturer gave the keto products (Fig. 6.7)[14–17].

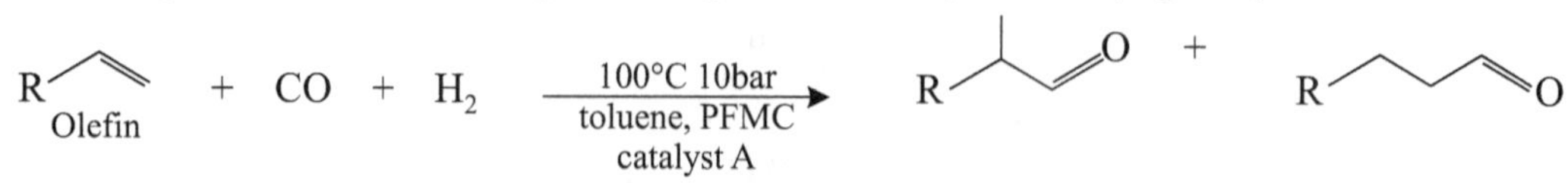

Fig. 6.7: Hydroformylation of olefins

(*ii*) **Oxidation of aldehydes** could be carried out due to affinity of fluorous compounds for molecular oxygen gave the corresponding aurboxylic acids (Fig. 6.8).

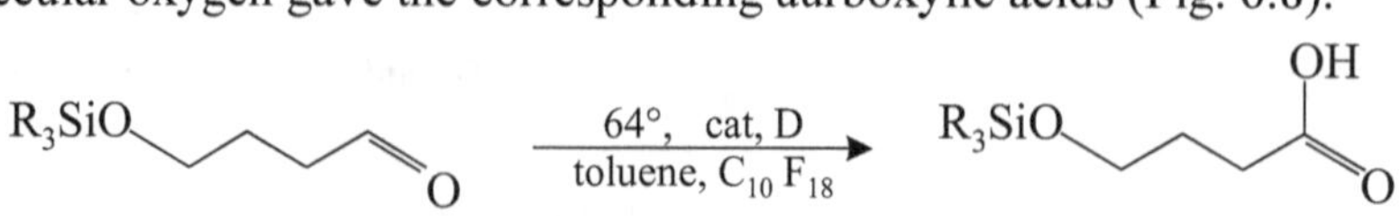

Fig. 6.8: Oxidation of aldehydes to carboxylic acids.

(*iii*) Palladium-catalysed carbon-carbon cross-coupling reactions like Heck, Still, Negishi and Sonogashira reactions could be carried out[18-22] under fluorous conditions. Recently, reaction Kits for Pd-catalysed C–C cross-coupling reactions in FBS have been commercially available.

(*iv*) Friedel-crafts acylation could be carried out in FBS using fluorous lanthanide methides[23-24], and non-flurorous $ZnCl_2$ used as catalyst in perfluorotriethylamine replacing toxic sym-tetrachloroethane[25].

(*v*) Enantioselective alkylation of benzaldehyde in presence of chiral binaphthols and arylzincthiolates FBC-conditions[26-28].

(*vi*) The Grignard reaction of an aldehyde using fluorous phase technique give secondary alcohols (Fig. 6.9). The grignard reaction was worked up by double triphase (fluorous/organic/aqueous) extraction comprising of the fluorous scavenging reaction of the alcohol followed by cleavege of the perfluoro moiety by the addition of CsF. The fluorinated silyl tag remains in the fluorous phase, the desired product is recovered from the organic phase (by evaporations). In this procudere (Fig. 6.9), the recycling of the fluorous tag is not possible – a disadvantage in this case.

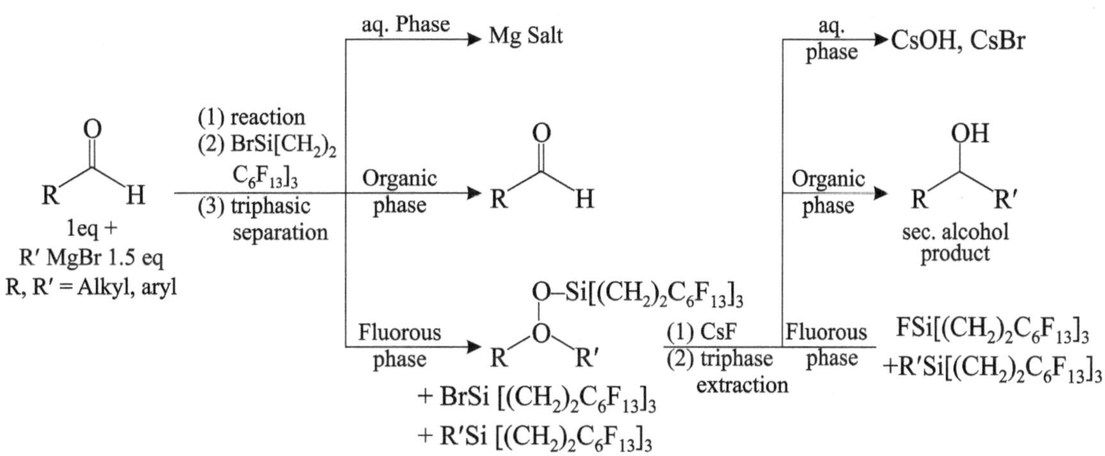

Fig. 6.9: Grignard. Reaction-perfluorotagging supported extractive work-up

(*vii*) Synthesis of tetrazoles by the reaction of nitriles with perfluoroazides. The desired tetrazole could be isolated by fluorobiphasic extraction. The procedure is highly chemoselective and is an excellent tool for the purification at the end of a multi step synthesis (Fig. 6.10)[29]

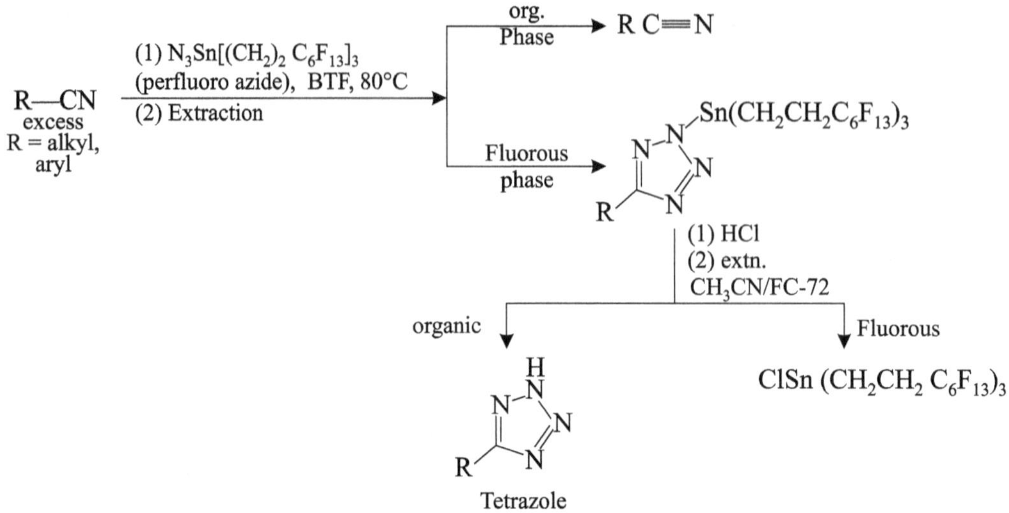

Fig. 6.10: Synthsis of tetrazoles

(*viii*) N-Alkylation of primary and secondary amines could be achieved by reaction with excess alkyl bromides; the remaining alky bromide was removed by filtration over FRPSG after scavenging with perfluoroalkyl thiol (Fig. 6.11).[30, 31]

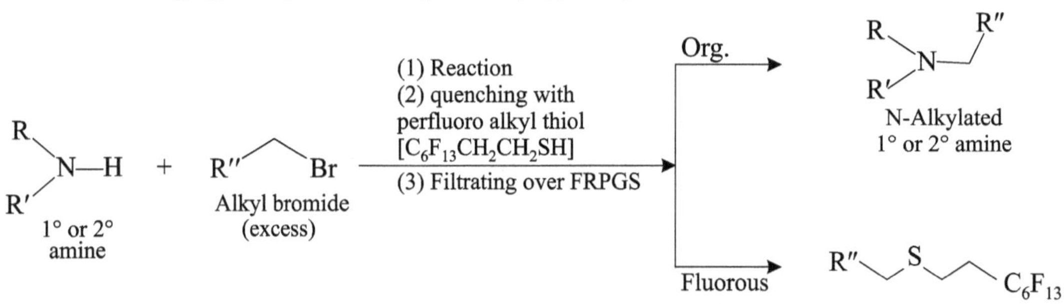

Fig. 6.11: Synthsis of N-Alkylated 1° or 2° amines.

(*ix*) **Still coupling** has been achieved (Fig. 6.12) with perfluoro-tagged tin compounds[32,33].

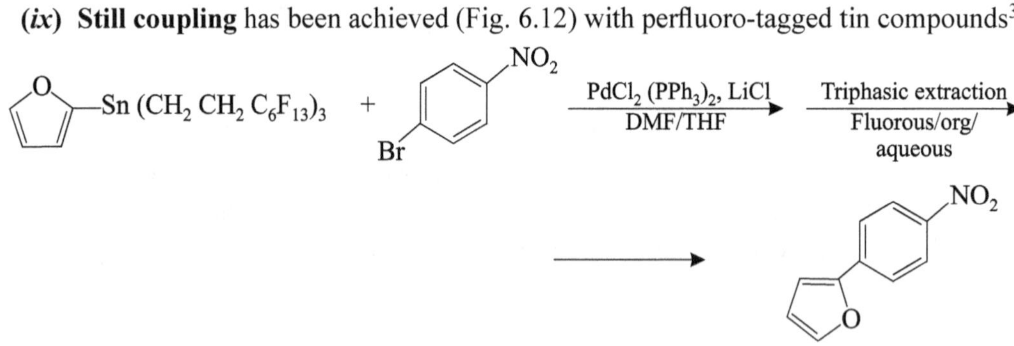

Fig. 6.12: Still coupling using perfluoro-tagged tin-compounds.

(*x*) 3H-Quinazoline-4-ones are prepared by **aza-Witting reaction** using fluorous triphenyl phosphine in order to remove the corresponding oxides by liquid-solid-extraction using FRPSG (Fig. 6.13)[34].

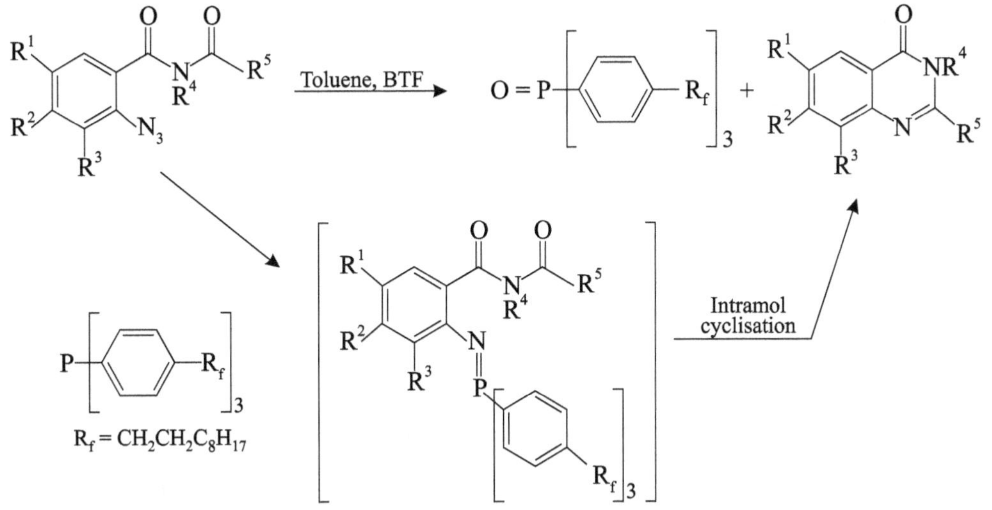

Fig. 6.13: Aza-Wittig reaction using trifluoro-triphenyl phosphine.

(*xi*) **Mitsunabu Reaction:** The reaction of 3,5-dinitrobenzoic acid with ethanol in presence of perfluorinated Mitsunabu reagents (A) and (B) gave the corresponding ester in 92% yield. The reagents (A) and (B) are helpful in removing triphenyl phosphine oxide and the hydrazide, which are formed as byproducts after reaction (Fig. 6.14)[35, 36]. The byproducts are retained on FRPSG after filtration and the non-fluorous products get directly eluted. The original reagents (A) and (B) could be regenerated by treatment with AlH_3 and Br_2.

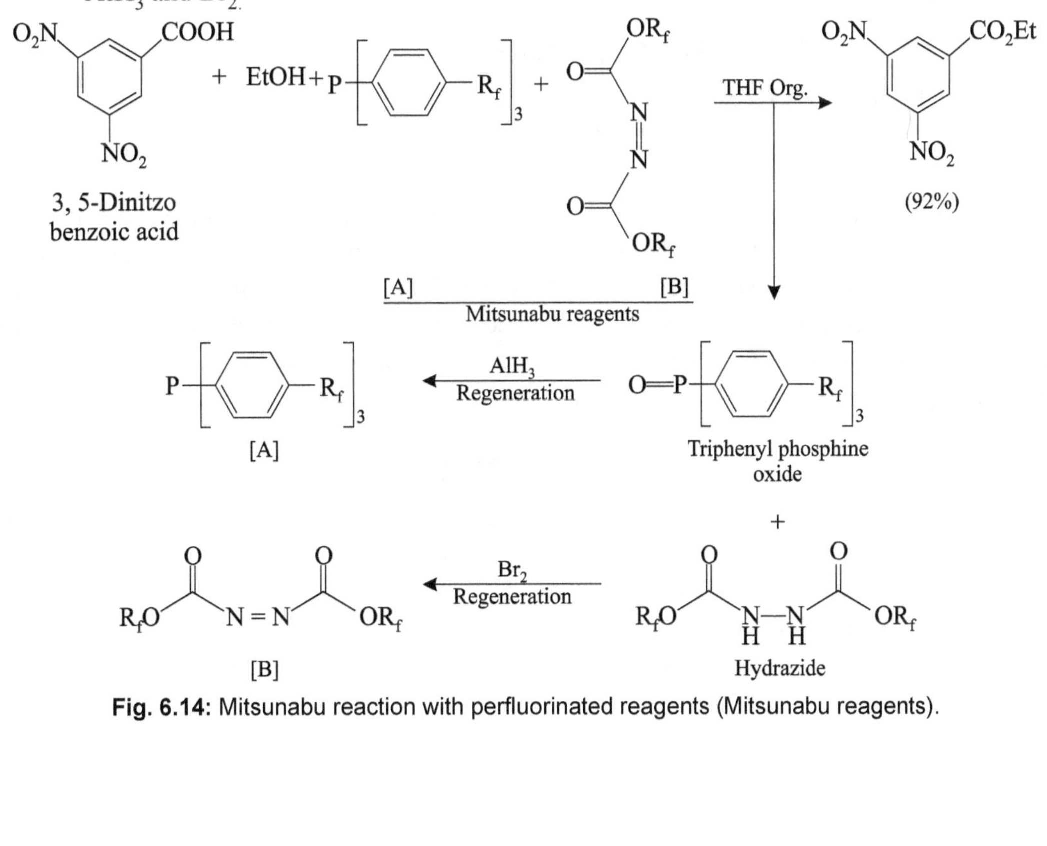

Fig. 6.14: Mitsunabu reaction with perfluorinated reagents (Mitsunabu reagents).

(*xii*) **Synthesis of perfluorinated sulfides and sulfoxides:** Perfluorinated sulfides and sulphoxides were synthesised[37] as given in Fig. 6.15.

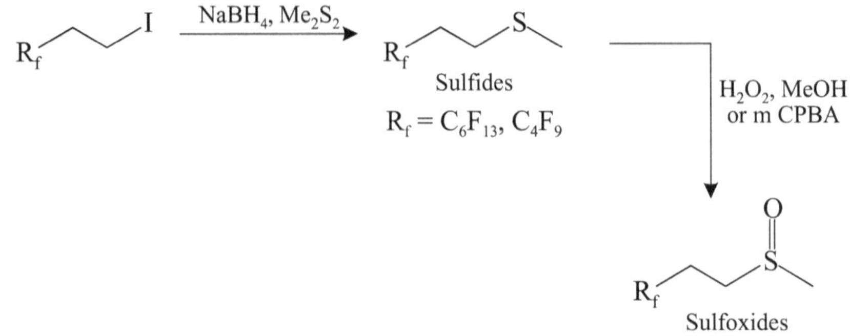

Fig. 6.15: Synthsis of perfluoroalkyl sulfides and sulfoxides.

Using the sulfoxides, a number of 1° and 2° alcohols could be converted into the corresponding aldehydes and ketones (**Swern oxidation and Corey Kim oxidation**).

Perfluorohexane has been used as a replacement of CCl_4 in bromination and as a 'universal' medium for suspension polymerisation[38]. It can be used with monomers which contains water – sensitive groups like an isocyante, chlorocarbonyl and trimethoxysilyl. Another fluorous solvent, perfluoroethylamine finds use as a solvent for Lewis acid catalysed reactions[39]. Benzotrifluoride has been used as a substitute for methylene chloride[40]. However, the potential of these fluorous solvents should be checked before using on a large scale. It has been found that tetrafluoromethane is a very potent greenhouse gas.

6.6 ENZYMATIC TRANSFORMATIONS

See Chapter 13, Section 13.2.5.

Conclusion

Perflucorinated hydrocarbons are unique solvents of the future. These have interesting applications in organic synthesis. Being immiscible with water and common organic solvents, they represent a third liquid phase–the fluorous phase. Since O_2 is soluble in these solvents, selective and efficient oxidations under mild conditions could be achieved. Individual components of reaction mixtures, which bear perfluoroalkyl substituents can be selectively extracted into the fluorous phase. This, in fact, is the basis of the so-called fluorous synthesis.

REFERENCES

1. Cited in D. Clarke, M.A. Ali, A.A. Clifford, A. Parratt, P. Rose, D. Schwinn, W. Bannwarth and C.M. Rayner, Current Topics in Medicinal Chemistry 2004, **4**, 729.
2. I.T. Hovath and J. Rabau, Science, 1994, **266,** 72.
3. P.L. Nostro, Adv. Colloid Interface Sci, 1995, **56**, 245.
4. W. Keim, M. Vogt, P. Wasserscheid, and B, Driessen-Holscher, J. Mol Catalysis A; Chemical 1999, 139, 171.

5. B. Betzemeier and P. Kochel, Top. Curr. Chem. 1999, **206**, 61.
6. L.P. Barthel-Rosa and J.A. Gladysz, Coord Chem, Rev., 1999, 109, 587.
7. C.C. Tzschucke, C. Market and H. Glaz, Argew. Chem. Int. Ed. 2002, **41** 4500.
8. S. Kairz, Z. Luo, D.P. Currian and W. Leitner, Synthsis, 1998, 1425.
9. D.P. Curran and Z. Luo, J. Am Chem. Soc., 1999, **121**, 9069.
10. Q. Zhan, L. LUO and D.P. Curran, J. Org. Chem. 2000, **65**, 8866.
11. D.P. Curran and S. Hadida, J. Am. Chem Soc., 1996, **118**, 2531.
12. D.P. Curran and Y. Oderaotoshi, Tetrahedron, 2001, **57**, 5253.
13. Z. Luo, Q. Zhang, Y. Oderatoshi and D.P. Curran, Science, 2001, **291**, 1766.
14. B. Richter, B.J. Deelman and G. V. Koten, J. Mol. Catalysis, A; Chemical 1995, **145**, 317.
15. J.J.J. Juliette, I.T. Horvath and J.A. Gladysz, Angew Chem. Int. Edn., 1997, **109**, 1610.
16. J.J.J Juliette, D. Rutherford, I.T. Horvath and J.A. Gladysz, J. Am Chem. Soc., 1999, **121**, 2696.
17. L.V. Dinh and J.A. Gladysz, Tetrahedron Lett., 1999, **40**, 8995.
18. J. Moineau, G. Pozzi, S. Quici and D. Sinou, Tetrahedran Lett., 1999, **40**, 7689.
19. B. Betzemeir and P. Knochl, Angew. Chem. Int. Edn, 1997, **36**, 2636.
20. S. Shneider and B. Bannwarth, Angew. Chem. Int. Edn., 2000, **39**, 4142.
21. S. Schneider and B. Bannwarth, Helv. Chim. Acta, 2001, **84**, 735.
22. C. Mancert and B. Bannwarth, Helv. Chim, Acta, 2002, **85**, 1877.
23. A. G. M. Barrett, D.C. Braddock, D. Catterick, D.Chandwick, J.P. Henschle and R.M. Mckinnell, Synlett, 2000, 847.
24. K.Mikami, Y. Mikami, Y. Matsumoto, J. Nishikido, F. Yamamoto and H. Nakajima, Tetrahedron Lett., 2001, **42**, 289.
25. T. Kitazume, J. Fluorinc Chem. 2000, **105**, 265.
26. H. Kleijn, E. Rijnberg, J. T. B.H. Jastrzebski and G. V. Koten Org. Lett., 1999, **1**, 853.
27. Y. Nakamura, S. Takeuchi, Y. Ohgo and D.P. Curran, Tetrahedron, 2000, **56**, 351.
28. Y. Tian and T.S. Chan, Tetrahedran Lett., 2000, **41**, 8813.
29. D.P. Curran, Angew. Chem Int. Ed., 1998, **37**, 1175.
30. R.J. Booth and J.C. Hodges, J.Am Chem. Soc. 1997, 119, 4882.
31. M.W. Creswell, G.L. Bolton, G.C. Hodges and M. Meppen, Tetrahedron, 1998, **54**, 3983.
32. M. Hoshiro, P. Degenkolb and D.P. Curran, J. Org. Chim., 1997, **62**, 8341.
33. J.K. Stille, Angew. Chem. Int. Ed., Engl. 1986, **25**, 508.
34. M. Lumbierres, M. Moreno-Manas and A. Vallribera, Tetrahedron, 2002, **58**, 4061.
35. O. Mitsunobu. M. Yamada and T. Mukaiyama, Bull. Chem. Soc. Jpn., 1967, **40**, 935.
36. S. Dandapani and D.P. Curran, Tetrahedron 2002, **58**, 3855.
37. D. Crich and S. Neelamkavi, Tetrahedron, 2002, **58**, 3865.
38. S.M. Pereia, G.P. Savage and G.W. Simpson, Synth. Commun, 1995, **25**, 1023; D.W. Zhu, Macro molecules, 1996, **29**, 2813.
39. H. Nakano and T. Kitazume, Green Chem., 1999, 1(1), 21.
40. M. Freenantle, Chem. Eng. News, March 30,1998, 32; M. J. Earle, P.B. Mc Cormac and K.R. Seddan Green Chem., 1991, 1(1), 23.

7. MISCELLANEOUS GREEN SOLVENTS

7

Miscellaneous Green Solvents

7.1 INTRODUCTION

We have seen organic transformations in green solvents or environmentally benign solvents. These solvents include water, ionic liquids, supercritical carbondioxide, polyethylene glycol and its solutions and Fluorous solvents. Besides these some other green solvents are ethyl lactate, gamma butyrolactone (GBL) and switchable solvents. A discussion on these form the subject matter of this chapter.

7.2 ETHYL LACATATE

It is a green solvent[1] and is derived from processing of corn. Ethyl lacate is the ester of lactic acid.

Ethyl lactate is noncorrosive, noncarcinogenic and 100% biodegradable. It is commonly used in paints and coating industry due to its high solvency power, high boiling point, low vapour pressure and low surface tension. It is used as a coating for wood, polystyrene and metals and also acts as an effective paint stripper. It has replaced solvents like toluene, acetone and xylene. It is also an effective cleaner for the polyurethane industry. Ethyl lactate has in fact eliminated the use of chlorinated solvents.

Due to low toxicity, ethyl lactate is used in pharmaceutical preparations, food additives[2] and fragrances. It is also used as solvent for nitrocellulose, cellulose acetate and cellulose ethers.[3]

Ethyl lactate is also used as tunable solvent for the synthesis of aryl aldimines[4].

As already stated, ethyl lactate is prepared by esterification of lactic acid which is prepared on a commercial scale by the fermentation of sugar by micro-organism Bacillus acidic lacatii (BAL).

$$\underset{\text{Sucrose}}{C_{12}H_{22}O_{11}} + H_2O \longrightarrow \underset{\text{Glu cos e}}{C_6H_{12}O_6} + \underset{\text{Fructose}}{C_6H_{12}O_6}$$

$$C_6H_{12}O_6 \xrightarrow{\text{BAL}} \underset{\text{Lactic acid}}{2CH_3CHOH\,COOH}$$

Depending with organism that is the source of Lactic acid, ethyl lactate can be either the levo (S) form or dextro (R) form. Most biologically sourced ethyl lactate is (–) – L–Lactate (ethyl (S) lactate). It is also produced from petrochemical stocks and this is in the form of racemic mixture.

Ethyl lacate can also be prepared by hydrogenation of ethyl pyruvate using heterogeneous catalytic system in supercritical fluids.

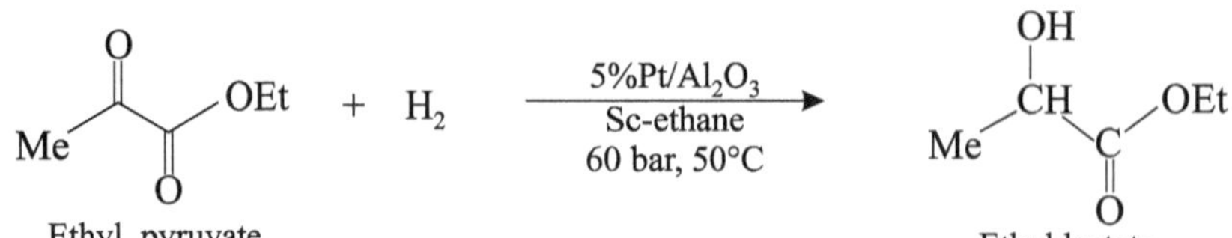

7.3 GAMMA-BUTYROLACTONE

γ-Butyrolactone (G BL), a hygroscopic colourless liquid having a weak characteristic odour is soluble in water. It is commonly used as a solvent and reagent in chemistry and also as a flavouring agent, cleaning solvent and as a superglue remover. It acts as a prodrug in humans for γ-hydroxybutyric acid (GHB) and is intoxicant having effects similar to alcohol.

γ-Butyrolactone is produced industrically[5] by dehydrogenation of γ-hydroxybutyric acid

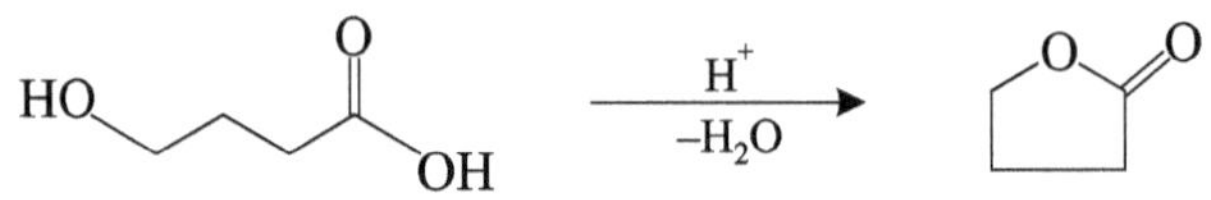

In the laboratory, it is obtained by the oxidation of tetrahydrofuran with aqueous sodium bromate.[6]

The solvent GBL is one of the most valuable alternative to the environmentally harmful chlorinated solvent. It has been widely used in polyner and paint industries. It is obtained by hydrogenation of maleic anhydride in an industrial important reaction.

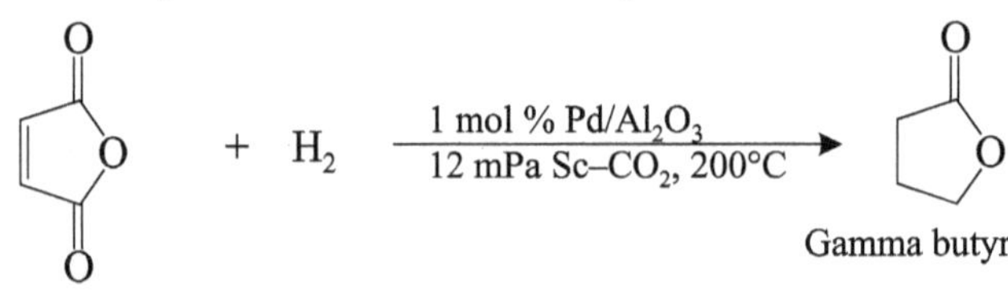

7.4 SOLVENTS WITH SWITCHABLE PROPERTIES

These solvents are believed to be new additions to green solvents and like the ionic liquids can be tailor-made for a particular process; their properties can be changed as desired to suit a reaction. In these solvents a gas such as carbon dioxide is introduced, the pressures involved are lower than used for supercritical carbondioxide. Thus these solvents are gas expanded liquids (GXLs) and particularly CO_2 expanded liquids (CXLs) and are believed to be alternative reaction media.[1]

The organic liquids like acetonitrile and hexane expand volumetrically as CO_2 dissolves in it. The advantages of CXLs over conventional reaction media and Sc–CO_2 are ease at removal of CO_2, milder process pressures (tens of bars) compared to Sc-CO_2 (about 100 bar). The compressed CO_2 in CXLs acts as an anti-solvent and so can be used to induce crystallisation of solutes. It is found that the solubility of gases such as hydrogen or carbon monoxide increases in CXLs compared to unexpanded liquids.

A large scale applications of gas expanded Liquids (GXLs) is enhanced oil recovery[8]. Using CO_2 helps in flushing out of oil reservoirs. Injection of CO_2 into the oil well displaces the remaining oil. This procedure is much better than conventional procedure using water. CO_2 lowers the viscosity of the crude oil and permeates the rock pores and this helps to flush out the remaining oil.

Some of the reactions performed in CXLs include homogeneous hydroformation, homogeneous hydrogenation, heterogeneous hydrogenation and oxidation[9]. An interesting reactions performed in CXLs in the formation of catalytic acetal formation. In this case the acid catalysts is generated in situ from the reaction of CO_2 with water or alcohol on pepressurization. The acid decamposed and there is no acid to dispose after acteal formation.

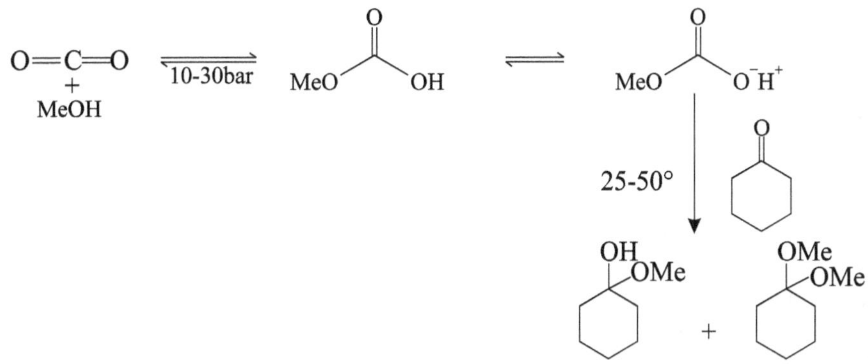

Besides solvents with switchable proporties (CXLs), there is another type of solvent. It is solvents of **switchable polarity**. As an example, CO_2 is used to reversibly form an ionic liquid from an alcohol and a organic base, *viz*., 1, 8-diazabicyclo-[5.4.0]-undec–7-ene (DBU)

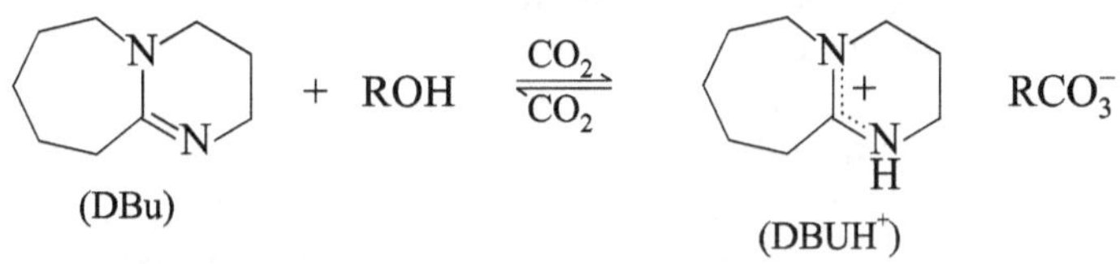

The ionic liquid ($DBUH^+$) can be switched back to its natural components (DBU + ROH) by bubbling nitrogen or argon through the mixture. Such a solvent is called switchable polarity solvent. Using such solvents there is no need for changing solvent after each step of a reaction. In this case, the carbonate based ionic liquid formed is polar and viscous compared to less viscous and less polar alcohol-base mixture.

Switchable polarity solvents (SPS) have been used for the polymerisation of styrene. Thus styrene is polymerised (in presence of an initiator) in DBU and Pr OH (use it PrOH gives a less viscous reaction mixture thereby facilitating the filtration step). After polymerisation the ionic liquid is formed. The formed polystyrene is filtered and the ionic liquid turned back into the less polar mixture using nitrogen and reused.[10]

A one-component switchable polarity solvent like dialkyl amines were found to be better than base-alcohol mixtures. As an illustrations N-ethyl-N-butyl amine was found to a good example. On treatment with CO_2, its polar form was obtained. The solubility of various solids and liquids was studied in both forms of SPS. It was subsequently used in the post-reaction separation of the polycarbonate formed in the catalytic Co-polymensation of CO_2 and cyclohexene oxide. Both the polymer and the catalyst are soluble in the amine. On exposure to CO_2, the solvent switches to its polar form and the polymer precipitates. The catalyst can be recycled.[11]

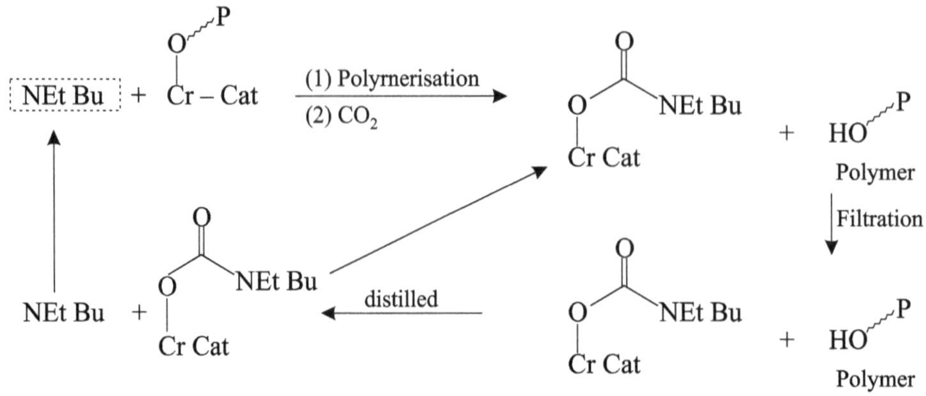

Diagramatic representation of copolymerisation of CO_2 in a switchable polarity solvent, ethyl butyl amine).

The same technology was used[8] for swtichable solvents to obtain switchable surfactants, which may have various applications. Their use reduces waste and solvent usage.

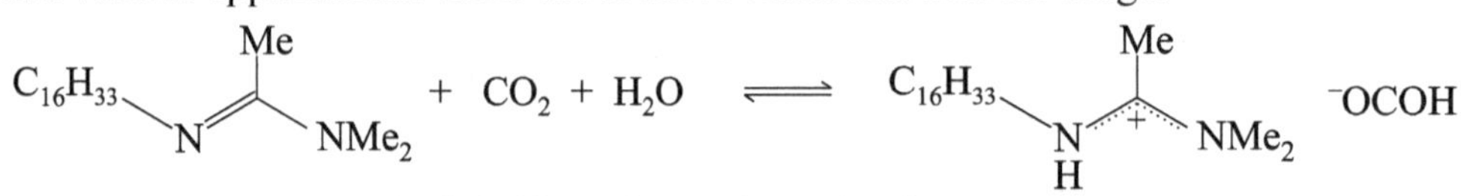

(Suitchable surfactant formation from long-chain alkyl amidine, CO_2 and H_2O)

Besdies solvents of switchable polarity, we also come across solvents of switchable volatility. It is well known that the volatility of solvent is of crucial important in their reaction. It has been found[12] that piperylene sulfone (PS) is a recyclable alternative to dimethylsulfoxide (DMSO). DMSO in addition to dimethylformamide (DMF) and hexamethyl phosphoramide (HMPA), is used as dipolar, aprotic solvent. The problem, however is that DMSO is difficult to remove from the product by distillation and is normally not recycled. However, PS decomposes at above 100°C to give trans -1,3-pentadiene and SO_2, which reform PS at room temperature.

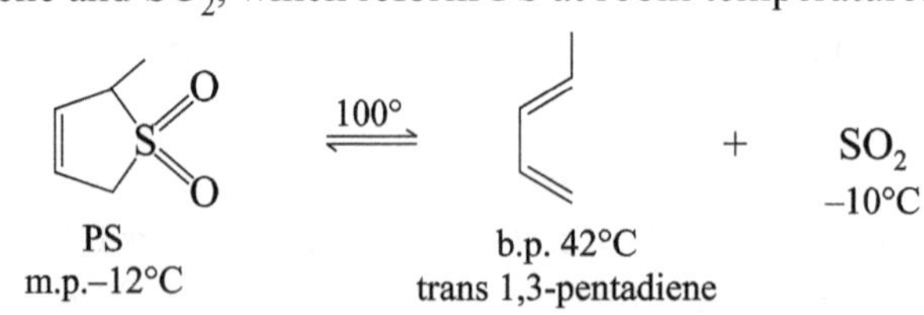

PS has the same dielectric constant as DMSO and in some anionic nucleophilic substitution reaction in PS is of the same order of magnitude as in DMSO.

PS has also been used in the copper catalysed aerobic oxidation primary alcohols.

Conclusion

Ethyl lactate, gamma butyrolactone and solvents with switchable properties are green solvents. However, their use in chemical transformation has not been exploited.

A number of green solvents have also been obtained from Biomass. For details see Chapter 16.

REFERENCES

1. Cited in Green Chmistry and Process M. Doble and A.K. Kruthiventi, Elsevier, Page 97.
2. [1] (http://www.food safety. gov/dms/eafus.html) U.S. Food and Drug Administration, Center for Food Safety and Applied Nutrition.
3. Industrial Solvents Handbook by Ernst W. Flick, 5th Edition. William Andrew Inc. 1998. ISBN 0-8155-1413-1, ISBN978-0-8155-1413-8.
4. J.S. Bennett, K.L. Charles M.R. Miner C.F. Heuberger, E.J. Spina, Green Chemistry 2009, **11(2)**, 166; Carla S.M. Pereira, Viviand M.T.M. Silva and Alirio E. Rodrigues, Green Chemistry, 2011 **13(10)**, 2658.
5. W. Schwarz, J. Schossig, R. Rossbocher and H. Höke, Butyralactone in Ullmann's Encyclopedia of Industrial Chemistry. Willey-VCH, 2000, Weinbein.
6. L. Metsger, S. Bittner, Tetrahedron 2000, **56(13)**, 1905.
7. Cited in D. Clarke, A.A. Mohammed, A.A. Clifford,A. Pavrutt, P. Rose, D. Schwenn, W. Bannwarth and C.M. Rayner, Current Topics in Medicinal Chemistry 2004, **4, 729**.
8. P.G. Jessop and B. Subrammaniam. Chem. Rev. 2007, **107**, 2666.
9. For details see RSC Green Chemistry Book Series, Alternative solvents for Green chemistry, by Francesca M. Kerton, Publish by the Royal Society of chemistry, www.rsc. org.
10. Ind. Eng. Chem. Res., 2008, **47**, 539.
11. J. Org. Chem., 2008, **73**, 127.
12. Liotta, Eckert and Jessop, Chem. Common. 2007, 1427.

8. SOLVENT-FREE ORGANIC TRANSFORMATIONS

8

Solvent-free Organic Transformations

8.1 INTRODUCTION

Solvent-free organic transformations or solid state organic transformations are the best alternatives in the contex of green chemistry. Such reactions are cheaper alternatives, and easy to operate and are devoid of environmental pollution. Solvent-free organic transformations can be conducted in following ways:

- Grinding the reactants using pestle mortar or ball mills
- Heating the reactants
- Heating the reactants using solid support
- Microwave assisted reactions.

8.2 SOLVENT-FREE ORGANIC TRANSFORMATIONS BY GRINDING THE REACTANTS USING MORTAR AND PESTLE

8.2.1 Aldol Condensation

The aldol condensation of aromatic aldehydes with lithium enolate of 3,3-dimethylbutanoate gives[1] 8 : 92 mixture of syn and anti products 70% yield (Fig. 8.1). The reaction is carried out by mixing freshly ground mixture of the reactants in vacuum for 3 days at room temperature.

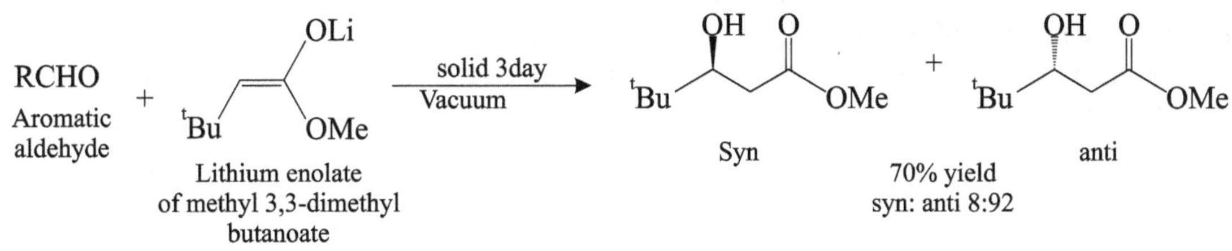

R = 2-$OCH_3C_6H_4$-, 4-Cl C_6H_4–, 4-$NO_2C_6H_4$–, 3-$NO_2C_6H_4$

Fig. 8.1: Aldol Condensation in Solid state.

The aldol condensation of an aldehyde and a ketone in presence solid NaOH by grinding the mixture in a pestle and mortar at room temperature for 5 min give[2] the chalcone (Fig. 8.2).

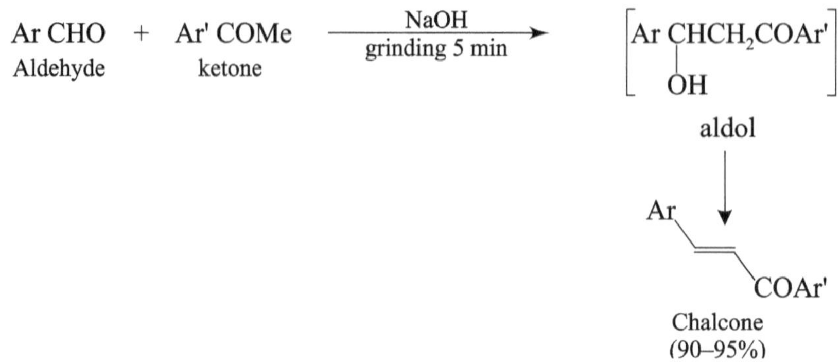

Fig. 8.2: Synthesis of chalcone.

See also section 8.3.3.

8.2.2 Reformatsky Reaction

Aromatic aldehydes on treatment with ethyl bromoacetate and Zn-NH_4Cl in the solid state gives[3] the corresponding Reformatsky reactions Product (Fig. 8.3).

$$\text{ArCHO} + \text{BrCH}_2\text{CO}_2\text{Et} \xrightarrow[\text{Solid State 3 hr.}]{\text{Zn}-\text{NH}_4\text{Cl}} \text{Ar CH(OH)CH}_2\text{CO}_2\text{Et} \atop (80\text{–}90\%)$$

Ar = Ph, *p*-BrC_6H_4, 3,4-methylenedioxyphenyl.

Fig. 8.3: Reformatsky reaction in solid state.

8.2.3 Synthesis of 1,4-dihydro-quinoxaline-2,3-dione

Quinoxaline diones and their derivatives and widely used in many fields as curatorial intermediates, bacteriocides and insecticides[4]. These moieties are also present in peptide antibiotics, which are used for the treatment at epilopsy, pain and other neurodegenerative disorders[5,6]. A number of synthetic methods of these compounds have been reported, which used catalysis and or some special techniques[5]. Most of these methods are associated with shortcommings like long reaction times, expensive reagents and low yields.

A one pot green synthesis of 1, 4-drihydroquinoxaline-2,3-dione has been reported[7]. It consists in grinding a mixture of oxalic acid dihydrate (1 m mol) and *o*-phenylene diamine (1 m mol) with a pestle in a mortar at room temperature in open atmosphere till a melt is obtained. The mixture is extracted will dilute alcohol (1 : 1) to get a pure product (Fig. 8.4).

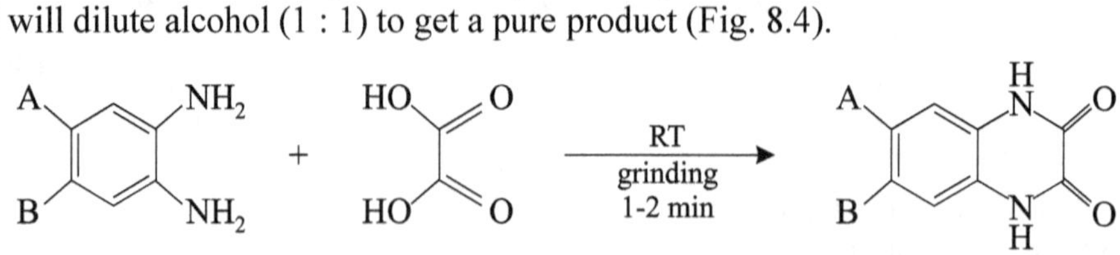

A	B	A	B
H	H	H	Me
H	NO_2	H	*n*-Pr
H	Cl	H	Ph

1,4-Dihydroquinoxaline -2,3-dione (90-98 %)

Fig. 8.4: Solid state synthesis of 1,4-dihydroquinoxaline -2,3-dione.

8.2.4 Synthesis of β-keto Sulfones from Ketones

β-keto sulfones are important intermediates for organic synthesis[8]. These are also precursors for optically active β-hydroxysulfones[9]. Some of the β-ketosulfones are found to possess fungicidal activity[10]. Earlier methods of preparation at β-ketosulfones involve oxidation β-ketosulfides[11], reaction of sulfonyl chlorides with silyl enol ethers[12], reaction of diazo sulfones with aldehydes[13], alkylation of arene sulfinate with α-haloketones[14], acylation of alkylsulfones, reaction of alkyl sulfones with N-acylbenzotriazoles[15] and reaction of sulfonyl chloride with arylacetylenes[16].

A facile one-pot synthesis of β-keto sulfones from ketones under solvent free conditions has been reported[17]. The method consist in reacting ketones with relatively benign [hydroxy (tosyloxy) iodo] benzene, PhI (OH) OTS and sodium arene sulfoniates in presence of PTC catalyst (tetrabutyl ammonium bromide, TBAB) under solvent free conditions at room temperature (Fig. 8.5).

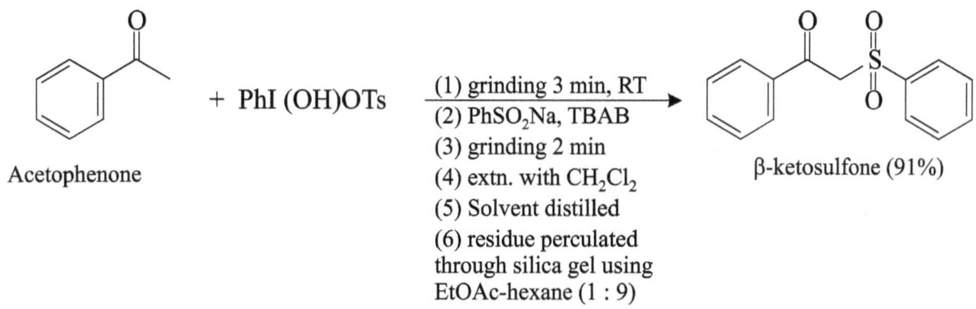

Fig. 8.5: One pot synthesis of β-ketosulfones from ketones under solvent free conditions.

Using the above method a number of ketones can be used for the preparation of the corresponding β-ketosulfones.

The above reaction (Fig. 8.5) is believed to proceed *via* the insitu generation of α-tosyloxyketones followed by nucleophilic substitution with sodium arene sulfinate under solvent free conditions.

8.2.5 Synthesis of α-Tosyloxy β-Ketosulfones

α-Tosyloxy β-ketosulfones are believed to be potentially useful precursors in organic synthesis. Most of the methods for α-functionalisation of β-ketosulfones required longer reaction times under refluxing conditions giving only moderate yields. A solvent-free facile synthesis[18] of novel α-tosyloxy-β-ketosulfones involves grinding of a neat mixture of α-benznesulfonyl-acetophenone (see section 8.2.3) and [hydroxy (tosyloxy) iodo] benzene [HTIB], commonly known as **Koser's reagent** at room temperature using a pestle mortar. (Fig. 8.6).

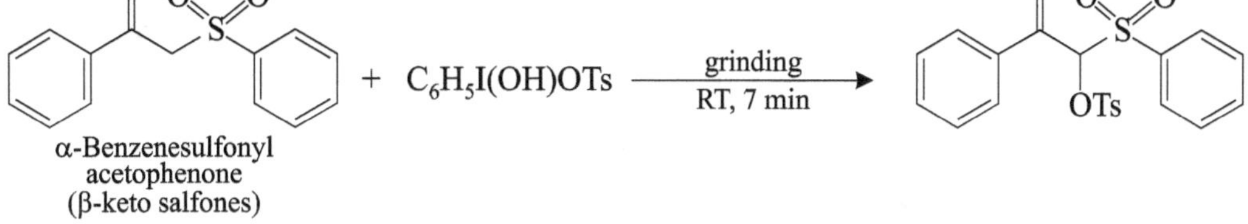

Fig. 8.6: Synthesis of α-tosyloxy-β-ketosulfones. Using this method a number of α-tosyloxy β-keto sulfone were synthesised[18].

8.2.6 Synthesis of 1-aryl-4-methyl-1, 2, 4-trizolo [4, 3-a]-quinoxalines

1,2,4-Triazol nucleus has been found to display antibacterial, CNS stimultative, sedative, antifungal and antitumor activities[19]. 1, 2, 4-Triazoles were synthesised earlier by the reaction 2-hydroxy quinoxaline with carboxylic acid at elevated temperature[20], 1-3-dipolar cycloaddition of aromatic nitriles in presence of base followed by elimination of hydrogen[21], photolysis of triazole-3-thiones[22] and oxidation of aryl hydrazones[23]. However, all these methods are multi step process and require toxic reagents and longer reaction times.

1-Aryl-4-methyl –1, 2, 4-trizolo [4, 3 –*a*] quinoxalines have been synthesised[24] by grinding a mixture of arenecarboxaldehyde, 3-methylquinoxalin–2*yl*–hydrazones[25] with iodobenzene diacetate. It is however, necessary in certain cases to warm the reaction mixture slightly. The hydrazones form a yellow eutectic mixture with iodobenzene diacetate before the occurence of mildly exothermic reaction. However, hydrazone derivatives of arenecarboxaldehyde containing an electron-donating group do not need warming (Fig. 8.7).

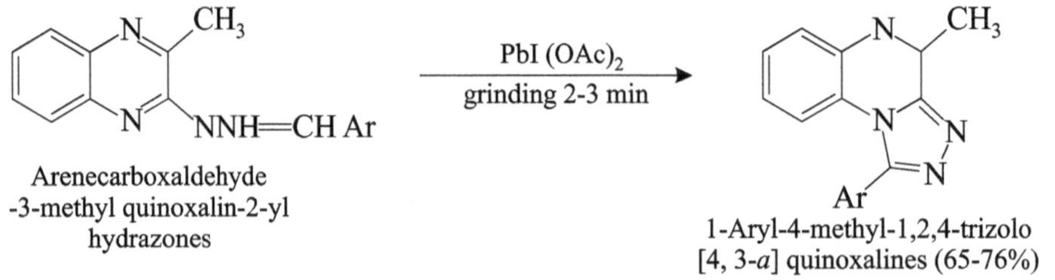

Ar = Ph, p-$CH_3C_6H_4$, p-ClC_6H_4, p-$OMeC_6H_4$, m-$OMeC_6H_5$, p-$NMe_2C_6H_4$.

Fig. 8.7: Solvent free synthesis of 1-aryl-4-methyl-1, 2, 4-trizolo [4,3-*a*] quinoxalines.

The required arene carboxaldehyde-3-methyl-quinoxalin-2*yl*-hydrazones were prepared by the literature method[25].

8.2.7 Synthesis of Oximes using Grindstone Chemistry

Conversion of carbonyl compounds into oximes is an important reaction in organic chemistry. Some of the applications of oximes include purification and characterisation of carbonyl compounds, conversion into nitriles, nitro compounds, nitrones, amines and synthesis of azahetrocycles. These are also useful for Beckmann rearrangement particularly during the manufacture of nylons. A number of procedure are available for the preparation of oximes, but most of them do not follow the concepts of green chemistry. These methods are associated with generation of pollutants, require high reaction temperature and low yields.

A rapid, convenient, solventless green method for the synthesis of oximes has been developed[26]. It consists in grinding together the ketone and hydroxylamine hydrochloride in presence of Bi_2O_3 (Fig. 8.8).

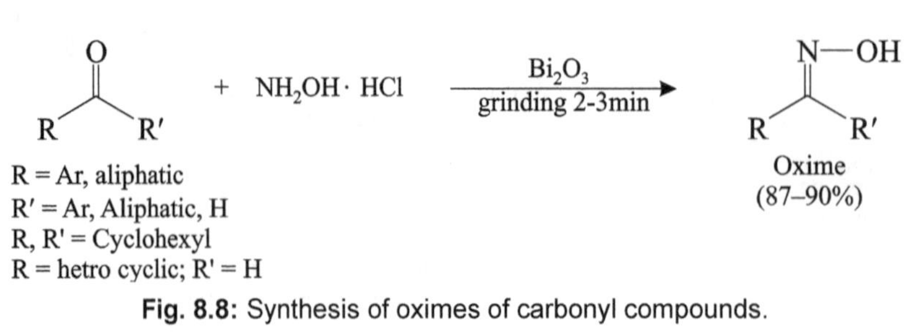

Fig. 8.8: Synthesis of oximes of carbonyl compounds.

8.2.8 Synthesis of Thionocarbamates

The thionocarbamates are prepared[27] by grinding together the ingredients with a mortar and pestle at room temperature. An examples is given below (Fig. 8.9).

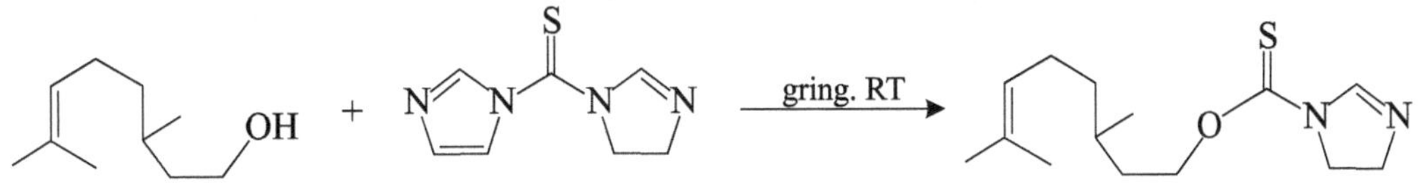

Thionocarbamates
(91%)

Fig. 8.9: Synthesis of Thionocarbamates.

8.2.9 Synthesis of Sym. Triisopropylbenzene

It is prepared by **Friedel-crafts alkylation**[28] of benzene with isopropyl benzene. The method consists in grinding the ingredients in a agate martar for about 45 min., pouring onto crushed ice and subsequent ether extraction (Fig. 8.10).

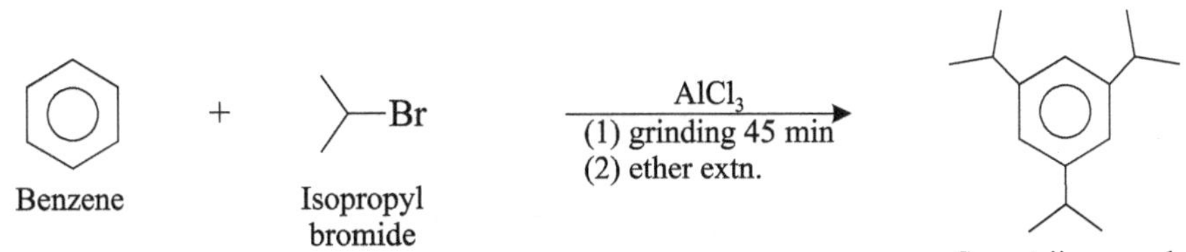

Benzene

Isopropyl bromide

Sym. triisopropyl benzene (95%)

Fig. 8.10: Synthesis of Sym. triisopropyl benzene.

8.2.10 Baker-Venkataraman Rearrangement

A facile Baker-Venkataraman Rearrangement can be performed by grinding a mixture of *o*-cinnamoyloxy acetophenone with KOH in a mortar pestle (Fig. 8.11)[29].

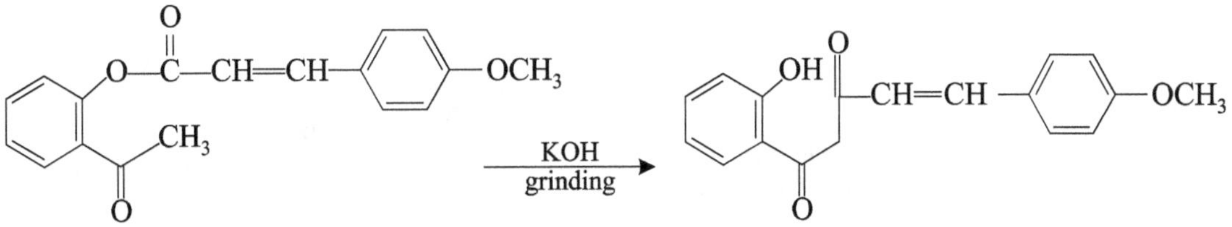

o-p-methoxycinnamoyloxy acetophenone

o-Hydroxy-β-diketone

Fig. 8.11: Baker-Venkataraman Rearrangement using a mortar and pestle.

So far solvent free reactions have been discussed which are performed using a mortar and pestle. It is found that only the reagents that form a eutectic melt by using mortar and pestle can react to give the desired product. However, such procedures are most economical methods for conducting a chemical reaction [see also Chapter 9, Section 9.5.24].

In cases when the reactants do not form electric melt or the reactants are high melting, such reactions can be performed by running them in ball mills[30]. The reactants are placed in a vessel along with a ball bearing. The vessel is sealed and placed in the milling apparatus where it is agitated. This high speed agitation provides enough kinetic energy to the ball in order to impact the reactants and start a chemical reaction. In fact, Ball mills are used for reactions on a commercial scale. The Ball mills, as we know are used in mining industry in grinding ores. The ball mill can be used repeatedly to make the same product, the problem of cleaning of between runs is eliminated. Following are given some of the transformations, which can be performed in Ball mills.

8.2.11 Synthesis of 1, 4-disubstituted–1, 2, 3-triazoles

The reaction of terminal alkynes with $Bn–N_3$ in presence of easy to prepare $Cu(PPh_3)_2\ NO_2$ complex as catalyst give the required 1, 4-disubstituted-1, 2, 3-triazoles. The reaction is conducted, by stirring in absence of any solvent and even in the absence of microwave irradiation. The reaction is known as copper-catalysed **Huisgen cycloaddtion** [Fig. 8.11(*a*)][30a]

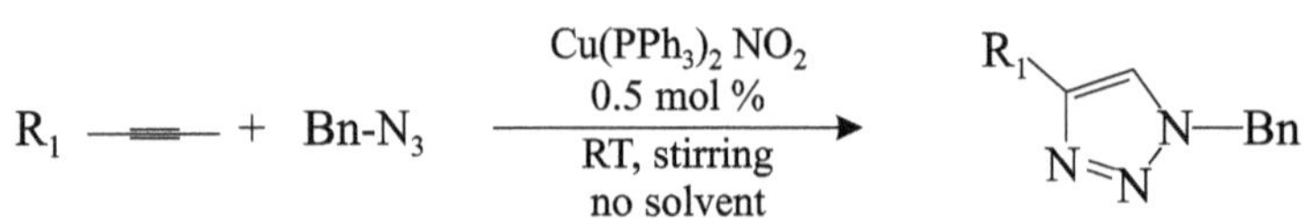

Fig. 8.11a: Copper-catalysed Huisgen cycloaddition reaction for the synthesis of 1, 4-disubstituted 1, 2, 3-triazoles.

8.3 Solvent Free Organic Transformations using Ball Milling

8.3.1 Wittig Reaction

The Wittig reaction of 4-bromobenzaldehyde with benzyltriphenylphosphonium bromide in a Ball mill in presence of K_2CO_3 in helium atmosphere gave the expected product, stilbene in 92% yield (Fig. 8.12)[31] with *cis: trans* ratio of 34 : 66 compared to reaction performed in solution where a yield of 98% with a *cis : trans* ratio of 56 : 44 reported.[32]

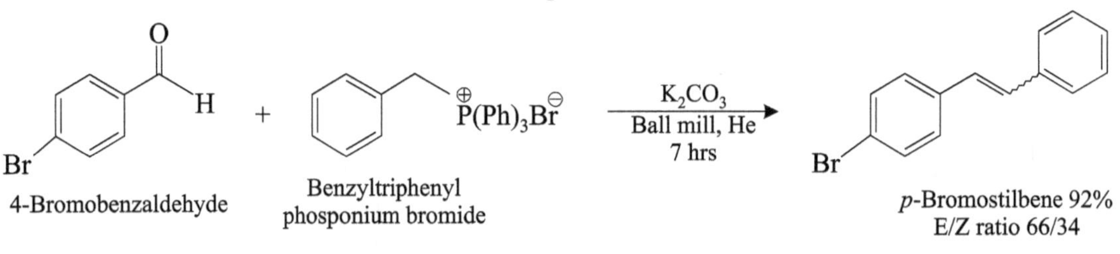

Fig. 8.12: Solvent Free Wittig reaction in ball mill.

As seen in the above reaction, a mild base K_2CO_3 is used compared to lithium enolate (to generate ylide) is necessary for the reaction conducted in solution.

8.3.2 Horner-Wadsworth-Emmons Reaction

The reaction of benzaldehyde with phosphonate ester in presence of K_2CO_3 using ball mill gave α, β-unsaturated ester in high diasteroselectivity (Fig. 8.13)[33, 34]. The product was isolated in pure state, the reaction product was treated with oxone to convert any remaining aldehyde into carboxylic acid) and finally the reaction mixture washed with $NaHCO_3$ solution and then filtered.

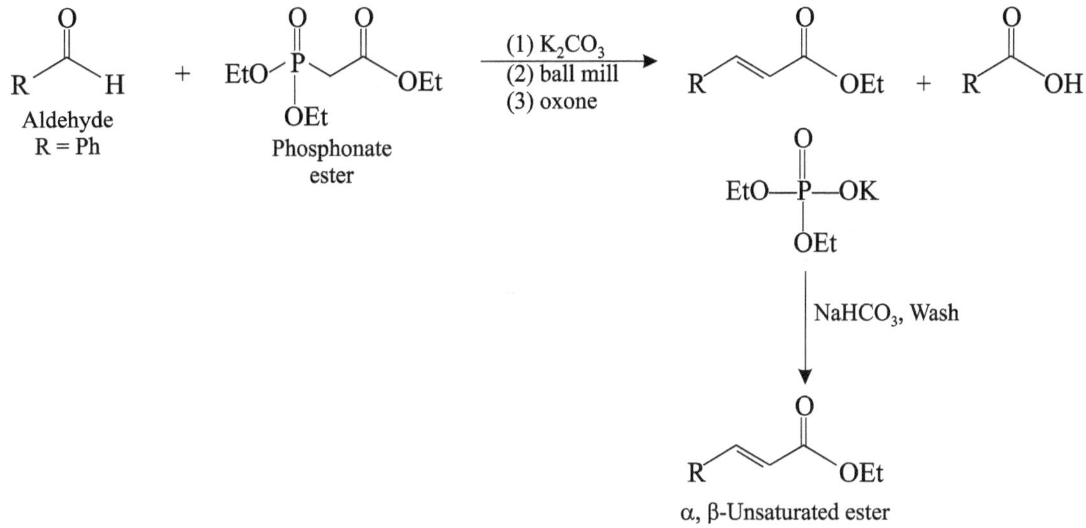

Fig. 8.13: Solvent free Horner-Wadsworth-Emmons reaction using ball mill.

8.3.3 Aldol Condensation

Stereoselective aldol condensation has been carried out under ball milling conditions[35, 36]. As an example aldol condensation of cyclohexanone and benzaldehyde derivative and 10 mol % (S)-proline using ball mill gave the aldol product in 99% yield and upto 99% enantioselectivity (Fig. 8.14)[37]

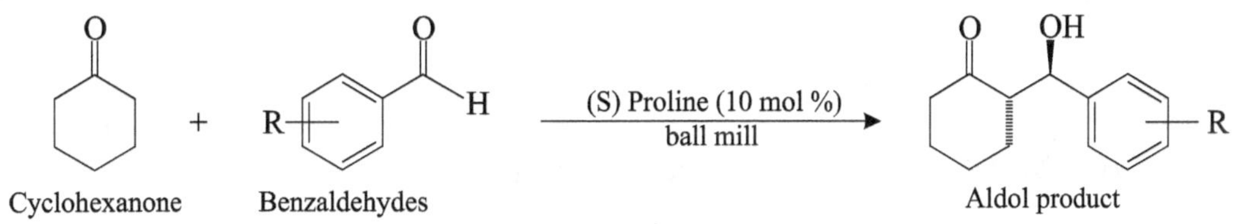

Fig. 8.14: Enantioselective aldol condensation under milling conditions.

8.3.4 Knoevenagel Condensation

It has been shown[38] that Knoevenagel condensation gives high yields of products under solvent free conditions using ball mill (Fig. 8.15)

$$\text{RCHO} + \text{CH}_2(\text{COOH})_2 \xrightarrow[\text{Ball mill}]{\text{K}_2\text{CO}_3} \text{RCH{=}C(COOH)}_2 \xrightarrow{-\text{CO}_2} \underset{\alpha,\ \beta\text{-unsat' d acid}}{\text{RCH{=}CHCOOH}}$$

Fig. 8.15: Knoevenagel condensation under solvent tree conditions using Ball Mill.

8.3.5 Sonogashira Coupling

The Sonogashira coupling reaction involving the reaction of an alkyl halide (or triflate) and a terminal alkyne in presence of copper iodide and a palladium catalyst is shown to be successful under solvent-free ball milling conditions is high yield (Fig. 8.16)[39].

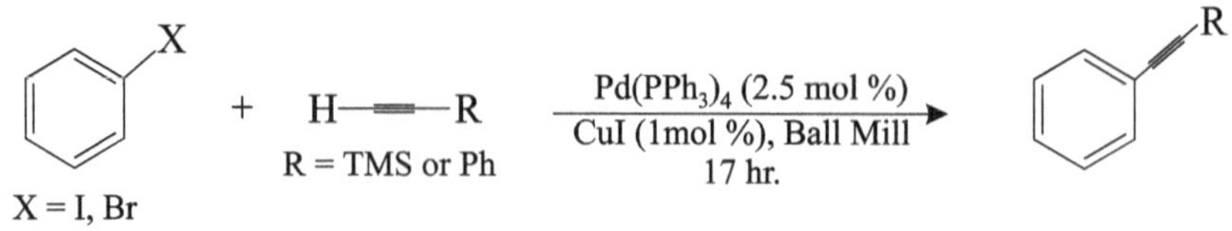

Fig. 8.16: Solvent free Sonogashira reaction under ball-milling conditions.

In the above reaction (Fig. 8.16), instead of using CuI, the reaction could be conducted in copper vial. Also it has been shown that Sonogashira reaction can be conducted[40] under ball milling conditions using palladium acetate and DABCO. The reaction was successful only with ary iodides (Fig. 8.17).

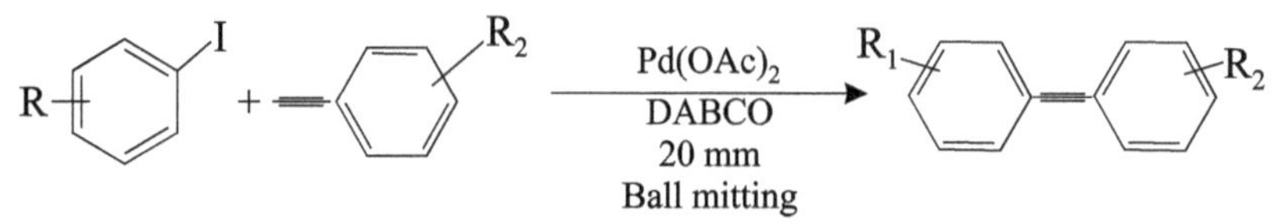

Fig. 8.17: Sonogashira reaction using Pd $(OAc)_2$ and DABCO under Ball milling conditions.

8.3.6 Heck Coupling

Heck coupling reaction has also been conducted under solvent-free ball milling conditions[41] Thus, the reaction of substituted aryl halides with olefins gave variety of organic compounds. The reaction was conducted under solvent free conditions using Pd $(OAc)_2$, $NaHCO_3$, HCOONa [reductant for Pd(II)], n Bu_4NCl and NaCl under ball milling conditions (Fig. 8.18). It was found that a combination of ball milling, heat and pressure gave best result.

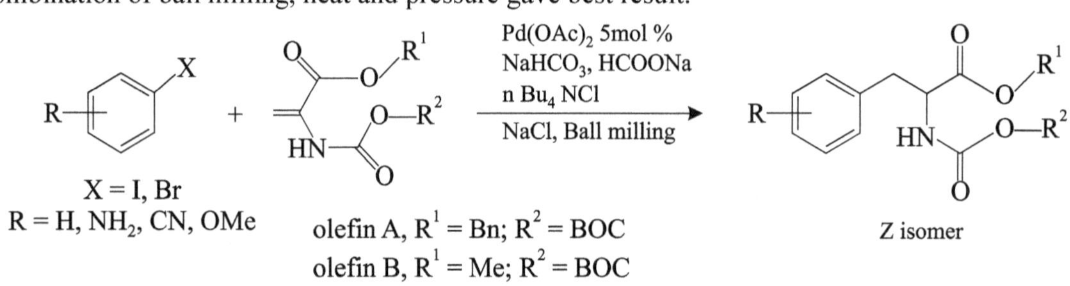

Fig. 8.18: Heck reaction-ball milled.

See also sections 9.5.6; 10.2.27.1

8.3.7 Suzuki Coupling

Suzuki coupling reaction between an aryl halide (iodide or bromide) and a boranic acid in presence of Pd catalyst and K_2CO_3 and NaCl under ball milling conditions[42] gave biphenyls (Fig. 8.19).

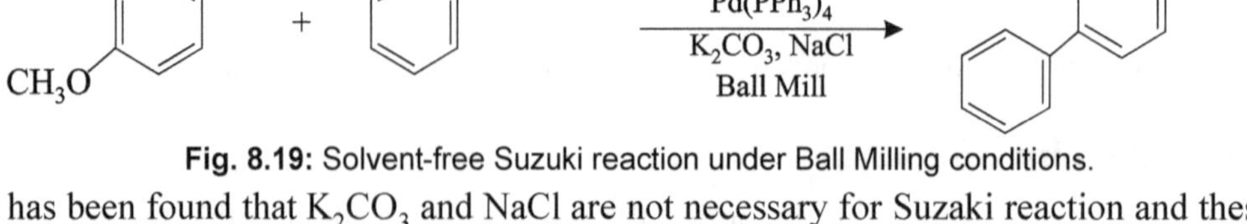

Fig. 8.19: Solvent-free Suzuki reaction under Ball Milling conditions.

It has been found that K_2CO_3 and NaCl are not necessary for Suzaki reaction and these could be replaced by solid-supported KF on basic alumina[43, 44]

See also secion 8.37; 10.2.27.2

8.4 SOLVENT-FREEORGANICTRANSFORMATIONSBYHEATINGTHEREACTANTS

These reactants are conducted by heating the reactants in an oil-bath or water bath.

8.4.1 Oxidation of Hydroxy Aldehydes and Ketones to Hydroxy Phenols using Urea-Hydrogen Peroxide Adduct (UHP)

Phenols and their derivatives are known to be extensively used in organic synthesis. The conversion of hydroxy benzaldehydes to hydroxy phenols is generally achieved by alkaline hydrogenperoxide[45] (Dakins Oxidation). Some other oxidants like peroxybonzoic acid[46] and peroxyacetic[47] acid have also been used. However, solid state solvent-free oxidation of hydroxy benzaldehydes with urea-hydrogenperoxide adduct is a much superior alternative in terms of shorter reaction time, cleaner product formation and ease of manipulation[48] (Fig. 8.20). The oxidation was conducted by heating the starting material and finely powdered urea-hydrogen peroxide adduct in an oil bath at 85°.

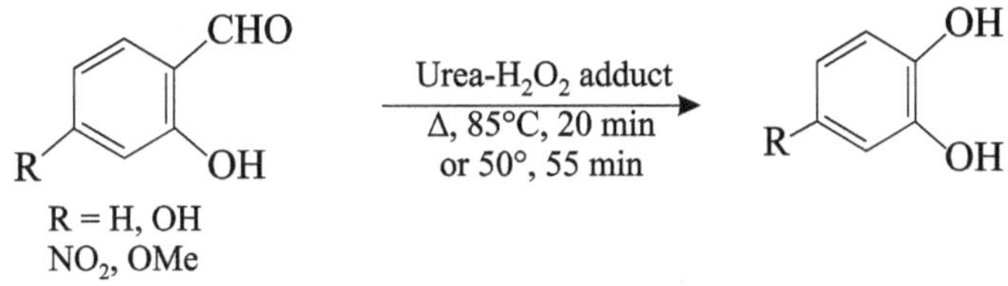

Fig. 8.20: Solid state oxidation of hydroxy aldehides to hydroxy phenols using UHP.

In a similar way hydroxy ketones on oxidation with UHP gave[48] the corresponding hydroxy phenols (Fig. 8.21).

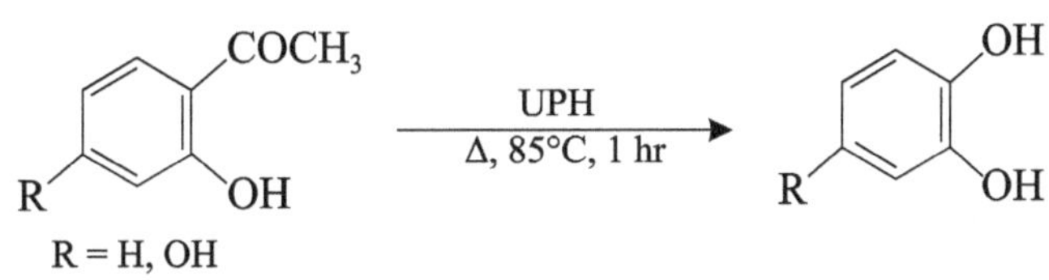

Fig. 8.21: Oxidation of hydroxy ketones to hydroxy phenols with UHP.

8.4.2 Oxidation of Nitriles to Amides Using UHP

The hydrolysis of nitriles is normally used for synthesis of carboxylic acid amides. The usual methods of hydrolysis of nitriles include using strong mineral acids[49], although some metals[50] and their oxides[50, 51] and complexes[50, 53] have also been used for conversion of carbonitriles to amides. Some other reagents like titanium tetrachloride in acetic acid[53], KF on Alumina[54], KOH in t-BuOH[55], sodium percarbonate in aq. MeOH[56] and some hydroxylamine derivatives[57] have also been used.

Eco-friendly hydrolysis of nitriles to amides has been achieved using UHP under solvent free conditions (Fig. 8.22)[48].

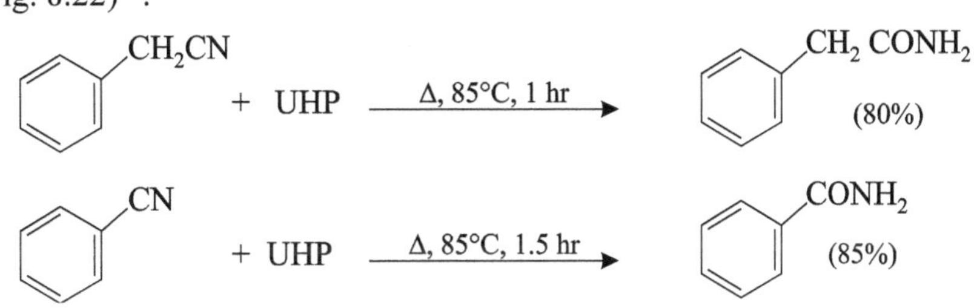

Fig. 8.22: Oxidation of nitriles to amides using UHP.

8.4.3 Selective Oxidation of Sulfides to Sulfoxides or Sulfones

Selective oxidation of sulfides to sulfoxides or sulfones is important for synthetic organic chemists. The conversion can be effected with hydrogen peroxide[58], chromic acid[59], nitric acid[60], manganese dioxide[59], ozone[61], peracids[62], selenium dioxide[63], sodium periodate[64], hypervalent iodine reagents[65], sodium perborate[66] and dinitrogen tetroxide[67]. Most of these procedures suffer from drawbacks like longer reaction times, use of corrosive acids, hazardous peracids and toxic metallic compounds. An eco-friendly method for the conversion of sulfides to sulfoxides and sulfones is by heating with UHP (Fig. 8.23 and 8.24).

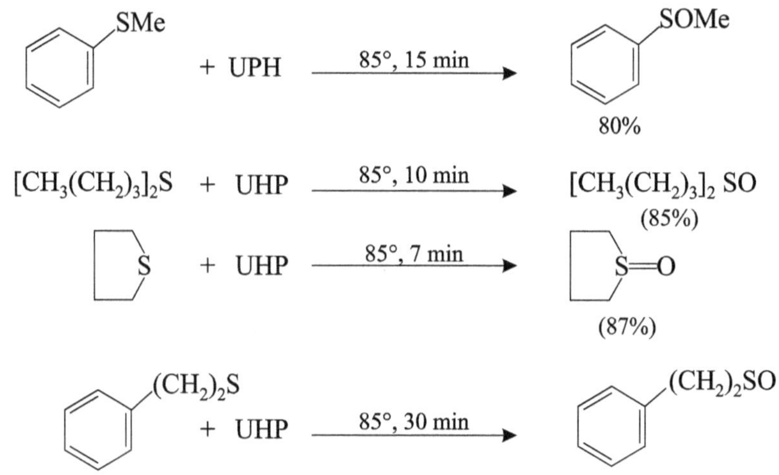

Fig. 8.23: Oxidation to Sulfides to Sulfoxides using UHP.

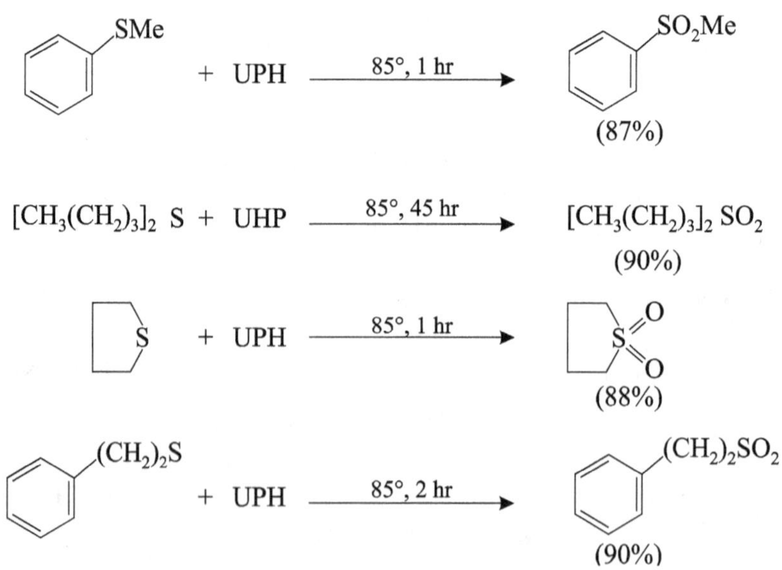

Fig. 8.24: Oxidation of Sulfides to Sulfones using UHP.

8.4.4 Oxidation of Nitrogen Heterocycles to N-oxides using UHP

Urea-hydrogen peroxide complex (UHP) has been used for the oxidation of Nitrogen heterocycles to their N-oxides (Fig. 8.25)[68].

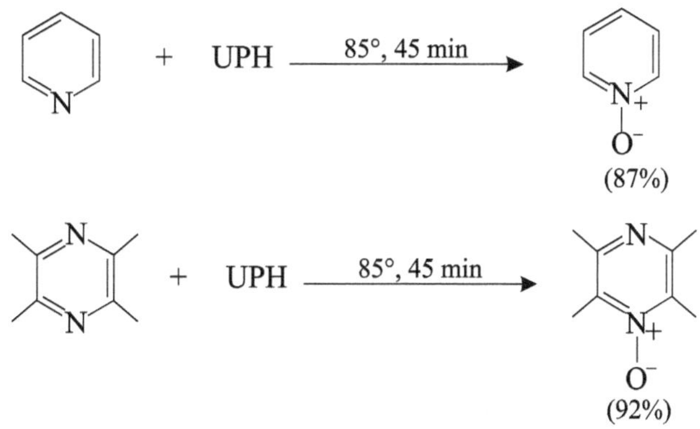

Fig. 8.25: Oxidation of N-heterocycles to N-oxides using UHP.

8.5 SOLVENT-FREE ORGANIC TRANSFORMATIONS USING CLAY-SUPPORTED REAGENTS

Following clay-supported reagent are commonly used.

(*i*) **Clayfen:** It is K-10 montmorillonite impregnated with Fe(III) nitrate[69]. It is used as an efficient oxidatnt for the oxidation of alcohols (about 90% yield of the aldehyde; there is no overoxidation to carboxylic acids in case of primary alcohols)[69], oxidation of benzoin to benzil[70] and oxidative coupling of thiols to disulfides via thionitrite intermediates[71].

(*ii*) **Claycop:** It is K-10 montmorillonite impregnated with Cu(II) nitrate[72]. It is used for the regeneration of carbonyls form their various protected forms such as tosylhydrazones, phenylhydrazones and semicarbazones[73]. It is also used to convert hydrazones into azides.

It is less reactive than clayfen but is more stable[74]. It oxidises dihydropyridines to pyridines[75]. It regenerates carbonyl groups from protective bisthioacetals, selenoacetals[76] and from thiocarbonyls[77].

(*iii*) **Clayzic:** It is aluminosilicate clay doped with zinc chloride[78]. It is a support reagent for **Friedel-crafts alkylation** of benzenoid compounds[79].

Following are given some solvent free organic transformations using clay-supported reagents.

8.5.1 Polymerisation of Styrene

Styrene on heating in presence acid-treated montmorillonite undergoes dimensation (Fig. 8.26)[80].

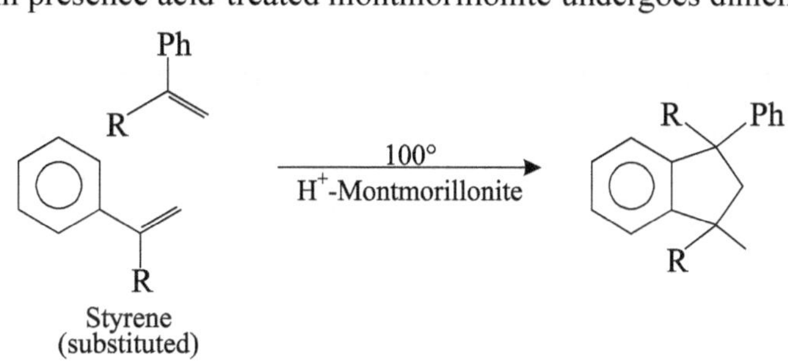

Fig. 8.26: Dimensation of Styrene.

8.5.2 Reaction of Triphenylamine with Na$^+$-montorillonite

Triphenylamine on heating with $\overset{+}{\text{Na}}$ montorillonite gives N, N′, N′, N′-tetraphenyl benzidine (Fig. 8.27)[81]

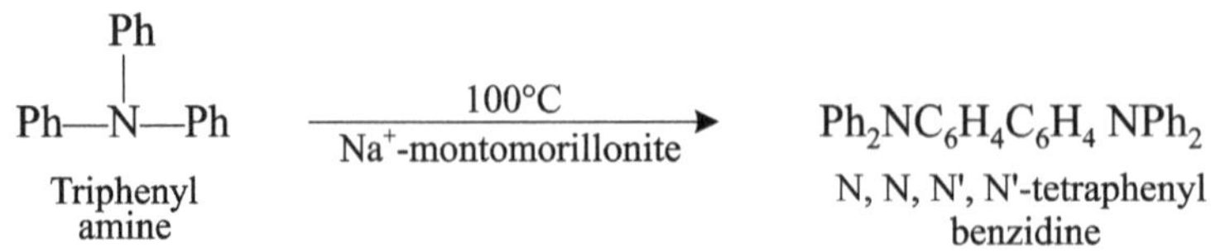

Fig. 8.27: Synthesis of N, N, N′, N′-tetraphenyl benzidine.

8.5.3 Dehydration of Alcohols

Primary alcohols on heating with Al (III) or H$^+$-exchanged clays gives mainly the dialkyl ethers with little olefin. However, 2° and 3° alcohols give exclusively alkenes (Fig. 8.28)[82].

$$\underset{1^\circ\text{ Alcohol}}{\text{R CH}_2\text{ CH}_2\text{ OH}} \xrightarrow[\text{Al(III)}^+-\text{montomorillonite}]{100^\circ} \underset{\text{dialkyl ethers}}{\text{RCH}_2\text{CH}_2\text{OCH}_2\text{CH}_2\text{R}} + \underset{\text{minor}}{\text{RCH}=\text{CH}_2}$$

Fig. 8.28: Dehydration of 1° alcohols to ethers.

8.5.4 Synthesis of Dialkyl Amines and Dialkyl Thioethers

Primary amines on treatment with Al (III)$^+$–montomorillonite loose a molecule of ammonia to give dialkylamines[83]. In a similar way thiols give dialkyl thioethers[84] by elimination of H_2S (Fig. 8.29)

$$\underset{1^\circ\text{ Amines}}{\text{R CH}_2\text{NH}_2} \xrightarrow[\text{Al(III)}^+-\text{montomorillonite}]{-\text{NH}_3,\,\Delta} \underset{\text{Dialkyl amines}}{\text{RCH}_2\text{NHCH}_2\text{R}}$$

$$\underset{\text{Thiols}}{\text{R CH}_2\text{ SH}} \xrightarrow[\text{Al (III)}^+-\text{montomorillonite}]{-\text{H}_2\text{S},\,\Delta} \underset{\text{Dialkyl thioethers}}{\text{R CH}_2\text{ S CH}_2\text{R}}$$

Fig. 8.28: Synthesis of dialkyl amines and dialkyl thioethers.

8.5.5 Addition of Terminal Alkenes to Alcohols and Thiols

Alcohols add on to terminal alkenes in presence of montomorillonite to give ethers[85, 86]. However, thiols add on to terminal alkenes in presence of acid-treated and Al (III)-exchanged montmorillonites at 200° to give dialkylsulphides is good yield (Fig. 8.29)[87].

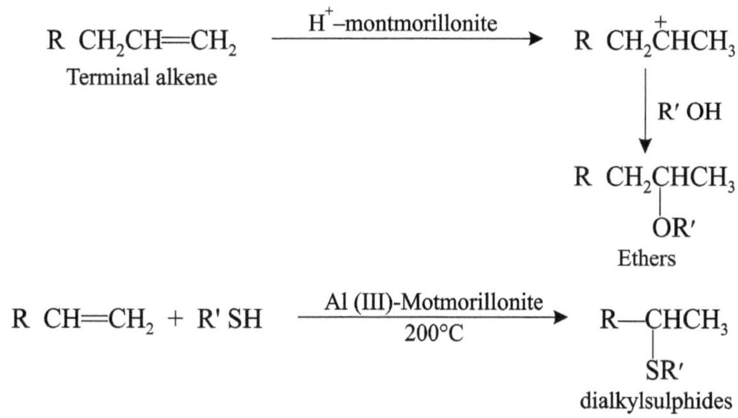

Fig. 8.29: Synthesis of ethers and thioethers.

8.5.6 Friedel-Crafts Reaction

Montmorillonite supported transition metals salts (zinc and nickel chlorides) are selective reagents for Friedel-Crafts alkylation. This catalytic system circumvents some of the drawbacks associated with using $AlCl_3$ (Fig. 8.30)[88,89].

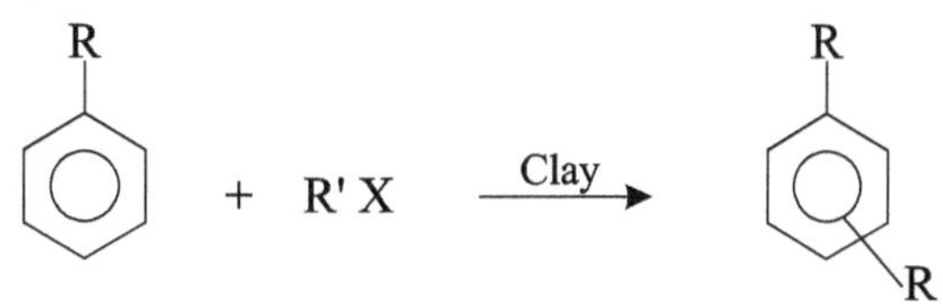

Fig. 8.30: Friedel-crafts alkylations.

It has been found[90] that Friedel-crafts acylation of anisole, mesitylene and xylene using Fe (III) modified clay gives quantitative yield of the acyl derivatives. Aromatic hydrocarbons can be nitrated[91] by montmorillonite supported cupric nitrate. Halobenzenes on nitration give the *p*-nitro compounds[92]. Electrophilic substitution of protonated formaldehyde to aromatic amines and phenol using acid treated montmorillonite give diarylmethane derivatives which are commercially important (Fig. 8.31).

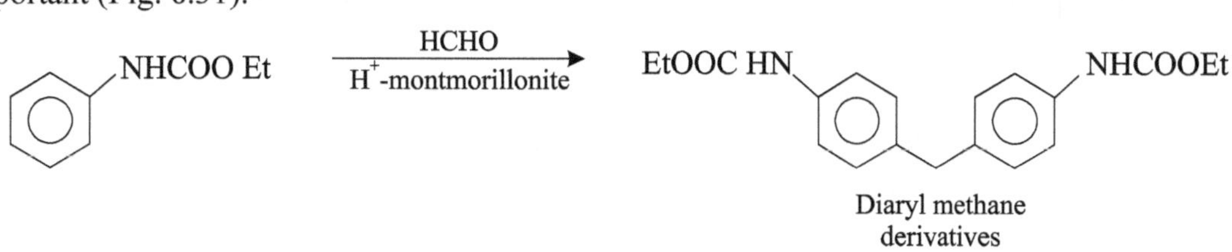

Fig. 8.31: Synthesis diarylmethane derivatives.

8.5.7 Diels-Alder Reaction

Fe (III) doped K10-montmorillonite along with 4-*tert*-butylphenol is an excellent catalytic system of unactivated dienophiles in Diels-Alder reaction[93]. It is found that cation-exchanged K10-montmorillonite on calcination and silylation (in order to eliminate acidic sites) give catalytic system which promotes Diels-Alder reaction of cyclopentadiene with trans-anethole[94] see also section 8.6.7 and 8.6.12.

8.5.8 Heck Reaction

It is an palladium-catalysed reaction of aryl halides with olefin acid is an important C–C bond forming reaction[95]. It has been found[96] that Pd Cl_2 or Pd $(OAc)_2$ in presence of tetraphenyl phosphonium salts exhibit much better catalitic activity. In the homogeneus reaction, the palladium salts are reduced to an active zerovalent palladium species. The catalyst activity and its reusability can be improved by a new clay-based catalyst system, palladium chloride and tetraphenylphosphonium bromide ($PdCl_2/Ph_4PBr$) intercalcalated clay (cat-I). This catalyst can be prepared[97] by refluxing sodium-exchanged clay with $PdCl_2$ and tetraphenylphosphonium bromide solution in water for about 48 hr. The clay material is finally filtered, washed with water and dried in an oven at 100-110°C for about 12 hrs. As an example, an equimolar amount (1 m mol) of aryl halide and styrene on refluxing with the cat-I and sodium acetate in DMF with continuous stirring gave the coupled product (Fig. 8.32).

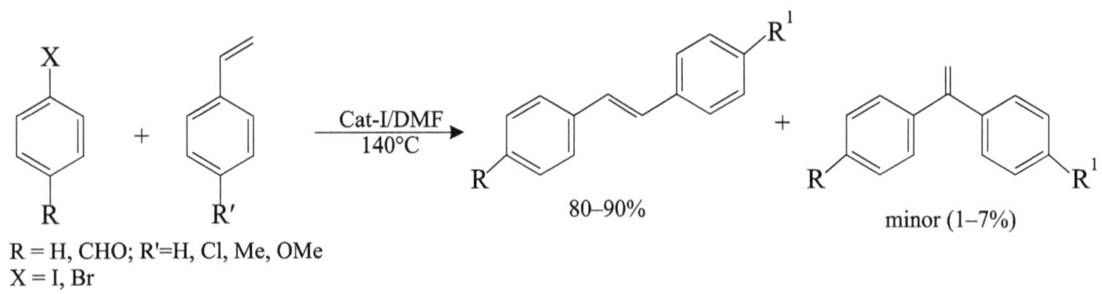

Fig. 8.32: Heck Reaction using a new clay-based catalytic system.

8.5.9 Suzki Coupling

The coupling of aryl halides with arylboronic acids or esters to give biaryl compounds is an important C–C bond forming reaction[98]. The use of cat. I system ($PdCl_2$, Ph_4 PBr)[97] gave much better yield of the biaryl compounds via cross-coupling reaction (Fig. 8.33)[99].

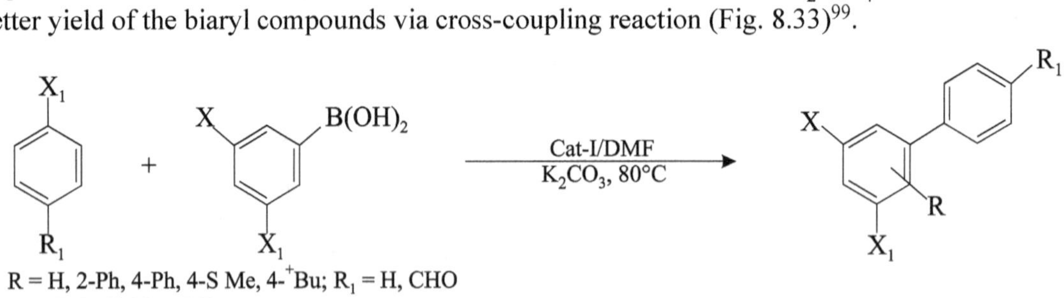

Fig. 8.33: Suzuki coupling using a new clay based catalytic system.

The above reaction (Fig. 8.33) can also be conducted in aqueous media using ultra sound[100].

8.6 MISCELLANEOUS SOLVENT-FREE ORGANIC TRANSFORMATIONS

8.6.1 Synthesis of Calix [4] resorcin arene

A supramolecular building block, Calix [4] resorcinarene is prepared by the condensation of equimolar quantities of benzaldehyde and resorcinol in presence of *p*-tolunesulfonic acid by grinding together in a mortar and pestle. A viscous paste is obtained in red colour after standing for 1 hr. The product is isolated by treating the mass with hot water (to remove PTS) and crystallised from methanol (Fig. 8.34)[101]

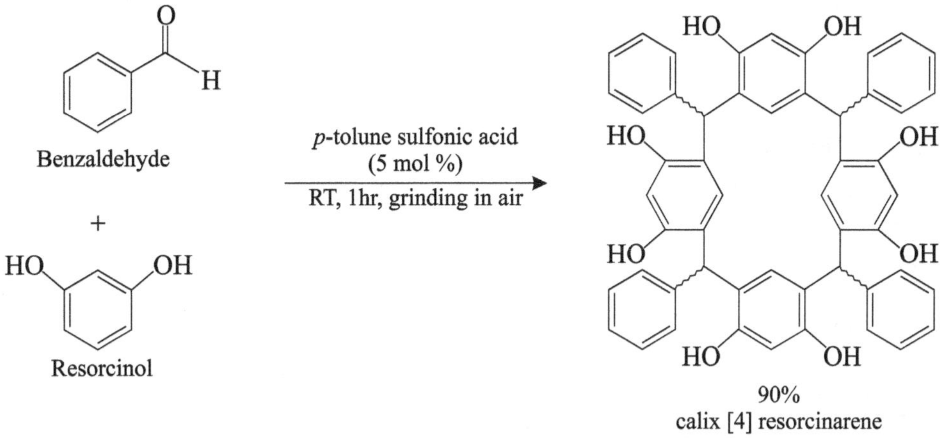

Fig. 8.34: Synthsis of calix[4] resorcinarene.

8.6.2. Oxidation of Benzophenone with m-CPBA

Benzophenone on oxidation with *m*-chloroperbenzoic and gives phenyl benzoate in solid state (Fig. 8.35)[102].

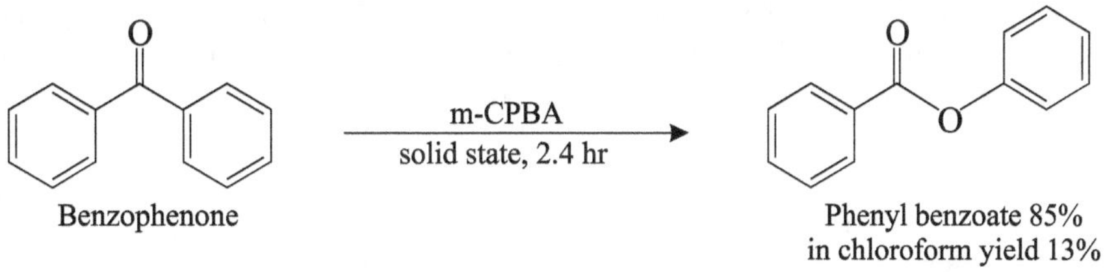

Fig. 8.35: Oxidation of benzophenone with m-CPBA in solid state.

8.6.3 Reduction of Ketones to Secondary Alcohols

Ketones, *e.g.* benzophenone on reduction with $NaBH_4$ in solid state gives the corresponding secondary alcohol (Fig. 8.36)[103].

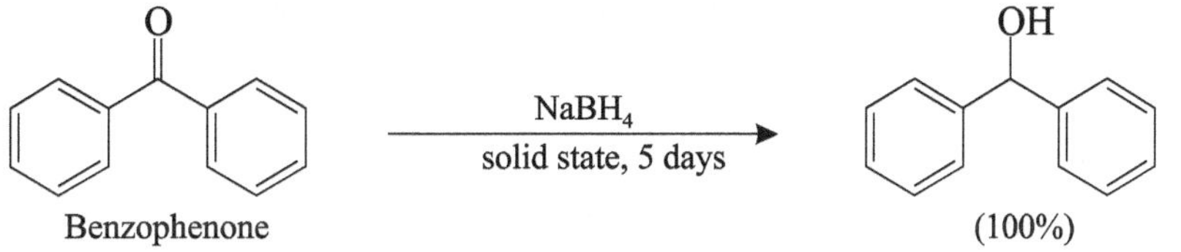

Fig. 8.36: Reduction of ketones with Na BH_4 in solid state.

see also section 8.6.10.

8.6.4 Michael Addition

The reaction of 2-phenylcyclohexanone with phenyl styryl ketone in presence of pot. *t*-butoxide and *n*-Bu_4 N Br under solvent free conditions give the Michael adduct (Fig. 8.37)[103].

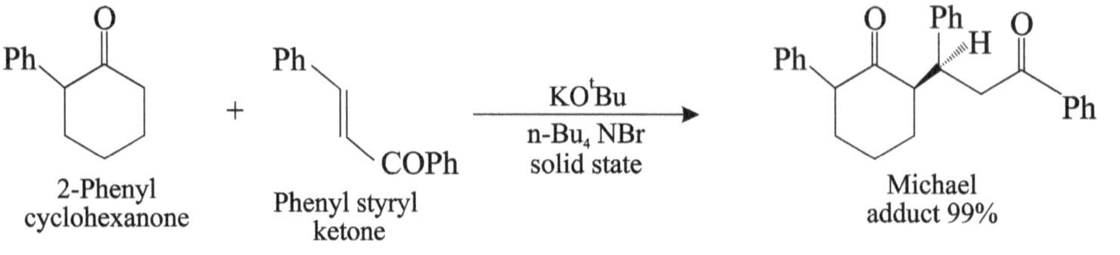

Fig. 8.37: Michael addition.

8.6.5 Photoaddition

The reaction of carbazole with diphenyl ethylene in solid state in presence of *uv* light gives the photoadduct (Fig. 8.38)[103].

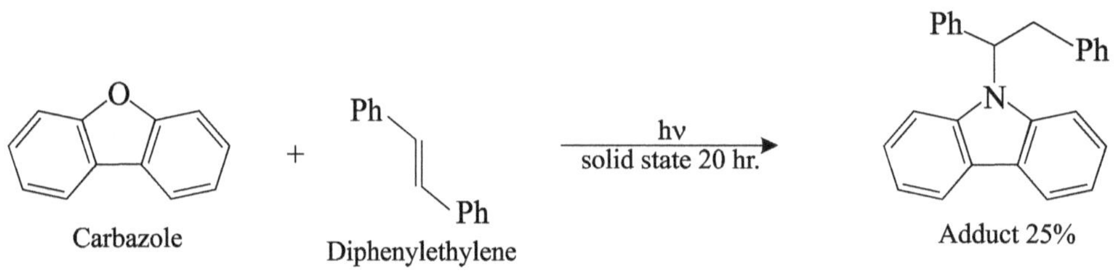

Fig. 8.38: Photoaddition.

8.6.6 Biginelli Condensation

Biginelli condensation of 1.3-dicarbonyl compounds with an aldehyde and urea by heating to 100° for 1 hr without solvent or catalyst give dihydropyrimidines, which have promising biological activites as antihypertensive and anticancer drugs (Fig. 8.39)[104].

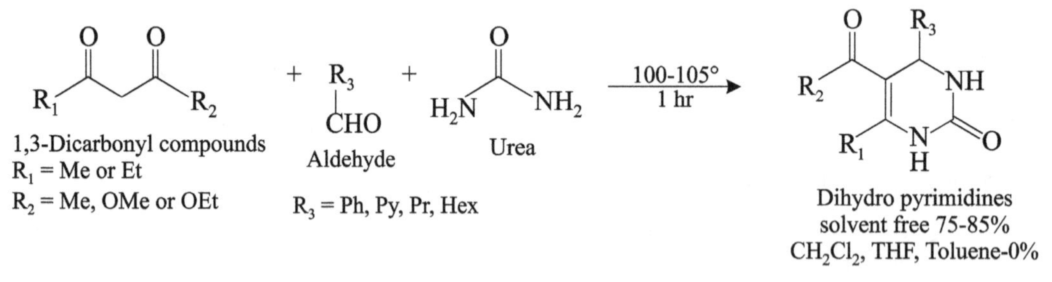

Fig. 8.39: Solvent free Biginelli condensation.

The above reaction can be scaled up to 1 kg.

8.6.7 Diels-Alder Cycloadditions

Diels-Alder cycloadditions can be catalysed by $HfCl_4$. 2THF in air and under solvent free conditions[105]. Thus 3-Ethoxycarbonyl coumarins on reaction with 1,3-butadines give cycloadduct in 95% yield. (Fig. 8.40).

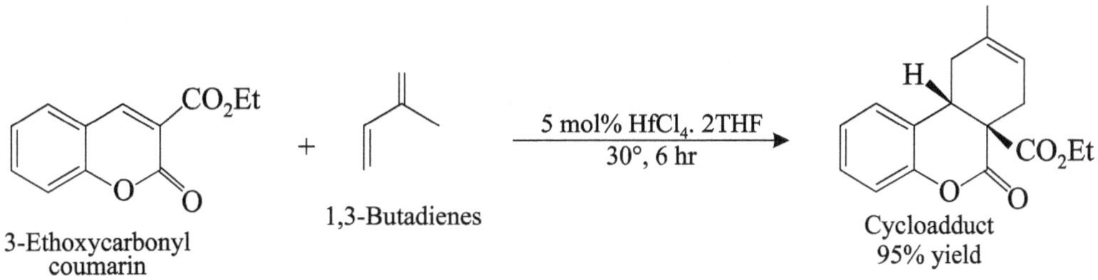

Fig. 8.40: Diels-Alder cycloadditions.

See also section 8.6.12.

8.6.8 Baylis-Hillman Reaction

The reaction of electrophile (generally an aldehyde) and an electro deficient olefin using high-speed ball mill (HSBM) gave the adduct under solvent free conditions (Fig. 8.41)[106]

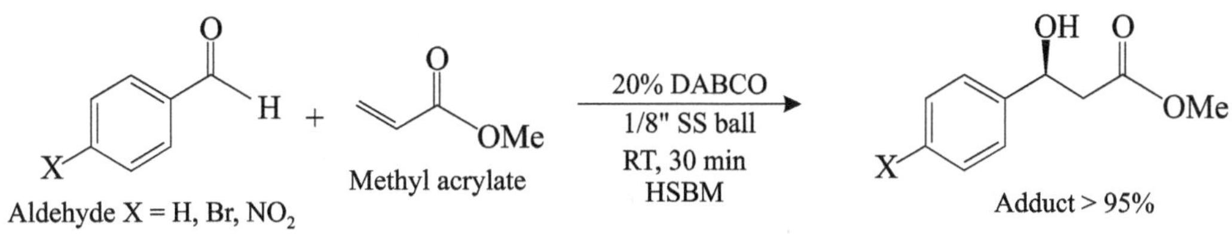

Fig. 8.41: Baylis-Hillman reaction Under Solvent free conditions and HSBM.

8.6.9 Cannizzaro, Tischenko and Meerwein-Ponndorf-verley Reactions

All the three reactions are promoted by lithium bromide and triethylamine under neat reaction conditions (Fig. 8.42)[107].

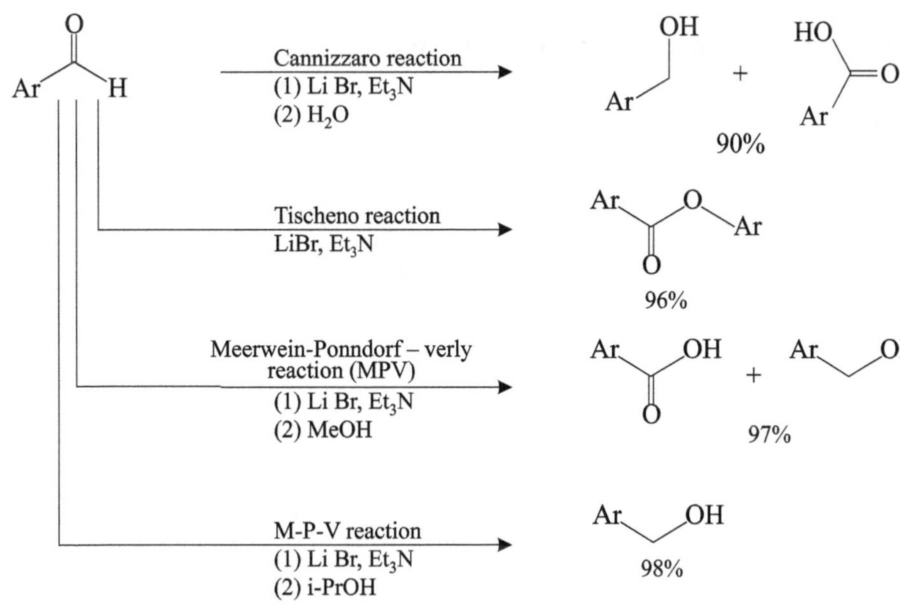

Fig. 8.42: Solvent free cannizzaro, Tischeno and Meerwein-Ponndorf-Verley reactions.

8.6.10 Reduction of Substituted Benzaldehydes, Acetophenones and Methyl Benzoates with $NaBH_4$ using HSBM

Substituted benzaldehydes, acetophenones and methyl benzoates could be reduced with $NaBH_4$ in air using an HSBM (Fig. 8.43)[108].

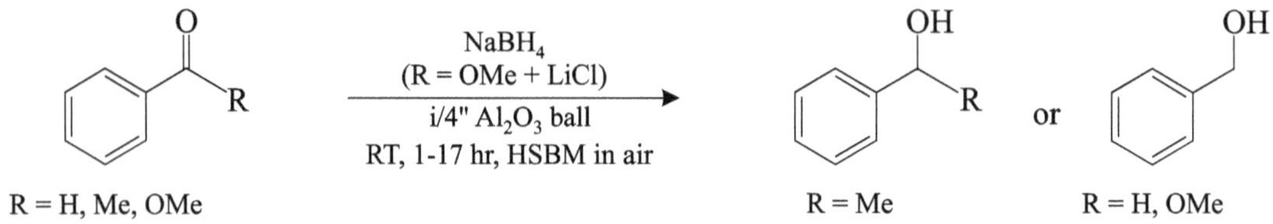

Fig. 8.43: Solvent free reductions of substituted benzaldehydes, acetophenones and methyl bonzoates with $NaBH_4$ using HSBM

8.6.11 Synthesis of S-alkyl Dithiocarbamates

A number of S-alkyl dithiocarbomates, useful for pharmaceutical and agrochemical industries are prepared by a one-pot reaction of an aliphatic primary and secondary amine, carbon disulfide and an alkyl-halide (Fig. 8.44)[109].

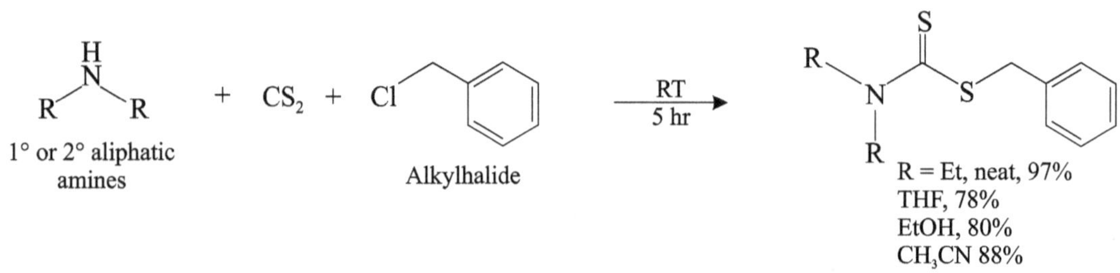

Fig. 8.44: Catalyst free one pot synthesis of alkyl dithiocarbonates.

8.6.12 Asymmetric Hetero-Diels-Alder Reaction

Asymmetric hetero-Diels-Alder reactions between Danishefsky's diene and benzaldehyde under solvent free conditions in presence of TADDOL-Cataylst gives the (S)-enantiomer adduct (Fig. 8.45).[110]

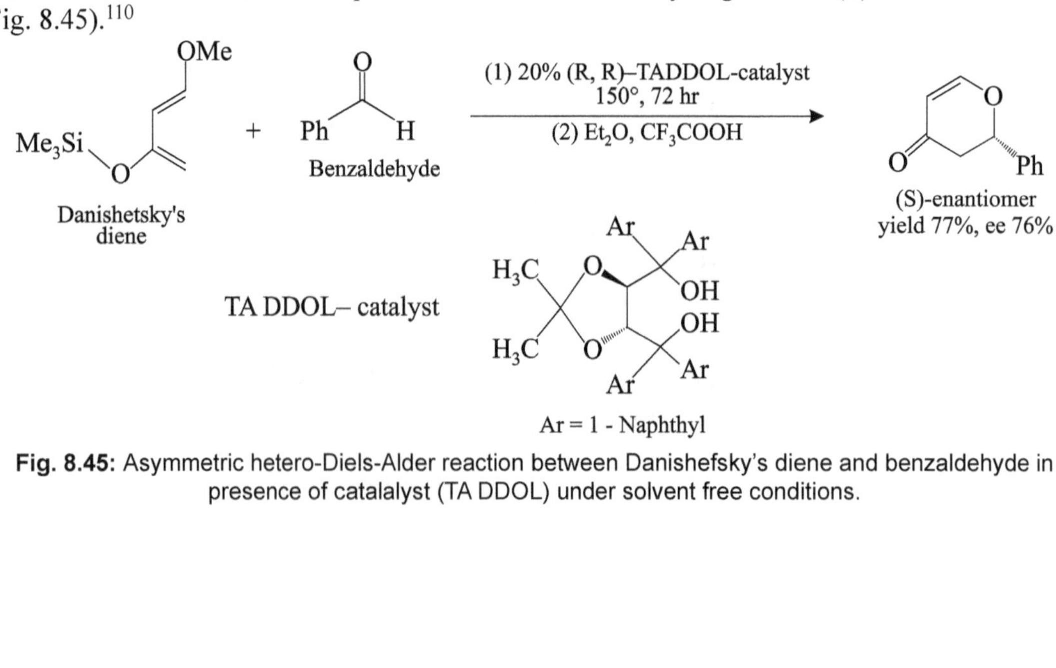

Fig. 8.45: Asymmetric hetero-Diels-Alder reaction between Danishefsky's diene and benzaldehyde in presence of catalalyst (TA DDOL) under solvent free conditions.

8.7 MICROWAVE ASSISTED SOLVENT FREE ORGANIC TRANSFORMATION

See Chapter 9 Section 9.4.

Conclusion

Solvent-free organic transformations or solventless reactions are the best choice in the context of Green Chemistry. It avoids the use of volatile organic solvents which is the main problem associated with environmental pollution. These reaction can be performed by grinding the reactants using a pestle mortar or by ball-milling. These can also be performed by heating the reactants and by heating using clay supported reagents. A number of such reactions can also be performed by heating the reactants using microwaves. All these reactions are important for use on commercial scale.

REFERENCES

1. W. Wef and R. Bakthavatechalam, Tetrahedron Lett., 1991, **32**, 1535.
2. F. Toda, K. Tanaka and K.Hamai, J. Chem. Soc., Perkin Trans 1990, 3207.
3. H. Tanka, S. Kishigami and F. Toda, J. Chem. Soc., 1991, **56**, 4333.
4. M.G. Moloney, Nat. Prod. Rep., 2002, **19**, 597.
5. F.F. Jian and P.S. Zhao, J. Mol. Struct, 2004, **705,** 133; H.C. Cheon, C.M. Lee, B.T. Kimb and K. J. Hwangb, Bioorg. Med. Chem. Lett., 2004, **14,** 2661.
6. H.T. Thakuria, A. Pramanik, B.M. Borah and G. Das, Tetrahedron Lett., 2006, **47,** 3135.
7. H. Thakuria and G. Das, J. Chem. Sci, 2006, **118,** 425.
8. N.S. Simpkins. In sulphones in Organic Synthesis; J.E. Baldwin, Ed., Pergamon Press, Oxford, 1993.
9. A. Svatos, Z. Hunkova, V. Kren, M. Hoskovec, D. Sÿaman, I. Valterova, J. Verloc and B. Koutek, Tetrahedron Asymmetry, 1996, 1, 1285; P. Bertus, P. Phansavath, V. Ratovelomanana-Vidal, J.–P. Genet, A.R. Touati, T. Homri and B.S. Hassine, Tetrahedram Asymmetry, 1999, **10,** 1369; V. Gotor, F. Rebolledo and R. Liz, Tetrahedron Asymmetry, 2001, **12,** 513.
10. W.M. Wolf, J. Mol. Struct., 1993, **474,** 113.
11. B.M. Trost and D.P. Curran, Tetrahedran Lett., 1981; **22**, 1287; A.–L. Fan, S. Cao, Z. Zhang, J. Heterocyl Chem., 1997, **34,** 1657.
12. N. Kamigata, K. Udodaira and T. Shimizu, J. Chem. Soc. Perkin, Trane. I, 1997, 783.
13. C.R. Holmquist and E.J. Roskamp, Tetrahedran Lett., 1992, **33**, 1131.
14. G.E. Vennstra and B. Zwaneburg, Synthesis, 1975, 519; J. Wildeman and A.M. Vanleusen, Synthesis, 1979, 733.
15. A.R. Katrizky, A.A.A. Abdel-Fattah and M.J. Wang, J. Org. Chem., 2003, **68,** 1443.
16. C. Lai, C. Yi, Y. Ziang and R. Hua, Tetrahedron Lett., 2005, **46,** 513.
17. D. Kumar, S. Sundaree, V.S. Rao and R.S. Varma, Tetrahedron Lett, 2006, **47**, 4197.
18. D. Kumar M.S. Sundaree. G. Patil, V.S. Rao and R.S. Varma, Tetrahedron Lett., 2006, **47,** 8 239.
19. N.D. Heindel and J.R. Reid, Hetrocycl Chem., 1980, **17,** 1087; B.S. Holla, B. Kalluaya, K.R. Sridhar, E. Drake, L.M. Thomas, K.K. Bhandary and M. Levne, Eur. J. Med. Chem., 1994, **29,** 301; M.A. Ghannoum, N.F. Eweiss, A.A. Bahajaj and M.A. Qureshi, Microbios, 1983, **37,** 151; S. Demirayak, K. Benkli and K. Guven, Eur. J. Med. Chem., 2000, **35,** 1037.

20. D.I. Shiho and S. Tagami, J. Am. Chem. Soc., 1960, **82,** 4044.
21. Z. Guolin and H. Yongzhou, J. Chem. Res (S), 2002, 560.
22. G. Jayanthi, S. Muthusamy, R. Paramasivam, V.I. Ramakrishanan, N. K. Ramasamy and P. Ramamurthy, J. Org. Chem., 1997, **62,** 57666.
23. O.V. Singh and V. George, Synth. Commun. 1994, **24,** 2627; D. Kumar, O. Prakash and S.P. Singh, J. Chem. Res (S), 1993, 244.
24. D. Kumar, V.G. Kondapalli, Chandra Sekhar H. Dhillon, V.S. Rao and R.S. Varma, Green Chem., 2004, **6,** 157.
25. M.I. Hearn, J. Schulz, P. Sinha, S. Collins, S. Hallenbeek and M. Kustin, J. Heterocycl. Chem., 1989, **26,** 581.
26. L. Sarikia, J.M. Baruah and A.J. Thakur, Organic and Medicinal Chemistry Letters. 2011, 1 : 12 http:// www. orgmed chemlett.com/content//1/1/12.
27. H. Hagiwara, S. Ohtsubo and M. Kato, Tetrahedron, 1997, **53,** 2415.
28. M. Ghiaci, J. Asghari, Synth. Commun. 1998, **28,** 2213; M. Ghiaci and G.H. Imanzadeh, Syn. Commum., 1998, **28,** 2275.
29. D. Sharma, S. Kumar and J.K. Makrande, Green Chem. Lett., 2009, **2,** 53.
30. TH Grindstaff, Chem. Eng, News Nov. 7, 1994, Page 3.
30a. D. Wang, N.Li, M. Zhao *et al*, Green Chem., 2010, **12,** 2120.
31. V.P. Balema, V.K. Percharsky *et al.*, J.Am. Chem. Soc., 2002, **124,** 6244.
32. R. Antonioletti, F. Bonadiels, A. Ciammaichella and A. Viglianti, Tetrahedron, 2008, **64,** 4644.
33. A. Baran, J. Martinez and F. Lamaty, Tetrahedron Lett. 2010, **51,** 6246.
34. W.C. Shearouse, C.M. Korte and J. Mach, Green, Chem., 2011, **13,** 598.
35. J.G. Hernandez and E. Juaristi, J. Org. Chem., 2011, **76,** 1464.
36. G. Guillena, M.d.C. Hita, C. Najera and S.F. Viozquez, Tetrahedron Asymmetry, 2007, **18,** 2300.
37. B. Rodriguez, T. Rantanen and C. Bolm, Angew. Chem., Int. Ed., 2006, **45,** 6924.
38. G. Kaupp. M.R. Naimi-Jamal and J. Schmeyers, Tetrahedron, 2003, **59,** 3753.
39. D.A. Fulmer, W.C. Shearouse, S.T. Medonza and J.Mack, Green Chem., 2009, **11,** 1821.
40. R. Thorwirth, A. Stolle and B. Ondruschka, Green Chem., 2010, **12,** 1985.
41. E. Tullberg, D. Peters and T. Frejd, J. Organomet. Chem, 2004, **689,** 3778.
42. S.F. Nelson, D. Peters and O. Axelsson, Synth. Commun., 2000, **30,** 350.
43. F. Schneider, and B. Ondruschka, Chem. Suschem, 2008, **1,** 622.
44. F. Schneider, A. Stolle, B. Ondruschka and H. Hopf, Org., Process Res. Dev., 2009, **13,** 44.
45. C.H. Hassall, Org. React., 1957, **9,** 73.
46. Y. Ogata and Y. Sawaki, J. Org. Chem., 1969, **34,** 3985.
47. J. Boeseken, W.D. Cohen and C.J. Kip, J. Recl. Trav. Chim. Pays –Bas, 1936, **55,** 875.
48. R.S. Varma and K.P. Naicker, Org, Lett., 1999, **1,** 189.
49. R.E. Steiger, Org. Synth, 1955, **3,** 663.
50. E.N. Zilberman, Usp. Khim, 1984, **53,** 1523; Chem. Abstr., 1984, **101,** 210082h.
51. K.T. Liv., M.H. Shih, H.W. Huang, and C. J. Hu, Synthesis, 1988, 715.
52. S.E. Diamond, B. Grant, G.M. Tom and H. Taube, Tetrahedron Lett, 1974, 4025.

53. T. Mukaigama, K. Kamino, S. Kobayashi and H. Takei, Chem. Lett, 1973, 357.
54. C.G. Rao, Synth. Commum, 1982, **12,** 17.
55. J.H. Hall and M.J. Gisler, J. Org. Chem 1976, **41,** 3769.
56. G.W. Kabalka, S.M. Deshpande, P.P. Wadgaonkar and N. Chatla, Synth. Commum, 1990, **20,** 1445.
57. T. Miyazawa, T. Endo and M. Okawarna, Synthesis 1984, 1034.
58. M. Gazdar and S.J. Smiles, J. Chem. Soc., 1908, **93,** 1833; D.A. Peak and T. I. Watkins, J. Chem. Soc., 1950, 445.
59. D. Edwards and J. B. Stenlaki. J. Chem. Soc., 1954, 3272; R. Knoll, J. Praki. Chem, 1926, **113,** 40.
60. C. Marcker, Justus Liebigs Ann. Chem., 1865, **136,** 75; F. G. Bordwell and P.J. Boutan, J. Am. Chem. Soc., 1957, **79,** 717.
61. L. Horner, H. Schaefer, W. Ludwig, Chem. Ber. 1958, **91,** 75; D. J. Barnard, J. Chem. Soc., 1957, 4547.
62. C.G. Overberger and R.W. Cummins, J. Am. Chem. Soc., 1953, **75,** 4250.
63. N.N. Melnikov, Usp. Khim, 1936, 5, 443.
64. R.G. Hiskey and M.A. Harpold, J. Org. Chem., 1967, **32**, 3191.
65. D.H.R. Barton, G.R.A. Godfrey, J.W. Morzycki, W.B. Motherwell and A. Stobie, Tetrahedron, 1982, 23, 457.
66. A. Mckillop and J.A. Tarbin, Tetrahedran Lett, 1983, **24,** 1505.
67. R.W. Whitaker and H.H. Sister, J. Org. Chem., 1960, **25,** 1038.
68. J. Muzart, Synthesis, 1995, 1325; A. Ohta and M. Ohta, Synthesis, 1985, 216; B. Klein and J. Berkowitz, J. Am. Chem. Soc., 1959, **81,** 5160.
69. A. Cornelis and P. Laszlo, Synthesis, 1980, 849.
70. M. Basemann, A. Cornelis and P. Laszlo, C.R. Hebd, Seances, Acad. Sci., 1984, **299,** 427.
71. A. Cornelis, N. Depaye, A. Gerstmans an P. Laszlo, Tetrahedron Lett., 1983, 24, 3103.
72. A. Cornelis, P.Y. Herze and P. Laszlo, Tetrahedran Lett., 1982, **23,** 5035.
73. P. Laszlo and E. Polla, Synthesis, 1985, 439.
74. A. Cornelis and P. Laszlo, Adrichim Acta 1988, 97.
75. M. Balogh, I. Hermecz, Z. Meszaros and P. Laszlo, Hel. Chim. Acta, 1984, **67,** 2270.
76. P. Laszlo, P. Pennetreau and A. Krief, Tetrahedron Lett., 1986, **27,** 3153.
77. S. Chalais, A. Cornelis, P. Laszlo and A. Mathy Tetrahedron Lett., 1985, **26,** 2327.
78. S. Barlow, J.H. Clark, M. R. Darby, A.P. Kybett, P. Landon and K. Marhn, J. Chem. Res. (S), 1991, 74.
79. S.J. Barlow, T.W. Bastock, J.H. Clark and S.R. Cullen. J. Chem Soc., Perkin Trans 2, 1994, 411.
80. C. Kato, K. Kuroda and H. Takahura, Clays Clay Miner. 1981, **29,** 294.
81. M.J. Tricker, D.T. B. Tennakoon, J.M. Thomas and J. Heald, Clays Clay Miner, 1975, **23,** 77.
82. J.A. Ballantine, K.J. Williams and J.M. Thomas, J. Mol. Catal, 1984, **26,** 37.
83. J.A. Ballantine, J.H. Purnell, M. Raganakorn, K.J. Williams and J.M. Thomas, J. Mol. Catal., 1985, 30, 373.
84. J.A. Ballantine, J.H. Purnell, M. Rayanakorn, K.J. Williams and J.M. Thomas, J. Mol. Catal., 1995, **30,** 373.

85. Gulf oil Canada, U.S. Patent, 4042633 (1980).
86. A. Bylina, J.M. Adam, S.H. Graham and J.M. Thomas, J. Chem. Soc. Chem. Cammun., 1980, 1003.
87. J.A. Ballantine, J.H. Purnell and J.M. Thomas, Clay Miner, 1983, **18,** 347.
88. J.H. Clark, A.P. Kybett, D.Macquarrie, S.L. Barlow and P. Landon, J. Chem., Soc., Chem. Commun. 1989, 1353.
89. P. Laszlo and A. Mathy., Helv. Chim. Acta., 1987, **70,** 559.
90. A. Cornelis, A. Gerstmans, P. Laszlo, A. Mathy and I. Ziebh, Catal. Lett., 1990, **6,** 103.
91. P. Laszlo and J. Vandormel Chem. Lett., 1988, 1843.
92. P. Laszlo and P. Pennetreau, J. Org. Chem., 1987, **52,** 2407.
93. P. Laszlo and J. Lucchette, Tetrahedron Lett., 1984, **25,** 1567.
94. J.M. Fraile, J.I. Garcia, J.A. Mayoral, T. Tarnai and P.J. Alonsu, J. Chem., Soc., Chem., commun., 1996, 1981.
95. R.F. Heck, Palladium reagention is Organic synthesis, Academic Press, London, 1985; A de Meijere and F.E. Meyer, Angew Chem. Int. Ed, Engl., 1994, **33,** 2379.
96. M.T. Reiz, G. Lohmer and R. Schwickardi, Angew Chem. Int. Ed. Engl., 1998, 37, 481.
97. R.S. Varma, K.P. Naicker and P.J. Liesen, Tetrahedron Lett., 1999, **40,** 2075.
98. S.P. Stanforth, Tetrahedron, 1998, **54,** 263; N. Miyaura and A. Suzuki, Chem., Rev., 1995, **95,** 2457; T. Ishigama, H. Kizaki, T. Hayashi, A. Suzuki and N. Miyaura, J. Org. Chem., 1998, **63,** 4726.
99. R.S. Varma and K.P. Naicker, Tetrahedron Lett., 1999, **40,** 439; R.S. Varma and K.P. Naicker, Green Chem., 1999, 1, 247.
100. R.S. Varma, K.P. Naicker, Tetrahedron Lett., 1999, **40,** 439; R.S. Varma, K.P. Naicker, Green Chem. 1999, **1,** 247.
101. B.A. Roberts, G.W.V. Cave, C.L. Raston and J.L. Scott., Green Chem., 2001, **3,** 280.
102. K. Tanka and F. Toda, Chem. Rev., 2000, **100,** 1025.
104. B.C. Ranu, A. Hajra and S.S. Dey, Org. Process. Res-Dev., 2002, **6,** 817.
105. F. Fringuelli, R. Girotti, F. Pizzo E. Zunino and L. Vaccaro, Adv. Synth. Catal., 2006, **348,** 297.
106. J. Mack and M. Shumba, Green Chem., 2007, **9,** 328.
107. M.M. Mojtabedi, E. Akbarzadeh, R. Sharifi and M.S. Abaee, Org. Lett, 2007, **9,** 2791.
108. J. Mack, D. Fulmer, S. Stofel and S. Santos, Green Chem., 2007, **9,** 1041.
109. N. Azizi, D. Fulmer, S. Stofel and N. Sastos, Green Chem., 2007, **9,** 1041.
110. X. Zhang, H. Du, Z. Wang, Y.D. Wu and K. Ding, J. Org. Chem., 2006, **71,** 2862.

9. MICROWAVE ASSISTED ORGANIC TRANSFORMATIONS

9

Microwave Assisted Organic Transformations

9.1 INTRODUCTION

The use of microwave energy normally result in good yields in a very short times. Using microwaves a number of organic transformations have been carried out. The technique provides rapid and relatively inexpensive access to high temperatures and pressures.

A microwave oven normally consists of a high power source, a wave guide and the oven cavity. The source is normally a magnetron in which microwaves are generated. A magnetron is a thermionic-diode having an anode and a directly heated cathode. As the cathode is heated, electrons are released and are attracted towards the anode. The andoe is made up of an even number of small cavities, each of which acting as a tuned circuit. The gap across the end of each cavity behaving as a capacitance.

The variable power available is a domestic microwave oven is produced by switching the magnetron on-and-off according to a duty cycle. As an example, a typical 600W oven with a 30 sec duty cycle can be made to deliver an average of 300W by switching the magnetron on and off every 15 sec.

For microwave induced reaction, the reaction vessel is a tall beaker with a capacity greater than the volume of the reaction mixture. Alternatively Telfon or polystyrene vessels can be sued. These materials are transparent to microwaves. Metallic containers cannot be used as a reaction vessel as it gets heated due to preferred absorption and reflection of rays.

The reactions in microwave oven can be conducted is a solvent medium or a solid support in which no solvent is used. For carrying the reaction in a solvent, the choice of the solvent is important. The solvent must have a dipole moment in order to absorb microwaves and a b.p. at least 20-30° higher then the desired reaction temperature. The solvents which can be used in a microwave oven are:

Solvent	Dielectric Constant	
	(ε)	B.P. (°C)
Ethanol	24	78
Dimethyl salfoxide	47	189
Isopopanol	0.799	82.5
I-propanol	0.757	97.2

Methanol	33	65
Water	80	100
Ethyl acetate	6	
Acetone	21	56.5
2-Butanone	18.4	79.6
Ethylene glycol	38	196
Acetic acid	6.1	118
Acetonitrile	38	81.6
Tetrahydrofuran	7.6	66

Solvents like dimethyl formanide, chloroform, dichloromethane, 1,4-dioxane and 1,2-dichloro ethane should not be used.

For solid state reactions, the reaction is performed on a solid support (no solvent) which couple effectively with microwaves. Examples of such solid support include phyllosilicate (M^{n+} – montomorillonite), silica and alumina.

The equipments, *viz.* microwave ovens are commercially available. For small scale laboratory preparation, domestic microwave ovens (Fig. 9.1) can be used. The microwave oven shown in Fig. 9.2 is Anton Paar Synthos 3000. It has two magnetrons and has a combined microwave output power from 0-1400 W. In this heavy-walled quartz reaction vessels (80 mL capacity) are used. These vessels can be used upto 80 bar pressure and high temperatures. The microwave oven shown in Fig. 9.3 has a continuous microwave power delivery system with power output 0-1600 W. In this, the reactions are performed in 1 or 3L RB flask. The temperature of the contents are monitered using a fibre-optic probe inserted into the reaction mixture by a glass thermowell. The contents of the vessel are stirred by a rotating magnetic plate, which is located below the floor of the microwave cavity and a Teflon-coated magnetic stirring bar in the vessel.

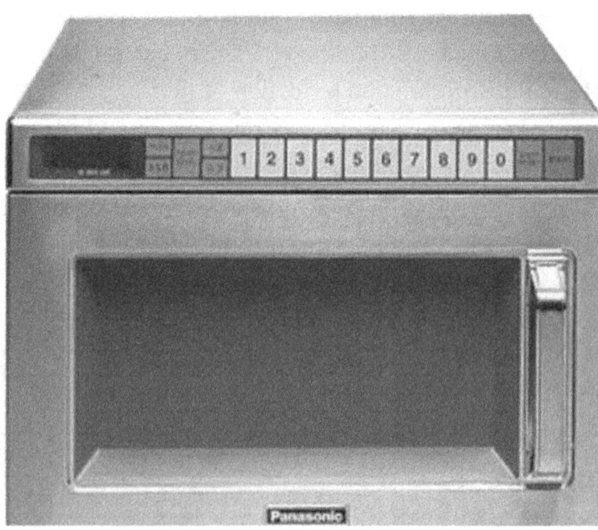

Fig. 9.1: Domestic MW oven.

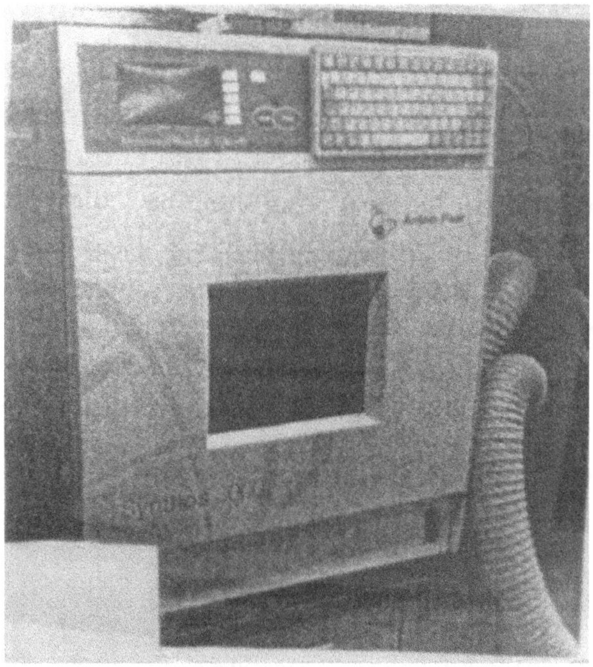

Fig. 9.2: MW oven with two magnetron (Anton Paar Synthos 3000).

Fig. 9.3: MW oven with continuous microwave-delivery system.

Broadly speaking, organic transformations using microwaves can be grouped in the following categories:

- M.W. assisted transformations in water.
- M.W. assisted transformations is organic solvent.
- M.W. assisted transformations in solid state.
- Scaling up of MW assisted transformations.

9.2 MICROWAVE ASSISTED ORGANIC TRANSFORMATIONS IN WATER

Water is the most abundant and cheapest green solvent available. It possesses high dielectric constant and is an excellent microwave absorbing material. Following are given some of the organic microwave assisted transformations in water.

9.2.1 Hofmann Elimination

The quaternary ammonium salts on heating at high temperatures yield the Hofmann elimination product is low yields. It is found that use of microwave irradiation in water/chloroform gave higher yields of thermally unstable Hofmann elimination product (Fig. 9.4).

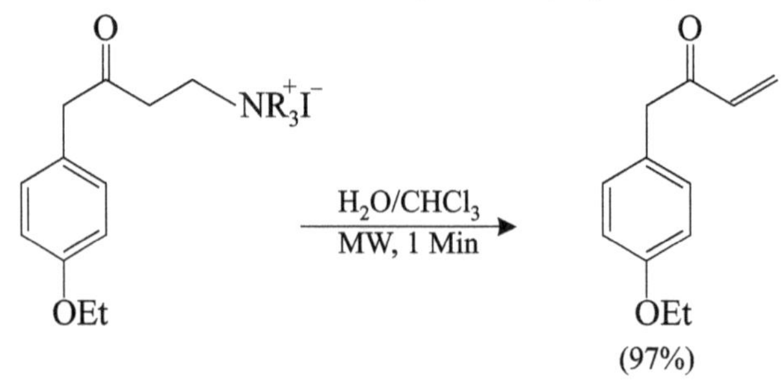

Fig. 9.4: Hofmann Elimination in H_2O-$CHCl_3$.

9.2.2 Hydrolysis of Benzyl Chloride

Benzyl chloride on hydrolysis by refluxing in water for about 35 min gives benzyl alcohol. However, hydrolysis with water in microwaves gives[1] 97% yield of benzyl alcohol (Fig. 9.5).

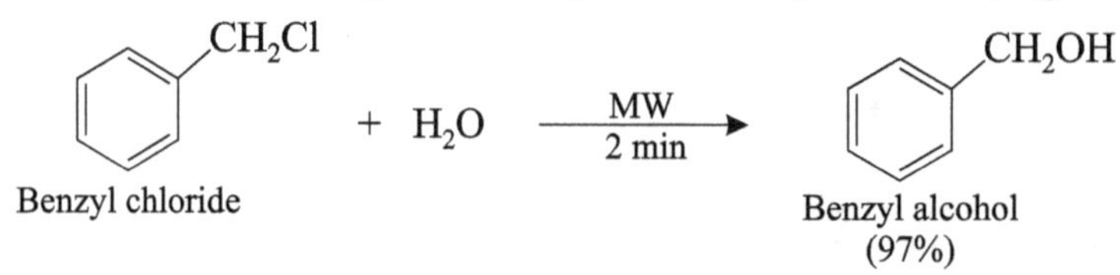

Fig. 9.5: Hydrolysis of benzyl chloride.

9.2.3 Hydrolysis of Benzamide

The hydrolysis of benzamide to benzoic acid by refluxing with dilute H_2SO_4 takes about 1 hr. However, under microwave irridation, the hydrolysis is complete in 7 min. giving 99% yield of benzoic and (Fig. 9.6).

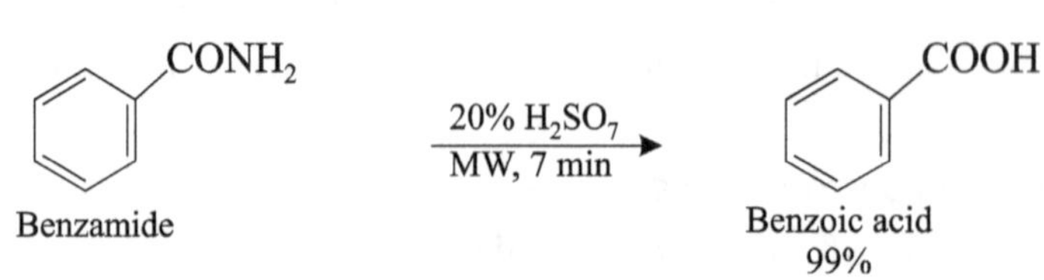

Fig. 9.6: Hydrolysis of Benzamide.

The reaction could be scaled up to 20 gms.

Similarly N-phenyl benzamide on heating with 20% H_2SO_4 is a microwave oven for 12 min. gives[2] benzoic acid is 74%, yield (Fig. 9.7). The conventional heating procedure takes about 18 hrs.

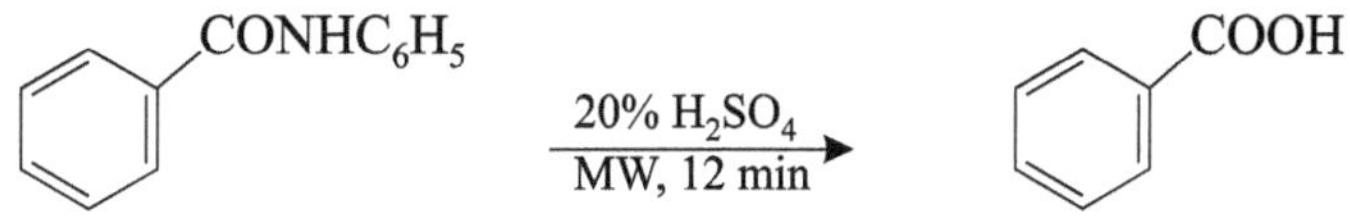

N-Phenyl benzamide Benzoic acid 74%

Fig. 9.7: Hydrolysis of N-phenyl benzamide.

9.2.4 Saponification

The esters on heating with aqueous sodium hydroxide under MW irradiation for 2.5 min give the corresponding acid (Fig. 9.8)[3]

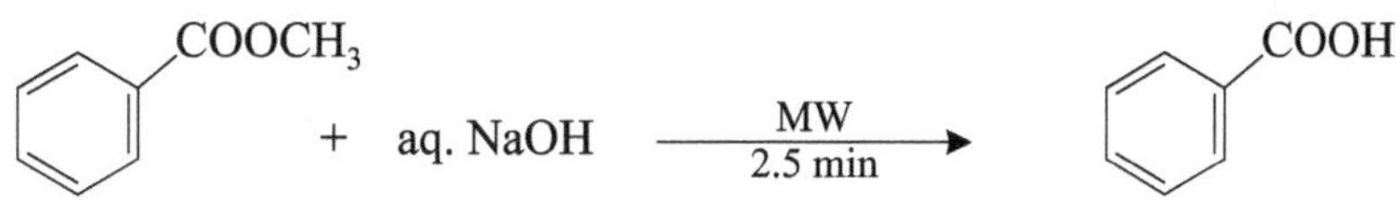

Methyl benzoate Benzoic acid

Fig. 9.8: Saponification.

See also chapter 10, section 10.2.2.

9.2.5 Demethylation

Phenolic ethers on heating with aqueous HBr in MW oven give the corresponding phenol (Fig. 9.9). The reaction takes about 70 hours on thermal heating.

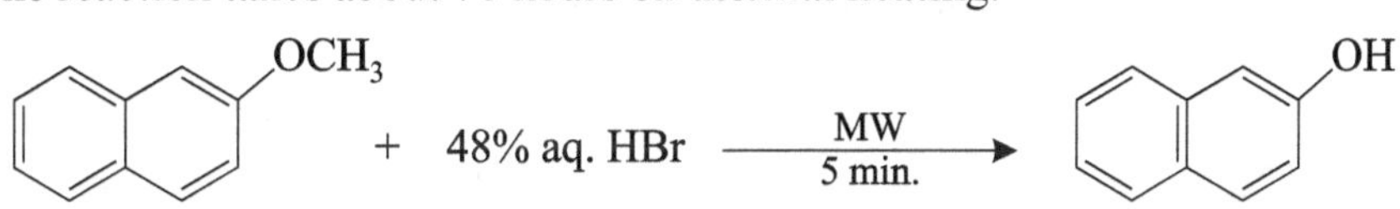

2-methoxy naphthalene β-Naphthol

Fig. 9.9: Demethylation.

9.2.6 Oxidation of Side Chains in Aromatic Compounds

Toluene on heating in MW oven with aq. $KMnO_4$ in presence of KOH give benzoic acid (Fig. 9.10)[3,4]. The reaction takes 5-6 hrs. on thermal refluxing.

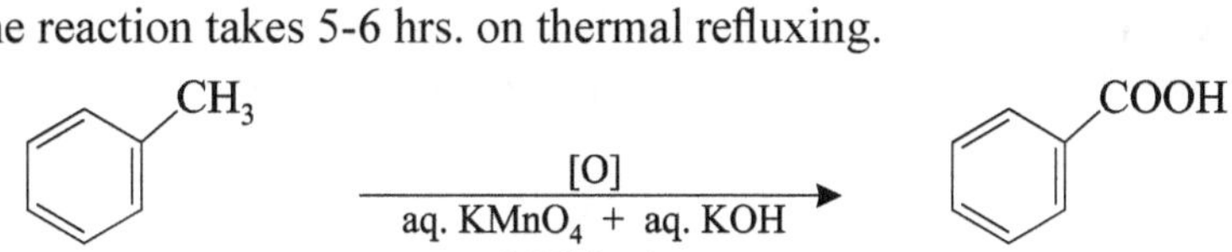

Toluene MW 5 min. Benzoic acid

Fig. 9.10: Oxidation of toluene.

9.2.7 Alkylation of Amines

Amines on heating in MW oven with alkyl halides in presence of aq. NaOH give the corresponding N-alkyl derivatives (Fig. 9.11)[5]. The reaction gives very poor yields under thermal conditions.

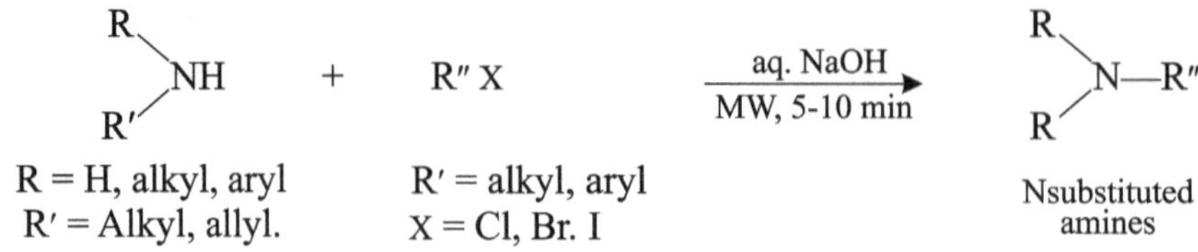

Fig. 9.11: Alkylation of Amines.

9.2.8 Synthesis of Azacycloalkanes

Aromatic amines on heating with 1,5-dibromopentane in aqueous K_2CO_3 for about 30 min. give azacycloalkanes (Fig. 9.12)[6]. The reaction takes several hours necessary for traditional synthesis[7].

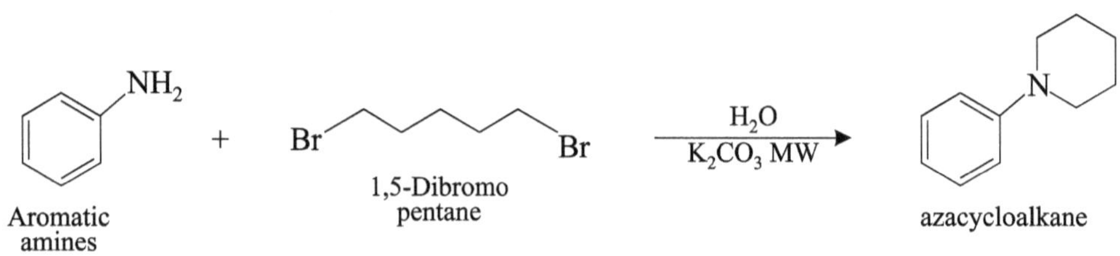

Fig. 9.12: Synthesis of aza cycloalkanes.

The above synthesis (Fig. 9.12) could be scaled up to 1 mol reactants.

9.2.9 Synthesis of Isoindolines

Aromatic amines on heating with o-dichloromethyl benzene in MW oven for about 30 min in presence of aqueous K_2CO_3 gave the corresponding isoindolines (Fig. 9.13).[6]

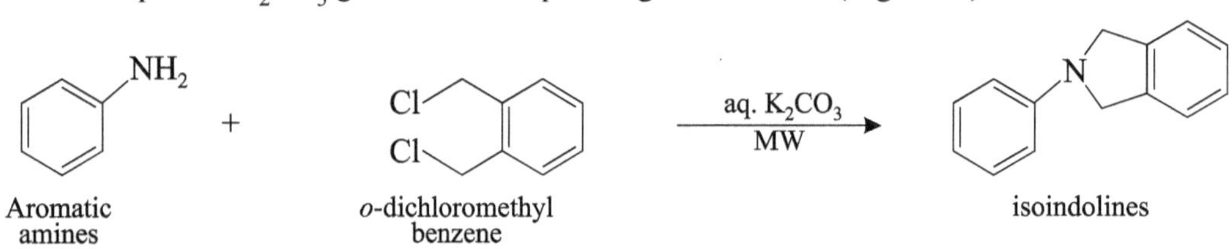

Fig. 9.13: Synthesis of isoindolines.

9.2.10 Sonogashira-type Coupling Reaction

The reaction between various aryl halides and arylalkenes in presence of aq. Na_2CO_3 and tetrabutyl ammonium bromide a PTC on MW heating gave the sonogashira-type coupling product (Fig. 9.14)[8]

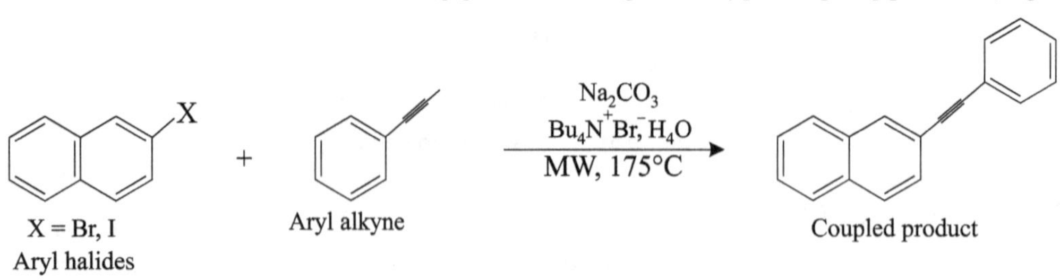

Fig. 9.14: Sonogashira-type coupling reaction.

9.2.11 Heck Coupling

Heck coupling of 4-bromoanisole (0.1 mol) and methyl acrylate (0.2 mol) in presence of tetrabutylammonium bromide as PTC, potassium carbonate (3.7 equiv.) as base and palladium loading of 0.002 mol% and water (as solvent) in microwave oven for 5 min. at 170°C and keeping the reaction mixture for 15 min more with continuous stirring using the MW oven (as shown in Fig. 9.3). Under these conditions a quantitative conversion of 4′-methoxy cinnamic acid was obtained (Fig. 9.15).[9]

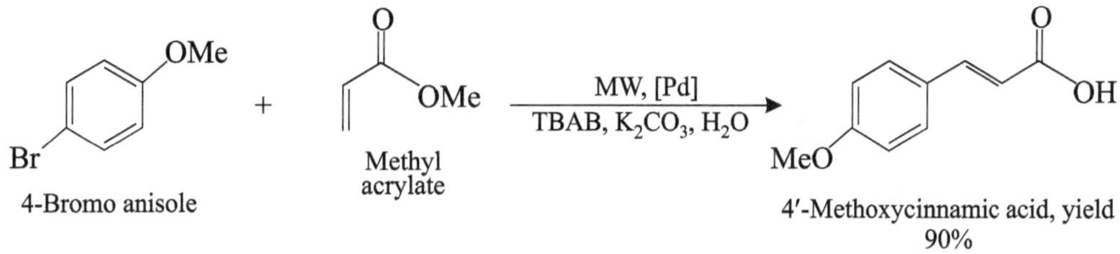

Fig. 9.15: MW assisted Heck coupling in water.

See also section 8.36, 9.5.6, 10.2.27.1.

9.2.12 Suzuki Reaction

Suzuki Reaction between aryl bromide and phenyl boronic acid in MW in presence of Pd (0.009 mol %), sodium carbonate in H_2O/EtOH gave the biphenyl in good yield. The reaction was carried out in open vessel at 86°C (Fig. 9.16).[9]

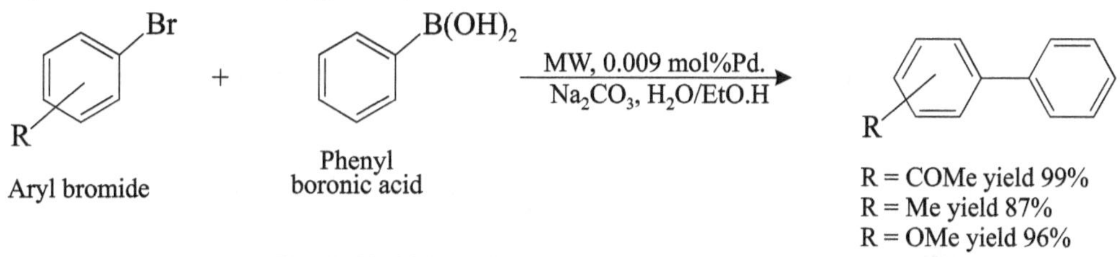

Fig. 9.16: MW assisted Suzuki Reaction in Water.

See also section 8.3.7; 10.2.27.2.

9.2.13 Hantzsch Synthesis of 1,4-dihydropyridines

The Hantzsch synthesis of 1, 4-dihydro pyridines involved the condensation of aldehyde with α, β-carbonyl compound and ammonia[10]. The 1,4-dihydropyridines have attracted considerable attension due to their biological properties[11]. It was found[9] that using aldehyde, α, β-carbonyl compound and NH_4OH in the ratio 1:3.4:10 and using water/ethanol as solvent and heating to reflux in a MW. oven and keeping the reaction mixture for 20 min at reflux temperature gave 96% yield of the 1,4 dihydropyridine (Fig. 9.17).

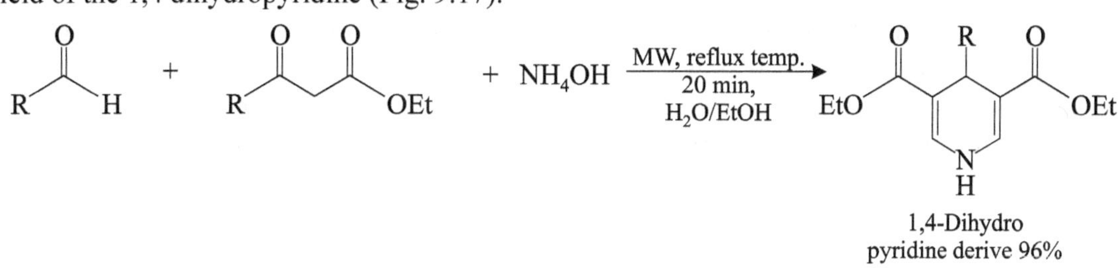

Fig. 9.17: MW assisted synthesis of 1,4-dihydropyridines in H_2O/EtOH.

9.2.14 Synthesis of Deferiprone

Deferiprone (3-hydroxy-1,2-dimethyl-4-pyridone) is synthesised from maltol by heating with 25% aqueous methylamine in MW oven at 160° for 1-3 min (Fig. 9.18).[12]

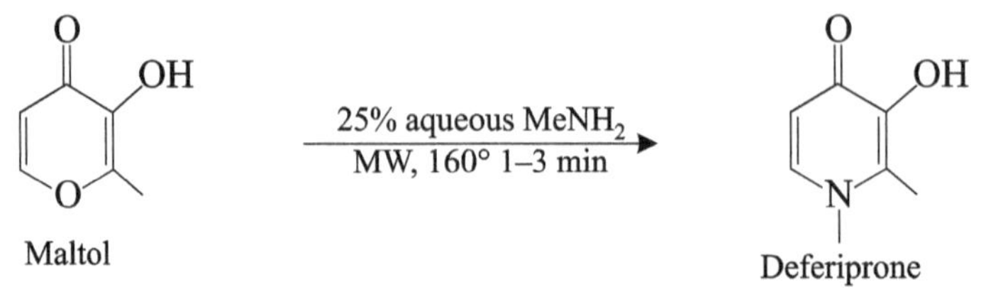

Fig. 9.18: Microwave assisted synthesis of Deferiprone in water.

9.2.15 Willgerodt Reaction

The Willgerodt reaction involves heating alkyl aryl ketones or styrene derivatives with sulfur and aqueous ammonia to form carboxylic acid amides possessing the same number of carbon atoms[13]. However, microwave heated reactions are completed within minutes at comparable temperatures[14] (Fig. 9.19).

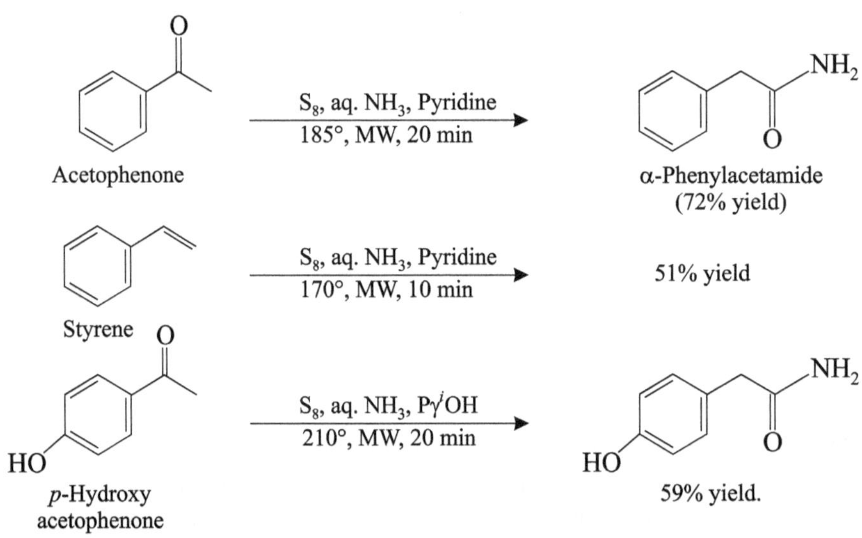

Fig. 9.19: Microwave-assisted Willgerodt reaction in water.

9.2.16 Miscellaneous MW Assisted Organic Transformations in Water

A number of organic transformations have been carried out in water using microwave heating. These transformations could be considered as being carried out using high temperature water (HTW), since microwave heating produces temperatures in the range 100-250°C (see also chapter 2, section 2.2.3). Some of the organic transformations in HTW are given in Table 9.1.

Table 9.1: Reactions in HTW[15,16]

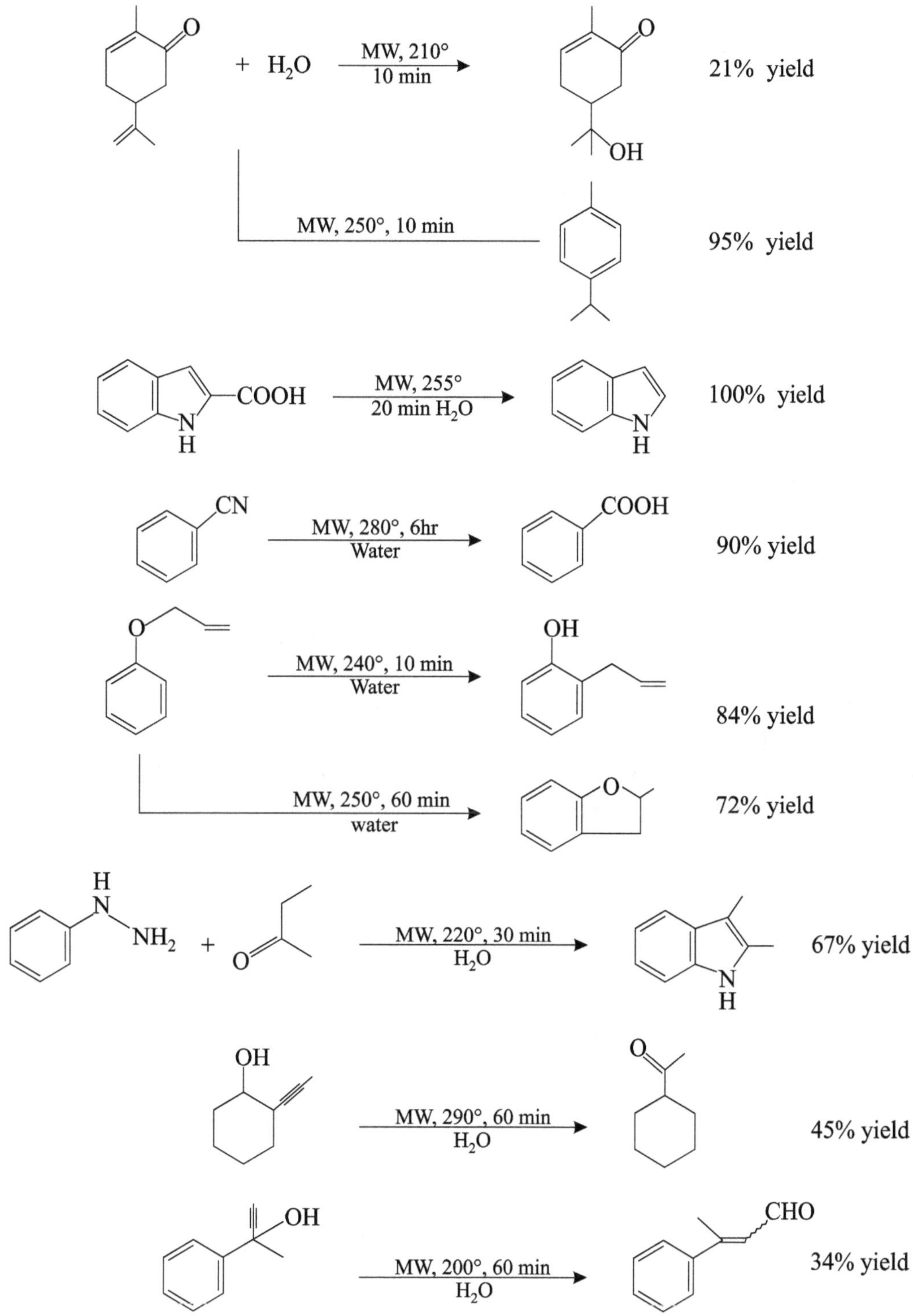

In the above transformations, water is believed to act as a catalyst and also as solvent. It is found that if the addition of acid or base was necessary, less reagent was required for high-temperature process than those at below 100°C and the reaction proceeds more selectively[16]. Following table gives reactions in dilute aqueous alkali or base at high temperatures[16,17,18].

Table 9.2: Reactions in dilute aq. acid or base at high temperature

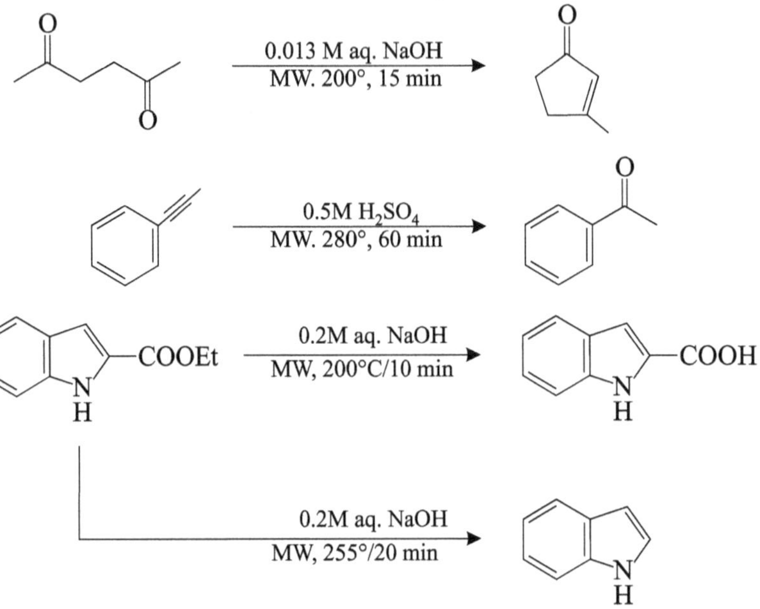

81% yield

90% yield

94% yield

93% yield

Besides the above miscellaneous reactions, microwaves have also been used in water for N-alkylations, synthesis of β-hydroxy sulfides and sulfoxides and hydrolysis of cellulose (Table 9.3 (cited in F.M. kerton, Alternative solvents for Green Chemistry, RSC Publishing, Page 57).

Table 9.3: Some MW Assisted Reactions using water as solvent

N-Alkylation

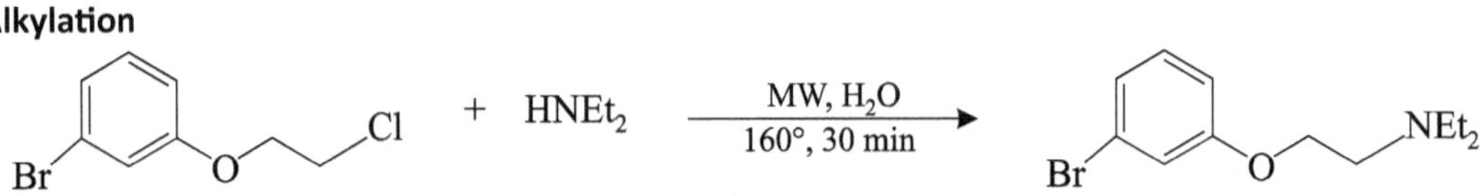

99%

Synthesis of β-hydroxy sulfides and sulfoxides

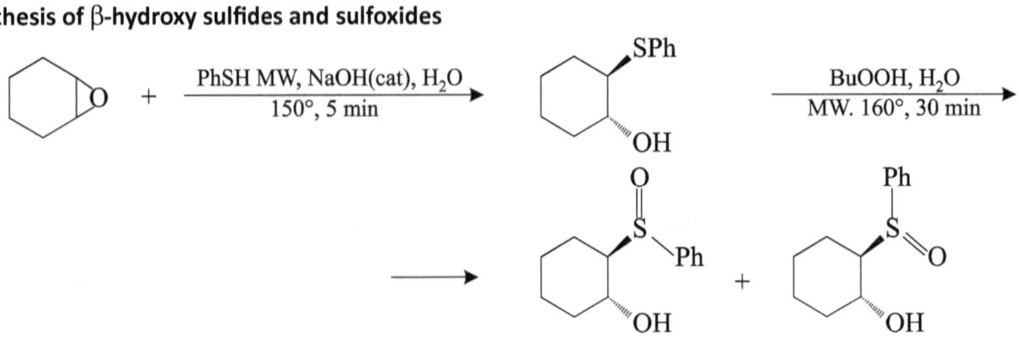

Hydrolysis of Celluose

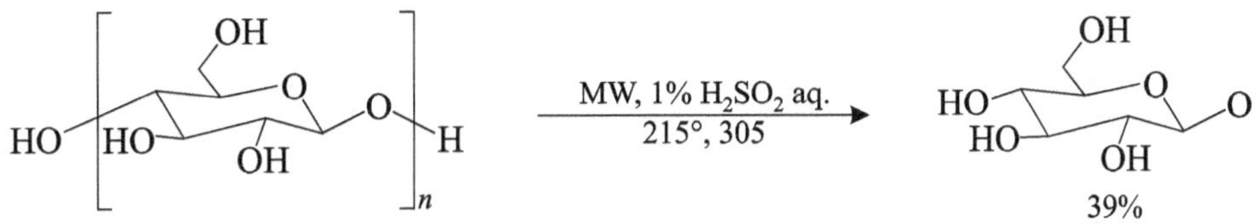

9.3 MICROWAVEASSISTEDORGANICTRANSFORMATIONSINORGANIC SOLVENTS

9.3.1 Esterification

A mixture of carboxylic acid and an alcohol on heating in a microwave oven for 6 min. in presence of catalytic amount of H_2SO_4 give[1, 2, 4] the corresponding ester (Fig. 9.20).

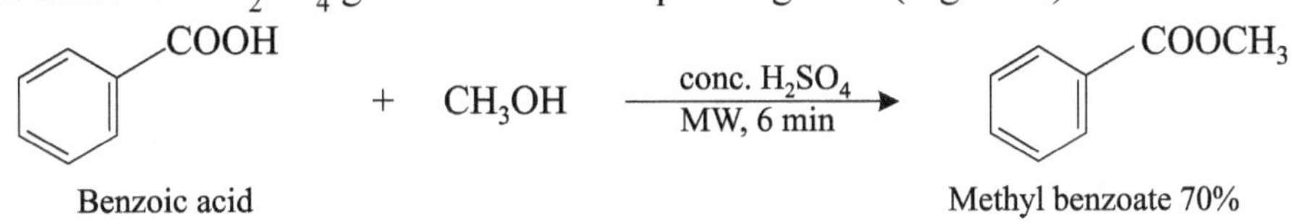

Fig. 9.20: Esterification.

The thermal method takes 1.5 hr. refluxing. Using microwave irridation, it is possible to esterify even hindered acids (like 2,4,6-trimethyl benzoic acid), which are difficult to esterify under thermal conditions. In the above cases, methanol acts as a reactant and also as a solvent. See also section 9.5.7 and 10.2.1.

9.3.2 Transesterification

Transesterification converts one ester into another by reaction with an alcohol in presence of acid catalyst. The reaction is much faster on microwave irridation (Fig. 9.21).

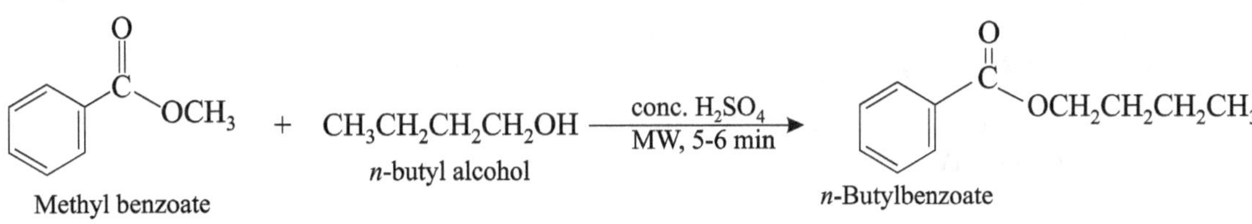

Fig. 9.21: Transesterification.

See also section 9.5.8.

9.3.3 Cycloaddition Reactions

(*i*) Cycloaddition reaction of 3-butenone with 2,3-dimethyl butadiene in xylene or dibutyl ether by heating in microwave give[19] 4-acetyl-1,2-dimethyl cyclohexene (Fig. 9.22).

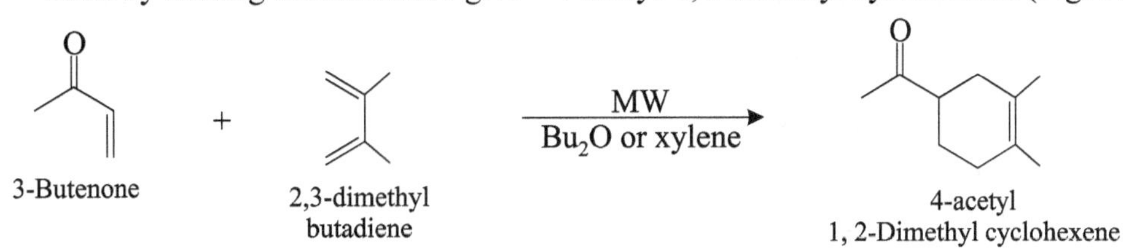

Fig. 9.22: Cycloaddition of 3-butenone with 2,3-dimethyl butadiene.

(*ii*) The cycloaddition reaction between cyclopentadiene and methyl acrylate (Fig. 9.23) under MW irradiation does not affect endo/exo selectivity[20]. The observed difference can be explained by the fact that under MW irradiation, the reaction occurs at higher temperatures then those under thermal conditions. Other investigators[21,22] have also found that the rates are identical in presence or absence of MW irradiation, the final yield depending on the temperature and not the mode of heating.

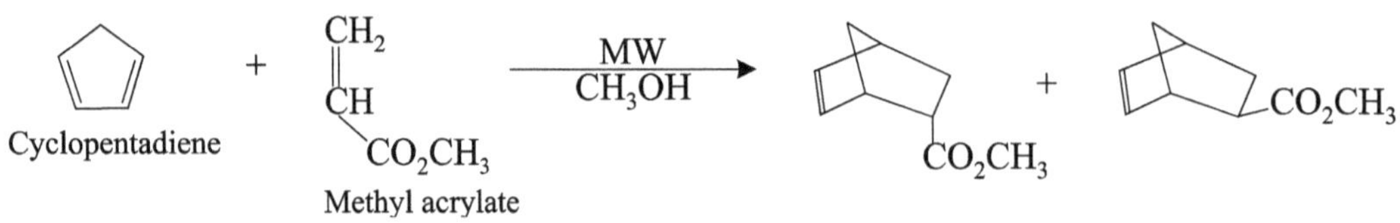

Fig. 9.23: Cycloaddition reaction of cyclopentadiene and methyl acrylate.

9.3.4 Synthesis of Benzodiazepin-2-Ones

Condensation of *o*-phenylene diamines with β-ketoesters in xylene under MW irridation gave benzodiazepin-2-ones (Fig. 9.24)[23,24]. It is found that the reaction did not proceed on thermal heating.

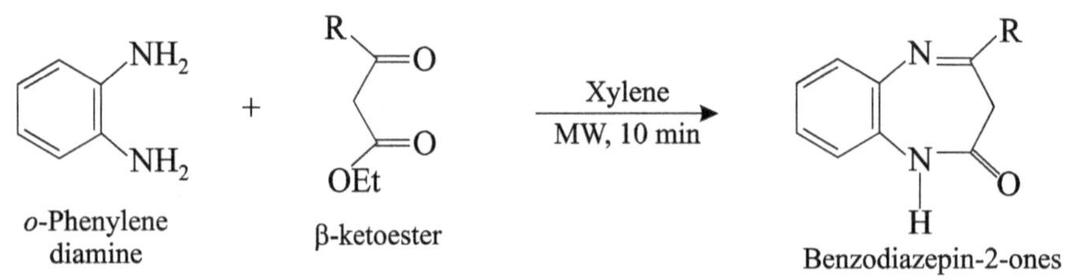

Fig. 9.24: Synthesis of benzodiazepin-2-ones.

See also Section 10.2.26.3.

9.3.5 Aromatic Nucleophilic Substitution Reaction

The reaction of *p*-chloronitrobenzene with ethyl alcohol in presence of NaOH and a phase transfer catalyst (PTC) under MW irridation gave *p*-ethoxynitrobenzene (Fig. 9.25)[25].

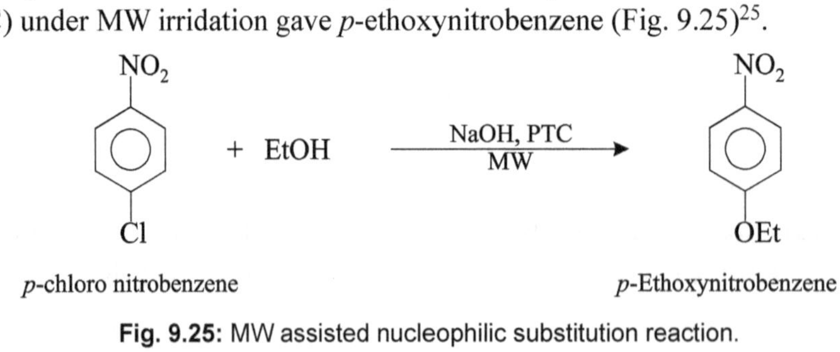

Fig. 9.25: MW assisted nucleophilic substitution reaction.

9.3.6 Pericyclic Reactions

(*i*) **Diels-Alder Reaction:** The reaction of anthracene with dimethyl fumarate in *p*-xylene on MW irridation gave[26] 87% yield of the adduct in 10 min. (Fig. 9.26). Conventional heating requires 4 hr. and the yield is only 67%.

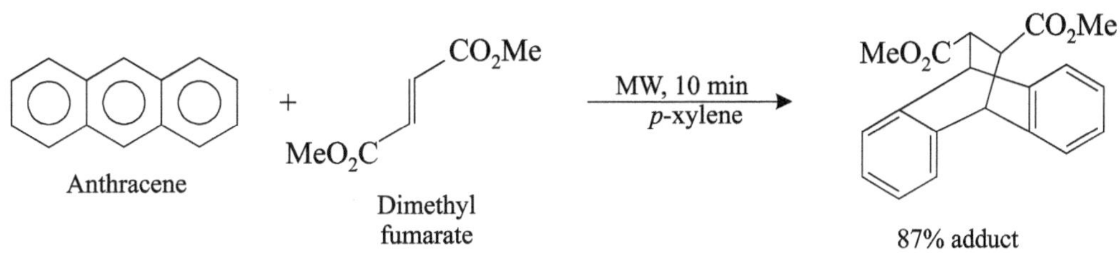

Fig. 9.26: Diels-Alder reaction of anthracene with dimethylfumarate.

(ii) Claisen Rearrangement: Claisen rearrangement using microwaves in presence of N-methyl formamide give 87% of the product. (Fig. 9.27)[27]

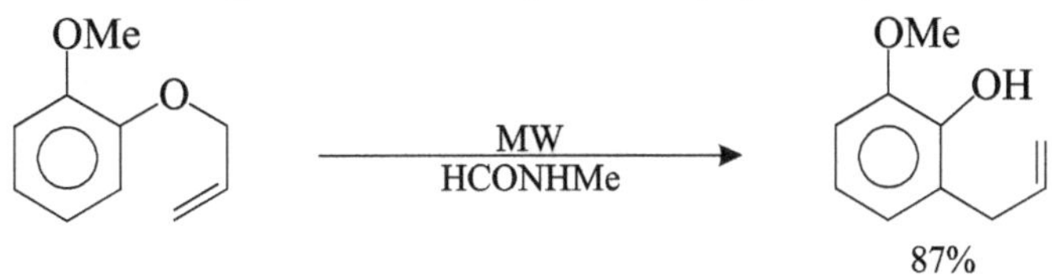

Fig. 9.27: MW assisted claisen rearrangement in HCONHMe.

The same rearrangement also occurs on heating without any solvent on heating in water at 240° in MW (see Section 9.2.16).

9.3.7 Reduction of Aldehydes to Alcohols

(*i*) Microwave-assisted reduction of benzaldehydes using the catalyst RuHCl (CO) $(PPh_3)_3$ in presence of formic acid gives the corresponding alcohols (Fig. 9.28)[28]

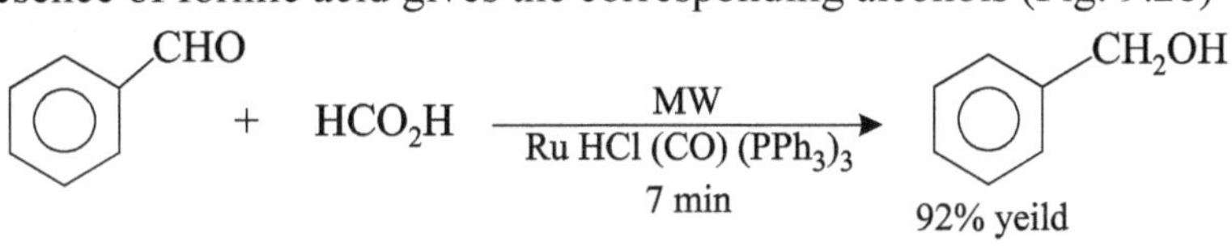

Fig. 9.28: Reduction of benzaldehydes.

(*ii*) The aldehydes can also be reduced by heating with iso propyl alcohol at about 225°C in a microwave (Fig. 9.29)[29]

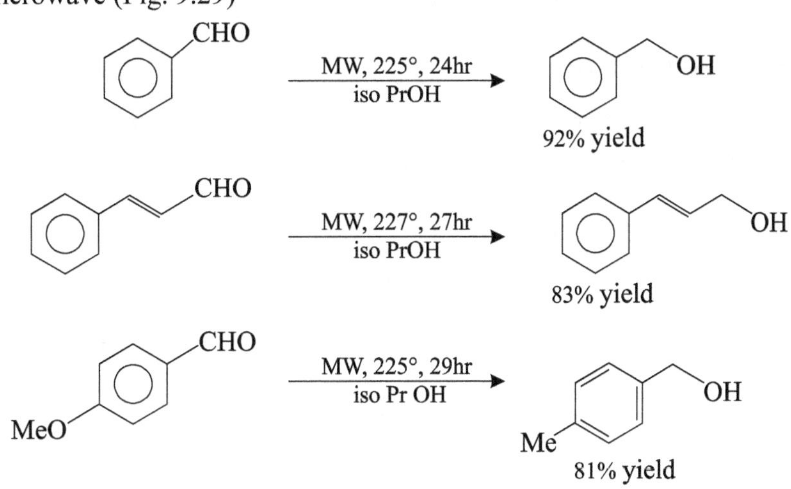

Fig. 9.29: Reduction of aldehydes (hydrogen-transfer reactions).

9.3.8 Catalytic Transfer Hydrogenation of β-lactam Derivatives

Substituted β-lactams can be reduced with 10% Raney Ni using ammonium formate in ethylene glycol in microwaves (Fig. 9.30)[30].

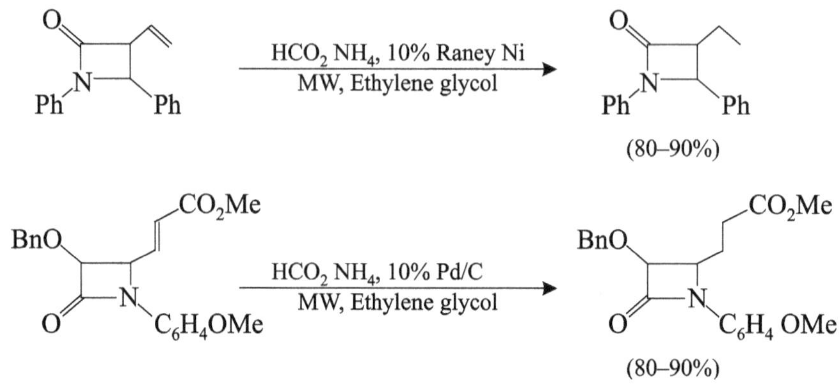

Fig. 9.30: Catalytic transfer hydrogenation of β-lactam derivatives.

9.3.9 Fries Rearrangement

Phenolic ketones are prepared by heating a mixture of *p*-cresyl acetate and $AlCl_3$ in chlorobenzene in a MW oven (Fig. 9.31)[31].

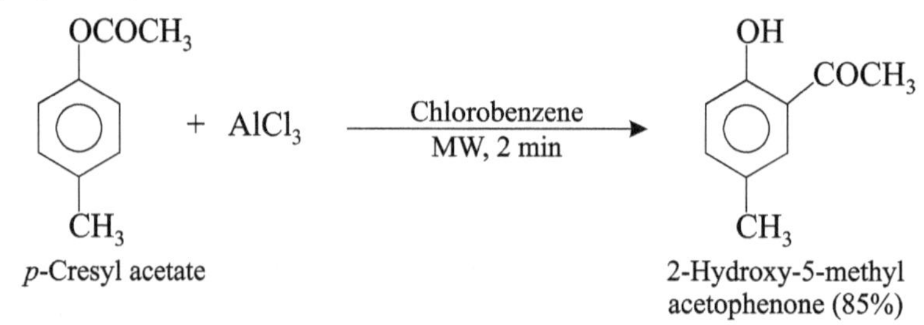

Fig. 9.31: Fries Rearrangement.

9.3.10 Synthesis of Chalcones

Aromatic ketones on heating with aromatic aldehydes in alcohol in a microwave oven for 30 sec gives the corresponding chalcone in quantitative yield (Fig. 9.32)[32].

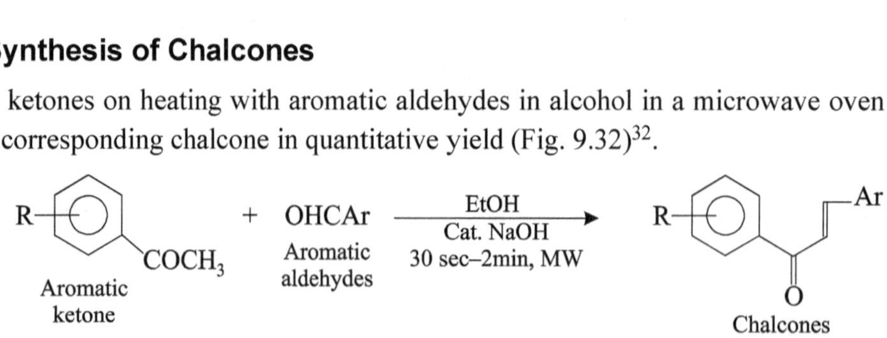

Fig. 9.32: Synthesis of Chalcones.

See also Chapter 10, section 10.2.25.2.

9.3.11 Decarboxylations

The decarboxylations of carboxylic acids can be conveniently effected by heating in quinoline in a MW oven (Fig. 9.33)[33].

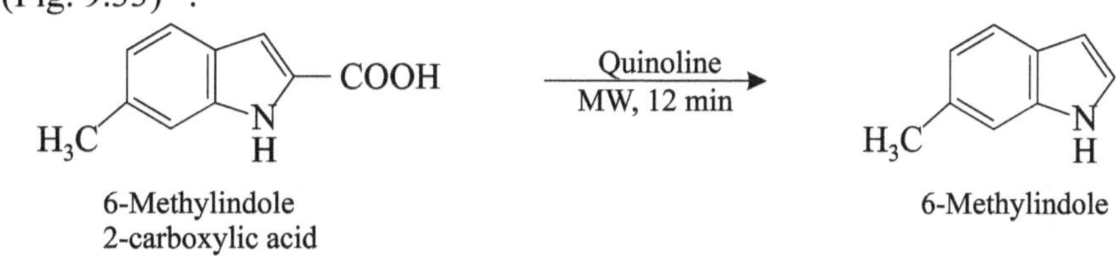

Fig. 9.33: Decarboxylations.

9.3.12 Suzuki Cross Coupling Reaction

MW-assisted Suzuki cross coupling of aryl halides with aryl boronic acid in presence of $PdCl_2$ as catalyst and KF using polyethylene glycol as solvent gives good yield of biaryls (Fig. 9.34)[34-37].

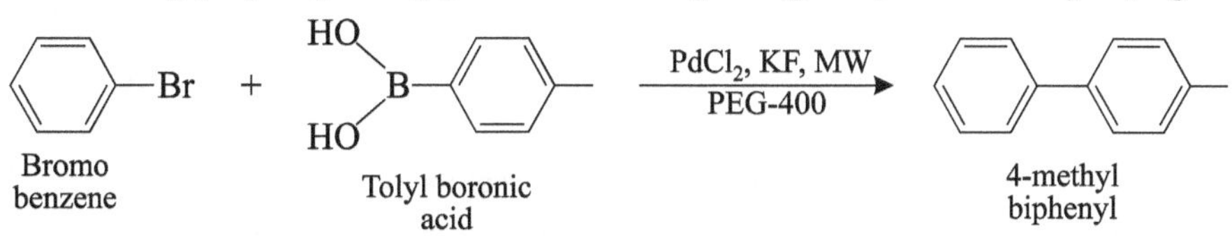

Fig. 9.34: Suzuki cross coupling reaction.

In the above procedure, the catalyst can be recycled[34].

9.3.13 Miscellaneous Organic Transformations in Organic Solvents

(*i*) **Racemization reactions:** The racemization of (–)-Vincadifformine has been achieved[38] under MW irradiation in DMF. The reaction involved two consecutive Diel-Alder cycloreversion and cycloaddition steps. The (+) isomer is an useful intermediate for the synthesis of an alkaloid vinamine (Fig. 9.35). Under classical conditions, much longer time is needed and the reaction is accompanied by considerable amount of decomposition products.

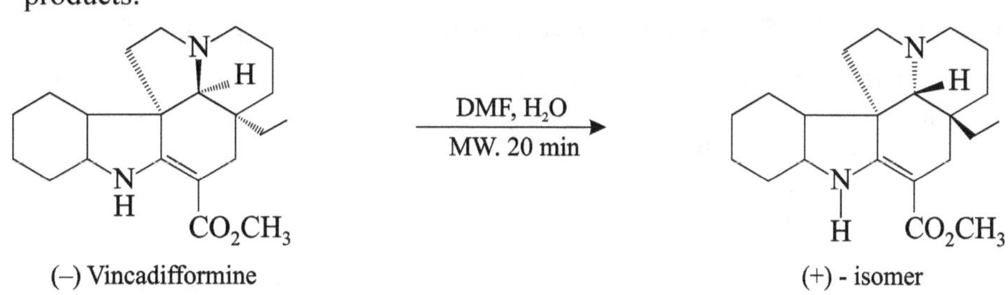

Fig. 9.35: Racemization Reaction.

(*ii*) **Stereoselective additions:** The reaction of tetrachlorophthaloyl glycine (TCPN) chloride with an imine under microwave irradiation in methylene chloride in presence of Et_3N gives exclusively the trans isomer (Fig. 9.36)[39]. However under classical conditions, a mixture of *cis* and *trans* isomers are obtained.

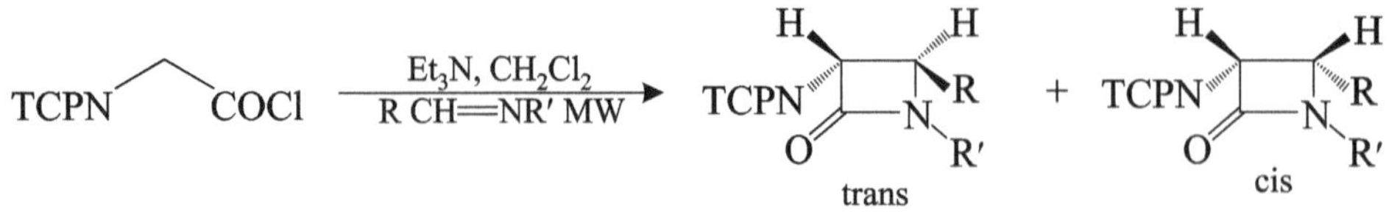

Fig. 9.36: A stereoselective reaction, synthesis of trans β-lactans.

Similarly addition of 2-aminothiophenol to glycidic ester under MW irridation is highly stereoselective. Using toluene as the solvent the cis product is obtained in major amount and using acetic acid, the trans product is obtained in major amount (Fig. 9.37)[40].

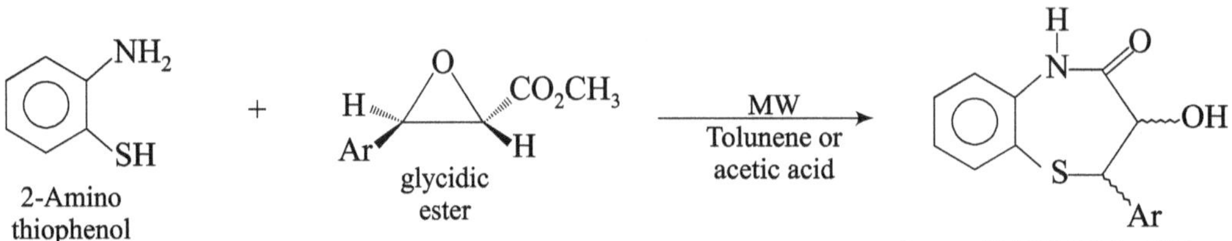

Power 390 W, PhMe 75%
cis: *trans* 9:1
Power 490 W, AcOH 84%
cis: *trans* 1:9

Fig. 9.37: MW-assisted stereoselective addition of thiophenols to glycidic esters.

(*iii*) **Williamson ether synthesis:** Alkyl halides on reaction with phenol in alkali in methanol give ethers (Fig. 9.38)[41].

$$\mathrm{BuCl + NaOPh \xrightarrow[\substack{150^\circ,\ 1000\ Kpa \\ MeOH}]{MW} \underset{67\%}{BuOPh}}$$

Fig. 9.38: Williamsons ether Synthesis

(*iv*) **Finkelstein Reaction:** The reaction of alkyl chlorides with NaI in acetone on MW irridation gives the corresponding iodides [Fig. 9.38(*a*)][42]

$$\mathrm{BuCl + NaI + Me_2CO \xrightarrow{MW} \underset{36\%}{BuI}}$$

Fig. 9.38(*a*): Finkelstein Reaction.

9.4 MICROWAVE ASSISTED REACTIONS IN SOLID STATE

Microwave assisted reactions in solid state have become extremely popular during the last two decades for the synthesis of several classes of organic molecules. Besides increase in yields, performing reactions in much shorter times these reactions are mostly environmentally benign. These reactions can be performed either by heating the substrates (in MW) in presence or absence of a catalyst or by performing reactions on solid supports.

9.4.1 Microwave Assisted Reactions in Solid State by Direct Heating of the Substrates

9.4.1.1 Reactions involving N-alkylations

Phthalimides or its potassium salt can be alkylated by reaction with alkyl halides in presence of potassium carbonate and a PTC (tetrabutyl ammonium bromide, TBAB) under MW irridation (Fig. 9.39)[43].

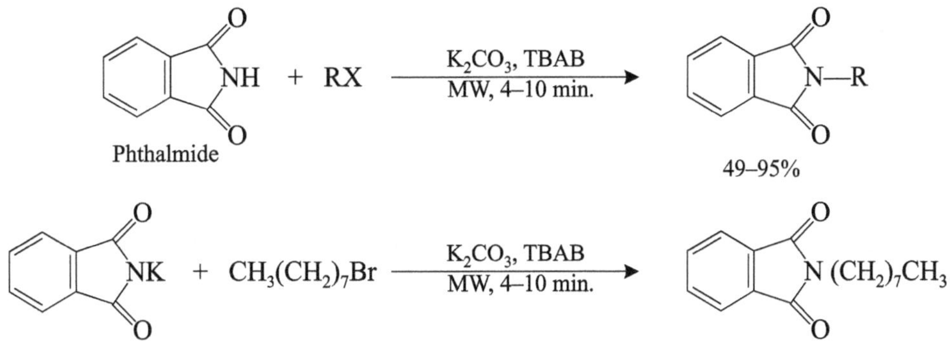

Fig. 9.39: N-Alkylation of phthalimide and its potassium salt.

In a similar way carbazole can be N-alkylated (Fig. 9.40)[44].

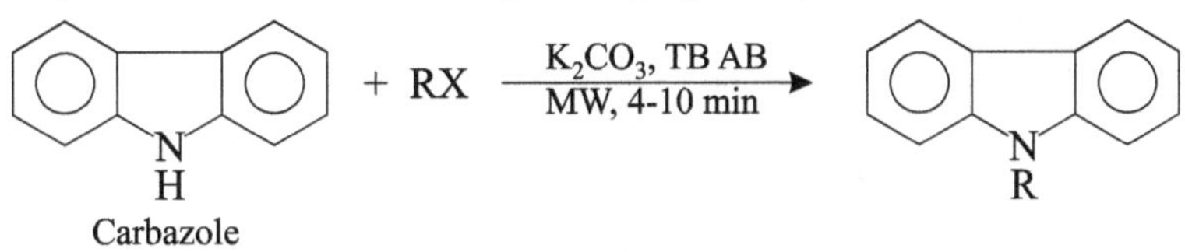

Fig. 9.40: N-Alkylation of Carbazole

9.4.1.2 Knoevenagel Condensation

The condensation of creatine with aldehydes is possible under MW irridation[45] (Fig. 9.41).

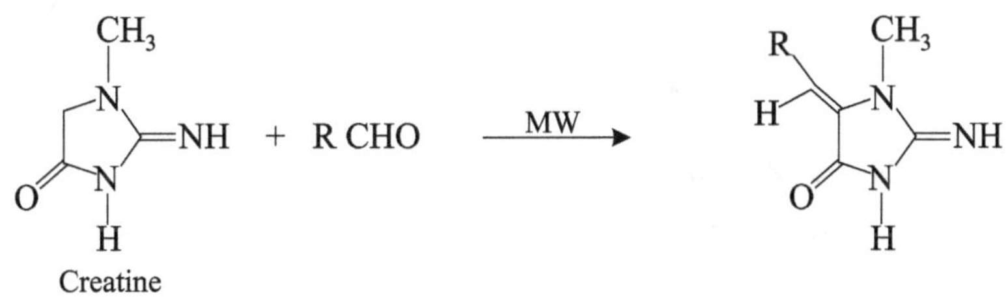

Fig. 9.41: Knoevenagel Condensation.

Salicylaldehydes also undergo Knoevenagel condensation with a variety of ethyl acetate derivatives under basic conditions (piperidine) under MW irridation to give coumarins (Fig. 9.42)[46]. The reaction is known as **pechmann reaction**[47].

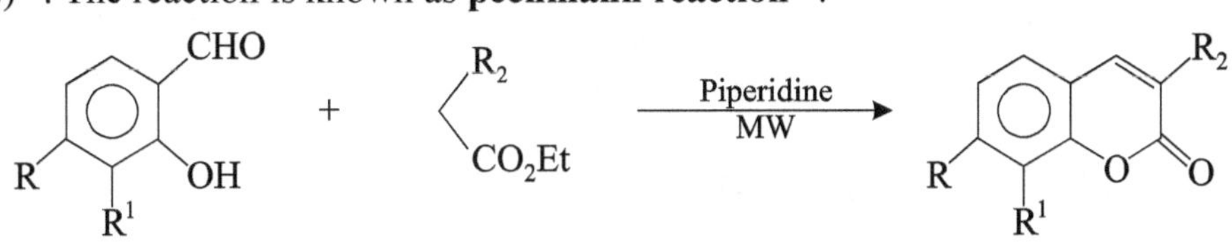

Fig. 9.42: Synthesis of Cumarins by Knoevenagel Condensations.

9.4.1.3 Synthesis of bridgehead Nitrogen Heterocycles

Pyrimidino [1, 6-*a*] benzimidazoles have been synthesised from N-acylimidates and activated 2-benzimidazoles on MW irridation (Fig. 9.43).[48]

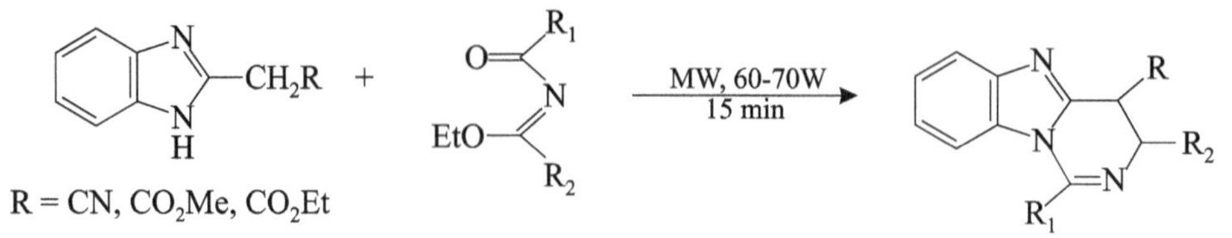

Fig. 9.43: Synthesis of pyridino [1, 6-*a*] benzimidazoles.

In a similar way 2,3-dihydroimidazolo [1, 2-c] pyrimidines were synthesised (Fig. 9.44)[48].

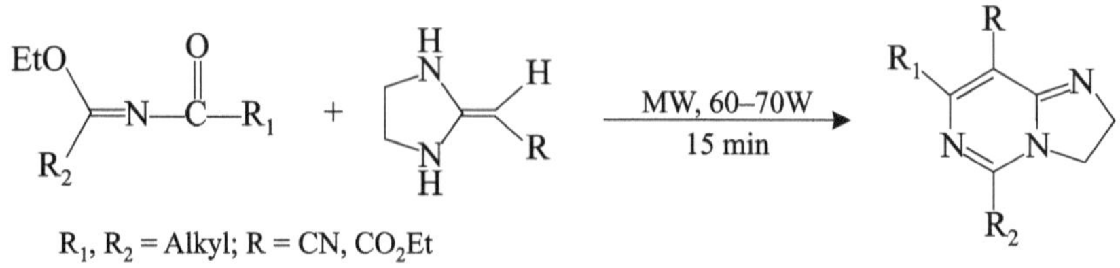

Fig. 9.44: Synthesis of 2,3-dihydroimidazolo [1,2-c] pyridines.

9.4.1.4 Synthesis of 2-oxazolines

The reaction of carboxylic acids with α, α, α-*tris* (hydroxymethyl) methylamine and microwave irridation give 2-oxazolines (Fig. 9.44).[49]

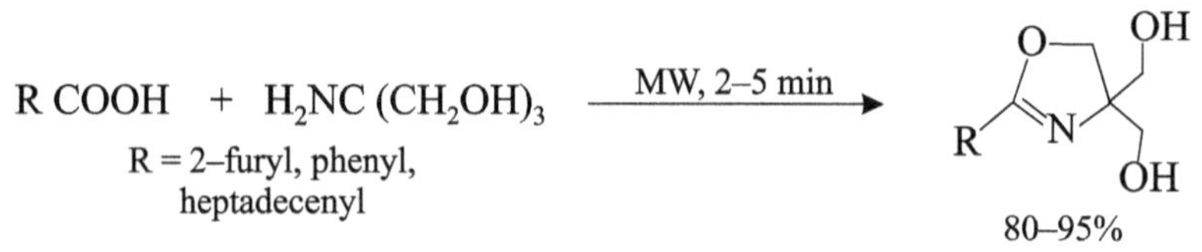

Fig. 9.45: Synthesis of 2-oxazolines.

9.4.1.5 Synthesis of Benzimidazole Derivatives

The reaction of neat succinic acid and homophthalic acid with substituted diamines under MW irradiation yielded benzoimidazole derivatives, that have anti cancer and anti-inflammatory activity (Fig. 9.46)[50].

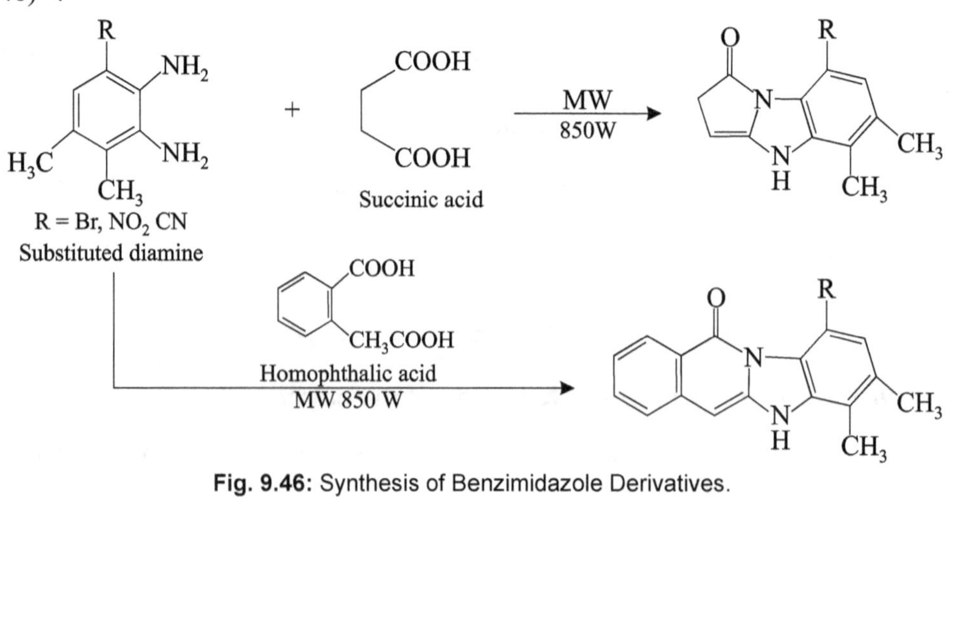

Fig. 9.46: Synthesis of Benzimidazole Derivatives.

9.4.1.6 Conversion of 3-nitrophthalic Acid into Nitrophthalimedo butyric Acid (NPB)

The reaction of 3-nitrophthalic acid with aminobutyric acid under MW irradiation gave nitrophthali midobutyric acid (NPB) (Fig. 9.47)[51]. NPB is used to protect the hydroxyl group in carbohydrates.

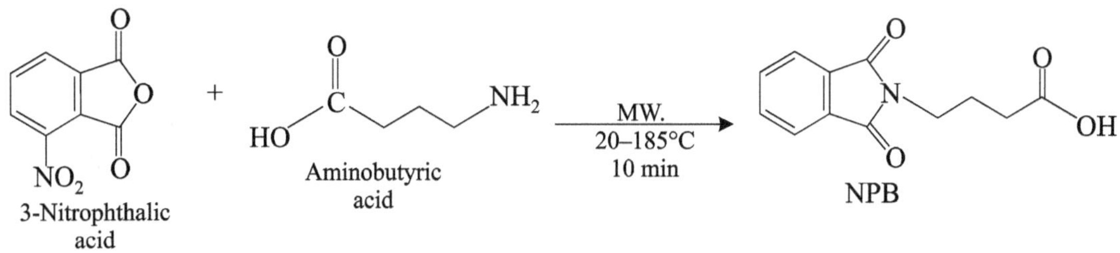

Fig. 9.47: Synthesis of NPB.

9.4.1.7 Peroxidative Oxidation of Secondary Alcohols to Ketones

Secondary alcohols could be oxidised to the corresponding ketones by microwave assisted solvent free peroxidative oxidation using a Cu(II) Complex as catalyst (Fig. 9.48)[52].

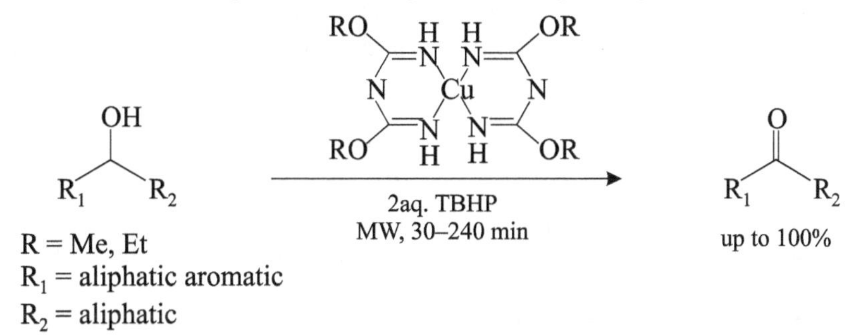

TBHP is tert-butylhydroperoxide (oxidant).

Fig. 9.48: Peroxidative Oxidation of Secondary Alcohols to Ketones

The Cu(II) complex, 2,4-alkoxy-1,3,5-Triazapentadeno complex was synthesised[52] from copper nitrate, sodium dicyanamide and methanol.

9.4.1.8 Synthesis of Esters from o-alkylisoureas

The *O*-alkylisoureas (prepared by the reaction of diisopropyl carbodimide with primary alcohol, catalysed by $Cu(OTf)_2$ under MW irridation) on treatment with a carboxylic acid solution in acetonitrile followed by MW irridation gave the esters in 70-90% yield (Fig. 9.49)[53]

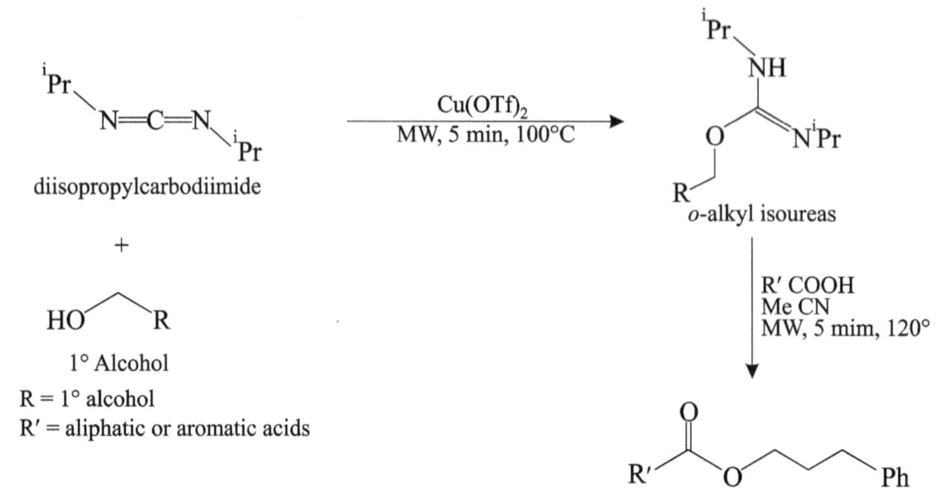

Fig. 9.49: MW assisted synthesis of esters from *o*-alkylisoureas.

9.4.1.9 Ritter Reaction

The reaction of nitriles with alcohols in presence of Nafion® catalyst on MW irradiation give amides (Fig. 9.50)[54]. The method is simple, fast (< 30 min) and convenient way of recovering the catalyst.

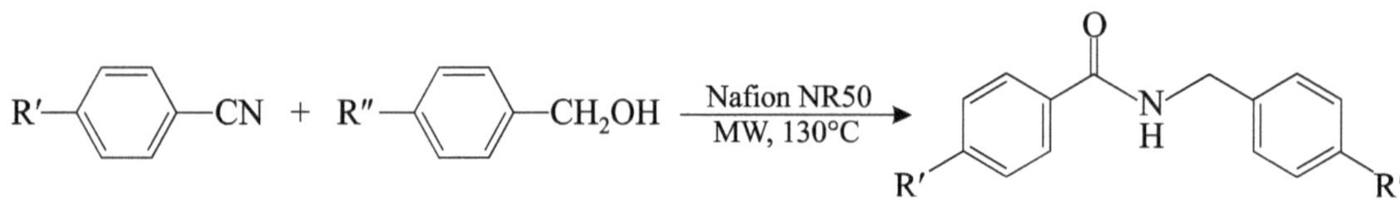

Fig. 9.50: Nafion® catalyst solvent-free Ritter reaction assisted by MW.

9.4.1.10 Synthesis of Indocyanine Dyes

Indocyanine dyes are obtained by MW irradiation of a mixture of indole quaternary salt (obtained by heating trimethyl indole with alkyl halides) and 2-methylthio quinoline quaternary salt (obtained by refluxing N-methyl-2-thioquinoline with methyl iodide) in presence of Et_3N (Fig. 9.51).[55]

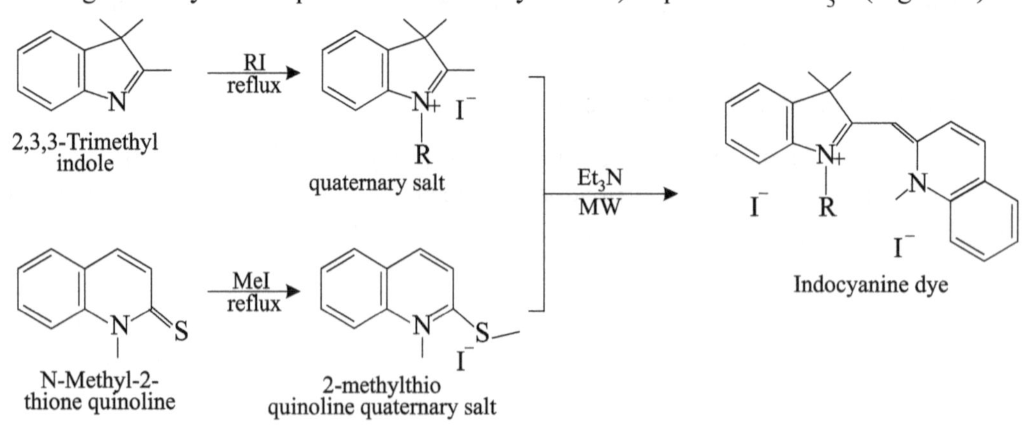

Fig. 9.51: Synthesis of Indocyanine Dyes.

9.4.1.11 Wittig Reaction

Phosphonium ylides undergo Wittig reaction with ketones in absence of solvent on MW irridiation giving good yields compared to conventional thermal heating[56] (Fig. 9.52).

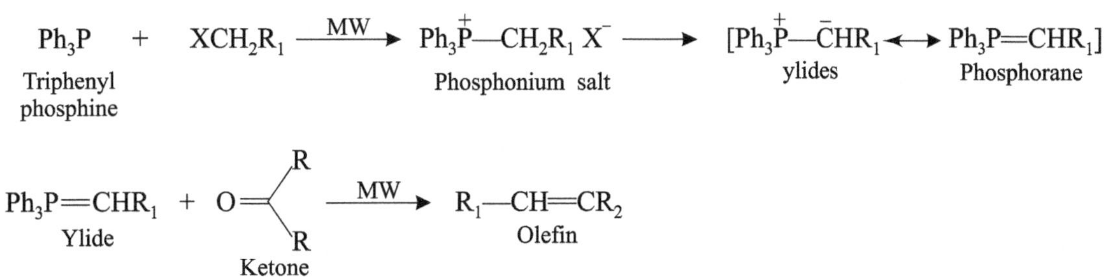

Fig. 9.52: The Wittig Reaction.

9.4.1.12 Henry Reaction

The condensation of nitroalkanes with carbonyl compounds under MW irridation in presence of catalytic amount of NH_4OAc gave[57] α,β-unsaturated nitroalkanes (Fig. 9.53).

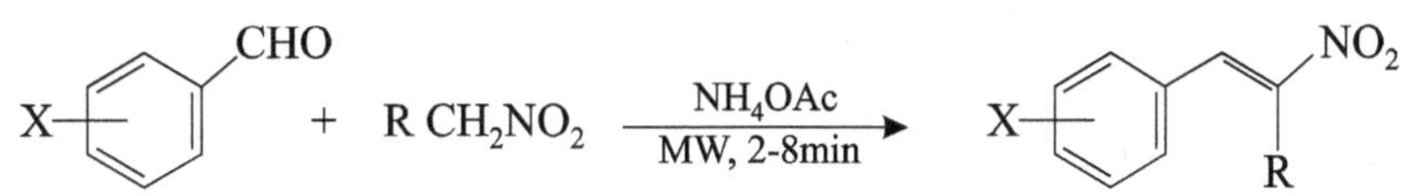

X = H; R = *p*-OH, *m, p*-$(OMe)_2$,
m-OMe, *p*-OH.
X = H, I

Fig. 9.53: Henry Reaction.

9.4.1.13 Synthesis of 2,3-disubstituted Quinoxalines

2,3-Disubstituted quinoxylines are synthesised from acyloins and *o*-phenylenediamine under MW irradiation (Fig. 9.54)[58].

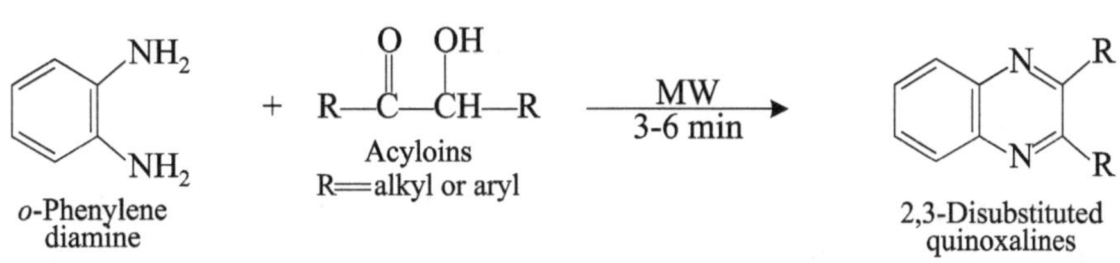

Fig. 9.54: Synthesis of 2,3-disubstituted quinoxalines.

9.4.1.14 Michael Addition

2′-Hydroxy-4′, 6′-dimethylchalcones undergo solid state intramolecular Michael type addition to yield[59] flavanones (Fig. 9.55).

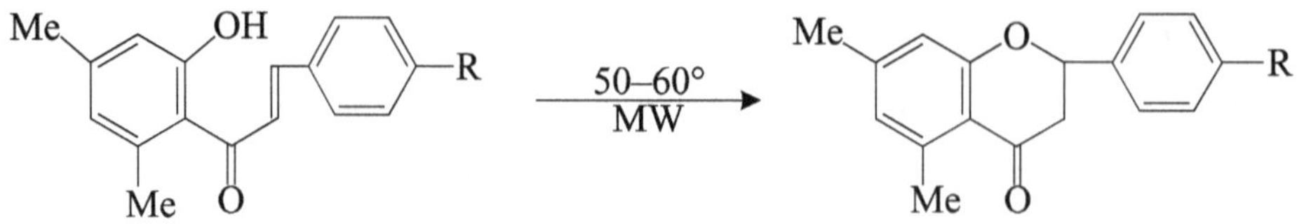

2′-Hydroxy-4′, 6′-dimethyl chalcones
R = H, Cl or Br

5, 7-Dimethyl flavanones

Fig. 9.55: Synthesis of Flavanones.

The Michael addition of chalcone to 2-phenyl Cyclohexanone under PTC conditions gave[60] 2,6-disubstituted cyclohexane derviative in high distereoselectivity (Fig. 9.56).

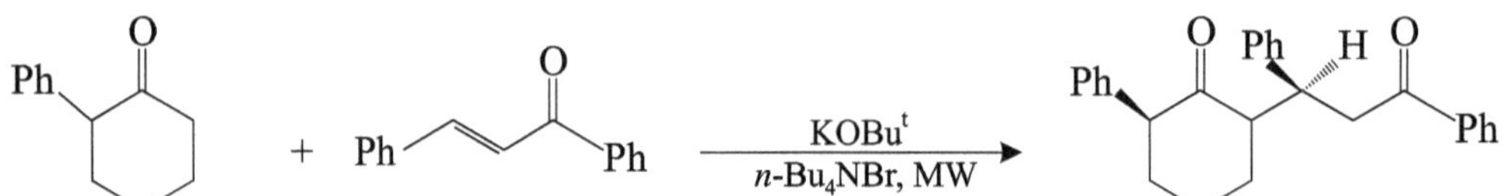

2-Phenyl cyclohexanone

Chalcone

99% ee

Fig. 9.56: Michael addition of 2-phenylcyclohexanone with Chalcone.

9.4.1.15 Synthesis of Aspirin

Aspirin is a single drug which is produced and used in maximum amount on a global scale. It is best prepared on several hundred grams by heating salicyclic acid in acetic anhydride in a MW oven (Fig. 9.57)[61].

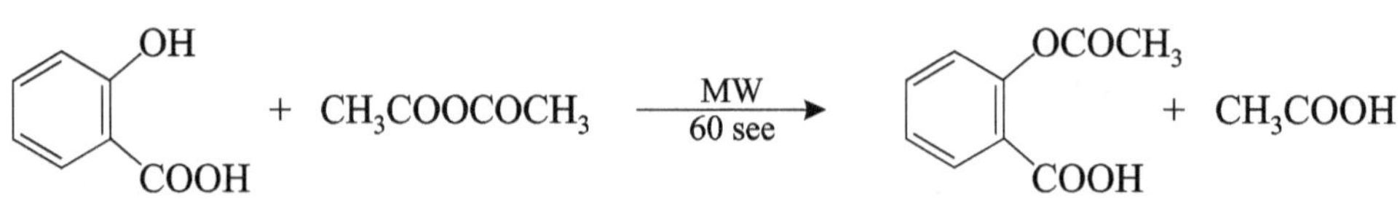

Aspirin

Fig. 9.57: Synthesis of aspirin.

See also Section 9.5.16.

9.4.1.16 Thionation Reactions: Synthesis of Thioketones, Thiolactones, Thioamides, thioesters and thioflavonoids

The conversion of carbonyl compounds to the corresponding thioanalogues involving the reaction of the substrates with Lawesson's reagent using MW irradiation under solvent free conditions is a convenient procedure (Fig. 9.58). The reaction avoids the use of dry hydrocarbon solvents like benzene, xylene, Et_3N or pyridine that are normally used by carrying out the reaction with P_2S_5 in presence of acids[62].

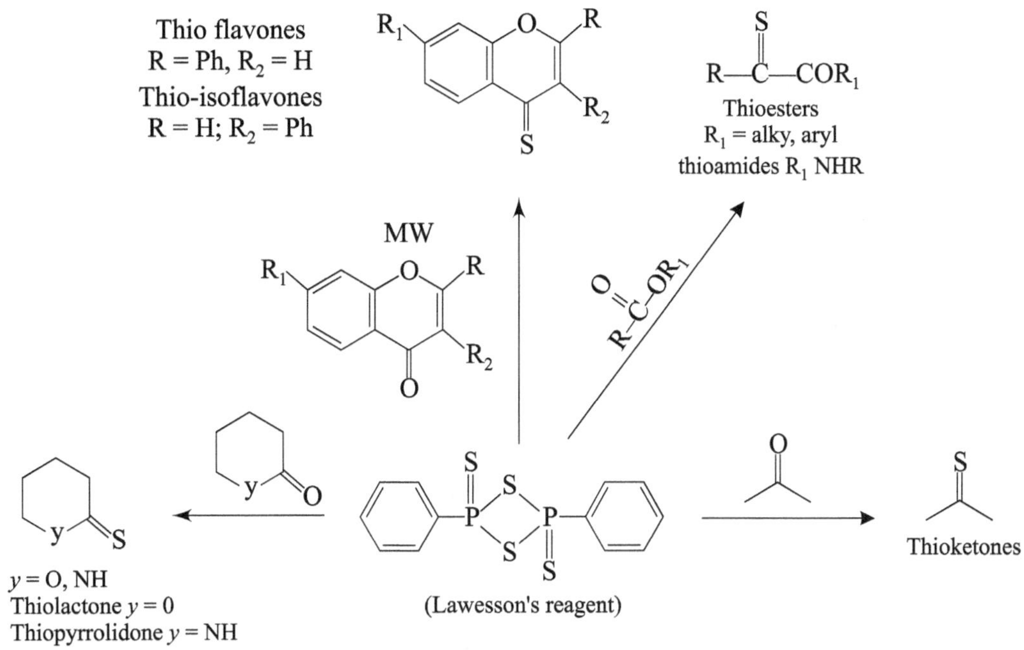

Fig. 9.58: Synthesis of Thioketones, thiolactones, thioamides and thioesters using Lawesson's reagent.

9.4.1.17 Beckmann Rearrangement

It involves rearrangement of oximes of ketones into analides by heating with acidic reagents like PCl_5, HCOOH, $SOCl_2$ etc. However, a solid-state MW assisted Beckmann rearrangement has been reported. The method consist in heating a mixture of oxime of a ketone with montorillonite K 10 clay in dry media for 7 min (Fig. 9.59)[62a].

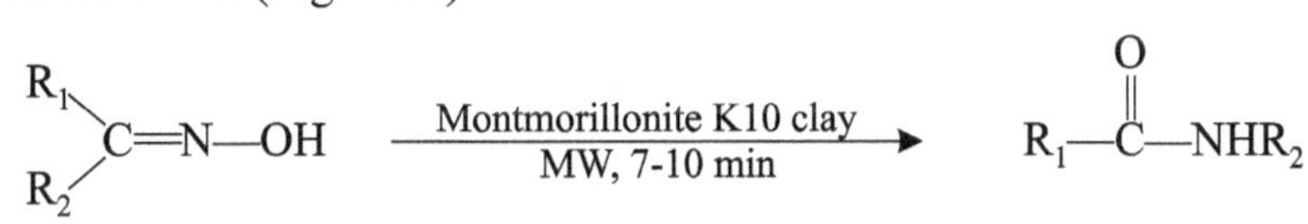

Fig. 9.59: Beckmann Rearrangement.

See also section 9.5.4.

9.4.2 Microwave Assisted Reactions in Solid State using Supported Reagents

In these reactions, the reactants are dissolved in a suitable solvent like water, alcohol, acetone, methylene chlorides etc. and the solution thus obtained is stirred with a suitable solid support or adsorbent like silica gel, alumina, phylliosilicate (M^{n+}-montmorillonite). Subsequently the solvent is removed in vacuo and the dried resultant solid support along with the reactants are used for carrying out the reaction under MW irridation. The work-up procedure involves extraction of the product from solid support. In some cases, the products can be isolated by simply distilling from the reaction vessel.

Most of the reactions are performed in open glass containers (Test tubes, beakers and RB flasks). Some of the supporting reagents, *viz.*, clay supported iron (III) nitrates (clayfen) and copper (II) nitrate (clay-top) are prepared by literature procedures (See also section 8.5, Chapter 8).

Following are given some important transformations using supported reagents.

9.4.2.1 Protection/deprotection Reactions

In a number of transformations, protection and deprotection are important steps. These normally involve use of acidic, basic and hazardous corrosive reagents. It is found that MW-accelerated protection/deprotection of functional groups can be carried out under solvent free conditions. As an example, acetals of 1-galactono-1,4-lactone can be prepared in excellent yield by adsorbing the lactone and the aldehyde on montmorillonite K10 or KSF clay followed by M.W. irradiation of the reaction mixtures (Fig. 9.60).[63]

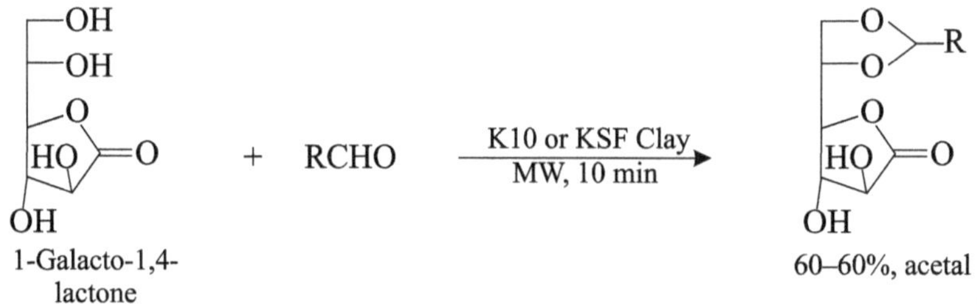

Fig. 9.60: Protection of OH groups of 1-galacto-1,4-lactone.

In a similar way, the C=O group of carbonyl compounds can be protected by treatment with ethylene glycol and MW irridation (Fig. 9.61)[64]

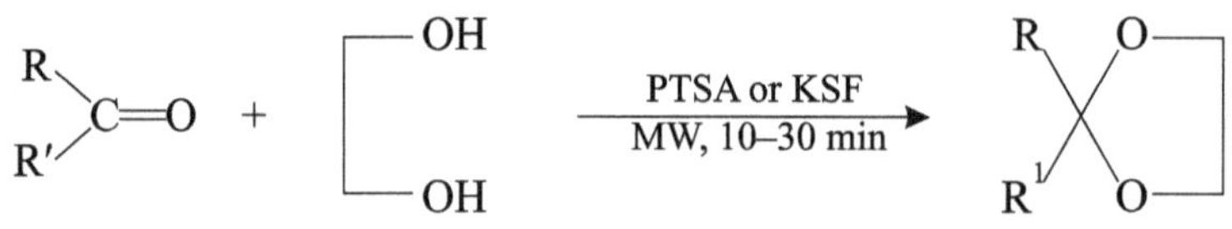

Fig. 9.61: Protection of Carbonyl Group.

9.4.2.2 Deprotection of Aldehyde Diacetates

The diacetates of aromatic aldehydes can be cleaved on a neutral surface by MW irridation (Fig. 9.62)[65].

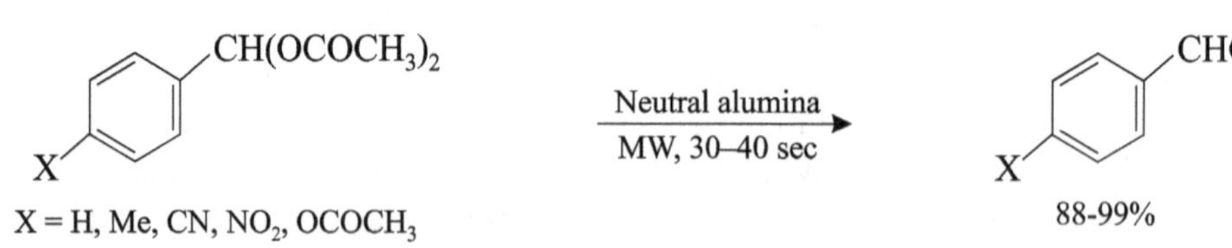

Fig. 9.62: Deprotection of aldehyde diacetates.

In case the substrate has an acetoxy function ($R–OCOCH_3$), microwave irridation for 2 min. is required to cleave both the diacetate and the ester group.

A similar reaction has been reported using zeolites (Fig. 9.63)[66].

$$R\,CH\,(OAc)_2 \xrightarrow[\text{Zeolite HSZ-360}]{\text{MW}} RCHO$$

$R = C_6H_5{-}C_6H_4$, p-$NO_2C_6H_4$, p-OMe C_6H_4, 4-Me.C_6H_4, Ph, $CH_3\,(CH_2)_{10}$

Fig. 9.63: Deprotection of aldehydes diacetates using Zeolites.

9.4.2.3 Debenzylation of Carboxylic Esters

Debenzylation of carboxylic esters is possible by microwave irradiation using acidic alumina (Fig. 9.64)[67].

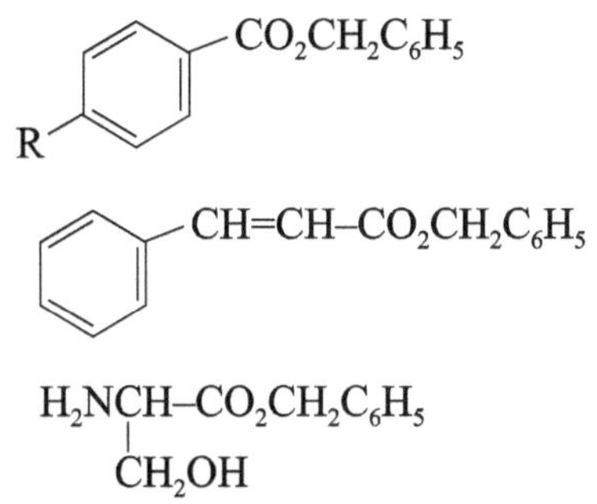

R = H, 7-min (20 min in oil bath), yield 92%
R = OMe, 7 min (20 min in oil bath), yield 89%
R = OH, 10 min, 92 % yield

10 min, yield 90% (20 min in oil bath)

3 min, neutral alumina, yield 92%
(10 min in oil bath)

Fig. 9.64: Debenzylation of Carboxylic Esters.

9.4.2.4 Selective Cleavage of N-tert-butoxycarbonyl Group

The selective cleavage of N-*tert* butoxycarbonyl (Nboc) group is readily, achieved by MW irridation in presence of $AlCl_3$ 'doped' neutral alumina (Fig. 9.65)[68].

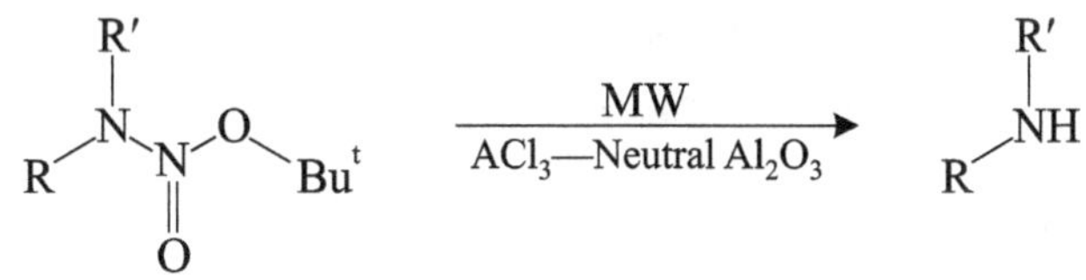

Fig. 9.65: Selective cleavage of N-*tert*-butoxycarbonyl group.

9.4.2.5 Deprotection of tert-butyldimethylsilyl Group (TBDMS)

The deprotection of TBDMS group attached to alcoholic OH group can be achieved by MW irridation on alumina surface (Fig. 9.66)[69]. This approach is very convenient compared to conventional approach using corrosive fluoride ions[69].

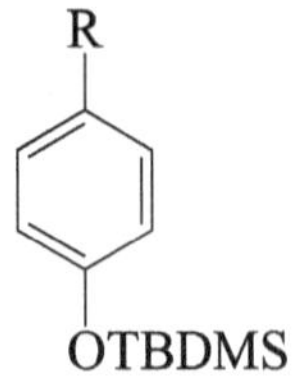

R = CHO, 10 min yield 91%
R = CO CH_3, 10 min yield 93%
R = $CH_2(CH_2)_2$ OTBDMS, 11min yield 78%

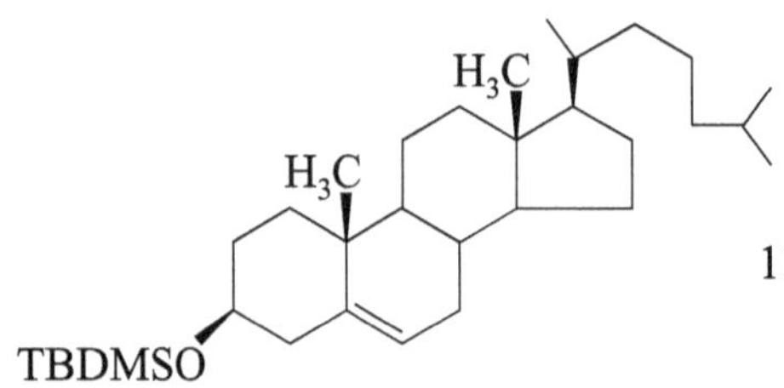

11 mim. yield 93%

Fig. 9.66: Deprotection of TBDS group.

9.4.2.6 Deprotection of Thioacetals

The deprotection of thioacetals can be achieved by MW irridation using clayfen (Fig. 9.67)[70]. The procedure is very convenient than using toxic heavy metals such as Hg^{2+}, Ag^{2+}, Ti^{4+}, Cd^{2+}, Tl^{3+} or reagents like benzeneseleninic anhydride.

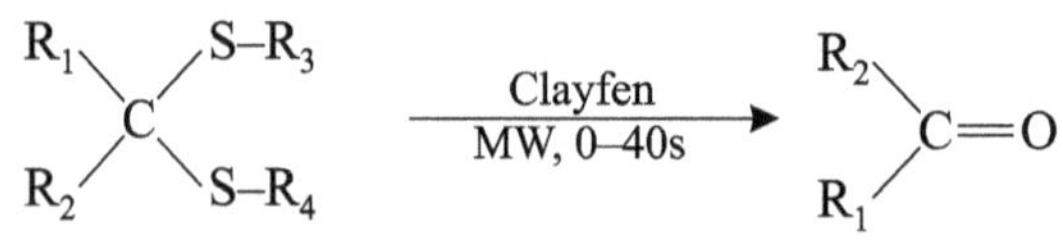

87-98%

R_1 = Ph, p-anisyl, 4-$NO_2C_6H_4$; R_2=H; R_3–R_4 = –$(CH_2)_2$–
R_1 = R_2 = Et; R_3–R_4 = –$(CH_2)_2$–; R_1 = R_2 = Ph; R_3 = R_4 = Et
R_1 = Ph; R_2 = Me; R_3–R_4 = –(CH_2)–

Fig. 9.67: Deprotection of thioacetals.

9.4.2.7 Deoximation of Oximes

Oximes are known to be used as protecting the C=O group. The reagents that have been used for deoximation include[71] Raney-nickel, pyridinium chloroformate, pyridinium chloroformate-H_2O_2, dinitrogen tetroxide, dimethyl dioxirane etc. A convenient method of deoximation by MW irradiation of oximes using ammonium persulphate on silica (Fig. 9.68)[72].

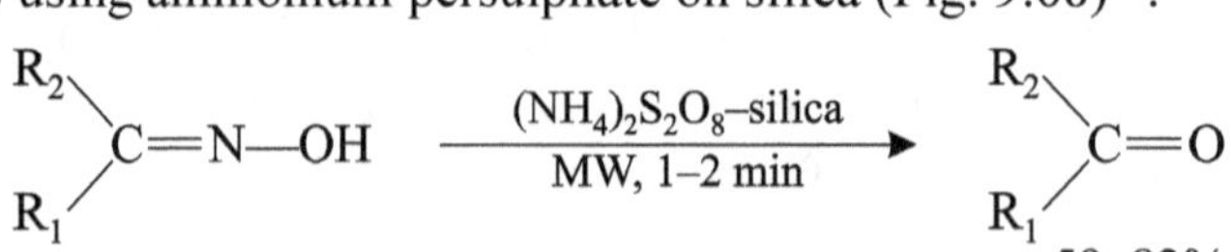

59–83%

R_1=CH_3; R_2 = Ph, *p*–Cl C_6H_4 *p*–Me C_6H_4, *p*-OMe C_6H_4
R_1 = H; R_2 = Ph, *p*-$NO_2C_6H_4$, *m*, *p*-$(OMe)_2$ C_6H_3.
R_1 – R_2 = 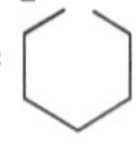

Fig. 9.68: Deoximation of Oximes.

The above reaction (Fig. 9.68) is applicable to both aldoximes and ketoximes. In this case, the role of the surface is critical. It should however, be noted that the same reagent supported on a clay gives predominately the Beckmann rearrangement product (*i.e.* amides)[72].

The ketoximes can be deoximated by MW irradiation using sodium periodate impregnated moist silica[73] (Fig. 9.69).

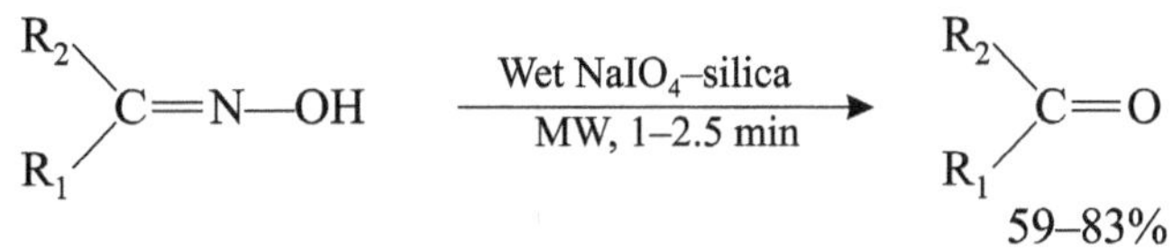

R_2=H; R_1 = *p*–Cl C_6H_4, *p*–Br C_6H_4, *p*-MeC_6H_4, *p*-OMe C_6H_4

R_1 = Ph; R_2 = Ph; R_1 = *n*-Bu; R_2 = Et,

$R_1 - R_2$ = 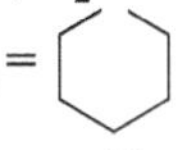

Fig. 9.69: Deoximation of ketoximes.

9.4.2.8 Cleavage of Semicarbazones and Phenyl Hydrazones

The cleavage of semicarbazones and phenylhydrazones of aldehydes and ketone can be effected by MW irradiation using ammonium persulfate impregnated on montmorillonite K10 clay (Fig. 9.70)[74].

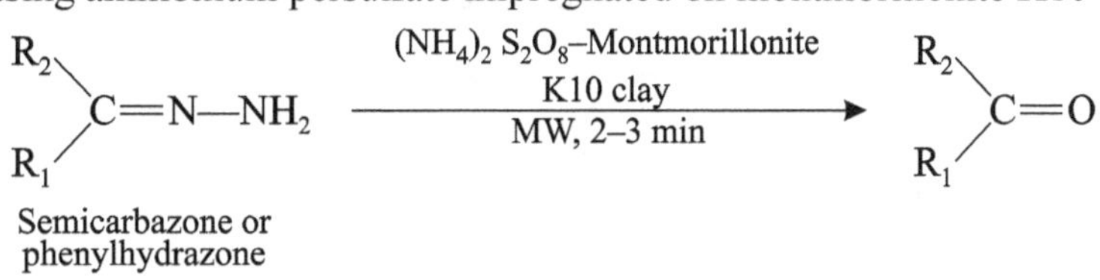

Fig. 9.70: Cleavage of semicarbazone and phenylhydrazones of carbonyl compounds.

9.4.2.9 Dethiocarbonylation

Dethiocarbonylation of thioketones or thioflavones can be conveniently achieved by MW irridation in presence of clayfen or clayan (Fig. 9.71)[75]. The method is much superior to using reagents like trifluoro acetic anhydride, thiophosgene, dimethyl selenoxide, benzene seleninic anhydride, benzoyl peroxide etc.

R_1–Me; R = H; R_1 = Ph; R=H, Br, Me

82–95%

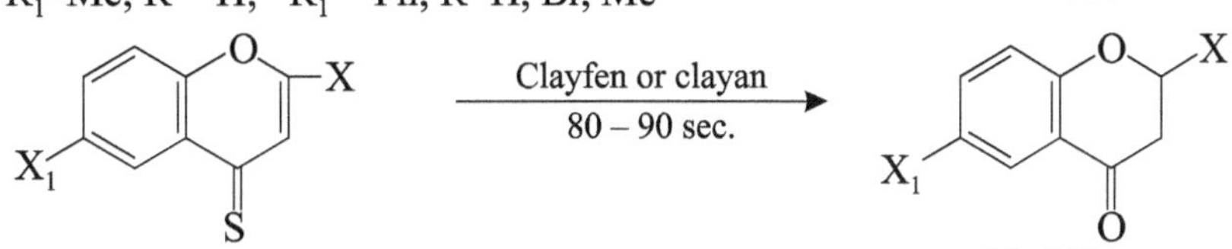

X = Ph, *p*-Me C_6H_4 ; X_1 = H

X = *p*-Me C_6H_4 ; X_1 = OMe

82–95%

Fig. 9.71: Dethiocarbonylation of thioketones and thioflavones.

9.4.3 Oxidations

9.4.3.1 Oxidation of Alcohols

The conventional oxidizing reagents[76] include $KMnO_4$, CrO_3, K_2CrO_4 and $K_2Cr_2O_7$. These reagents have their own limitations. It is most appropriate to perform oxidation with supported reagents using MW irradiation[77]. Following are given some of the reagents used for the oxidation of alcohols.

***(a)* Oxidation with Clayfen:** Montmorillonite K10 clay-supported iron (III) nitrate (clayfen) has been used to oxidise primary and secondary alcohols to carbonyl compounds on MW irridation (Fig. 9.72)[78].

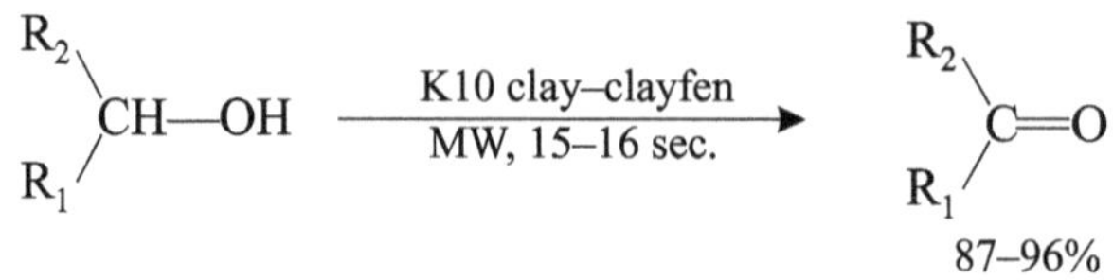

R_1 = Ph, *p*-Me C_6H_4, *p*-OMe C_6H_4; R_2 = H
R_1 = Ph; R_2 = Et, Ph CO
R_1 = *p*-MeO C_6H_4; R_2 = *p*-OMeC_6H_4CO

Fig. 9.72: Oxidation of alcohols with Clayfen.

***(b)* Oxidation with manganese dioxide-silica:** Using MnO_2-silica and MW irridation, alcohols are conventiently oxidised to carbonyl compounds (Fig. 9.73)[79].

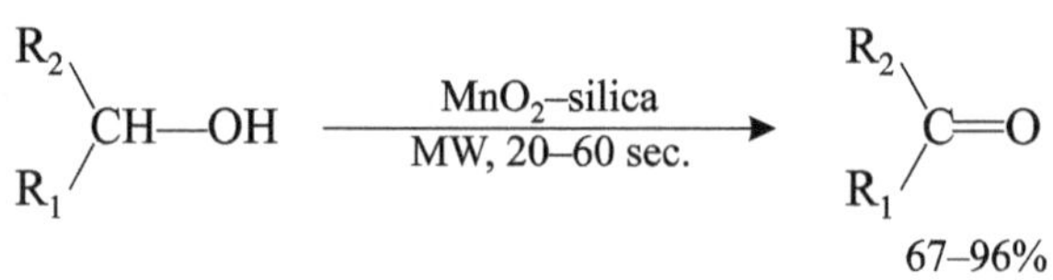

R_1 = Ph, *p*-MeC_6H_4, *p*-OMeC_6H_4, Ph CH = CH; R_2 = H
R_1 = Ph; R_2 = Et, PhCO
R_1 = *p*-MeO C_6H_4; R_2 = *p*-OMeC_6H_5CO

Fig. 9.73: Oxidation of alcohols with MnO_2-Silica.

***(c)* Oxidation with Claycop-hydrogen peroxide:** Copper (II) nitrate impregnated on K10 clay (claycop)-hydrogen peroxide is an effective oxidising reagent. It oxidises a variety of substrates and gives carbonyl compounds in excellent yield (Fig. 9.74).[80]

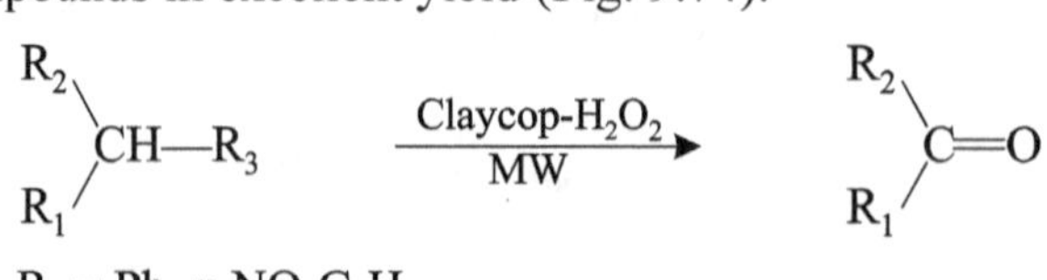

R_1 = Ph, *p*-$NO_2C_6H_4$
R_2 = H, Ph
R_3 = H, Br, COOH, CN, NH_2

Fig. 9.74: Oxidation of a variety of substrates with claycop-H_2O_2.

(*d*) **Oxidation with chromium trioxide impregnated wet alumina:** Chromium trioxide impregnated pre-moistened alumina is an effecient oxidising agent and converts benzyl alcohols to carbonyl compounds on MW irridation (Fig. 9.75)[81]. In these cases no overoxidtion is observed.

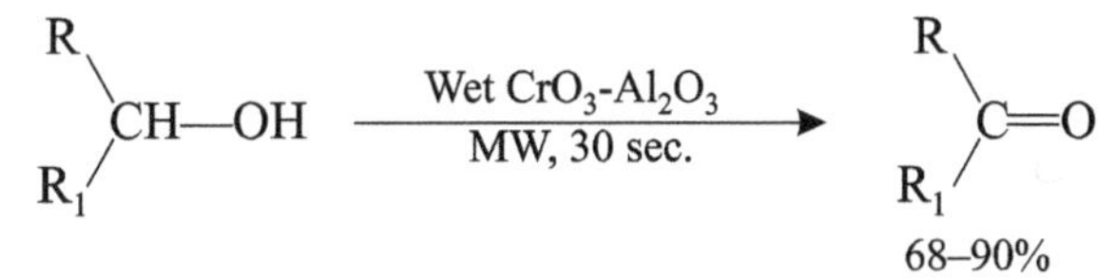

R = Ph, *p*-MeC_6H_4, *p*-$OMeC_6H_4$, *p*-$NO_2C_6H_4$; R_1 = H
R = Ph; R_1 = Me, Ph, PhCO

Fig. 9.75: Oxidation of alcohols with wet CrO_3-Al_2O_3.

(*e*) **Oxidation with iodobenzene diacetate (IBD) 'doped' alumina:** Oxidation of alcohols to carbonyl compound can conducted with alumina supported-IBD and MW irridation in quantitative yields (Fig. 9.76)[82].

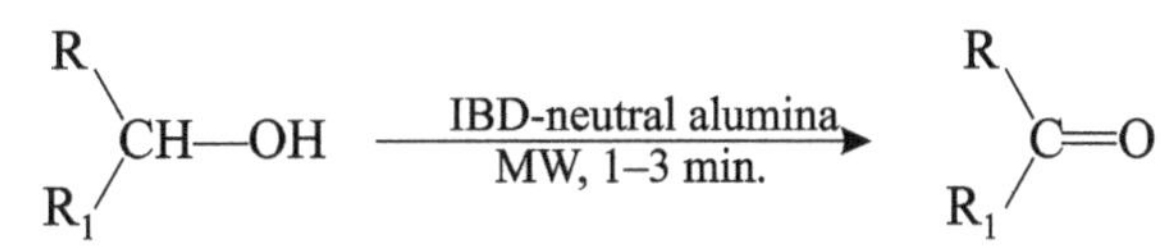

Fig. 9.76: Oxidation of alchols with IBD-Al_2O_3.

(*f*) **Oxidation with Copper (II) sulphate-alumina:** The reagent $CuSO_4$-Al_2O_3 oxideses both symmetrical and unsymmetrical benzoins to benzils on MW irridation (Fig. 9.77)[83].

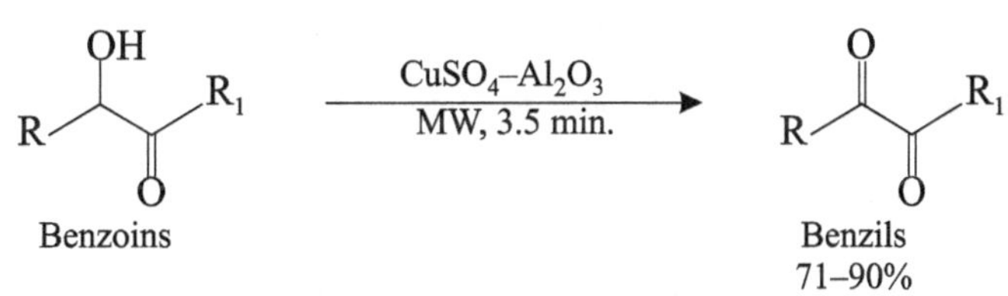

R = R_1 = C_6H_5, *p*-MeC_6H_4, *p*-$OMeC_6H_4$, *p*-Cl, C_6H_4
R = C_6H_5, R_1 = *p*-MeC_6H_4, *p*-$OMeC_6H_4$
R = Me, R_1 = C_6H_5

Fig. 9.77: Oxidation of benzoins to benzils using $CuSO_4$-Al_2O_3

The same oxidation can also be effected with Oxone® wet alumina[84].

9.4.3.2 Oxidation of Sulfides

(*a*) **Sodium periodate-silica oxidation:** Sulfides are usually oxidised using strong oxidising agents like HNO_3, H_2O_2, CrO_3, per acides and per iodate[85]. However, using MW irridation the oxidation of sulfides to sulfoxides or sulfones can be achieved in presence of silica doped with 10% sodium periodate using pulse technique[85] (Fig. 7.78)

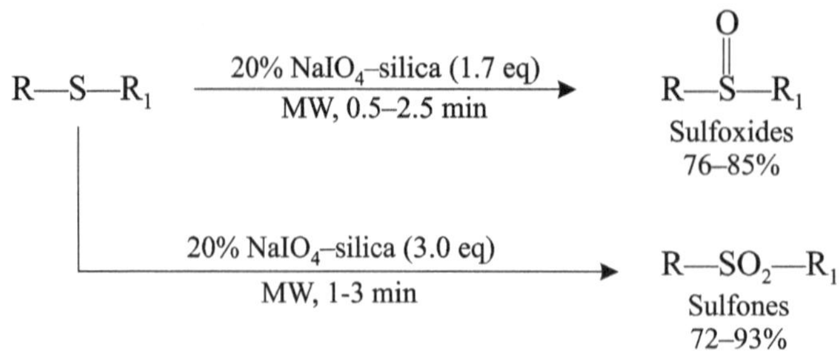

$R = R_1$ = Ph, Ph CH_2; *n*-Bu

R = Ph CH_2; R_1 = Ph; R = Ph, *n*-C_{12} H_{25}; R_1 = Me

Fig. 9.78: Oxidation of Sulfides to Sulfoxides or Sulfones.

Even benzothiophenes are oxidised to the corresponding sulfoxides and sulfones using ultrasonic and microwave irradiation respectively in presence of $NaIO_4$-silica[85].

***(b)* Iodobenzene diacetate-alumina:** IBD-alumina, which was shown to be used for oxidation of alcohols to ketones [See section 9.4.3.1 (*e*)] can also be used for the selective oxidation of alkyl, aryl and cyclic sulphides to the corresponding sulfoxides on MW activation (Fig. 9.79)[86].

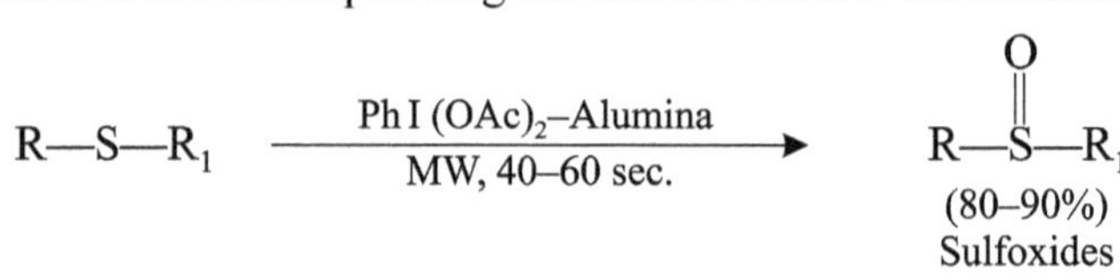

$R = R_1$ = i-Pr, Ph, Ph CH_2; R = Ph; R_1 = Me, Ph CH_2

R = n-$C_{12}H_{25}$; R_1 = Me; $R = R_1$ =

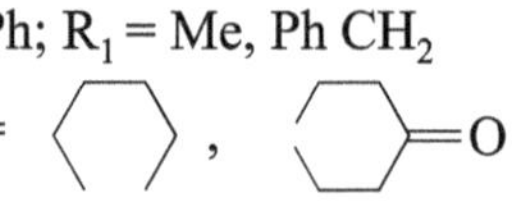

Fig. 9.79: Oxidation of Sulfides to Sulfones by Ph I $(OAc)_2$-Alumina.

9.4.3.3 Oxidation of Enamines

β, β-Disubstituted enamines can be oxidised into carbonyl compounds using $KMnO_4$-Al_2O_3 in domestic (255 W, 82°) or in focussed (330 W, 140°) MW oven. The yields are much better in the latter. Ketones and are formed by heating in an oil bath at 140°. (Fig. 9.80)[87].

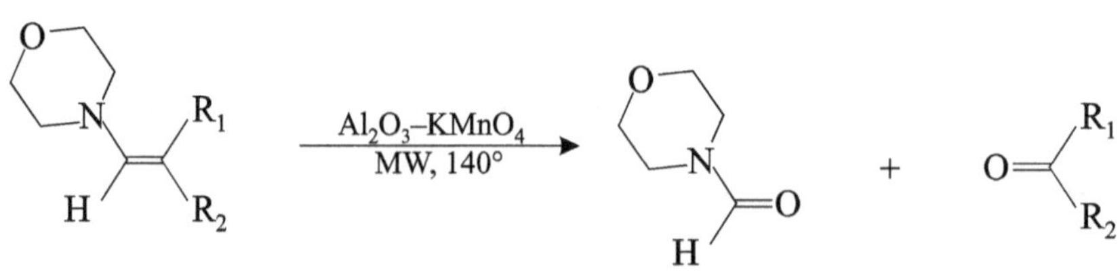

Fig. 9.80: Oxidation of Enamines.

9.4.3.4 Oxidation of Arenes

Arenes can be oxidised to ketones using focussed microwaves in presence of $KMnO_4$ impregnated alumina (Fig. 9.81)[88].

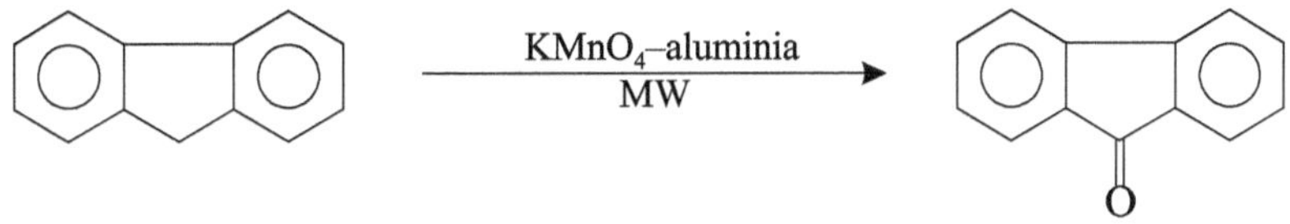

Fig. 9.81: Oxidation of Arenes.

9.4.4 Reductions

9.4.4.1 Reduction of Carbonyl Compounds

Sodium borohydride has been used for the reduction of carbonyl compounds. The solid state reduction of ketones has been possible by mixing with $NaBH_4$ and storing the mixture for 5 days[89]. However, the reduction proceeds much faster by using 10% $NaBH_4$-alumina and irridating the mixture for 0.5-2 min in a MW oven (Fig. 9.82)[89].

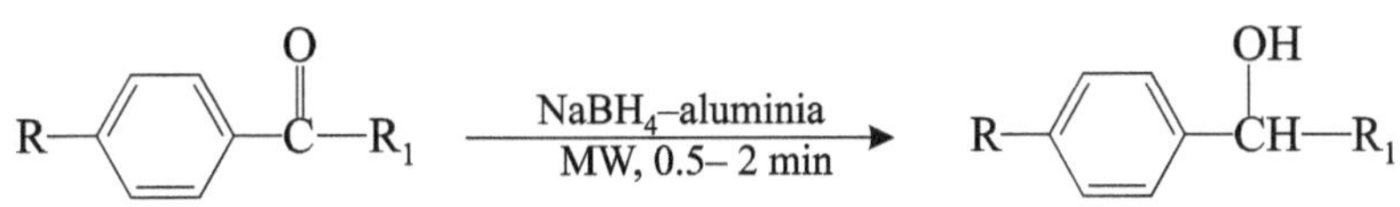

R = Cl,Me, NO_2; R_1 = H; R = H; R_1 = Me, Ph

R = Ph; R_1 = PhCH(OH); R_1 = R_2 = Me

Fig. 9.82: Reduction of Ketones with $NaBH_4$-alumina.

The above procedure can also be used for the reduction of trans-cinnamaldehyde; in this case the olefinic moiety remains intact and only the aldehydic moiety, is reduced at room temperature.

The reduction of carbonyl compounds using isopropyl alcohol in presence of aluminium alkoxides under MW irradiation give the corresponding alcohols (Fig. 9.83)[90].

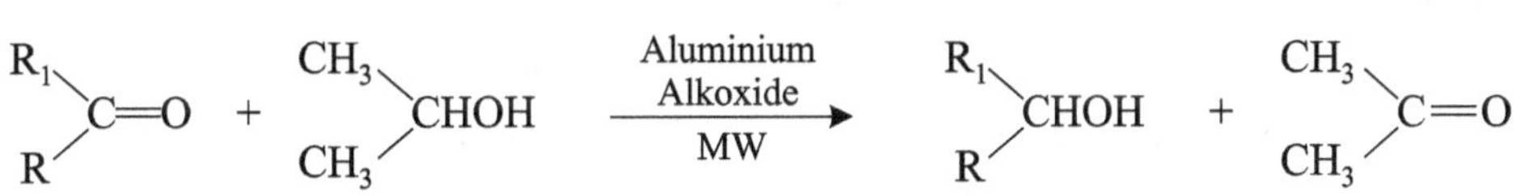

Fig. 9.83: Reduction of carbonyl compounds with aluminium alkoxide.

The reduction of carbonyl compounds using isopropyl alcohol and alumina was earlier recorded[91].

9.9.4.2 Reductive Alkylation of Amines

The reductive amination of carbonyl compounds has been usually achieved by using sodium cyanoborohydride, sodium triacetoxyborohydride or $NaBH_4$ coupled with H_2SO_4. Most of these reagent produce waste. A environmentally benign method involves the solvent-free reductive amination of carbonyl compounds by using wet montmorillonite K10 clay supported $NaBH_4$ and subsequent irradiation by microwaves (Fig. 9.84)[92].

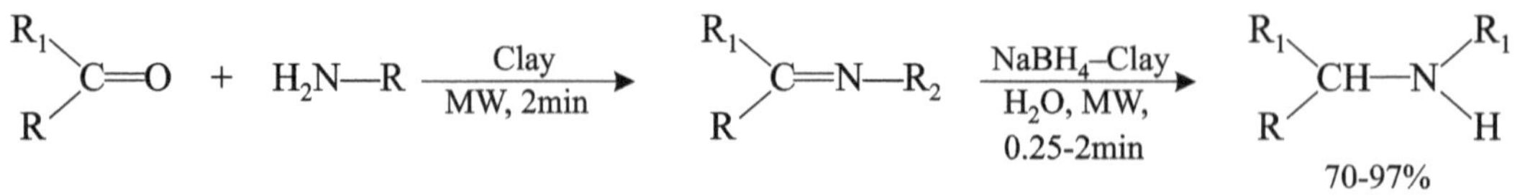

Fig. 9.84: Reductive amination of amines.

9.4.5 Pinacol-Pinacolone Rearrangement

Pinacol-Pinacolone rearrangement has been carried out by microwave irridation of gem-diols (Pinacols) with Al^{3+}-Montmorillonite K10 clay for 15 min (Fig. 9.85)[93].

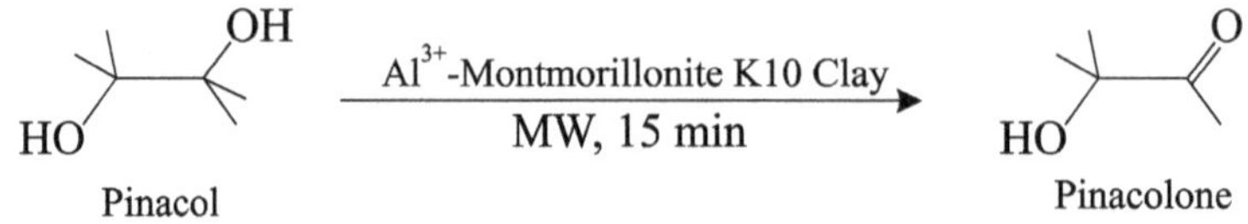

Fig. 9.85: Pinacol-Pinacolone rearrangement.

9.4.6 Beckmann Rearrangement

Ketoximes on MW irridation in presence of Montmorillonite K10 clay gives the Beckmann rearrangement product (Fig. 9.86)[94]

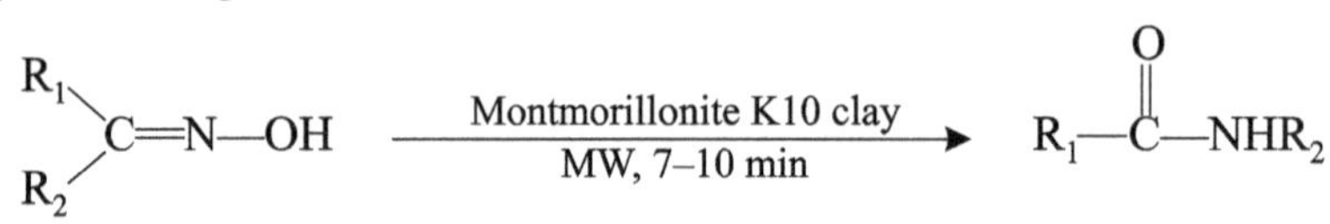

Fig. 9.86: Beckmann rearrangement.

9.4.7 Synthesis of Substituted Thiazoles

Substituted thiazoles are obtained by microwave irradiation of α-tosyloxyketones [which are generated in situ from aryl methyl ketones and [hydroxy (tosyloxy) iodo] benzene (HTIB)] with thioamides in presence of K10 clay (Fig. 9.87)[95]

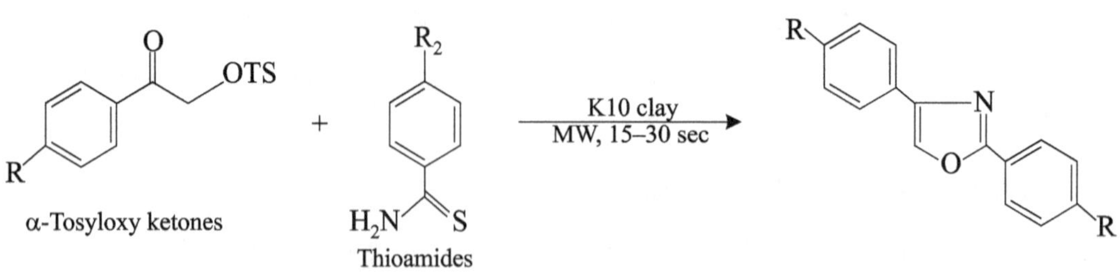

Fig. 9.87: Synthesis of substituted Thiazoles.

It is found that the reaction of α-tosyloxyketones with ethylenethioureas (MW irridation) gave bridgehead heterocydes (Fig. 9.88)[95].

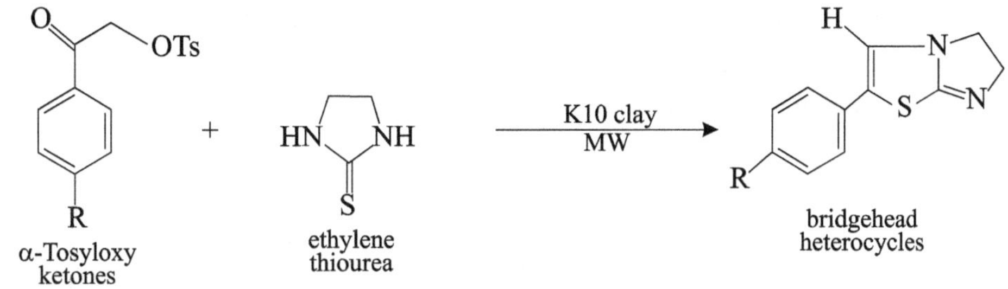

Fig. 9.88: Synthesis of bridgehead heterocycles.

See also Section 9.5.5.

9.4.8 Synthesis of 2-aroyl benzofurans

The naturally occurring and pharmacologically important 2-arylbenzofurans are obtained by the solid state reaction α-tosyloxyketones and salicyladehydes in the presence of a base like KF doped alumina using M.W. irradiation[95] (Fig. 9.89).

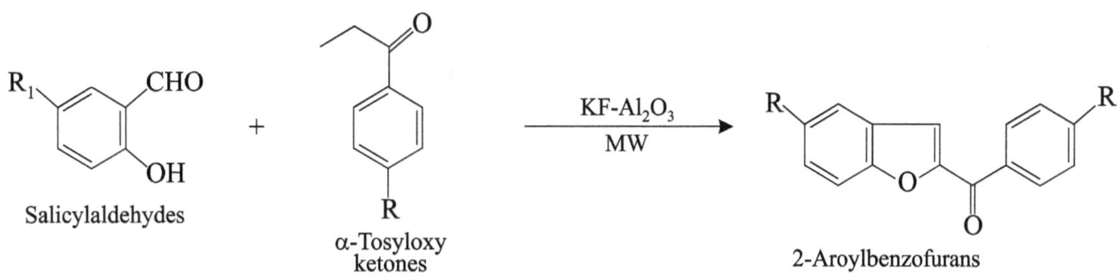

Fig. 9.89: Synthesis of 2-Aroylbenzofurans.

9.4.9 Synthesis of Flavones

Flavones belong to the class of naturally occurring phenolic compounds, which are found to occur in plant kingdom and are known to display a variety of biological activities. Flavones are prepared by the well known Allan-Robinson synthesis or by the cyclisation of 1, 3-diketones[96]. These are best prepared by MW irridation of *o*-hydroxydibenzoyl methanes adsorbed on montmorillonite K10 clay[96] (Fig. 9.90)

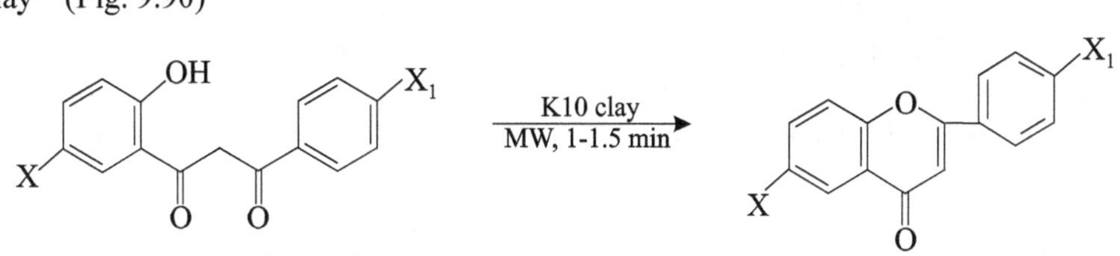

X = H; X_1 = H, Me, OMe, NO_2
X = OMe; X_1 = H, Me, OMe

Fig. 9.90: Synthesis of Flavones.

9.4.10 Synthesis of 2-aryl-1,2,3,4-tetrahydro-4-quinolones

The 2-aryl-1, 2, 3, 4-tetrahydro-4-quinolones are medicinally important and are prepared by MW irridation of 2′-aminochalcones in presence of montmorillonite K10 clay[97] (Fig. 9.91).

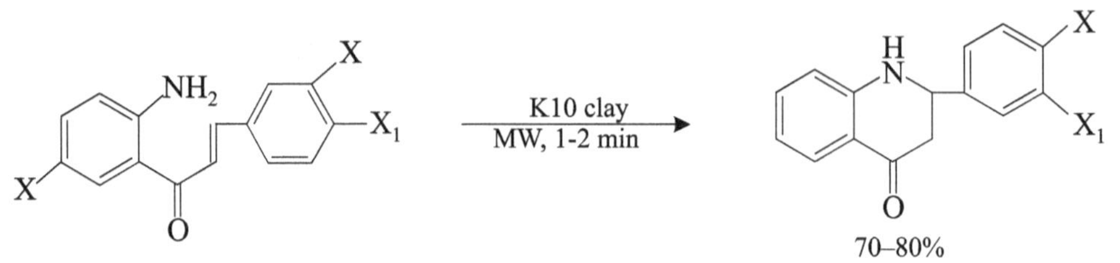

X = Cl, Br, Me, OMe, NO_2; X_1 = H
X = X_1 = H, OMe

Fig. 9.91: Synthesis of 2-aryl-1, 2, 3, 4-tetrahydro quinolones.

9.4.11 Transformation of Aromatic Aldehydes to Nitriles

Aromatic aldehydes are known to be converted into corresponding nitriles by dehydration of the oximes using a variety of reagents. Arylaldehydes are easily converted into nitrile in a one step process by the reaction of aryl aldehydes with hydroxylamine hydrochloride supported on montmorillonite K10 clay followed by MW irradiation (Fig. 9.92)[98].

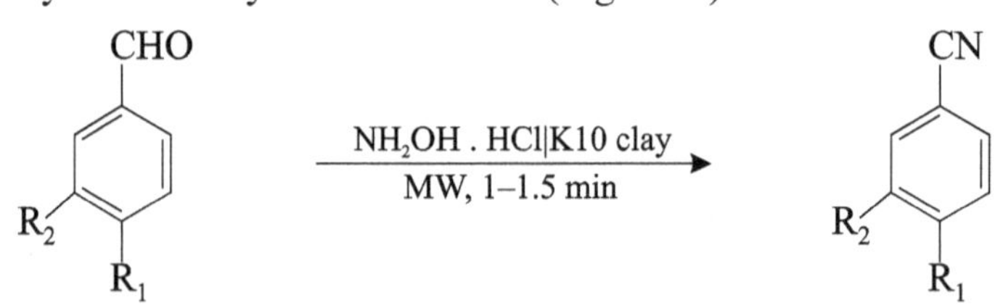

89–95%

R_1 = H; R_2 = H, OMe, NO_2, OH, Br, Me
R_1 = R_2 = OMe

Fig. 9.92: Conversion of aromatic aldehydes into corresponding nitriles.

The aliphatic aldehydes give only poor yields (10-15%) of nitriles.

9.4.12 Conversion of Aldehydes to Alcohols

The aldehydes which lack α-hydrogens are converted into a mixture of alcoholc and carboxylic acids by the well known **cannizzaro reaction.**[99] The conversion of aldehydes to alcohols can also be effected by $LiAlH_4$, but the yields are low. However, crossed cannizzaro reaction[100] using para formaldehyde give higher yield of alcohol. It is found[101] that the reaction of aldehydes lacking α-hydrogens with para formaldehyde in presence of $Ba(OH)_2.8H_2O$ on MW irridation gave 80-99% of the corresponding alcohol. The reaction is known as **solid state cannizzaro reaction** (Fig. 9.93).

$$RCHO + (CH_2O)_n \xrightarrow[Ba(OH)_2.8H_2O]{MW} \underset{80\text{-}90\%}{RCH_2OH} + \underset{1\text{-}20\%}{RCOOH}$$

Fig. 9.93: Conversion of aldehydes into alcohols.

The aldehydes having an electron withdrawing substitution undergo reaction at a much faster rate.

9.5 SCALING UP OF MICROWAVE ASSISTED TRANSFORMATIONS

9.5.1 Introduction

It is now well known that microwave has been routinely used for organic transformations on a laboratory scale. However, in industries it is necessary to scale up the transformations. In general two possible scale-up operations are available[102]. These are continuous-flow microwave cell and batch-type process. The former, *viz.* continuous-flow microcell has been successfully used for a number of different reactions[103]. This technology is difficult to process solids, highly viscous liquids or reaction mixtures which are heterogeneous. The bach-type process can involve either one large vessel or parallel batch reactors. This technology is believed to be suitable for scaling up of organic transformations.[102]

Different types of Microwave instruments are now commercially available. Different instruments have one or more of the following options:

- Having continuous microwave delivery system.
- A fibre-optic probe inserted directly into the reaction mixture by means of a glass thermowell. By this the temperature of the contents of the reaction mixture in reaction vessel can be measured.
- Fitted by a rotating magnetic plate located below the floor of the microwave cavity and Teflon-coated magnetic stirring bar in the vessel. This arrangement helps to stir the reaction mixture.
- Fitted with a reflux condenser which is attached to the reaction flask. This arrangement is helpful to carryout transformation in volatile solvents.
- Fitted with a Dean-Stark apparatus which is attached to the reaction flask. This arrangement is helpful for carrying out transformation in which water or alcohol is eliminated. Examples include esterification, transestrification.
- Some instruments are equipped with two magnetrons, which combined continuous microwave output power from 0 to 1400 W.

Following are given some of the organic transformation which can be scaled up. The extent of scaling up is also mentioned.

9.5.2 Suzuki Coupling Reaction

The reaction could be scaled up to 1 mol scale[104] (Fig. 9.94).

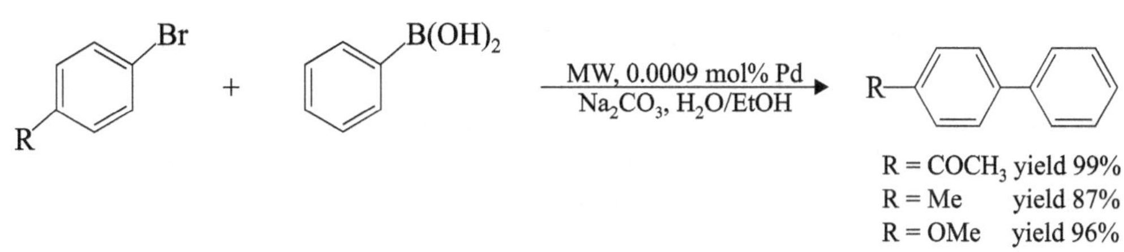

Fig. 9.94: Suzuki reaction on 1 mol. scale.

The above reaction was performed in a CEM Discover microwave unit. This instrument is equipped with an access port in the top of the cavity through which a reflux condenser could be attached. The temperature of the reaction mixture was measured using a fibre-optic temperature probe inserted into a thermal well that was itself-inserted into the reaction mixture.

9.5.3 Hentzsch Synthesis of 1,4-dihydropyridines

The reaction could be scaled up to 1 mol scale. The 1,4-dihydropyridines are known to have considerable biological activities (see section 9.2.13). These are obtained by the condensation of an aldehyde with α, β-dicarbonyl compound and ammonia. It was found that using, aldehyde, α, β-dicarbonyl compound and ammonium hydroxide in the ratio 1:3.4:10 gave on yield of 76% on a 1 mol scale. The reaction was conducted by heating the reaction mixture to reflux, keeping at this temperature for 20 min (Fig. 9.95)[102].

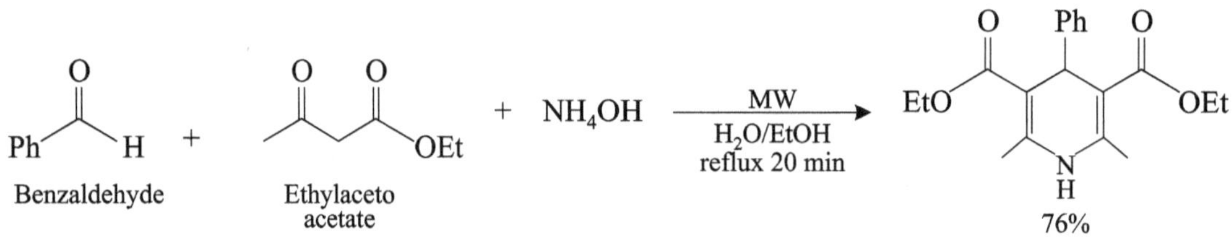

Fig. 9.95: Hentzsch Synthesis of 1,4-dihydro pyridines on one mole scale.

9.5.4 Beckmann Rearrangement

This reaction could be scaled up to 1 mol scale. The Beckmann rearrangement is particularly important for the manufacture of γ-caprolactam. The reaction consists in heating a mixture of a ketone, *e.g.*, benzophenone and hydroxylamine hydrochloride in acetic acid/sulphuric acid as solvent at 120° in a microwave oven. The product, N-phenyl benzamide was obtained in 95% yield (Fig. 9.96)[103,104].

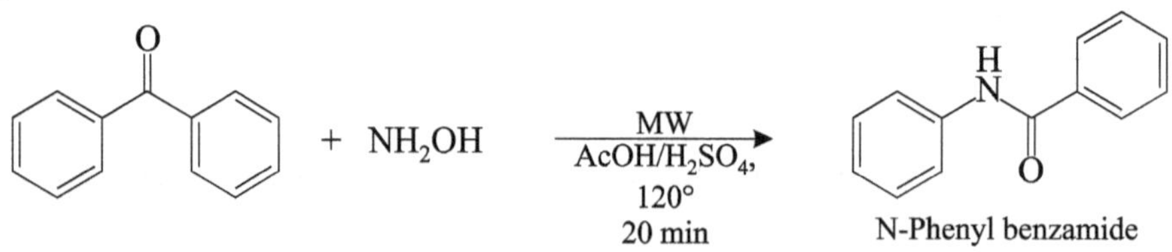

Fig. 9.96: Beckmann rearrangement on molar scale.

9.5.5 Synthesis of 2-amino-4-phenyl Thiazole

Compounds containing 2-aminothiazole moiety have application for the treatment of a number of health conditions[105]. 2-Amino-4-phenyl thiazole is best prepared by the condensation of α-haloketones with thiourea[106]. The use of MW heating facilitates the reaction[107]. It was found that using a 1 : 1.5 stoichiometric ratio of phenacyl chloride to thiourea as reagents and ethanol as solvent and heating in MW oven, the reaction was complete in 5 min at 140°C. The reaction could be scaled to 12.5 m mol scale (Fig. 9.97)[102].

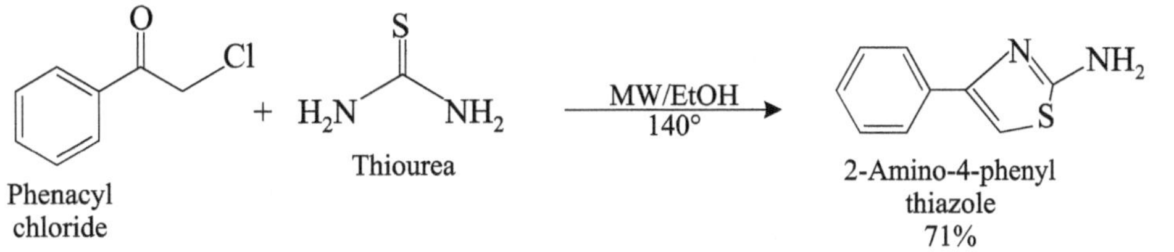

Fig. 9.97: Synthesis to 2-amino-4-phenyl thiazole on 12.5 m mol scale.

Using continuous flow microwave reaction[108, 109] the preparation of 2-amino-4-phenyl thiazole could be scaled up to 0.5 mol giving 97% yield.

9.5.6 Heck Coupling Reaction

The reaction could be scaled up to 0.1 mol. Thus, the reaction aryl bromide (*e.g.*, 4-bromoanisole) (0.1 mol) and methyl acrylate (0.2 mol) in presence of tetrabutylammonium bromide (1 equiv) as phase-transfer agent, potassium carbonate (3.7 equiv.) as base and palladium loading of 0.002 mol% and water as solvent was heated in a MW oven to 170° during 5 minutes and held at this temperature for 15 minutes. The mixture was continuously stirred during the reaction. The product obtained was 4′-methoxycinnamic acid is 93% yield (Fig. 9.98)[102].

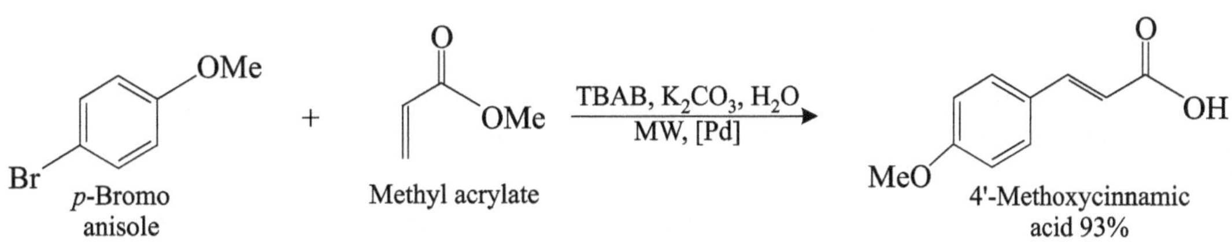

Fig. 9.98: Heck coupling reaction on 0.1 mol scale.

Using continuous flow microwave reaction[108,109], the preparation of 4′-methoxy cinnamic acid could be scaled up to 1 mol.

9.5.7 Esterification

The esters are used on small and large scales in chemical industries, particularly fine chemicals, flavour and fragrance. The esterification of acetic acid with butanol in presence of H_2SO_4 was performed by heating the reaction mixture in a MW oven, which was fitted with a Dean-stark trap in order to remove water formed during esterification. With this arrangement the esterification proceeds to completion. The reagents (acetic acid and *n*-butanol) were used in 1 : 1 molar ratio and H_2SO_4 (1 wt%) as catalyst. The reaction mixture was heated till no more water collected in Dean-stark trap. By this process 82% conversion to butyl acetate was obtained. The reaction could be sealed to 6 mol. In place of *n*-butanol, secondary alcohols like cyclohexanol or 2-octanol could also be used (Fig. 9.99)[102].

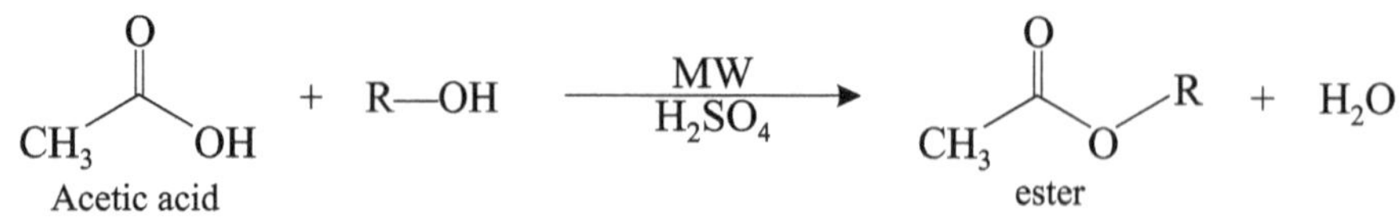

Alcohol = butanol (2 moles) 82% Conversion
Alcohol = butanol (6 Moles) 98% Conversion
Alcohol = cyclohexanol (2 moles) 80% Conversion
Alcohol = 2-octanol (2 moles) 86% Conversion

Fig. 9.99: Esterification of acetic acid with 1° and 2° alcohols.

Using continuous flow microwave reaction[108,109], the esterification could be scaled up to 22 moles.

The esterification can also be done by heating potassium acetate and alkylhalide in presence of a PTC in a MW oven (Fig. 9.100)[110].

$$CH_3COOK + n\,C_8H_{17}Br \xrightarrow[MW]{PTC} \underset{98\%}{CH_3COOnC_8H_{17}}$$

Fig. 9.100: Esterification in presence of PTC.

The esterification could be scaled up to using 2 mol of each of the reactants.

9.5.8 Transesterification

The transesterification is of special interest in connection with the synthesis of biofuels[111]. As an example diethyl adipate is heated in MW oven with butanol (diethyl adipate: butanol, 1 : 2 molar ratio) and using H_2SO_4 (1 wt %) as catalyst. The reaction could be driven towards completion by removal of ethanol produced (using Dean-Stark trap). The reaction mixture was initially heated to 105° and then held at this temperature for 10 min. Subsequently the temperature was raised to 115° and holding the reaction mixture at 115° for 10 min. By this procedure all the ethanol produced in transesterification is removed. Finally the temperature was raised to 145° to remove any unreacted butanol. Dibutyl adipate was obtained in 59% yield (Fig. 9.101)[102].

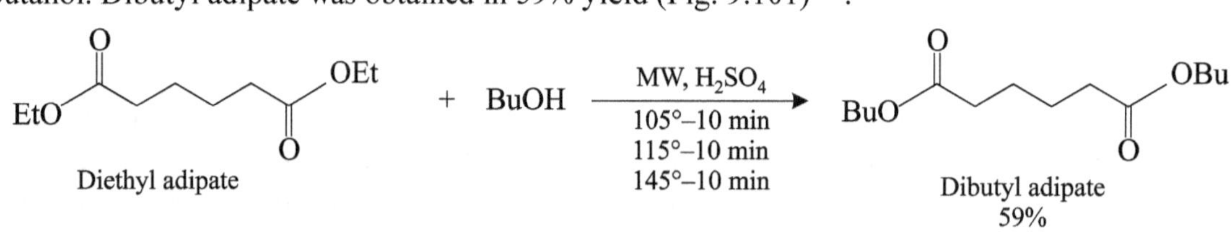

Fig. 9.101: Transesterification.

9.5.9 Synthesis of 2-octene

It is obtained by the dehydration of 2-octanol using 1 wt % H_2SO_4 as catalyst and heating in a MW oven fitted with a Dean-Stark trap. The formation of 2-octene commenced as the reaction mixture reached a temperature of 145°. The formed 2-octene and water gets collected in the Dean-Stark trap. The 2-octene was a mixture of *cis*- and *trans*-isomer along with a trace of 1-octene. The reaction could be scaled up adding more of 2-octanol at a rate the 2-octene distils. Using this technique a total of 460 mL of 2-octanol could be dehydrated (Fig. 9.102)[102].

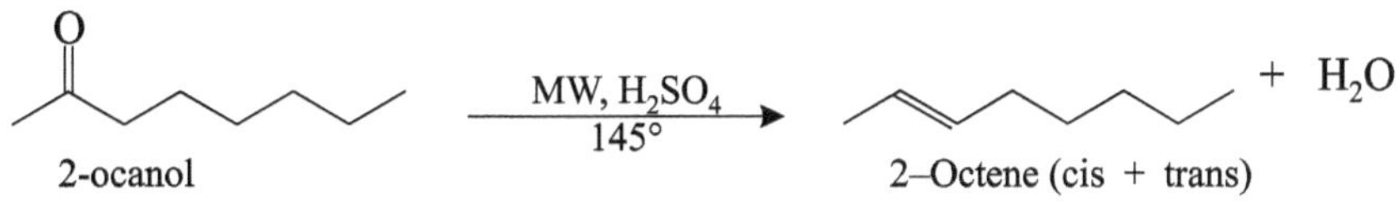

Fig. 9.102: Synthesis of 2-octene.

9.5.10 Synthesis of Fluorescein

Fluorescein is prepared from phthalic anhydride and resorcinol by heating in a MW oven. The starting materials (2 mol of resorcinol and 1 mol of phthatic anhydride) are ground in a mortar and pestle. The mixture is loaded in a reaction pot. Using an initial microwave power of 800 W, the mixture is heated to 100° during 5 min. Subsequently using a microwave power of 1600 W, the melted mixture is heated from 100 to 220° and kept at this temperature for 10 min. The reaction mixture is cooled, dissolved in aqueous KOH and the clear basic solution poured into dilute H_2SO_4. The precipitated fluorescein is filtered, added to water and boiled 10 min using a MW oven; this treatment dissolves any remaining starting material. The fluorescein thus obtained is quite pure yield 80%. (Fig. 9.103)[102].

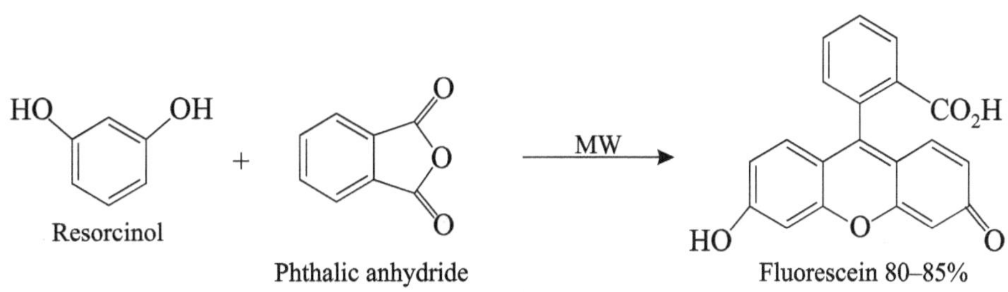

Fig. 9.103: Preparation of Fluorescein.

The reaction could be scaled up to using 3 mol of resorcinol and 1.5 mol phthalic anhydride. During the reaction, the rotation of the reaction vessel ensures good mixing of the reactants.

9.5.11 Du Pont HCN Industrial Process[112]

It involves the condensation of methane and ammonia over alumina, maintained at 1200° by MW heating (Fig. 9.104)

$$CH_4 + NH_3 \xrightarrow[\text{Catalyst}]{1200°C} HCN + 3H_2$$

Fig. 9.104: Du Pont HCN Industrial Process.

9.5.12 Phenylacylation of 1, 2, 4-triazole

The phenylacylation of 1,2,4-triazole could be scaled up to 540 m mol. In this case there was enhanced selectivity to the N-1 product under MW irridation (Fig. 9.105)[113]. Classical heating gave a mixture of N-1 and N-4 alkylated product as well as quaternary sats[114].

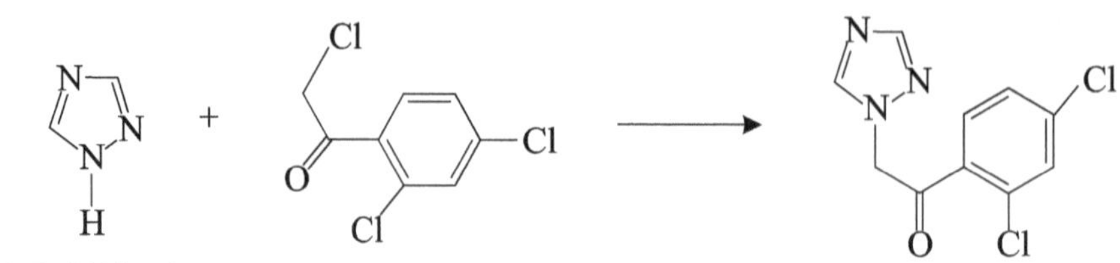

Fig. 9.105: Phenylacylation of 1,2,4-triazole.

9.5.13 Synthesis of Ionic Liquids

Solvent-free synthesis and scale-up of ionic liquids under MW irridation has been reported[115]. Using a commercial MW reactor, ionic liquids based upon the 1-alkylpyridinium, 1-alkyl-3-methylimidazolium, 1-alkyl-2-methylpyrazolium and 3-alkyl-4-methyl thiazolium cations on scales from 50 m mol to 2 mol were prepared (Fig. 9.106).

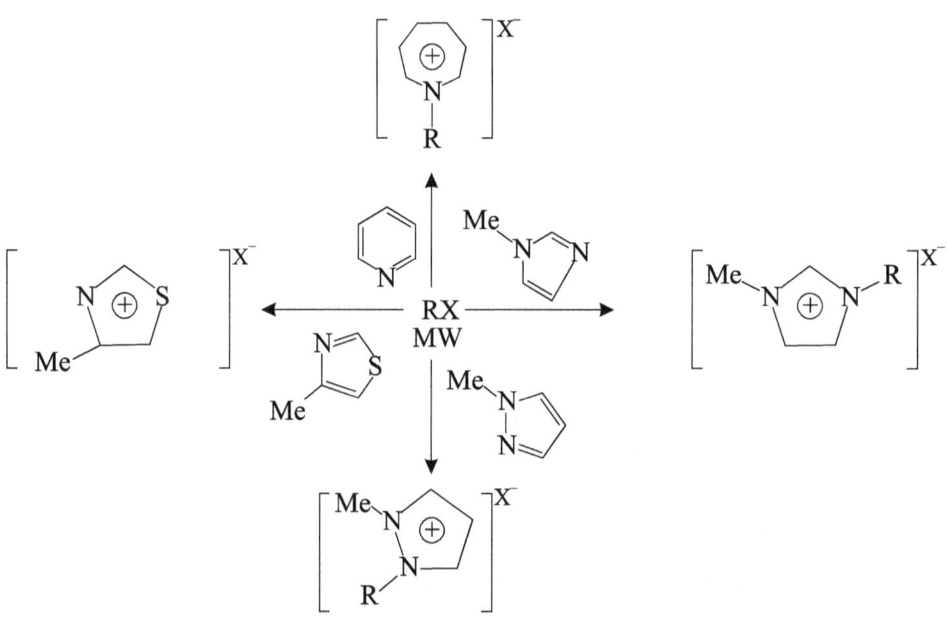

Fig. 9.106: MW assisted synthesis of ionic liquids.

9.5.14 Scaling up of MW assisted Reactions using Solvent-Swollen Polymer Beads

Solvent-swollen polymer beads have been used for scaling up MW assisted reactions[116]. It is found that the swelling behaviour of beads did not alter after 20 minutes MW irradiation (700 W). This technique has been used for peptide hydrolysis[117] peptide coupling[118], Suziki and Stille coupling[119-121] and fluorous still coupling[122].

9.5.15 Microwave-assisted Deuterium Exchange

MW-assisted deuterium exchange has been effected in reactions like hydrogenations, borohydride reductions, decarboxylations etc. (Fig. 9.107)[123]. In addition to rate acceleration, environmentally friendly routes have been developed, especially solventless reactions that minimise waste minimisation[123-126].

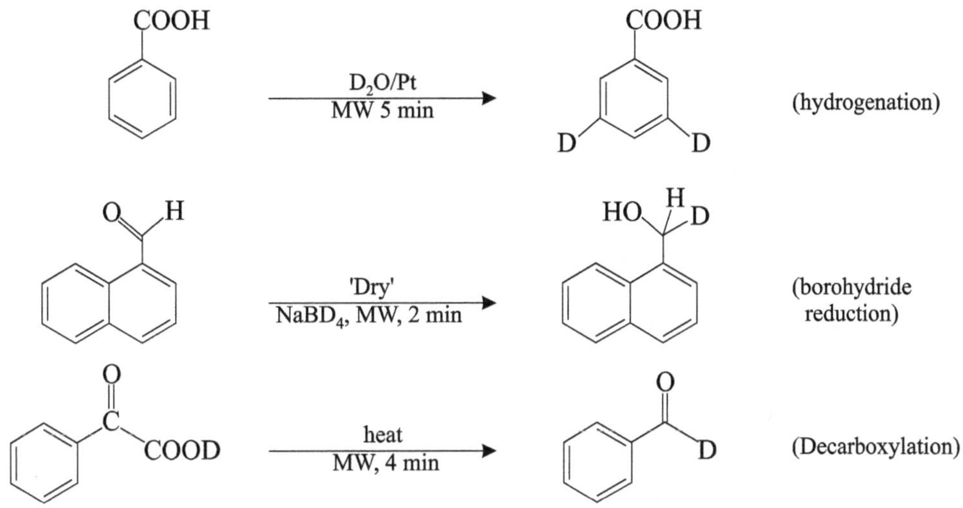

Fig. 9.107: M-W assisted deuterium exchange.

9.5.16 Synthesis of Aspirin

It is synthesised from salicylic acid and acetic anhydride using commercial microwave reactors (Fig. 9.108)[127]. By this process high purity aspirin is obtained is 80% yield. The process could be scaled up to 500-800 g scale.

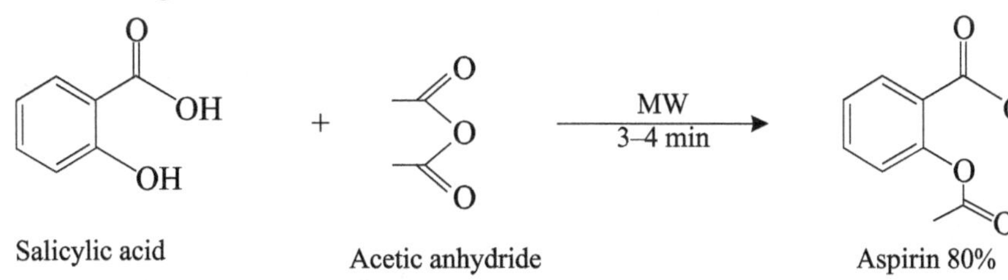

Fig. 9.108: Microwave-assisted synthesis of asparin (500-800 g scale).

9.5.17 Synthesis of Paracetamol

N-Acetylation of *p*-aminophenol (50g) with acetic acid (200 mL) on heating in MW oven gave 55% yield of paracetamol (Fig. 9.109)[127].

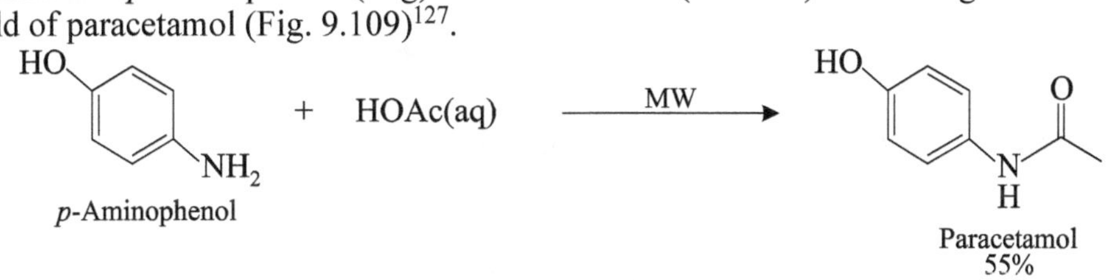

Fig. 9.109: Synthesis of Paracetamol.

9.5.18 Synthesis of Tetrachlorophthalimidoacetic Acid

It was synthesised from tetrachloro-phthalic anhydride and glycine using DMF as solvent and heating in a MW oven for 8 min. (Fig. 9.110)[127]. The reaction was scaled up to molar scale.

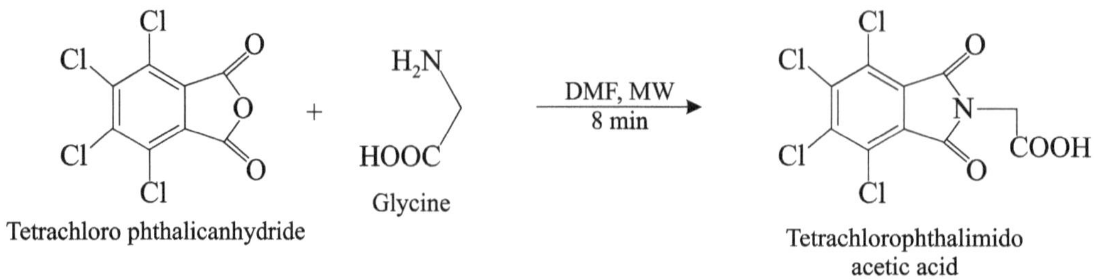

Fig. 9.110: Synthesis of tetrachlorophthalimido acetic acid.

9.5.19 Synthesis of 3-carboxycoumarin

3-Carboxycoumarin was obtained by the reaction of salicylaldehyde with malonic acid is presence of clay/KSF (aq) by MW irridation for 12 min. (Fig. 9.111)[127].

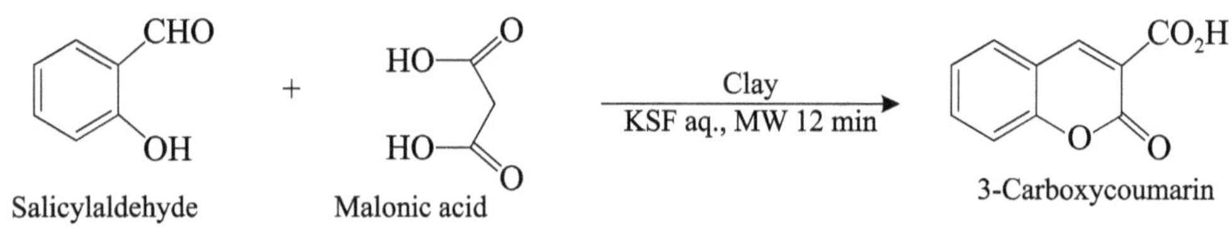

Fig. 9.111: Synthesis of 3-carboxycoumarin.

9.5.20 Synthesis of 2, 3-Dimethyl Indole

2, 3-Dimethyl indole is obtained by heating a mixture of phenyl hydrazine and butan-2-one by a one-pot reaction in water at 220° for 30 min. The reaction known as **Fischer-indole synthesis**. The product formed is 2, 3-dimethyl indole (Fig. 9.112)[128].

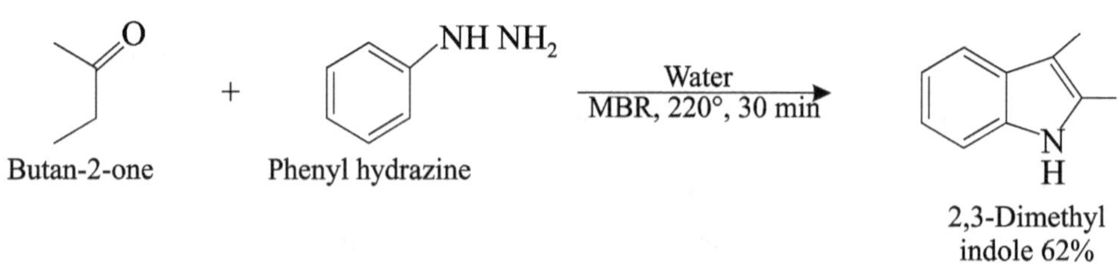

Fig. 9.112: Synthesis of 2,3-dimethyl indole.

9.5.21 Decarboxylation of Indole-2-carboxylic Acid

The decarboxylation of indole-2-carboxylic acid can be effected by heating in a MW oven a 255° for 20 min (Fig. 9.113)[128].

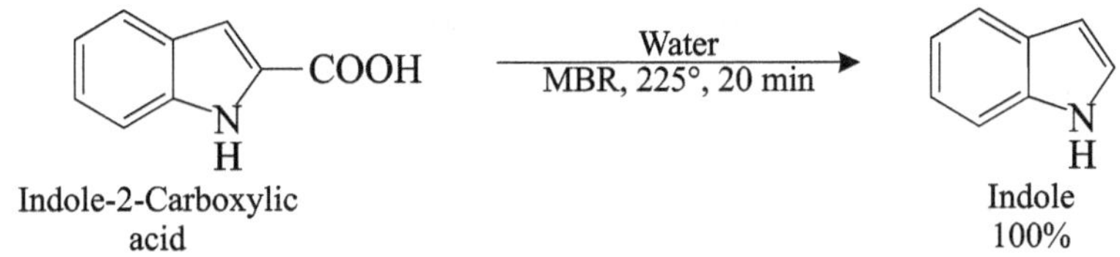

Fig. 9.113: Decarboxylation of indole-2-carboxylic acid.

9.5.22 Friedel-Crafts Reaction

Friedel-crafts reactions could be conducted on a large laboratory scale via a continuous flow process[129]. This, acylation of aromatic ethers gave the corresponding ketones (Fig. 9.114)[129].

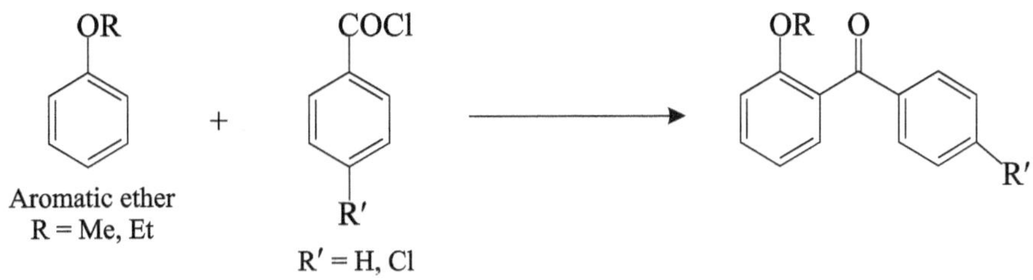

Fig. 9.114: Friedel-crafts reaction.

In a similar way mesitylene could be sulphonylated (Fig. 9.115).

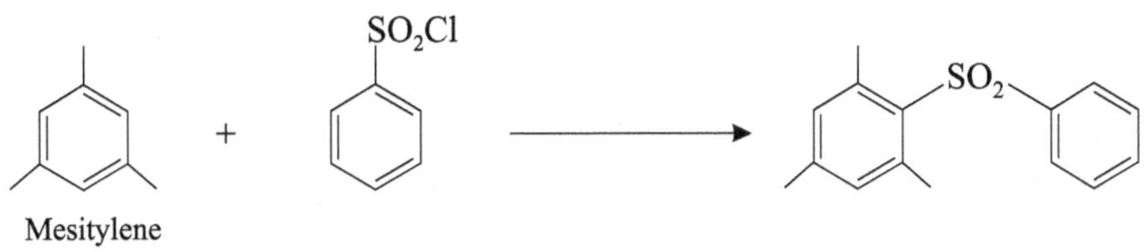

Fig. 9.115: Sulphonylation of Mesetylene.

9.5.23 Methylation of Phenols, Indoles and Benzamidazoles

Methylation of phenols, indoles and benzamidazoles could be effected in a much shorter time with dimethyl carbonate and DBU as catalyst[130] using a commercial (CMR microwave unit Fig. 9.116). The procedure could be scaled up by passing 5g of the substrate through the reactor every 6 min and monitering with HPLC until the reactions were complete.

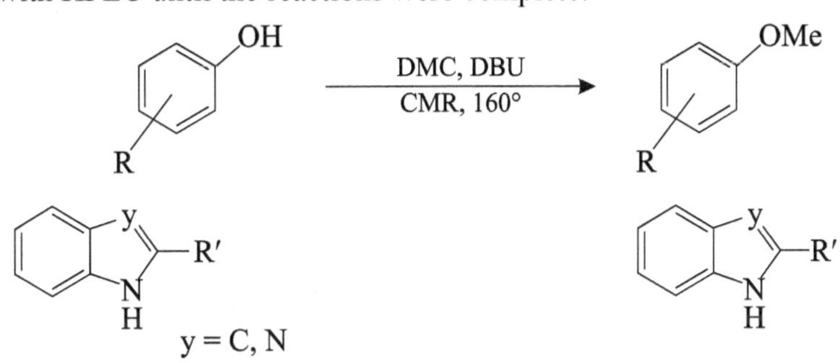

Fig. 9.116: Methylation of phenols, indoles and benzamidazoles.

9.5.24 Synthesis of alkyl-2-(hydroxymethyl) Acrylates

These were prepared by a modified **Baylis-Hillman reaction**. The method involves the reaction of acrylate derivatives with formalin in presence of 1,4-diazobicyclo [2.2.2] octane (DABCO) using a CMR microwave unit at 160-180° (Fig. 9.117)[131].

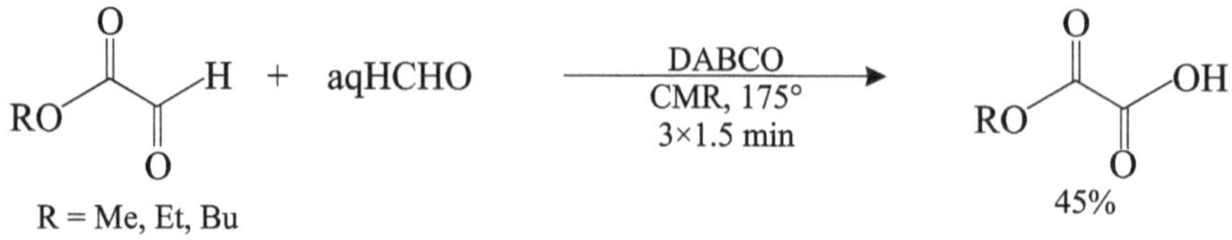

Fig. 9.117: Synthesis of alkyl 2-(hydroxymethyl) acrylates.

9.5.25 Depolymerisation of Cellulose

The depolymerisation of cellulose into glucose could be effected by heating with 1% aqueous H_2SO_4 in a MBR microwave unit[131]. The method involved raising the temperature to 215° within 2 min (in the MBR), maintaining this temperature for 30 sec. and cooling (Fig. 9.118)[131].

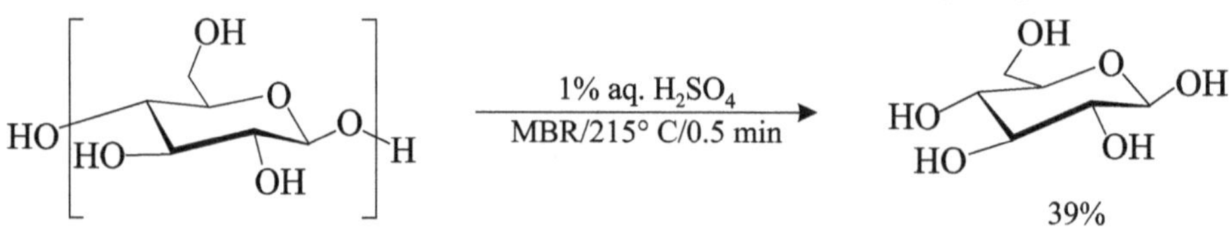

Fig. 9.118: MW assisted depolymerisation of Cellulose.

9.5.26 Rearrangement of 2-benzoyloxyacetophenone

The rearrangement of 2-benzoyloxy-acetophenone to 1-(2′-hydroxyphenyl)-3-phenylpropan-1,3-dione (β-diketone) is known as **Baker-venkatoraman rearrangement**. It is normally a base catalysed rearrangement[132]. The rearrangement can be easily performed by a combination of MW and uv radiation (MW-UV) first reported[133] in 1999. Using this technique 2-benzyloxyacetophenone could be rearranged to 1-(2′-hydroxyphenyl)-3-phenylpropan-1,3-dione (Fig. 9.119)[133].

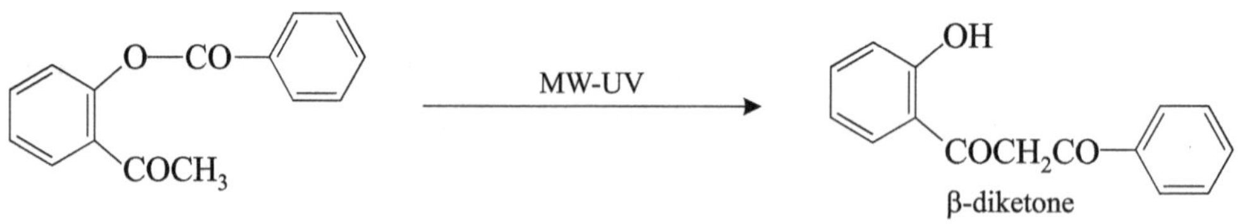

Fig. 9.119: Rearrangement of 2-benzyloxyacetophenone.

BIBLIOGRAPHY

1. A. Loupy, Modern Solvents in Organic Synthesis, Topics Curr. Chem., 1999, **206**, 153.
2. L. Perrux and A. Loupy, Tetrahedron, 2001, **57**, 9194.
3. R.S. Varma, in 'Microwaves in organic synthesis' (A. Loupy, Ed) Chapter 6, PP, 181-218, Wiley-VCH, New York, 2002.

4. R.S. Varma, Green Chemistry, 1949, **1**, 43-55.
5. R.S. Varma, Green Chemistry, Environment Friendly Alternatives, R. Sandhi and M.M. Srivastava (Eds.), Narosa, 2012, Chapter 6, Page 93-105 and references cited there in.
6. S. Caddick, Microwave Assisted Organic Reactions, Tetrahedron, 1995, **51**, 10103-10432.
7. D. Dallinger and C.O. Kappe, Microwave Assistant Synthesis in Water as Solvent, Chem. Rev., 2007, **107**, 2563-2591.
8. R.S. Varma, Chemical Synthesis using Microwave Irradiation, Astra Zeneca Research Foundation India, Bangalore.

Conclusion

The use of microwaves for organic transformations prevents formation of wastes and is energy efficient. In a number of cases, MW heating enhances catalytic reactions. In fact, microwave heating significantly improves the practice the organic synthesis. A large number of transformation can be scaled up by using microwaves.

REFERENCES

1. R.N. Gedye, W. Rank and K.C. Westaway, Can. J. Chem., 1991, **69**, 706.
2. R.N. Gedge, F.E. Smith and K.C. Westaway, Can. J. Chem., 1988, **66**, 17.
3. R.N. Gedge, W. Rank and K.C. Westaway, Can. J. Chem., 1988, **66**, 700.
4. R.N. Gedge, F. Smith, K. Westaway, L. Baldisera, Laberge and Rousell, Tetrahedron Lett., 1986, **26**, 279.
5. Y. Ju and R.S. Varma, Green Chem., 2004, **6**, 219.
6. T.M. Barnard, G.S. Vanier and M.J. Collins, Org. Process Res. Dev., 2006, **10**, 1233.
7. C.R. Strauss, Aust. J. Chem., 1999, **52**, 83.
8. P. Appukkuttan, W. Dehaen and E. van der Eycken, Eur. J. Org. Chem., 2003, 4713.
9. M.D. Bowman, J.L. Holcomb, C.M. Kormos, N.E. Leadbeater and V.A. Williams, Organic Process Research and Development, 2008, **7**, 41.
10. A. Hantzch, Justus Liebigs Ann. Chem., 1882, **215**, 1.
11. Cited in Ref 9, Pages 43.
12. T. Cablewski, A.F. Fourx and C.R. Strauss, J. Org. Chem., 1994, **59**, 3408.
13. E.V. Brown, Synthesis, 1975, 358.
14. C.R. Strauss and R.W. Trainor, Aust. J. Chem., 1998, **51**, 703.
15. J. Bagnell, T. Cablewski, C.R. Strauss and R.W. Trainor, J. Org. Chem., 1996, **61**, 7355 and references cited therein.
16. J. An., L. Bagenell, T. Cablewski, C.R. Strauss and R.W. Trainor, J. Org. Chem., 1997, **62**, 2505 and references cited therein.
17. C.R. Strauss, K.D. Raner, R.W.Trainer and J.S. Thorn, Aust. Pat. 677876 (1977).
18. T. Cablewski, P.A. Curr, P.J. Pajalic and C.R. Strauss, Unpublished data.
19. J. Berlan, S. Giboreau, S. Lefeuvre and C. Merchand, Tetrahedron Lett., 1991, **32**, 2363.

20. R.N. Gedye, W. Rank and K.C. Westaway, Can. J. Chem., 1991, **69**, 706.
21. S.D.Pollington, G.Bond, R.B. Moyes, D.A. Whan J.P. Candlin and J.R. Jennings, J. Org. Chem., 1991, **56**, 1313.
22. K.D. Raner and C.R. Strauss, J. Org. Chem., 1992, **57**, 6231; D. Constable, K.D. Raner, P. Somto and C.R. Strauss, J. Microwave Power Electromagnetic Energy, 1992, **27**, 195.
23. K. Bougrin, A.K. Bennani, S.F. Tetouani and M. Soufiaouri, Tetrahedron Lett., 1994, **35**, 8373.
24. A.C.S. Reddy, P.S. Rao and R. Venkataratnam, Tetrahedron Lett., 1996, **37**, 2845.
25. Y. Yuncheng, G. Dabin and J. Yulin, Synth. Commun., 1992, **22**, 2117.
26. R.J. Giguere, T.L. Bray, S.M. Duncan and G. Majetich, Tetrahedron Lett., 1986, **27**, 4945.
27. R.J. Giguere, A.M. Namen, B.O. Lopez, A. Arepally, D.E. Ramos, G. Majetich and J. Defauw, Tetrahedron Lett., 1987, **28**, 6553.
28. E.M. Gordon, D.C. Gaba, K.A. Jebber and D.M. Zacharias, Organometallics, 1993, **12**, 5020.
29. L. Bagnell and C.R. Strauss, Chem. Commun. in Press, cited in C.R. Strauss, Aust. J. Chem., 1999, **52**, 93.
30. A.K. Bose, B.K. Banik, K.J. Barakat and M.S. Manhas, Synlett., 1993, 575.
31. V. Sridar and V.S. Sundara Rao, Indian J. Chem., 1941, **63**, 3232.
32. G. Optiz and E. Tempel, Liebigs Ann. Chem., 1966, **699**, 68.
33. G.B. Jones and B.J. Chapman, J. Org. Chem., 1993, **58**, 5558.
34. V.V. Namboodri and R.S. Varma, Green Chem., 2001, **3**, 146.
35. N. Miyaura and A. Suzuki's, Chem. Rev., 1995, **95**, 2457.
36. N.E. Leadbeatex and M. Marco, Org. Lett., 2002, **4**, 2973.
37. N.E. Leadbeater and M. Marco, J. Org. Chem., 2003, **68**, 888.
38. S. Takano, A. Kijima, T. Sugihara, S. Satoh and K. Ogasawara, Chem., Lett., 1989, 87.
39. A.K. Bose, B.K. Banik and M.S. Manhas, Tetrahedron Lett., 1995, **36**, 213.
40. J.A. Vega, S. Cueto, A. Ramos, J.J. Vaquero, J.L. Garcia-Navio, J. Alvarez-Builla and J. Ezquerra, Tetrahedron Lett., 1996, **37**, 6413.
41. S.T. Chem., S.-H. Chiou and K.T. Wang, J. Chem. Commun, 1990, 807.
42. B.S. Furnes, A.H. Hannaford, V. Rogers, F.W.G. Smith and A.R. Tatchell, Vogal's Textbook of Practical Organic Chemistry, 4th Edn. Longman, New York, 1978.
43. D. Bogdal, J. Pielichowski and A. Boran, Synlett, 1996, 873.
44. D. Bogdal, J. Pielichowski and K. Jaskot, Synth. Commun., 1997, **27**, 1553.
45. D. Villemin and B. Martin, Synth. Commun, 1995, **25**, 3135.
46. D. Bogdal, J. Chem. Res. (s), 1998, 468.
47. V. Singh, J. Singh, P. Kaur and G.L. Kad, J. Chem. Res. (S), 1997, 58.
48. M. Rahmouni, A. Derdour, J.-P. Bazureau and J. Hamelin, Tetrahedron Lett., 1994, **35**, 4563; Syn. Commun., 1996, **26**, 453.
49. A.L. Marrero-Terrero and A. Lospy, Synlett, 1996, 245.
50. S.M. Sondhi, R. Rani, J. Singh *et al*, Bioorg. Med. Chem. Lett., 2010, **20**, 2306.
51. K.S. Ko, G. Park, Y. Yu and N.L. Pohl, Org. Lett., 2008, **10**, 5381.
52. P.J. Figiel, M.N. Kopylovich, M.N. Larsi *et al*, Chem. Commun., 2010, **46**, 2766.

53. A. Chighine, S. Crosignani, M.-C. Arnal *et al*., J. Org., Chem., 2009, **74**, 4753.
54. V. Polshettiwar and R.S. Varma, Tetrahedron Lett., 2008, **49**, 2661.
55. Y.-L. Fu, W. Huang and C.-L. Li, Dyes Pigm., 2009, **82**, 409.
56. V. Spinella, T. Fortunati and R. Sorieni, Synlett., 1997, 93.
57. R.S. Varma, R. Dahiya and S. Kumar, Tetrahedron Lett., 1997, **38**, 5131.
58. F. Juncai, L. Yang, M. Quinghua and L. Bin, Synth. Commun., 1998, **28**, 193.
59. B. Satish, K. Panneersel-Vam, D. Zacharids and G.R. Desivaju, J. Chem. Soc. Perkin Trans., 1955, **2**, 325.
60. E. Diez-Barra, A. de la Hoz, S. Merino and P. Sanchez-Verdu, Tetrahedron Lett., 1997, **38**, 2359.
61. A.K. Bose, B.K. Banik, N. Lavlnskaria, M. Jayraman and M.S. Manhas, Chemtech., Sept. 1977, 18.
62. R.S. Varma and D. Kumar, Org. Lett. 1999, **1**, 647.
62a. I. Almena, A. Diaz-Ortiz, E. Diez-Barra, A. Hos and A. Loupy, Chem. Lett., 1996, 333; S. Caddick, Tetrahedron, 1995, 10400.
63. D. Villemin and A.B. Alloum, Synth Commun., 1991, **21**, 63.
64. B. Perio, M.J. Dozias, P. Jacquault and J. Hamelin, Tetrahedron Lett., 1997, **38**, 7867.
65. R.S. Varma, A.K. Chatterjee and M. Varma, Tetrahedron Lett., 1993, **34**, 3207.
66. R. Ballini, M. Bordoni, G. Bosica, R. Maggi and G. Sartori, Tetrahedron Lett., 1998, **39**, 7587; R. Ballini, G. Bosica and M. Parrini, Tetrahedron Lett., 1998, **39**, 7963.
67. R.S. Varma, A.K. Chatterjee and M. Varma, Tetrahedron Lett., 1993, **34**, 4603.
68. D. Subhas Bose and V. Lakshminarayana, Tetrahedron Lett., 1998, **39**, 5631.
69. R.S. Varma, J.B. Lamture and M. Varma, Tetrahedron Lett., 1993, **34**, 3029.
70. R.S. Varma and R.K. Saini, Tetrahedron Lett., 1997, **38**, 2623.
71. R.S. Varma and H.M. Meshram, Tetrahedron Lett., 1997, **38**, 5427.
72. A.I. Bosch, P. de la Cruez, E. Diez-Barra, A. Loupy and F. Langa, Synlett, 1995, 1259.
73. R.S. Varma, R. Dahiya and R.K. Saini, Tetrahedron lett., 1997, **38**, 8819.
74. R.S. Varma and D. Kumar, Synth. Commun., 1999, 29, xxx.
75. R.S. Varma and D. Kumar, Tetrahedron Lett., 1997, **38**, 2043.
76. Comprehensive Organic Synthesis (Oxidation), ed. B.M. Trost, Pergamon Press, New York, 1991, Vol. 7.
77. R.S. Varma, Green Chem., 1999, 47.
78. R.S. Varma and R. Dahiya, Tetrahedron Lett., 1997, **38**, 2043.
79. R.S. Varma R.K. Saini and R. Dahiya, Tetrahedron Lett., 1997, **38**, 7823.
80. R.S. Varma and R. Dahiya, Tetrahedron Lett., 1998, **39**, 1307.
81. R.S. Varma and R.K. Saini, Tetrahedron Lett., 1998, **39**, 1481.
82. R.S. Varma, R. Dahiya and R.K. Saini, Tetrahedron Lett., 1997, **38**, 7029.
83. R.S. Varma, D. Kumar and R. Dahiya, J. Chem. Res. (S), 1998, 324.
84. R.S. Varma, R. Dahiya and D. Kumar, Molecules online, 1998, **2**, 82.
85. R.S. Varma, R.K. Saini and H.M. Meshram, Tetrahedron Lett., 1997, **38**, 6525.
86. R.S. Varma, R.K. Saini and R. Dahiya, J. Chem. Res. (S), 1998, 120.
87. H. Benhaliliba, A. Derdour, J.-P. Bezureau, F. Texier-Baullet and J. Hamelin, Tetrahedron Lett., 1998, **39**, 541.

88. A. Qussaid and A. Loupy, J. Chem. Res. (S), 1997, 342.
89. R.S. Varma and R.K. Saini, Tetrahedron Lett., 1997, **38**, 4337.
90. D. Barby and S. Torchy, Tetrahedron Lett., 1997, **38**, 2959.
91. G.H. Posner, A.W. Runquist and M.J. Chapdelaine J. Org. Chem., 1977, **42**, 1202 and references cited therein.
92. R.S. Varma and R. Dahiya, Tetrahedron, 1998, **54**, 6243.
93. A. Gutierrez, A. Loupy, G. Bram and E. Ruiz-Hitzky, Tetrahedron Lett., 1989, **30**, 945.
94. A.I. Bosch, P. de la Cruez, E. Diez-Barra, A. Loupy and F. Langa, Synlett, 1995, 1259.
95. R.S. Varma, D. Kumar and P.J. Liesen, J. Chem. Soc., Perkin Trans. I, 1998, 4093.
96. R.S. Varma and R. Dahiya, J. Org. Chem., 1998, **63**, 8038.
97. R.S. Varma and R.K. Saini, Synlett., 1997, 857.
98. R.S. Varma and K.P. Naicker, Molecules on line, 1998, **2**, 94; R.S. Varma, K.P. Naicker, D. Kumar, R. Dahiya and P.I. Liesen, J. Microwave Power Electromagnetic Energy, 1993, **34**, 113.
99. S. Cannizzaro, Ann., 1853, **88**, 129; T.A. Geissaman, Org. React., (NY), 1944, II, 94.
100. C.G. Swain, A.L. Powell, W.A. Sheppard and C.R. Morgon, J. Am. Chem. Soc., 1979, **101**, 3571.
101. R.S. Varma, K.P. Naicker and P.J. Liesen, Tetrahedron Lett., 1998, **39**, 8437.
102. Cited in M.D. Bowman, J.L. Holocomb, C.M. Kormos, N.E. Leadbeater and V.A. Williams, Organic Process Research and Development, 2008, **12**, 41-57 and the references cited therein.
103. For a recent review, see C.O. Kappe and T.N. Glasnov, Macromol. Rapid Commun., 2007, 28, 395.
104. N.E. Leadbeater, V.A. Williams, T.M. Barnard and M.J. Collins, Org. Process Res. Dev., 2006, **10**, 833.
105. S. Bondock, W. Khalifa and A.A. Fadda, Eur. J. Med. Chem., 2007, **42**, 948.
106. S.P. Watson, N. Baily, A.W. Dean, D.B. Judd, D. Middlemiss and R. Storer, Bioorg. Med. Chem. Lett., 1996, **6**, 1409.
107. G.W. Kabalka and A.R. Mereddy, Tetrahedron Lett., 2006, **47**, 5171.
108. M.D. Bowman, J.L. Holcomb, C.M. Kormas et al., Org. Process Res. Dev., 2007, **12**, 41.
109. F. Bergamelli, M. Iannelli, J.A. Marafic and J.D. Moselly, Org. Process Res. Dev., 2010, **14**, 926.
110. F. Chemat, M. Poux, J.L. DiMartino and J. Berlan, Chem. Eng. Tech., 1996, **19(5),** 420.
111. N.E. Leadbeater and L.M. Stencel, Energy fuels, 2006, **20**, 2281.
112. J.K.S. Wan and T.A.Koch, Res. Chem. Internal, 1994, 20, 29.
113. J. Cléophax, M. Liagre, A. Loupy and A. Petit, Org. Process. Res. Dev., 2000, **4**, 498.
114. D. Abenhaim, E. Diez-Barra, A. De la Hoz., A.Loupy and A. Sanchez-Migallon, Heterocycles, 1994, **38**, 973.
115. M. Deetlefs and K. Seddon, Green Chem., 2003, **5**, 181.
116. A. Lew, P.O. Krutzik, M.E. Mart and A.R. Chamberlin, J. Comb. Chem., 2004, **4**, 95 and references cited therein.
117. H.M. Yu, S.T. Chen., S.H. Chiou and K.T. Wang, J. Chromatogr, 1988, **456**, 357.
118. H.M. Yu, S.T. Chen and K.T. Wang, J. Org. Chem., 1992, **57**, 4781.
119. A. Stadler and C.O. Kappe, Eur. J. Org. Chem., 2001, 919.
120. M. Larhed, G. Lindeberg and A. Hallberg, 1996, **37**, 8219.

121. M. Larhed, C. Moberg and A. Hallberg., Acc. Chem. Res., 2002, 35, 717.

122. M. Larhed, M. Hoshino, S. Hadida, D.P. Curran and A. Hallberg, J. Org. Chem., 1997, **62**, 5583.

123. J.R. Jones and S.Y. Lu in Proc. Int. Conf. on Microwave Chemistry, 4-7 Sept., 2000, Institute National Polytechnique de Toulouse, P. 51.

124. J.M. Barthez, A.V. Filikov, L.B. Frederiksen, M.L. Huguet, J.R. Jones and S.-Y. Lu, Can. J. Chem., 1998, **76**, 726.

125. W.Th. Erb., J.R. Jones and S.-y. Lu, J. Chem. Res. (S), 1999, 728.

126. L.B. Frederiksen, T.H. Grobosch, J.R. Jones, S.-Y. Lu and C.C. Zhau, J. Chem. Res. (S), 2000, 42.

127. A.K. Bose, M.S. Manhas, S.N. Ganguly, A. Sharmat, M. Huarotte, S. Rumthao, M. Jayaraman and B.K. Banik, MORE Chemistry Techniques for process development. In: Fifth International Electronic Conference on Synthetic Organic Chemistry (ECSOC-5) E0047.

128. J. An., L. Bagnell, T. Cablewski, C.R. Strauss and R.W. Trainor, J. Org. Chem., 1997, **62**, 2505.

129. J. Marque, G. Salmoria, M. Poux, A. Laporterie, L. Dubac and N. Rogues, Ind. Eng. Chem. Res., 2001, **40**, 4485.

130. W.-C. Shieh, S. Dell and O. Ripic, Org. Lett., 2001, **3**, 4279.

131. C.R. Strauss and R.W. Trainer, Aust. J. Chem., 1995, **48**, 1665.

132. W. Baker, J. Chem. Soc., 1933, 1381; H.S. Mahal, K. Venkataraman, J. Chem. Soc., 1934, 1767.

133. S. Chemat, A. Aouabed, P.V. Bartels, D.C. Esveld and F. Chemat, J. Microwave Power Electromag. Energy, 1999, **34**, 55.

10. ULTRASOUND ASSISTED ORGANIC TRANSFORMATIONS

10

Ultrasound Assisted Organic Transformations

10.1 INTRODUCTION

Ultrasound are sound waves having frequencies higher than to which human ear can respond (*i.e.*, > 16 kHz). Ultrasound has been used to promote chemical reactions; this is called sonochemistry. During organic transformations, the effects of ultrasound are believed to be due to cavitation. In fact, it is cavitation that induces very high local temperatures and pressures inside the bubbles (cavities). This aspect is subsequently discussed.

Ultrasound has been used to accelerate a number of organic transformations. A number of reviews on chemical applications of ultrasound have been published[1].

The instrumentation for generating ultrasound requires an ultrasonic transducer, which is a device which converts electrical or mechanical energy into sound energy. The common used are the electromechanical transducers, which convert energy into sound. These are commonly based on piezo-electric effect. The piezo-electric effect is the production of a potential difference across opposite faces of a crystal of a material when it is subjected to sudden compression and is found in some crystalline materials like quartz. The inverse effect is produced when a rapidly alternating potential is placed across the faces of a piezo-electric crystal. This induces dimensional changes in the crystal resulting in the generation of vibrational (sound) energy[1-8].

As already stated cavitation induces very high local temperatures and pressures inside the bubbles (cavities). The ultrasound (waves of compression and expansion) is generated by a piezoelectric ceramic in a probe or a cleaning bath. It passes through a liquid with expansion cycles exerting negative pressures on the liquid. This applied negative pressure is quite strong to breakdown the intramolecular ven der Waals force of the liquid. This results in the formation of small cavities or gas-filled microbubbles. These micro bubbles are formed at pre-existing weak points in the liquid. The micro bubbles, as they are formed absorb energy from ultrasound waves and grow. However, a stage is reached when the microbubbles cannot absorb energy as efficiently. In the absence of energy input, the cavity cannot sustain itself and implodes. In fact, it is this implosion of the cavity that creates environment for chemical transformations[4].

The efficiency of the bubble collapse depends[9] on the vapour pressure, temperature, thermal conductivity and viscosity, the ultrasonic frequency and acoustic intensity.

The wavelength of ultrasound between successive compression waves is approximately 10 to 10^{-3} cm; it does not interect directly with molecules to start a chemical change. It is believed[10,11] that the bubble collapse in the liquid produce large amount of energy. It has been shown that the bubbles have temperatures around 5000K and pressures of approximately 1000 atm.

Different types of ultrasonic equipments are available. Two of these are commonly used. These are ultrasonic cleaning bath and direct immersion sonic Horm. The first one, *i.e.*, ultrasonic cleaning bath (Fig. 10.1) is most accessible and simple instrument. In this, the amount of power reaching the 'reaction' immersed in the bath is not readily quantifiable since it will depend on the size of the bath and the type of the reaction vessel. Also the control of temperature is not easy. At times it does not give reproducible results. The second type, *viz*., direct immersion sonic horn is very efficient for transmitting ultrasonic energy into a reaction. The method consisting in placing an ultrasonic probe directly into the reaction system. In this method much higher ultrasonic power can be used as there is no energy loss during the transfer of ultrasound through the bath meadia and the reaction vessel.

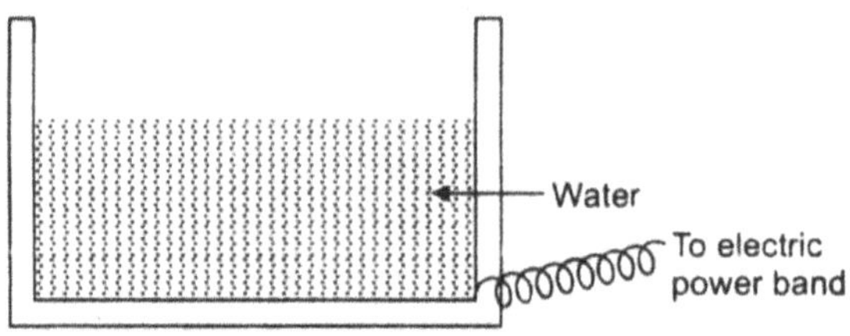

Fig. 10.1: An ultrasonic cleaning bath.

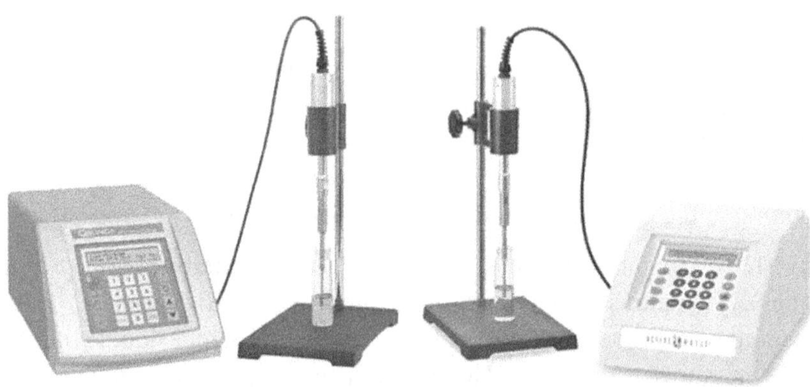

Fig. 10.2: Direct immersion sonic Horn.

Following three types of sonochemical reactions are possible:

(*a*) **Homogeneous Reactions:** The reactions proceed via radical or radical-ion intermediate. In these reactions, sonication effects reactions proceeding through radicals. Ionic reactions are not affected. In case of volatile molecules, the bubbles or cavities act as a microreactor, as in this case, the molecules enter the microbubbles and the high temperature and pressure produced during cavitation break their chemical bonds.

(*b*) **Heterogeneous Liquid-Liquid Reactions:** The use of PTC in organic aqueous biphase system is well known to catalyse these reactions. However, ultrasound is much more effective in these reactions since ultrasonic waves are known to generate extremely fine emulsions which result in large interfacial contact between the liquids; the result is increase in the reactivity between species dissolved in separate liquids and so react much faster than the conventional PTC conditions. In certain case, a combination of sonication and PTC gives much better results.

(*c*) **Heterogeneous Solid-Liquid Reactions:** These are of two types. Type I are those in which the solid serves as ane of the reagents and is consumed during the reaction. The type II are those in which the solid often a metal functions as a catalyst or is consumed. Type I reactions have been used to improve yields.

In type II reactions, the cavotational erosion is the major effect which is observed when ultrasonic waves propagate towards or in the vicinity of a solid. The erosion of metals follows the sequence Pb > Mg > Zn > Sn > Cu.

Ultrasound has been used to accelerate a large number of synthetically useful organic transformations, some of which are given below.

10.2 SYNTHETIC APPLICATIONS

All reactions are carried out at room temperature unless otherwise specified. The symbol)))) is used for reactions carried out on exposure to ultrasound.

10.2.1 Esterification

The esterification of carboxylic acids with alcohols is generally carried out in presence of catalysts like H_2SO_4, *p*-toluene sulfonic acid, dicyclohexylcarbodiimide etc. These reactions are slow and give low yields. A simple procedure consist is the reaction of carboxylic acid and alcohol in presence of catalytic amount of H_2SO_4 using ultrasound[6] (Fig. 10.3).

$$\mathrm{RCOOH + R'OH \xrightarrow[))))]{H_2SO_4,\ RT} RCOOR'}$$

Fig. 10.3: Esterification.

Low intensity ultrasound (cleaning bath) has also been used[12] for esterification.

10.2.2 Saponification

Normally saponification need aggressive conditions. However, it can be conducted under mild conditions using sonication[13] (Fig. 10.4).

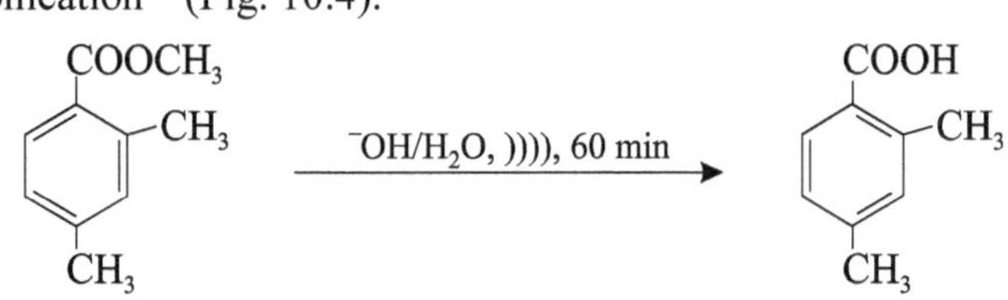

Methyl 2,4-dimethyl benzoate

2,4-Dimethyl benzoic acid 94%

Fig. 10.4: Saponification.

The rate of increase of saponification in the above case is attributed to emulsifying effect.

A number of 4-nitrophenyl esters can also be saponified using sonication[14].

10.2.3 Solvolysis/Hydrolysis

The solvolysis of *tert*-butyl chloride using ultrasound results in rate enhancement of two (Fig. 10.5)[15].

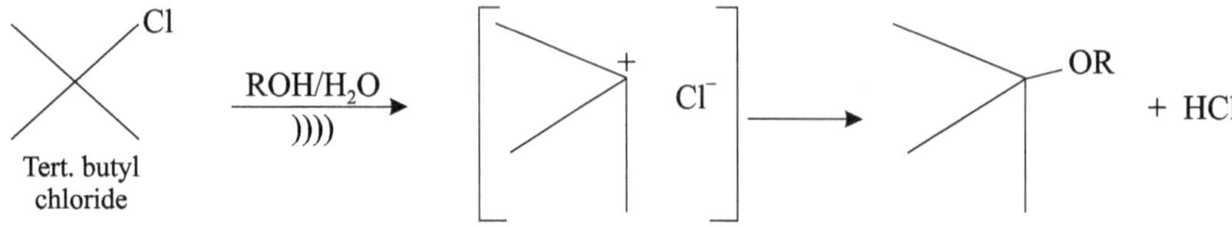

Fig. 10.5: Solvolysis of *tert.* butyl chloride

Basic hydrolysis of nitriles is also improved by sonication[16] (Fig. 10.6).

$$\text{ArCN} \xrightarrow[))))]{OH^-/H_2O} \underset{33\%}{\text{ArCOOH}}$$

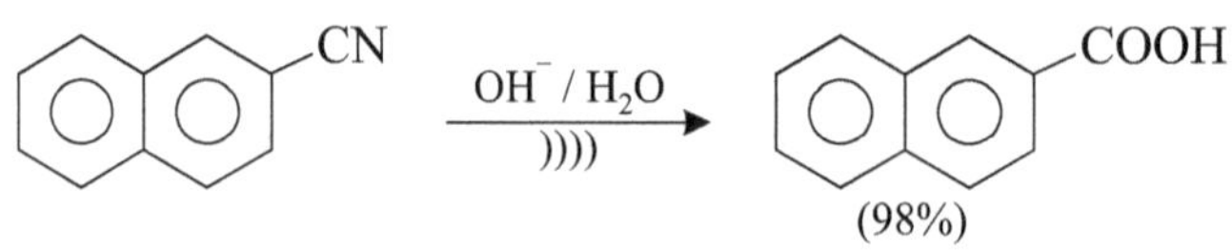

Fig. 10.6: Hydrolysis of nitriles

10.2.4 Friedel Crafts Reaction

Friedel crafts reaction in facilitated[17,18] by ultrasound (Fig. 10.7).

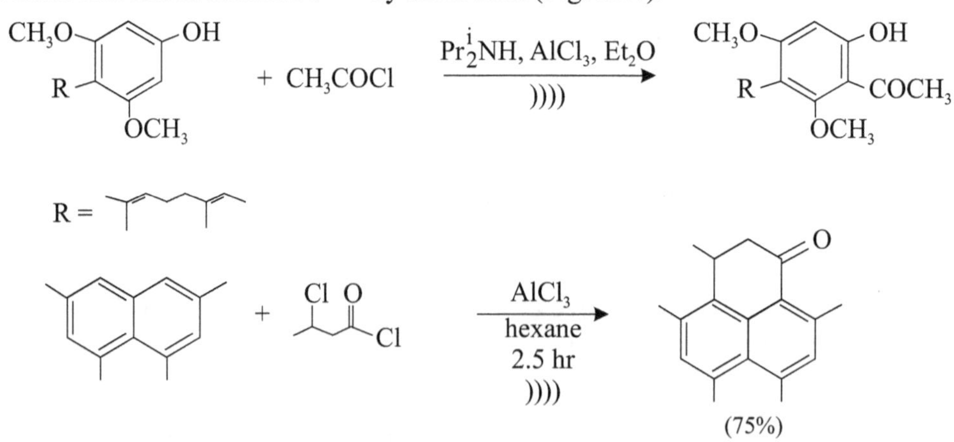

Fig. 10.7: Friedel Crafts Reaction.

10.2.5 Substitutions

Halides can be converted into cyanides. As an example benzyl bromide in toluene on reaction with potassium cyanide in presence of alumina (catalyst) gives the corresponding cyanides in 76%

yield. However, in the absence of ultrasound 83% of Friedel-crafts product is obtained (Fig. 10.7). It is believed that the ultrasonic dispersion of KCN on alumina surface decreases the Friedel-crafts activity and promotes the nucleophilic displacement of CN^- on its surface.

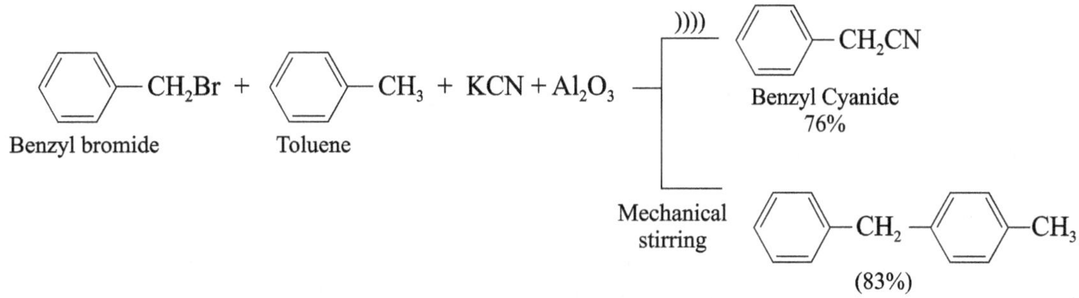

Fig. 10.7: Conversion of benzyl bromide into benzyl cyanide

Other substitution reactions include conversion of acylchlorides into the corresponding cyanide[19], alkyl halides into corresponding sulphocyanide and azides[20] (Fig. 10.8).

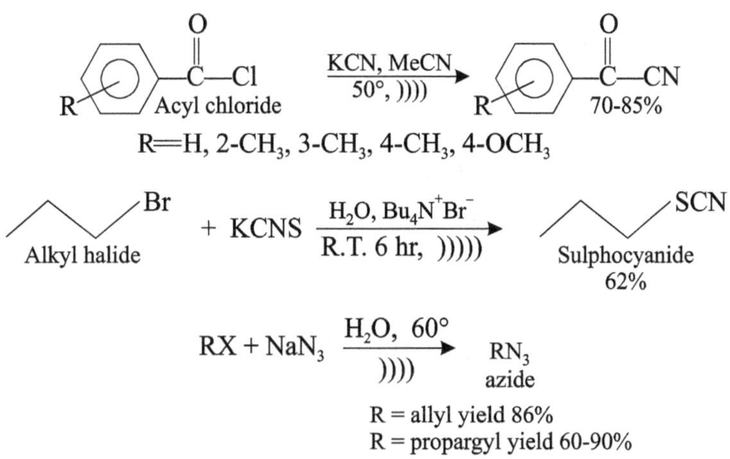

Fig. 10.8: Some substitution reactions.

10.2.6 1,4-Addition to α, β-unsaturated Carbonyl Compounds

Normally organocopper reagents are used for 1,4 addition to α, β-carbonyl compounds[21]. It has, however, been observed that there is significant improvements in yields, rates and ease of experimental technique if orgocopper compounds are generated in situ by sonication (bath 50 kHz) of copper (I) compounds, organic halides and lithium sand in diethylether-THF at 0° are allowed to react with α-enones (Fig. 10.9).

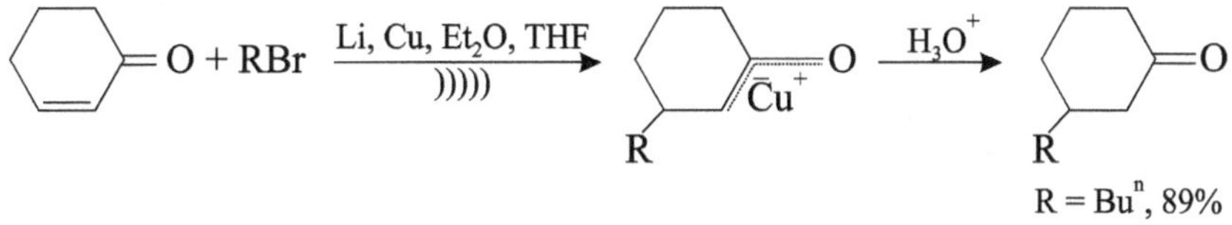

Fig. 10.9: 1,4-Addition to α, β-Unsaturated Carbonyl Compounds.

Even arylzinc compounds (generated ultrasonically using cleaning bath) are excellent reagents for the 1,4-addition to α-enones and α-enals in presence of catalytic quantities of Ni (acac)$_2$ (Fig. 10.10)[22]

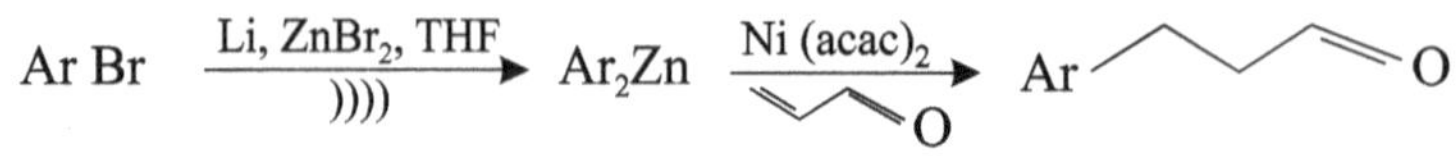

Fig. 10.10: 1,4-Addition of aryl zinc compounds to α, β-unsaturated carbonyl compounds.

See also Michal addition, section 10.2.25.4[22].

10.2.7 Cycloadditions

Cycloaddition of α, α′-dibromoketones to 1,3-dienes protomated by zinc is facilitated by ultrasound. This method is useful for the synthesis of highly hindered bicyclo [3.2.1] Oct-6-en-3 ones[23] (Fig. 10.11). In the absence of ultrasound the yields are low and requires much more times (24 hr).

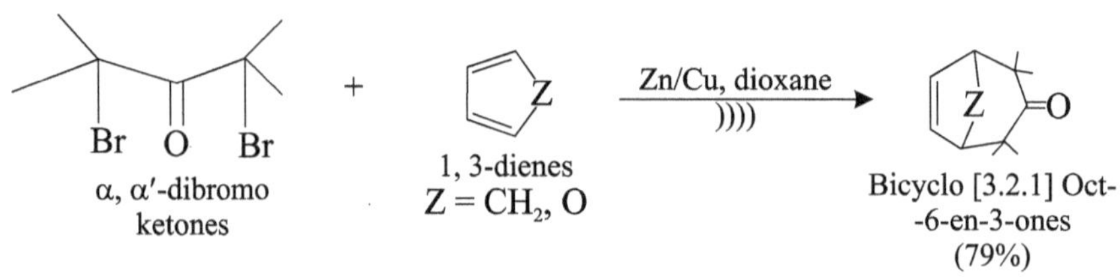

Fig. 10.11: Cycloaddition of α, α′-dibromoketones to 1,3-dienes

The [2 + 2] cycloaddition of dichloroketene to alkene is also improved by sonication (Fig. 10.12)[24].

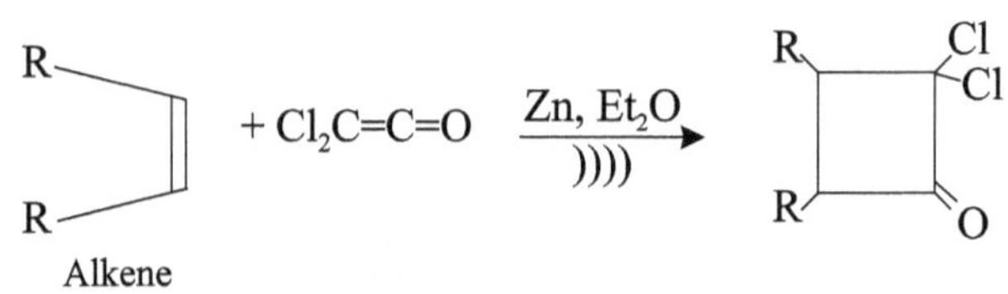

Fig. 10.12: [2+2] cycloaddition of dichloroketene to alkenes.

An interesting application of cycloaddition reaction is the cyclopropanation of methyl oleate with diodomethane giving 82% yield of the adduct (Fig. 10.13)[25].

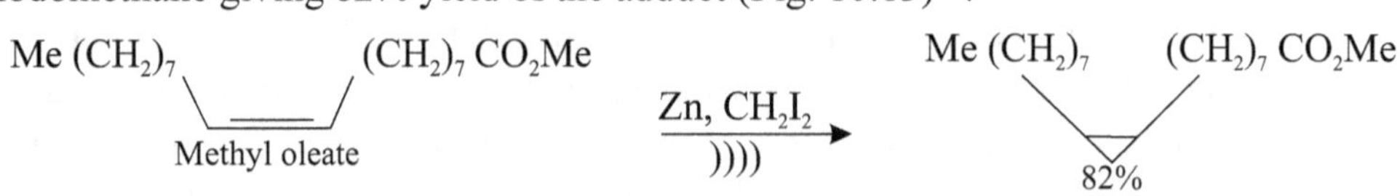

Fig. 10.13: Cyclopropanation of methyl oleate.

The above cyclopropanation has several advantages and can also be scaled up[26].

Another example of cyclopropanation is the reaction of styrene with dichlorocarbene[27] (generated in situ by the sonication of chloroform in presence of NaOH) (Fig. 10.14). In this case both sonication and stirring are necessary. Only sonication reduces the yield to 38% (20 hr sonication) and only stirring reduces the yield 31% compared to 96% yield in combined sonication and stirring.

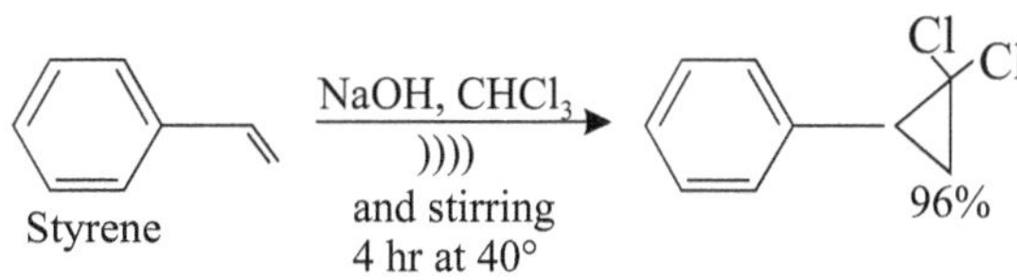

Fig. 10.14: Cyclopropanation of Styrene.

The above reaction is best conducted in a cleaning bath.

See also section 10.2.7.

10.2.8 Hydrogenations

Hydrogenations of alkenes can be effected by formic acid and Pd/C at room temperature in presence of low intensity ultrasound (cleaning bath, 50 kHz) (Fig. 10.15)[28]

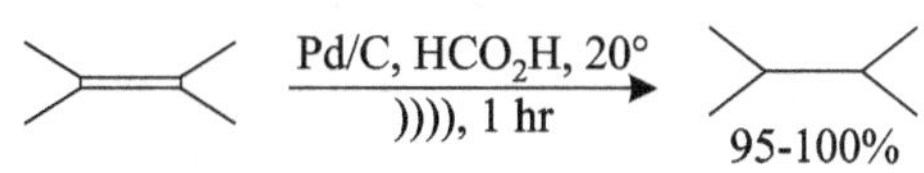

Fig. 10.15: Hydrogenation alkenes

In a similar way, the hydrazine-palladium-on carbon couple is also useful for hydrogenation of alkenes at room temperature using ultrasonic bath[29].

This procedure has been commercially used for ultrasonic hydrogenation of soyabean oil[30] using hydrogen gas and a solid catalyst (either 1% copper chromite or 0.1% Nysel (25% nickel) at 115 psig and 180°.

10.2.9 Coupling Reactions

Sonication is effective in promoting homocoupling of organometallic intermediates[31,32] obtained by the reaction of alkyl, aryl or vinyl halides with lithium wire in THF (ultrasonic bath, 117W, 50 kHz). (Fig. 10.16). No reaction takes place in the absence of ultrasound.

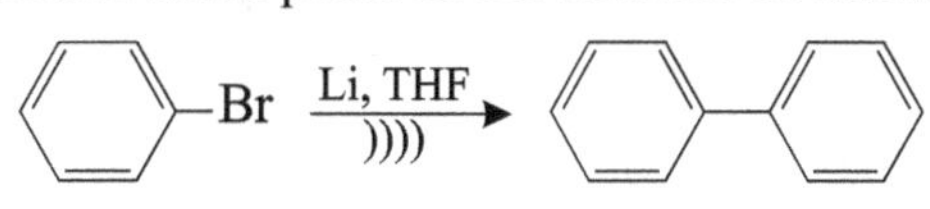

Fig. 10.16: Coupling of alkyl, aryl or vinyl halides.

In a similar way, coupling of benzyl halides in presence of copper or nickel powder (generated by lithium reduction of the corresponding halides in presence of ultrasound) gives high yields of dibenzyl (Fig. 10.17)[33].

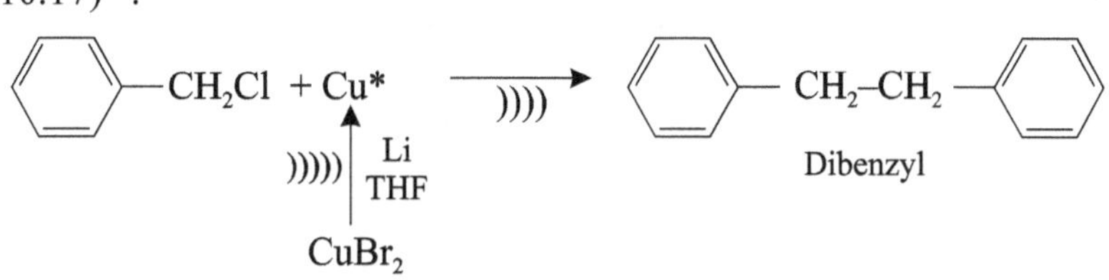

Fig. 10.17: Coupling of benzyl halides.

Ullmann Coupling of activated aryl halides like 2-iodonitrobenzene in DMF using high intensity ultrasound from a sonic horn gave 64-fold increase in rate over a mechanically stirred reaction (Fig. 10.18)[34].

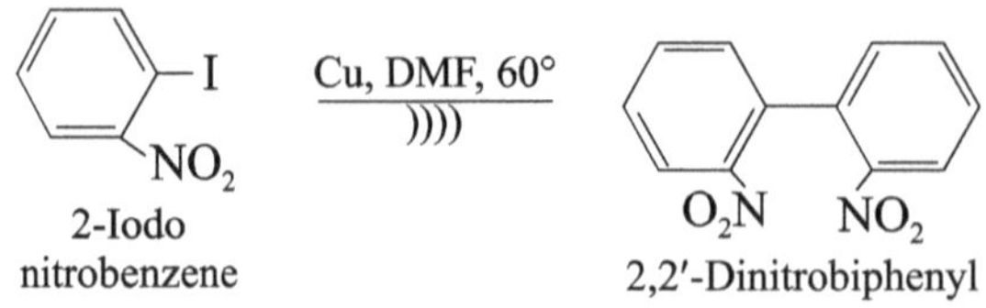

Fig. 10.18: Ullmann coupling of activated aryl halides.

It is found that **Ullmann-type coupling** of aryl sulphonates promoted by in situ generated nickel (II) complex is facilitated by ultrasound (Fig. 10.19)[35]. Best yields are obtained for triflates ($R = CF_3$) but tosylates ($R = 4CH_3C_6H_4$) give much lower yields.

$$ArOSO_2R \xrightarrow[\text{DMF, 60°,))))}]{\text{NiCl}_2\text{, Zn, PPh}_3\text{, NaI}} Ar - Ar$$

Fig. 10.19: Ulmann-type coupling of aryl sulphonates.

Cross-coupling reactions of perfluoroalkylzinc reagents with vinyl, allyl or aryl halides can also be effected using cleaning bath (35 or 45 kHz) (Fig. 10.20)[36-38].

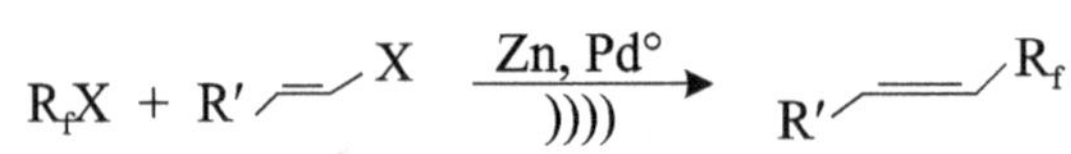

Fig. 10.20: Cross Coupling reactions of perfluoroalkylzinc reagents with vinyl, allyl or aryl halides

10.2.10 Dehalogenations

α, α-Dibromoorthoxylene in dioxane on ultrasonic irridation in presence of zinc gives a xylylene intermediates, which adds on to any dienophile in the reaction mixture giving high yields of the adducts (Fig. 10.21)[39]. In the absence of dienophile, a dimer is obtained.

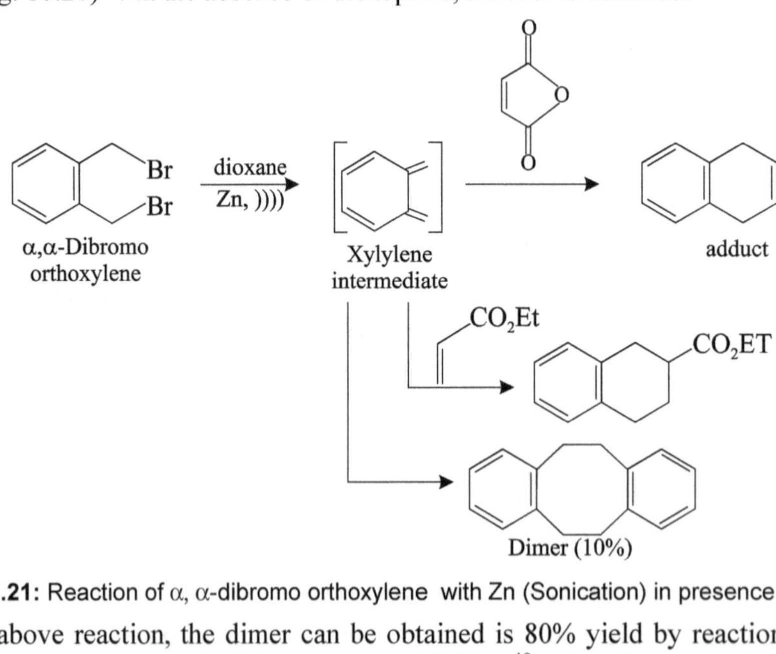

Fig. 10.21: Reaction of α, α-dibromo orthoxylene with Zn (Sonication) in presence of dienophiles.

In the above reaction, the dimer can be obtained is 80% yield by reaction of α, α-dibromo orthoxylene in an ultrasonic bath in presence of lithium[40]. There is no reaction in the absence of ultrasound.

A number of gem-dihalogenocyclopropanes on ultrasound irridation in presence of lithium, sodium or magnesium in THF gave the rearranged products (Fig. 10.22)[41].

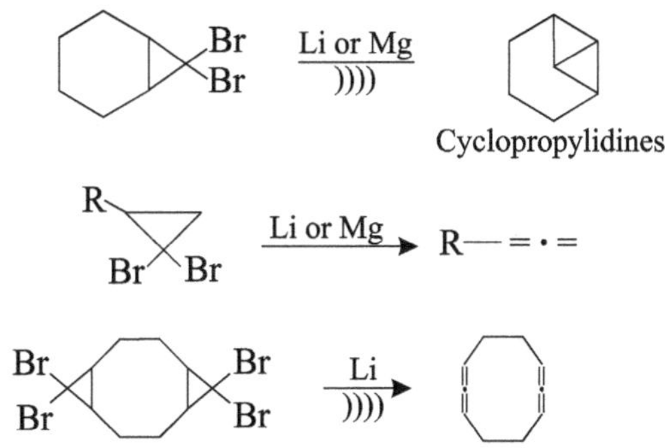

Fig. 10.22: Rearrangement of some gem dihalogenocyclopropanes.

Sonochemical debromination of 6-bromopencillinate ester with zinc in dioxane followed by treatment with aq. NH_4Cl gave penicillinate esters in good yields (Fig. 10.23)[42].

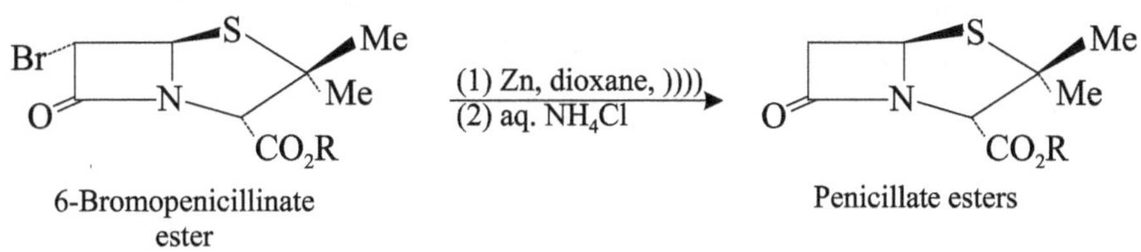

Fig. 10.23: Sonochemical debromination of 6-bromopenicillate esters.

The above method (Fig. 10.23) is efficient and cleaner than the usual debrominating agents Bu_3^nSnH or Pd-C/H_2 used for this purpose.

Reductive debromination of α, α′-dibromoketone in acetic acid using dispersed mercury emulsions gave[43] α-acetoxyketones (A) and (B). In case ketones are used as solvent (in place of acetic acid), the products are isopropylidene-1,3-dioxolan derivatives[44] (Fig. 10.24).

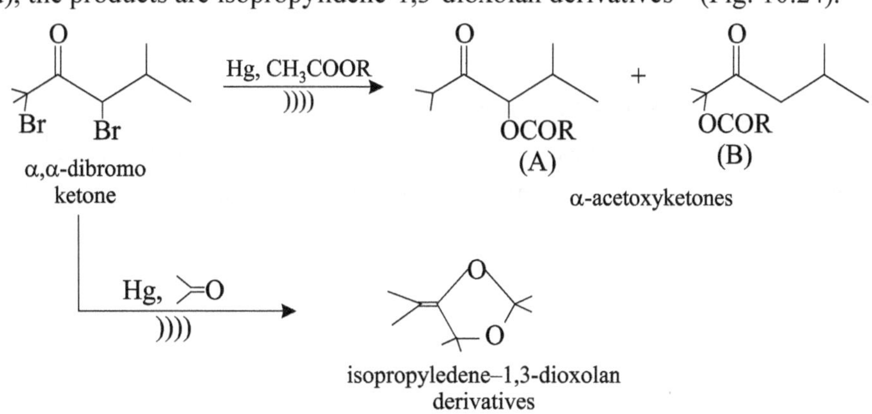

Fig. 10.24: Reductive debromination of α, α′-dibromoketone.

The reductive dehalogenation of aryl halides with nickel (II) chloride and Zinc in aq. HMPA is facilitated by low intensity ultrasound[45] (equation 1, Fig. 10.25). In place of nickel (II) chloride and zinc in aq. HMPA, $LiAlH_4$ and 1,2-dimethyoxyethane can also be used (equation 2, Fig. 10.25)[46].

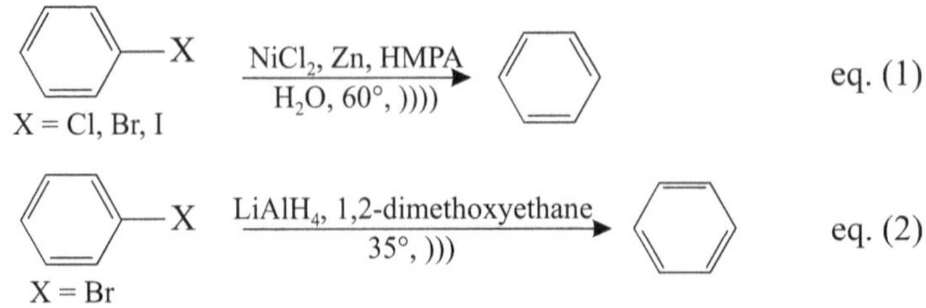

Fig. 10.25: Reductive dehalogenation of aryl halides.

10.2.11 Alkylations

10.2.11.1 N-Alkylations

N-Alkylation of secondary amines is possible under sonication in presence of a PTC, polyethylene glycol monoethyl ether. As an example, benzopyrrole on methylation with methyl iodide in toluene in presence of KOH and PTC gives[47] the corresponding N-methyl ether is 65% yield compared to 60% yield under normal conditions (20°, 5 hr) (Fig. 10.26).

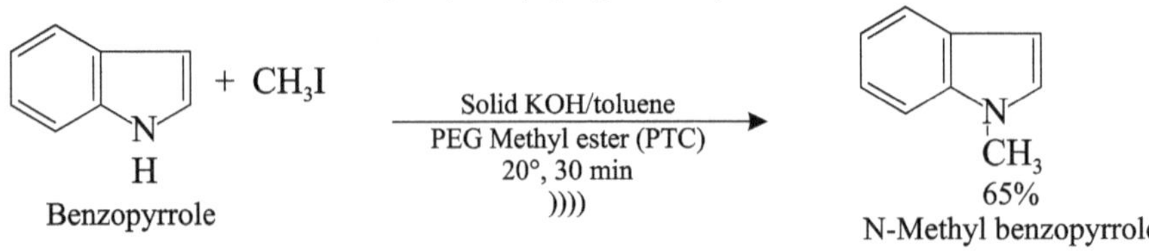

Fig. 10.26: N-methylation of benzopyrrole.

Similarly, diphenylamine on sonication with benzyl bromide gives 98% yield of the corresponding N-benzylated product in 1 hr compared to 70% yield in 48 hr under normal reflux conditions (Fig. 10.27)[47].

Ph_2NH	+	$PhCH_2Br$	Solid KOH/toluene → PEG Methyl ether 20°, 1 hr,))))	Ph_2NCH_2Ph
Diphenyl amine		Benzyl bromide		98% N-Benzyl diphenylamine

Fig. 10.27: N benzylation of diphenylamine.

In the absence of PTC, the above reaction does not take place under sonication.

N-Alkylation of crown compounds can also be effected under sonication (Fig. 10.28)[48].

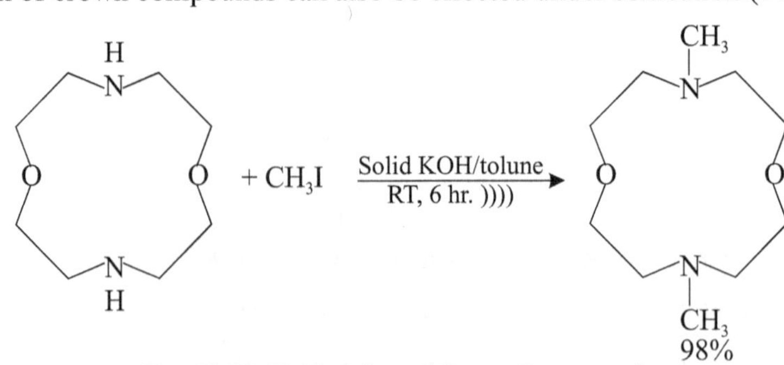

Fig. 10.28: N-Alkylation of Crown Compounds.

10.2.11.2 C-Alkylations

Alkylation of dihydroisoquinoline derivative has been effected[49] by sonication with an aryl halide using solid NaOH (Fig. 10.29).

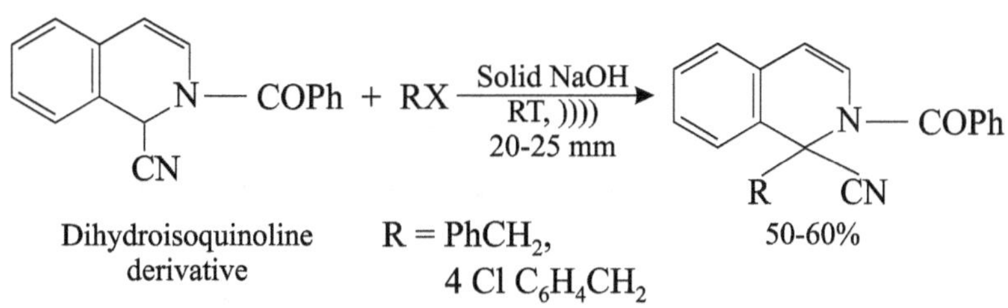

Fig. 10.29: C-Alkylation of dihydroisoquinoline derivative.

An interesting case of C-alkylation involves the synthesis of spiroketones from cycloalkanones by reaction with X $(CH_2)_4$X in benzene in presence of *t*-BuOK (Fig. 10.30)[50].

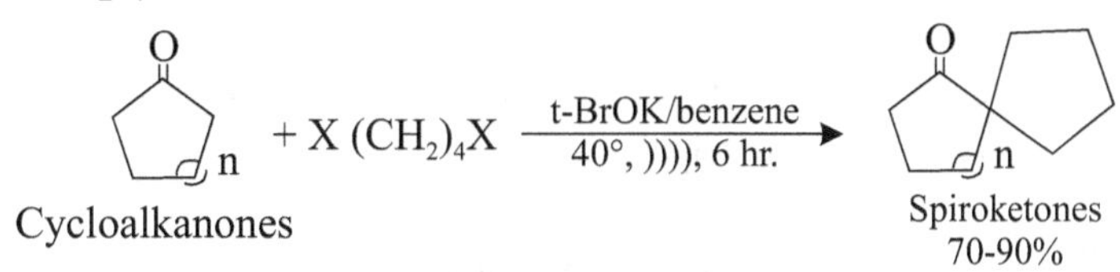

Fig. 10.30: Synthesis of Spiroketones.

In the case of cyclopentanone (n = 1), the reaction takes place under stirring only.

10.2.11.3 S-Alkylation

Aryl and alkylthiols can be alkylated[51] in presence of K_2CO_3/DMF at room temperature under sonication (Fig. 10.31).

$$\text{RSH} + \text{R'X} \xrightarrow[\text{))))}]{\text{K}_2\text{CO}_3\,/\,\text{DMF}} \text{RSR'}$$

Fig. 10.31: Alkylation of thiols.

10.2.11.4 O-Alkylation

The reaction of primary alcohols with benzyl bromide in presence of silver oxide gives 72% yield of the *O*-benzylated product under sonication[52]. However, without sonication, under normal conditions the yields are low (Fig. 10.32).

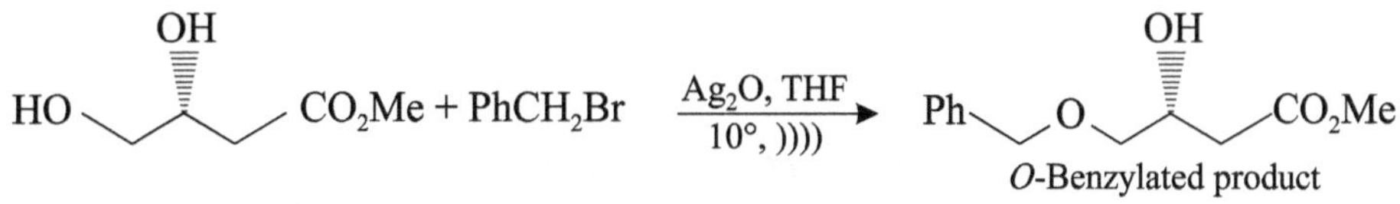

Fig. 10.32: Benzylation of primary alcohols.

Similarly phenolic hydroxy groups on sonication with alkyl halides using a probe generator[53] give the corresponding ethers (Fig. 10.33).

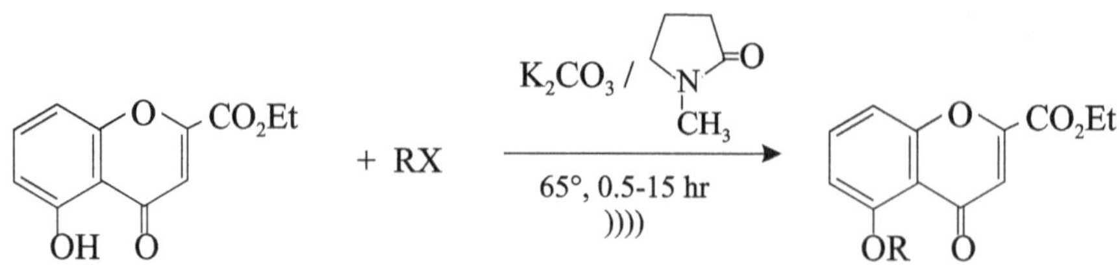

Fig. 10.33: Synthesis of ethers.

The hydroxyl groups in alcohol can be alkylated with aryl halides using sonication in presence of KOH, PEG (Fig. 10.34)[54].

$$\underset{\text{Ethyl alcohol}}{C_2H_5OH} + C_6H_5X \xrightarrow[))))]{\text{KOH, PEG}} \underset{\substack{\text{Ethyl phenyl elther} \\ 80\%}}{C_2H_5\ OC_6H_5}$$

Fig. 10.34: Synthesis of ethyl phenyl ether.

10.2.12 Oxidation

The oxidation of alcohols with solid $KMnO_4$ in hexane or benzene is considerably enhanced by sonication[55] in an ultrasonic bath (Fig. 10.35).

$$R^1R^2CH\ OH \xrightarrow[\text{RT,))))}]{KMnO_4\text{/hexane or benzene}} R^1R^2C=O$$

Fig. 10.35: Oxidation of alcohols with solid $KMnO_4$.

Thus, oxidation of octan-2-ol (R^1 = C_6H_{13}; R^2 = H) gives the corresponding aldehyde, $C_6H_{13}CHO$ in 92.8% yield (5 hr sonication) compared to 2% yield by mechanical stirring. In a similar way, cyclohexanol gives 53% yield of cyclohexanone (5 hr sonication) compared to 4% under usual conditions. Similarly, cinnanyl aldehyde gives cinnamic, acid (3 hr sonication) in 82% yield compared to 4% under usual conditions.

It has been shown[56] that crystalline MnO_2, which has low reactivity is activated by sonication. Thus sonication of cinnanyl alcohol, $C_6H_5CH = CH\text{–}CH_2OH$, geraniol or 1-phenylethanol with MnO_2 gives the corresponding aldehydes. In these oxidations, best results are obtained by using octane as solvent.

Benzylic halides have been oxidised[57] with aqueous sodium hydrochlorite at room temperature on sonication. It is believed that the oxidations proceed via benzylic hypohalides (Fig. 10.36).

$$\text{Ar R CHX} + \text{NaOCl} \xrightarrow[))))]{CH_3CN,\ RT} \text{Ar RCHOCl}$$

$$\text{Ar R CHOCl} + \text{NaOCl} \xrightarrow[))))]{CH_3CN,\ RT} \text{Ar R CHO}$$

Fig. 10.36: Oxidation of benzylic halides with aq. NaOCl.

10.2.13 Reduction

It has already been stated (see section 10.2.8) that sonication plays an important role in the reduction of alkenes by Pd/C, HCO_2H.

Sonication is known to increase the activation of nickel powder[58], which is turn has also been used for the reduction of alkenes.

Catalytically active nickel can be obtained by the sonochemical reduction of its salt such as chloride with zinc powder. Under these conditions excess of metallic zinc is also activated and reduces the water present in the medium producing hydrogen gas[59]. Thus, not only the catalyst (Ni) but also the reagent (H_2) is produced with maximal efficiency and safety. This procedure has been used for the selective reduction of C–C double bonds in α, β-unsat'd carbonyl compounds which is reduced much faster than the carbonyl group (Fig. 10.37).

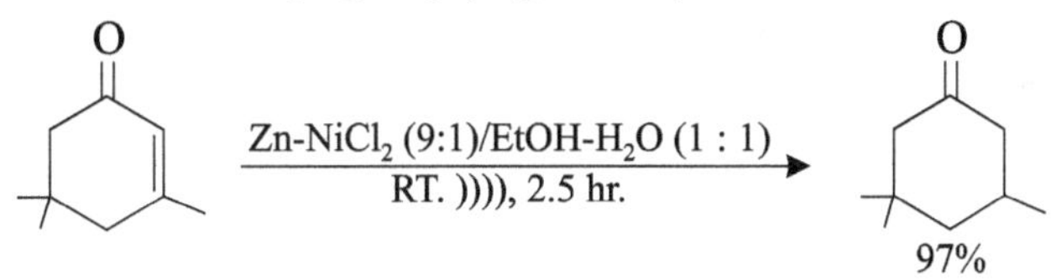

Fig. 10.37: Selective reduction of C=C in presence of carbonyl group.

In the absence of sonication, the above reaction takes 48 hrs.

Another example of selective reduction of C=C in presence of C=O is given below. As seen, the product obtained depends on the pH of the medium (Fig. 10.38)[60].

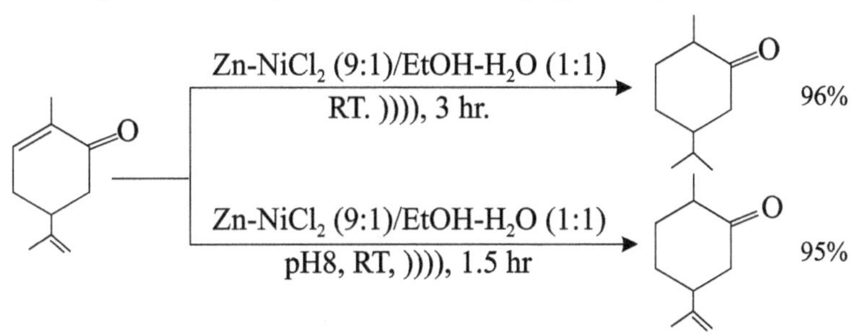

Fig. 10.38: Selective reduction of C=C in presence of C=O depends on the pH of the medium.

Sonication has also been used to reduce carbonyl group in presence of metal (eq., Li, Na or *K*) in THF. Thus, camphor on reduction gives a mixture of endo and exo borneol (Fig. 10.39)[61].

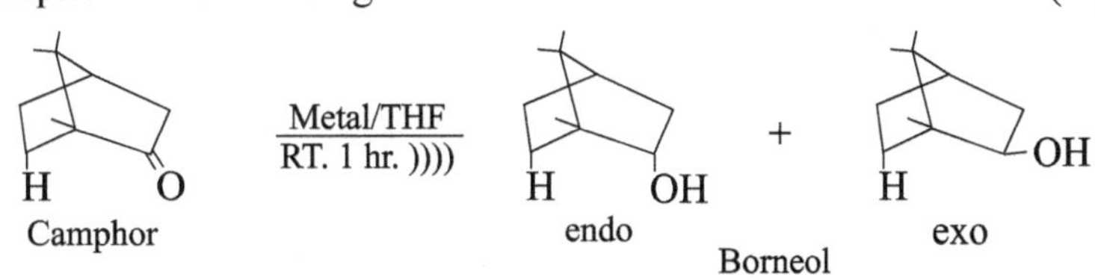

Fig. 10.39: Reduction of Camphor

Carbonyl groups which have less negative redox potential such as quinones or α-diketones can be reduced by sonication with zinc in presence of trimethylchlorosilane[62].

The ketones can also be reduced by Clemmenson's reduction by sonication in improved yields[63].

10.2.14 Hydroboration

Hydroboration is considerably enhanced[64] by low intensity ultrasound, particularly in a heterogeneous system. Thus, the reaction of cyclohexene with $BH_3.SMe_2$ in THF in presence of ultrasound gives tricyclohexylborane (Fig. 10.40).

Traditionally the above procedure require 24 hrs at 25°C compared to 1 hr. on sonication.

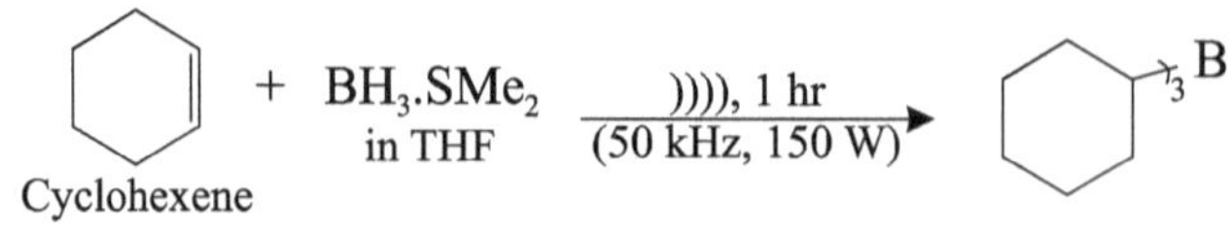

Fig. 10.40: Hydroboration of Cyclohexene

Thus, it has been shown that sonication provides an interesting alternative for the synthesis of organic boranes in excellent yields and in much shorter time. The method has also been used for the preparation of chiral reagents, Alpine-borane reagent can also be effected conveniently from pinine and 9-borabicyclononane (9.BBN) (Fig. 10.41).

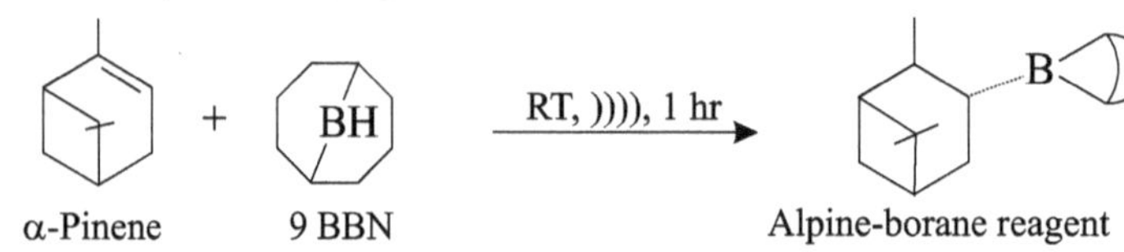

Fig. 10.41: Synthesis of Alpine-borone reagent.

The above reaction takes 5 hrs. for normal preparation.

10.2.15 Hydrosilation and Hydroalkylation

Alkene can be hydrosilated[65, 66] by sonication is a cleaning bath at much lower temperature in presence of Pt–C catalyst (Fig. 10.42).

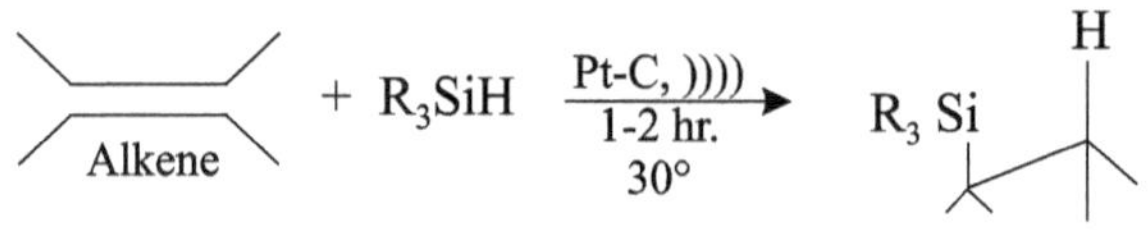

Fig. 10.42: Hydrosilation of alkenes.

In a similar way alkynes can be hydroperfluoroalkylated by using perfluoroalkylcurprates (generated in situ by the reaction of perfluoroalkyl halides (Rf) with zinc and copper (I) iodide) in THF. The reaction is regiospecific. Similar results are obtained[67] for perfluoroalkylation of dienes catalysed by Cp_2TiCl_2 (Fig. 10.43).

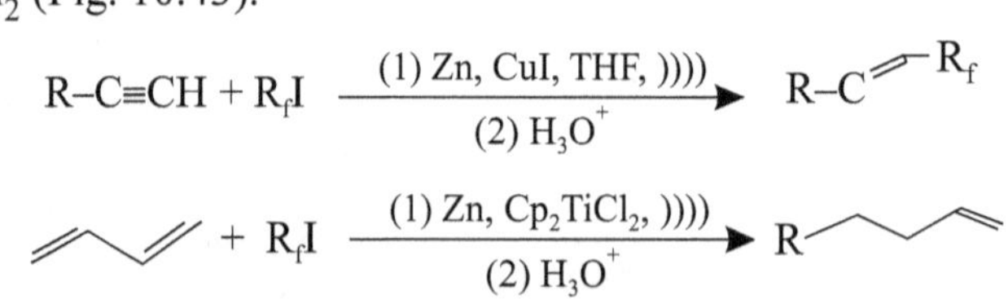

Fig. 10.43: Hydroperfluoroalkylation of alkynes and dienes

It has been found that dienes can be prepared[68] by the reaction of alkynes with sonochemically generated highly functionalized alkyl zinc intermediates (Fig. 10.44).

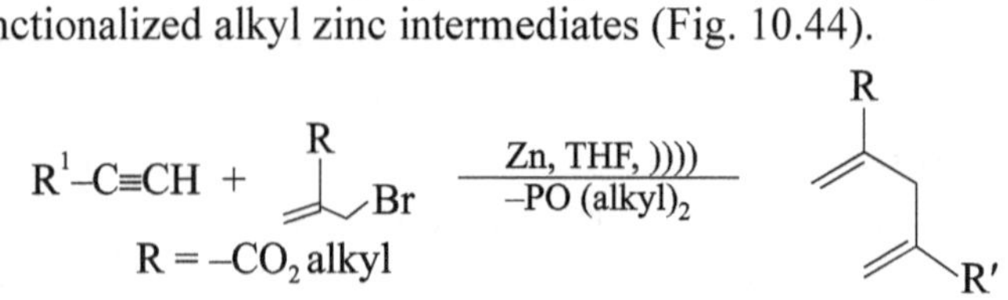

Fig. 10.44: Synthesis of dienes.

10.2.16 Grignard Reagents

Grignard reagents can be prepared by sonication of alkyl halides with magnesium in ether. Under these conditions magnesium gets activated (Fig. 10.45)[69]. No activators are needed.

$$\mathrm{R-X+Mg \xrightarrow[))))]{\text{ether}} \underset{(90\%)}{RMgX}}$$

Fig. 10.45: Preparation of Grignard reagents.

Grignard reagents have also been prepared[70] from olefins as shown below (Fig. 10.46)

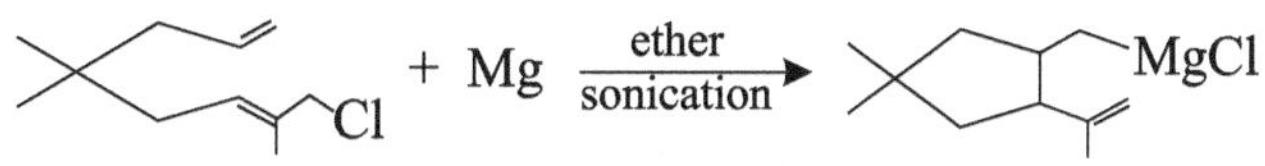

Fig. 10.46: Grignard reagents from olefins.

Similarly, *n*-propyl, *n*-butyl and phenyl lithium are prepared in > 90% yield by the sonication of appropriate bromide with lithium wire (Fig. 10.47)[69,71]

$$\underset{\mathrm{R\,=\,Pr,\,Bu,\,Ph}}{\mathrm{R-X+Li}} \xrightarrow{))))} \underset{90\%}{\mathrm{R-Li}}$$

Fig. 10.47: Preparation of n-Pr, n-Bu, n-Ph lithium.

10.2.17 Diels-Alder Cycloaddition Reaction

Diels-Alder Cycloaddition reaction of various dienes (mostly belonging to 1-vinyl cyclohexene series) with *o*-quinonones give the expected adducts on sonication having better regioselectivity and higher yields. It is found that best results are obtained by irradiation of neat mixture of reagents (Fig. 10.48)[72].

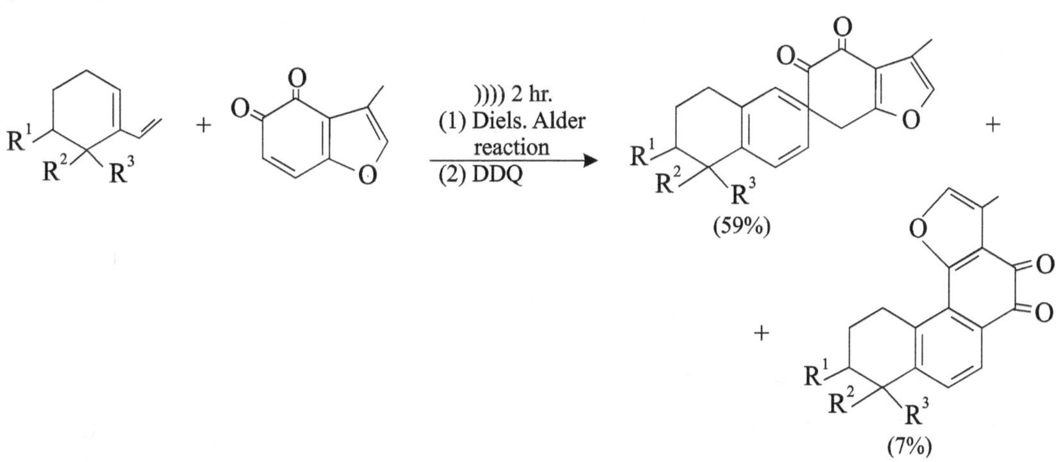

Fig. 10.48: Sonochemical Diels Alder Reaction.

10.2.18 Curtius Rearrangement

Sonication of benzoyl azide in benzene solution gives much better yield of phenyl isocyanate compared to reaction carried out on stirring (Fig. 10.49)[73].

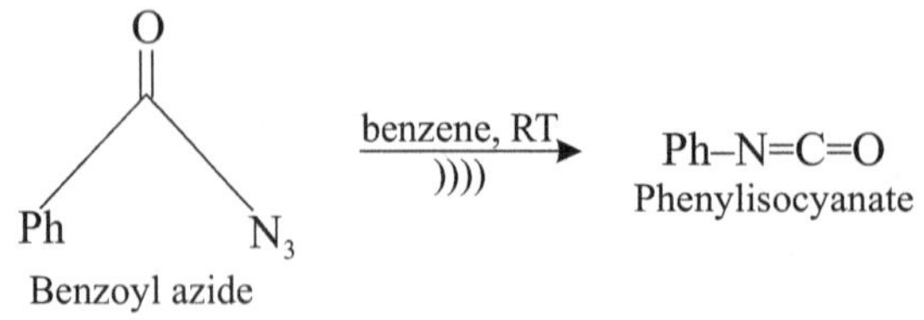

Fig. 10.49: Sonochemical Curtius reaction.

12.2.19 Strecker Synthesis

Strecker synthesis of aminonitriles is possible using ultrasonic acceleration (Fig. 10.50)[74].

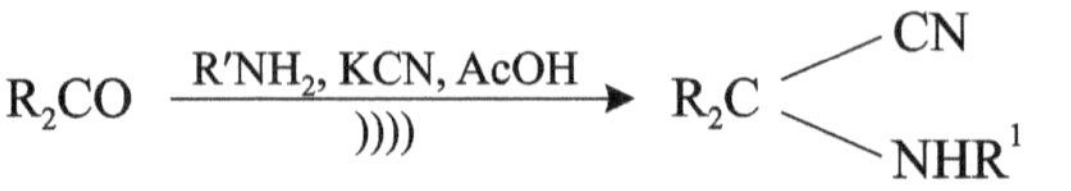

Fig. 10.50: Strecker synthesis.

A modified strecker synthesis for the preparation of α-aminonitrile in excellent yield consist in adsorption of the reagents on the surface of the catalyst before the reaction. This technique permits a combination of 'supported reagents' with sonochemical activation, the side reactions are suppressed (Fig. 10.51)[75].

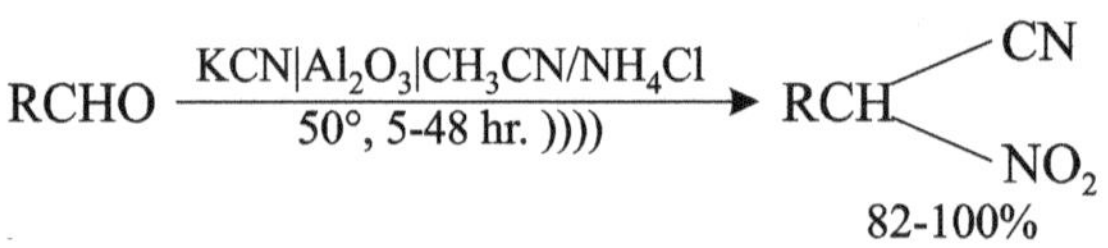

Fig. 10.51: A modified strecker synthesis.

In a homogeneous acetic acid solution, a cyclic amine ketone undergoes easy strecker reaction (Fig. 10.52)[74].

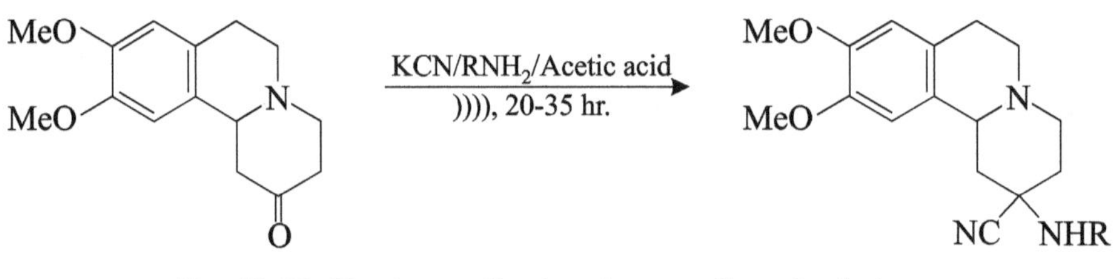

Fig. 10.52: Strecker synthesis using a cyclic amine ketone.

10.2.20 Cannizzaro Reaction

The reaction of aromatic aldehydes using $Ba(OH)_2$ as catalyst is greatly accelerated by low intensity ultrasound (cleaning bath) giving 100% yield of disproportionation products after 10 min of sonication. No reaction is observed during this period is absence of sonication (Fig. 10.53)[76].

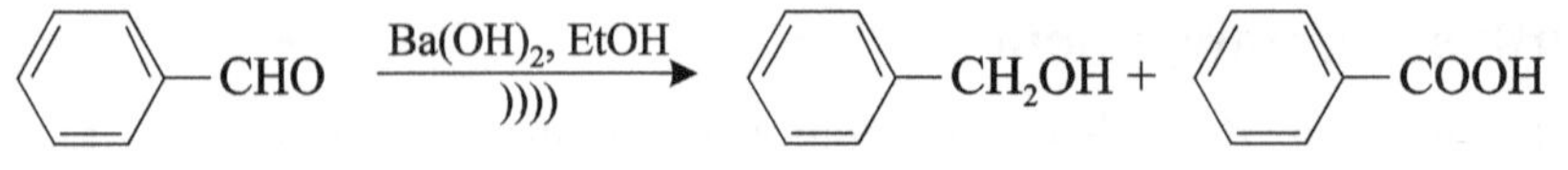

Fig. 10.53: Sonochemical Connazzaro Reaction

10.2.21 Reformatsky Reaction

The Reformatsky reaction under sonication[77] gives excellent yields compared to traditional methods such as those employing activated zinc or trimethyl borate as cosolvent. It is however necessary to activate the metal with iodine and run the reaction in dioxane instead of traditional aromatic solvents. (Fig. 10.54).

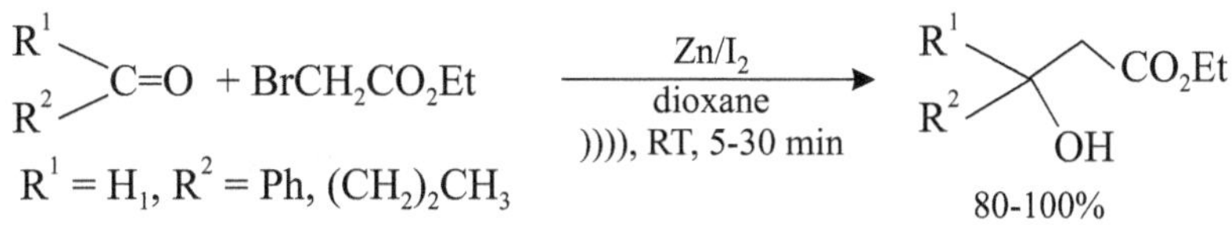

Fig. 10.54: Sonochemical Reformatsky Reaction.

The Reformatsky reaction using Schiffs bases gives β-lactams (Fig. 10.55)[78].

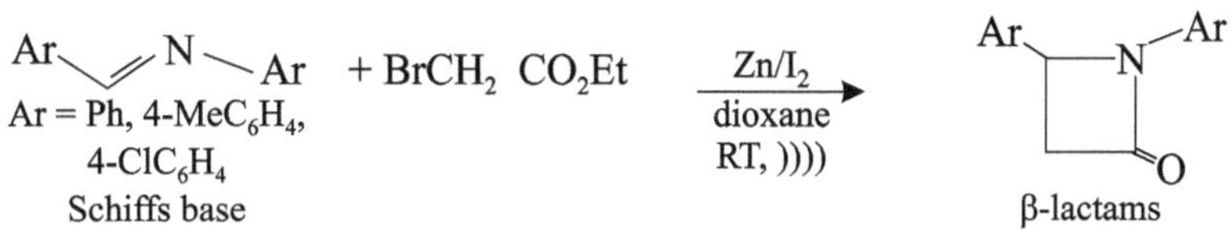

Fig. 10.55: Synthesis of β-lactams

10.2.22 Barbier Reaction

The reaction of a ketone with an organometallic reagent (obtained in situ) to give the corresponding alcohol is known as Barbier reaction. Same products are obtained by using a ketone and a Grignard reagent, which is separately obtained. Thus, the Barbier reaction is time saving and even organometallics which are unstable can also be used. However, the main disadvantage in the normal Barbier reaction is that only reactive alkyl halides can be used. This difficulty is overcome if the reaction is carried out in presence of ultrasound[79]. The reaction is carried out in THF and even imperfectly dried alkyl halides give excellent yield of the expected alcohol by reaction with lithium and various aldehydes and ketones. Even reactive halides, *e.g.*, allyl or benzyl halides which normally give Wurtz coupling can also be used. Some examples are given in Fig. 10.56.

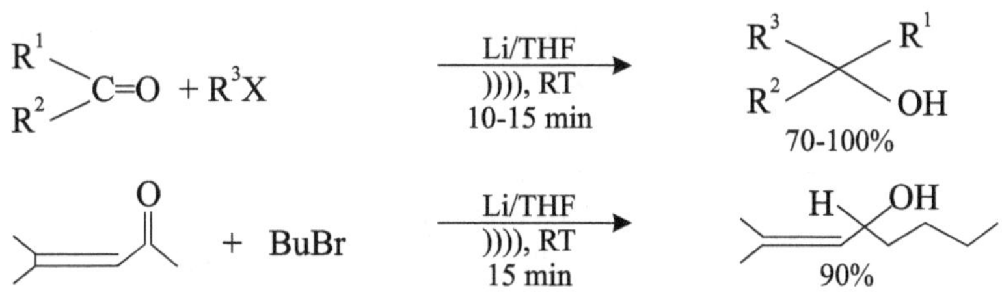

Fig. 10.56: Sonuchimical Barbier Reaction.

Sonochemical Barbier reaction has been used for a total synthesis of pentalenic acid (Fig. 10.57)[79].

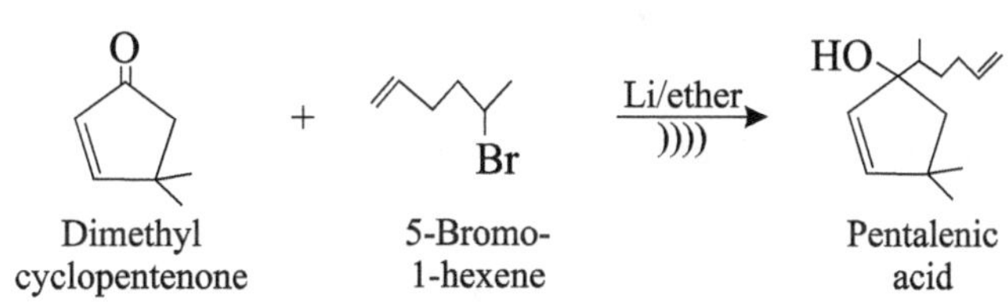

Fig. 10.57: Synthesis of Pentalenic Acid.

The Barbier reaction can also be applied to amides[80] and isocyanates[81] (Fig. 10.58)

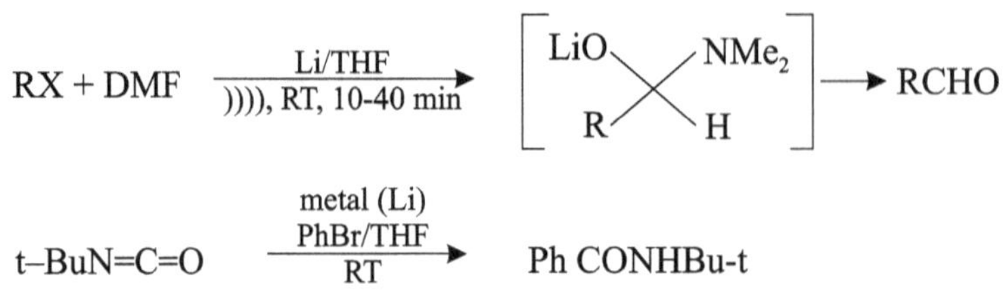

Fig. 10.58: Barbier reaction as applied to amides and isocyanates.

10.2.23 Dieckmann Cyclisation

The reaction of diethyl adipate with potassium in presence of ultrasound gives ethyl cyclopentanone 2-carboxylate (Fig. 10.59)[82].

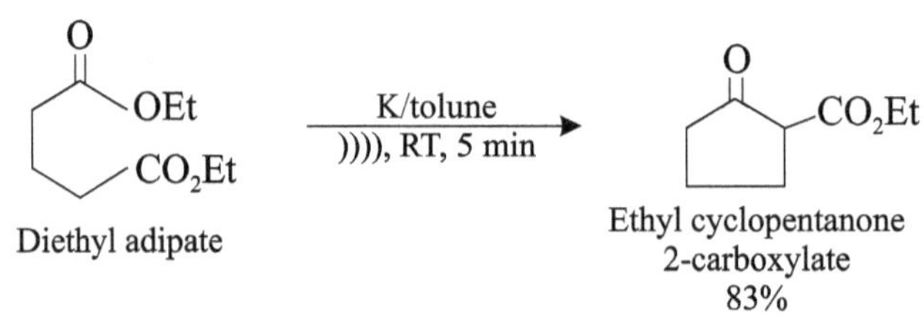

Fig. 10.59: Dieckmann Cyclization.

In presence of ultrasound, potassium is easily transformed to a silvery blue suspension in toluene.

10.2.24 Oxymercuration of Olefins

The oxymercuration of olefins in a well known procedure for C–O bond formation. It is generally performed using mercuric acetate or trifluoroacetate. Since other salts of mercury are not easily available, it has now been shown that any mercury salt can be prepared from mercuric oxide and an organic acid under sonication[83]. The oxylation of olefins can be carried out as one pot reaction under sonication and excellent yields are obtained. A typical example is given below (Fig. 10.60).

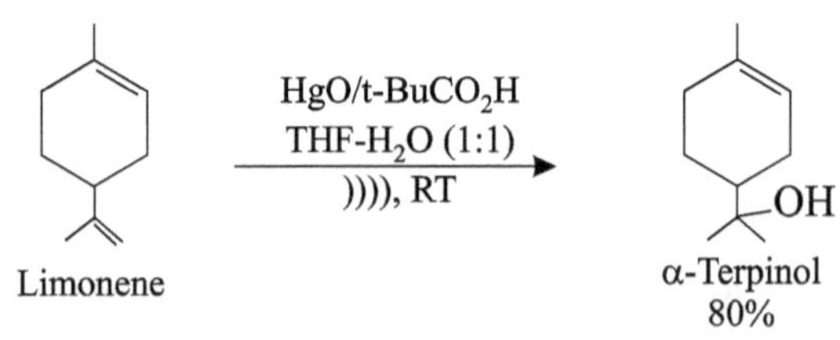

Fig. 10.60: Conversion of Limonene into α-terpinol.

10.2.25 Condensation Reaction

10.2.25.1 Synthesis of 3-nitro-2H-Chromenes

3-Nitro-2H-chromenes are synthesised by the condensation of *o*-hydroxy benzaldehydes with β-nitrostyrene derivatives in presence of ultrasound using a 60 kHz cleaning bath (Fig. 10.61)[84] in presence of basic alumina catalyst.

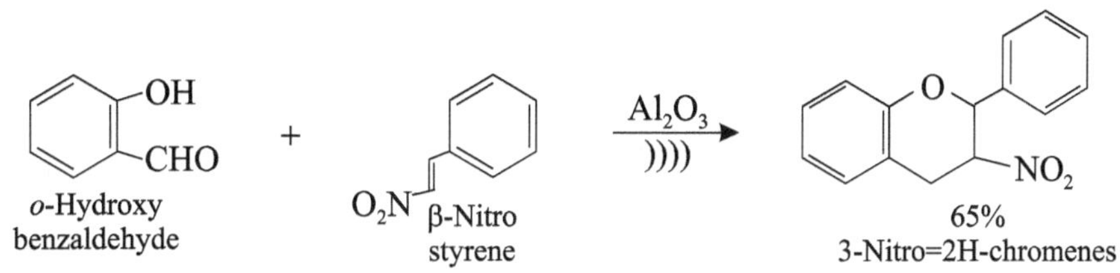

Fig. 10.61: Synthesis of 3-nitro-2H-chromenes.

10.2.25.2 Knoevenagel Condensation

The Knoevenagel condensation involves the condensation of aldehydes or ketones having no α-hydrogen atom with an active methylene compound. This condensation is usually carried out by heating using an organic base, Lewis base or acid as catalyst. The reaction takes long reaction time, is difficult to work up and gives low yields. It has been shown that using ultrasound the Knoevenagel condensation can be carried out in mild conditions, much shorter time and higher yields. Thus, pyridine catalysed knoevenagel condensation of ethyl cyanoacetate with a number of aromatic aldehydes in presence of ultrasound gives α-cyanocinnamate is excellent yields (90-95%) (Fig. 10.62)[85] compared to 50-60% yield in reactions without ultrasound.

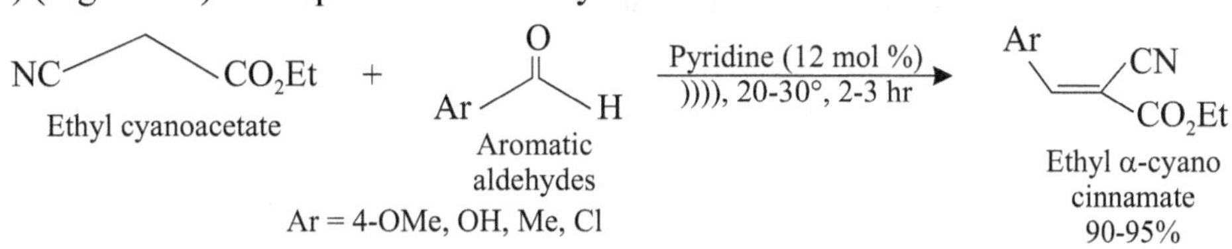

Fig. 10.62: Sonochemical Knoevenagel condensation in presence of pyridine.

The above reaction (Fig. 10.62) can also be conducted by sonication in presence of KF-Al_2O_3[86]. The reaction was carried out in ethanol and KF-Al_2O_3 could be recycled without any loss in yield.

Knoevenagel condensation of malononitrile with aromatic aldehydes in presence of ultrasound and in the absence of any catalyst gives aryl methylene malononitrile (Fig. 10.63)[87].

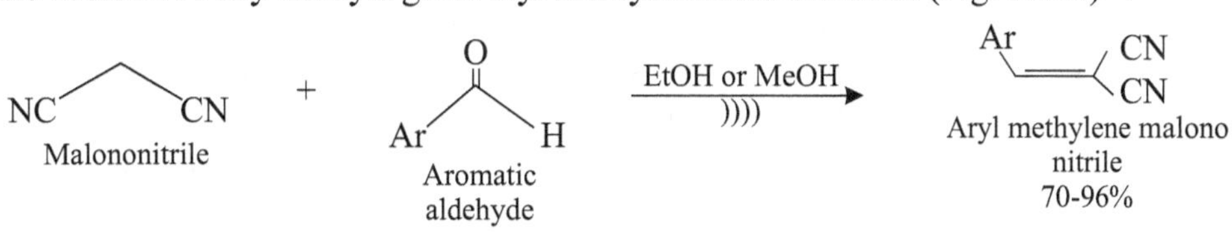

Fig. 10.63: Knoevenagel condensation of malononitrile with aromatic aldehydes.

10.2.25.3 Synthesis of Chalcones

Chalcones are important intermediates for the synthesis of a number of biologically active compounds. These are generally prepared by the **Claisen-Schmidt Condensation** of aromatic aldehydes with ketones in presence of a base like NaOH, KOH and $Ba(OH)_2$. The reaction takes longer times and difficult workup procedure. The chalkones are conveniently synthesised by sonochemical procedure. Thus, the reaction of an aromatic aldehyde with a ketone is presence of $Ba(OH)_2$ catalyst in presence of ultrasound[88] at room temperature gives the chalcone in good yields in much shorter time (25 min) compared to 5 hr. stirring in the conventional procedure (Fig. 10.64). The above condensation can also be carried out using KOH as catalyst in presence of ultrasound[89,90].

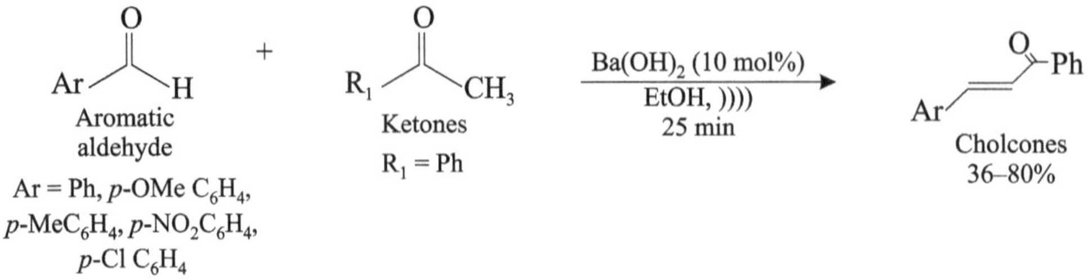

Fig. 10.64: Sonochemical Synthesis of chalcone.

The above condensation (Fig. 10.64) can also be carried out in presence of NaOH (16 mol%), EtOH (at 20-40°, 4-5 min sonication, yield 80-90%). A reusable catalyst, KF-Al_2O_3 can also catalyse the above condensation (KF-Al_2O_3, MeOH, 25-40°, 5-240 min sonication, yield 84-90%). An acid-montmorillonite[91] (KSF) can also be used as catalyst (KSF, RT, 30-240 min sonication, yield 80-95%) can also be used in the above Claisen-schmidt reaction.

10.2.25.4 Synthesis of Nitroalkenes

Nitroalkenes can be synthesised[92] by ultrasound promoted Knoevenagel condensation of aromatic aldehydes with nitromethane. The reaction is catalyzed by NH_4OAc-HOAc (Fig. 10.65).

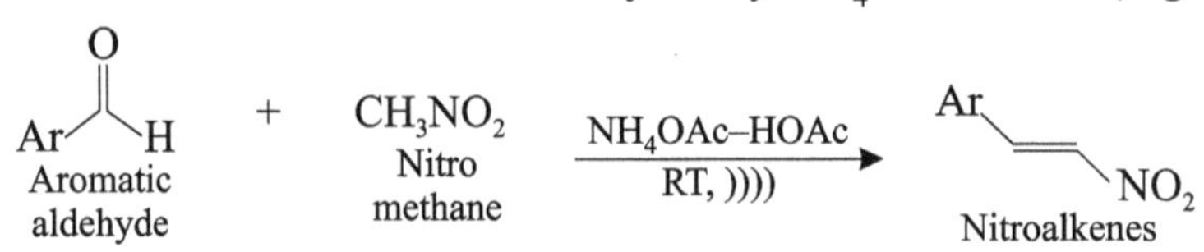

Fig. 10.65: Synthesis of nitroalkenes.

10.2.25.5 Michael Addition

It is an addition reaction between an α, β-unsaturated carbonyl compound and a compound with an active methyl group in presence of a base. The reaction takes long time, low yields and difficult to workup. It is found that the use of ultrasound gives much better results. As an example, potassium hydroxide catalysed Michael addition of ethyl acetoacetate to chalcone in presence of ultrasound[93] despite mild conditions gives product (a cyclic product) which is different under thermal conditions (an acyclic compound (B) is obtained) (Fig. 10.66).

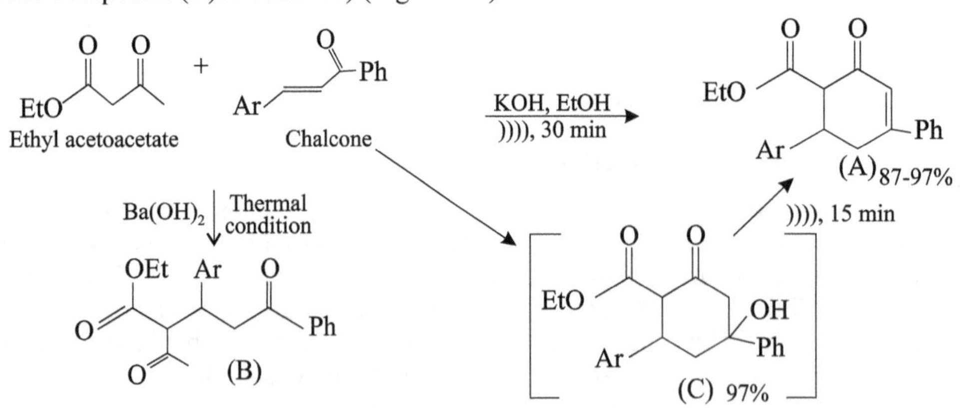

Fig. 10.66: Sonochemical Michael addition of ethylacetoacetate with Chalcone.

In the above reaction, the cyclic product (A) is likely to be obtained from the intermediate (C), which was isolated in shorter time. The acyclic product (B) could also be formed under ultrasonic conditions[94]; in this case basic KF-Al_2O_3 is used as a catalyst.

10.2.25.6 Mannich Reaction

Compounds containing at least one active hydrogen atom condense with an aldehyde and a primary or secondary amine (in the form of its hydrochloride) to give the product known as Mannich base (α β-carbonyl compound). It has been found that the one-pot reaction can be conducted in presence of ultrasound using sulfamic acid as an efficient, inexpensive and recyclable green catalyst. As an example Mannic reaction of acetophenone with an aldehyde (benzaldehyde) and an amine (aniline) on sonication in presence of sulfamic acid give excellent yield of the Mannic base (Fig. 10.67)[95].

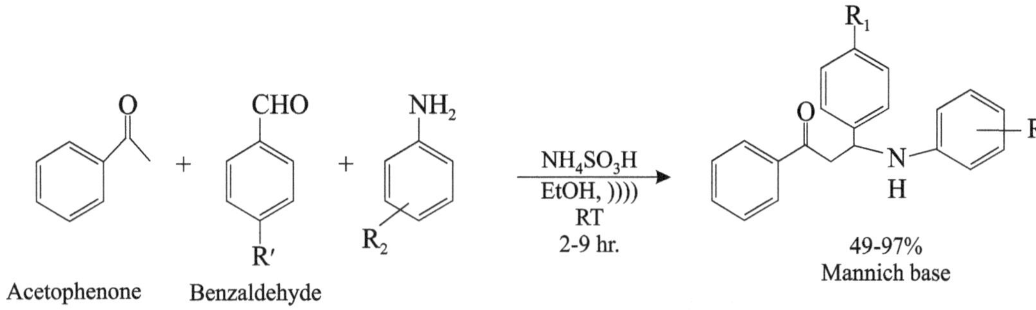

Fig. 10.67: Ulsound assisted Mannich Reaction.

Under conventional conditions, the Mannich Reaction took much longer times (18-72 hrs.) and the yields were 24-90%.

The Mannic reaction of cycloketones with aromatic aldehydes and aromatic amines, promoted by ultrasound in water using the catalyst Ga $(OTf)_3$ give good yield of the Mannich base with good anti-selectives (Fig. 10.68)[96].

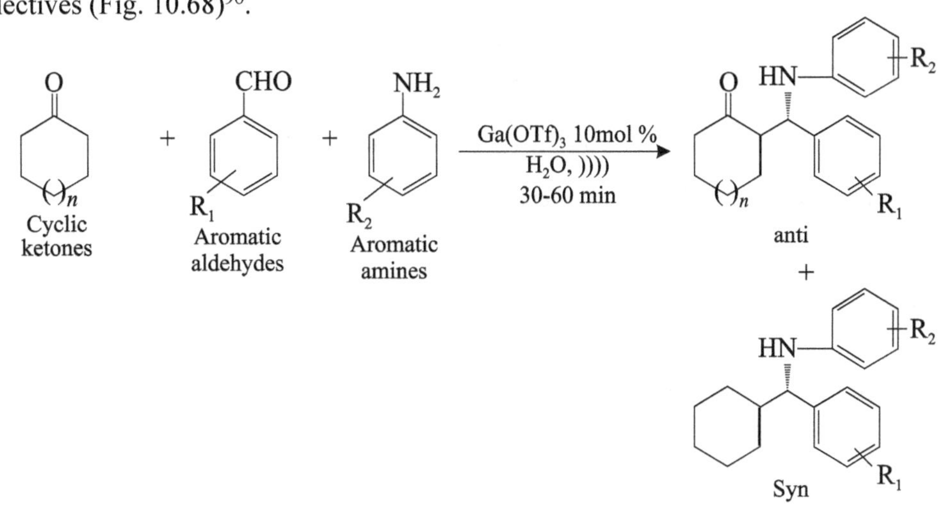

Fig. 10.68: Sonochemical Mannic Reaction in Water in presence of Ga $(OTf)_3$.

Mannic Reaction can also be conducted in presence of (S)-proline as catalyst. Thus the reaction between hydroxyacetone, aromatic aldehyde and *p*-anisidine in presence of (S)-proline on ultrasonic irridation gave the corresponding Mannic products in good yields with good diasteo and enantio selectives (Fig. 10.69)[97].

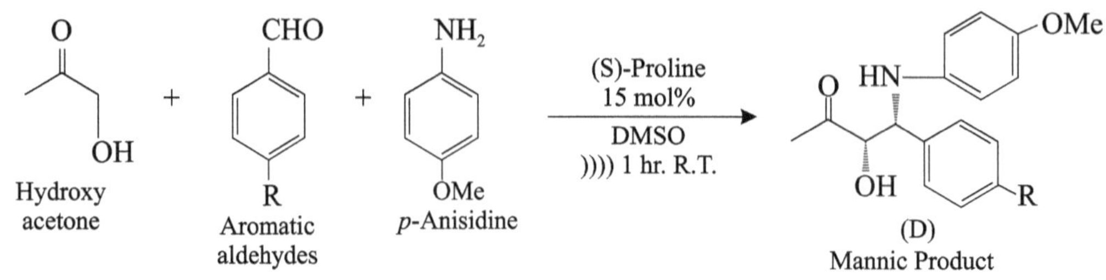

Fig. 10.69: Sonochemical (S)-Proline catalysed Mannich reaction.

The Mannich Product (D) (Fig. 10.69) could also be obtained by Mannich reaction between the hydroxyacetone and preformed imine (E) (Fig. 10.69)[97].

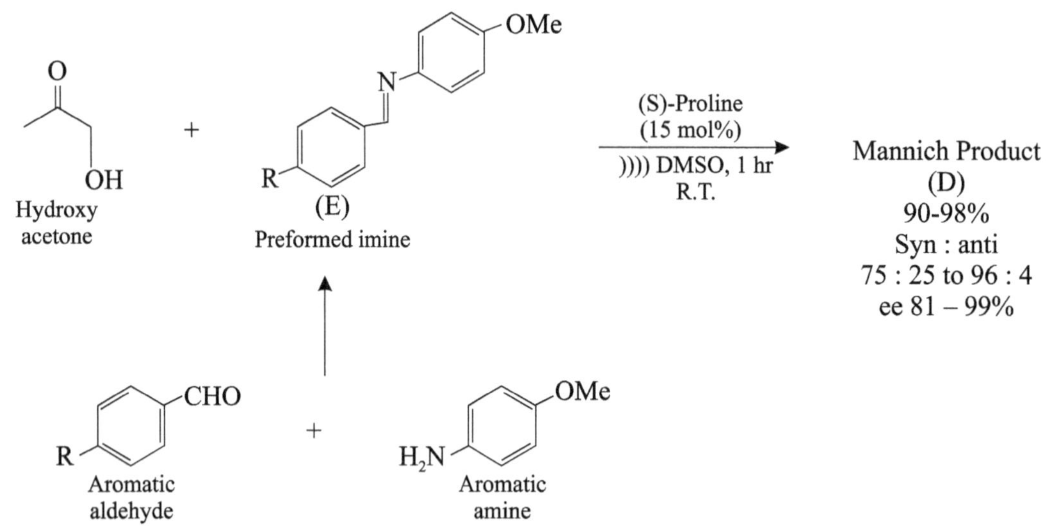

Fig. 10.70: Mannic Reaction between hydroxyacetone and preformed imine.

The above reaction (Fig. 10.70) on thermal heating [(S)-Proline (15 mol %), DMSO, Δ, 1hr.) gives 85-93% of the Mannich product (D), syn. anti 75 : 25 to 92 : 10; ee 65-93%.

10.2.26 Synthesis of Heterocycles

The heterocycles are important in a number of fields like organic, agricultural, pharmaceutical and medicinal chemistry. These are conveniently prepared by ultrasound promoted heterocyclisation of appropriately functionalized substrates using readily available starting materials under mild conditions[98].

10.2.26.1 Synthesis of 1H-benzotriazoles

These are prepared by the cyclisation of *o*-phenelelenediamine with sodium nitrite in acetic acid under ultrasound irridation. The 1H-benzotriazoles thus obtained on acylation under a ultrasound irridation give 1-acylbenzotriazoles (Fig. 10.71)[99].

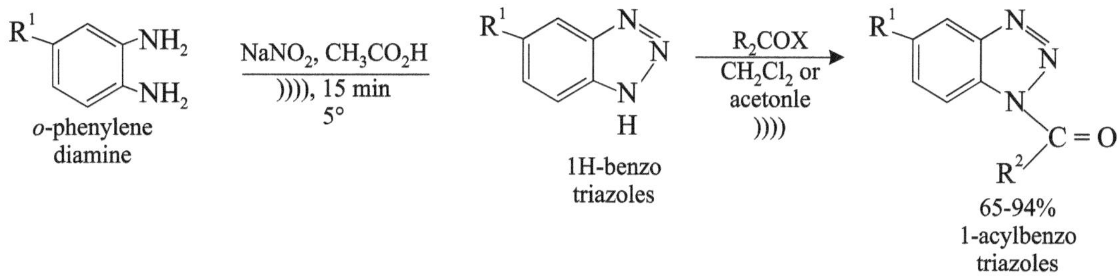

Fig. 10.71: Synthesis of 1H-benzotriazoles.

1,4-Disubstituted 1,2,3-triazoles were prepared by **Huisgen 1,3-dipolar cyclization** of terminal alkynes and alkyl/aryl halides in presence of sodium azide using 10 mol% CuI in aqueous medium under ultrasound irradiation (Fig. 10.72)[100].

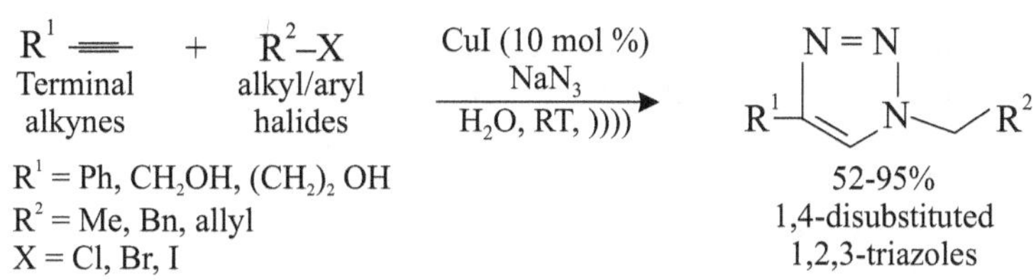

Fig. 10.72: Synthesis of 1,4-disubstituted 1,2,3-triazoles.

10.2.26.2 Synthesis of Pyrazoles

Substituted pyrazoles are obtained by ultrasound irridation of α-OXO thiooxoester or from α-oxo ketene, O,N-acetals and hydrazine derivatives using montmorillonite K-10 as a solid support (Fig. 10.73)[101].

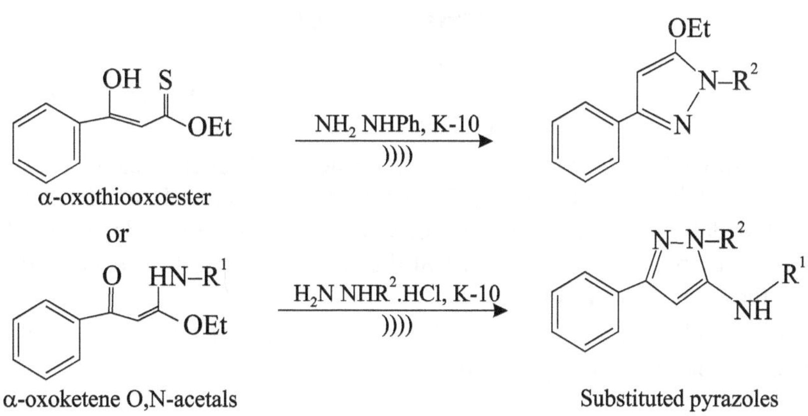

Fig. 10.73: Synthesis of substituted pyrazoles.

3,5-Dimethylpyrazoles on sonication in presence of N-halosuccinimide (NXSX), ICl or molecular iodine or bromine gave the halogenated pyrazoles in excellent yield (Fig. 10.74)[102] in a shorter time and convenient easily workup.

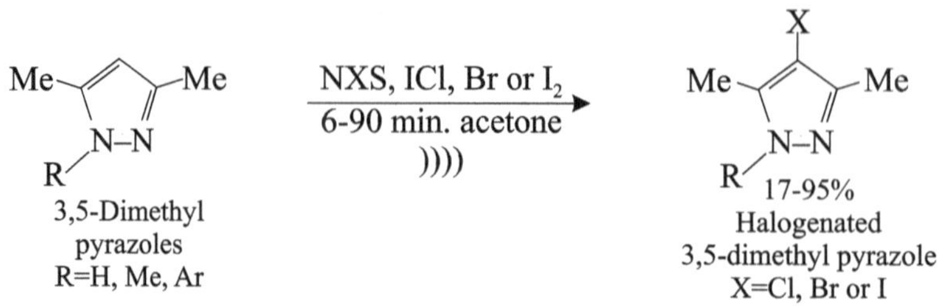

Fig. 10.74: Synthesis of halogenated 3,5-dimethyl pyrazoles.

10.2.26.3 Biginelli Reaction

It involves the reaction of ethyl acetoacetate, an aldehyde and urea to give 3,4-dihydropyrimidine 2 (1H) ones in presence of Brosted acids and/or Lewis acid catalyst. The dihydropyrimidones find use in pharmaceutical industry as calcium channel blocker, antihypersive agents.

It has been shown that Biginelli reaction could be carried out under milder conditions; the reaction is accelerated by a factor of 40 times or more under sonication[103]. The Biginelli reaction, a three component reaction between ethyl acetoacetate, aldehyde and urea (or thiourea) in presence of catalytic amount of HCl was completed in 2-5 min at room temperature to give dihydro pyrimidinones in excellent yield (Fig. 10.75)

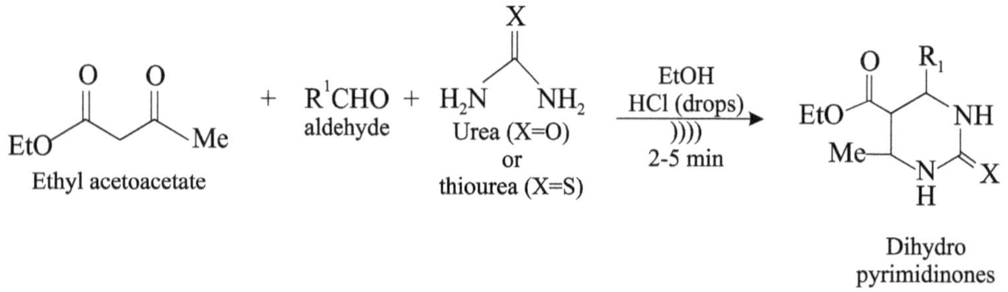

Fig. 10.75: Ultrasonic Biginelli Reaction

The Biginnelli reaction could also be conducted in solvent free condition[104]. It is catalyzed with HCl (1 mol%) or trifluoroacetic acid and is completed in 15-45 min in case urea is used and 60-90 min. in case thiourea is used. The yields is 72-97%.

The above reaction (Fig. 10.75) can also be performed in absence of catalyst[105]. In this case, the reaction is carried out in 1-*n*-butylimidazolium tetrafluoroborate ([Hbim]BF_4) (an ionic liquid) under ultrasound irridation, the reaction is complete in 40-90 min giving 85-97% of the product and the ionic liquid can be recovered and reused in the same reaction. In this case the ionic liquid is believed to play the role of Bronsted acid.

The Beginelli product can also be produced by using NH_4Cl in methanol and ultrasonic irradiation using a cleaning bath (Fig. 10.76)[106]

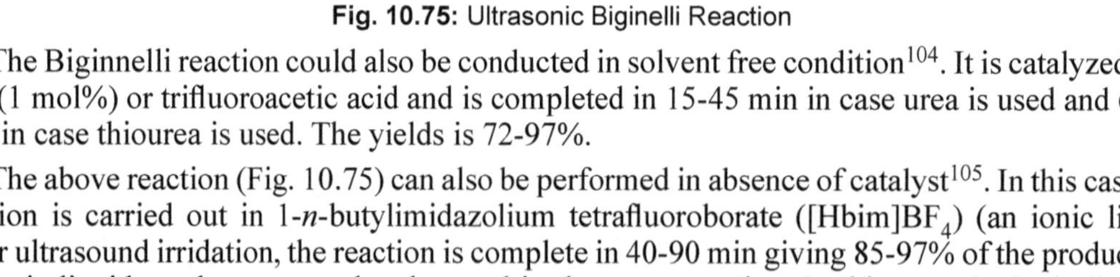

Fig. 10.76: Biginelli reaction using NH_4Cl

10.2.26.4 Synthesis of 1,5-benzodiazepines

The reaction of *o*-phenylene diamine derivatives with 2, 4-pantadione or a ketone under ultrasound irridation gives 1, 5-benzodiazepines. The reaction is catalysted by 10 mole % of *p*-toluenesulfonic acid (PTSA) (Fig. 10.77)[107].

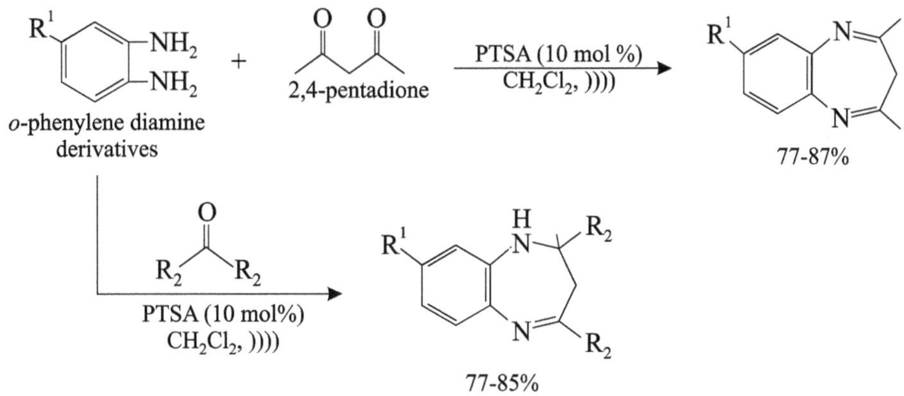

Fig. 10.77: Synthesis of 1,5-benzodiazepines.

10.2.27 Coupling Reactions

10.2.27.1 Heck Reaction

It involves the reaction of iodobenzenes with activated alkenes in presence of 2 mol % Pd $(OAc)_2$ using ionic liquids as solvent. The reaction is considerably enhanced in presence of ultrasound irradiation (Fig. 10.78)[108].

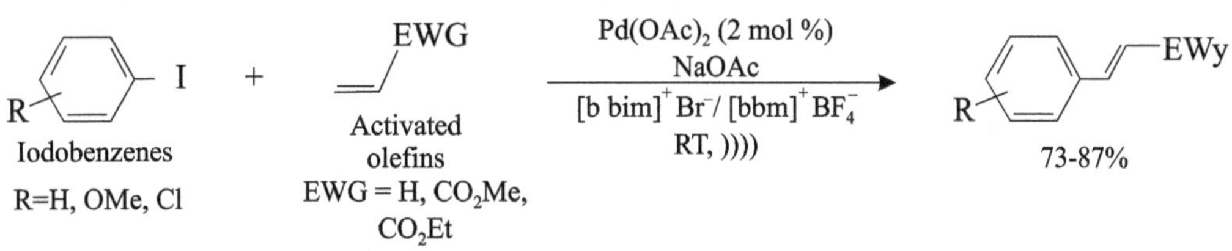

Fig. 10.78: Sonochemical Heck Reaction.

The above reaction could also be performed[109] in presence of heterogeneous Pd/C. Solubilized Pd was redeposited onto the support by using sodium formate as reducing agent [10% Pd/C, Et_3N, NMP, RT,)))), 70% yield). The reaction did not take place in absence of ultrasound.

Under conventional condition, the reaction requires high temperature and longer reaction times. The catalyst Pd/C is recyclable.

A **regioselective Heck reaction** of 2,4-diiodobenzenes is catalysed by palladium nanoparticles (obtained in situ from $PdCl_2$ in aqueous solution under sonication) has been reported[110]. The Pd nanoparticles could be reused. The regioselectivity was para-over ortho substitution (Fig. 10.79).

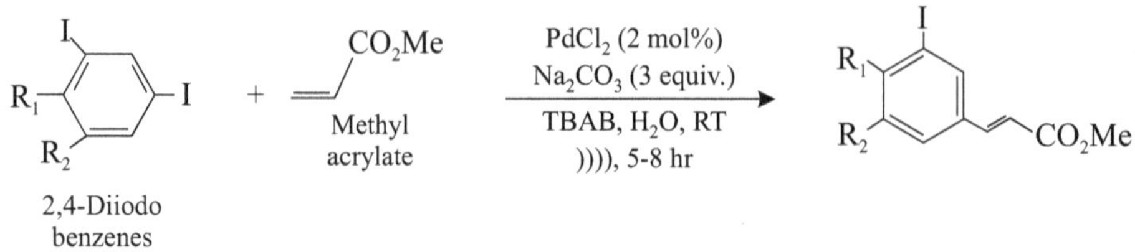

Fig. 10.79: A regioselective Heck Reaction.

10.2.27.2 Suzuki Reaction

The cross-coupling of halobenzenes with aryl boronic acids in ionic liquid in presence of Pd/C under sonochemical conditions gave the coupled products (Fig. 10.80)[111].

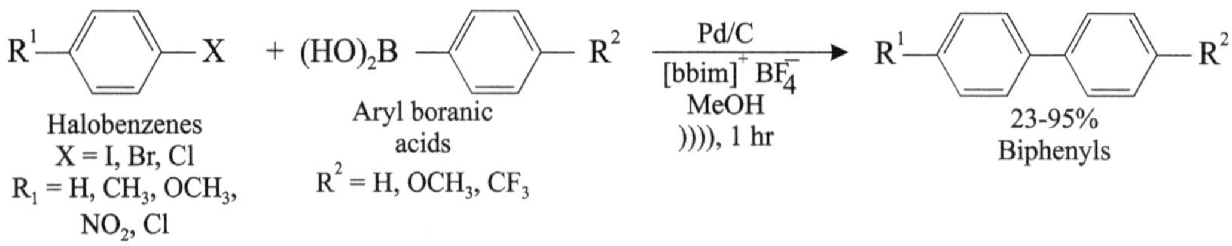

Fig. 10.80: Sonochemical Suzuki Reaction.

In the above reaction, in place of ionic liquid KF in methanol-water could also be used[112].

An **aqueous Suzuki-Miyaura** reaction mediated by ultrasound irridation using very low loading of palladium nanoparticles stabilized in polyvinylpyrrolidone as a catalyst has been reported[113]. It is found that ultrasonic irridation can considerably accelerate heterogeneous Suzuki coupling[114] of phenylboronic acid with aryl halide in neat water in the presence of TBAB and ligand-free cyclopalladated ferrocenylimes (A) and (B) (Fig. 10.81). In this case even aryl chlorides could cross-couple with boronic acid.

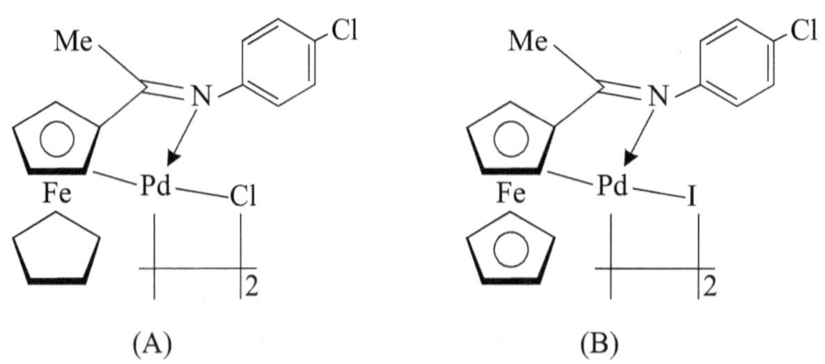

Fig. 10.81: Cyclopalladated ferrocenylimes (A) and (B).

A cross-coupling reaction[115] under sonication has been used to prepare intermediates of Vitamin D_3 with modified side chain. The method involves reaction of an iodide containing vitamin D triene with α, β-unsaturated compound under sonochemical conditions (Fig. 10.82).

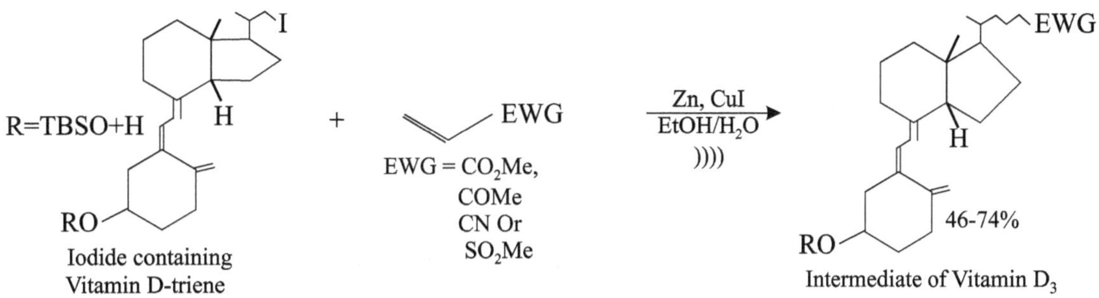

Fig. 10.82: Synthesis of an intermediate of Vitamin D_3.

10.2.28 Synthesis of Functionalized Aryl Acetylenes

The reaction of aryl acetylenes with metallic lithium with electrophiles in presence of ultrasound gives the corresponding functionalized aryl acetylenes (Fig. 10.83)[116].

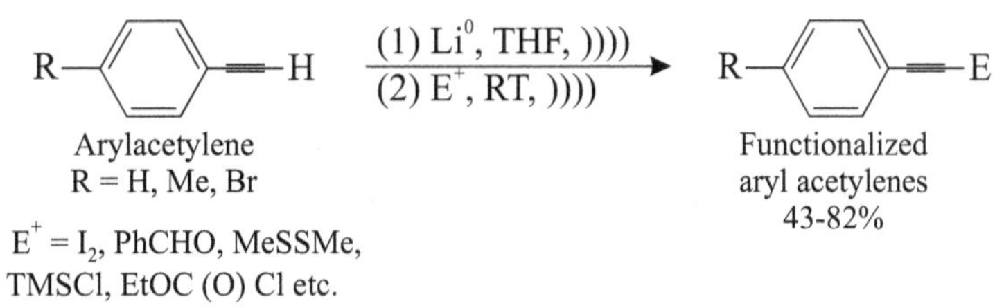

Fig. 10.83: Synthesis of functionalized aryl acetylene.

10.2.29 Synthesis of Imines

The reaction of aldehydes with primary amines catalyzed by silica in presence of ultrasound gives imines in excellent yields (Fig. 10.84)[117].

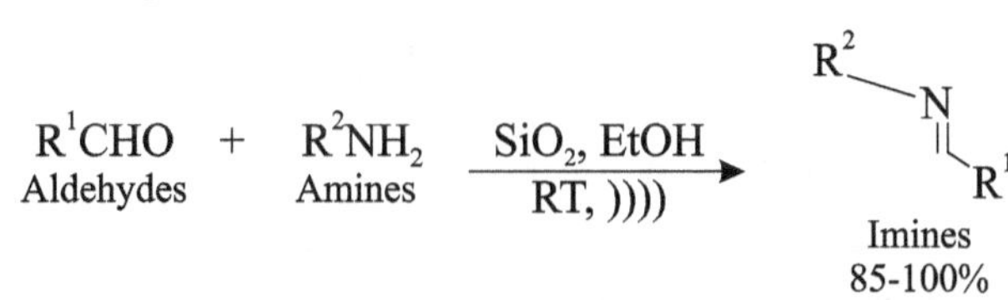

Fig. 10.84: Synthesis of Imines.

Conclusion

It has been found that ultrasound has tremendous application in organic synthesis. It dramatically effects the rates of chemical reactions and organic transformations. In fact, ultrasound should be used in all organic transformation to find out whether the rate of the transformation increases.

REFERENCES

1. J.P. Lorimer and T.J. Mason, Chem. Soc. Rev., 1987, **16**, 239-274.
2. J.L. Lindley and T.J. Mason, Chem. Soc. Rev., 1987, **16**, 275-311.

3. K.S. Suslick, Science, 1990, **247**, 1439.
4. G. Cavotto and P. Cintas, Chem. Soc. Rev., 2006, **35**, 180.
5. H Fillion and J.L. Luche, Synthetic Organic Sonochemistry, 1998, Plenum Press, New York.
6. J.M. Khurana, P.K. Sahoo and G.C. Maikap, Synth. Commun., 1990, 2267.
7. K.S. Suslick, Modern Synthetic methods, 1986, **4**, 1.
8. J.M. Khurana, Chemical Education, 1990, 24-29.
9. C.M. Sehgal and S.Y. Wang, J. Am. Chem. Soc., 1981, **103**, 6606.
10. K.S. Susleck, D.A. Hammerton and R.E. Cline, Jr., J. Am. Chem. Soc., 1986, **108**, 5641.
11. F.B. Flint and K.S. Susleck, Science, 1991, **253**, 1397.
12. R.S. Davidson, A Safdar, J.D. spencer and D.W. Lewis, Ultrasonics, 1987, **25**, 35.
13. S. Moon, L. Duclin and J.V. Croney, Tetrahedron Lett., 1979, 3917.
14. D.S. Kristol, H. Klotz and R.C. Parker, Tetrahedron Lett., 1991, **22**, 907.
15. J.P. Lorimer, T.M. Mason and B.P. Mistry, Ultrasonics,1987, **25**, 23.
16. J. Elguero, P. Goya, J. Lissavetzky and A.M. Valdeomillos, C.R. Acid Sci. Paris, 1984, **248**, 877.
17. D.M. Trost and B.P. Coppola, J. Am. Chem. Soc., 1982, **104**, 6879.
18. P. Boudjout, W.H. Ohrbom and J.B. Woell, Synth. Commun., 1986, **16**, 401.
19. T. Ando, J. Kawale and T. Hanafusa, Synthesis, 1983, 637.
20. H. Priebe, Acta. Chem. Scand., Ser. B., 1984, **38**, 895.
21. G.H. Posner, Org. React. (N.Y.), 1972, **19**, 1.
22. J.C.S. Barboza, C. Petrier and J.L.Luchi, Tetrahedron Lett., 1985, **26**, 829.
23. N. Joshi and H.M.R. Hoffmann, Tetrahedron Lett., 1986, **27**, 687.
24. G. Mehta and H.S.P. Rao, Synth. Commun., 1985, **15**, 991.
25. O. Repic and S. Vogt, Tetrahedron Lett., 1982, **23**, 2729.
26. O. Repic, P.G. Lee and N. Giger, Org. Prepn. Proc. Int., 1984, **16**, 25.
27. S.L. Regan and A. Singh, J. Org. Chem., 1982, **47**, 1587.
28. P. Boudjouk and B.H. Han, J. Catal., 1983, **79**, 489.
29. D.H. Shin and B.H. Han, Bull. Korean. Chem. Soc., 1985, **6**, 247.
30. K.J. Moulton, S. Koritala and E.N. Frankel, J. Am. Oil. Chem. Soc., 1983, **60**, 1257.
31. B.H. Han and P. Boudjouk, Tetrahedron Lett., 1981, **22**, 2757.
32. T.D. Lash and D. Berry, J. Chem. Educ., 1985, **62**, 85.
33. P. Boudjouk, D.P. Thompson, W.H. Ohrbom and B.H. Han., Organometallics, 1986, **5**, 1257.
34. J. Lindley, J.P. Lorimer and T.J. Mason, Ultrasonics, 1986, **24**, 292.
35. T. Yamashita, Y. Inoue, T. Kondo and H. Hashimoto, Chem. Lett., 1986, 407.
36. T. Kitazume and N.Ishikawa, J. Am. Chem. Soc. 1985, **107**, 5186.
37. N. Ishikawa and T. Kitazume, European Patent, 0082252 Al, 1982.
38. T. Kitazume and N. Ishikawa, Chem. Lett., 1981, 1679.
39. B.H. Han and P. Boudjouk, J. Org. Chem., 1982, **47**, 751.
40. P. Boudjouk, R. Sooriyakumaran and B.H. Han. J. Org. Chem., 1986, **51**, 2818.
41. L. Xu, F. Tao andT. Yu, Tetrahedron Lett., 1985, **8**, 749.

42. J. Brennan and F.H.S. Hussain, Synthesis, 1985, **8**, 749.
43. A.J. Fry and D. Herr, Tetrahedron Lett., 1978, **19**, 1721.
44. A.J. Fry, G.S. Ginsberg and R.A. Parante, J. Chem. Soc. Chem. Commun., 1978, 1040.
45. T. Yamashita, Y. Inoue, T. Kondo and H. Hashimoto, Bull. Chem. Soc. Jpn., 1985, **58**, 2709.
46. B.H. Han and P. Boudjouk, Tetrahedron Lett., 1982, **23**, 1643.
47. R.S. Davidson, A.M. Patel, A. Safdar and D. Thornthwaite, Tetrahedron Lett., 1983, **24**, 5907.
48. J. Jurizak and R. Ostaszewki, Tetrahedron Lett., 1998, **29**, 959.
49. J. Ezquerra and J. Alvarcz-Builla, J. Chem. Soc., Chem. Commun., 1984, 54.
50. T. Fujita, S. Watanabe, S. Sakamoto and H. Hashimoto, Chem. Ind. (London),1986, 427.
51. J.M. Khurana and P.K. Sahoo, Synth Commun., 1992, 1691.
52. R.D. Walkup and R.T. Cunningham, Tetrahedron Lett., 1987, **28**, 4019.
53. T.J. Mason, J.P. Lorimer, A.T. Turner and A.R. Harris J. Chem. Res. (S), 1988, 80.
54. R.S. Davidson, A. Safdar, J.D. Spencer and D.W. Lewis, Ultrasonics, 1987, **25**, 35.
55. J. Yamakawi, S. Sumi, T. Ando and J. Hanafusa, Chemistry Lett., 1983, 379.
56. T. Kimura, M. Fujita and T. Ando, Chemistry Lett., 1988, 137.
57. J.M. Khurana, P.K. Sahoo, S.S. Titus and G.C. Maikap, Synthetic Communications, 1990, **20(9)**, 1357.
58. K.S. Suslick and D.J. Casadonte. J. Am. Chem. Soc., 1987, **109**, 3459.
59. C. Petrier and J.L. Luchi, Tetrahedron Lett., 1989, **28**, 2347.
60. C. Petrier and J.L. Luci, Tetrahedron Lett., 1989, **28**, 2351.
61. J.W. Huffman, W. Liao and R.H. Wallace, Tetrahedron Lett., 1987, **28**, 3315.
62. P. Boudjouk and J. So, Synthetic Commun., 1986, **16**, 775.
63. W.P. Reeves, J.A. Murry, D.W. Willoughby and W.J. Friedrich, Synthetic Commun., 1988, **18**, 1961.
64. H.C. Brown and U.S. Racherla, Tetrahedron Lett., 1985, **26**, 2187.
65. B.H. Han and P. Boudjouk, Organometallics, 1983, **2**, 769.
66. B.H. Han and P. Boudjouk, Tetrahedron Lett., 1981, **22**, 2757.
67. T. Kitazumi and N. Tshikawa, J. Am. Chem. Soc., 1985, **107**, 5186.
68. P. Knochel and J.F. Normant, Tetrahedron Lett., 1984, **25**, 1475.
69. J.L. Luche and J.C. Damiano, J. Am. Chem. Soc., 1980, **102**, 7926.
70. W. Oppolzer and A. Nakao, Tetrahedron Lett., 1986, **27**, 5471.
71. G.C. Markap, Ph.D. Thesis, Delhi University, 1991.
72. J. Lee and J.K. Snyder, J. Am. Chem. Soc., 1989, **111**, 1522
73. C.W. Porter and L. Young, J. Am. Chem. Soc., 1938, **60**, 1497.
74. J.C.Menediz, G.G. Trigo and M.M. Solhuber, Tetrahedron Lett., 1986, **27**, 3283.
75. T. Hanafusa, J. Ichihara and T. Ashida, Chemistry Lett., 1987, 687.
76. A. Fuentes and J.V. Sinisterra, Tetrahedron Lett., 1986, **27**, 2967.
77. B. Han and P.J. Boudjouk, J. Org. Chem., 1982, **47**, 5030.
78. A.K. Bose, K. Gupta and M.S. Manhas, J. Chem. Soc. Chem. Commun., 1984, 86.
79. M. Ihara, M. Katogi, K.Fukumoto and T. Kametani J. Chem. Soc. Chem. Commun, 1987, 721.

80. C. Petrier, A.L. Gemal, J.L. Luche, Tetrahedron Lett., 1982, **23**, 3361.
81. J. Einhorn and J.L. Luche, Tetrahedron Lett., 1986, **27**, 501.
82. J.L.Luche, C. Petrier and C. Duputy, Tetrahedron Lett., 1985, **26**, 753.
83. E. Einhorn, C. Einhorn and J.L.Luchi, J. Org. Chem. in press.
84. R.S. Varma and G.W. Kabalka, Heterocycles, 1985, **23**, 139.
85. J.T. Li, T.S. Li, L.J. Li and X. Cheng, Ultrason. Sonochem., 1999, **6**, 199.
86. S.X. Wang, J.T. Li, W.Z. Yang and T.S.Li, Ultrason. Sonochem., 2002, **9**, 159.
87. W.Z. Xu, G.F. Chen., J.F. Han *et al*, Chem. Ind., 2003, **1**, 109.
88. A. Fuentes, J.M. Marinas and V.N. Sinistera Tetrahedron Lett., 1987, **28**, 4541.
89. J.T. Li, G.F. Chen, J.X. Wang and T.S. Li, Synth. Commun., 1999, **29**, 965.
90. J.T. Li, W.Z. Yang, S.X. Wang *et al*, Ultrason. Sonochem., 2002, **9**, 237.
91. M. Chtourou, R. Abdelhedi, M.H. Frikha and M. Traibelsi, Ultrason. Sonochem., 2010, **17**, 246.
92. J. McNulty, J.A. Steere and S. Wolf, Tetrahedron Lett., 1998, **39**, 8013.
93. A. Gorcia-Raso, J. Garcia-Raso, R. Companer *et al*, Synthesis, 1982, **12**, 1037.
94. J.T. Li, G.F. Chen, W.Z. Xu and T.S. Li, Ultrason. Sonochem., 2003, **10**, 115.
95. H. Zeng, H. Li and H. Shao, Ultrason. Sonochem., 2009, **16**, 758.
96. G. Zhang, Z. Huang and J. Zhou, Chin. J. Chem., 2009, **27**, 1967.
97. M.L. Kantam, C.V. Rajasekhar, G. Gopikrishna *et el*, Tetrahedron Lett., 2006, **47**, 5965.
98. R. Cella and H.A. Stefani, Tetrahedron Lett., 2009, **65**, 2619.
99. C.M.P. Pereira, H.A. Stefani, K.P. Guzen and A.T.G. Orfao, Lett. Org. Chem., 2007, **4**, 43.
100. R. Sheedhar and P.S. Reddy, Synth. Commun., 2 007, **37**, 805.
101. M.E.F. Braibante, H.T. Braibante, L.C. Tavares *et al* Synthesis, 2007, 2485.
102. H.A. Stefani, C.M.P. Pereira, R.B. Almeida et al. Tetrahedron Lett., 2005, **46**, 6833.
103. M.S. Zhidovinova, O.V. Fedorova, G.L. Rusinov and I.G. Ovchinnikova, Russ. Chem. Bull., 2003, **52**, 2527.
104. K. Singh, S. Singh and P. Kaur, Lett. Org. Chem., 2006, **3**, 201.
105. A.R. Gholap, K. Venkatesan, T. Daniel *et al*, Green Chem., 2004, **6**, 147.
106. H.A. Stefani, C.B. Oliveira, C.B. Almeida *et al*., Eur. J. Med. Chem., 2006, **41**, 513.
107. K.P. Guzen, R. Cella and H.A. Stefani, Tetrahedron Lett., 2006, **47**, 8133.
108. R.R. Deshmukh, H. Rajagopal and K.V. Srinivasan, Chem. Commun., 2001, 1544.
109. G.V. Ambulgekar, B.M., Bhanage and S.D. Samant, Tetrahedron Lett., 2005, **46**, 2403.
110. Z. Zhang, Z. Zha, C. Gan *et al*., J. Org. Chem., 2006, **71**, 4339.
111. R. Rajagopal, D.V. Jarikote and K.V. Srinivasan, Chem. Commun. 2002, 616.
112. V. Polackova, M. Hut'ka and S. Toma, Ultrason. Sonochem., 2005, **12**, 99.
113. A.L.F. De Souza, L.C. Da Silva, B.L. Oliveria and O.A.C. Antunes, Tetrahedron Lett., 2008, **49**, 3895.
114. J. Zhang, F. Yang, G. Ren *et al*, Ultrason. Sonochem., 2008, **15**, 115.
115. J. Perez-Sestelo, J.L. Mascareñas, L. Castedo and A. Mourino, Tetrahedron Lett., 1994, **35**, 275.
116. H.A. Stefani, R. Cella, F.A. Dörr *et al*, Tetrahedron Lett., 2005, **46**, 2001.
117. K.P. Guzen, A.S. Guarezemini, A.T.G. Orfão, Tetrahedron Lett., 2007, **48**, 1845.

11. PHOTOCHEMICAL ORGANIC TRANSFORMATIONS

11

Photochemical Organic Transformations

11.1 INTRODUCTION

Most of the photochemical transformations are carried out using light with shorter wavelengths than those found in sunlight. In fact photochemistry is believed to be a key approach to green chemistry. This is because the photochemical transformations are carried out under mild conditions. The reagent is these transformations is the photon, which can be considered to be the best green solvent; it is either absorbed causing the reaction or not absorbed in which case it travels out of the reaction vessel. No residue is left in either of the cases in contrast to the chemical reagents which are added and have to be eliminated after the reaction.

Even with the green advantage associated with photochemical transformations, these are generally not used in industrial applications. This is because in photochemical transformations a mixture of product are normally obtained. Also for photochemical transformations an expensive experimental apparatus is required. However, with all the limitations, it is expected that photochemical transformation may find a place in industrial application in times to come.

In photochemical organic transformations mostly the following lamp are used:

- Low pressure mercury arc, emitting at 254 nm, typically 8-15 W. These are long tubes, 10-20 cm, and are suited for external illumination of the reaction vessels. Cooling is not required but if required a fan is used when lamps are in a confined space. The same lamp fitted with an outside phosphor converts light to longer wavelength at 310 or 365 nm.
- Medium pressure mercury arcs, emitting intensively in the near UV, particularly 100 W or more. These are compact/sources and are suitable for internal illumination. These require water cooling.

11.2 PHOTOCHEMICAL TRANSFORMATIONS

The earliest photochemical transformation known to mankind is **photosynthesis of sugars** by plants using sun-light, CO_2 and H_2O in presence of chlorophyll. Some examples of photochemical transformations are given below:

11.2.1 Synthesis of Benzopinacol

It is obtained by photolysis of benzophenone using sunlight in presence of an alcoholic solvent (particularly isopropyl alcohol) (Fig. 11.1).

$$2(C_6H_5)_2C=O + (CH_3)_2CHOH \xrightarrow{hv} C_6H_5-\underset{OH}{\overset{C_6H_5}{C}}-\underset{OH}{\overset{C_6H_5}{C}}-C_6H_5 + CH_3COCH_3$$

Benzophenone — Isopropyl alcohol — Benzopinacol

Fig. 11.1: Synthesis of benzopinacol.

The above reaction (Fig. 11.1) is more efficient when hydrogen donor is benhydrol [$(C_6H_5)_2CHO$], the reduction product of benzophenone.

11.2.2 Photochemical Transformations of Olefins

Photochemical reactions in case of olefins involve two types of electronic absorption:

(*i*) $\sigma \rightarrow \pi^*$ excitation (requires more energy and is available only from light of wavelengths lower than 150 nm and is difficult to achieve under usual experimental conditions) and

(*ii*) $\pi \rightarrow \pi^*$ excitation (requires the absorption of light of about 180-210 nm for nonconjugated olefins and above 220 nm for conjugated olefins). In case of olefins, most of the photochemical reactions involve $\pi \rightarrow \pi^*$ excitation.

11.2.3 Isomerization of Olefins

Olefins undergo isomerisation on irradiation with UV light. In case of simple olefins, E-isomers absorb, energy more effectively and at a slightly different wavelength than the Z-isomer. Usually the E-isomer is partially converted into its thermodynamically less stable Z isomer. An example is the interconversion of fumaric and maleic acids (Fig. 11.2).

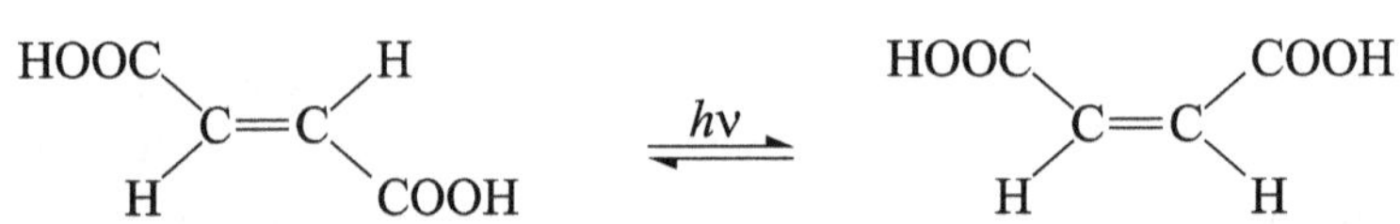

E-isomer Fumaric acid — Z-isomer Maleic acid

Fig. 11.2: Interconversion of fumaric and maleic acid

11.2.4 Conversion of Maleic Acid into Terebic Acid

The photocatalyst alkylation of maleic acid in isopropanol-water (1 : 1) mixture in presence of a water soluble benzophenone derivative (A) *e.g.*, disodium benzophenondisulfonate as a mixture of isomers gave naturally occurring terebic acid (Fig. 11.3)[1]. The terebic acid is believed to have formed by the attack of dimethyl hydroxy methyl radical (generated from isopropanol) onto maleic acid followed by lactonization.

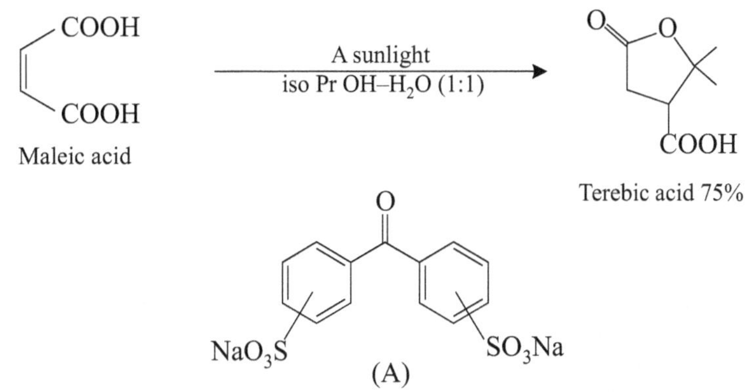

Fig. 11.3: Conversion to maleic acid into terebic acid.

It is appropriate to mention that the use of benzophenone as an initator is normally accompanied by some practical limitations; in particular, byproducts derived from benzophenone (*e.g.* benzopinacol) makes the isolation of the formed product (terebic acid) difficult. The use of water-soluble benzophenone derivative (A) has been proposed to overcome this problem. The desired product can be removed from the pyrolysed mixture by extraction with an organic solvent; the by-products formed from (A) remains in aqueous phase.

11.2.5 Photoisomerisation of *cis* and *trans* stilbenes

The photochemical *cis-trans* isomerism of stilbenes represents the simplest case of light induced geometrical isomerisation. As an example, it is found that irridation of trans stilbene in hexane in UV light results in the formation of the *cis*-isomer. The *cis-trans* ratio becomes constant after some time of irradiation. This is called photostationary state and is also attained on irridation of the *cis* isomer. The equilibrium favours the formation of thermodynamically less stable *cis* form. At photostationary state the *cis* and *trans* isomers are in relative amount of 10 : 1 (Fig. 11.4)

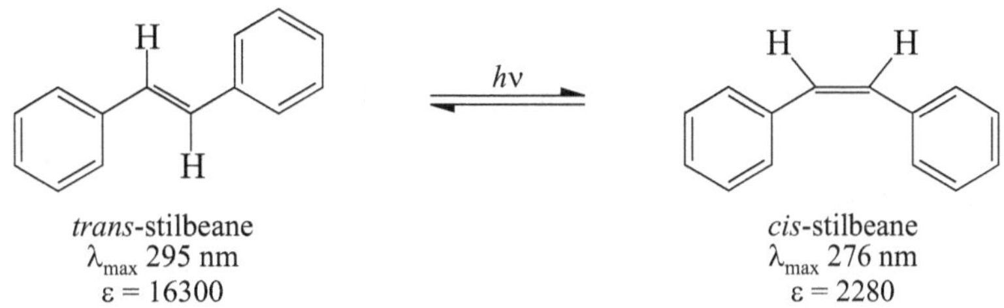

Fig. 11.4: Photochemical *cis-trans* isomerism.

See also section 11.2.9.

11.2.6 Photochemical Cycloaddition Reactions

2 + 2 and 4 + 2 cycloadditions occur photochemically with or without sensitizers. Examples of both types are given below (Fig. 11.5).

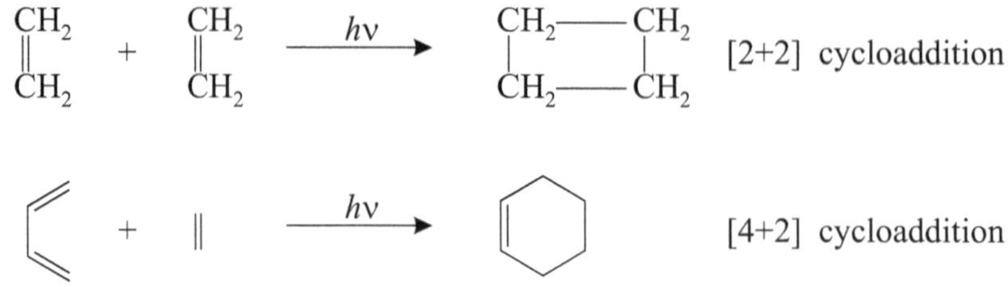

Fig. 11.5: 2 + 2 and 4 + 2 cycloadditions.

The cycloaddition of olefins giving four membered ring is a synthetically useful process. A typical example is the dimerization of cyclopentenone on irridation with light to give a mixture of 'head to head' and 'head to tail' dimers; these are formed via an excited dimer from the ($\pi \rightarrow \pi^*$) cyclopentenone and a molecule of ground state cyclopentenone (Fig. 11.6).

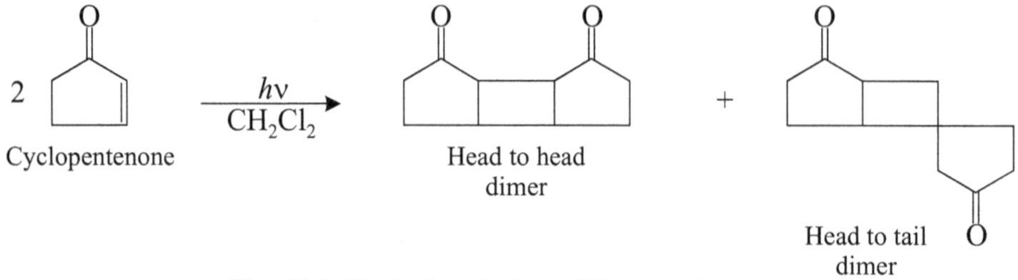

Fig. 11.6: Photodimerisation of Cyclopentenone.

See also section 11.2.5.

A very interesting photochemical cycloaddition reaction is **Paterno-Büchi reaction.** It involves the cycloaddition of the triplet state carbonyl compounds with the ground state of olifins to give oxatanes (oxa-cyclobutanes) (Fig. 11.7)

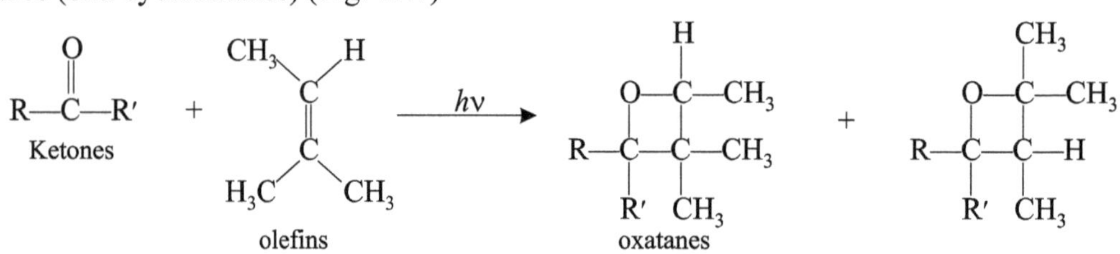

Fig. 11.7: Photochemical Paterno-Büchi reaction.

Another example of Paterno-Büchi reaction is the photocycloaddition of butyraldehyde to 2-methyl-2-butene (Fig. 11.8)

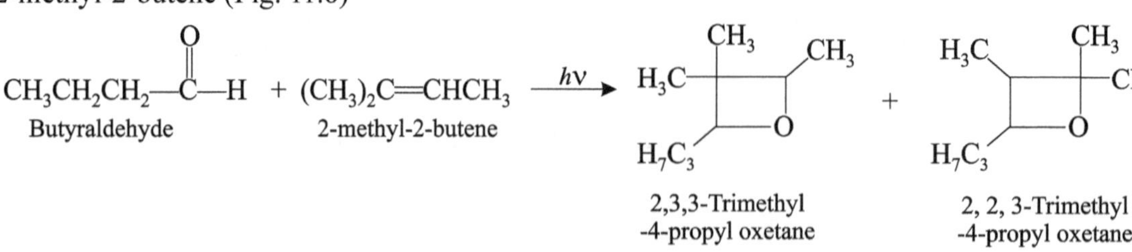

Fig. 11.8: Photocycloaddition of Butyraldehyde to 2-methyl-2-butene.

The photocycloaddition is not stereospecific. This is well understood by the observation that photocycloaddition of benzophenone with *cis* and *trans*-2-butene give the same mixture of *cis*-and *trans*-oxetanes (Fig. 11.9).

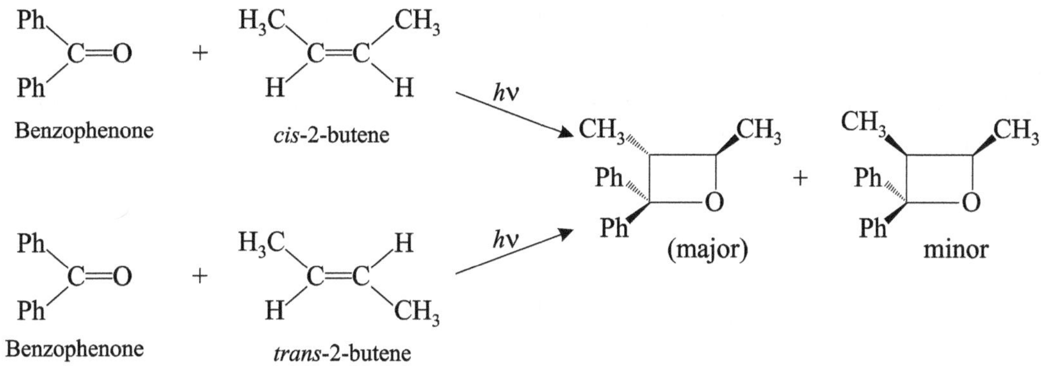

Fig. 11.9: Photocycloaddition of benzophenone with *cis*- and *trans*- 2-butene.

Photochemical addition of acrylonitrile and methyl acetylene dicarboxylate with benzene is given below (Fig. 11.10). In the later case, the formed cyclobutane being unstable undergoes spontaneous ring opening.

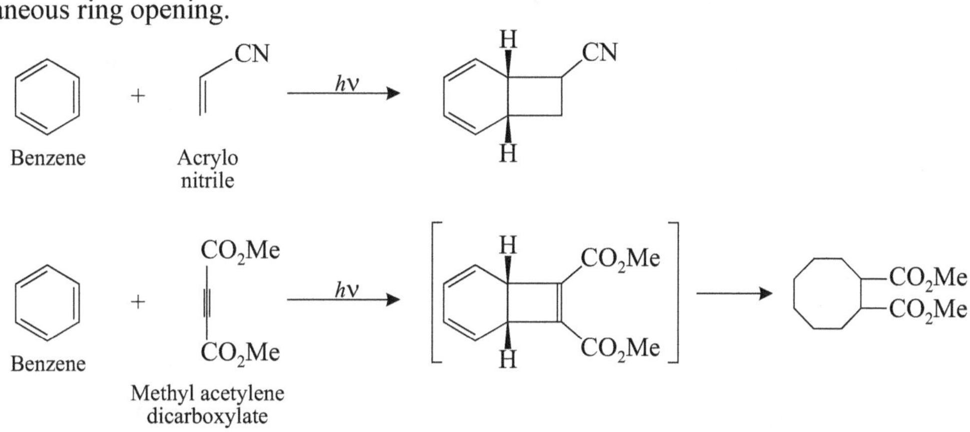

Fig. 11.10: Photochemical addition of acrylonitrile and methyl acetylene dicarboxylate with benzene

The cycloaddition could also proceed in an intramolecular fashion. Some examples are given below (Fig. 11.11).

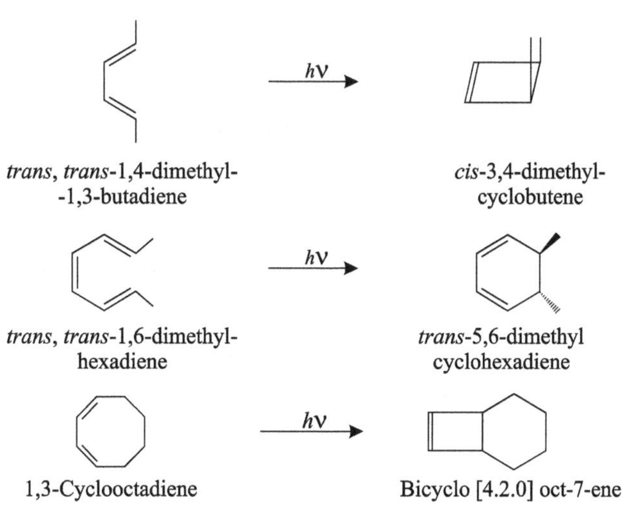

Fig. 11.11: Examples of intramolecular cycloadditions.

10.2.7 Photoinduced Substitution of Aromatic Compounds

Photoexcitation[2] of an aqueous solution of 4-methoxy-1-nitronaphthalene in presence of CN^- ions give 4-methoxy-1-naphthalene carbonitrile (Fig. 11.12). The yields are much better if the reaction is performed in presence of a PTC [hexadecyl (trimethyl) ammonium chloride (CTAC)].

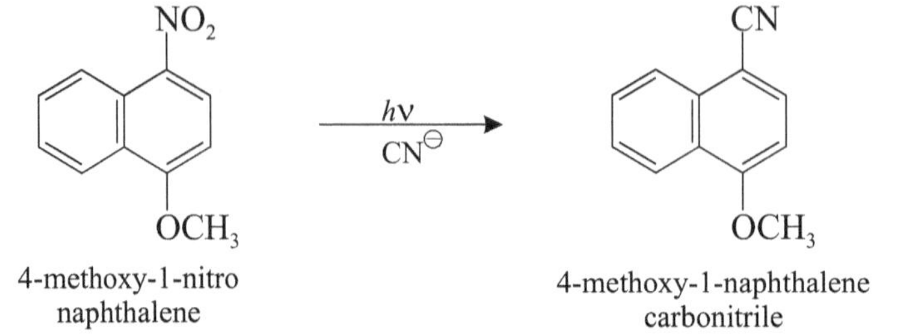

Fig. 11.12: Photoinduced substitution of organic compounds.

The irridation of *o*-fluoroanisole in presence of aqueous KCN gives catechol monomethyl ether as the major product. On the other hand irridation of *p*-fluoroanisole in presence of aq. KCN gives *p*-cyanoanisole as the major product (Fig. 11.13)[3].

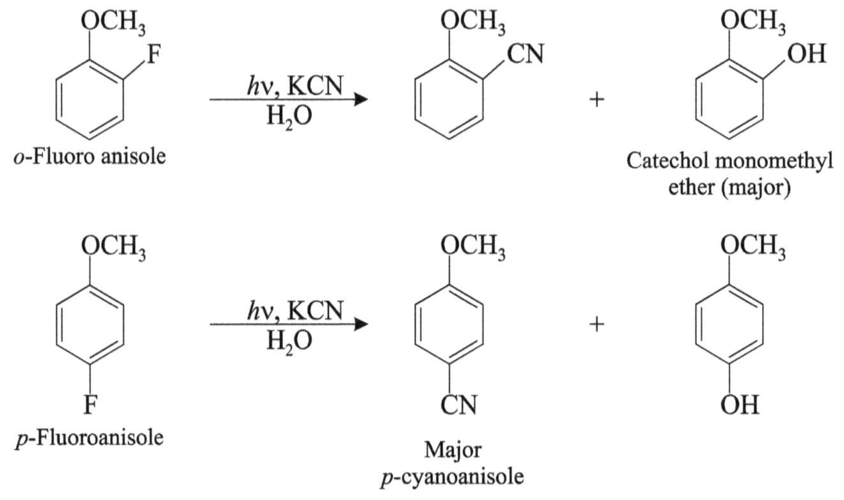

Fig. 11.13: Photoinduced substitution in *o*- and *p*- fluoroanisoles.

11.2.8 Photoirridation of dibenzoyldiazomethane in Presence of Amino Acid Derivative

Dibenzoyldiazomethane on photoirridation in presence of amino acid derivative in aqueous solution gave the addition product via the formation of photogenerated carbene which rearranged to benzoyl ketene (Fig. 11.14)[4].

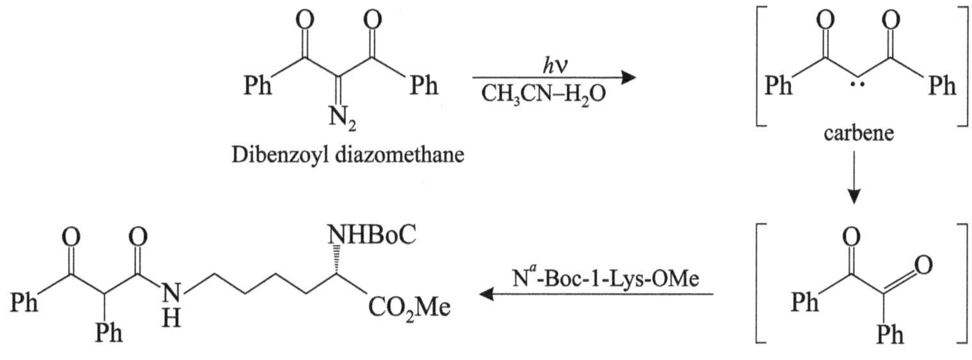

(55%)

Fig. 11.14: Photo irridation of dibenzoyldiazomethane in presence of amino acid derivative.

11.2.9 Photochemical Cycloadditions in Water

Organic substrates, such as stilbenes and alkyl cinnamates which have poor solubilities in water, photodimerize in water. Thus, as an example, stilbene on photochemical reaction give a mixture of dimeric products (Fig. 11.15)[5]. There was no dimerisation in benzene, only *cis-trans* isomerism was observed (See Section 10.2.4). However, addition of LiCl (increasing hydrophobic effect) increases the yield of dimerisation. The addition of guanidine chloride (decreasing hydrophobic effect) lowers the yield.

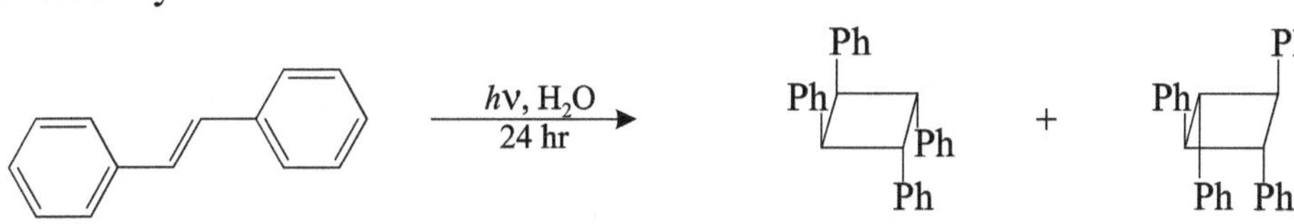

Benzene	0%	0%
H_2O	12%	10%
H_2O + LiCl	25%	17%
H_2O + guanidium chloride	8%	6%

Fig. 11.15: Photodimerisation of stilbene.

As in the case of stilbene (Fig. 11.15), similar results were obtained with alkyl cinnamates[5].

In a similar way, coumarin dimerished more efficiently in water than in organic solvents[6].

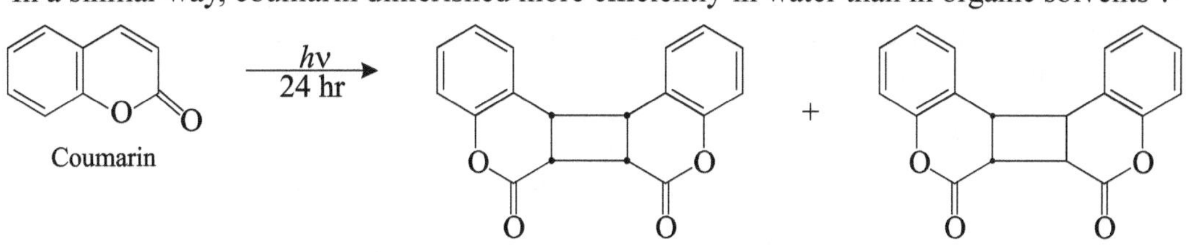

Solvent	yield
benzene	$< 10^{-5}$
Methanol	$< 10^{-5}$
Water	2×10^{-3}

Fig. 11.15: Photodimerisation of Coumarin.

11.2.10 Photochemical Reactions in Solid State

There are only few reports of photochemical reactions in solid state:

(*i*) Cinnamic acid (single crystal) on photoirridation gives[7] tuxillic acid (Fig. 11.16).

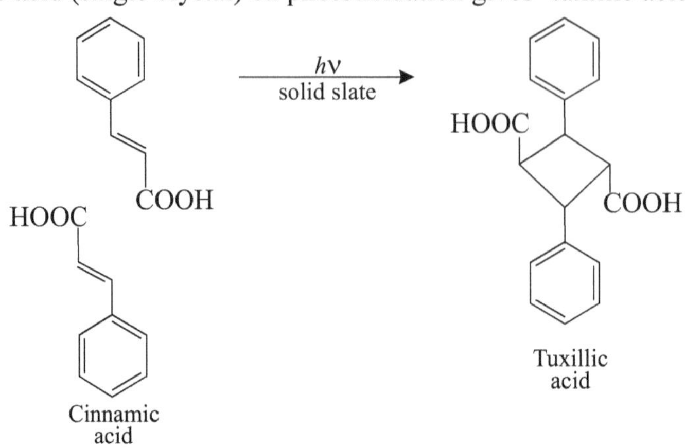

Fig. 11.16: Photochemical reaction of Cinnamic Acid in Solid state.

Cinnamic acid can also be photodimerised by irridation of its double salts with certain diamines in the solid state. Thus, the double salt of cinnamic acid and *o*-diamino cyclohexane on irridation in solid state gives[8] β-truxinic acid as the major product (Fig. 11.17).

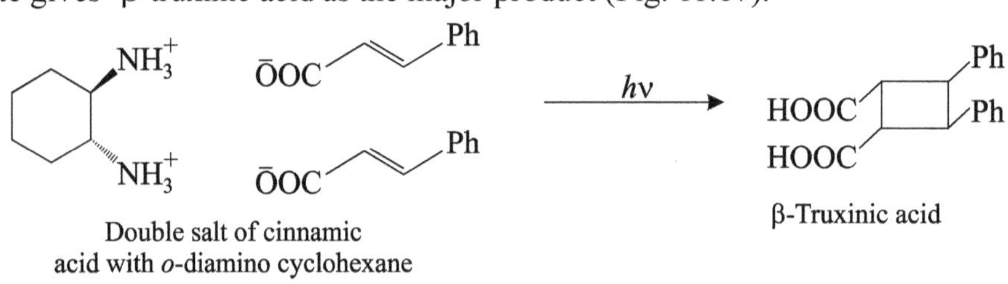

Fig. 11.17: Photoirridation of double salt of cinnamic acid with o-diamino-cyclohexane in solid state.

11.2.11 Synthesis of Allylphenols and allylanisoles

The allyl phenols and allyl anisoles have nematicial and antimicrobial molluscicidal activity and are also used in the manufacture of perfumes and fragrances. These are prepared by photoirridation *p*-chlorophenol or *p*-chloro anisole in MeCN-H_2O in presence of allyl trimethyl silane. The reaction proceeds via the formation of aryl cation (obtained from *p*-chlorophenol or *p*-chloro anisoles by photoirridation) followed by reaction with allyl trimethylsilane and subsequent elimination of Me_3Si^+ from the adduct carbocation (Fig. 11.18)[9].

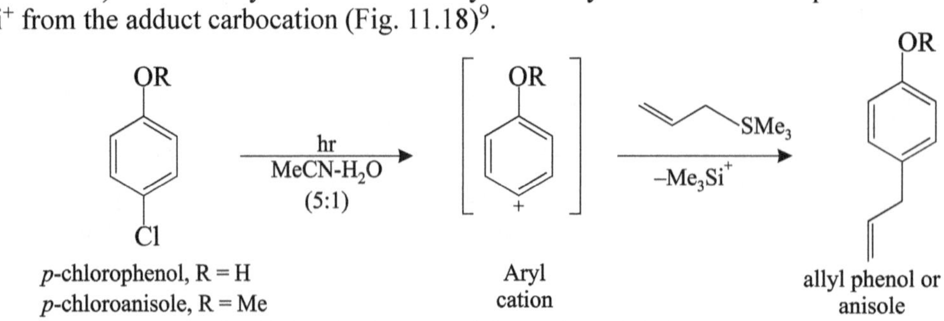

Fig. 11.18: Synthesis of allylphenols and allyl anisoles.

11.2.12 Synthesis of α-arylpropionic Acid Derivatives

The α-arylpropionic acid derivatives are employed as nonsteroidal anti-inflammatory drugs and are prepared from 4-chloroanilines as the phenyl cation precursors and Keten silyl acetals as nucleophiles as given below (Fig. 11.19)[9].

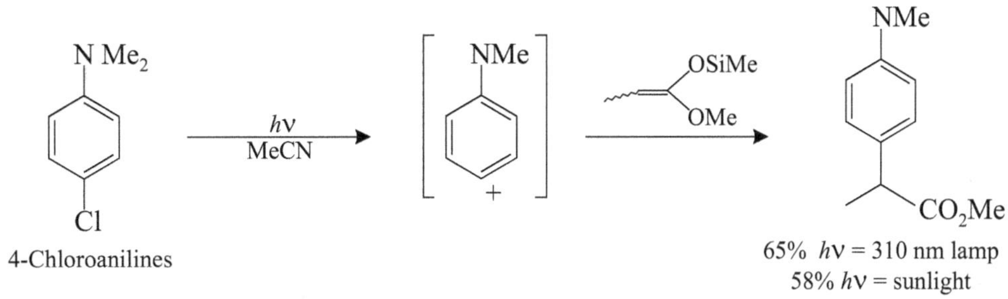

Fig. 11.19: Synthesis of α-arylpropionic acid derivatives.

11.2.13 Synthesis of 13-cis-retionic Acid

The 13-*cis* retionic acid, an active principle that finds practical application, in particular for treatment of acne. It was prepared by a patented procedure[10] involving a **Wittig condensation**. A mixture of the desired acid, 13-*cis* retionic acid (A) along with two isomers, *viz*., 11-*cis*, 13-*cis* retionic acid (B) and 11-*trans*, 13-*trans* retionic acid (C) was formed in the ratio 23 : 64 : 13. The mixture of retionic acids (A, B and C) was irradiated in water in presence of KOH and a sensitizer, Rose Bengal for 2 hr (Fig. 11.20). On the basis of HPLC of the mixture, it was found that the required 13-*cis* retionic acid (A) was present to the extent of 65.7%. The cooled mixture on acidification (H_2SO_4) precipitated (A), which was dried and crystallised (99% purity, yield 53%.[11])

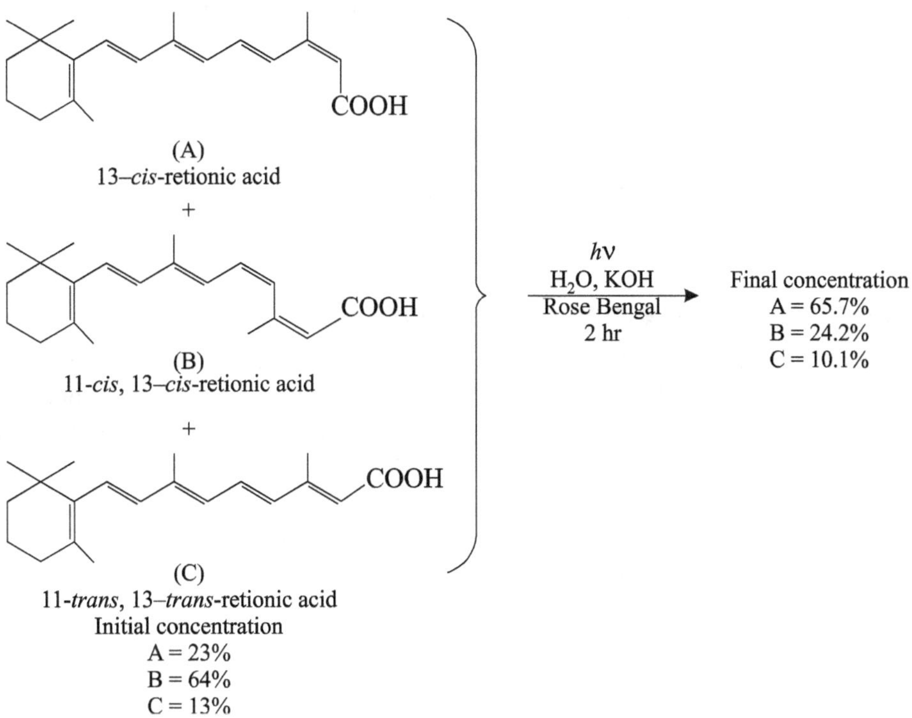

Fig. 11.20: Synthesis of 13-*cis*-retionic acid.

11.2.14 Synthesis of Previtamin D_3

It is prepared by a photochemical reaction starting from 7-dehydrocholesterol (7-DHC) (provitamin D_3) and subjecting to UV light. The reaction involves cleavage of the 9, 10-bond of 7-DHC to give (Z)-triene previtamin D_3, which can be easily converted by thermal rearrangement into vitamin D_3. The main problem is that the (Z) triene previtamin D_3 can easily absorb the UV light giving undesired photochemical byproducts such as Tachisterol and lumisterol. This problem has been overcome by carrying the reaction at a low conversion of 7-DHC. The yield could be improved by using quasi monochromatic excimer or exciplex lamps. It is found that the conversion of 7-DHC less than 10% permitted the preparation of previtamin D_3 in a selective fashion (Fig. 11.21)[12]

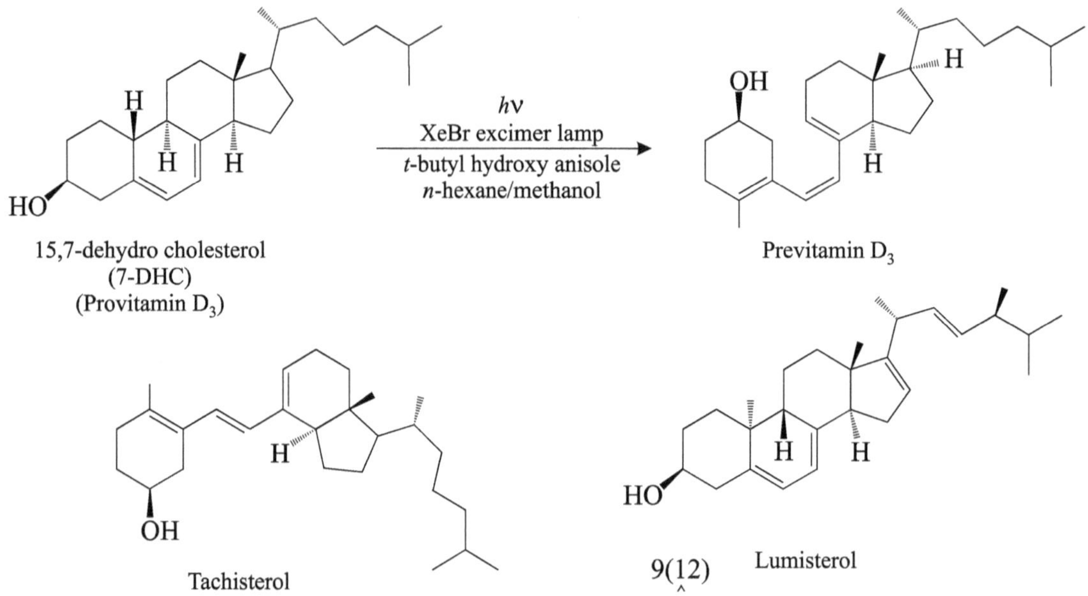

Fig. 11.21: Synthesis of Previtamin D_3

See also section 11.2.20.2 (part *b*).

11.2.15 Synthesis of Three-membered Rings

The presence of strained three membered (*e.g.*, cyclopropane) ring is quite common in natural and man made molecules having biological activities.[13,14] Natural compounds like pheromones and terpenes are known to contain three member rings. These rings are mostly prepared by a methylene carbon addition on to a double bond[15]. However, in photochemical methods the starting materials, *viz.*, 1,4-dienes are quite different. The pyrethroid derivatives are neurotoxic insecticides based on cyclopropanecarboxylic acid skeleton (*e.g.* esters of chrysanthemic acid). These are prepared photochemically as given below (Fig. 11.22)[16,17].

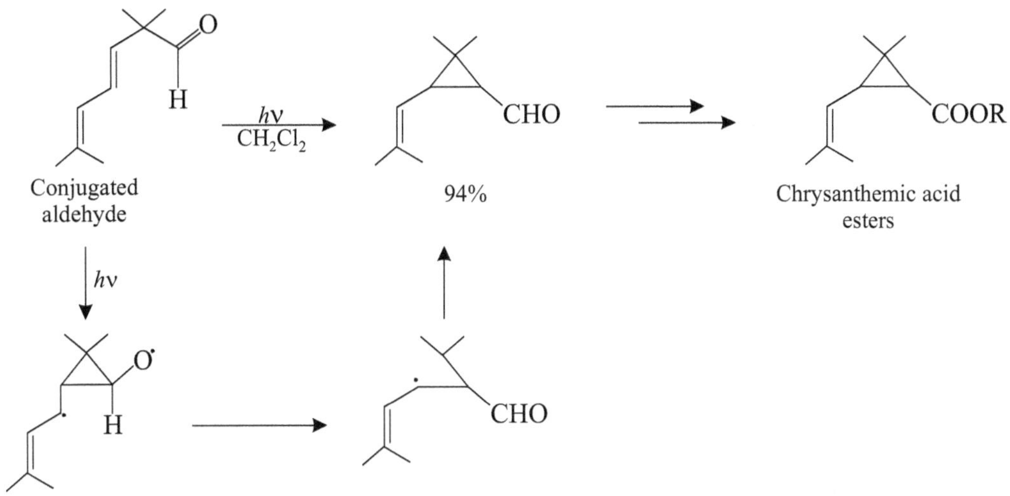

Fig. 11.22: Synthesis of chrysanthemic acid esters.

A number of other compounds, *viz.* (±)-hirsutene (a polyquinone sesquiterpenoids)[18], (–)-complicatic acid[19], (±) – $\Delta^{9(12)}$ – capnellene[20] were also synthesised from precursors containing cyclopropane rings.

A promising route for the synthesis of three membered ring involves the sterecontrolled synthesis of cyclopropane rings pramoted by irridation of benzoin esters as described below (Fig. 11.23)[19].

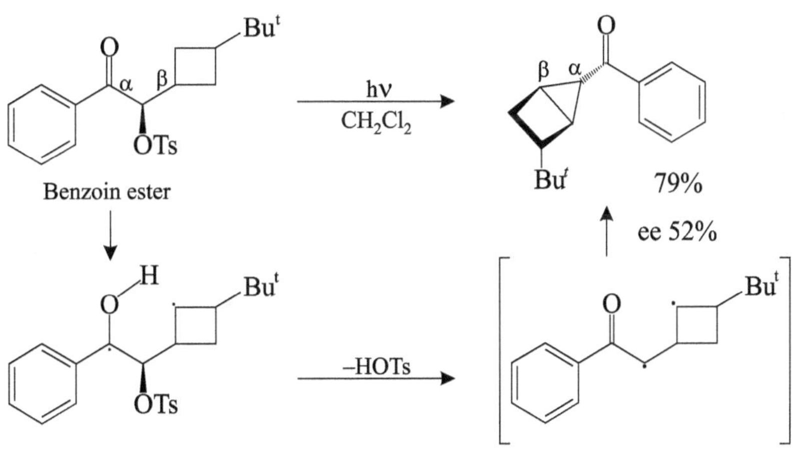

Fig. 11.23: A new route for the synthesis of three membered ring.

11.2.16 Synthesis of Cyclobutane Ring

The cyclobutane moiety is present in both natural and man-made molecules and have interesting biological and pharmaceutical properties. The four membered ring is mostly produced photochemically by (2+2) cycloaddition processes[21] (See also section 11.2.6, Fig. 11.5).

Interesting natural compounds containing a four membered ring are **Rhodopeptins**, which are cyclic lipopeptides isolated from Gram-positive bacteria Rhodococcus sp. Mer-N1033 and exhibit antifungal activity. The rhodopeptin analog is synthesised in high-yield by [2+2] cycloaddition reaction between ethylene and uracil derivative. Various steps involved are given below (Fig. 11.24)[22].

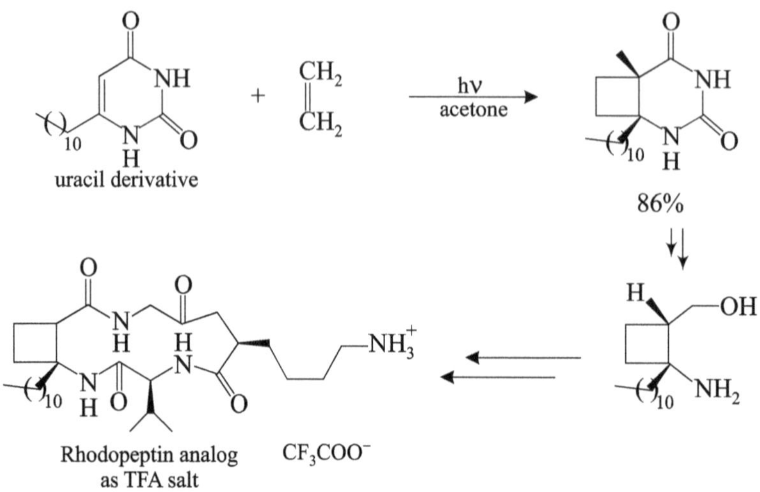

Fig. 11.24: Synthesis of Rhodopeptin analog.

A sesquiterpene, **Merilactone A** having neurotropic activity is a potential drug for the treatments of neurodegeneration associated with Alzheimer's and Parkinsons diseases is prepared by photochemical cycloaddition between 1,2-dichloroethylene and 2,3-dimethylmaleic anhydride in acetone. Various steps involved in the synthesis are given below (Fig. 11.25)[23]

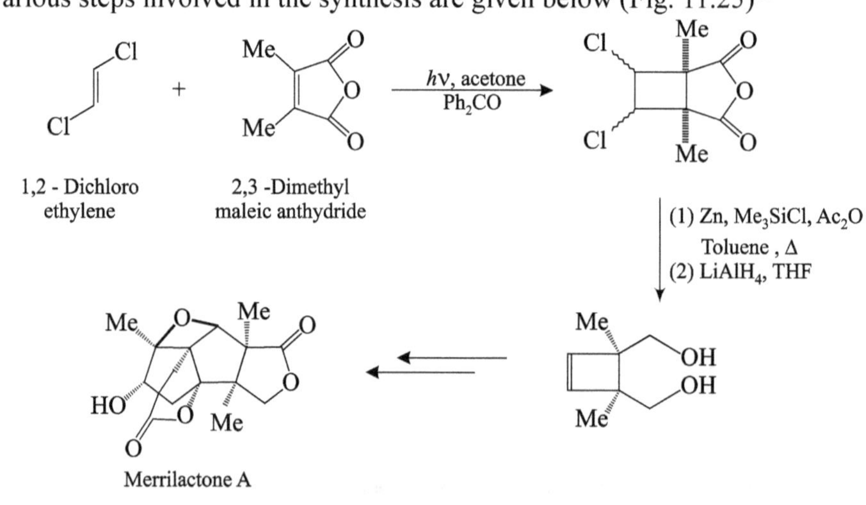

Fig. 11.25: Synthesis of Merrilactone A.

A number of other compounds containing cyclobutane ring have been synthesised. These include (+)-lineatin[24], Biyouyanagin[25], the alkaloid (+)-meloscine[26] and Grandisol[27].

The cyclobutane ring can also be formed in a intramolecular fashion. Some cases have already been included earlier (see section 11.2.5, Fig. 11.11). Other example, include synthesis of sesquiterpene (±) **punctaporonin C**. Various steps involved in its synthesis are given below (Fig. 11.26)[28].

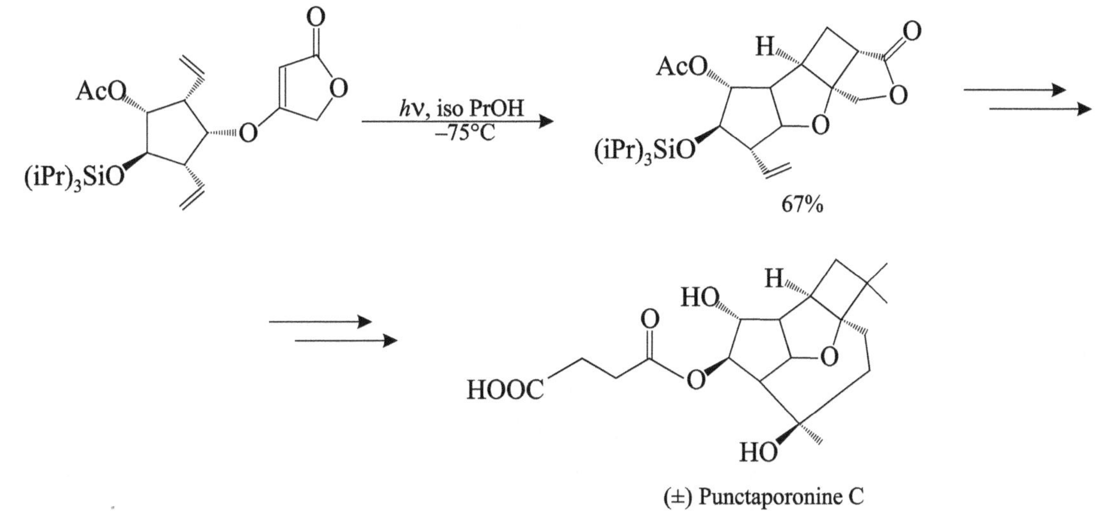

Fig. 11.26: Synthesis of (±) Punctaporonin C

(±) **Ginkgolite** B, the active component of Ginkgo biloba extract, commonly used as herbal medicine for treatment of respiratory and circulatory disorders has been synthesised by intramolecular photochemical reaction (Fig. 11.27)[29].

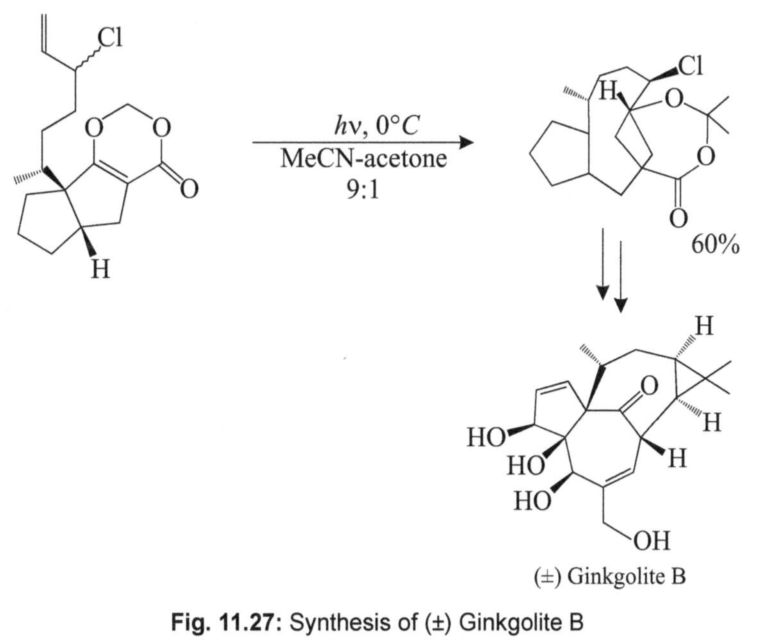

Fig. 11.27: Synthesis of (±) Ginkgolite B

The synthesis of cyclobutane ring containing oxygen as an heteroatom are known as **oxetanes**. These are synthesised by **Paterno-Buchi reaction** involving photochemical [2+2] cycloaddition between an olefin and a carbonyl group (see section 11.2.5, Figs. 11.7, 11.8 and 11.9)

The synthesis of **merrilactone** A, containing an oxctane ring system has been synthesised as given below (Fig. 11.28)[30].

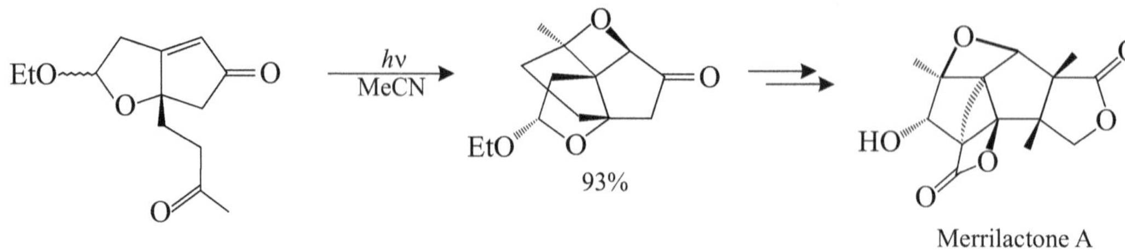

Fig. 11.28: Synthesis of Merrilactone A

11.2.17 Synthesis of Five-membered Ring

A number of Aglaia species are found to contain **Rocaglamides**, a group of cyclopenta [b] benzofurans and are known to have several biological effects, *e.g.,* anticancer and anti-leukemic activity. For example, **silvestrol** shows cytotoxic activity towards human cancer cells. It is obtained by the irridation of a flavonol. Various steps involved are given below (Fig. 11.29)[31].

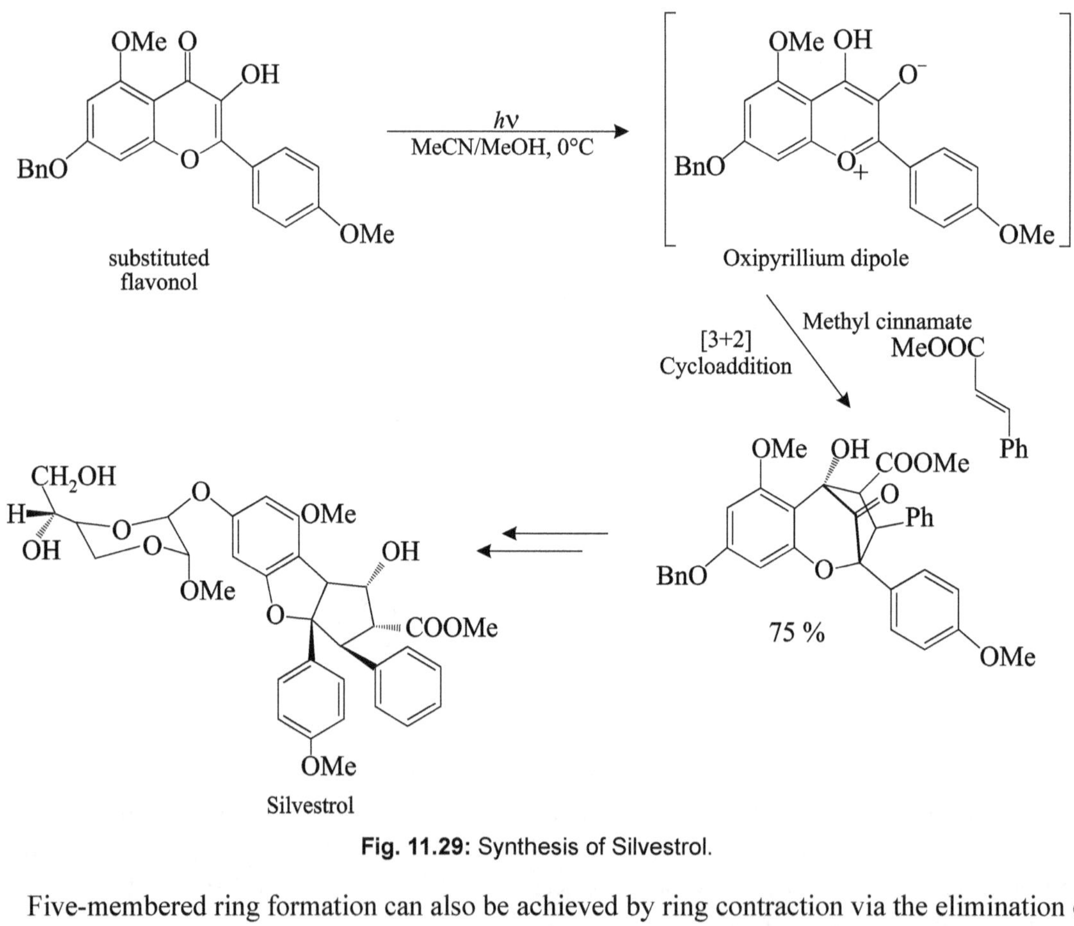

Fig. 11.29: Synthesis of Silvestrol.

Five-membered ring formation can also be achieved by ring contraction via the elimination of small stable molecules like CO, N_2, CO_2. As an example, the antifungal **herbertenolide** is obtained by a two step process from β-ketoester (Fig. 11.30)[32].

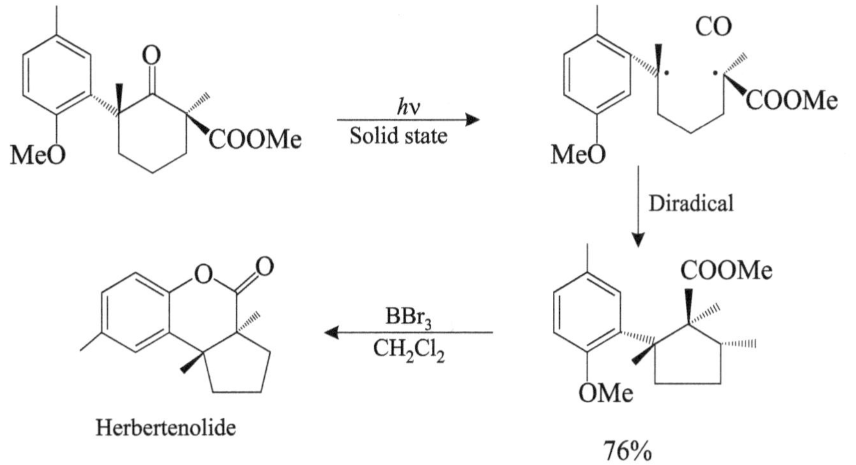

Fig. 11.30: Synthesis of herbertenolide.

A bioactive molecule **palominol** has been synthesised by photoirridation of an α,β-unsaturated diazoketone followed by heating in neat 1, 8-diazobicyclo [5.4.0] undec-7ene (DBU). The thermal treatment caused loss of nitrogen and formed ring-contracted ester. Final methylation using MeLi gave palominol (Fig. 11.31)[33].

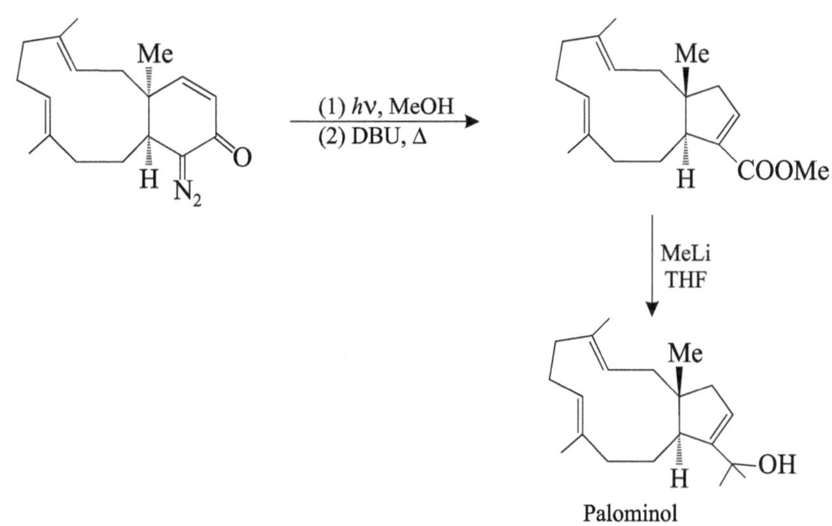

Fig. 11.31: Synthesis of Palominol.

The five-membered ring can also be formed by photoirridation of pyridinium salts. As an example (+)-**mannostatin** has been synthesised by irridation of pyridinium perchlorate. The formed amino diol was transformed into (+)-mannostatin. Various step involved are given below (Fig. 11.32)[34].

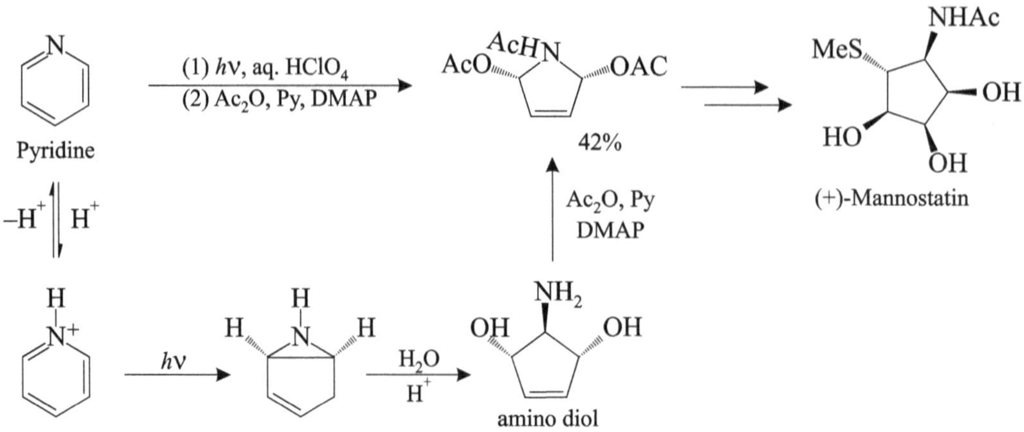

Fig. 11.32: Synthesis of (+)-Mannostatin.

11.2.18 Synthesis of Six-membered Rings

As already stated six-membered rings are formed by photochemical [4 + 2] cycloadditions (see section 11.2.6, Fig. 11.5). Six-membered rings are also formed by photochemical cyclisations based on diverse reactions. As an example photolysis of 5-chloro-N-phenyl-1H-pyrazole-4-carboxamide derivative under acetone sensitization gave pyrolozo [4, 3-c] quinalin-4-one in quantitative yield. The formed quinolin-4-ones have been studied for their antihypertensive, antitumar, antimalarial and antipyretic activity. The mechanism of the reaction is as shown below (Fig. 11.33)[35].

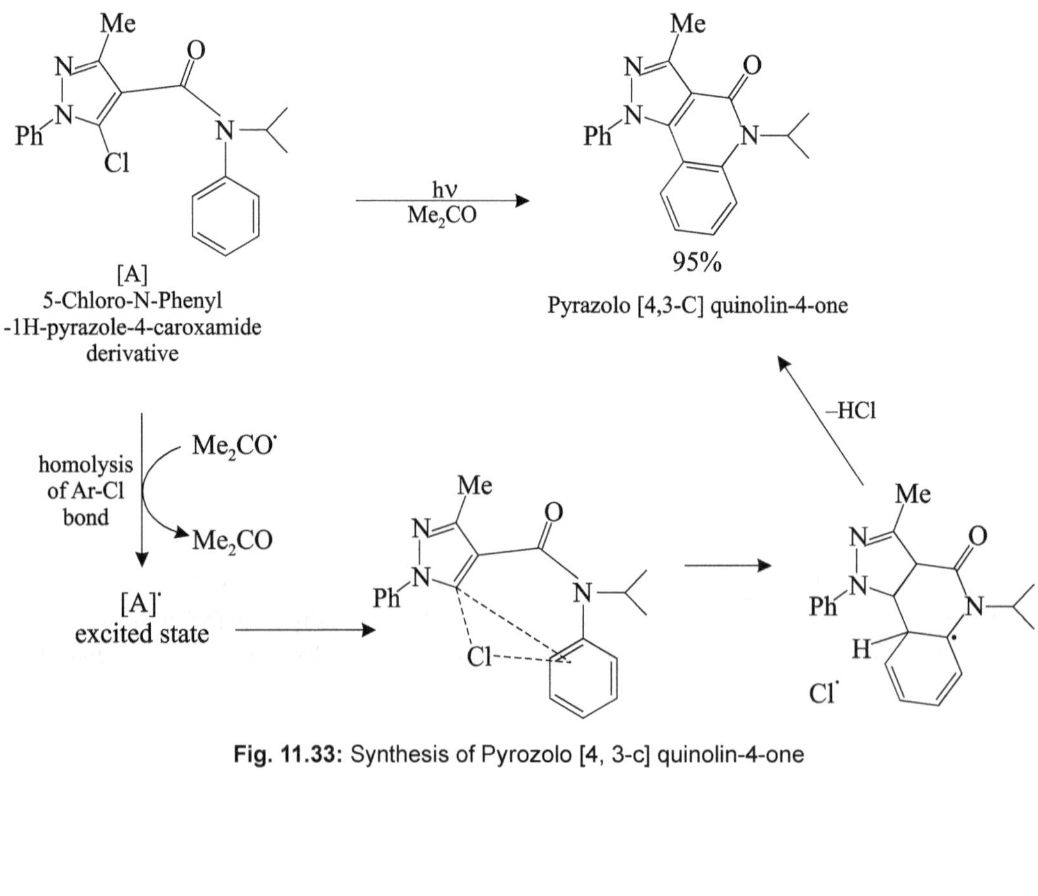

Fig. 11.33: Synthesis of Pyrozolo [4, 3-c] quinolin-4-one

Hamigeran A belongs to a class of natural compounds called Hamigerans and are isolated from some sponges in New Zealand and have a range of biological activities including antiviral and cytotoxic activity. Hamigeran A is prepared from substituted benzaldehyde. Various steps involved are shown below (Fig. 11.34)[36].

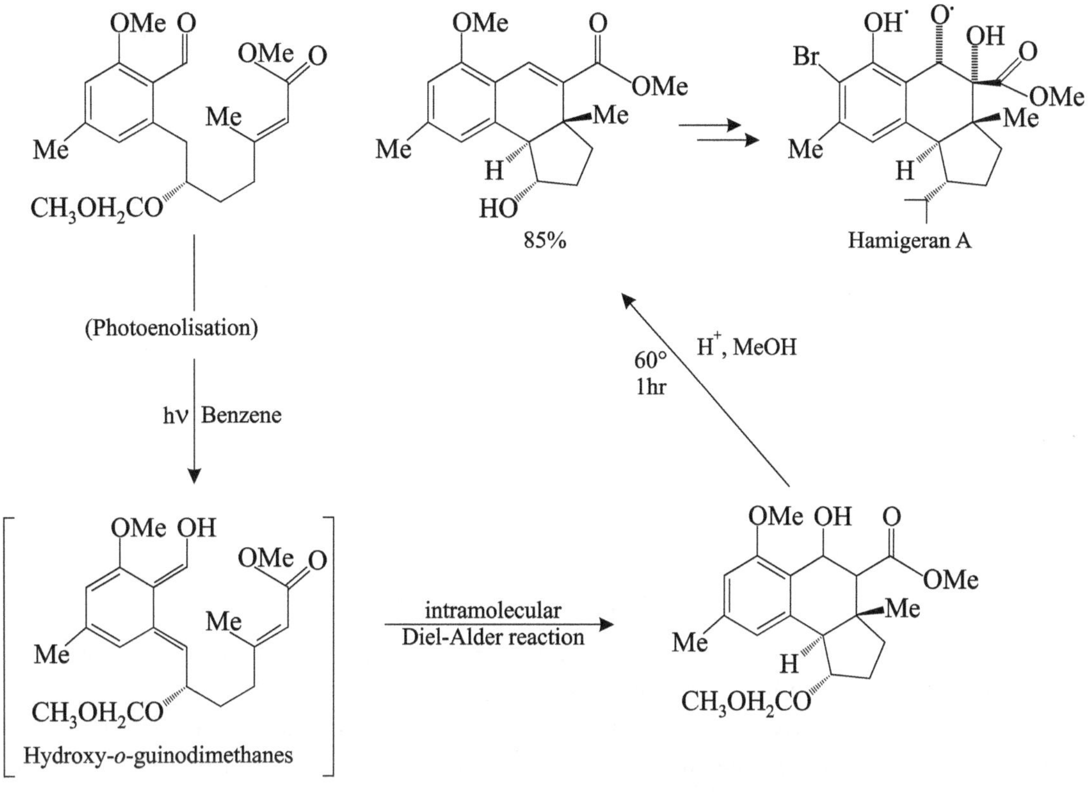

Fig. 11.34: Synthesis of Hamigeran A.

11.2.19 Synthesis of Larger Rings

N-Alkenyl-substituted maleimides in MeCN on irridation underwent a [2 + 2] intramolecular photocycloaddition followed by a rearrangement to form an azepine ring, which is considered to be a [5 + 2] phytocycloaddition. The skeleton of **cephalotoxine** a drug for myelogenous leukaemia is accessible by the above procedure (Fig. 11.35)[37]. The isomer (A) is considered a promising precursor for the synthesis of (–)-cephalotoxine.

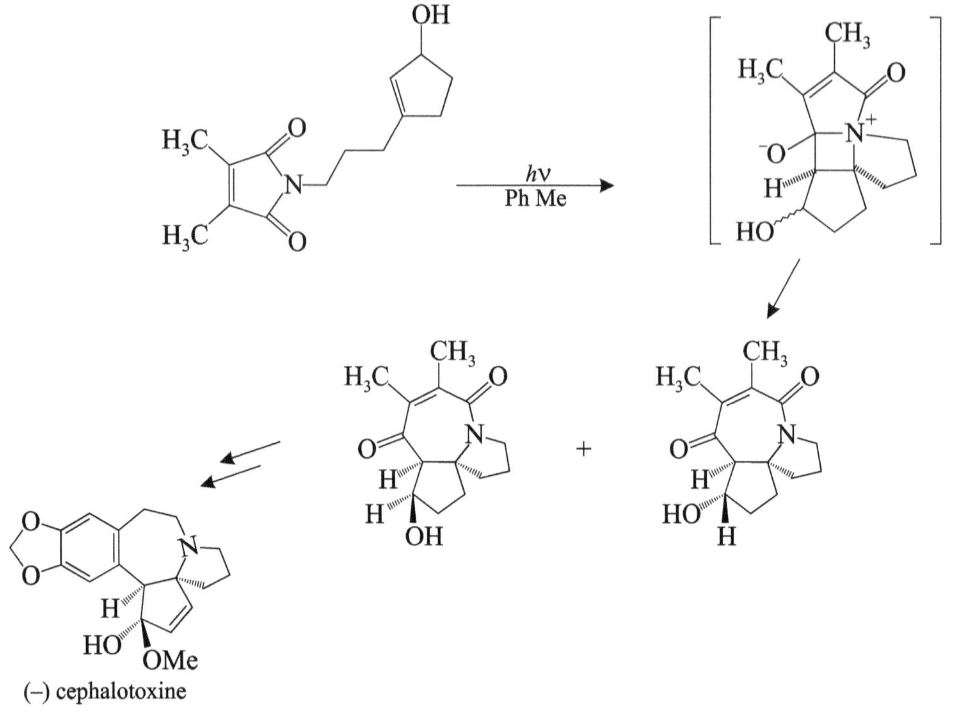

Fig. 11.35: Synthesis of (–) cephalotoxine.

An interesting photochemical reaction is the macroclisation of ω-phthalimidoalkylcarboxylate salt to give 12-membered annelated amide via elimination[38] of CO_2. The reaction involves an intramolecular photo induced electron transfer (PET) reaction where the carboxylate group is the donor and one of the group is phthalimide is the acceptor. The final product is formed by radical-radical coupling after the loss of CO_2 (Fig. 11.36).

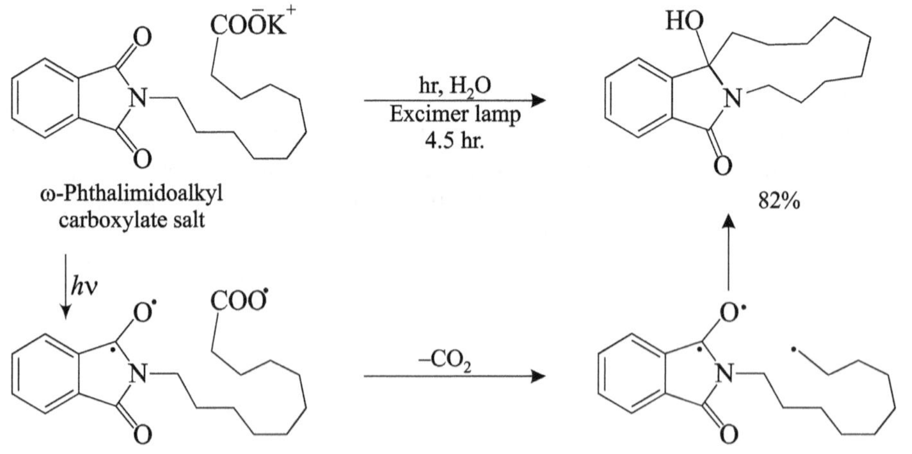

Fig. 11.36: Synthesis of 12-membered annelated amide.

11.2.20 Photooxygenation

It involves the use of photogenarated singlet oxygen for the oxidation of terpenes and conjugated dienes. Using this procedure the synthesis of anti-helmintic **ascaridol** has been achieved by photooxygenation of α-terpene. Best results have been obtained by using Rose Bengal photosensitized oxygenation of a solution of α-tepentine is methanol. The ascaridol is obtained in 85% conversion using low intensity light source (a 20 W lamp) (Fig. 11.37)[39].

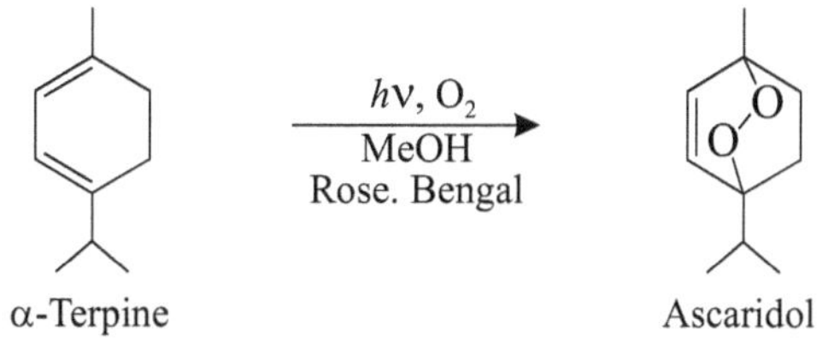

Fig. 11.37: Synthesis of ascaridol.

Using a similar procedure, an anthraquinone derivative, **Juglone** (5-hydroxy-1,4-naphthoquinone), a biologically active compound has been synthesised by dye photooxidation of commercially available 1,5-dihydroxynaphthalene via an endoperoxide. Various steps involved are given below (Fig. 11.38)[40].

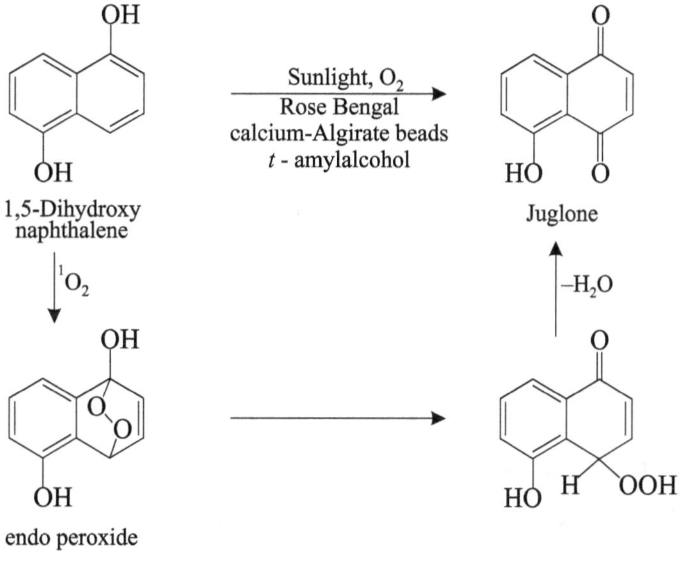

Fig. 11.38: Synthesis of Juglone.

A promising field for industrial photochemistry is the synthesis of fragrances. As an example rose oxide is synthesised by photooxygenation of citronellol using Rose Bengal as sensitizer. A mixture of two hydroperoxides is obtained, which on reduction with sulphite gives the alcohols. The main product undergoes allylic rearrangement in acid solution followed by cyclization with elimination of water to give rose oxide (Fig. 11.39)[41].

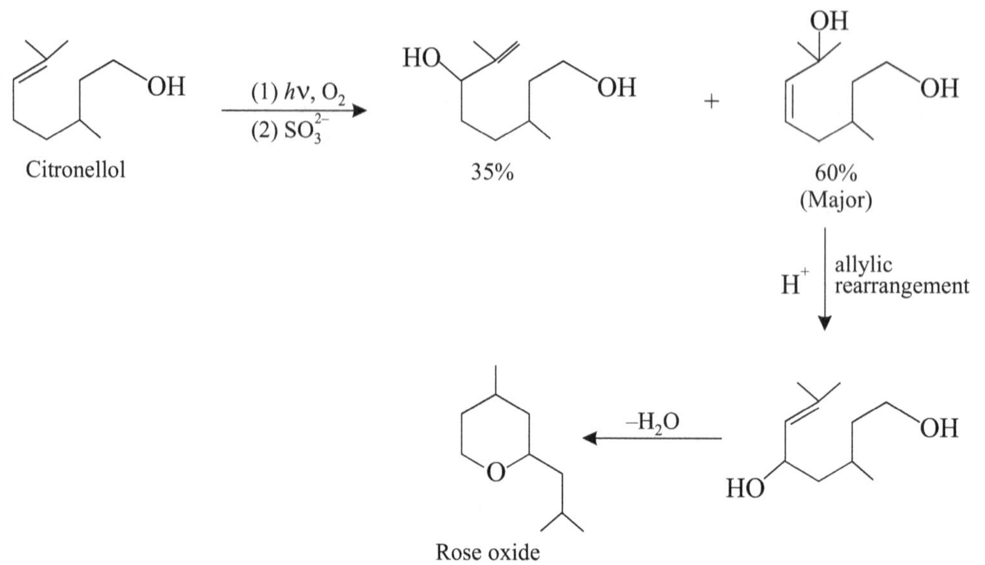

Fig. 11.39: Synthesis of Rose Oxide.

Photooxidation is also useful for the conversion of α-pinene into trans-pinocarveol (Fig. 11.39*a*).

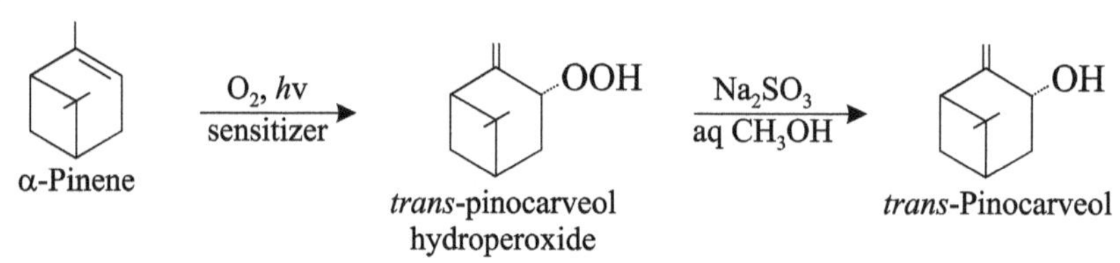

Fig. 11.39a: Conversion of α-pinene *into trans*-pinocareal.

11.2.21 Some Industrial Applications of Photochemistry

Some industrial applications of photochemistry include free-radical chlorination, sulfochlorination and nitrosation. Besides photochemical reactions have been used for the synthesis of vitamins, drugs and fragrances (For fragrances, see section 11.2.20)

11.2.21.1 Free-Radical Reactions

(a) Free-Radical Chlorination: The most of significant application of photochemical synthesis is in the area of radical chain reactions, particularly free radical chlorination. As an example chlorination in externally irriated glass tube was described by Philips Corporation[42], *viz.*, the production of monochloroalkanes from a $C_{11\text{-}14}$ *n*-paraffin cut (Fig. 11.40).

$$RH + Cl_2 \xrightarrow[<40°C]{hv} RCl + HCl$$

$$RH = \text{mixture of } n\text{-alkanes}$$

Fig. 11.40: Synthesis of monochloroalkanes.

The chloroalkanes are used for the alkylation of benzene to give alkyl benzenes, which are starting materials for the synthesis of alkylbenzene sulfones, an important class of detergents (Fig. 11.41)[43].

Fig. 11.41: Synthesis of Alkyl benzene Sulfones.

Photochlorination of benzene gives hexachlorocyclohexane. An apparatus for the preparation of hexachlorocyclohexane was developed by Ethyl corporation[44]. This method is used to produce γ-isomer of hexachlorocyclohexane, an insecticide known as Lindane or gammexane (γ-BHC), a biodegradable product. However, the yield of 15% is not satisfactory (Fig. 11.42).

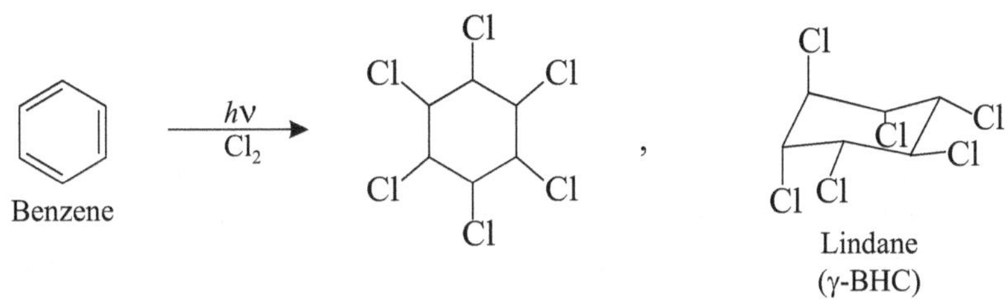

Fig. 11.42: Synthesis of γ-BHC (Lindane)

Photochlorination of toluene is known to give a mixture of benzyl chloride, benzylidene dichloride and benzotrichloride[45]. Mixture of the first two compounds *viz.* benzylchloride and benzylidine dichloride are produced via a series of four lead-lined reactors at 80-110°C[45] (Fig. 11.43).

$$\underset{\text{Toluene}}{C_6H_5-CH_3}\xrightarrow[Cl_2]{h\nu}\underset{\text{Benzyl chloride}}{C_6H_5CH_2Cl}\xrightarrow[Cl_2]{h\nu}\underset{\text{Benzylidene dichloride}}{C_6H_5-CHCl_2}\xrightarrow[Cl_2]{h\nu}\underset{\text{Benzotrichloride}}{C_6H_5-CCl_3}$$

Fig. 11.43: Photochemical chlorination of toluene.

It is well known that thermal chlorination of toluene gives mainly benzyl alcohol, a well known fragrance. It also finds use in the synthesis of drugs, disinfactants and emulsifiers[46]. On hydrolysis or on reaction with benzoic acid, benzylidene dichloride gives benzaldehyde; in the latter case an valuable intermediate benzoylchloride is also obtained (Fig. 11.44).

$$\underset{\text{Benzylidene dichloride}}{C_6H_5-CHCl_2}+\underset{\text{Benzoic acid}}{C_6H_5-CO_2H}\longrightarrow\underset{\text{Benzaldehyde}}{C_6H_5CHO}+\underset{\text{Benzoyl chloride}}{C_6H_5-COCl}+HCl$$

Fig. 11.44: Conversion of benzylidene dichloride into benzaldehyde and benzoyl chloride.

Benzotrichloride is an intermediate for dyes of the triphenylmethane xanthene and anthraquinone series and also for plant protection agents.

***(b)* Free radical sulfochlorination:** Photochemical sulfochlorination of paraffins is of industrial importance[43,47]. In sulfochlorination the light is responsible for the formation of chlorine atoms from chlorine. The sulfonyl chloride group is found to be distributed almost randomly over all C atoms of the hydrocarbon chain[48] (Fig. 11.45).

$$\underset{\text{mixture of } n\text{-alkenes}}{RH} + SO_2 + Cl_2 \xrightarrow{h\nu} RSO_2Cl + HCl$$

Fig. 11.45: Sulfochlorination of paraffins.

Sulfochlorination does not stop at the monosulfonyl chlorides. As di- and polysulfonyl chlorides have undesirable properties, the reaction is stopped at conversion of 30-50%. Yields of 80 to 90% are then obtained.

Most of the alkanesulfonyl chlorides produced industrially[49] on hydrolysis by NaOH give water-soluble alkanesulfonates which find use mainly as emulsifiers for polymerisations[43]. (Fig. 11.46)

$$\underset{\text{Alkanesulfonyl chlorides}}{RSO_2Cl} + 2NaOH \longrightarrow \underset{\text{Alkanesulfonates}}{RSO_3Na} + NaCl + H_2O$$

Fig. 11.46: Synthesis of alkanesulfonates.

Alkane sulfonyl chlorides on reaction which ammonia gives the corresponding sulfonamides, which are used as textile auxiliaries[50]. The sulfonamides on reaction with chloroacetic acid give a mixture of sulfonylamino acetic acids, which are used as emulsifiers and as anticorrosion agents for mineral oils[50,51] (Fig. 11.47).

$$\underset{\text{Alkanesulfonyl chlorides}}{RSO_2Cl} + NH_3 \longrightarrow \underset{\text{Sulfonamides}}{RSO_2NH_2} \xrightarrow[\text{NaOH}]{ClCH_2CO_2H} \underset{\text{Sulfonyl aminoacetic acids}}{RSO_2NHCH_2CO_2H}$$

Fig. 11.47: Conversion of alkane sulfonyl chlorides into sulfonamides and sulfonylamino acetic acids.

***(c)* Photochemical Sulfoxidation:** Alkane sulfonates are prepared by photochemical sulfoxidation. In this procedure, oxygen is used as an oxidizing agent instead of chlorine used earlier (Fig. 11.48).

$$\underset{C_{14-18}\text{-}n\text{-alkanes}}{RH} + SO_2 + \frac{1}{2}O_2 \xrightarrow{h\nu} RSO_3H$$

$$RSO_3H + NaOH \longrightarrow RSO_3Na + H_2O$$

Fig. 11.48: Photochemical Sulfoxidation.

Photochemical sulfoxidation has been developed as in industrial process at Hoechst[43,52,53].

***(d)* Photonitrosation:** Light induced reaction of nitrosyl chloride with cyclohexane gives cyclohexane oxime, which is used for the synthesis of **caprolactam**, the monomer of **nylon 6** (Fig. 11.49).

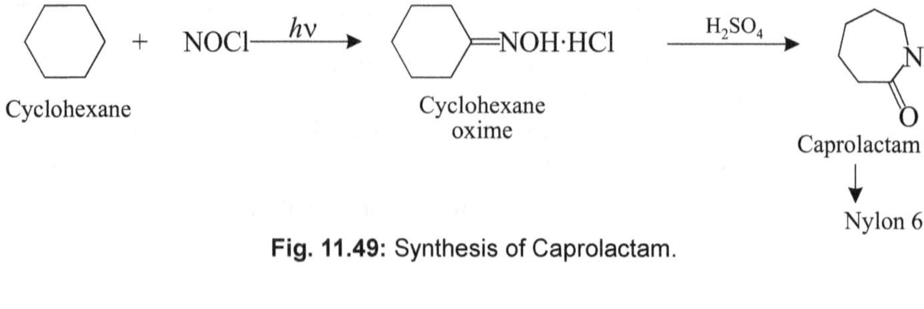

Fig. 11.49: Synthesis of Caprolactam.

Using the above methol, a Japanese chemical company Toray started large scale production[54] of caprolactum.

Another industrial application of photonitrosation is the production of **laurylllactam,** the monomer of Nylon12[55]. The starting material was cyododecane which could be prepared from butadiene in two steps. The cyclododecane was converted into the oxime by NOCl and light[56] (Fig. 11.50).

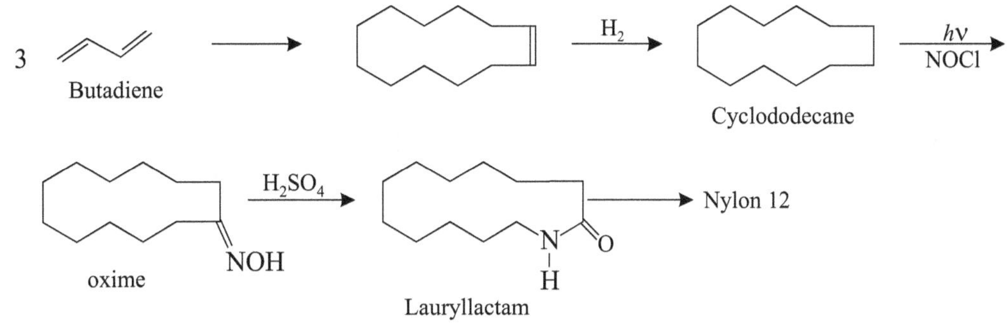

Fig. 11.50: Synthesis of Lauryllactam.

The photochemical Lauryllactam was developed by ATO in France.[57].

11.2.21.2 Photochemical Synthesis of Vitamin D and Related Compounds

(*a*) **Vitamin D_2:** Vitamin D2, also known as **ergocalciferol** was the earliest vitamin to be synthesised by photoirridation of ergosterol. Various steps involved are given below (Fig. 11.51).

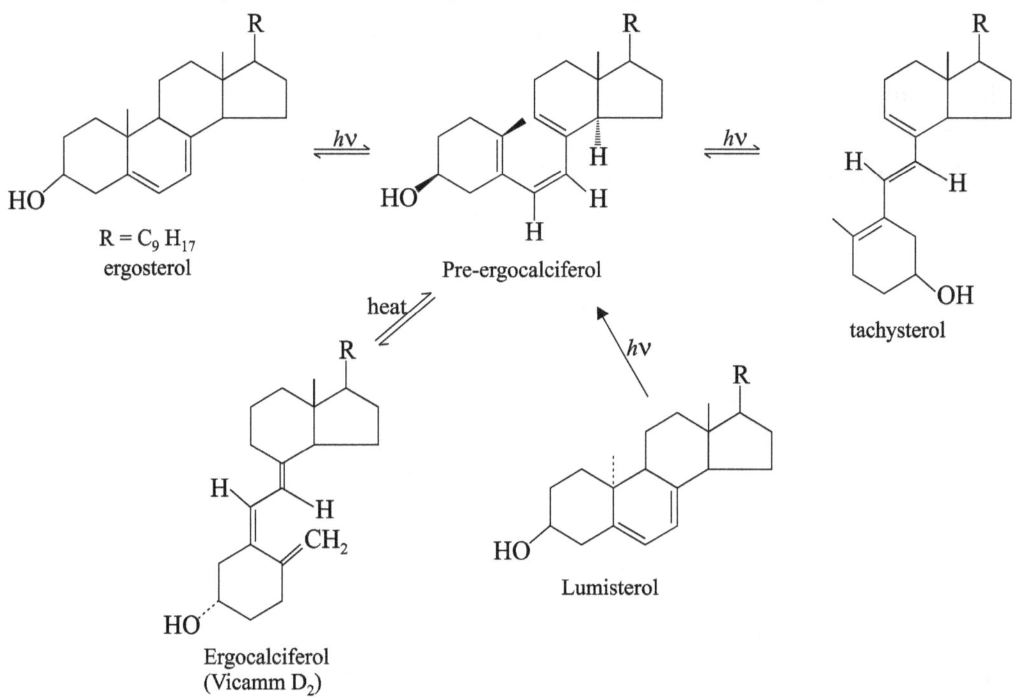

Fig. 11.51: Synthesis of Vitamin D_2 .

***(b)* Vitamin D_3:** It is obtained by photochemical synthesis of 7-dehydrocholesterol, which is obtained in four steps from cholesterol[58]. Photochemical ring opening gives previtamin D_3, which on heating gives the thermodynamically more stable Vitamin D_3 (Fig. 11.52). The photolysis of 7-dehydrocholesterol is performed in dilute solution (about 1%) using ethanol[58,59].

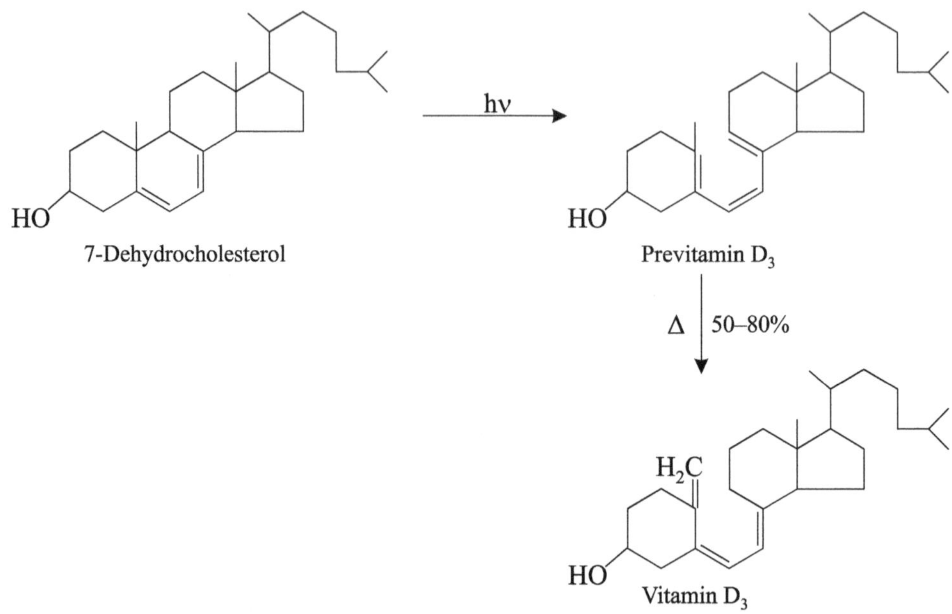

Fig. 11.52: Synthesis of Vitamin D_3.

The Vitamin D_3 is not used to prevent or cure rickets in children but is used as an additive in animal nutrition.

***(c)* Hydroxy derivatives of Vitamin D_3:** It has been established that 1α, 25-dihydrovitamin D_3 (a metabolite of vitamin D_3) formed in liver and kidney is the actual active form, which is responsible for regulating calcium metabolism.

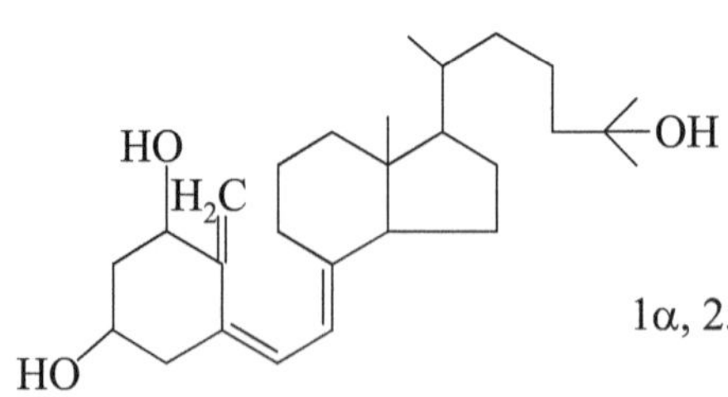

1α, 25-dihydrovitamin D3

It is found that 1α-hydroxy-vitamin D_3 is as active as 1α, 25-dihydrovitamin D_3. It is synthesised as follows: (Fig. 10.53). The intermediate (A) is obtained from cholesterol in six steps[56,60]. Its photochemical ring opening give the previtamin D_3 derivative, which on heating gives the diacetate[56,60], which on alkaline hydrolysis gives 1α-hydroxy-vitaminD_3[61].

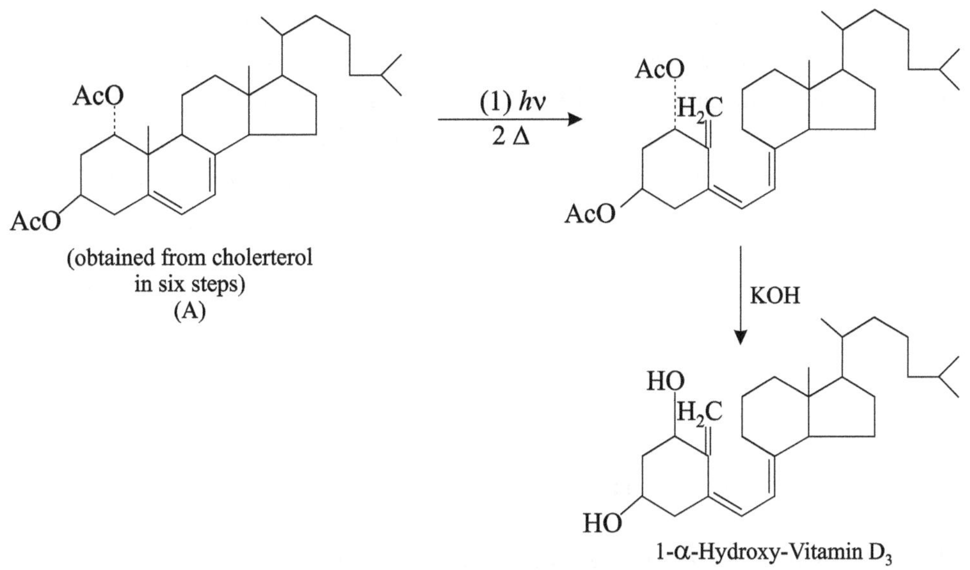

Fig. 11.53: Synthesis of 1-α-hydroxy-vitamin D_3.

(*d*) Dydrogesterone: The sex hormone, dydrogesterone is obtained by photoirridation of the pregnadiene derivative to give retrosteroid, which on cleaving with HCl in alcohol give dydrogesterone (Fig. 10.54)[62].

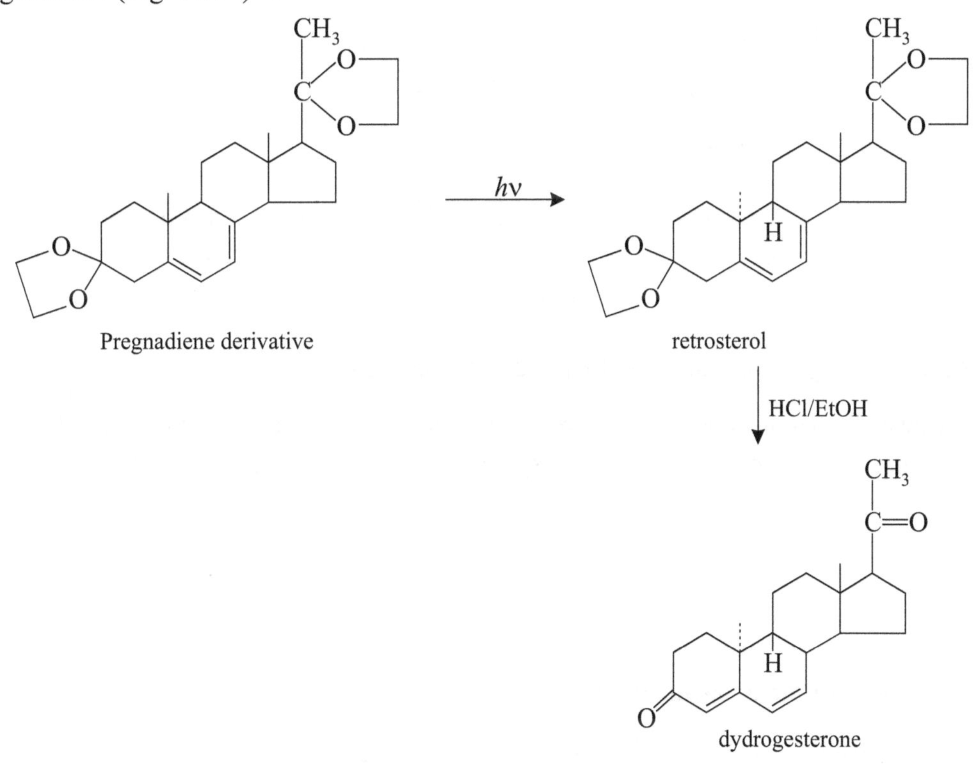

Fig. 11.54: Synthesis of dydrogesterone.

***(e)* Photoisomerisation of Vitamin A acetate:** The Wittig synthesis of vitamin A acetate performed on industrial scale as developed by BASF[63] resulted in a mixture of stereoisomers, the all-*trans* and 11-*cis* form. Only, the all trans isomer is suitable for use in pharmaceuticals and animal feeds. A photochemical method has been developed by BASF[64] for converting 11-cis into all-trans-vitamin A acetate. The method consists in irridating the stereoisomeric mixture with visible light in presence of a sensitizer like chlorophyll A or tetraphenyl porphinatozinc (Fig. 11.55).

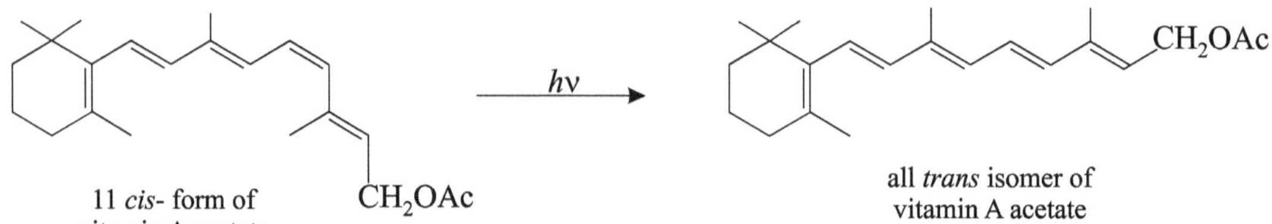

11 *cis*- form of vitamin A acetate

all *trans* isomer of vitamin A acetate

Fig. 11.55: Photoisomerisation of 11 cis form into all trans isomer of vitamin A acetate

11.2.22 Miscellaneous Photochemical Applications

11.2.22.1 Cleavage of Dithianes and Benzyl Ethers

Dithianes and benzyl ethers can be cleaved by the use of visible light in presence of a dye (Fig. 11.56)[65].

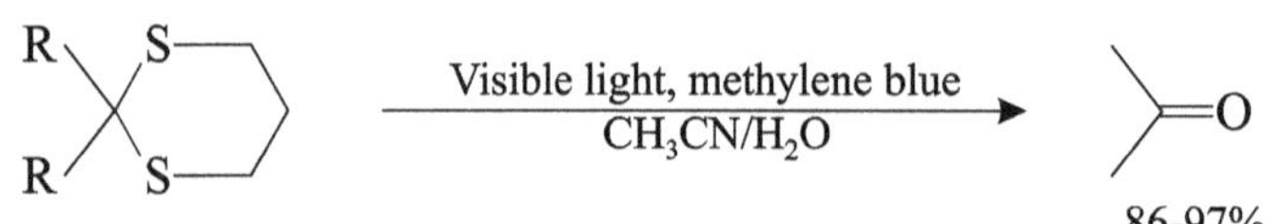

Dithianes 86-97%

$$ArCH_2OR \xrightarrow[CH_3CN/H_2O]{\text{Visible light, methylene blue}} ArCH_2OH$$

Benzyl ethers 70–100%

Fig. 11.56: Cleavage of dithianes and benzyl ethers.

11.2.22.2 Hydrolysis of Benzonitrile to Benzamide

Benzonitrile in methanolic potassium hydroxide can be hydrolysed to benzamide in 96% yield with > 420 nm light at 20° in presence of oxophosphoporphrin catalyst (Fig. 11.57)[66].

$$C_6H_5CN \xrightarrow[\text{methamolic KOH, } 20^\circ]{\text{light} > 420 \text{ nm}} C_6H_5CONH_2$$

Benzonitrile — Oxophosphoporphin catalyst — Benzamide 96%

Fig. 11.57: Hydrolysis of benzonitrile to benzamide.

11.2.22.3 Synthesis of 2-substituted pyridines

2-substituted pyridines are prepared by the reaction of acetylene with a nitrile using 400 nm light in presence of Co(I) catalyst in H_2O (Fig. 11.58)[67].

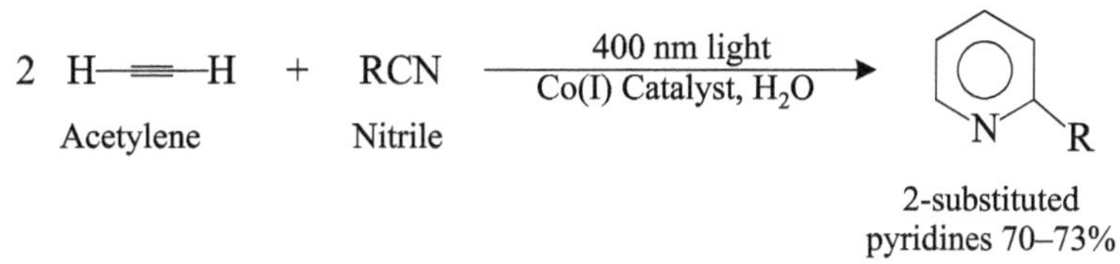

Fig. 11.58: Synthesis of 2-substituted pyridines.

11.2.22.4 Conversion of nitrobenzene into phenyl hydroxylamine

Irradiation of nitrobenzene in presence of zinc oxide in alcohol gives phenyl hydroxylamine is 73% yield (Fig. 11.59)[68].

$$\underset{\text{Nitrobenzone}}{C_6H_5NO_2} \xrightarrow[\text{ZnO}]{\text{hv}} \underset{\substack{\text{Phenyl hydroxylamine} \\ 73\%}}{C_6H_5NHOH}$$

Fig. 11.59: Conversion of nitrobenzene into phenyl hydroxylamine.

12.2.22.5 Conversion of 4-cyanophenylazide into 4-cyanoaniline

Irradiation of 4-cyanophenylazide in ethanol in presence of titanium dioxide gave 4-cyanoaniline is 90% yield (Fig. 11.60).

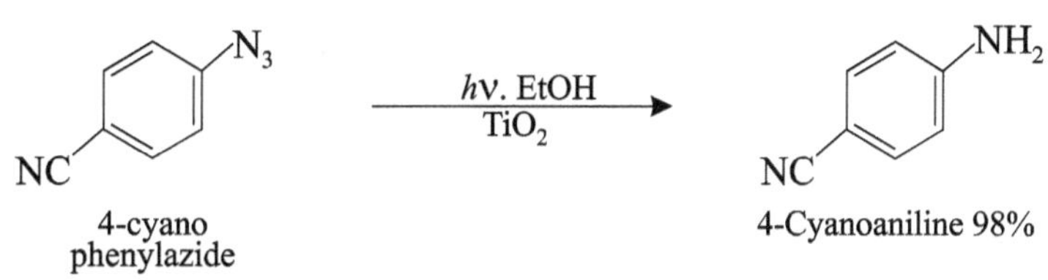

Fig. 11.60: Conversion of 4-cyanophenylazide into 4-cyanoaniline.

11.2.22.6 Addition of aldehydes to α, β-unsaturated carbonyl compounds

The addition of aldehydes to quinone in presence of light gives acylhydroguinones[69]. The method provides a viable environmentally alternative to the **Friedel-crafts acylation reaction** (Fig. 11.61), which normally requires the use of corrosive acid chlorides, use of strong Lewis acids ($AlCl_3$, $TiCl_4$ etc.) and use of hazardous solvents.

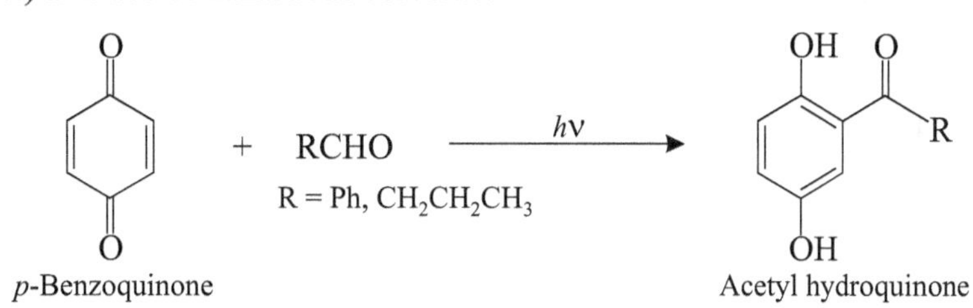

Fig. 11.61: Synthesis of acetyl hydroquinone.

The reaction is believed to take place as follows (Fig. 11.62)[70].

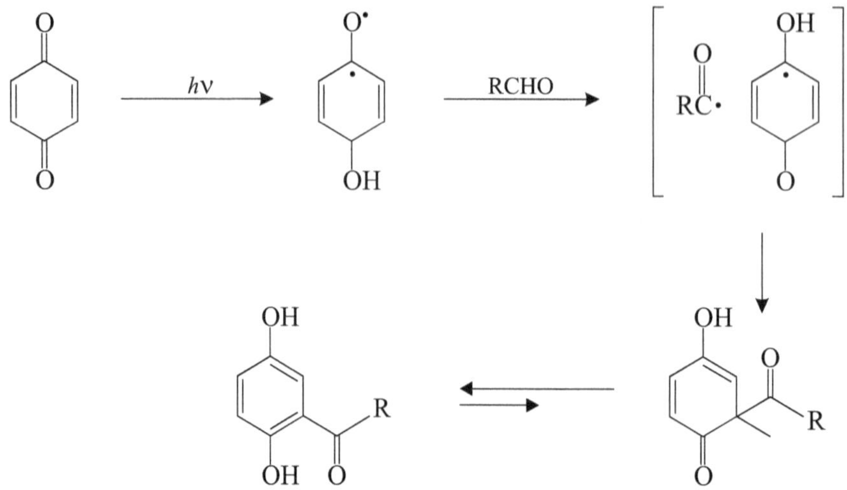

Fig. 11.62: Mechanism of the formation of acetyl hydroquinones.

In cases, where the yields are poor, addition of benzophenone (catalytic quantity) improves the yield[71].

The above process has been improved by eliminating benzene as solvent and replacing with supercritical carbon dioxide[72].

11.2.22.7 Synthesis of 1,4-naphthoquinone Photomer

1, 4-Naphthoquinone on irridation in benzene solution with UV light gives 1,4-naphthoquinone dimer (Fig. 11.63)

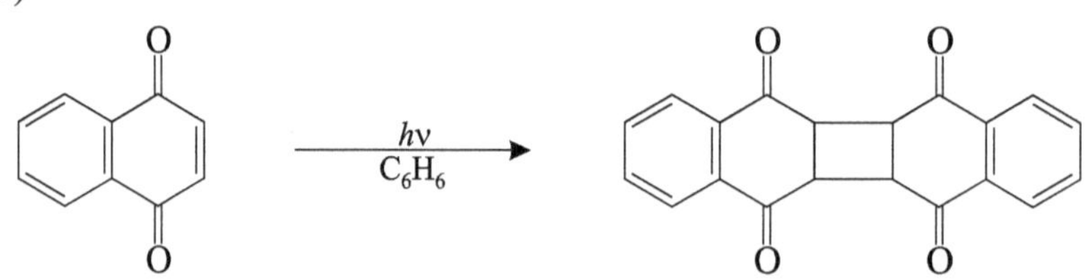

1,4-Naphthoquinone 1,4-Naphthoquinone dimer

Fig. 11.63: Synthesis of 1,4-naphthoquinone dimer

12.2.22.8 9-Phenyl Phenanthrene

Triphenyl ethylene in cyclohexane on irridation in presence of iodine gives 9-phenyl phenanthrene (Fig. 11.64).

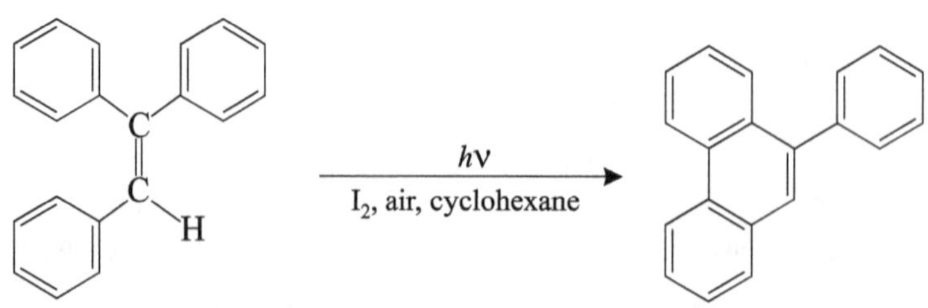

Triphenyl ethylene 9-Phenyl phenanthrene

Fig. 11.64: Synthesis of 9-Phenyl phenanthrene.

12.2.22.9 Photochemical Arndt-Eistert Synthesis

Photochemical reaction of α-naphthoyl chloride with diazomethane in presence of methanol gives α-naphthyl acetate (11.65)[73].

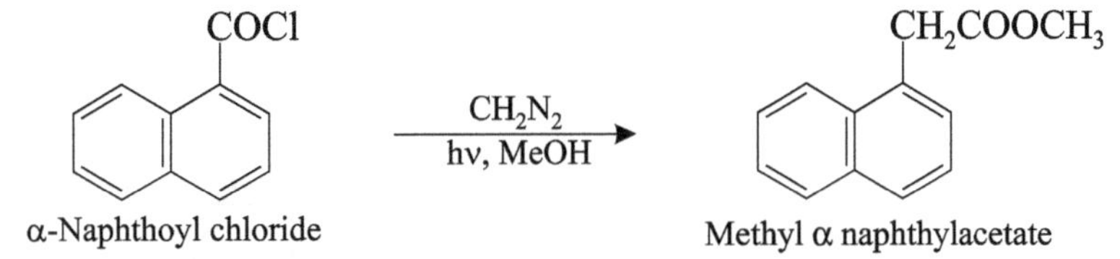

Fig. 11.65: Photochemical Arndt-Eistert Reaction.

12.2.22.10 Barton Reaction

In Barton reaction[74], a methyl group in the δ-position to an OH group is converted into an oxime group. The reaction provides an excellent synthetic route for the conversion of 18-methyl group to 18-substituted steroids. As an example β-acetoxy-5α-preganan-20β-ol is converted into the oxime derivative (Fig. 11.66).

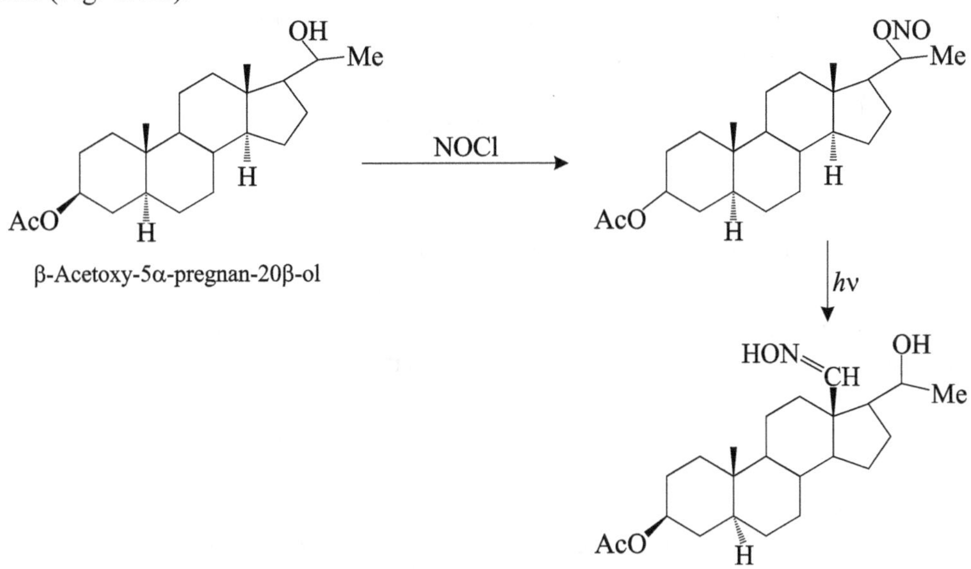

Fig. 11.66: Barton Reaction.

The mechanism of Barton reaction is as given below[75] (Fig. 11.67).

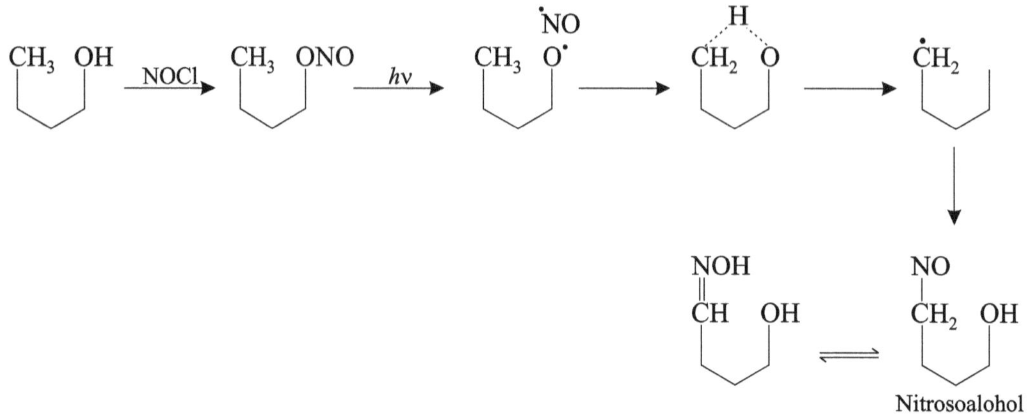

Fig. 11.67: Mechanism of Barton reaction.

Conclusion

Photochemistry is believed to be a key approach to Green Chemistry. The transformations can be carried out under mild conditions. No byproducts are formed. It is also useful for a large number of organic transformations including industrial scale preparations.

REFERENCES

1. D. Dondi, S.Protti, A. Albini *et al.*, Green Chem., 2009, **10**, 1653.
2. R.R. Hautala and R.L. Letsinger, J. Org. Chem., 1991, **36**, 3762.
3. J.H. Liu and R.G. Weiss, J. Org. Chem., 1985, **50**, 3655.
4. K. Nakatani, J. Shirai, R. Tomaki and I. Saito, Tetrahedron Lett., 1995, **36**, 5363.
5. Y. Ito, T. Kajita, K. Kunimoto and T. Matsuura, J. Org. Chem., 1989, **54**, 587.
6. K. Muthuramu and V. Ramamurthy, J. Org.Chem., 1982, **47**, 3976.
7. V. Ramamurthy, Ed. Photochemistry of Organized and Constraint Media, VCH, Weinheimn, Germany, 1991; CRC Handbook of Organic Photochemistry and Photobiology, W.H. Horspool (Ed.), CRC Press, Boca Raton, Fl., 1995.
8. Y. Ito, B. Borecka, J. Trotter and J.R. Scheffer, Tetrahedron Lett., 1995, **36**, 6083; Y. Ito, B. Borecka, G.Olovasson, J. Trotter and J.R. Scheffer, Tetrahedron Lett., 1995, **36**, 6087; Y. Ito and B. Olovassan, J. Chem. Soc; Perkin Trans., I, 1997, 127.
9. S. Protti, M. Fagnoni and A. Albini, J. Org. Chem., 2005, **3**, 2868.
10. U.S. Patent 20010018533 (A1).
11. A.G. Magnone, European Patent EP-0850925 (A1).
12. R.Reintjens and A. Puhl, International Patent CN-101663269 (A).
13. S. Protti, M. Fagnoni and A. Albeni, Org. Biorg. Chem., 2005, **3**, 2868.
14. A. Fraboni, M. Fagnoni and A. Albini, J. Org. Chem., 2003, **68**, 4886; V. Dichiarante, M. Fagnoni and A. Albini, J. Org. Chem., 2010, **75**, 1271.
15. W.A. Donaldson, Tetrahedron, 2001, **57**, 8589.
16. D.V. Armesto, M.R. Ortiz, A.R. Agarrabeita and N. El-Boulifi, Tetrahedron, 2010, **66**, 8690.

17. D.V. Armesto, G.M. Gomez, W.M. Horspool and T.F. Bermejo, International Patent WO9216499 (A1).
18. G. Mehta and A. Srikrishna, Chem. Rev., 2004, **97**, 671.
19. P. Wessig and O. Mühling, Angew.Chem. Int. Ed., 2005, **44**, 6778.
20. D.-S. Hsu, Y.-y. Chou, Y.-S. Tung and C-C. Liao, Chem. Eur, J., 2010, **16**, 3121.
21. J. Iriondo-Alberdi and M.F. Greaney, Eur J. Org. Chem., 2007, 4801 and references cited therein.
22. O. Roy, S. Faure and D.J. Aitken, Tetrahedron Lett., 2006, **47**, 5981.
23. M. Inoue, T. Sato and M. Hirama, J. Am. Chem. Soc., 2003, **125,** 10772.
24. R. Alibes', P. de March, M. Figueredo *et al.*, Org. Lett., 2004, **6**, 1449.
25. C. Du, L. Li, Y. Li and Z. Xie, Angew. Chem. Int. Ed., 2009, **48**, 7853.
26. P. Selig, E. Herdtweck and T. Bach, Chem. Eur. J., 2009, **15**, 3509.
27. P. Esser, B.Pohlmann and H.-D. Schart, Angew.Chem. Int. Ed., 1994, **33**, 2009.
28. M. Fleck and T. back, Angew.Chem.Int. Ed., 2008, **47**, 6189.
29. M.T. Crimmins, J.M. Pace, P.G. Nantermet *et al.*, J. Am. Chem. Soc., 2002, **122**, 8453.
30. J. Irondo-Alberti, J. Perea-Buscela and M. Greaney, Org. Lett., 2005, **7**, 3969.
31. T.E. Adams, M. El. Sous, B.C. Hawkins et al, J. Am. Chem. Soc. 2009, **131**, 1607.
32. D. Ng, Z. Yang and M.A. Garcia-Garibay, Org. Lett., 2004, **6**, 645.
33. S.A. Snyder and E.J. Cocey, J. Am. Chem. Soc., 2006, **128**, 740.
34. R. Ling and P.S. Mariano, J. Org. Chem., 1998, **63**, 6072.
35. S. Lu, W. Zhang, J. Pan and J. Zhang, Synthesis, 2008, 1517.
36. K.C. Nicolaou, D. Gray and J. Tae, J. Am. Chem. Soc., 2004, **126**, 613; K.C. Nicolaou, D.L.E. Gray and J. Tae, Angew. Chem. Int. Ed., 2001, **40**, 3675.
37. K.I. Booker-Milburn, L.F. Dudin, C.E. Anson and S.D. Guile, Org. Lett., 2001, **3**, 3005.
38. A.G. Griesbeck, N. Hoffmann and K.D. Warzecha, Acc. Chem. Res., 2007, **40**, 128.
39. R.C.R. Wootton, R. Fortt and A.J. de Mellow, Org. Process Res. Dev., 2002, **6**, 187.
40. D.T. Santos, J.Q. Albarelli. K. Joyce and M. Oelgemoller, J. Chem. Tech. Biotech., 2009, **84**, 1026.
41. O. Ohloff, E. Klein and G.O. Schenck, Angew. Chem., 1961, **73**, 578.
42. T. Hutson, Jr. and R.S. Logan, Chem. Eng. Prog. 1972, **68(5),** 76.
43. F. Broich, Fette, Seifen, Anstrichm, 1970, **72**, 17.
44. L.J. Governale and J.T. Clarke, Chem. Eng. Prog., 1956, **52(7)**, 281.
45. H.G. Haring and H.W. Knol, Chem. Process Eng., 1964, **45**, 560, 619, 690.
46. K.A. Lipper in Ullmanns Encyclopadie der Technischen Chemie 4th Edn. Verlagchemie, Weinheim, 1975, Vol. 9, P. 525.
47. H.I. Joschek, Chem. Ztg., 1969, **93**, 655.
48. F. Asinger, Ber. Dtsch. Chem. Ges., 1944, **77**, 191.
49. K. Lindnr: Tenside, Textilhilfsmittel, Waschrohstoffe. Wissenschaftl, Verlagsges, Stuttgart, 1964, Vol. 1, P. 705.
50. Reference 49, P. 717.
51. F. Asinger: Die Petrochemische Industrie, Akademic-verlag, Borlin, 1971, P. 731.
52. L. Orthner, Angew. Chem., 1950, **62**, 302; R. Graf, Justus Liebigs Ann. Chem., 1452, **578**, 50.

53. C. Beermann, Eur. Chem. News, Normal Paraffins Supplement, Dec. 2, 1966, P. 36; H. Hartig, Chem.-ztg, 1975, **99**, 179.
54. P. Turner, Int. Chem. 1970, **9(5,6)**, 51; P. Huime, and P.E. Turner, Chem. Process Eng., 1967, **11,** 96; Y. Ito and Y. Hara, DAS 1468737 (1962). Toray.
55. Inf. Chim, 1970, **8(3,4)**, 47.
56. D.H.R. Barton, R.H. Hesse, M.M. Pecher; and E. Rizzardo, J. Am. Chem. Soc., 1973, **95**, 2748; N. Müller, DOS 2400189, BASF Ag.
57. Chem. Ind. (Düsseldorf', 1975, **27**, 735.
58. J. Fragner: Vitamine. Gustav Fisher Verlag, Jena 1964, Vol. I, P. 613.
59. S.B. Greenbaum in Kirk-Othmer: Enclopedia of Chemical Technology, 2nd Ed. Interscience, New York 1970, Vol. 21, P. 549.
60. A. Fürst, L. Labler, W. Meier and K.-A. Pfoetner, Helv. Chim. Acta, 1973, **56**, 1708.
61. Eur. Chem. News, August 22/29, 1975.
62. O.A. deBruin, H.F.L. Schöler and J.N. Walop, Philips Tech. Rundsch. 1967, **28(3,4)**, 72; M.P. Rappoldt and T.R. Rix, Rec. Trav. Chim. Pays-Bos, 1971; **90**, 27.
63. W.Reif, H. Grossner, Chem. Ing. Tech., 1973, **45**, 646; H. Pommer, Angew Chem., 1977, **89**, 437; Angew.Chem. Int. Ed. Engl., 1977**, 16**, 423.
64. M. Fischer, W.-W. Wiersdorff, D. Nürrenbach, D. Horn and F. Feichtmayr, DBP 2210800, BASF AG.
65. G.A. Epling and Q. Wang In: P.T. Anastas, and C.T. Farrts, Benign by Design, A.C.S. Symp. 577, Washingon, DC, 1994; 64.
66. T. Igarashi, K. Konishi and T. Aida, Chem. Lett., 1998, 1039.
67. B. Heller and G. Oehme, J. Chem. Soc. Chem. Commun, 1995, 179.
68. R. Bao, W.S. Dodson, T.C. Bruton and Y.Li, Preprints Paper Natl Meet ACS Div. Environ. Chem., 1994, 34(2), 411.
69. G.A. Kraus and M. Kirihara, J. Org. Chem., 1992, **57**, 3256.
70. G.A. Kraus, M. Kirihara and Y. Wu. In Benign by Design: Alternative Synthetic Design for Pollution Prevention; P.T. Anatas and C.A. Farris, Eds., ACS Symposium Series 577; American Chemical Society, Washington, D.C.,1994, pp. 76-83.
71. G.A. Kraus and P. Lius, Tetrahedron Lett., 1994, **35**, 7723.
72. R. Pauct, M.L. Grimm, G.A. Kraus and J.M. Tanko, Tetrahedron Lett., 2001, **42**, 1415.
73. A.B. Smith, Chem. Commun., 1974, 695.
74. D.H.R. Barton, J.M. Beaton, L.E. Geller and M.M. Pechet, J. Am. Chem. Soc., 1960, **82**, 2640; 1961, **83**, 4076.
75. D.H.R. Barton *et al*, J. Chem. Soc. Perkin trans I, 1979, 1159.

12. ORGANIC TRANSFORMATIONS USING PHASE TRANSFER CATALYSTS

12

Organic Transformations using Phase Transfer Catalysts

12.1 INTRODUCTION

In organic transformations difficulties are normally encountered in case the organic compound is soluble in an organic solvent and the reagent in water. In such cases, the organic compound and the reagent will react very slowly. The result is negligible reaction. The problem can be solved by using a catalyst which is soluble in water as well as organic solvents. Two types of catalysts are generally available. These are phase transfer catalysts and **crown ethers.**

The **phase transfer catalysts** (PTC) are ionic substances, usually quarternary ammonium salts, where the size of the hydrocarbon group in the cation in large enough to confer good solubility of the salt in organic solvents. In fact, the PTC reaction describes a methodology for accelerating reaction between water insoluble organic compounds and water soluble reagents. The basic function of the catalyst (PTC) is to transfer the anion from aqueous phase to organic phase. As an example heating and stirring a mixture of 1-chlorooctane with sodium cyanide for several days give practically no yield of 1-cycanooctane. However, if a small quantity of PTC is added, the product is formed in 2 hrs. (Fig. 12.1)[1].

$$\underset{\text{1-Chlorooctane}}{CH_3(CH_2)_6CH_2Cl} + NaCN + H_2O \xrightarrow[\text{PTC}]{\text{decane}} \underset{\substack{95\% \\ \text{1-Cyano octane}}}{CH_3(CH_2)_6CH_2CN}$$

$$(CH_3CH_2)_{15}P^+(n\text{–Bu})_3 \quad \Delta,\ 105°,\ 2\text{ hr}$$

Fig. 12.1: Reaction of 1-chlorooctane with NaCN in presence of a PTC.

In the presence of PTC (Q^+CN^-), following three equilibrium are set up (Fig. 12.2).

Organic phase: $$\overset{+}{Q}\ \overset{-}{CN} + R'Cl \xrightarrow{4} R'CN + \overset{+}{Q}\overset{-}{Cl}$$

$1 \Updownarrow$ $\qquad\qquad\qquad\qquad$ $\Updownarrow 2$

Aqueous phase: $$\overset{+}{Q}\ \overset{-}{CN} + \overset{+}{Na}\overset{-}{Cl} \overset{3}{\rightleftharpoons} \overset{+}{Na}\overset{-}{CN} + \overset{+}{Q}\overset{-}{Cl}$$

$$[Q = R_4N^+ \text{ or } R_4P^+]$$

Fig. 12.2: Mechanism of PTC Reaction

The Na^+ ions remain in aqueous phase since they cannot pass into organic phase. The Q^+ ions do cross the interface and carry the anion with them. In the beginning of the reaction CN^- is the chief anion present in aqueous phase. This gets carried into organic phase as Q^+CN^- (equilibrium 1), where it reacts with R′Cl to form R′CN and Cl^-. The Cl^- (as Q^+Cl^-) gets carried into the aqueous phase (equilibrium 2). The equilibrium takes place entirely in the aqueous phase; it alloys Q^+CN^- to be regenerated. All the three equilibriums (1, 2 and 3) are normally reached much faster than the actual reaction 4 and so the later is the rate determining step.

As already stated phase transfer catalysts are quaternary 'onium' salts such as ammonium, phosphonium and tertiary sulfonium salts. However, in practice, only a limited number of ammonium and phosphonium salts are used. These include:

- Aliquat 336: $N^+CH_3(C_8H_{17})_3\ Cl^-$
  Methyl trioctylammonium chloride
- Benzyl trimethylammonium chloride or bromide
  $N^+ (CH_3)_3\ CH_2C_6H_5X^-$ (X = Cl or Br)
- Benzyltriethylammonium chloride or bromide
  $^+N\ (C_2H_5)_3\ CH_2C_6H_5X^-$ (X = Cl or Br)
- Tetra-*n*-butyl ammonium chloride, bromide, chlorate or hydroxide
  $N^+\ (n\text{-Bu})_4\ X^-$ (X = Cl, Br, ClO_4, OH)
- Cetyl trimethylammonium chloride or bromide
  $N^+\ (CH_3)_3(CH_2)_{15}\ CH_3\ X^-$ (X = Cl or Br)
- Benzyl tributyl ammonium chloride
  $C_6H_5CH_2\ (n\text{-}C_4H_9)_3\ N^+\ Cl^-$
- Benzyl triphenyl phosphonium iodide
  $C_6H_5\ CH_2(C_6H_5)_3\ P^+I^-$

The PTC reactions have considerable advantages over conventional procedures. The PTC reactions:

- Do not require vigorous conditions and the reactions are much fast.
- There is no need to use expensive aprotic solvents.
- The reactions occur at low temperatures.
- There is no need to use anhydrous conditions as water is used as one of the phases.
- PTC makes available anions in organic phase resulting in increased nucleophilicity.
- A number of reactions which require strong bases like alkoxide, sodamide, sodium hydride etc. can proceed even in presence of OH^- as it becomes strongly nucleophilic in presence of PTC.

As already stated the **crown ethers** can also act as phase transfer catalysts. The crown ethers are a family of cyclic polyethers and are large ring compounds containing several oxygen atoms. Some typical examples of crown ethers include the following:

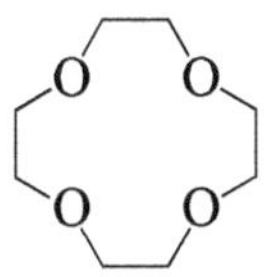

[12]-crown-4

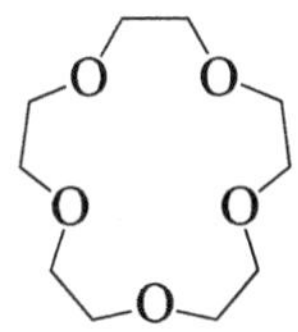

[15]-crown-5

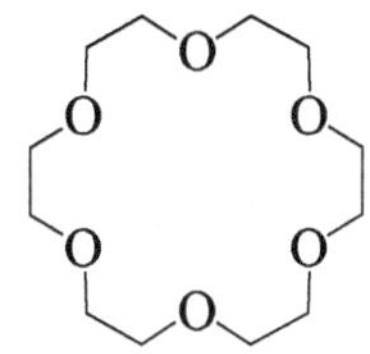

[18]-crown-6

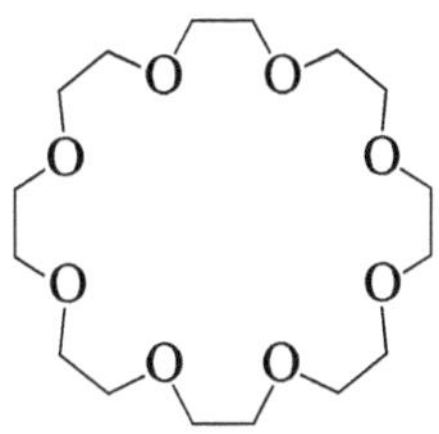

[24]-crown-8

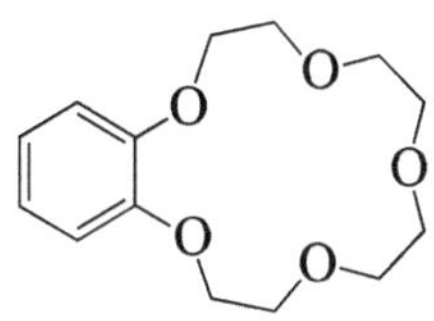

Benzo-[15] crown-5

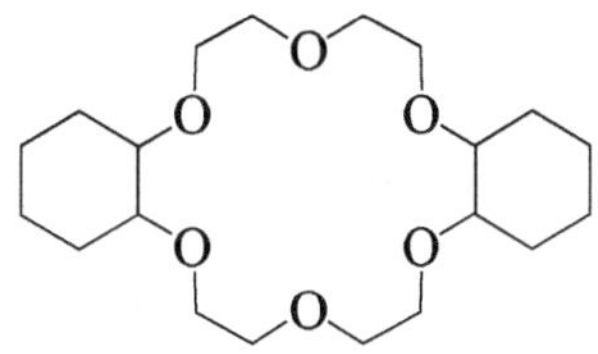

Dicyclohexyl-[18]–crown-6

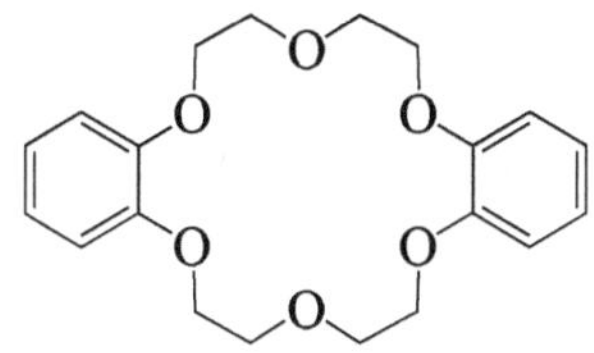

Dibenzo-[18]-crown-6

It is found that crown ethers, *viz.*, [18]-crown-6, NaOH dissolves more than expected amount and that it forms a stable alkali and alkaline earth metal cation complexes which are soluble in organic solvents. In fact, the alkali metal complexation is one of the crown ethers most important property.

The [18]-crown-6 molecule is shaped like a daughnut and has a cavity in the middle. The cavity of [18]-crown-6 has a diameter of 2.7Å. In this cavity, K^+ ions, having a diameter 2.66Å just fits into the cavity of crown, where it is held by unshared pair of electrons on six oxygen atoms. Thus, the crown ether holds K^+ by some ion-dipole forces, which are much stronger.

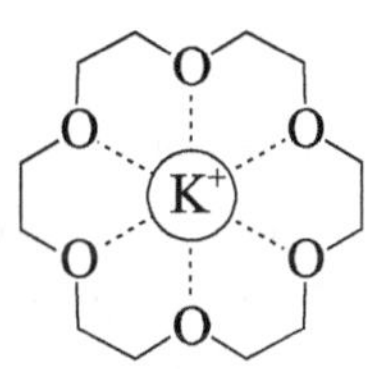

It is found that a particular crown ether is more specific for a particular metal ion. This is well understood by the data given in the following Table 12.1:

Table 12.1

Cation	Cation diameter (Å)	Crown ether	Cavity diameter [Å]
Li^+	1.36	[12] C–4	1.2–1.5
Na^+	1.90	[15] C–5	1.7–2.2
K^+	2.66	[18] C–6	2.6–3.2
Cs^+	3.38	[21] C–7	3.4–4.3

The croun ethers due to their behaviour of complexation through metal ions, act as powerful phase transfer catalyst. Thus, they are used to transfer ionic compounds into organic phase from aqueous phase.

Unlike quaternary ions like PTC, the crown ethers are neutral molecules and the cation is held together by ion-dipole bond instead of covalent bonds as in quaternary ions.

12.2 APPLICATIONS OF PHASE TRANSFER CATALYSIS IN ORGANIC SYNTHESIS

Following are given some of the important applications of Phase Transfer Catalysts.

12.2.1 Synthesis of Nitriles

As already state nitriles can be obtained from alkyl chloride in decane by heating with aqueous NaCN in presence of a PTC (see Fig. 12.1)

Carboxylic acid chlorides on reaction with aqueous NaCN in presence of $Bu_4N^+X^-$ gives[2] the corresponding benzoyl cyanide (Fig. 12.3).

Fig. 12.3: Synthesis of benzoyl cyanides from benzoyl chlorides.

In place of PTC, crown ethers can also be used. Thus, using crown ethers, benzyl chloride, *p*-nitrobenzyl chloride, *p*-chloro benzyl chloride and 3, 4-dimethoxy benzyl chloride can be converted into the corresponding nitriles in 85-90% yield.[2a] This method has also been used for the conversion of trimethyl silyl chloride which is hydrolytically sensitive, into the corresponding cyanide (Fig. 12.3*a*)

$$(CH_3)_3\ SiCl \xrightarrow[\substack{CH_2Cl_2 \\ 40,\ 24\ hr.}]{\substack{KCN \\ [18]crown\ 6}} \underset{45\%}{(CH_3)_3\ Si\ CN}$$

Fig. 12.3a: Synthesis of trimethylsilyl cyanide.

Crown ethers have also been used for the nucleophilic displacement of chloride by cyanide at hindered position. Thus the reaction[2b] of 2-chloro-2-methylcyclohexanone with KCN in CH_3CN in presence of [18]-crown-6 gives excellent yield of the cyanide at room temperature. However, at higher temperature **Favorskii rearrangement** takes place giving 5 membered compound (Fig. 12.3*b*)

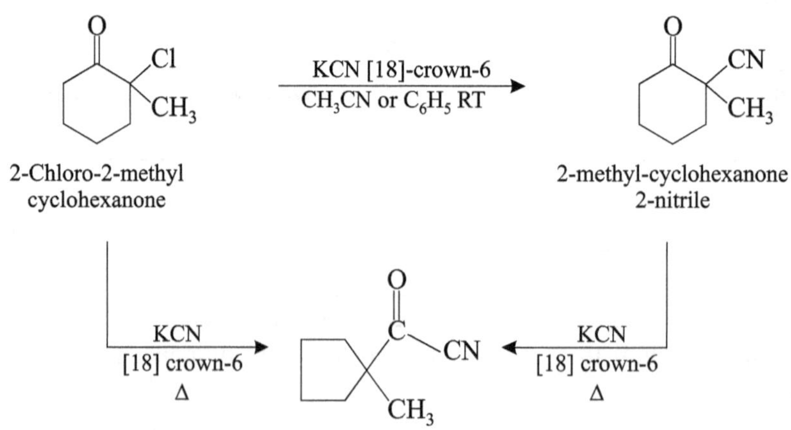

Fig. 12.3b: Nucleophilic displacement of chloride by cyanide at hindered position.

Similar reaction[2b] of 2-chloro-4-methylcyclohexanone gives the corresponding cyanide. Even sulfonyl chlorides RSO_2Cl are readily converted[2C] to the corresponding fluorides by [18]-crown 6 in CH_3CN at 25°C. Yield 60-96%.

Nitriles can also be prepared by the reaction amides[3], thioamides[3], aldoximes[4] and amidines by reaction with chloroform and aqueous NaOH in presence of a PTC, $C_6H_5\,CH_2N^+Et_3Cl^-$. The reaction takes place via the formational of dichlorocarbene (Fig. 12.4).

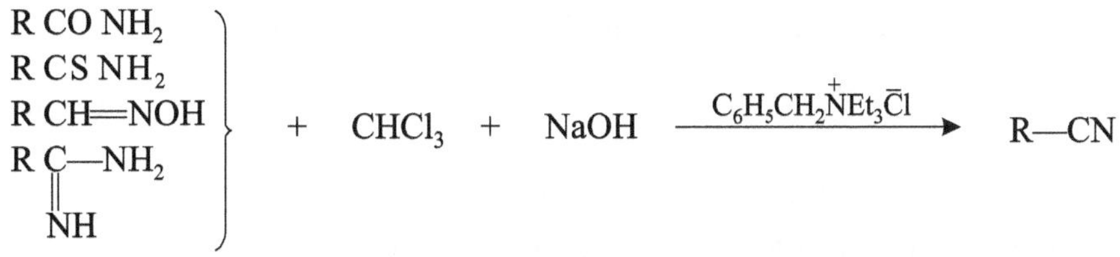

Fig. 12.4: A convenient synthesis of nitriles.

Using the above method (Fig. 12.4) benzonitrile is prepared[3] from benzamide. The formation of nitriles from amides is shown below (Fig. 12.5).

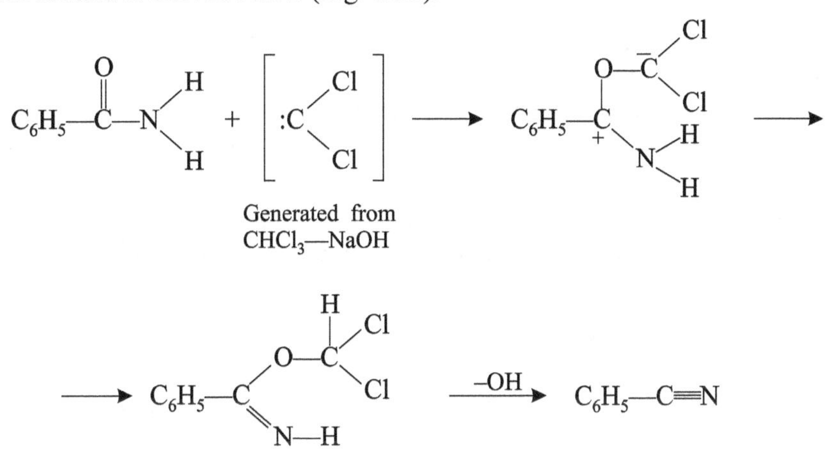

Fig. 12.5: Mechanism of the formation of nitrile from amide.

For the mechanism of the formation of dichlorocarbene from $CHCl_3$/NaOH see section 12.2.11.

12.2.2 Synthesis of Isonitriles

Isonitriles are prepared by **phase transfer Hofmann carbylamine reaction.** The method consists in the reaction of primary amines with dichlorocarbene-generated by PTC method (Fig. 12.6)[5].

$$RNH_2 + CHCl_3 + \underset{\text{aq.}}{NaOH} \xrightarrow{C_6H_5-CH_2N^+Et_3Cl^-} \underset{40\text{-}60\%}{R-N\equiv C}$$

Fig. 12.6: Synthesis of isonitriles.

12.2.3 Synthesis of Alkyl Flruorides

Alkyl fluorides can be obtained from alkyl halides, *viz.*, alkyl chlorides or bromides with potassium fluoride in presence of phase transfer catalyst (Fig. 12.7)[6].

$$\underset{X=Cl,\,Br}{RX} + KF \xrightarrow{PTC} R-F + KX$$

Fig. 12.7: Conversion of alkyl chlorides or bromides into alkyl fluorides.

Using the above method 1-chlorooctane is converted into 1-fluorooctane[6]. The procedure is much superior to conventional methods[7,8] using anhydrous KF and high boiling polar solvents.

Alkyl fluorides can also be prepared from alkyl-methane sulphonates, which are obtained from the corresponding alcohols by treatment with methylsulphonyl chloride in presence of PTC (Fig. 12.8).

$$R-OH + CH_3SO_2Cl \xrightarrow{C_6H_5N} R\,OSO_2CH_3 \xrightarrow[\substack{KF \\ 2\text{ hr. }120°C}]{PTC} R-F + K^+O_3^- SCH_3$$

Fig. 12.8: Conversion of alcohols into alkyl fluorides.

12.2.4 Conversion of Alkyl Halides into Alcohols

The displacement of halogen by hydroxyl group under PTC conditions is defficult. This is because the hydroxide ion being highly hydrated compared to chloride or bromide is poorly transferred into organic phase. Even if alcohol is once formed, it is converted into alkoxide resulting in the formation of ethers. The alkyl *halidely, e.g.,* $1\text{–}C_8H_{17}$ Br may give elimination product, *viz.,* $CH_3(CH_2)_5\text{–}CH{=}CH_2$.

It is found[9] that 1-bromooctane can be converted into the corresponding alcohol is 90-100% yield by using n-$Bu_4\,N^+\,OH^-$, $C_{16}H_{33}\,Et_3\,N^+OH^-$ or $(C_8H_{17})_3C_2H_5\,N^+\,OH^-$ as PTC. The reaction is carried out at 80° for 48 hrs.

The displacement of halogen by OH is best accomplished by using 'betanine' quaternay salts, $R_3N^+CH_2CO_2^-$; these are 10-50 times more active than tetraalkyl ammonium salts in the reaction of 1-haloalkanes with aq. NaOH (Fig. 12.9)[2b]

$$\underset{\text{1-Bromooctane}}{1\text{-}C_8H_{17}Br} + (C_{12}H_{25})_3\overset{+}{N}CH_2COO^- \longrightarrow 1\text{-}C_8H_{17}OC\,CH_2\overset{+}{N}\,(C_{12}H_{25})_3Br^-$$

$$1\text{-}C_8H_{17}OC\,CH_2\overset{+}{N}\,(C_{12}H_{25})_3Br^- \underset{Br}{\overset{^-OH}{\rightleftharpoons}} 1\text{-}C_8H_{17}OC\,CH_2\overset{+}{N}\,(C_{12}H_{25})_3\,{}^-OH$$

$$1\text{-}C_8H_{17}OC\,CH_2\overset{+}{N}\,(C_{12}H_{25})_3\,{}^-OH \longrightarrow 1\text{-}C_8H_{17}\,OH + (C_{12}H_{25})_3\overset{+}{N}CH_2COO^-$$

Fig. 12.9: Conversion of 1-bromooctane into 1-octanol using betanine quaternary salt as PTC

The betanine catalyst is also effective the conversion[2b] of 1, 4-dibromobutane to tetrahydrofuran.

In the displacement reactions, the use of crown ethers is enormous even in those cases where there is possibility of elimination reaction taking place[10]. Thus, the reaction of 4-bromocyclohexenone with potassium acetate in presence of crown ether gives 4-acetoxy cyclohexenone (Fig. 12.10).

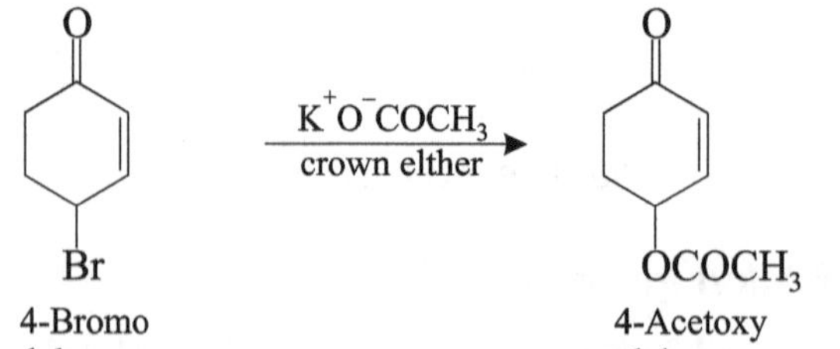

4-Bromo cyclohexenone 4-Acetoxy cyclohexenone

Fig. 12.10: Conversion to 4-bromocyclohexenone into the corresponding acetoxy compound.

12.2.5 Azides from Alkyl Halides

Alkyl halides can be converted into azides by treatment with sodium azide in presence of a PTC[11,12]. The utility of this technique is illustrated by the conversion of 3-methyl-5-chloro-1H-pyrazol-4-carboxaldehyde into 3-methyl-5-azido-1H-pyrozol-4-carboxaldehyde.

12.2.6 Sodium Alkyl Sulphonates from Alkyl Halides

Alkyl halides on treatment with sodium bisulfite in presence of a PTC gives the corresponding alkyl sulphonates. (Fig. 12.11)[13].

$$\underset{R=C_6H_5-CH_2}{RCl} + Na_2SO_3 \xrightarrow[2\text{ hr, }100°]{(C_2H_5)_4N^+Cl^-} \underset{99\%}{RSO_3\ Na}$$

Fig. 12.11: Conversion of alkyl halide into sodium alkyl sulphonate.

12.2.7 Alkyl Nitrates, Thiocyanates, Cyanates and *p*-toluene Sulphonates from Alkyl Halides

The reaction of alkyl halides with sodium nitrite gives the corresponding alkyl nitrates (Fig. 12.12)[14].

$$R-Br + \underset{aq.}{NaNO_2} \xrightarrow{R_4N^+X^-} R-NO_2 + NaBr$$

Fig. 12.12: Alkyl nitrates from alkyl halides.

In a similar way, alkyl bromides on reaction with sodium thiocyanate (NaSCN), sodium cyanate (NaCNO) and sodium *p*-toluene sulphonate ($NaO_2SC_6H_4$ p-CH_3) is presence of quaternary ammonium nitrates give the corresponding thiocynates[9,28], cyanates[29] and *p*-toluene sulphonates[30] respectively (Fig. 12.13).

$$RBr + Nax \xrightarrow{R_4N^+X^-} RX + NaBr$$
$$X = SCN, CNO, O_2SC_6H_4\ p\text{-}CH_3$$

Fig. 12.13: Conversion of alkyl halides into the corresponding thiocyanates, cyanates and *p*-toluene sulphonates.

12.2.8 Aryl Ethers and Thioethers

The reaction of phenols in aqueous NaOH with an alkylating agent in methylene chloride in presence of a PTC, *viz.*, C_6H_5–$CH_2(C_4H_9)_3$ N^+Cl^- gives aryl ethers (Fig. 12.14).

$$ArOH + RX \xrightarrow[\text{aq. NaOH}]{PTC/CH_2Cl_2} ArOR + HX$$
$$X = Cl \quad \text{Stirring, 3-10 hr.}$$

Fig. 12.14: Synthesis of aryl ethers.

The above PTC method is useful[14] for the methylation of nitrophenols (*o*-, *m*-, *p*) with methyl iodide (80% yield); these phenols are alkylated only by special techniques. Also, *o*-dihydroxy groups can be methylenated[15] with methylene bromide in 70-80% yield.

The PTC method is also useful for the alkylation of chelated hydroxyl group as in salicylaldehyde.

The reaction of 3 or 5-nitro-2-chloro pyridine with substituted phenols in aqueous NaOH in presence of PTC (TBAS) gives heteroanalogs of biphenyl ethers[16] is 70-90% yield (Fig. 12.15). This method has been used for the preparation of agrochemicals.

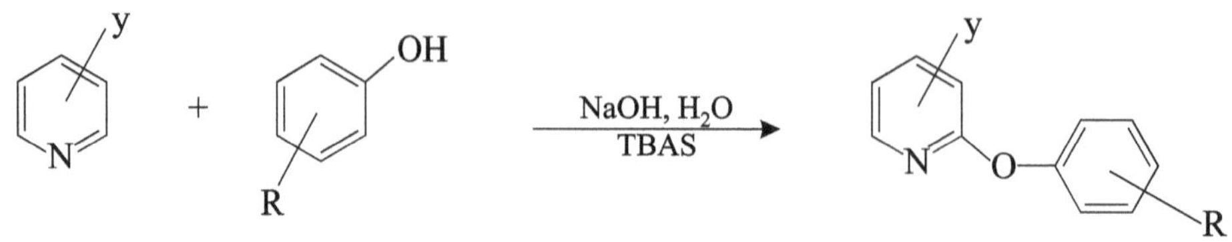

Fig. 12.15: Synthesis of heteroanalogs of biphenyl ethers.

Heteroaryl thioethers have been prepared by the reaction of thiophenol with chloro and bromo heterocycles under PTC conditions[17,18]. Thus, the reaction of 5-nitro-2-chloropyridine with sodium sulphide in presence of PTC gives[18] bipyridyl sulphide (Fig. 12.16).

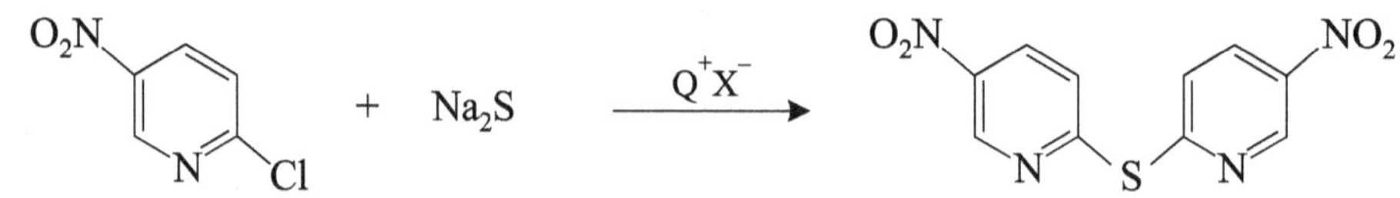

5-Nitro-2- Chloropyridine *bis* pyridylsulphide

Fig. 12.16: Synthesis bispyridyl sulphides

Dialkyl sulphides are obtained[19] by the reaction of alkyl halides and Na_2S in presence of PTC (Fig. 12.17).

$$\underset{\text{aq.}}{RX + Na_2S} \xrightarrow[70°,\ 40\ \text{hr.}]{C_{16}H_{35}P^{T}Bu_3Br^-} R-S-R$$

Fig. 12.17: Synthesis of disulphides.

12.2.9 Esterification

Carboxylic acids can be esterified[20,21] with alkyl halides in presence of triethylamine. In this case, phase transfer catalyst is obtained in situ by the reaction of triethylamine and alkyl halide (Fig. 12.18).

$$\underset{\text{Triethyl amine}}{Et_3N} + \underset{\text{Alkyl halide}}{RX} \longrightarrow \underset{\text{PTC (generated in situ)}}{Et_3N^+RX^-}$$

$$\underset{\text{Carboxylic acid as Sodium salt (aq. soln.)}}{R'CO_2Na} + \underset{\text{Alkyl halide}}{RX} \xrightarrow{\text{PTC}} \underset{\text{Ester}}{R'COOR} + NaX$$

Fig. 12.18: Esterification of carboxylic acids.

The above procedure is applicable only to reactive alkyl halides, *e.g.,* benzyl chloride, which readily form quaternary salts. Alternatively, quaternary ammonium or phosphonium salts can be used in the esterification of carboxylic acid with alkyl halides.

Crown ethers have also been used for esterification. Thus, *p*-bromophenacyl esters can be prepared[23] in 92% yield by the reaction of *p*-bromophenacyl bromide with potassium salt of carboxylic acid by using [18]-crown-6 (Fig. 12.19).

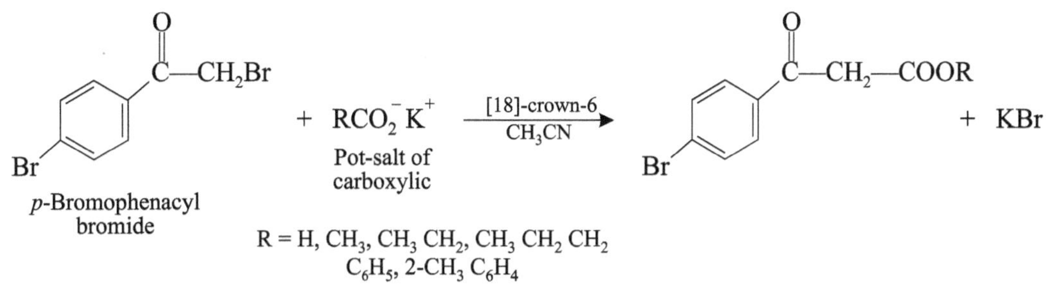

Fig. 12.19: Esterification using crown esters.

The phenacyl esters of fatty acids can also be obtained[24,25] in quantitative yield using crown ethers as catalysts.

12.2.10 Saponification

Saponification is normally carried out with KOH. The main problem is the insolubility of KOH in organic solvents like toluene. Using crown ethers, *e.g.* dicyclohexyl [18]-crown-6, it has been possible to transport KOH into the organic phase. This procedure finds application in the saponification[26,27] of hindered esters (Fig. 12.20).

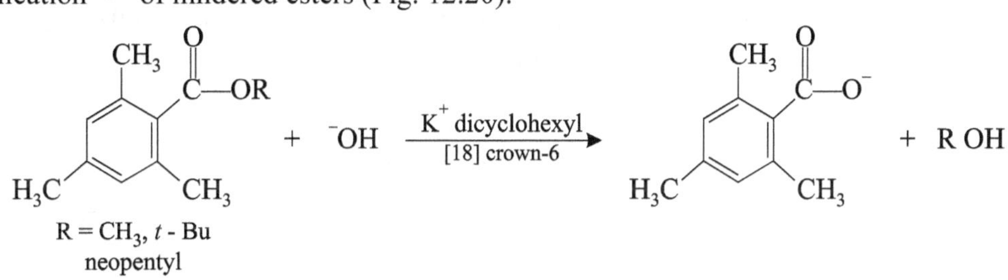

Fig. 12.20: Saponification of hindered esters.

Saponification of esters by aq. NaOH is also facilitated by PTC[27a].

12.2.11 Dichlorocarbene

Dichlorocarbene, synthetically useful intermediates[28-30] are normally generated by the reaction of chloroform with a base. These reactions require anhydrous conditions as the dichlorocarbene in aqueous solution undergo hydrolysis (Fig. 12.21).

$$CHCl_3 + Base \rightarrow CCl_3^- \xrightarrow{-Cl^-} :CCl_2 \xrightarrow{H_2O} CO + HCOO^- + Cl^-$$

Fig. 12.21: Generation of dichlorocarbene in anhydrous conditions.

The hydrolysis of dichlorocarbene in aqueous medium is supported by the fact that addition of chloroform to a mixture of cyclohexene and aq. NaOH (25%) gives[31] less than 5% yield (based on $CHCl_3$) of the adduct, *viz.*, 2,2-dichloro-bicyclo (4.1.0) heptane. However, using sodium ethoxide or potassium tert. butoxide in anhydrous solvent, the adduct is obtained[32] in 60-70% yield. It is found that using a PTC, *viz.* benzyl triethyl ammonium chloride, the adduct is obtained in 60-70% yield[33] (Fig. 12.22)

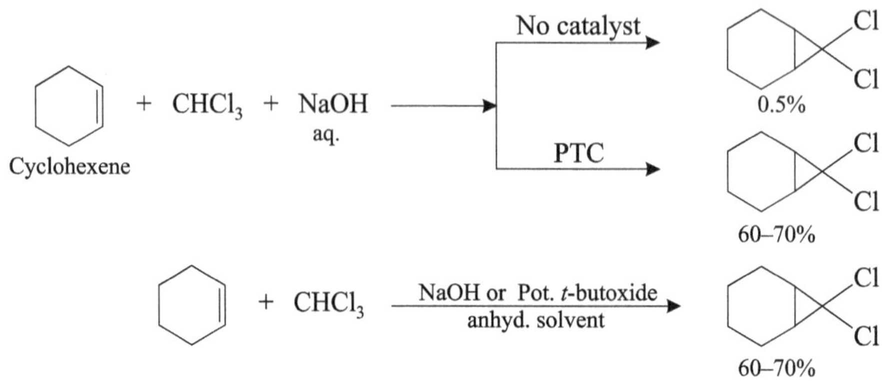

Fig. 12.22: Generation of dichloro carbene in situ and reacting with cyclohexene using PTC in aq. NaOH.

In the above reaction, using PTC, the generated dichlorocarbene is transferred to organic phase and reacts with cyclohexene than reacting with water.

Crown ethers, *e.g.*, dicyclohexyl-18-crown-6 have also been used to in situ generation of dichlorocarbene. Thus the reaction of olefins with chloroform in 50% aq. NaOH in presence of dibenzo[18] crown-6 gives the corresponding dichlorocyclopropanes (Fig. 12.23)[34]

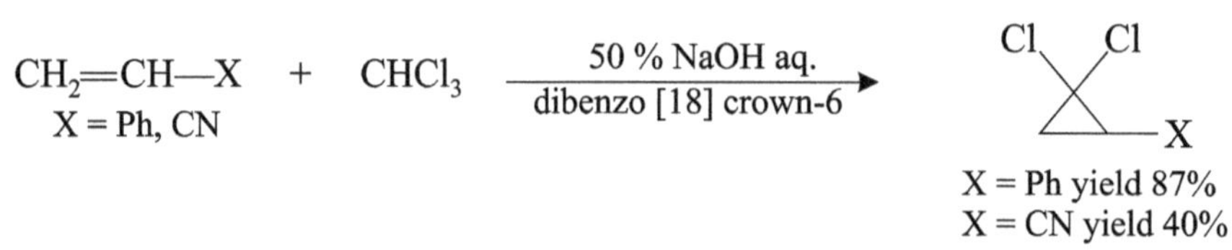

Fig. 12.23: Generation of insitu dichlorocarbene in presence of crown ether.

Using the above method cyclohexene give the corresponding gem-dichlorocyclopropane[34a].

The dichlorocarbene generated insitu by PTC method has been used for addition to C = N bonds (Schiffs base)[34*b*,35] and insersion into C–H bonds (as in adamantanes[36] and THF[37]) (Fig. 12.24)

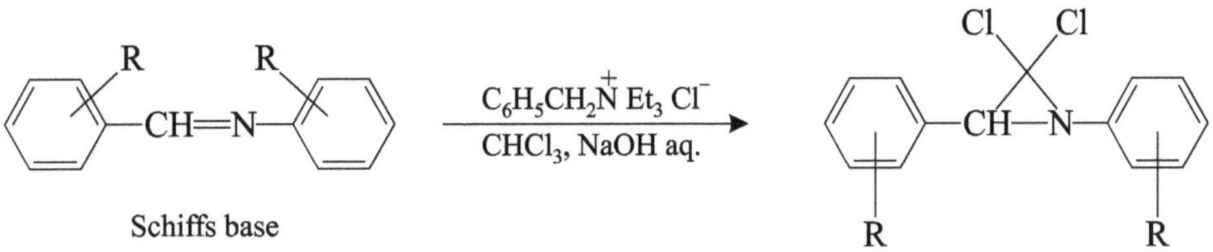

Schiffs base

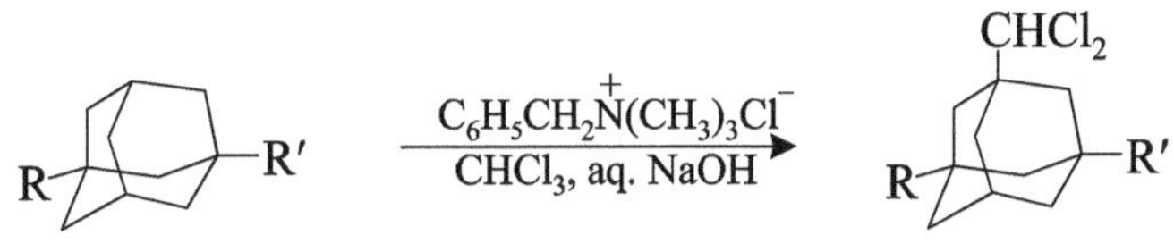

Adamantanes

R = R′ = H; yield 54%
R = CH_3, R′ = H; yield 100%

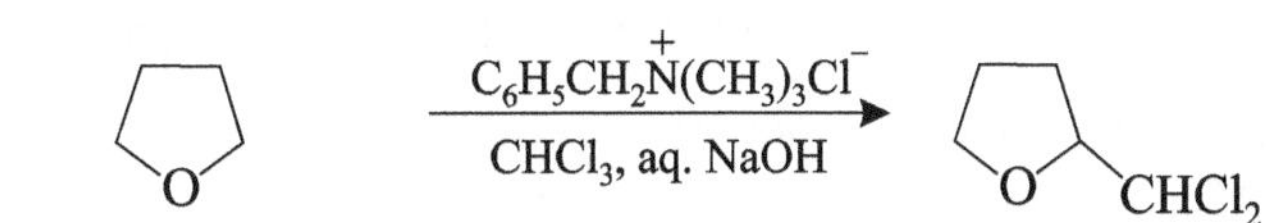

THF 18%

Fig. 12.24: Addition of dichlorocarbene into C=N bonds and C–H bonds.

12.2.12 Diazomethane

It is prepared by the reaction of hydrazine with dichlorocarbene generated in situ by the PTC method[38,39] (Fig. 12.25).

$$NH_2NH_2 + CHCl_3 + NaOH \xrightarrow[\text{ether or } CH_2Cl_2]{\text{PTC}} CH_2N_2 \text{ in ether or } CH_2Cl_2$$

Fig. 12.25: Generation of diazomethane

In this reaction the yields are low (35%). However, use of crown ether, *viz*. [18] crown-6 gives[38] better yields (50%).

12.2.13 Reaction of Dichlorocarbene with Amines

Dichlorocarbene generated in situ by the PTC method reacts with primary amines gives isonitriles (For details see section 12.2.2).

Secondary amines on reaction with dichlorocarbene (generated by PTC method) gives N,N-disubstituted formamides (Fig. 12.26)[40,41].

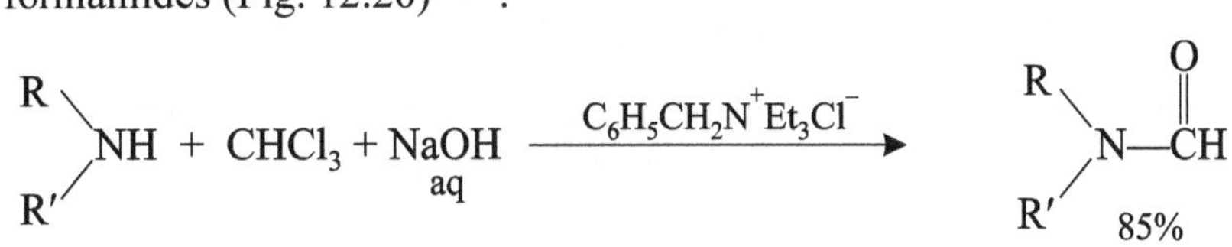

R, R′ = Ethyl, 2-butyl, allyl, cyclohexyl

Fig. 12.26: Reaction of dichlorocarbene with secondary amines.

Tertiary amines on reaction with chloroform and aq. NaOH in presence of $C_6H_5CH_2N^+Et_3Cl^-$ results[42] in elimination of carbon bridge (Fig. 12.27).

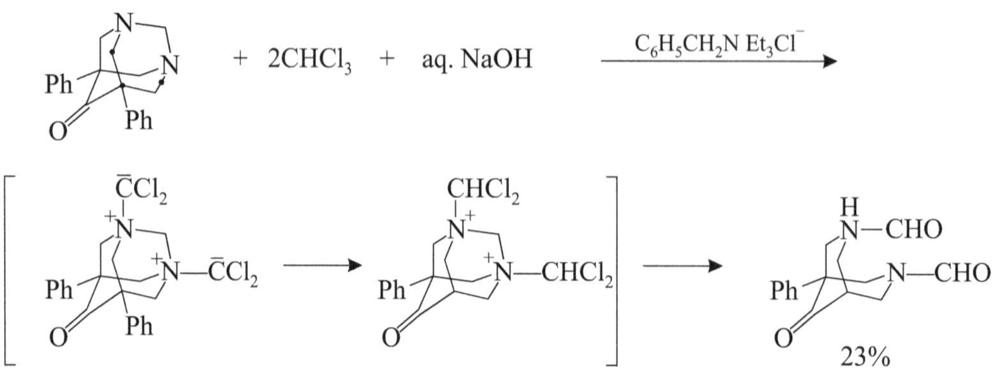

Fig. 12.27: Reaction of tertiary amines with dichlorocarbene.

12.2.14 Reaction of Dichlorobenzene with Substituted Indoles

Substituted indoles on treatment with dihalocarbenes (generated in situ by the reaction of haloform in presence of aq. NaOH and a PTC) results in ring expansion giving quinolines (Fig. 12.28)[43]

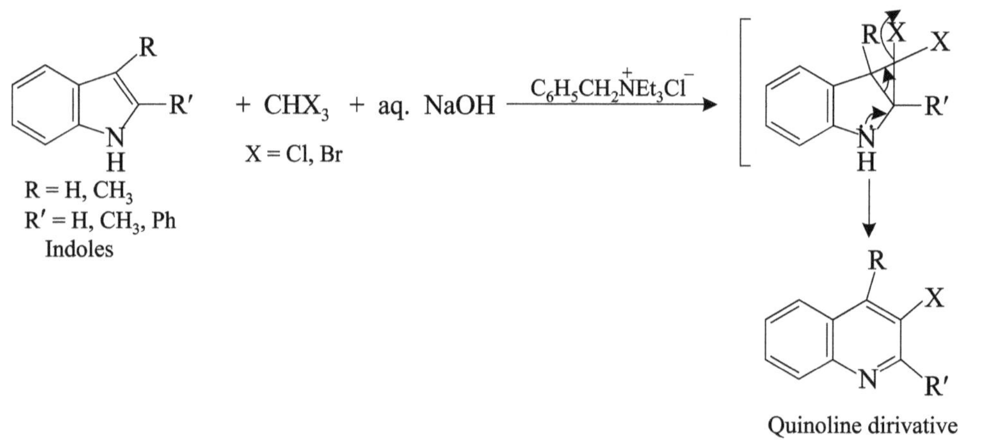

Fig. 12.28: Reaction of dihalocarbenes with substituted indoles.

The ring expansion technique (Fig. 12.28) has been applied[44] to a variety of heterocyclic compounds.

12.2.15 Reaction of Dichlorocarbene with Alcohols

Alcohols (*e.g.*, cyclohexanol) react with dichlorocarbene to give[45] a mixture of corresponding chlorides, olefins and their corresponding dichlorocyclopropanes. Following mechanism has been proposed[46] (Fig. 12.29).

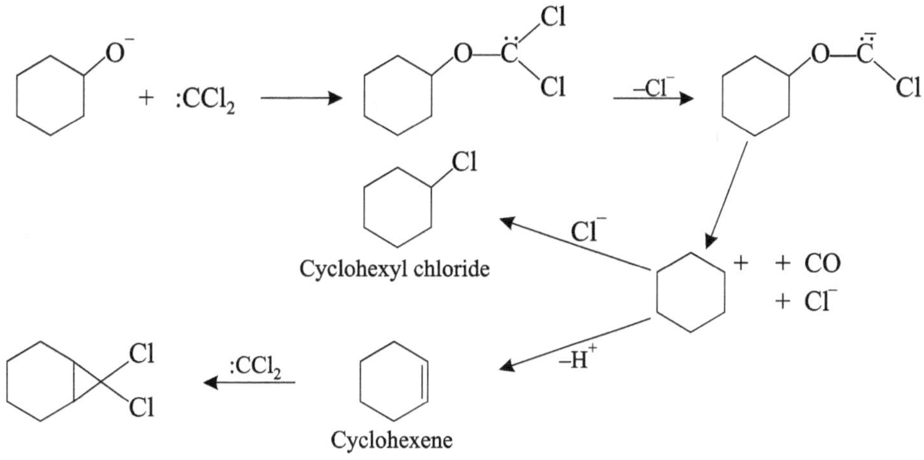

Fig. 12.29: Reaction of dichlorocarbene with cyclohexanol.

It has been found that alcohols react with dichlorocarbene (generated by PTC method) to give the corresponding chlorides (Fig. 12.30)[47].

$$ROH + CHCl_3 + \underset{aq}{NaOH} \xrightarrow{C_6H_5CH_2N^+Et_3Cl^-} RCl + NaCl + H_2O$$

Fig. 12.30: Reaction of alcohols with dichlorocarbene give alkyl chlorides.

Using the above procedure, a number of alcohols like 1-adamantyl alcohol, benzyl alcohols, 1-menthyl alcohol, 2 exo and 2-endo norbornyl alcohols and steroidal alcohols could be converted into the corresponding chlorides. The steroid chlorides are obtained with original configuration.

1,2-Diols on treatment with dichlorocarbene undergo pinacol rearrangement to give ketones (Fig. 12.31)[48].

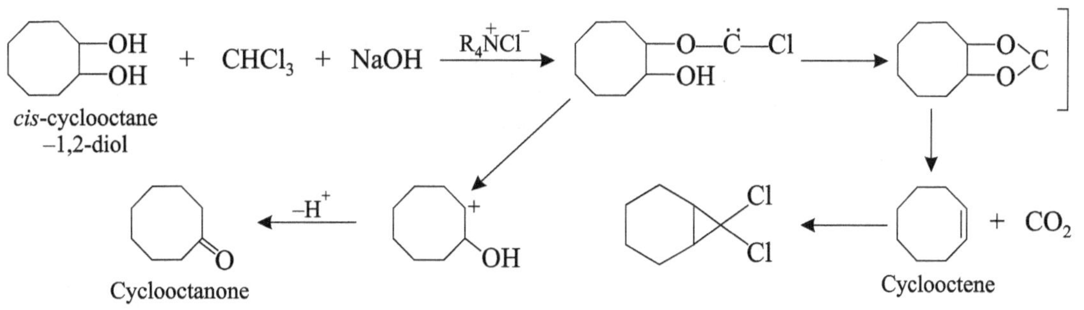

Fig. 12.31: Reaction of *cis* cyclooctane 1,2-diol with dichlorocarbene.

12.2.16 Reaction of Dichlorocarbene with Aldehydes

Aromatic aldehydes on reaction with dichlorocarbene give mandelic acids. The reaction proceeds via the formation of a three membered cyclic mechanism (Fig. 12.32)[49].

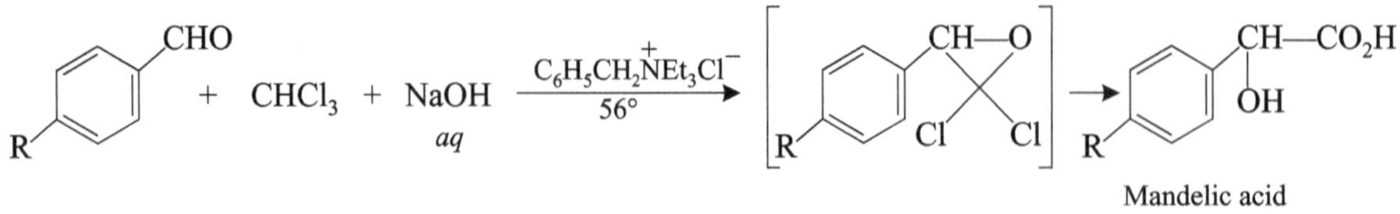

Fig. 12.32: The reaction of aromatic aldehydes with dichlorocarbene give mandelic acids.

The above reaction is temperature dependent and is best carried out at 56°. At lower temperature α-trichlorocarbinols are obtained[50]. However, at moderately high temperature[51], carboxylic acids are the main product (Fig. 12.33).

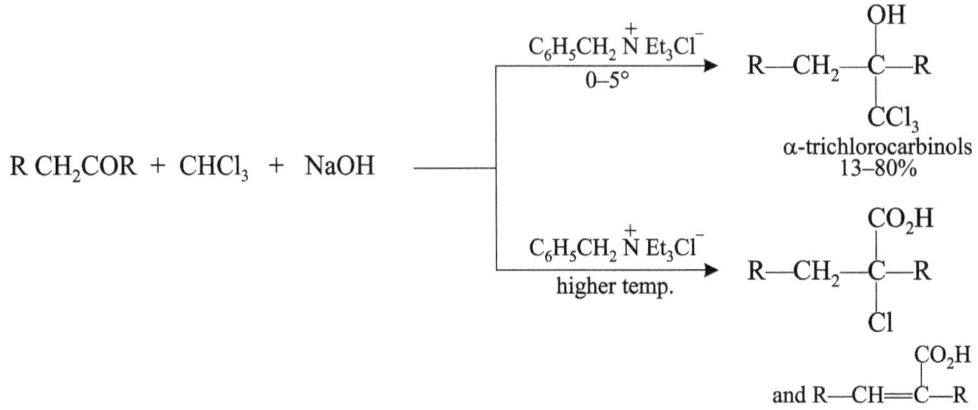

Fig. 12.33: Reaction of carbonyl compound at low or higher temp. with dichlorocarbene

Unsaturated aldehydes add : CCl_2 at the C=C bond and not at the C=O bond[52].

12.2.17 Vinylidene Carbenes

Dimethylvinylidene carbenes can be generated[53] from 3-chloro-3-methyl-1-butyne by reacting with base under vigorously anhydrous conditions. These add on to olefins in situ to give dimethylvinylidenecyclopropanes (Fig. 12.34).

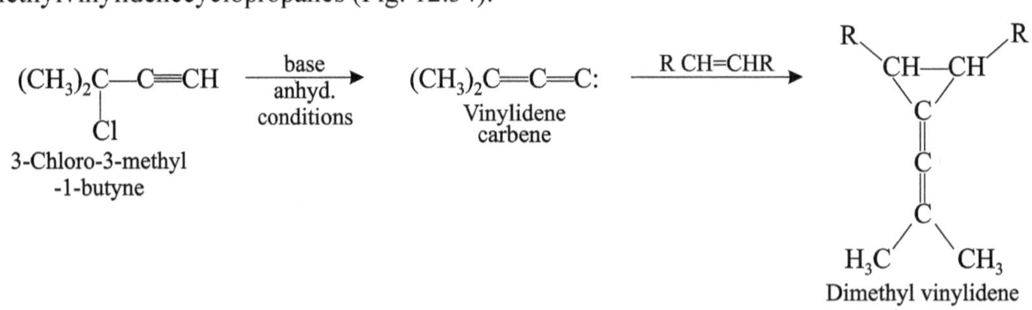

Fig. 12.34: Generation of Divinylidene carbenes.

It has been found[54-57] that the phase transfer technique with aq. NaOH is more convenient and also gave better yields. Using this procedure a number olefins add on to vinylidene carbenes to give the corresponding dimethylvinylidenecyclopropanes. The crown ethers, *viz.*, dibenzo-18-crown-6, dicyclohexyl-18-crown-6 and 18-crown-6 and more effective[58] than quaternary ammonium salts (PTC).

12.2.18 Vinyl Carbenes

Vinyl carbenes are normally obtained[59-61] by the reaction of N-nitrosoacetylaminomethyl substituted carbinols with strong base (Fig. 12.35).

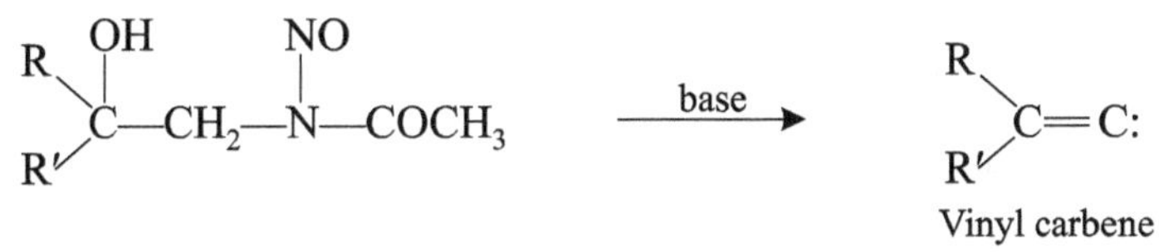

Fig. 12.35: Generation of Vinyl Carbenes.

The vinyl carbenes are highly relative carbenes and the reaction proceeds especially well in presence of PTC and aq. NaOH. The vinyl carbenes add on to olefins, aldehydes, ketones, azides, thiocyanate, iodide and triphenyl phosphate. Many of the products obtained by this technique are difficult to prepare by other methods (Fig. 12.36).

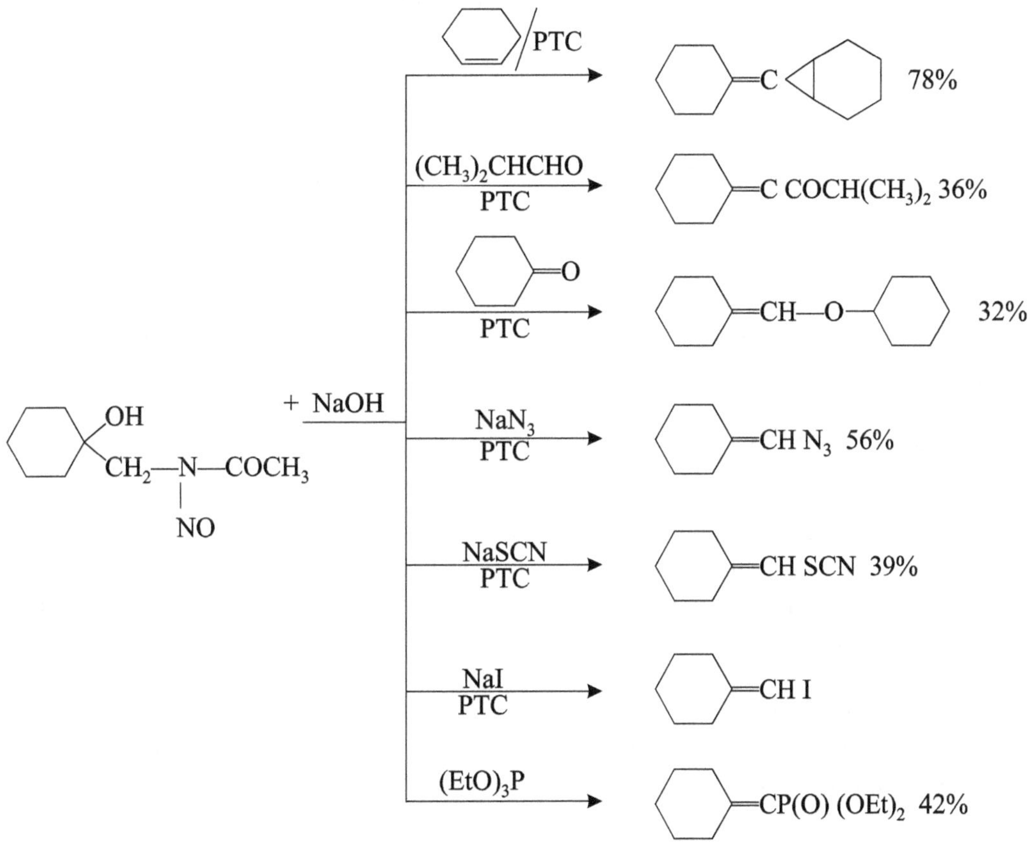

Fig. 12.36: Addition of Vinyl carbenes generated in situ to various substrates.

12.2.19 Elimination Reactions

Phase transfer catalysts have been used for dehydrohalogenation of alkyl halides with aq. NaOH (Fig. 12.37)[62].

$$\underset{\substack{|\\ \text{Br}}}{\text{R CH}}\text{—CH}_2\text{—R}' + \text{aq. NaOH} \xrightarrow{\text{PTC}} \text{R CH=CH R}'$$

PTC = tetra butyl ammonium hydrogen sulfate

Fig. 12.37: Dehydrohalogenation of alkyl halides.

Vic-dibromoalkanes can be neatly debrominated to the corresponding olefins by a phase transfer catalyst using sodium thiosulphate with a catalytic amount of sodium iodide[63] (Fig. 12.38).

$$\text{R—}\underset{\substack{|\\ \text{Br}}}{\text{CH}}\text{—}\underset{\substack{|\\ \text{Br}}}{\text{CH}}\text{—R}' + Na_2S_2O_3 \xrightarrow[\text{NaI}]{C_{16}H_{33}Bu_3\overset{+}{P}\overset{-}{Br}} \text{R CH=CH R}'$$

Fig. 12.38: Debromination of Vic-dibromoalkanes.

12.2.20 Carbonylation of Aryl and Vinyl Halides

Aryl and Vinyl halides can be carbonylated[64] by using PTC as catalyst under photostimulation in presence of CO_2Co_8 and CO. Thus aryl halides can be converted into the corresponding carboxylic acids using tetrabutyl ammonium bromide as PTC (Fig. 12.39).

$$\text{Ar X} + \underset{\text{aq}}{\text{NaOH}} \xrightarrow[\text{CO, } h\nu \text{ (350 nm)}]{\text{benzene, } CO_2Co_8,\ Bu_4\overset{+}{N}Br} \text{Ar COONa} + \text{Nax}^-\ H_2O$$

Fig. 12.39: Carbonylation of aryl halides.

The reaction is applicable to aryl bromides only. The method is also applicable to vinyl bromides. In place of 350 nm UV lamp, the irradiation can also be achieved[65] by a commercial sun lamp.

12.2.21 Alkylations

12.2.21.1 O-Alkylations

Phenols can be *O*-alkylated by reacting with alkyl halides in presence of K_2CO_3. In presence of crown ethers, the product is obtained in much better yield. In fact crown ethers can enhance the solubility of K_2CO_3 which in turn helps to scavenge protons. An example is the conversion of phenol into benzyl phenyl ether in quantitative yield by using K_2CO_3 in presence of catalytic amount of 18-crown-6.

12.2.20.2 C-Alkylations

Use of Phase Transfer Catalysts permits a simple method for alkylation of a variety of compounds. As an example, nitriles can be monoalkylated (Fig. 12.40)[66].

$$C_6H_5CH_2CN + C_2H_5 - Cl + \underset{\text{aq}}{\text{NaOH}} \xrightarrow{C_6H_5CH_2N^+Et_3Cl^-} C_6H_5 - \underset{\substack{|\\ C_2H_5}}{CHCN}$$

Fig. 12.40: Monoalkylation of phenylacetonitrile

The above technique is useful for most types of alkylations except those compounds which are easily hydrolysed.

The reaction requires a molar quantity of quaternary ammonium bisulphate rather than a catalytic amount. All steps are carried out in the same pot. Various steps involved are given below (Fig. 12.41).

$$\underset{(CH_2Cl_2\ \text{Solution})}{(C_4H_9)_4\overset{+}{N}HSO_4^- } + \underset{\text{aq.}}{NaOH} \longrightarrow (C_4H_9)_4\overset{+}{N}\bar{O}H \rightarrow NaHSO_4$$

$$(C_4H_9)_4\overset{+}{N}O\bar{H} + CH_2(CO_2Et)_2 \longrightarrow (C_4H_9)_4\overset{+}{N}\bar{C}H(CO_2Et)_2 + H_2O$$

$$(C_4H_9)_4\overset{+}{N}\bar{C}H(CO_2Et)_2 + RX \longrightarrow RCH(CO_2Et)_2 + (C_4H_9)_2\overset{+}{N}X^-$$

Fig. 12.41: Mechanism of mono alkylation.

It has been shown that phenyl acetonitrile can be alkylated with di-(2-chloroethyl) ether or the corresponding thioanalogue[67], Cl $(CH_2)_2$-S-$(CH_2)_2$Cl or 1,2-ethylenedithiocyanate[68] (Fig. 12.42).

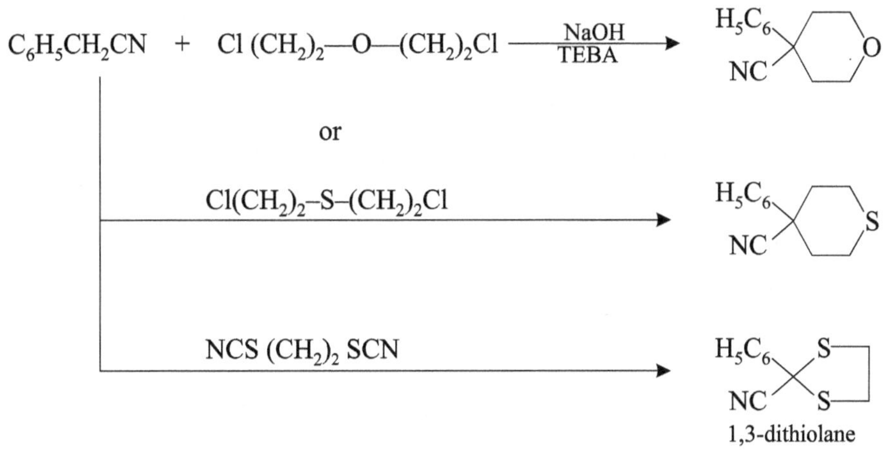

Fig. 12.42: Phenylacctonitrile on reaction with various Synthones gives cyclic compounds

Reactive methylene compounds on alkylation with 1,2-dibromoethane gives excellent yield of the corresponding 1,1-cyclopropane dicarboxylic acids (Fig. 12.43)[69].

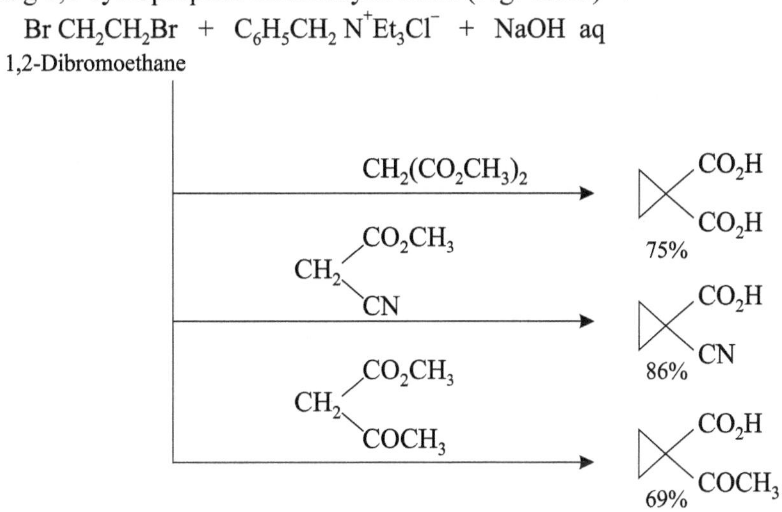

Fig. 12.43: Alkylation of reactive methylene compounds with 1,2-dibromoethane.

In a similar way oxindole[70] reacts at positions 3 with 2 moles of alkyl halide and undergoes cycloaddition with an α, ω dihalide (Fig. 12.44).

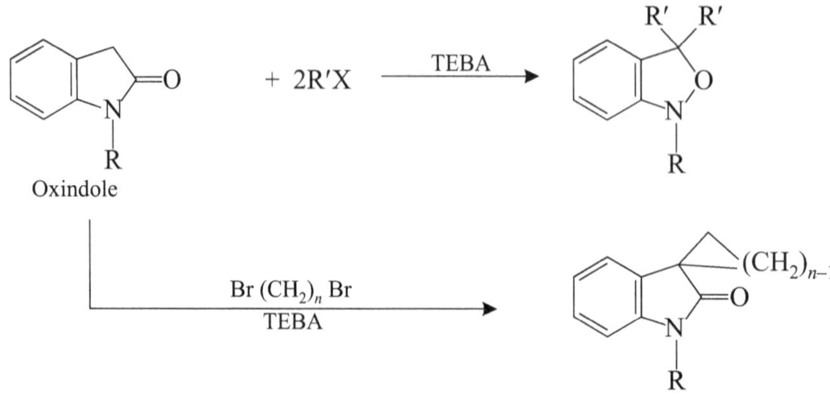

Fig. 12.44: Alkylation of oxindoles.

N-Benzoyl-1,2-dihydroisoquinaldenitrile can be alkylated in presence of NaOH and TEBA (Fig. 12.45)[71]

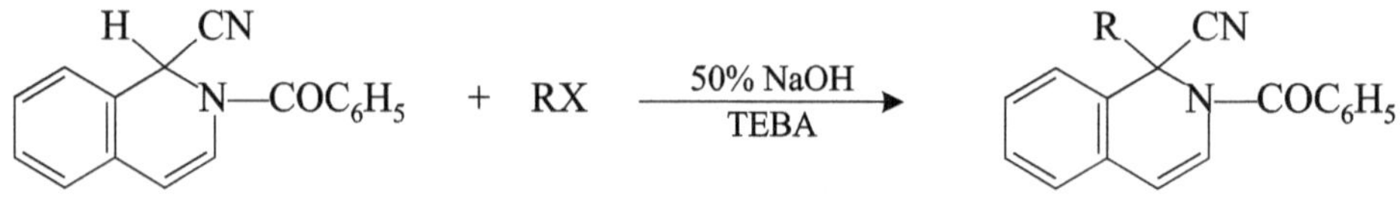

Fig. 12.45: Alkylation of N-benzoyl-1,2-dihydroisoquinaldenitrile.

Pyridine cannot be alkylated in positions 2,4 and 6 unless the nitrogen atom is first quaternised. After alkylation[72] by PTC, the N atom can be dequaternised by 4-methyl thiophenol to yield the pyridine moiety (Fig. 12.46).

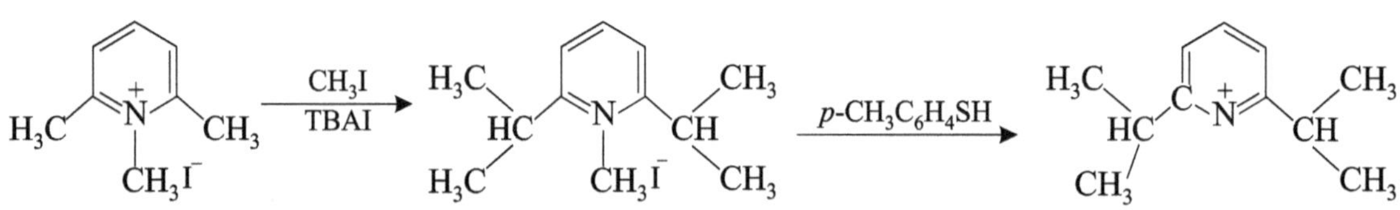

Fig. 12.46: Alkylation pyridine at 2 and 6 positions.

C-Alkylations can also be effected in presence of crown ethers. Some of such reactions are given below (Fig. 12.47)[73].

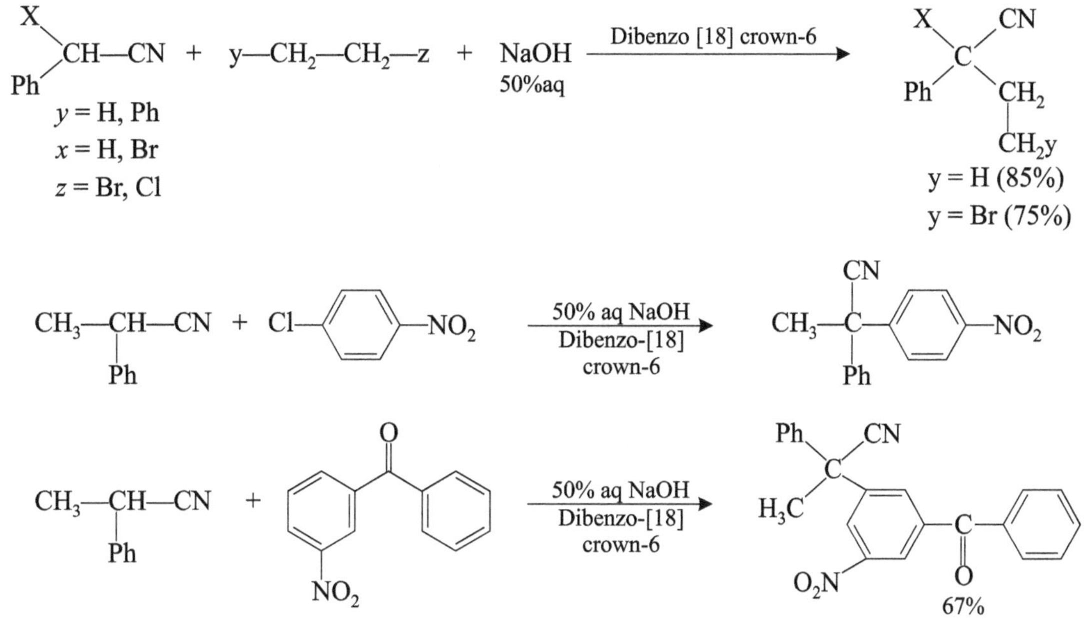

Fig. 12.47: Alkylations in presence of crown ethers.

Activated ketones having on aromatic substitution at the α-CH_2 group (*e.g.* $C_6H_5CH_2COCH_3$) can be alkylated[74-76] using PTC (Fig. 12.48).

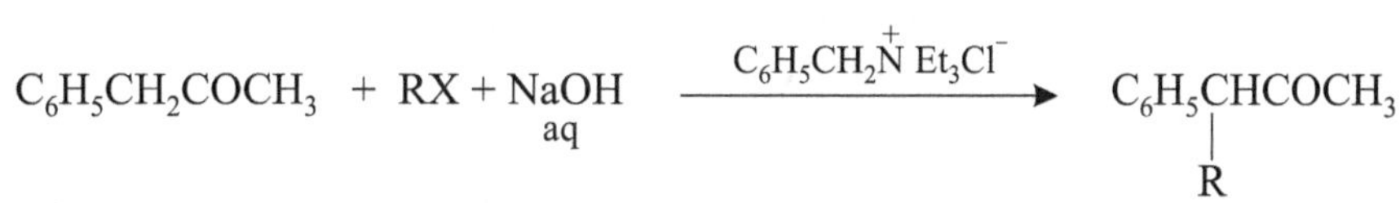

Fig. 12.48: Alkylation of activated ketones.

In case the reacting alkyl chloride is reactive, *e.g.*, benzyl or allyl chloride, dialkyl products are obtained.

Crown ethers have also been used for alkylation of ketones. Thus, benzyl methyl ketone on reaction with n-butylbromide in 50% aq. NaOH in presence of dicyclohexyl-[18]-crown-6 gives 90% of the alkylated product (Fig. 12.49)[77].

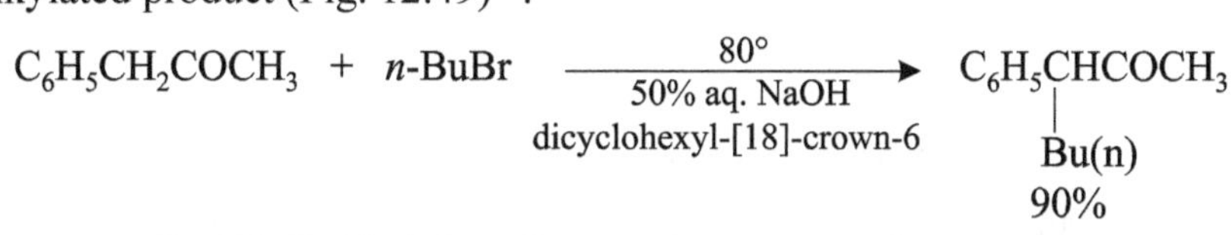

Fig. 12.49: Alkylation of ketones in presence of crown ether

Aldehydes containing only an α-hydrogen atom (*e.g.*, isobutraldehyde) can be alkylated[78] with alkyl halides in presence of 50% aq. NaOH and catalytic amount of PTC (Fig. 12.50)

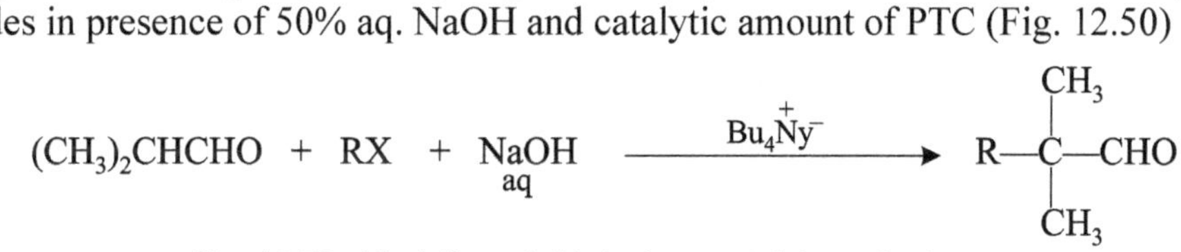

Fig. 12.50: Alkylation of aldehydes containing α-hydrogen.

In the above reaction, only reactive alkyl halides, *e.g.*, allyl and benzyl chlorides are used. In case the alkyl halide is less active, the alkylation is conducted by dropwise addition of isobutyraldehyde to a mixture of aq NaOH, benzene and PTC at 70°.

It has been found that many aldehydes and ketones condense with acetonitrile in presence of solid KOH using [18]-crown-6 as cataysed to give unsaturated nitriles (Fig. 12.51)[79].

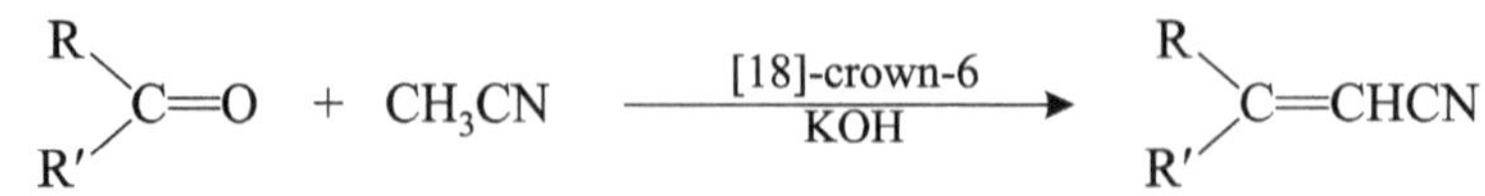

Fig. 12.51: Condensation of aldehydes or ketones with CH_3CN in presence of solid KOH and crown ether.

12.2.21.3 N-Alkylations

Aziridines cannot be N-alkylated under conventional conditions due to rapid decomposition to give open chain compounds. However, N-alkylation can be carried out under PTC conditions[80] (Fig. 12.52).

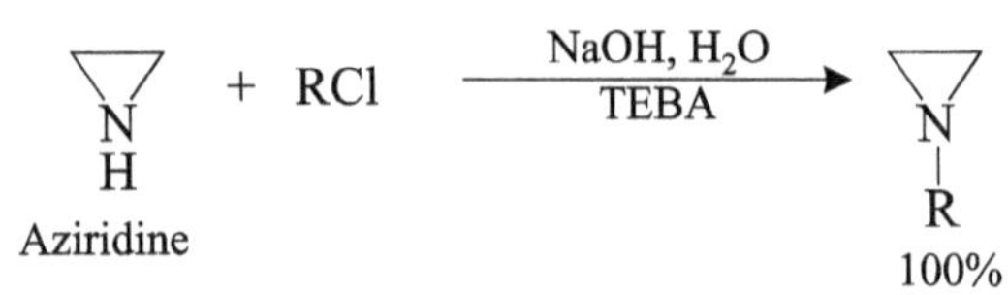

Fig. 12.52: N-Alkylation of aziridine.

Similarly, N-alkylation of pyrrole under conventional conditions occurs at 2 and 3 positions. However, under PTC conditions, N-alkylation occurs[81]. The ratio of N-alkylation versus C-alkylation of pyrrole and other heterocycles increased considerably if the alkylation is carried out[82] with *t*-Bu OK in presence of [18]-crown-6.

N-alkylation of substituted benzoindoles with alkyl bromides in presence of $TBAHSO_4$ gives the N-substituted product (Fig. 12.53)[83]

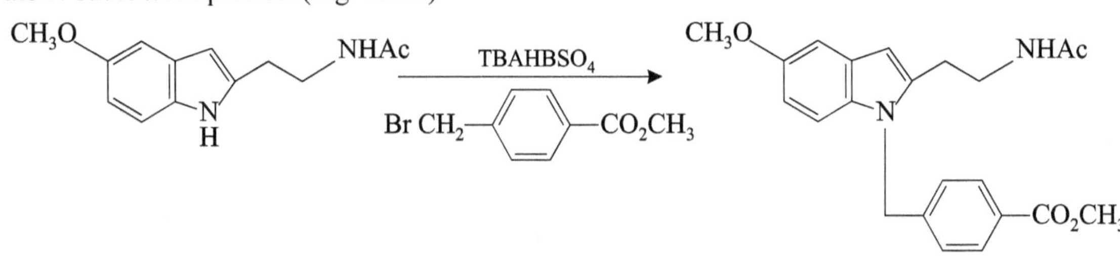

Fig. 12.53: N-alkylation of substituted benzoindoles.

Benzimidazoles can also be N-alkylated in presence of 18-crown-6. In this case two isomers are obtained[84] (Fig. 10.54)

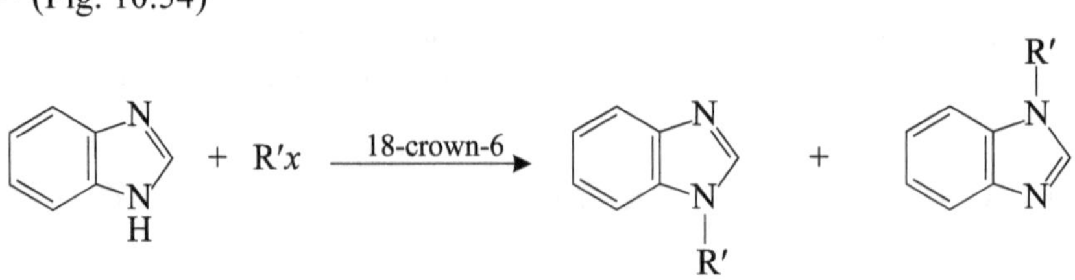

Fig. 12.54: N-Alkylation of benzimidazoles.

The N-alkylation of pyrrole and indole has also been carried[85] out in presence of crown ethers. In case of benzoindole alkylation occurs at N and C–3 (Fig. 12.55).

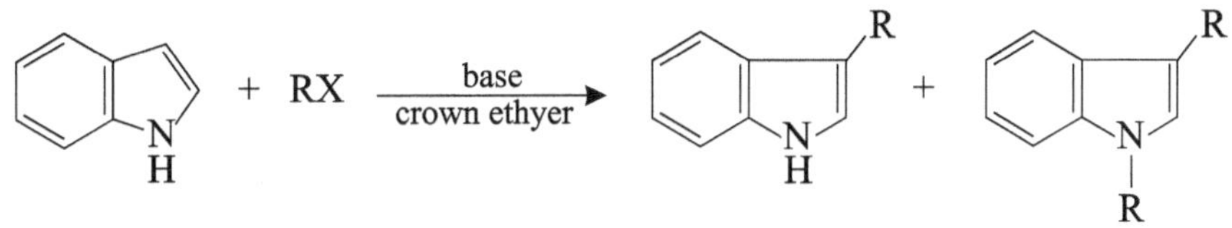

Fig. 12.55: Alkylation of benzoindole in presence of crown ether.

12.2.21.4 S-Alkylations

S-Alkylated products are obtained in major amount by the alkylation of S, N-nucleophilic heterocycles. In some cases, however, N-alkylation follows S-alkylation to give N,S-dialkylated products (Fig. 12.56)[86].

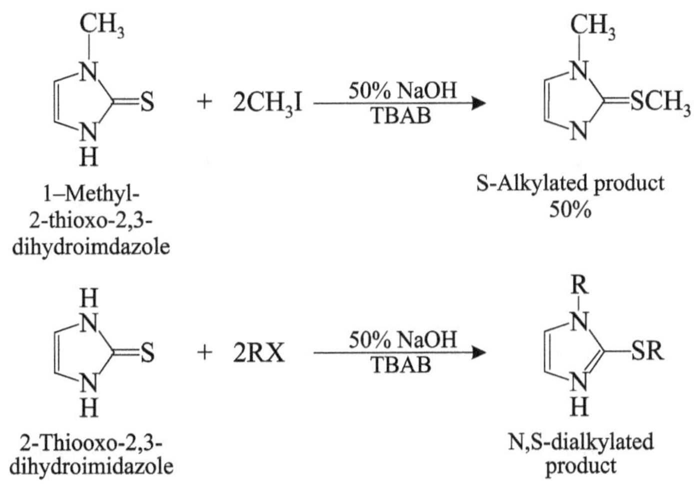

Fig. 12.56: S and S, N alkylation.

12.2.21.5 Alkylation of Mercaptans and Thiophenols

Alkylation of mercaptans with alkyl halides in presence of quaternary ammonium salts give the corresponding alkyl sulphides (Fig. 12.57).

$$RSH + R'-X \xrightarrow{PTC} \underset{70\text{-}75\%}{RSR'}$$

$$R = C_2H_5, n-C_4H_9-, (CH_3)_2CHCH_2^-$$

$$R'X = 1-C_8H_{17}Cl(\text{or Br}), 2-C_8H_{17}Cl\ (\text{or Br})$$

$$PTC = C_{16}H_{33}P^+Bu_3Br^- \text{ or } (C_8H_{17})_3N^+CH_3Cl^-$$

Fig. 12.57: Alkylation of mercaptans.

Thiophenols can be alkylated with methyl iodide, ethyl bromide, dimethyl sulphate and 1-C_8H_{17} Br to give[87,88] the corresponding alkyl derivatives.

12.2.22 Benzoin Condensation

Benzoin condensation of aldehydes, which do not contain α-hydrogen is strongly catalysed by quaternary ammonium cyanide[89] (Fig. 12.58).

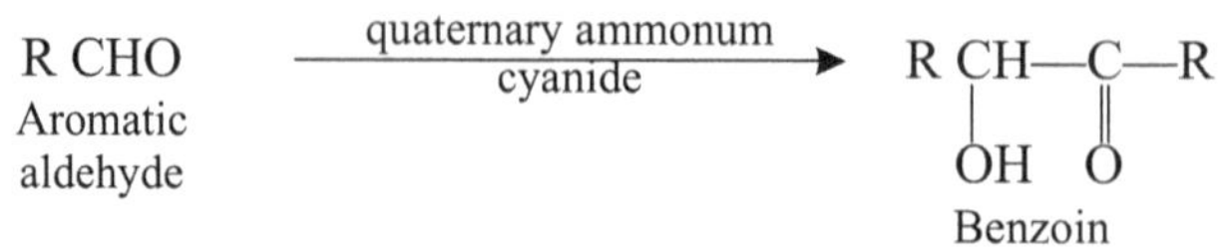

Fig. 12.58: Benzoin Condensation

It is known that lower aliphatic aldehydes containing an α-hydrogen undergo **aldol condensation** with aq. NaOH even in the absence of a PTC. However, as the chain length increases, the solubility of the aldehyde decreases in aqueous phase and the condensation becomes slower. In such cases, PTC becomes more effective. It has, however been found[90] that with solid KOH and using 3-benzyl-4-methyl thiazolium chloride as catalyst, acetoin is formed from acetaldehyde (Fig. 12.59)

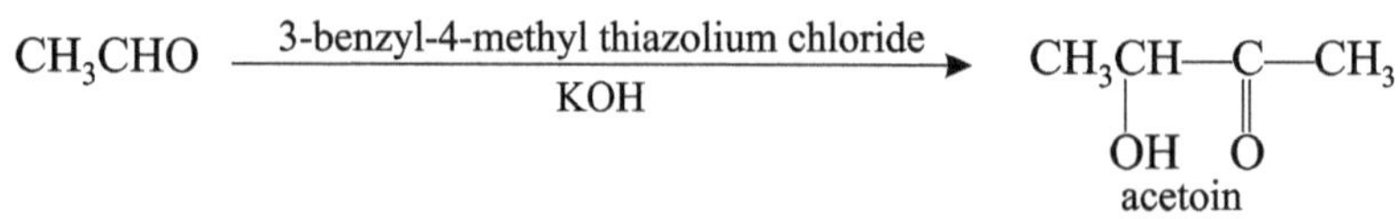

Fig. 12.59: Formation of acetoin from acetaldehyde.

The above procedure (Fig. 12.59) is useful for converting benzaldehyde to benzoin.

The Benzoin condensation of aromatic aldehydes with aq. NaCN are strongly catalysed by quaternary ammonium salts[89] (Fig. 12.60)

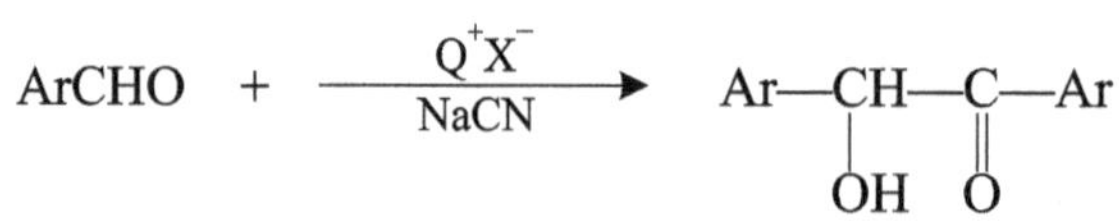

Fig. 12.60: Benzoin condensation in presence of quaternary ammonium salts.

In a similar way, acyloin condensation with both aliphatic or aromatic aldehydes is easily effected by N-laurylthiazolium bromide as catalyst with an aqueous phosphate solution[91] (Fig. 12.61)

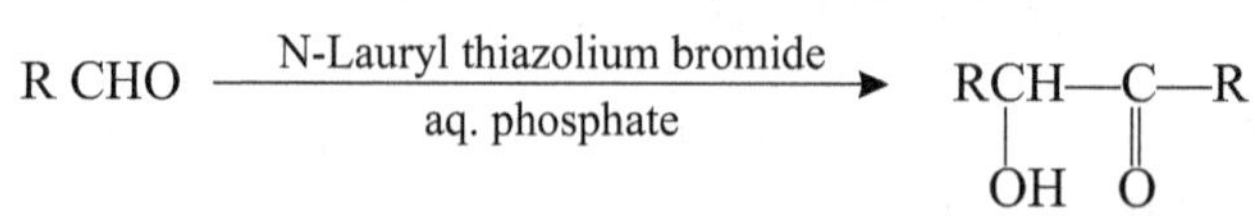

Fig. 12.61: Acyloin condensation of aliphatic or aromatic aldehydes.

It is found that aromatic aldehydes react in a few minutes but aliphatic aldehydes require 5-10 hr.

Benzoin condensation can also be carried out with either aq. KCN/neat aromatic aldehyde or solid KOH/aldehyde dissolved in benzene or acetonitrile at 25-60° using 18-crown-6 or dibenzo-18-crown-6 as catalyst.

12.2.23 Darzen Reaction

The Darzen reaction consists in the condensation of aldehydes and ketones with chloroacetonitrile in presence of a base to give α, β-epoxy esters called glycidic ester (Fig. 12.62)[92].

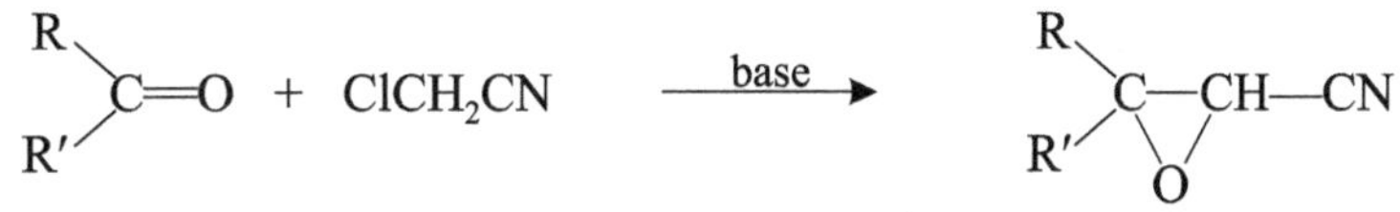

Fig. 12.62: Darzen reaction.

The use of PTC has made Darzen reaction very simple and the yields are 70-80%. (Fig. 12.63)[93]

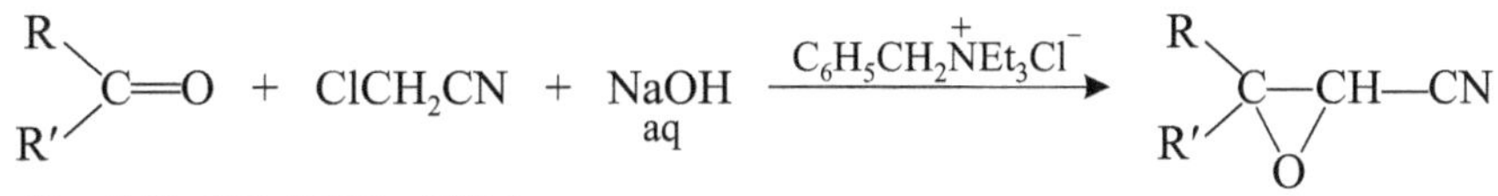

$R = C_6H_5, CH_3 (CH_2)_5, (CH_2)_4,$
$R' = H, CH_3.$

Fig. 12.63: PTC catalysed Darzen Reaction.

With aldehydes and unsymmetrical ketones both possible stereoisomers are obtained. However, with more acidic ketones, *e.g.*, phenylacetone, the ketone carbanion is formed leading to alkylation of the ketone (Fig. 12.64).

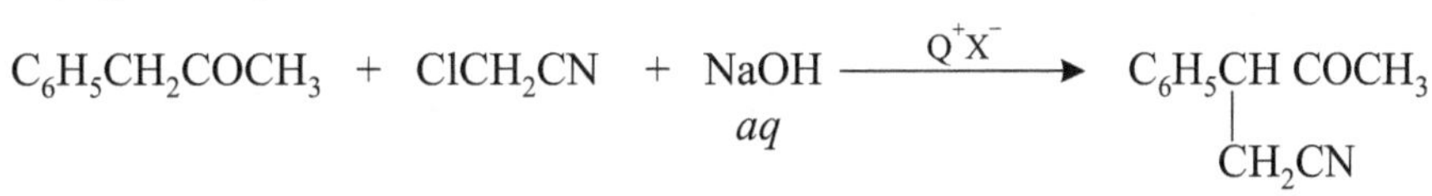

Fig. 12.64: Acidic ketone on Darzen reaction give alkylation rather than the expected glycidic ester.

With aromatic aldehydes, crown ethers have been used for Darzen reaction.

The PTC catalysed reaction of α-thiocarbonyl compounds with 2-chloroacrylonitrile gave 2-cyano-2,3-epoxytetrahydrothiophenes; in this case most likely the reaction involves thiol addition followed by an **intramolecular Darzen reaction** (Fig. 12.65)[94]

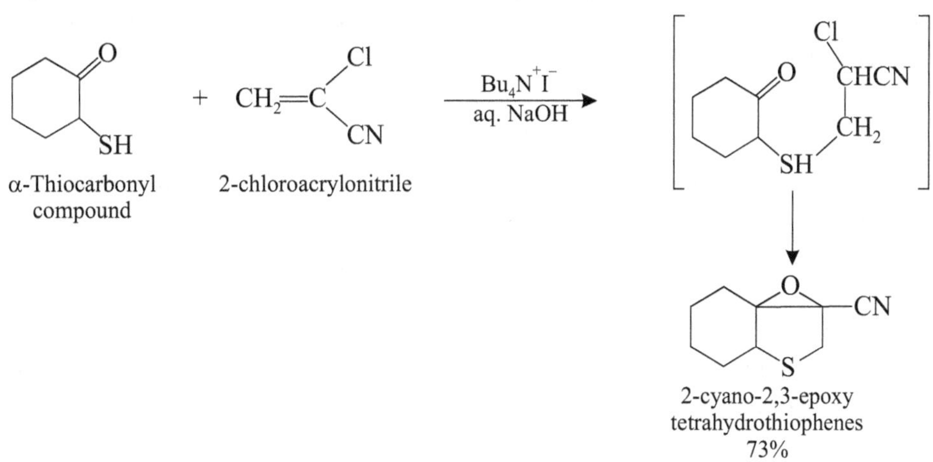

Fig. 12.65: Intramolecular Darzen reaction.

3.2.24 Michael Reaction

It involves the reaction of α, β-unsaturated carbonyl compounds with compounds containing reactive methylene group in presence of a base to give an adduct (Fig. 12.66).

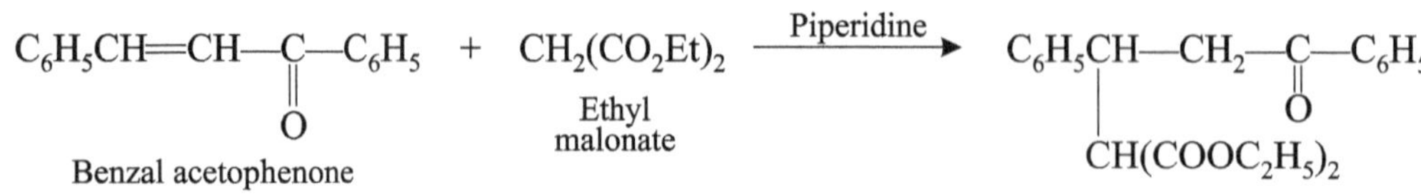

Fig. 12.66: Michael reaction.

Michael Reaction involving reaction of active nitriles to acetylenes is catalysed by quaternary ammonium salt (Fig. 12.67)[95].

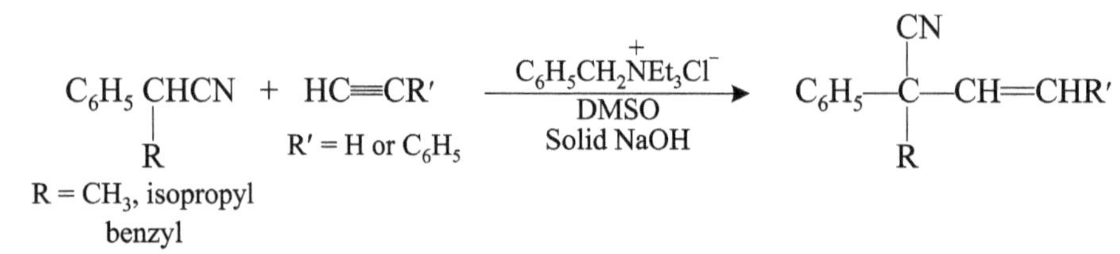

Fig. 12.67: Michael reaction of activated nitriles with acetylenes in presence of PTC.

12.2.25 Williamsons Ether Synthesis

Phase transfer technique provides a simple and convenient procedure for conducting Williamsons ether synthesis. It is found that using excess alcohol or excess alkyl halide, lower temperature and larger alcohol (*e.g.*, $C_8H_{17}OH$) gives higher yields of ether (Fig. 12.68)[96,97].

$$C_8H_{17}OH + C_4H_9Cl \xrightarrow[\text{aq. NaOH}]{\text{PTC}} C_8H_{17}OC_4H_9 + \underset{\text{Byproduct}}{C_8H_{17}OC_8H_{17}}$$

Fig. 12.68: PTC catalysed Williamsons ether Synthesis.

Maximum yields are obtained[97] by using 5 fold excess of aq. NaOH (50%) over alcohol, excess of alkyl halide (also used as solvent) and tetrabutyl ammonium bisulfate (1-5 mol) as catalyst at 25-27°. Primary alcohols require greater time or greater amount of catalyst.

Dimethyl sulphate is known not to react with most alcohols in presence of aq. NaOH or using alkali metal alkoxide, the reaction proceeds[98] with tetrabutylammonium salts as catalyst. Activated primary alcohols give much higher yields of ethers. However, secondary alcohols react very slowly and tertiary alcohols do not react.

Ethers from phenols can also be obtained by using crown ethers as PTC.

12.2.26 Wittig Reaction

The Wittig reaction involves a reaction between a phosphorane or phosphonium ylid and an aldehyde or ketone to give an alkene (Fig. 12.69)[99].

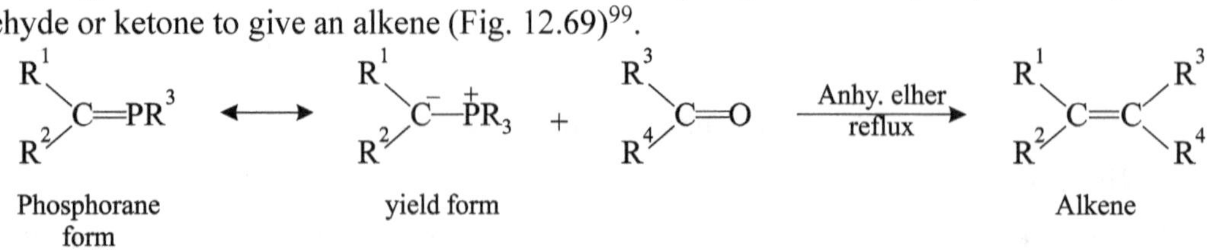

Fig. 12.69: The Wittig reaction.

The reaction provides a procedure for introducing C = C in place of C = O.

It has been found[100,101,97] that PTC, *viz.*, alkyl triphenylphosphorium salts react with aq. NaOH to generate the ylides, which combine with organic phase aldehydes to give olefins (Fig. 12.70).

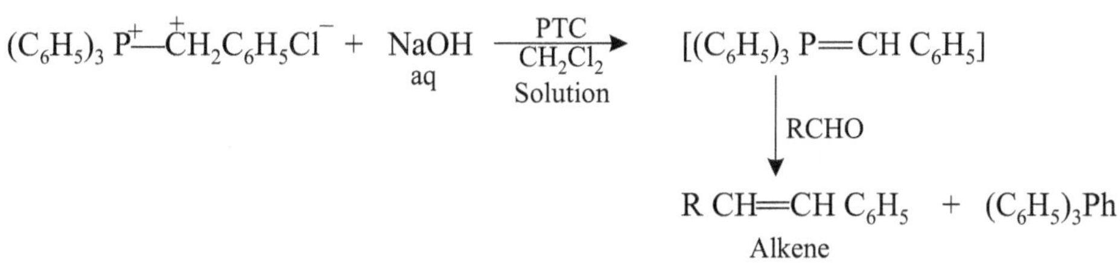

Fig. 12.70: PTC catalysed Wittig reaction in presence of aq NaOH.

The PTC catalysed Wittig reaction is limited to aldehydes only. In certain cases, the phosphonium halides on treatment with aq. NaOH gives crystalline phosphonium ylides[102] (Fig. 12.71). Such crystalline ylides are also obtained when the phosphonium salt contains a $COOCH_3$[103] or a CHO[103a] group rather than a cyano group.

$$\bar{X}(C_6H_5)_3\overset{+}{P}CH_2CH{=}CH\ CN + \underset{aq}{NaOH} \longrightarrow (C_6H_5)_3\overset{+}{P}\bar{C}H{-}CH{=}CHCN + NaCl$$

Fig. 12.71: Formation of crystalline phosphonium ylides

2.2.27 Wittig-Horner Reaction

Wittig-Horner reaction, a modification of Wittig reaction, a readily available phosphine oxide, $Ph_2\overset{O}{\overset{\|}{P}}CH_2R$ is used. Its lithio derivative on reaction with aldehydes or ketones gives α-hydroxy phosphene oxides, which on treatment with NaOH eliminates water to give the corresponding alkene. This step is stereospecific, eryltro hydroxy phosphine oxide gives the z-alkene and threo compound give E-alkene by Syn elimination. Various steps are given in Fig. 12.72.

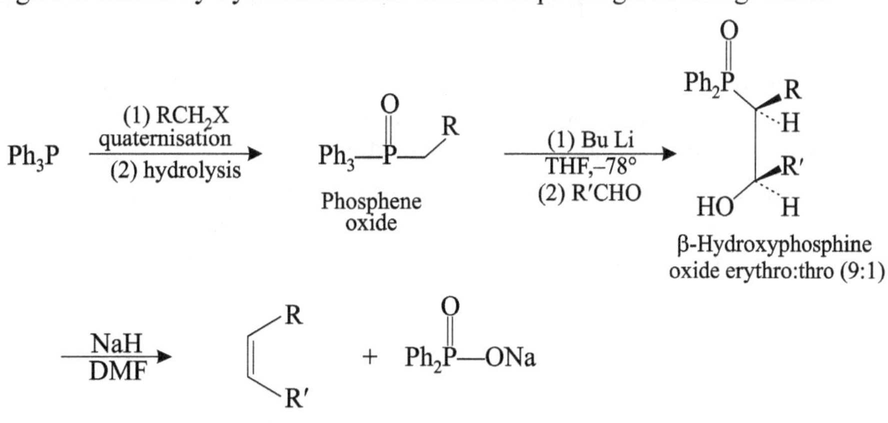

Fig. 12.72: Wittig-Horner Reaction.

The PTC catalysed Wittig-Horner reaction using aqueous NaOH and either tetraalkyl ammonium salts or crown ether as catalyst has also been carried out (Fig. 12.73)[104-107].

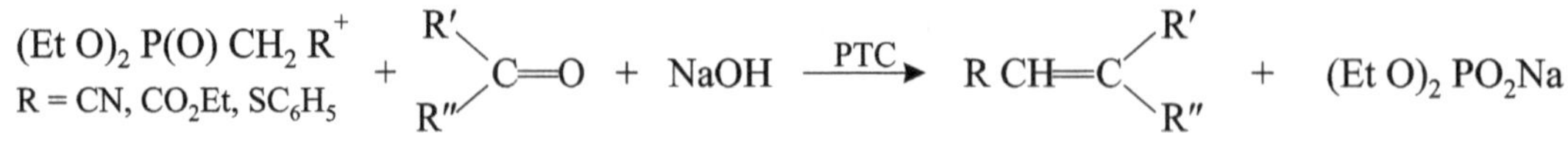

Fig. 12.73: PTC catalysed Wittig-Horner reaction.

12.2.28 Sulphur Ylids

These are obtained from commercially available trimethylsulphonium salts. These sulphur ylids on treatment with aldehydes or ketones give the corresponding oxaranes (Fig. 12.74)[108]

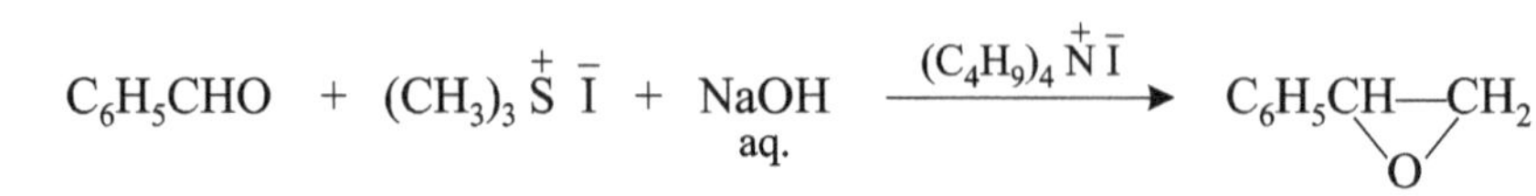

Fig. 12.74: Synthesis of oxiranes from trimethyl sulphonium salts.

12.2.29 Oxidation

12.2.29.1 Potassium Permanganate Oxidation

Of a wide variety of agents available for the oxidation of aromatic compounds, probably potassium permanganate is most widely used. $KMnO_4$ is a derivative of hexavalent manganese and is a powerful oxidant. Its reactivity depends on whether it is used in acidic, neutral or basic conditions. $KMnO_4$ is generally used in aqueous solution and this restricts its usefulness since many compounds are not sufficiently soluble in water. Some of solvents used for $KMnO_4$ oxidation are acetic acid, *tert*-butanol, dry acetone and pyridine.

It is found that oxidation with $KMnO_4$ can be effected in presence of a PTC (*e.g.*, tetraalkyl ammonium or phosphonium salts). The permanganate anion strongly associates[109,110] with quaternary cations and is easily extracted into organic solvents by these reagents. The $Q^+MnO_4^-$ reacts much faster in organic solvents. A very small amount of PTC is sufficient to transfer all permanganate from aqueous to organic phase. The permanganate ion transferred into organic phase oxidies[110,111] olefins, alcohols, nitroalkanes, nitriles and 1,3,5-trimethylbenzene.

The solution of $KMnO_4$ in benzene, thus obtained by the action of quaternary salt to a suspension of benzene and $KMnO_4$ in water is known[110] as 'Purple benzene'. The benzene solution can be separated and used for oxidation under anhydrous conditions.

Alternative oxidation with aqueous solution of $KMnO_4$ can also be effected is presence of a crown ether, dicyclohexano-18-crown 6, which forms a complex. Under these conditions, permanganate becomes soluble in benzene and the resulting solutions are excellent reagents for the oxidation of a variety of organic substrates in organic solvents (Fig. 12.75).

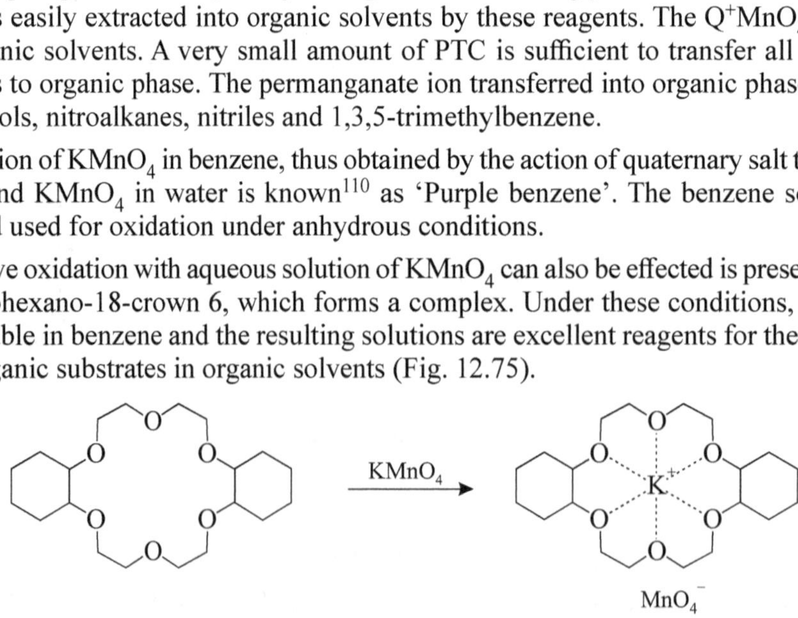

Fig. 12.75: Formation of $KMnO_4$ Complex with dicyclohexano-18-crown-6.

Using this procedure, substituted catechols are converted into the corresponding *o*-qinones in high yields (Fig. 12.76)[112]

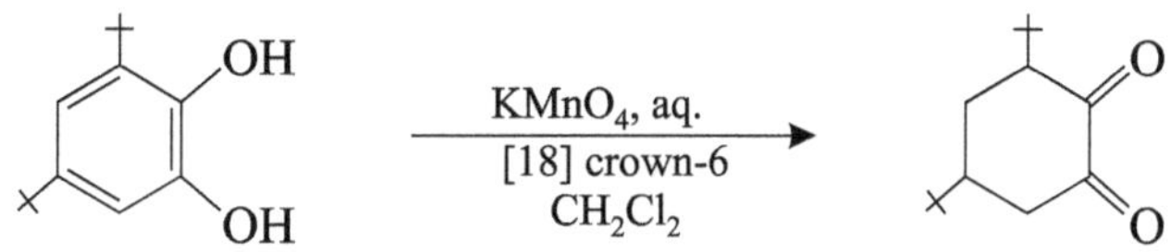

Fig. 12.76: Oxidation of catechols into *o*-quinones.

$KMnO_4$ solubilized in benzene by croun ether has been used[113] for the oxidation of benzyl alcohol, trans stilbene, toluene and benzaldehyde into benzoic acid in quantitative yield. Oxidation of α-pinene by the same reagent gives[113] pinonic acid in 90% yield (Fig. 12.77).

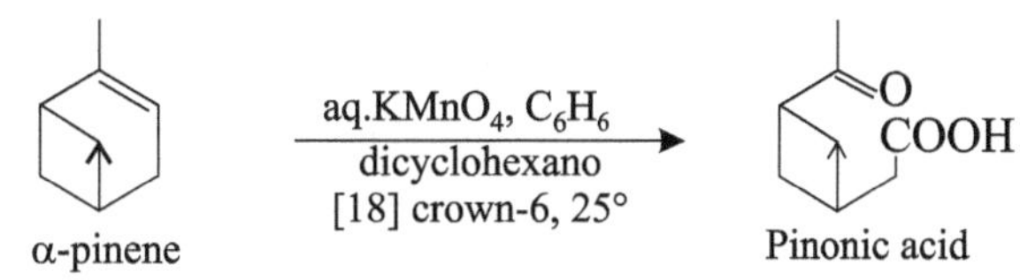

Fig. 12.77: Oxidation of α-pinene into pinonic acid.

12.2.29.2 Chromate Oxidation

Not much work has been reported on phase transferred chromate ions as oxidants for organic compounds. It has been shown[114-116] that quaternary bichromate salts in organic solvents are of low oxidizing power but have higher selectivity. Amberlyst A-26, a commercial resin can be converted into $HCrO_4^-$ form by stirring the chloride form of the resign (35g) into a solution of CrO_3 (15g) in water (100 mL). The resin thus obtained (3.8 mole to CrO_3 per g of resin) is stable at room temperature. This resin has been used for the oxidation of alcohol[115], benzylic or alkylic halides[116] in solvents like hexane, benzene chloroform and THF and give the corresponding aldehydes or ketones in high yields. In these oxidations carboxylic acids are not obtained.

It is found that commercially available quaternary salt ADOGEN 464 (a mixture of trialkyl ($C_{6\text{-}10}$) methyl ammonium chloride) is effective in transferring dichromate into methylene chloride, Chloroform, carbon tetrachloride etc. using powered $K_2Cr_2O_7$. These solutions can oxidise benzyl alcohol, 1-phenylethanol, cinnamyl alcohol and cyclodecanol to the corresponding aldehydes.[117]

12.2.29.3 Hypochlorite Oxidation

Hypochlorite anions can be transferred[117] to organic phase by quaternary cations. This technique finds a number of application in the oxidation of alcohols and amines. Some examples are given in Fig. 12.78.

$$\underset{\text{Benzyl alcohol}}{C_6H_5CH_2OH} + \underset{(aq)}{NaOCl} \xrightarrow[\substack{CH_2Cl_2 \\ 75\ mm}]{Bu_4N^+X^-} \underset{76\%}{C_6H_5CHO}$$

$$\underset{\text{aliphatic alcohol}}{RCH_2OH} + NaOCl \xrightarrow[\text{Slow reaction}]{Bu_4N^+X^-} (R\ CHO) \rightarrow RCO_2H$$

$$\underset{\alpha\text{-Methyl benzyl amine}}{C_6H_5\underset{|}{CH}(CH_3) - NH_2} + NaOCl \xrightarrow[\substack{EtOAc \\ (140\ min)}]{Bu_4N^+X^-} \underset{98\%}{C_6H_5COCH_3}$$

$$\underset{\text{1-octylamine}}{n\text{-}C_7H_{15}CH_2NH_2} + NaOCl \xrightarrow[\substack{EtOAc \\ (35\ min)}]{Bu_4N^+X^-} \underset{\substack{\text{1-Cyanoheptane} \\ 60\%}}{n\text{-}C_7H_{15}CN}$$

Fig. 12.78: Oxidation by hypochlorite anion in presence of quaternary cations.

A special application of phase transfer catalysed hypochlorite oxidation is the conversion of some polycyclic aromatic compounds to arene oxide (Fig. 12.79)[118]

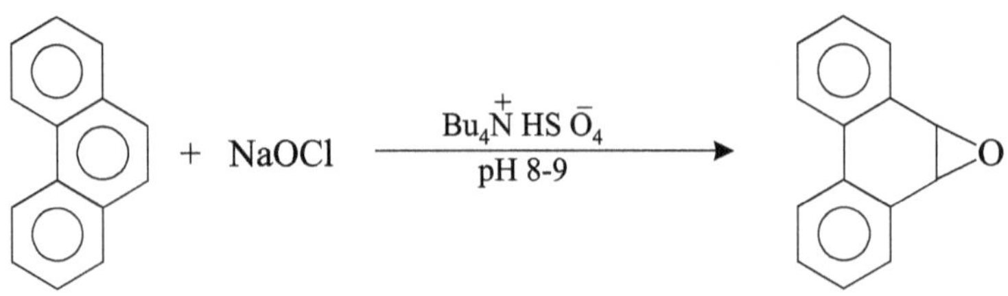

90%

Fig. 12.79: Oxidation of polycyclic aromatic compounds to arene oxide.

12.2.29.4 *Oxidation with Osmium or Ruthenium Oxides in Presence of Periodic Acid*

Osmium tetraoxide in catalytic amount with periodiate has been used[119] for the oxidation of olefins to aldehydes via the formation of 1, 2-diols (Fig. 12.80).

$$RCH = CHR' + OsO_4 + NaIO_4 \longrightarrow RCHO + R'CHO$$

Fig. 12.80: Oxidation of olefins with $OsO_4/NaIO_4$

In homogeneous solutions, H_2O_2 in *t*-butanol, and acetone have been used[120,121] as oxidants. Heterogeneous systems involving chlorate[122], hypochlorite[123], ferrocyanite[124] and periodate[125] have also been used. It is found that quaternary ammonium salts have been used[126] as phase transfer catalysts for heterogeneous osmium catalysed oxidation of olefins with periodic acid.

Ruthenium oxide like Osmium tetroxide is expensive and is used[127,128] in catalytic amount with oxidising agents like HIO_4, MnO_4^-, BrO_3^-, Cl_2, OCl^- in order to maintain Ru is RuO_4 state. Quaternary ammonium and phosphonium salts are used[126] as PTC for the oxidation of olefins to carboxylic acids. In these reactions, aldehydes can be obtained as major products in case the reaction is carefully controlled.

12.2.29.5 Potassium Ferricyanide Oxidation

1,2-Disubstituted hydrazines on oxidation with $K_3Fe(CN)_6$ in presence of 2,4,6-triphenylphenol (TPP) as PTC in presence of NaOH[129] to give 1,2-disubstituted azo compounds (Fig. 12.81)

$$\underset{\text{1,2-Disubstituted hydrazines}}{\text{RNH NHR}'} + K_3Fe(CN)_6 \xrightarrow[\text{NaOH}]{\text{TPP}} \underset{\substack{\text{63-98\%} \\ \text{1,2-Disubstituted} \\ \text{azo compounds}}}{\text{RN} = \text{NR}}$$

Fig. 12.81: Dehydrogenation of 1,2-disubstituted hydrazines to the corresponding azo compounds.

In the above reaction 2, 4, 6-triphenylphenol (TPP) acts as catalyst by transfer to the aqueous phase as anion which is oxidized to triaryl phenoxyradical, which in turn move into the organic phase where it dehydrogenate the hydrazine and simultaneously regenerating triarylphenol (Fig. 12.82).

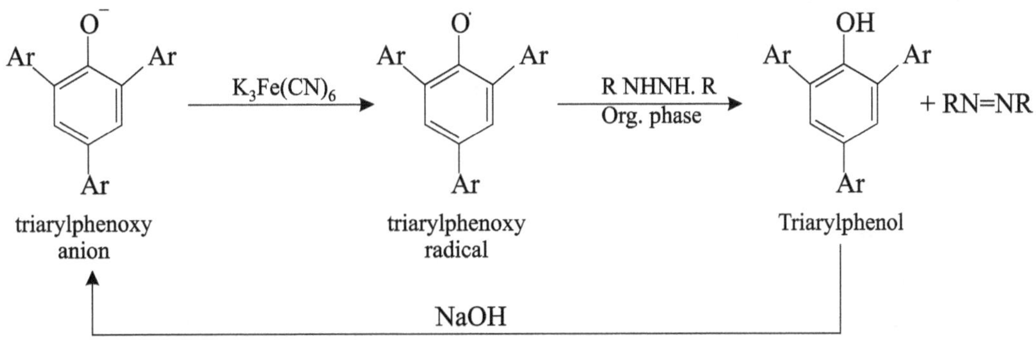

Fig. 12.82: Mechanism of dehydrogenation of disubstituted hydrazines.

12.2.29.6 Air Oxidations

Fluorene is oxidised to fluorenone by using benzene as solvent in presence of PTC and air as oxidant (Fig. 12.83)[216]

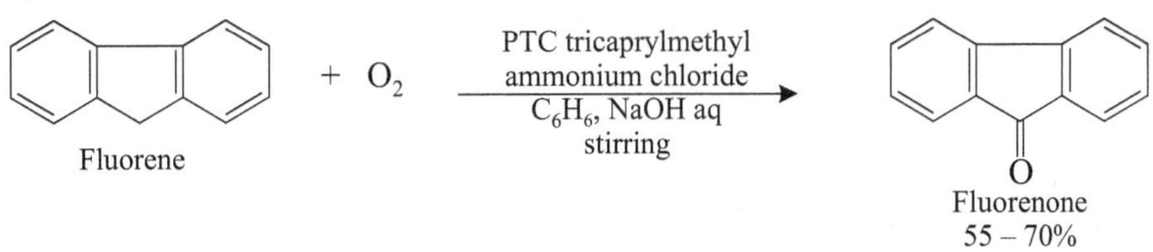

Fig. 12.83: Oxidation of fluorene to fluorenone.

12.2.29.7 Oxidation with Peroxides

PTC transfer the peroxide anions to the organic phase as salts of percarboxylic acids. In this form, these are weak oxidant and this has been used[130] for the PTC oxidation of hydrazones to diazo compounds (Fig. 12.84). Iodine is found to be an effective co-catalyst at a level 10^{-4} mole per mole of hydrazone

$$\underset{\text{in } CH_2Cl_2}{(C_6H_5)_2C = NNH_2} + CH_3CO_3H + \underset{\text{aq.}}{NaOH} \xrightarrow[\substack{I_2\ \text{(CO catalyst)} \\ 0°,\ 45\ \text{min}}]{(C_8H_{17})_3N^+C_3H_7Cl^-} (C_6H_5)_2C = N = N$$

Fig. 12.84: Oxidation of hydrazones to diazo compounds.

12.2.30 Reduction

12.2.30.1 Hydride Reduction

Tetraalkyl ammonium borohydrides[131,132] have been used as reducing agents in systems which have solubility problems. These salts transfer BH_4^- anion from aqueous solution to organic phase. It has been found that quaternary salts having a hydroxyl group on the β-carbon are much superior catalysts for borohydride reductions. The tetraalkylammonium borohydrides are obtained in situ by the reaction of tetraalkyl ammonium halides with sodium borohydride in aqueous medium. The compound, in general is reduced in organic solvent (*e.g.*, benzene) with sodium or potassium borohydride, water and appropriate quaternary ammonium halide (stirring at RT for 0.5-7 hr.)

The quaternary salts, *viz.*, (–)- N-Dodecyl-N, N-dimethyl ephedrinium bromide is used[133] for the reduction of aldehydes and ketones

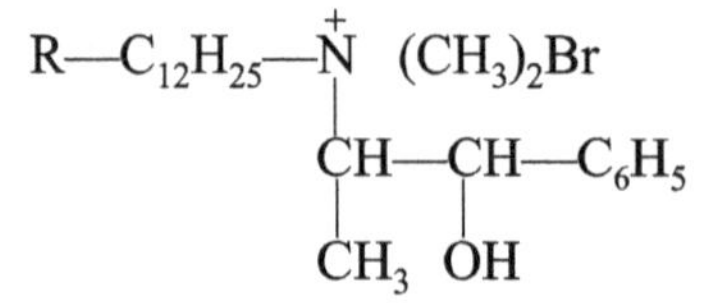

Tetraalkyl ammonium cyanoborohydrides prepared by the ion extraction technique have highly selective reducing activity in homogeneous solution[134].

Crown ethers have also been used[135] for the sodium borohydrile reductions.

12.2.30.2 Reduction by Diborane

Borohydride anion is easily transferred into methylene chloride solution by ion pair extraction with tetrabutylammonium bisulphate R_4N^+ BH_4^- salts can also be prepared from $R_4N^+I^-$ by using ion exchange resins[136]. It has been found that the treatment of the borohydride solution in CH_2Cl with alkyl halides like CH_3I, C_2H_5Br liberates diborane, which may be used for the usual reduction hydroboration reactions[137].

The diborane solutions can also be used for the usual olefin hydroboration reaction, followed by oxidation of the formed adduct with 30% aq H_2O_2 to give alcohols. Using this procedure cyclohexene and phenylalkyl ether gives alcohols in 90% and 70% yields respectively.

12.2.30.3 Reduction by Formamidine Sulfinic Acid

Formamidine sulfinic acid is a strong reducing agent and is prepared[138] by the oxidation of thiourea with H_2O_2 (Fig. 12.85)

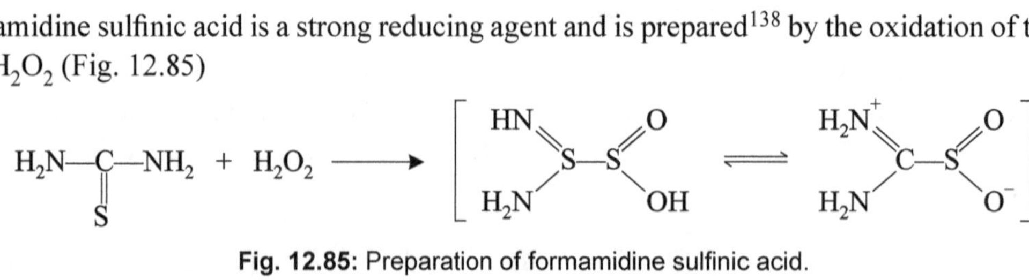

Fig. 12.85: Preparation of formamidine sulfinic acid.

The reagent can be transferred[139] by quaternary ammonium salts into organic solutions for reaction with organic substrates. It reduces disulphides to thiols (Fig. 12.86).

$$R-S-S-R + CN_2SO_2H_4 \xrightarrow[\text{NaOH aq.}]{C_{16}H_{33}Bu_3P^+Br^-} 2RSH + H_2NCSNH_2$$
$$\text{60-70\%}$$
$$R = C_6H_5-, C_6H_5-CH_2-, \text{n-}C_4H_9-, \text{Sec.}C_4H_9-$$

Fig. 12.86: Reduction of disulphides to thiols.

12.2.31 Superoxide Anion

The superoxide radical anion is available in the form cheap potassium superoxide (KO_2) or sodium superoxide (NaO_2). Its use has been restricted for synthetic transformations due to solubility problems associated with K^+ and Na^+ salts. It is found[140] that crown ethers, *viz.*, 18-crown-6, dicyclohexyl-18-crown-6 and dibenzo-18-crown-6 can help to dissolve superoxides in benzene, tetrahydrofuran and dimethylformamide. The resulting solution of naked superoxide on treatment with primary and secondary alkyl bromides give peroxides as the major product. The reaction is believed to proceed as follows (Fig. 12.87).

$$R.Br + O_2^{\dot{-}} \longrightarrow R-O-O^{\bullet} + Br^-$$
$$R-O-O^{\bullet} + O_2^- \longrightarrow R-O-O^- + O_2$$
$$R-O-O^- + R-Br \longrightarrow R-O-O-R + Br^-$$

Fig. 12.87: Reaction of Superoxide with alkyl halide gives peroxides.

The use of crown ethers has provided[141] a simple process for the conversion of bromides to alcohols (Fig. 12.88). In this procedure 4 equivalent of KO_2 and 2 equivalent of crown ether are employed.

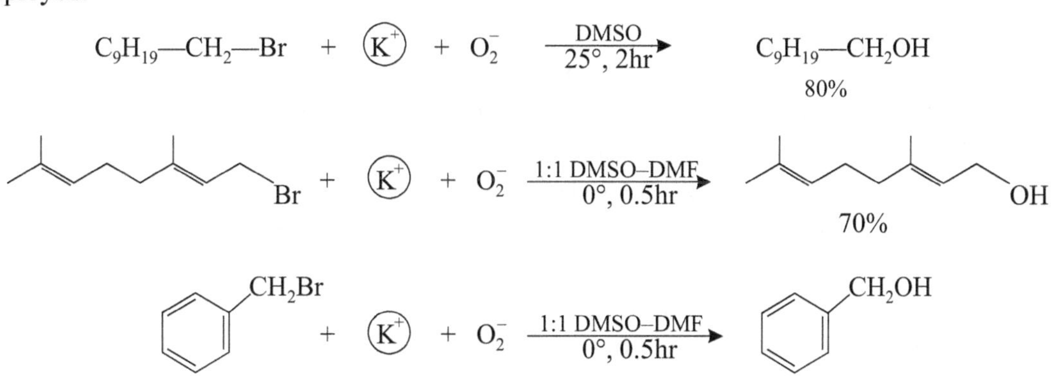

Fig. 12.88: Conversion of bromides to alcohols.

12.2.32 Phase Transfer Catalysts in Pharmaceutical Industry

Phase transfer catalysts have been used to obtain kilogram quantities of various intermediates which find use for the synthesis of a large number of pharmaceutical chemicals including drugs. For details see chapter 18.

Conclusion

The phase transfer catalyst have gained wide applications in organic synthesis. Using this procedure a variety of reactions can be performed in excellent yields and purity. Even reactions which did not proceed satisfactorily under usual conditions could be performed satisfactorily under PTC conditions. This method has been used for the preparation of kilogram quantities of a large number of intermediates which find use in pharmaceutical Industry. The methodology is versatile and the scope of applications is tremendous.

REFERENCES

1. C.M. Starks, J.Am. Chem. Soc., 1971, **93**, 195.
2. K.E. Koening and W.P. Weber, Tetrahedron Lett., 1974, 2275; H. Habner, Ann. Chem. Pharm., 1861, **120**, 330.
2*a*. J.W. Zubrick, B.I. Dunbar and H.C. Durst, Tetrahedron Lett., 1975, 71.
2*b*. C.M. Starks and C. Liotta, Phase Transfer Catalysts, Academic Press, N.Y. P. 127.
2*c*. T.A. Bianchi and L.A. Catt., J. Org. Chem., 1977, **42**, 2039.
3. T. Saraie, K. Ishiguro, K. Kawashima and K. Morita, Tetrahedron Lett., 1971, 2121.
4. J. Graefe, Z. Chem., 1975, **15**, 301.
5. W.B. Weber, G.W. Gokil, Angew. Chem. Int. Ed. Eng., 1972, **11**, 530.
6. D. Landini, F. Moutanari and F. Rolla, Synthesis, 1974, 428.
7. H. Hudlicky, Organic Fluorine Compounds, Plenum, New York, 1971.
8. C.M. sparks, J. Chem. Educ., 1968, **45**, 185.
9. P.W. Herriott and D. Picker, Tetrahedron Lett., 1972, 4521.
10. H.D. Durst, Tetrahedron Lett., 1974, 2421.
11. Reeves and Bahr, Synthesis, 1976, 823.
12. Nakajima, Oda amer Inouge, Tetrahedron Lett., 1978, 3107.
13. R. Lantzsch, A. Markhold and K.F. Lehmint (to Bayer AG), German Patent, 2, 545, 644 (1977).
14. A. McKillop, J.C. Fiaud and R.P. Hug, Tetrahedron, 1974, 30, 1379.
15. A.P. Bashail and J.F. Collins, Tetrahedron Lett., 1075, 3489.
16. H. Alsaidi, R. Gallo and J. Metzner, Synthesis, 1980, 921.
17. H. Dou, P. Hassanby, G. Vermin and J. Metzger, Helv. Chim. Acta, 1978, **61**, 3143.
18. H. Alsaidi, R. Gallo and J. Metzger, C.R. Acad. Sci., Ser C, 1979, **289**, 203.
19. D. Landini and F. Rulla, Synthesis, 1974, 565.
20. H.E. Hennis, L.R. Thompson and J.P. Long, Ind. Eng. Chem. Prod. Res. Dev., 1968, **7,** 96.
21. H.E. Hennis, J.P. Esterly, L.J. Collins and L.R. Thompson, Ind. Eng. Chem. Prod. Res. Dev., 1967, **6**, 193.
22. B.W. Peace, F. Carman and D.S. Wulfman, Synthesis, 1971, 658.
23. H.D. Durst, Tetrahedron Lett., 1974, 2421.
24. H.D. Durst, M.Milano, E.J. Kihta, S.A. Connelly and E. Grushka, Anal. Chem., 1975, **47**, 1797.
25. P.J. Pedersen, J. Am. Chem. Soc., 1967, **89**, 2485, 7017; 1970, **92**, 386, 391.
26. C.J. Pederson and H.K. Friensdorff, Angew. Chem. Int. Ed. Engl, 1972, **11**, 16.
27. C.J. Pederson, J. Am. Chem. Soc., 1967, **89**, 2485, 7017; 1970, **92**, 386, 391.
27*a*. C.M. Starks, J. Am. Chem. Soc., 1971, **93**, 195.
28. W. Kirmse, Carbene Chemistry, Academic Press, Ny, 1964, P. 152.

29. C.A. Buehler, J. Chem. Educ., 1972, **49**, 239.
30. E. Chinoporos, Chem. Rev., 1963, **63**, 235.
31. W. Von E. Doering and A.K. Hoffmann, J. Am. Chem. Soc. 1954, **76**, 6162.
32. W.E. Parham and E.E. Schweizer., Org. React., 1963, **13**, 55.
33. M. Makosza, A. Kacprowicz and M. Fedorynski, Tetrahedron Lett., 1975, 2119.
34. M. Makosza and M. Ludwikow, Angew. Chem. Int. Ed. Engl, 1974, **13**, 665.
34*a*. R.R. Kostikov and A.P. Molchanov, Zn. Org. Khima 1975, **11**, 1767.
34*b*. M. Makosza and A. Kacprowicz, Rocz. Chem., 1974, **48**, 2129; C.A., 1975, **83**, 78798.
35. J. Craefe, M. Alder and M. Muehlstaedt, Z. Chem., 1975, **15**, 14.
36. I. Tabuski, Z. Yoshida and N. Takahashi, J. Am. Chem. Soc., 1970, **92**, 6670.
37. S. H. Goh, K.C. Chan, T.S. Khan and H.L. Chong, Aust. J. Chem., 1975, **28**, 381.
38. D.T. Sepp, K.V. Scherer and W.B. Weber, Tetrahedron Lett., 1974, 2983.
39. H. Staudinger and O. Kupfer, Ber., 1912, **45**, 501.
40. J. Graefe, I. Forehlick and M. Muehlstoedt, Z. Chem., 1974, **14**, 434.
41. M. Makosza and A. Kacprowicz, Rocz. Chem., 1975, **49**, 1627.
42. T. Sasaki, S. Eguchi, T. Kiriyama and Y. Sakito, J. Org. Chem., 1973, **38**, 1648.
43. S. Kwon, Y. Nishimura, M. Ikeda and Y. Tamura, Synthesis, 1976, 249.
44. F. De Angelis, A. Gambacorta and R. Nicolitti, Synthesis, 1976, 798.
45. J. Hine, E.L.Pollitzer and H. Wagner, J. Am. Chem. Soc., 1953, **75**, 5607.
46. P.S. Skell and I. Starer, J. Am. Chem. Soc., 1959, **81**, 4117.
47. I. Tabushi, Z. Yoshida and N. Takahashi, J. Am. Chem. Soc., 1971, **93**, 1820.
48. P. Stronquist, M. Radcliffe and W.P. Weber, Tetrahedron Lett., 1973, 4523.
49. A. Merz, Synthesis, 1974, 724.
50. A. Merz and R. Tomohogh, Chem. Ber., 1977, **110**, 96.
51. P. Kuhl, M. Muhlstadt and J. Graefe, Synthesis, 1976, 825.
52. A. Khusid, G.V. Kryshtal, V.A. Dombrovsky V.F. Kucherov, L.A. Yanovskaya, V.P. Radentseb and O.S. Chizhov, Tetrahedron 1977, **33**, 77.
53. H.D. Hartzler, J.Am. Chem. Soc., 1961, **83**, 4490; 1961, **83**, 4997; J. Org. Chem., 1964, **29**, 1311.
54. T. Sasaki, S. Eguchi and T. Ogawa, J. Org. Chem.., 1974, **39**, 1927.
55. T.B. Patrick, Tetrahedron Lett., 1974, 1407.
56. S. Julia, D. Michelot and G. Linstrumelle, C.R. Hebd, Seances Acad. Sci, Ser. C, 1974, **278**, 1523.
57. G.F. Hennison, J.J. Sheehan and D.E. Maloney, J.Am. Chem. Soc., 1950, **72**, 3542; G.F. Hennison and A.P. Boisselle, J. Org. Chem., 1961, **26**, 725.
58. T. Sasaki, S. Egushi, M. Ono and F. Nakata, J. Org. Chem., 1976, **41**, 2308.
59. M.S. Newmann and S.J. Gromelski, J. Org. Chem., 1972, **37**, 3220.
60. M.S. Newmann and Z.U. Din, J. Org. Chem., 1973, **38**, 547.
61. M.S. Newmann and W.C. Liang, J. Org. Chem., 1973, **38,** 2438.
62. For a review see E.V. Dehmlov, Angew. Chem. Int. Ed. Engl, 1977, **16**, 493; W.B. Weber and G.W. Gokel, Phase Transfer catalyst in Organic Synthesis, Springer-Verlag, Berlin, 1977.

63. D. Landini, S. Quici and F. Rolla, Synthesis, 1975, 397.
64. J.J. Brunet, C. Sidot and P. Caubere, Tetrahedron Lett., 1981, 1013.
65. J.J. Brunet, C. Sidol and P. Caubere, J. Org. Chem., 1983, **48**, 1166.
66. M. Makosza, Tetrahedron, 1968, **24**, 175.
67. I. Rylski, F. Gajewski and M. Dabrowaka, Acta Pol. Pharm., 1973, **30**, 21.
68. M. Makosza and M. Fedorynski, Synthesis, 1974, 274.
69. R.K. Singh and S. Danishefsky, J. Org. Chem., 1975, 40, 2969.
70. M. Makosza and M. Fedorynski, Roct. Chim., 1971, **45**, 1861.
71. M. Makosza, Tetrahedron Lett., 1969, 67.
72. L.S. Hart, C.R. Killen and K.J. Nauders, J. Chem. Soc. Chem. Commun., 1979, 24.
73. M. Makosza and M. Ludwikow, Anglo. Chem. Int. Ed. Engl., 1974, **13**, 655.
74. J. Jarrouse and C.R. Hebd, Scances Acad Sci, Ser. C., 1951, **232,** 1424.
75. J. Jonczyk, B. Serafin and M. Makosza, Roz. Chem., 1971, 45, 1027.
76. J. Jonczyk, B. Serafin and E. Skulimowska, Rocz. Chem., 1971, **45**, 2097.
77. M. Cinquini, F. Montanori and P. Tundo, Chem. Commun., 1974, 878.
78. H. Dietl and K.C. Brannock, Tetrahedron Lett., 1973, 1273.
79. G.W. Gokel, S.A. DiBiase and B.A. Lipiska, Tetrahedron Lett., 1976, 3498.
80. M. Maurette, A. Lopez, R. Martino and A. Lattes, C.R. Acad. Sci. Ser. C., 1976, **282**, 599.
81. A. Jonczyk and M. Makosza, Rocz. Chem., 1975, **49**, 1203.
82. W.C. Guisda and J. Mathre, J. Org. Chem., 1980, **45,** 3172.
83. S.O. Desilva and V. Snieckus, Can. J. Chem., 1978, **56,** 1621.
84. I.J. Mathias and D. Bukett, Tetrahedron Lett 1970, 4709.
85. E. Santaniello, C. Farachi and P. Pouti, Synthesis, 1970, 617; J. Sundberg, The Chemistryof Indoles, Academic Press, New York, 1970.
86. A. Mahamoud, J.P. Galy, E.J. Vincent and J. Barbe, Synthesis, 1981, 917.
87. D. Nasipuri, S. Banerjee, B.V. Alak and N.P. Daw, Synthesis, 1980, 850.
88. A.W. Harriott and D. Picker, Synthesis, 1997, 447.
89. J. Solodar, Tetrahedron Lett., 1971, 287.
90. A.F. Babicheva, A.I. Vovk and A.A. Yasnickov, Ukr. Khim. Zh., 1976, **42**, 259.
91. W. Tagaki and H. Hara, J. Chem. Soc. Chem. Commun., 1973, 891.
92. C. Darzen, Campt. Rend., 1904, **139**, 1214; 1905, **141**, 766; 1906, **142**, 214; M. Ballester, Chem. Rev., 1955, **55**, 283.
93. A. Jonczyk, M. Fedorynski and M. Makosa, Tetrahedron Lett., 1972, 2395.
94. J.M. McIntosh and H. Khalih, J. Org. Chem., 1977, **42**, 2123.
95. M. Makosza, Tetrahedron Lett., 1966, 5489.
96. J. Jarrouse, C.R. Hebd. Scieces Acad. Sci. Ser. C., 1951, **45**, 1027.
97. W. Tagaki, I. Inouse, Y. Yano and T. Okonogi, Tetrahedron Lett., 1974, 2587.
98. A. Merz, Angew Chem. Int. Ed. Engl., 1973, **12**, 846.

99. For a general Treatose, see Cadogan: Organophosphorous reagents in organic synthesis. Academic Press, N.Y., 1970; For a monograph see Johnson, 'Ylid Chemistry', Academic Press, N.Y., 1966; Reviews-Bestmann and Vostrowsky, Top Curr. Chem., 1983, **109,** 85, 164.
100. G. Markl and A. Merz, Synthesis, 1975, 295.
101. S. Hung and I. Stemmler, Tetrahedron Lett., 1974, 3151.
102. P.D. Croce, J. Chem. Soc., Perkin Trans I, 1976, 619.
103. F. Bohlmann and C. Zdero, Ber., 1973, 106, 3779.
103a. M.J. Berenquer, J. Castello, J. Fermandez and R.M. Galard, Tetrahedron Lett., 1976, 396.
104. M.Mikolajczky, S. Grzejszczyk, W. Midura and A. Zatorski, Synthesis, 1976, 396.
105. R.C. Piechucki, Synthesis, 1974, 869.
106. E. D'Incan and J. Seyden-Penne, Synthesis, 1975, 516.
107. M. Mikokajczuk, S. Grzejszczak, W. Mridura and A. Zatorski, Synthesis, 1975, 278.
108. A. Merz. and G. Märkl, Angw. Chem. Int. Ed. Eng., 1973, **12**, 845.
109. W.A. Gibson and R.B. White, Anal. Chim. Acta, 1955, **12**, 413.
110. A.N. Herriott and D. Picker, Tetrahedron Lett., 1974, 1511.
111. A.P. Krapcho, J.R. Larsen and J.M. Eldridge, J. Org. Chem., 1971, **42**, 3749.
112. C.M. Starks and C. Liotta, Phase Transfer catalysis, Academic Press, N.Y.
113. D. Sam and H.E. Simmons, J. Am. Chem. Soc., 1972, 94, 4024.
114. I.A. Shevtshuk and T.N. Simonova, Ukr. Khim. Zr, 1964, 30, 983.
115. G. Cainelli, G. Cardillo, M. Orena and S. Sandri, J. Am. Chemical Soc., 1976, **98**, 6737.
116. G. Cawdillo, G. Giamecian and S. Sandri, Tetrahedron Lett., 1976, 3985.
117. G.A. Lee and H.H. Feedman, Tetrahedron Lett., 1976, 1641.
118. S. Krishnan, D.G. Kuhn and G.A. Hamilton, J.Am. Chem. Soc., 1977, **99**, 8121.
119. F.D. Gunstone, Adv. Org. Chem., 1960, **1**, 103.
120. N.A. Milas and S. Sussman, J.Am. Chem. Soc., 1936, **58**, 1302; 1937, **59**, 2345.
121. J.W. Cook and R. Schoental, J. Chem. Soc., 1950, 47.
122. G. Braun, J.Am. Chem. Soc., 1929, **51**, 228.
123. R.W. Cummins (to FMS corporation), US patent, 3, 846, 478(1974).
124. J.S. Mayell, Ind. Eng. Chem. Prod. Res. Dev., 1968, **7**, 129.
125. R. Pappo, D.S. Allen, R.U. Lemieux and W.S. Johnson, J. Org. Chem., 1956, **21**, 478.
126. C.M. Starks and C. Liotta, Phase Transfer catalysts, Principles and Techniques, Academic Press, Inc., Ny, P. 310.
127. J.L. Courtney and K.F. Swanborough, Rev. Pure. Appl. Chem., 1972, **22**, 47.
128. K.A. Keblys and M. Dubck (to Ethyl Corpn), US Patent, 3, 309, 649 (1968)
129. K. Dimroth and W. Tuncher, Synthesis, 1977, 339.
130. J.R. Adamson, R. Bywood, D.T. Eastlick, G. Gallagler, D. Walker and E.W. Wilson, J. Chem. Soc. Perkin Trans., 1975, 2050.
131. E.A. Sullivan and A.A. Hinckley, J. Org. Chem., 1962, **27**, 3731.
132. A. Barandstom, U. Janggren and B. Lamn., Tetrahedron Lett., 1972, 3173.
133. S. Colonna and R. Fornasier, Synthesis, 1975, 531.

134. R.O. Hutchins and D. Kandasamy, J. Am. Chem. Soc., 1973, **95**, 6131.
135. T. Matsuda and K. Koida, Bull. Chem. Soc. Jpn., 1973, **46**, 2259.
136. K.M. Semenko, O.V. Kravchenko and S.P. Shikin, Zh. Neorg. Khim, 1975, **20**, 2334; C.A. 1975, **83**, 201303.
137. K. Brändström, U-Junggren and B. Lamm., Tetrahedron Lett., 1972, 3173.
138. E.E. Reid, Org. Chem. Bivalent Sulf., 1963, **5**, 39.
139. G. Borgogono, S. Colonna and Rornasier, Synthesis, 1975, 529.
140. J.S. Valentine and A.B. Curtis, J. Am. Chem. Soc., 1975, **97**, 224.
141. M. Makosza and M. Ludwikow, Angew. Chem. Int. Ed. Eng. 1974, **13**, 655.

13. BIOCATALYTIC TRANSFORMATIONS

13

Biocatalytic Transformations

13.1 INTRODUCTION

Biocatalysts are the well known enzymes, which are proteins in nature and are capable of catalytic activity. The biocatalysts can prevent waste generation by performing the process with high stereo- and regioselectivity, prevent the generation of waste by using water as the green solvent. In fact biocatalysis has become one of the greenest technologies.

The earliest biocatalytic conversion known to mankind is the manufacture of ethylalcohol from molasses the mother liquor left after crystallisation of cane sugar from concentrated Juice. This conversion is brought about by the enzyme 'invertase' (present in yeast) which converts sucrose into glucose and fructose and finally the enzyme 'zymase' (present in yeast) converts glucose and frucstose into ethyl alcohol.

In nature, biocatalysts are responsible for numerous chemical reactions to occur with extraordinary ease in each of our body cells. A typical reaction that takes place in the lungs is the dissociation of carbonic acid into CO_2 and H_2O with the help of enzyme carbonic anhydrase.

$$H_2CO_3 \xrightarrow{\text{Carbonic anhydrase}} CO_2 + H_2O$$

Another example is the decomposition of H_2O_2, a toxic byproduct of metabolism by the enzyme catalase. It catalyses the decomposition of about 10^7 molecules of H_2O_2 in one second.

A characteristic property of biocatalysts is their specificity of action, which can be in any of the following ways:

- The biocatalysts are reaction specific, since they catalyse a particular type of reaction, *e.g.*, estarases hydrolyses only esters.
- The biocatalysts are substrate specific, since they react only with a specific substrate. For example, urease and phosphate hydrolyse only urea and phosphate esters respectively.
- A number of biocatalysts exhibit kinetic specificity. For example, though esterases hydrolyse all esters but they hydrolyse different esters at different rates.
- Many biocatalysts are stereospecific. For example maltase hydrolyse α-glucosides but not β-glucosides which are hydrolysed by emulsion.

An biocatalyst may exhibit more than one specificities. For example, estarase hydrolyse only esters but can hydrolyse one enantiomer of an optically active ester more rapidly than the other.

The activity of the biocatalysts is influenced by a number of factors. These include, temperature, pH of the medium and concentration of the biocatalyst and the substrate.

The biocatalytic transformations have distinct advantages. These are generally carried out in single step and in aqueous medium at ambient temperature and pressure. Also, in these transformation, protection and deprotection of functional groups is not necessary and the transformations are stereospecific. Finally much less energy is needed in these transformations.

In recent times, the biocatalysts have revolutionised the area of synthetic organic chemistry. The importance of biocatalysts is evident from the facts that some of the chemicals have been produced on much larger global scale. These include[1] ethanol (15 million metric tons/year), monosodium glutamate (1 MT/year), citric acid (400,000 MT/year), Lysine (115,000 MT/year), gluconic acid (50,000 MT/year) and penicillin[2] (15,000 MT/year). In addition to these, many valuable compounds are obtained in lesser amounts[3]. These include lactic acid, malic acid, L-phenylalanine, L-aspartic acid, other amino acids[4], flavours and fragancies[5], steroids, vitamins, antibiotics, human growth hormone, erythropoietin and a number of enzymes[6].

Biocatalysts can catalyse reaction, which are otherwise difficult by chemical procedures. Biocatalysts can be broadly classified into six major categories (Table 13.1).

Table 13.1: Reactions catalysed by biocatalysts

Class of biocatalyst	Examples	Reactions catalysed
Hydrolases	Lipase, protease, esterase, nitrilase, nitrilehydratase, glycosidase, phosphatase	Hydrolysis reactions (in water)
Oxidoreductases	Dehydrogenase, oxidase, oxygenase, peroxidase	Oxidation, Reduction
Transferases	Transaminase, glycosyltrasfease, tansaldose	Transfer of a group from one molecule to another
Lyasis	Decarboxylase, dehydrase, deoxyribosephosphate aldolase	Nonhydroxylic bond cleavage
Isomerases	Racemase, mutase	Rearrangement (intramolecular)
Ligases	DNA ligae	Bond formation requiring triphosphate

13.2 BIOCATALYSTS IN ORGANIC SYNTHESIS

Biocatalysts are considered to be important tools in organic synthesis. Following are given some of the application of biocatalysts in organic synthesis.

13.2.1 Oxidation

13.2.1.1 Hydroxylation of Aromatic Rings

The reaction of benzene with *Pseudomonas putida* gives[7,8] cyclohexadiene 1,2-diols. The reaction involves conversion of a π bond of benzene into racemic and chiral, nonracemic cyclohexadiene (Fig. 13.1).

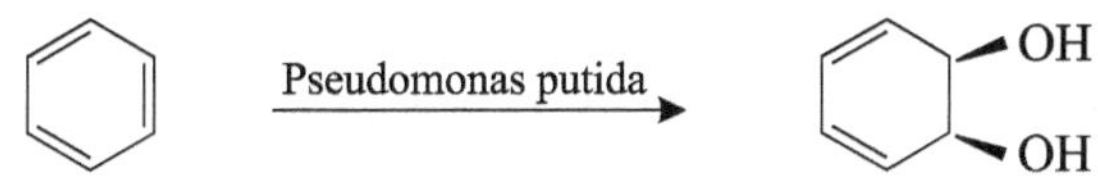

Benzene Cyclohexadiene 1,2-diol

Fig. 13.1: Hydroxylation of benzene.

The above reaction is not restricted to benzene. It is found that substituted benzenes like toluene, chlorobenzene, styrene and phenyl acetylenes also undergo *cis* hydroxylations in positions 2 and 3 (Fig. 13.2)[9].

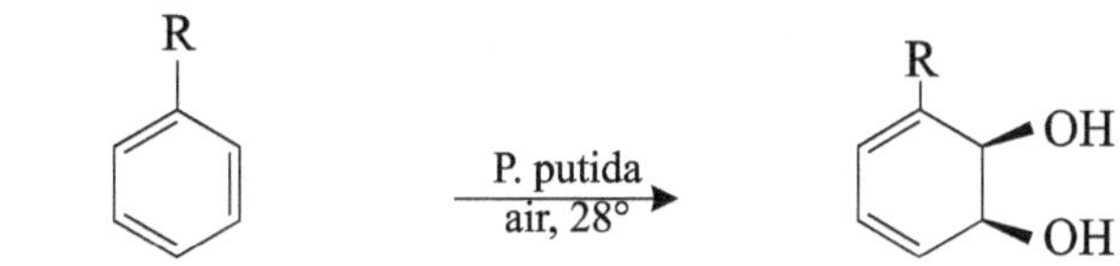

Toluene, R = CH_3
Chlorobenzene, R=Cl
Styrene, R = –CH=CH_2
Phenylacetylene R = –C≡CH

1-Substituted
Cyclohexadiene 2,3-diols

Fig. 13.2: Hydroxylation of substituted benzenes.

See also Fig. 13.37.

It is found that some aromatic compounds can be hydroxylated without loss of aromaticity.[10] Phenols can be hydroxylated selectively; such reaction are difficult to accomplish chemically. (Fig. 13.3)[11].

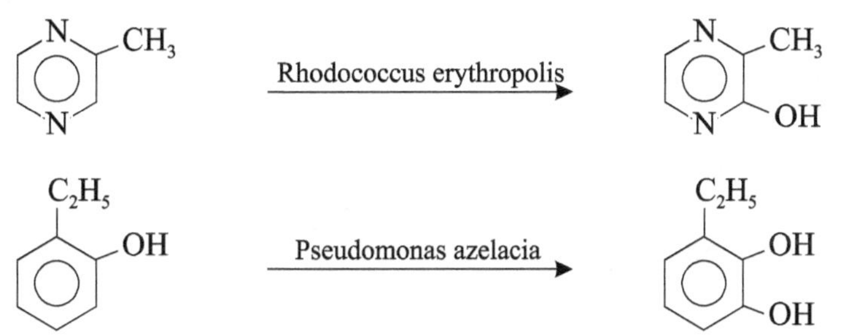

95%

Fig. 13.3: Hydroxylation of aromatic compounds without loss of aromaticity

The diols are important synthons for asymmetric synthesis of important natural products.

13.2.1.2 Biocatalytic Oxidation of Side Chain in Aromatic Nucleus

Compared to chemical oxidations (*e.g.*, oxidation with $KMnO_4$), which oxidises side chain attached to aromatic nucleus to the corresponding carboxylic acids, biocatalytic oxidations proceed without degration or only with limited degradation. Same examples are given below (Fig. 13.4)[12].

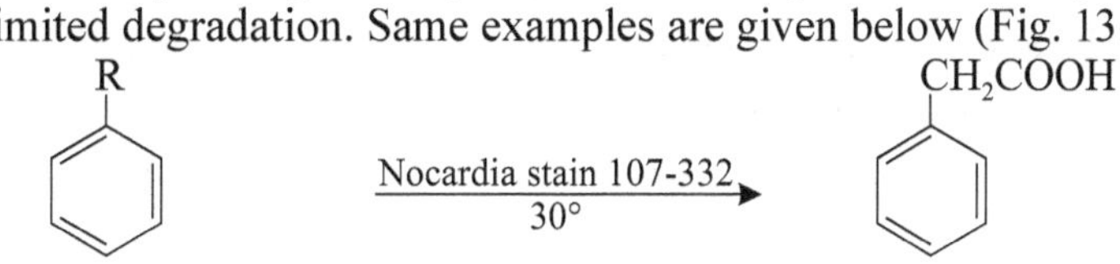

Ethylbenzene, R = CH_2CH_3
n-Butyl benzene, R = $CH_2CH_2CH_2CH_3$
Dedecylbenzene, R = $CH_2CH_2C_{10}H_{21}$

Phenylacetic
acid

Fig. 13.4: Oxidation of side chain in aromatic nucleus.

However, amylbenzene is oxidised by cellulomonas galba initially to 5-phenylvaleric acid, which is converted into trans cinnamic acid by β-degradation and dehydrogenation (Fig. 13.5)[13]

$$\underset{\text{Amyl benzene}}{C_6H_5(CH_2)_4CH_3} \xrightarrow{\text{Cellulomonas galba}} \underset{\text{5-Phenylvaleric acid}}{C_6H_5(CH_2)_4COOH} \rightarrow \underset{\substack{\text{trans}\\ \text{cinnamtic acid}\\ \text{80-100\%}}}{C_6H_5CH=CHCO_2H}$$

Fig. 13.5: Oxidation of amyl benzene to give cinnamic acid.

It is found that side chains can be selectively oxidised (Fig. 13.6)[10]

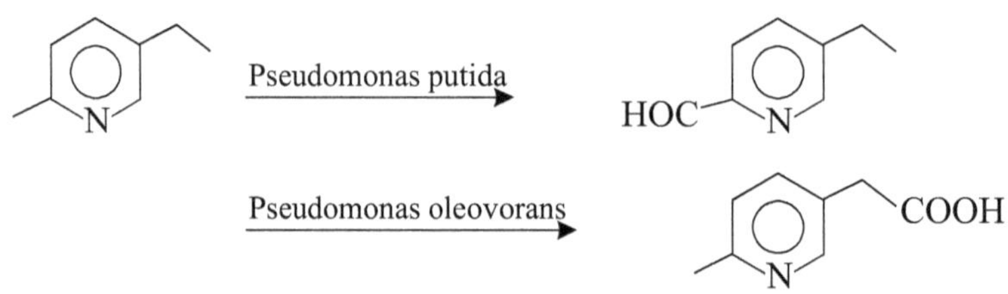

Fig. 13.6: Selective oxidation of side chain in aromatic compounds.

Oxidation of aromatic methyl group to the corresponding aldehyde by Luccase in presence of oxygen in presence of benzothiazoline compound (Fig. 13.7)[14].

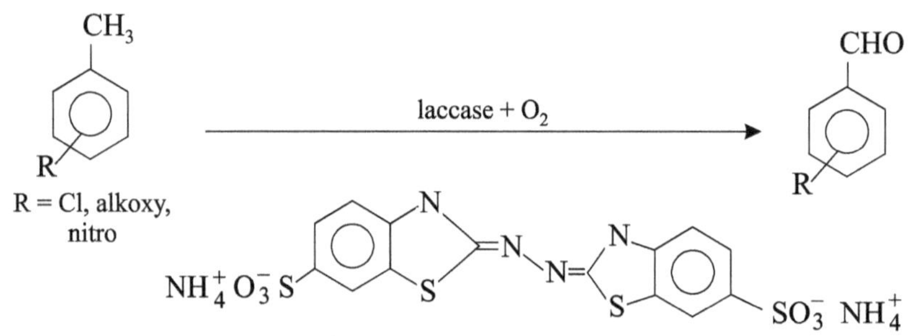

Fig. 13.7: Oxidation of aromatic methyl group to the corresponding aldehyde.

Such transformations (Fig. 13.7) can also use 1-hydroxy-1-H-benzotriazole in place of benzothiazole[15].

13.2.1.3 Biochemical Oxidation of Secondary Alcohol to Ketones

It is well known that[16] a number of chemical oxidising agents are used to oxidise secondary alcohols into ketones. Such conversion can be effected biochemically. Thus using horse liver dehydrogenase, (±)-trans-3-methylcyclohexanol is converted[17] into (–)-(S)-3-methyl cyclohexanone (50% yield, 100% ee). Similarly, (±))-cis-2-methylcyclopentanol is converted into (+)-(S)-2-methylcyclopentanone (55% yield, 96% ee).

The oxidation of secondary alcohol in presence of primary alcoholic group is difficult by chemical means. However, such oxidations can be effected by Acetobacter suboxydans. Thus, aldonit is converted into adinose (Fig. 13.8)[18].

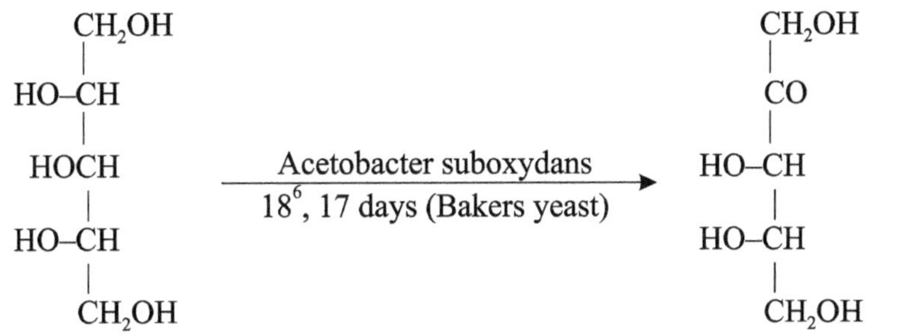

L-Aldonit L-Adinose (95%)

Fig. 13.8: Oxidation of Secondary alcohol in presence of primary alcoholic group

13.2.1.4 Biocatalytic Baeyer-Villiger Oxidation

The oxidation of ketones to esters with H_2O_2 or with peracids is known as Baeyer-Villiger Oxidation[19]. Baeyer-Villiger oxidation can be brought about biocatalytically. For example, phenyl acetone on treatment with cyclohexanone oxygenase in presence of O_2 gives benzyl acetate (Fig. 13.9).

$$\underset{\text{Phenyl acetone}}{C_6H_5CH_2COCH_3} \xrightarrow[\text{EnZ-FAD, NADPH, H}^+]{\text{Cyclohexanone oxygenase, O}_2} \underset{\text{Benzyl acetate}}{C_6H_5CH_2OCOCH_3}$$

Fig. 13.9: Biocatalytic Baeyer-Villager oxidation of phenyl acetone to benzyl acetate.

A typical Baeyer-villiger oxidation is the oxidation of cyclohexanone using cyclohexanone oxygenase from Acinobacter strain NCIB 9871 to give ε-caprolactone (Fig. 13.10)[20].

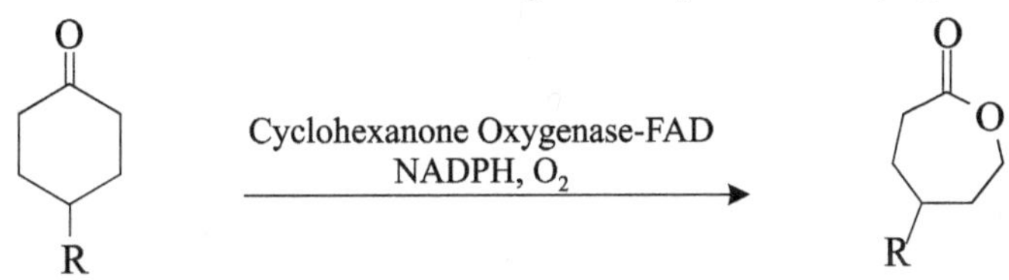

Cyclohexanone
R=H
ε-caprolactone

Fig. 13.10: Baeyer-Villiger oxidation of cyclohexanone.

Similarly 4-methyl cyclohexanone (Fig. 13.10; R = Me) is converted into the corresponding lactone with > 98% ee[21].

The Baeyer-Villiger oxidation of cyclic ketones (*e.g.* cyclobutanone) can also be carried out with monooxygenases from P. putida or A. calcoaceticus, the former is preferred as the latter is pathogenic (Fig. 13.11)[22].

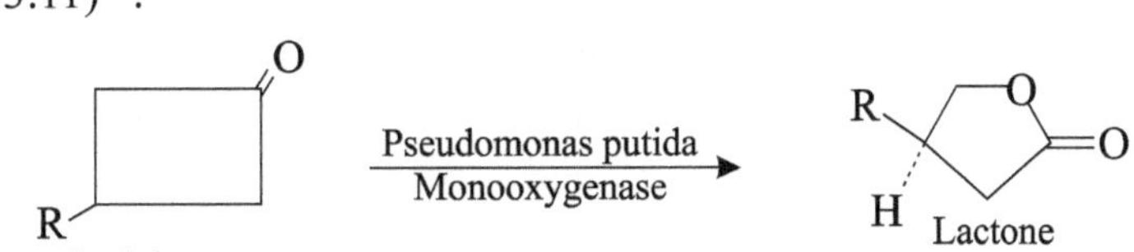

Cyclobutanone

Fig. 13.11: Baeyer-Villiger oxidation of cyclobutanones.

A number of Baeyer-Villiger oxidations of steroids have been accomplished biochemically. Thus, 19-nortestosterone on treatment with Aspergillus tamarii give 19-nortestololactone (Fig. 13.12)[23]

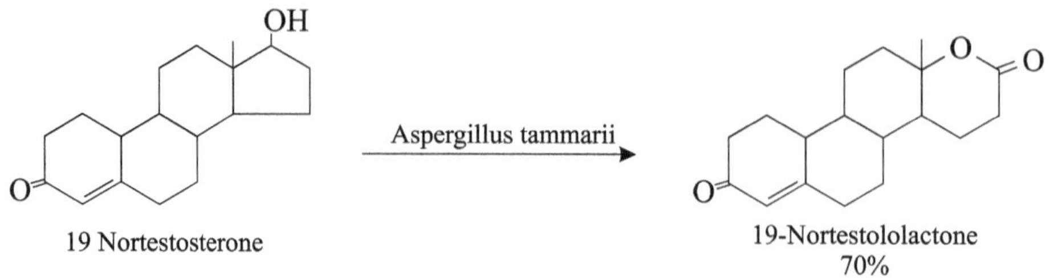

Fig. 13.12: Biochemical oxidation in Steroids.

Progesterone and testosterone can be converted into Δ^1-dehydrotestoloctone by fermentation with cylindrocarpan radiciola. Testololactone is obtained from progesterone by biochemical oxidation with penicillium chrysogenum and from 4-androstene-3,17-dione by treating with penicillium lilacinum[24] (Fig. 13.13)[25]

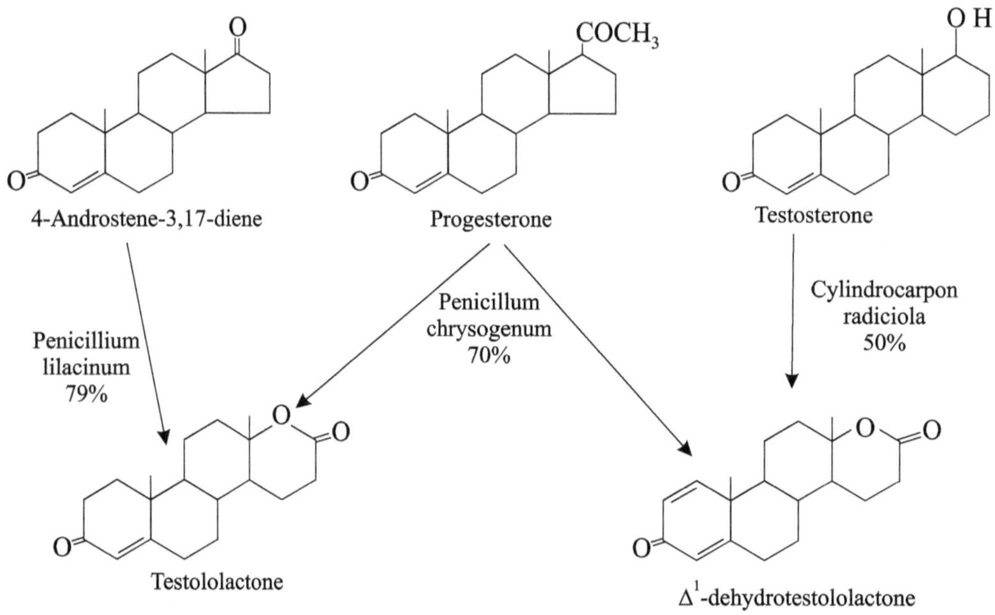

Fig. 13.13: Biochemical Baeyer-Villiger oxidation of 4-androstene-3,17-dione, progesterone and testosterone.

13.2.1.5 Hydroxylation in Steroids

Hydroxylation in steroids can be accomplished in different positions by microorganisms. The most favoured position of hydroxylation is 11α[26-33] followed by 11β[34,35], 6β[34,36], 7α[37], 10β[34], 12α[34], 12β[36], 13β[38], 14α[34], 15α[37,38], 15β[36,37,39], 17α[27] and 17β[34].

Hydroxylations have been accomplished in case of progesterone[29,30,36], dioxycorticosterone[26,37], pregnane dione[32], androstenedione[18] and their derivatives.

Progesterone yields[29] 11α-hydroxy together with 6β, 11α-dihydroxy derivative with Rhizopus arrhizus. On the other hand progesterone give[36] 12β, 15β-dihydroxy derivative with calonectria decora (Fig. 13.14).

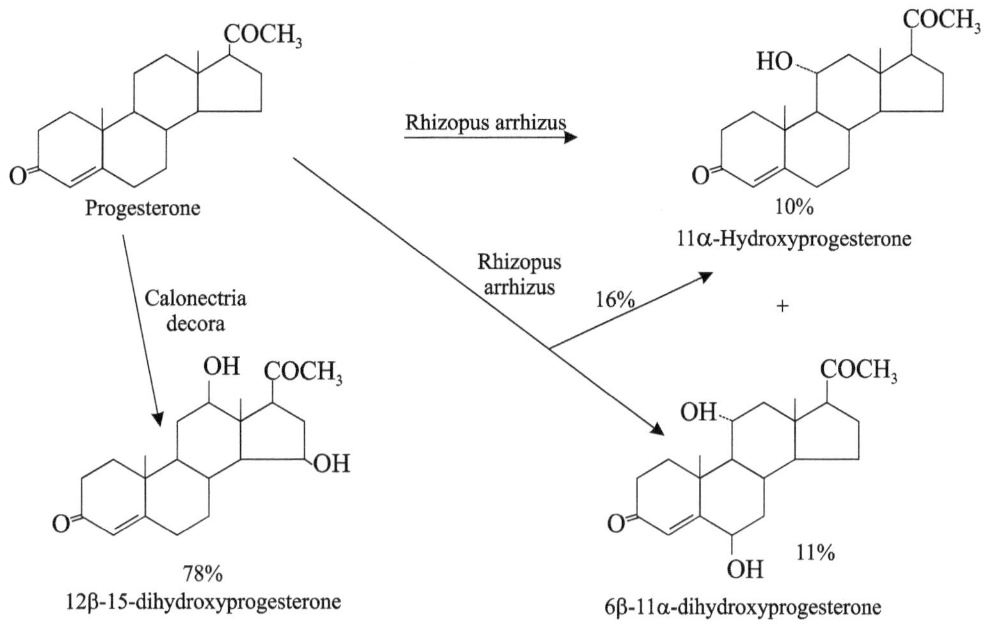

Fig. 13.14: Hydroxylation of Progesterone.

Rhizopus nigricans hydroxylates[31,32] 6-dehydroprogesterone and 3,17-pregnanedione in 50-60 and 39% yields respectively. The same organism brings about 11α-hydroxylation[33] on 16α-17α-epoxyprogesterone.

Using the enzyme Aspergillus niger, deoxycorticosterone (4-pregnen-21-ol-3,20 dione) is converted into Δ^4-pregene-11α, 17α-diol-3,20 dione and to 17α-hydroxydesoxycorticosterone by Peziza species (Fig. 13.15)[26].

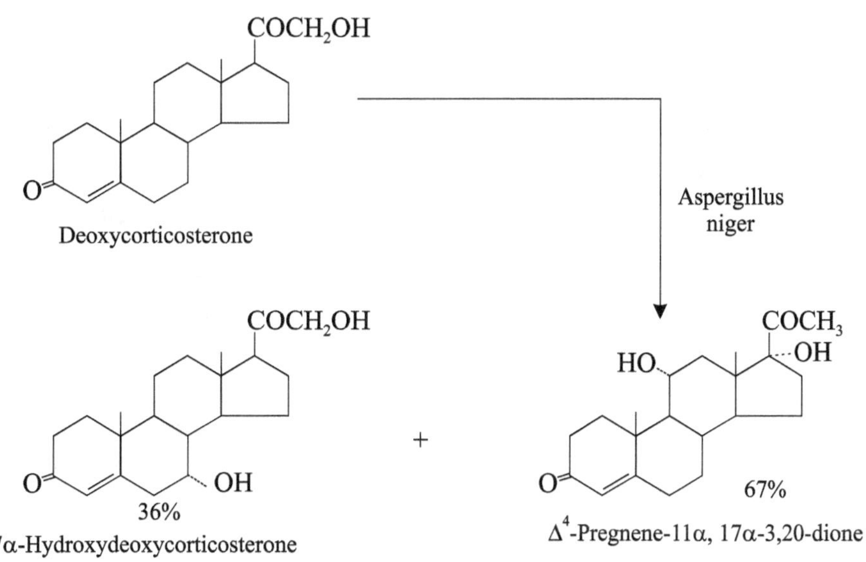

Fig. 13.15: Hydroxylation of deoxycorticosterone

17α-Hydroxydeoxycorticosterone is hydroxylated in the 11α-position by cunninghamella blakesleena H-344 and in 15-position by Bacillus megaterium (Fig. 13.16)[39].

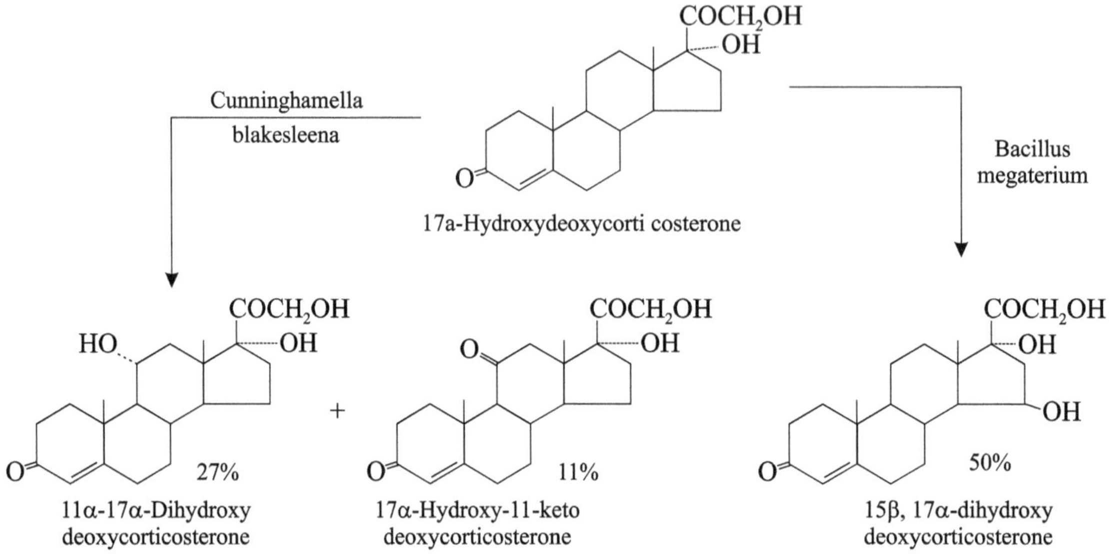

Fig. 13.16: Hydroxylation in 17α-hydroxdeoxycorticosterone

Δ^4-Andostene-3,17-dione is converted[38] into 15α-hydroxy-Δ^4-androstene-3, 17-dione by fusarium lini and into 6β-hydroxy-Δ^4-andosten-17 dione along with 15α-hydroxy-Δ^4-androstene-3, 17-dione with the microorganism Gibberella Saubinetii[35,38] (Fig. 13.17).

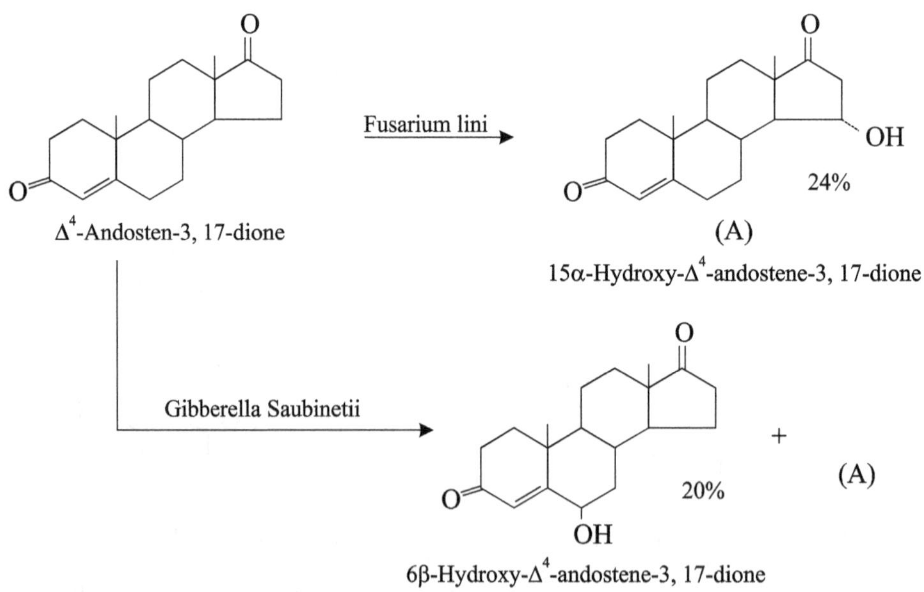

Fig. 13.17: Hydroxylation in Δ^4-Andostene-3, 17-dione.

13.2.1.6 Oxidation of Hydroxy Ketones to Diones in Steroidal Alcohols

In steroids, if the alcoholic group (usually secondary) is remote enough from the keto group, its oxidation to diketones can be usually effected by reagents like chromium trioxide in aq. H_2SO_4; in these in addition to oxidation, isomerisation of double bonds, dehydrogenations and hydroxylation may also occur. The above transformation can best be carried out biochemically. Thus, dehydroepiandosterone is converted[40] into 4-andostene-3,17-dione by corynebacterium simplex. The same product is also obtained from dehydroandostrone using bacterium-infected yeast[41] (Fig. 13.18).

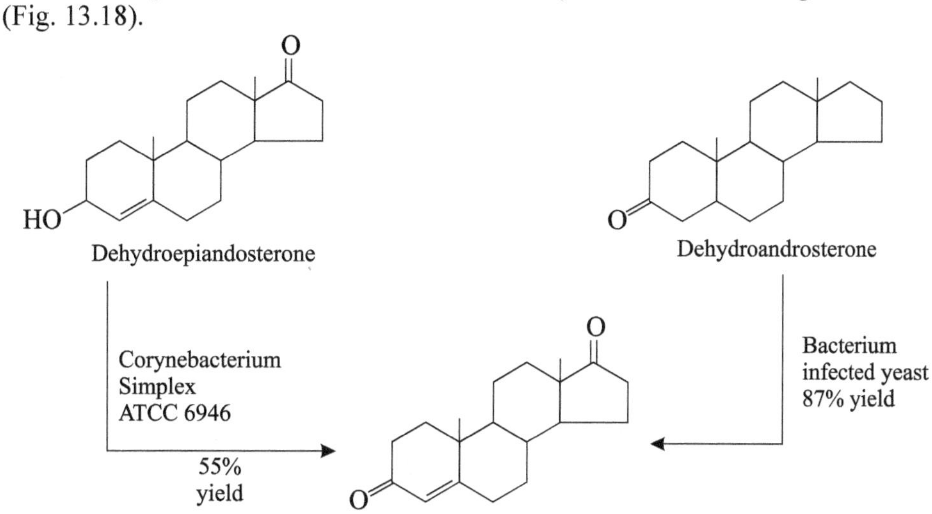

Fig. 13.18: Oxidation of dehydroepiandosterone.

11β-Hydroxysteroids can be oxidised[42] to 11-ketosteroids with Rhizopus arrhizus; the reaction is also accompanied by β-hydroxylation in position 6. The same reaction can also be achieved by Helicostylum piriforme or cunninghamella blakesleeana, but in this case there is α-hydroxylation in 9 position in addition to oxidation. The primary hydroxyl group is not effected in the above transformations (Fig. 13.19).

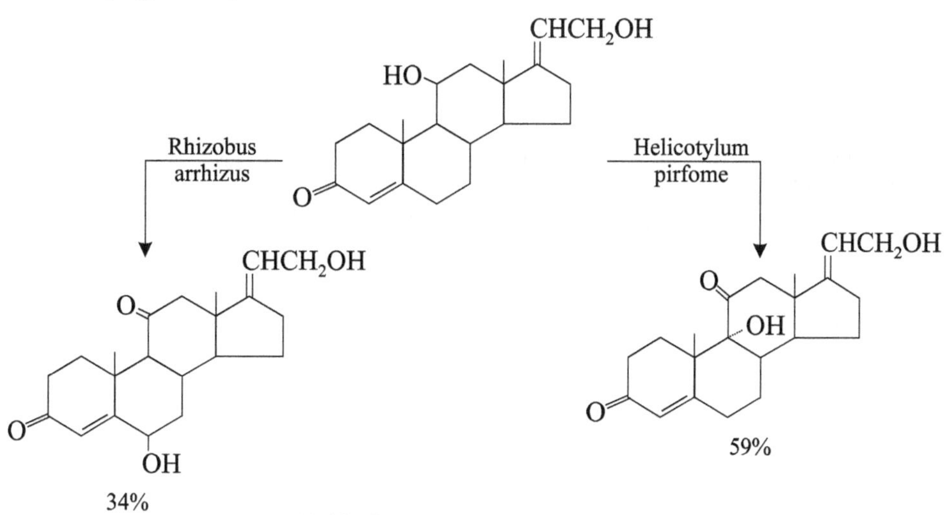

Fig. 13.19: Oxidation of 11β-hydroxysteroids.

13.2.1.7 Biochemical Oxidation of Amines

Amino groups are oxidised to nitro groups with streptomyces thioluteus. A typical example is given below (Fig. 13.20)[43].

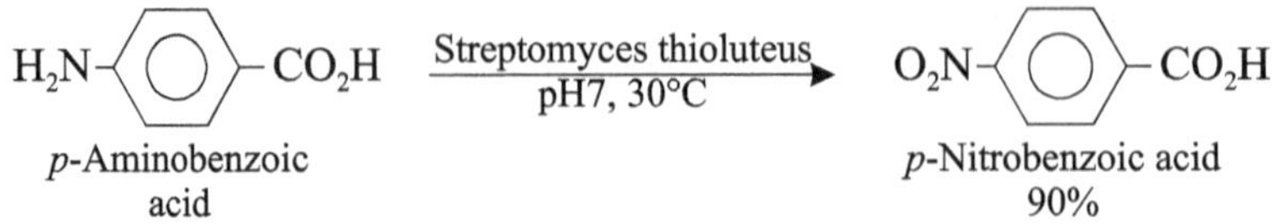

Fig. 13.20: Biochemical oxidation of primary amines.

Tryptamine on biochemical oxidation with Hygrophorus conicus gives[44] 2-indolene-3-acetic acid (Fig. 13.21).

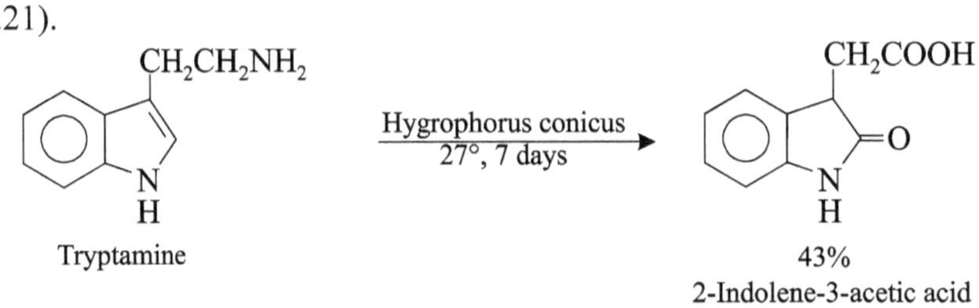

Fig. 13.21: Biochemical oxidation of Tryptamine

N-Benzoylated cyclic amines *e.g.* N-benzoylated piperidine on treatment with sporotrichum sulfurescens give the corresponding 4-hydroxy compound. However, N-benzoyl hexamethyleneimine gives 4-hydroxy-4-keto and 3-keto derivatives (Fig. 13.22)[45].

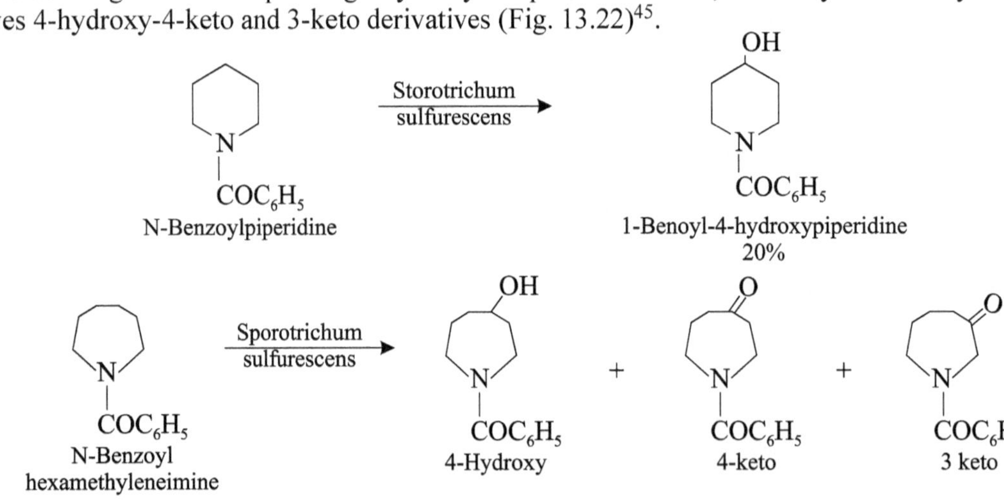

Fig. 13.22: Oxidation of N-benzoylated cyclic amines.

N-Acetyl-1-adamantamine can be hydroxylated on the tertiary and also on secondary carbons to yield N-acetyl-1-adamantamine-4α-ol and N-acetyl-1-adamantamine-3-ol (Fig. 13.23)[46].

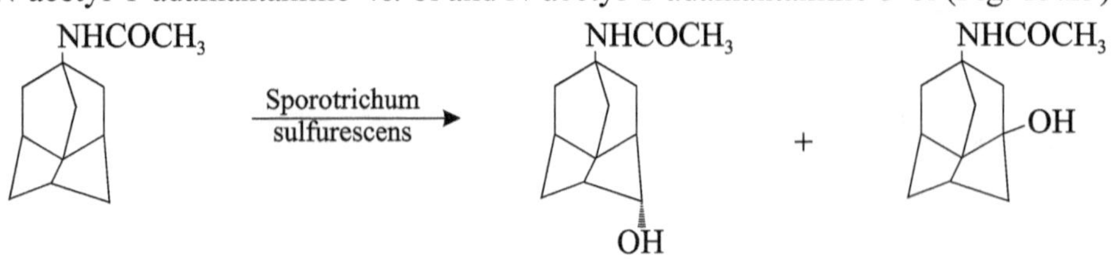

Fig. 13.23: Hydroxylation of N-acetyl-1-adamantamine.

13.2.1.8 Miscellaneous Oxidations

(*a*) Cholesterol can be oxidised to the ketone, cholestenone with cholesterol oxidase in reversed micelles of water is isooctane in 100 yield[47].

(*b*) E-coniferyl alcohol is converted into (+) pinoresinol using an oxidase in the presence of a guiding protein. In the absence of protein, racemic product is obtained[48].

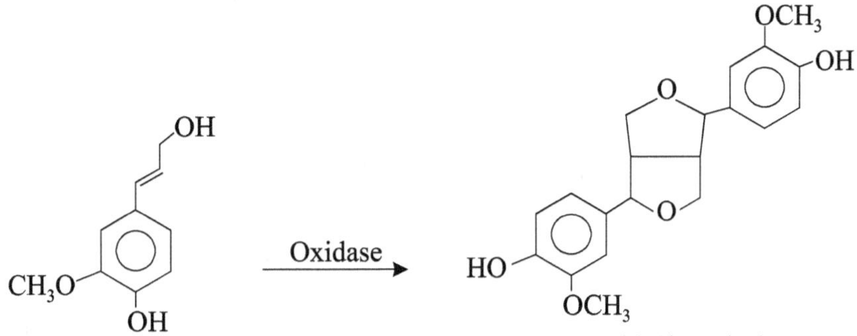

E-coniferyl alcohol (+)-Pinoresinol

(*c*) Lipase has been used to convert carboxylic acids to peracids with H_2O_2. The peracids are used in situ for the oxidation of olefins like oleic acid, oleyl alcohol etc.[49]

(*d*) Diazepam is hydroxylated, the OH group entering the heterocyclic ring with Pelloioularia filamentosa[50].

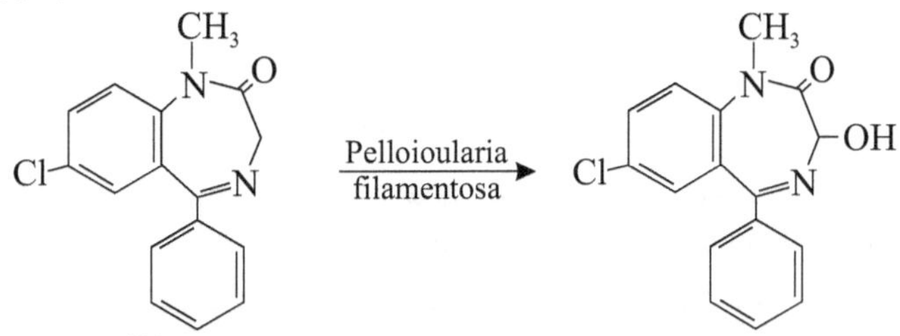

Diazepam

(*e*) Oleic acid on oxidation with Pseudomonas sp gives[51] 10-hydroxysteric acid

$$\underset{\text{Oleic acid}}{CH_3(CH_2)_7CH=CH(CH_2)_7COOH} \xrightarrow{\text{Pseudomonas sp}} \underset{\text{10-Hydroxystearic acid}}{CH_3(CH_2)_7\underset{\displaystyle OH}{\underset{|}{C}H}-CH_2(CH_2)_7COOH}$$

(*f*) Meliotoic acid on oxidation with Arthrobacter species gives 3-hydroxy derivative[52].

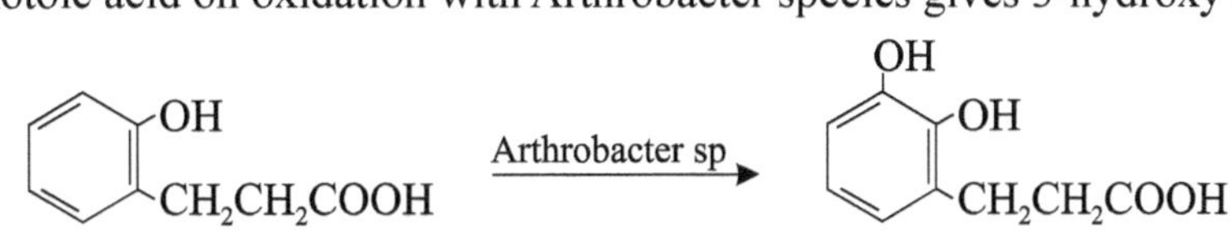

(*g*) Nicotine is hydroxylated at position 6 by Arthrobacter sp.[53]

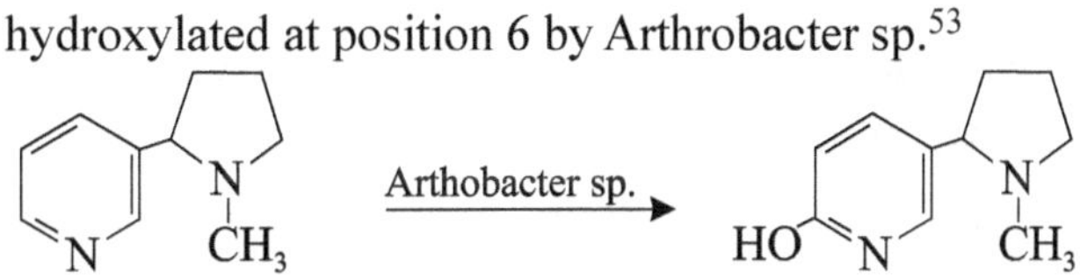

Nicoline

(*h*) Benzene is oxidised into trans, trans muconic acid by Micrococcus spheroids[54].

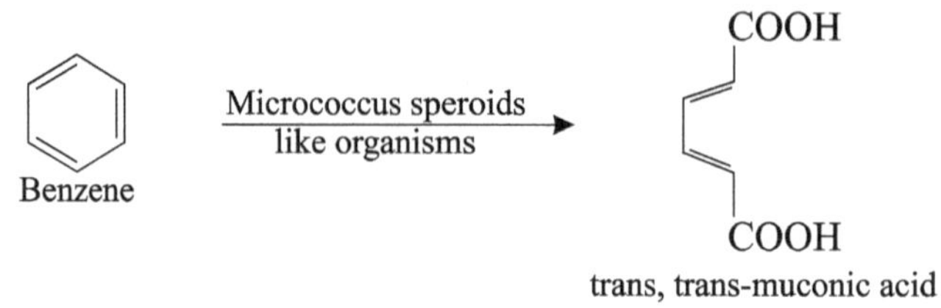

13.2.2 Reduction

Like biocatalytic oxidations, the biocatalytic reductions are in general straight forward and stereoselective. Prelog was the first to study the reduction of carbonyl group. Thus, the reduction of ketones with currularia fulcata, gave predictable stereochemical induction depending on the groups [Large (L) or small (S)] present around the carbonyl. This procedure is referred to as **Prelog's rule**[55]. According to this rule, if the steric difference between large and small groups attached to carbonyl group is large enough, the enzyme delivers hydrogen from the less hindered face (over small group) to give the corresponding alcohol (Fig. 13.24).

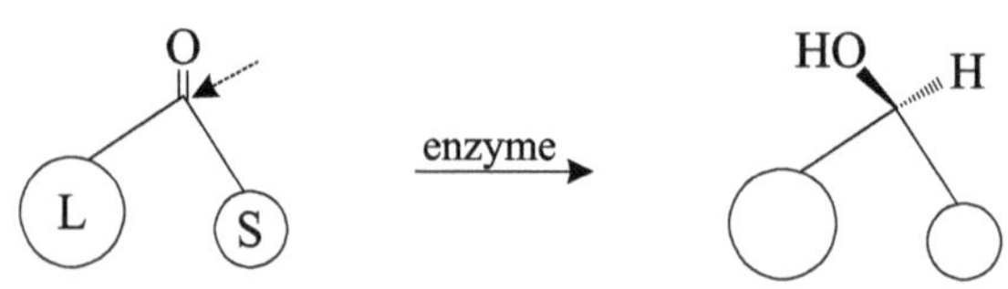

Fig. 13.24: Prelog's rule.

Two common enzymatic systems are yeast alcohol dehydrogenate (YAD) and horse liver alcohol dehydrogenate (HLADH). A very common 'reagent' is **Bakers yeast** (Saccharomyces Cerevisial). It is for a selective reduction of β-ketoesters and β-diketones. Thus, reduction of ethyl cacetoacetate with Baker's yeast gives the (S)-alcohol. However, the reduction of ethyl β-ketovalerate gives the (R)-alcohol (Fig. 13.25)[56].

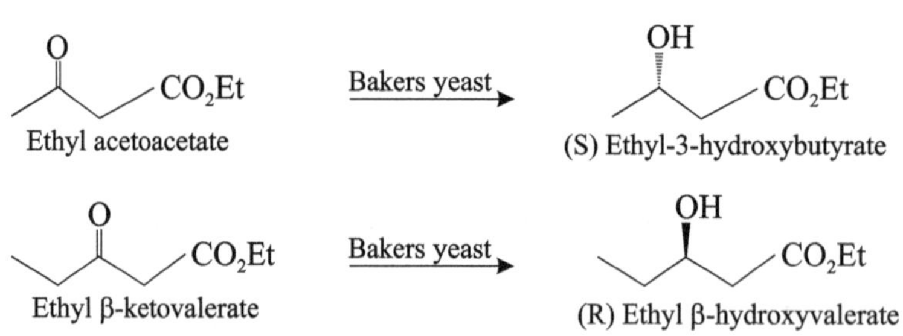

Fig. 13.25: Reduction of ethyl acetoacetate and ethyl β-ketovalerate.

It was shown that the selectivity changed from S selectivity (with small chain esters) to R-selectivity (with long chain esters)[56].

Reduction of β-diketones, *e.g.*, 2,4-hexanedione gave the β-ketoalcohol, *viz.*, (S)-5-hydroxy-3-hexanone (Fig. 13.26)[57].

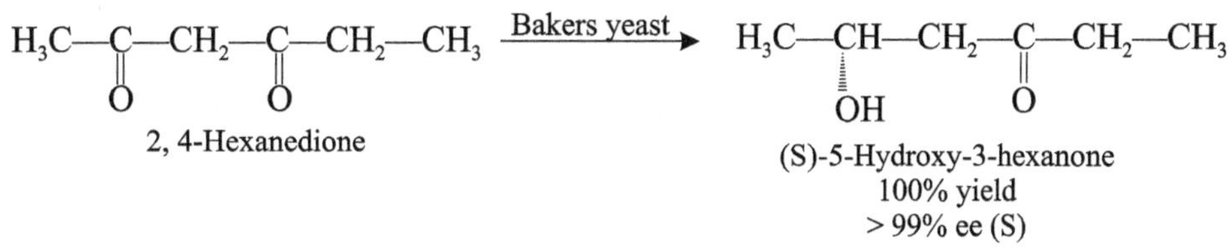

Fig. 13.26: Reduction of β-diketones

Some β-ketoesters like furan derivatives (Fig. 13.27) could not be reduced by Baker's yeast. However, other yeast, *e.g.*, Kloectera Saturmus could reduce the ketonic moiety[58] in 47% to yield a 47 : 53 mixture of syn and anti diasteromers in 87% *ee* and 53% *ee* respectively[58].

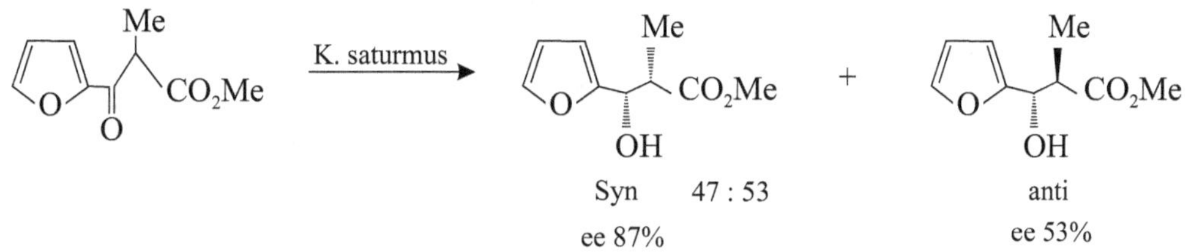

Fig. 13.27: Reduction of furan derivatives with K. Saturmus.

Another yeast, Saccharomyces delbrueckii in the above reduction (Fig. 13.27) gave[59] 99% *ee* syn and 94% ee of anti diasteromers.

Simple ketones can also be selectively reduced by Bakers yeast[60]. Thus, the ketonic moiety in the side chain of cyclopentadione can be reduced to R alcohol with Baker's yeast (Fig. 13.28). The R-alcohol is used in the synthesis of norgestral[61].

Fig. 13.28: Reduction of ketonic moiety to the corresponding R alcohol.

Conjugated carbonyls can also be reduced by Beauveria sulfurescens. Thus, trans-croton aldehyde on reduction gives[62] 2-buten-1-ol in 80% yield. However, 2-methyl-2-pentenal gives[62] a mixture of 31% conjugated alcohol and 69% of completely reduced product (Fig. 13.29).

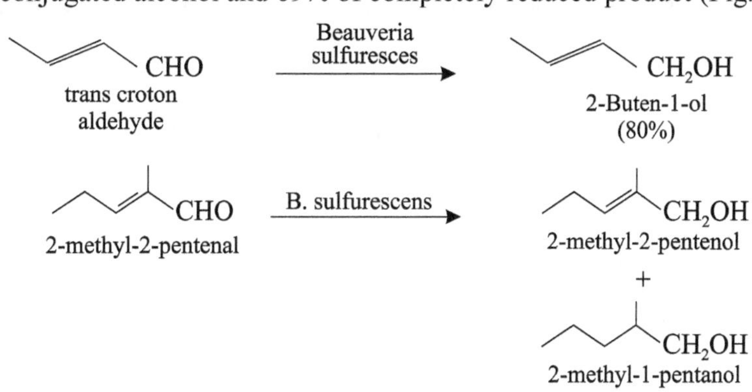

Fig. 13.29: Reduction of conjugated carbonyls.

It has been shown that the reduction of 2-methyl-2-pentenal (Fig. 13.29) gave the S-alcohol by a stepwise process[63] involving first reducing the conjugated double bond (establishing the S-stereochemistry, followed by reduction of the aldehyde moiety to the alcohol. On the basis of extensive work, Veschambre formulated[62] a rule for this type of reduction. According to this rule (now referred to as a modification of Prelog's rule), the hydrogen is delivered to the double bond α to the carbonyl from the front or back side depending on the relative bulk of the groups attached to the carbonyl group. Thus, reduction of geranial with Baker's yeast gives R-citronellol in agreement with the prediction. The Z-isomers (neral) was predicted to give *S*-isomer, but reduction with Baker's yeast gave 6 : 4 mixture, possibly due to isomerisation of the double bond in neral prior to the delivery of hydrogen[63].

It has been found that small ketones, *e.g.*, ethyl methylketone on reduction with Thermoanaerobium brockii gave[64] the R alcohol (2-butanol) is 12% yield (48% ee.). However, reduction of larger ketones, *e.g.*, 2-hexanone gave the S-alcohol (85% yield and 96% *ee.*) Fig. 13.30.

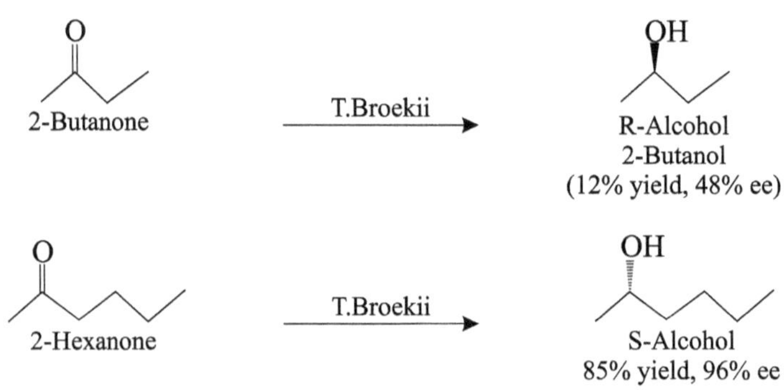

Fig. 13.30: Comparative reduction of small and larger ketones.

On the basis of the above (Fig. 13.30) it is established that the selectivity of reduction depended on the nature and the size of the groups attached to the carbonyl.

An interesting biocatalytic reduction is the conversion of sodium salts of carboxylic acids to the corresponding alcohols with colletotrichum gloeosporioxides in 70% yield[65]. This reduction is difficult to carryout with usual inorganic catalysts.

Reduction of quinoline oxide with Baker's yeast in alkaline medium give quinoline (Fig. 13.31)[66].

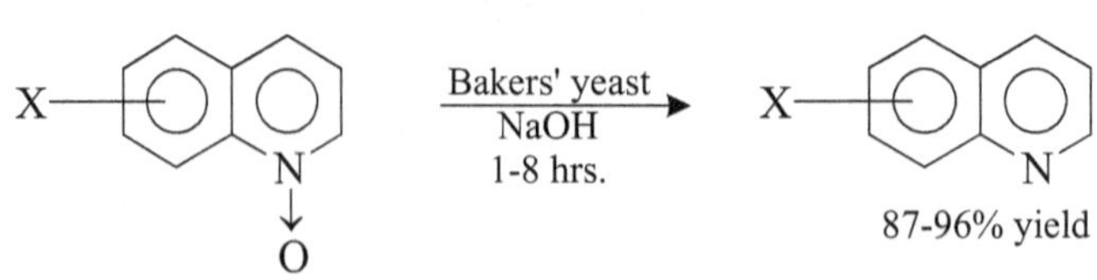

Fig. 13.31: Reduction of quinoline oxide to quinoline

13.2.3 Transesterification

Biocatalysts can be used for hydrolysis of oils and fats and subsequent reesterification[67] with alcohol of oil and rapeseed oils using lipases[68]. The process is referred to as transestrification and is much simpler to the conventional method in which fat or oil is hydrolysed in water at 150-260° for 3-24 hrs. and subsequent reesterification of the formed carboxylic acid.

The method, is fact can be used for making biodiesel fuel by the lipase-catalysed alcoholysis of oils such as rapeseed oil is more than 90% conversion[69]. As an example, the reaction of triolein and ethanol occurs as shown below (Fig. 13.32).

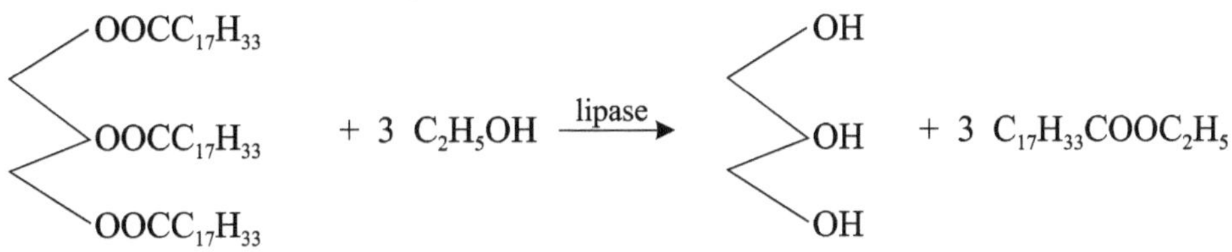

Triolein

Fig. 13.32: Transesterification of triolein.

13.2.4 Selective Hydrolysis of Diesters

Biocatalysts have been used to selective hydrolyse only one of the two esters groups in a molecule. Thus, an esterase has been used to hydrolyse dimethyl adipate to more than 99% half ester[70]. Pig liver esterase has been used to hydrolyse dialkyl phthales to the monoesters in 84-93% in 1-13 hrs.[71]. Also, a porcine pancreatic lipase hydrolyses linear diacetates to the mono acetates in 79-95% yields[72].

13.2.5 Enzymatic Synthesis of Carboxylic Acids, Amides and Nitriles

(*a*) Synthesis of Carboxylic Acids: It is well known that acetic acid, formic acid, lactic acid, citric acid and malic acid are synthesised on a large scale using biocatalysts (Fig. 13.33).

$$CH_3CH_2OH \xrightarrow[\text{alcohol dehydrogenase}]{[O]} CH_3CHO \xrightarrow[\text{alcohol dehydrogenase}]{[O]} CH_3COOH$$

$$CH_3OH \xrightarrow[\text{alcohol dehydrogenase}]{[O]} HCHO \xrightarrow[\text{alcohol dehydrogenase}]{[O]} HCOOH$$

$$\underset{\text{Lactose}}{C_{12}H_{22}O_{11}} + H_2O \xrightarrow[\text{acidic lacti}]{\text{Bacillus}} \underset{\text{Lactic acid}}{CH_3CH(OH)COOH}$$

Fig. 13.33: Large scale production of some simple acids.

Limed straw and bassage (the residue left after the juice is obtained from sugarcane) on treatment with rumen bacteria in an anaerobic fermentor produces acetic acid, propionic acid and butyric acid[73]. These acids can be converted into alcohols with hydrogen. Such synthesis permits the conversion of unwanted wastes to valuable chemicals. Propionic acid can also be made from hemicelluloses in wood[74].

Besides these, a number of other carboxylic acids can also be produced by biocatalytic methods. As an example succinic acid is prepared from glucose by genetically engineered E. Coli[75]. The succinic acid thus obtained is converted into commercial chemicals by known methods into 1,4-butanediol, tetrahydrofuran and *n*-methylpyrrolidiones.

Glyoxylic acid is obtained from glycolic acid by treatment with microbes in presence of O_2.[76] The catalyst can be recovered and reused 25-30 times. Glycoxylic acid has been used to prepare an herbicide, N.(phosphono) methyl glycine (Fig. 13.34).

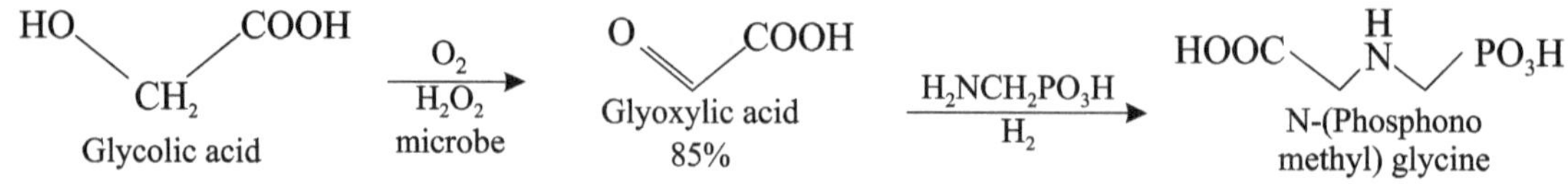

Fig. 13.34: Preparation of glyoxylic acid.

Glycine is converted into L-serine by reacting with formaldehyde using E. Coli[77] (Fig. 13.35).

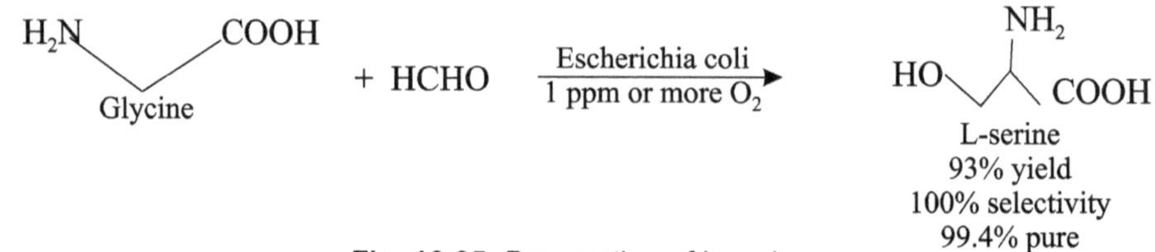

Fig. 13.35: Preparation of L-serine.

Long-chain unsaturated fatty acids are of interest for good health[78] can also be produced biocatalycilly[79]. Thus, archidonic acid is produced by growing fungus, Mortierella alpina, on a mixture of glucose and potatoes[80]. Gamma-Linolenic acid is produced by growing cunninghamella echinulatum on starch plus potassium nitrate or urea[81]. The eicosapentaenoic acid and doccosahexaenoic acids from fish oils help prevent heart ailments. Attempts have been made to prepare them more conveniently from algal, Antarctic sea ice bacteria and fungal cultures[82]. The structures of the four long-chain unsaturated fatty acids are given below (Fig. 13.36).

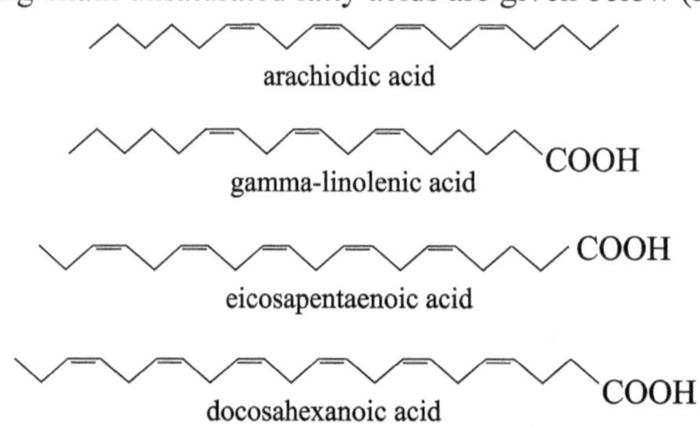

Fig. 13.36: Structures of some long-chain unsaturated fatty acids.

Linoleic acid can be hydroxylated to give 10-hydroxy-12(z) octadecenoic acid (an analogue of ricinoleic acid) by A. Flavobacterium sp[83]. Ricinoleic acid (12-hydroxy-9(z) octadecenoic acid) from caster oil is used in the preparation of paints and varnishes[84]. *n*-Pentadecane on treatment with cryptocous neoformans gives the corresponding α, ω-dicarboxylic acid $[HOOC(CH_2)_{13}COOH]$[85].

4-Hydroxybenzoic acid, a monomer used for the preparation of polyesters for use at high temperature is obtained from toluene by treatment with P. putida[86] (Fig. 13.37).

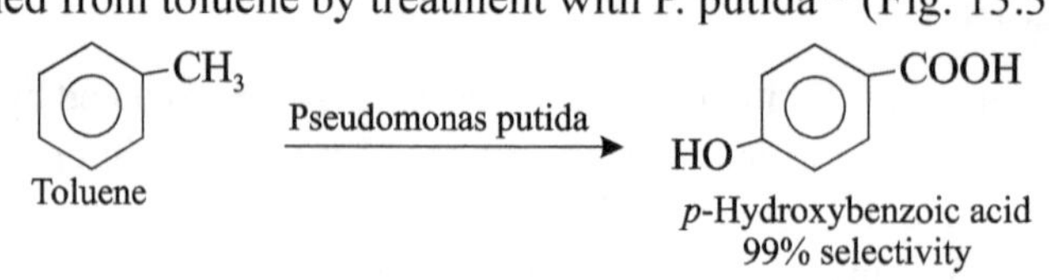

Fig. 13.37: Conversion of toluene into *p*-hydroxybenzoic acid

Pyrrole-2-carboxylic acid can be prepared from pyrrole and CO_2 using Bacillus megaterium (Fig. 13.38)[87].

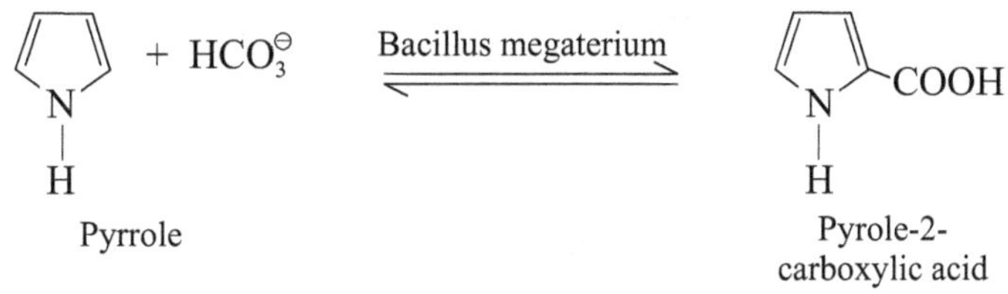

Fig. 13.38: Preparation of pyrrole-2-carboxylic acid.

***(b)* Synthesis of amides:** Triglycerides are converted into the corresponding amides by treatment with C. Antarctica and ammonia in *tert*-butyl alcohol[88]. The enzyme can also be used to convert oleic acid to olemide in 90% yield in *tert.* amyl alcohol containing *n*-butanol (2 equivalents). Similarly, dimethyl sucinate is converted into monoesteramides by candida antarctica lipase[89] (Fig. 13.39).

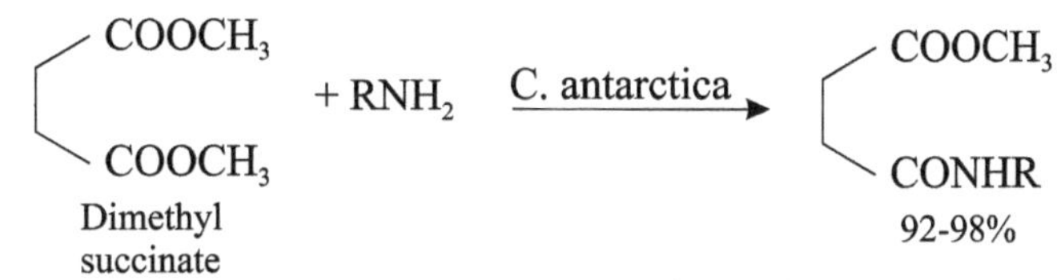

Fig. 13.39: Conversion of diesters into monoesteramides.

Amides can also be prepared by the enzymic hydrolysis of nitriles using the enzyme Rhodoccus rhodocrous (Fig. 13.40)[90].

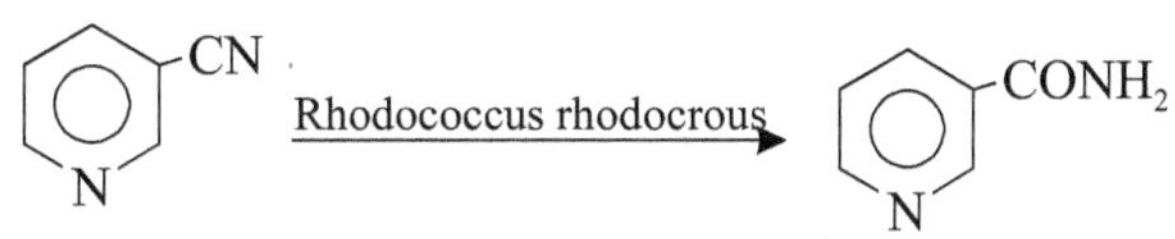

Fig. 13.40: Conversion of nitriles into amides.

The product of hydrolysis of nitrile may be an amide or a carboxylic acid depending on the microorganism (Fig. 13.41)[91].

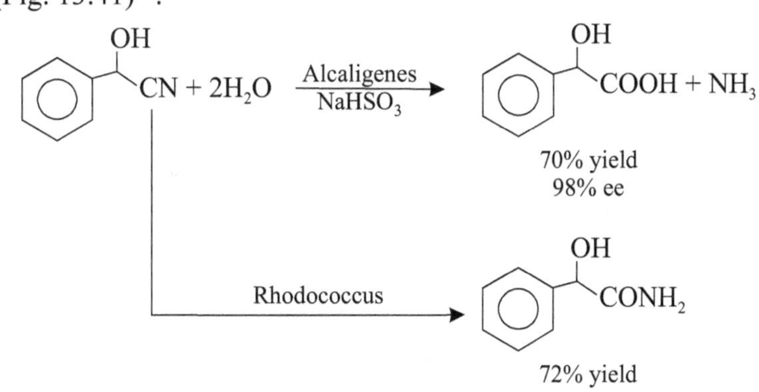

Fig. 13.41: Hydrolysis of a nitrile into carboxylic acid or a amide.

13.2.6 Use of Biocatalysts in the Manufacture of Aromatic Compounds

Methods have been developed[90] to produce adipic acid, catechol and DHS from glucose using genetically altered E Coli-bacteria[91]. The E. Coli bacteria used in this process are genetically

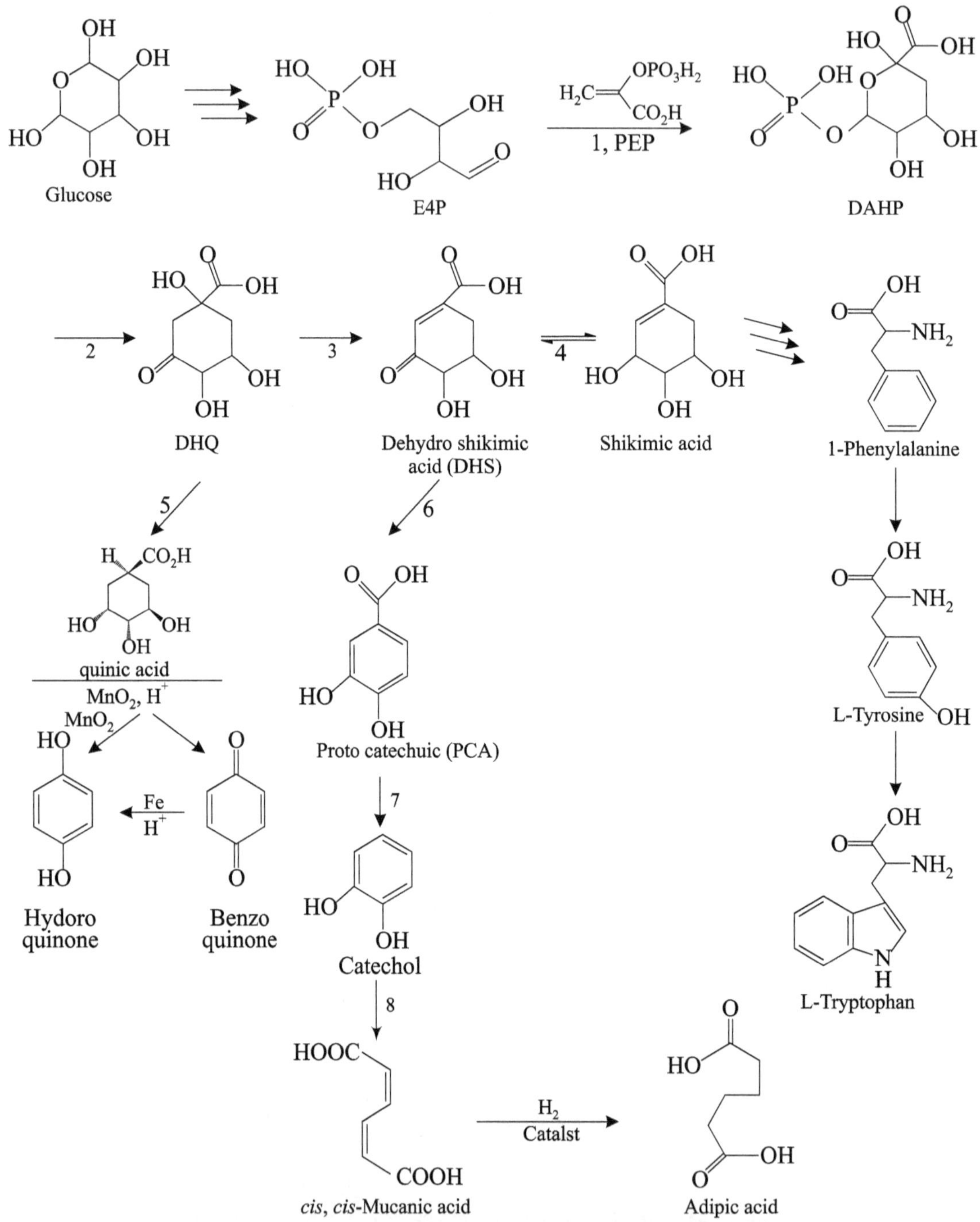

Fig. 13.42: Biocatalytic Synthesis of aromatics, shikimic acid, L-phenylalanine, L-Tyrosine and L-Tryptophane are synthesised by Unmodified E. Coli and the others are synthesised by modified E. Coli. Various enzymes used for various steps are given in Fig. 13.42:

(1) DAHP synthase, (2) DHQ synthase, (3) DHQ dehydratase, (4) Shikimate dehydrogenase, (5) quinate dehydrogenase (6) DHS dehydrate, (7) Protocatechuate decarboxylase (8) Catechol 1,2-deoxygenase.

engineered using recombinent DNA technology. The process consist in spicing out a segment of DNA out of one organism and subsequently adding into the DNA of another organism, giving the second organism the desired quality that was present in the first organism. This technology has been used to alter the metabolic pathways of glucose in E. Coli (Fig. 13.42).

One strain of the modified E. Coli can convert glucose into cis, cis-muconic acid, which is hydrogenated to give adipic acid (Fig. 13.42). In unmodified E. Coli, glucose is transformed into the amino acids L-phenylalanine, L-tryptophane. The intermediates in the production of amino acids include E4P, DAHP, 3-dehydroshikimic acid (DHS) and Shikimic acid. However, modified E. coli converts DHS into protocatechuic acid (PCA), which is converted into catechol and subsequently catechol is converted into *cis*, *cis*-muconic acid. Finally, *cis*, *cis*-mucanic acid is catalytically hydrogenated to produce adipic acid[92].

As seen (Fig. 13.42), adipic acid has been synthesised by environmentally benign bicocatalytic process from glucose[93], which is obtained from biomass.

Indigo, the dye commonly used is synthesised enzymatically by removing enzymatcally the side chain of L-tryptophan, which on dihydroxylation using another enzyme gave a dihydroxylated product. Final oxidation with oxygen gave indigo (Fig. 13.43).

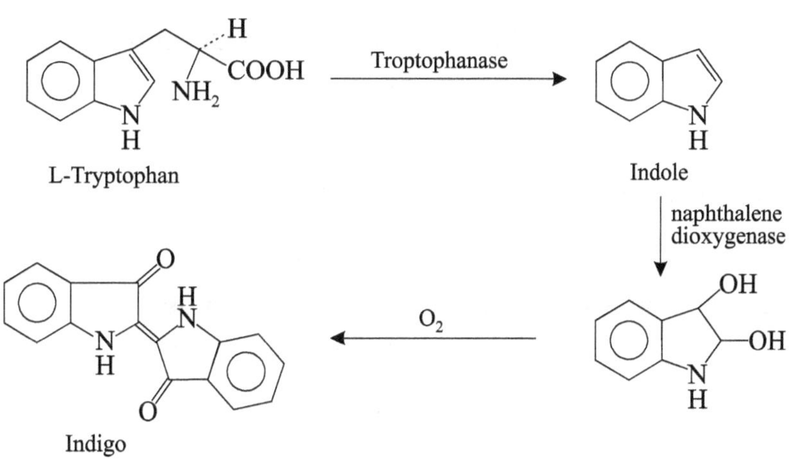

Fig. 13.43: Green Synthesis of Indigo.

It is well known that the present commercial route uses highly toxic aniline produces waste salts (Fig. 13.44).

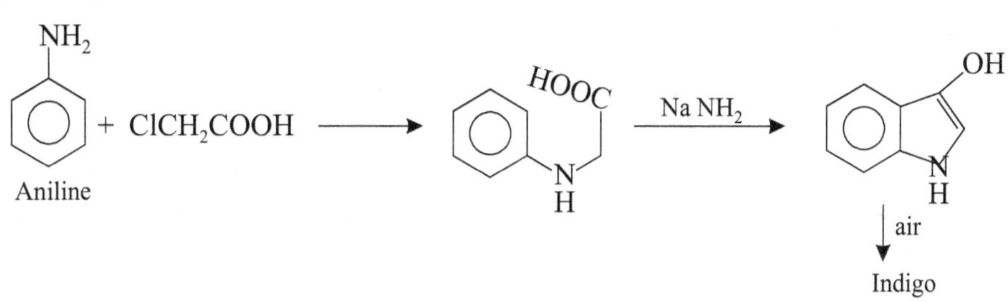

Fig. 13.44: Present day commercial synthesis of Indigo.

An even greener way to make indigo is to raise it as a crop using a new commercial strain producing five times more indigo than the usual plant[94].

13.3 BIOCATALYTIC TRANSFORMATIONS IN GREEN SOLVENTS

Mostly the biocatalytic transformations were earlier conducted in aqueous medium at ambient temperature and pressure. However, some biocatalysts solutions are unstable in pure water and so these transformations are conducted in phosphate or citrate buffers. The use of biocatalysts in organic solvents suffer from the disadvantage of reduced activity, selectivity or stability of the biocatalyst as compared to aqueous media. Also the organic solvents are volatile organic compounds and have considerable environmental problems. In view of all these, attempts have been much to use green solvents for biocatalytic transformations. The green solvents include ionic liquids, super critical carbon dioxide, polyethylene glycol and fluorous solvents.

13.3.1 Biocatalytic Transformations in Ionic Liquids

The Ionic liquids are becoming popular for their use in organic synthesis (for details see Chapter 3) are biochemical[95]. For a discussion on biotransformations in ionic liquids see Chapter 3, Section 3.5.

13.3.2 Biocatalytic Transformation in Supercritical Carbon dioxide

A number of examples have been reported using biocatalysts in SC–CO_2 for organic transformations. For details see Chapter 4, Section 4.10.

13.3.3 Biocatalytic Transformations in Polyethylene Glycol

For details see Chapter 5, Section 5.11.

13.3.4 Biocatalytic Transformation in Fluorous Solvents

Following are given some of the biocatalyst catalysed reactions in fluorous solvents.

13.3.4.1 Kinetic Resolution of Rac. 1-phenylethanol

The kinetic resolution of rac.1-phenyl ethanol using the catalyst nonozym 435 in fluorous solvent, *e.g.*, hydrofluorocarbons (HFCs) increases the rate and also the yield compared to those in hexane and methyl tert. butyl ether[96]. The HFCs used are 1,1,1,2-tetrafluoroethane and 1,1,1,2,3,3,3-heptafluoropropane.

13.3.4.2 Resolution of 1-(2-naphthyl) Ethanol

The reaction of rac-1-(2-naphthylalcohol) with a fluorinated carboxylic acid ($C_8F_{17}CH_2CH_2COOH$) in presence of Chirazyme L-2 (CABL) (CZL-2) gave a mixture of (S)-alcohol, fluorinated carboxylic acid and the corresponding (S) fluorinated ester. The resulting mixture was transferred to the source phase (methanol) of a U-tube. The nonfluorinated (R)-alcohol remained in the source phase (methanol) and both the fluorinated (S)-ester and the catalystic fluorinated acid passed to the fluorous phase (FC-72). The (S)-ester on reaction with methanol undergoes alcoholysis, the fluorous tag is removed and the resulting (S)-ester passes to the receiving phase (methanol) on the opposite side of the U-tube to the source phases. In this way both the enantiomers of 1-(2-naphthyl) ethanol are obtained in high yields and enantioselectives (Fig. 13.45). The same procedure is used for the kinetic resolution of rac. 1-phenylethanol using the chirazyme L-2 (CZL-2) and a highly fluorinated carboxylic acid in acetonitrile[97,98].

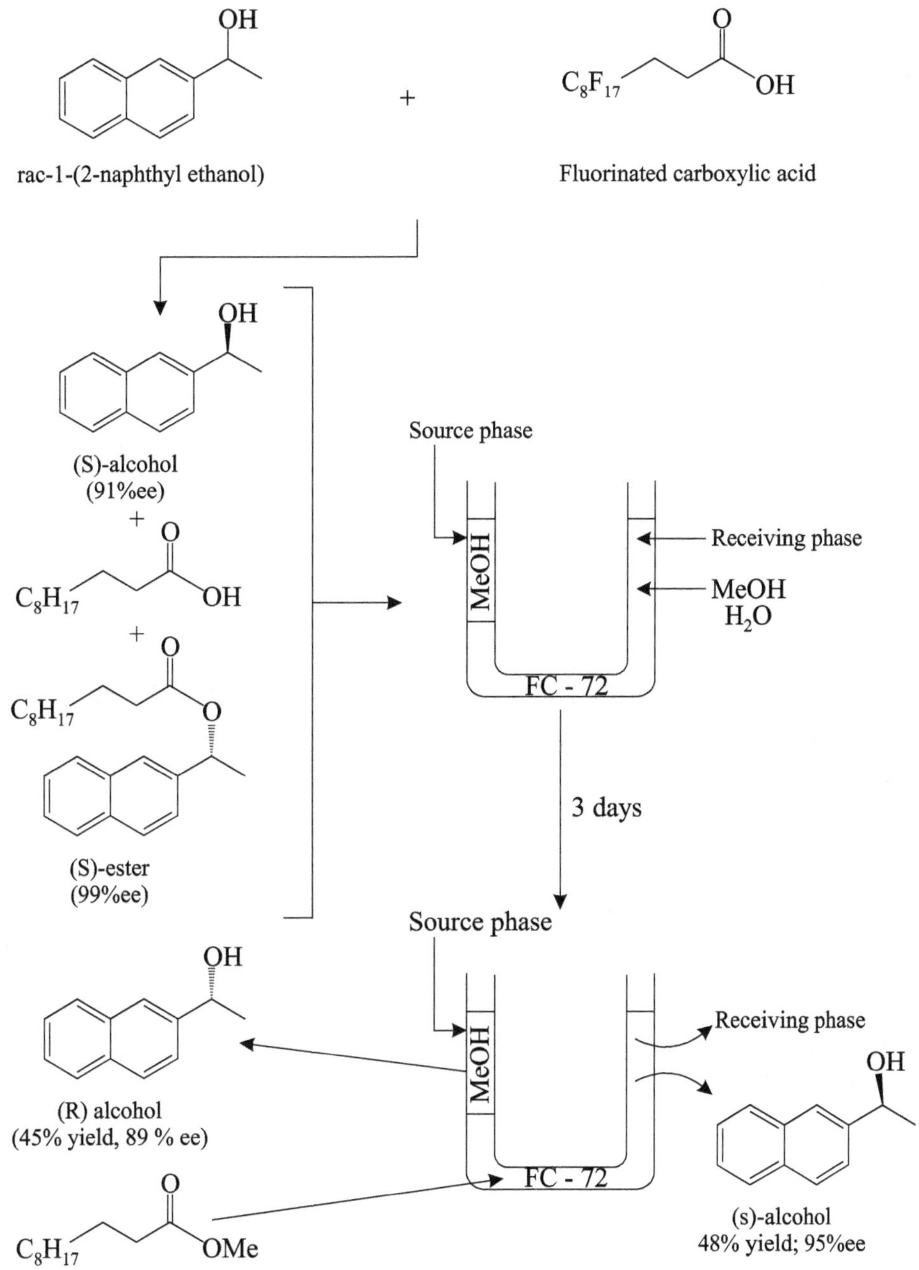

Fig. 13.45: Resolution of 1-(2-naphthyl) alcohol.

13.3.4.3 Enantioselective Esterification of Rac. α-methylpentanoic acid

α-Methylpentanoic acid is dissolved in hexane and alcohol and fluorinated ester (both dissolved in perfluoro hexane). The catalyst used in candida rutoga lipase (CRL), which being insoluble was used as suspension. The catalyst (CRL) selectively catalysed the esterification of (S) α-methyl pentanoic acid to the corresponding (S)-fluorinated ester, which remained in the fluorous solvent. The remaining (R)-α-methylpentanoic acid is removed by phase separation on cooling the system (Fig. 13.46)[99,100].

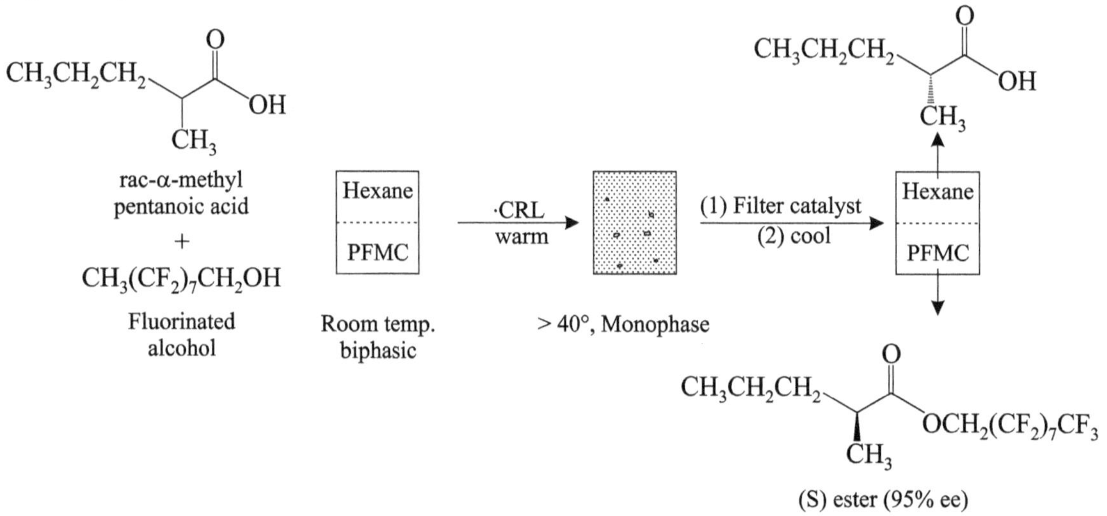

Fig. 13.46: Enantioselective esterification of Rac. α-methyl pentanoic acid.

13.4 MISCELLANEOUS BIOCATALYTIC TRANSFORMATIONS

(*a*) The Biocatalytic hydrolysis of amide in pencillin G has replaced the multistep chemical synthesis that produces waste salts[101] (Fig. 13.47)

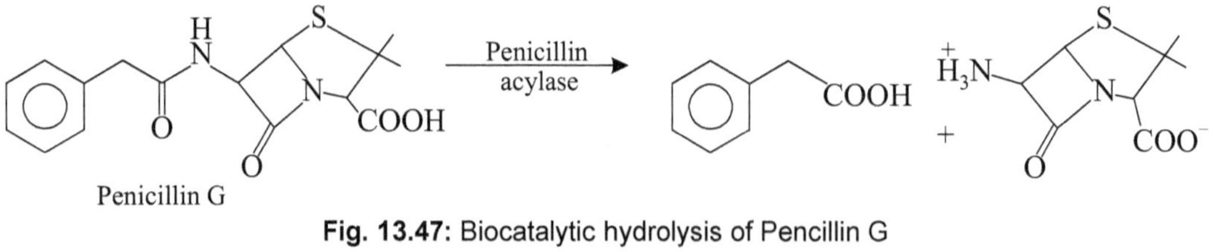

Fig. 13.47: Biocatalytic hydrolysis of Pencillin G

(*b*) The use of expandase to convert penicillins to cephalosporins has replaced a costly multistep chemical synthesis[102] (Fig. 13.48).

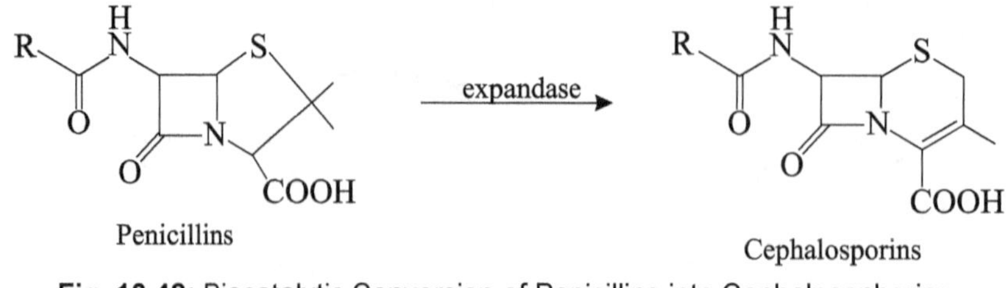

Fig. 13.48: Biocatalytic Conversion of Penicillins into Cephalnosphorins.

(*c*) Penicillin acylase is also used to remove phenylacetyl protecting group from amines in the synthesis of oligonucleotides[103].

(*d*) Cyanohydrins from aldehydes can be made by using hydroxynitrile lyases[104] (Fig. 13.49)

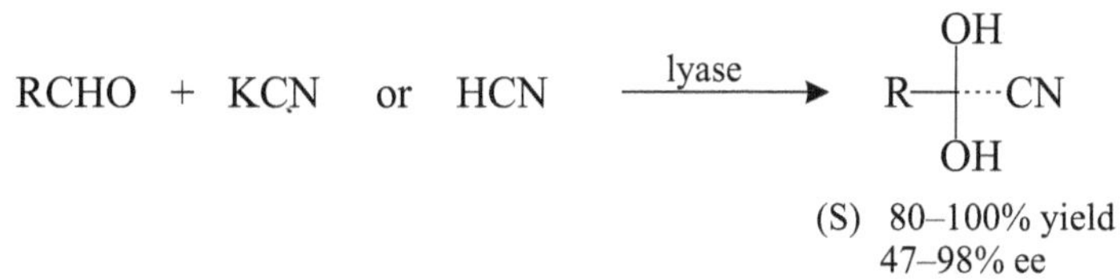

Fig. 13.49: Conversion of aldehyes into cyanohydrins

(*e*) Glycerol (which is obtained as a byproduct during the preparation of biofuel by alcoholysis of rapeseed oil) can be converted into propionic acid with a propinobacterium in a membrane reactor without the formation of acetic acid[105]. Glyceral can also be converted to 3-hydroxypropinaldehyde by Klebsiella oxytica[106]. This product, *viz*. 3-hydroxy-propinaldehyde can be converted into acrylic acid by dehydration and oxidation.

Glycerol can also be converted to 1,3-propanediol which can be used to prepare poly (trimethylene terephthalate)[107] using K.pneumoniate.

(*f*) Genetically engineered bacteria is used to convert glucose to 2-keto-L-gulonic acid, thus elimating four steps in the synthesis of Vitamin C[108].

(*g*) Prednisolone, a cortisone analog used as a drug against rehumatoid arthritis is obtained on a commercial scale from a steroid percussor, Reichstein's compound, which is passed through a series of two different columns each containing a specific enzyme attached to a polyacramide polymer support (Fig. 13.50)[109]

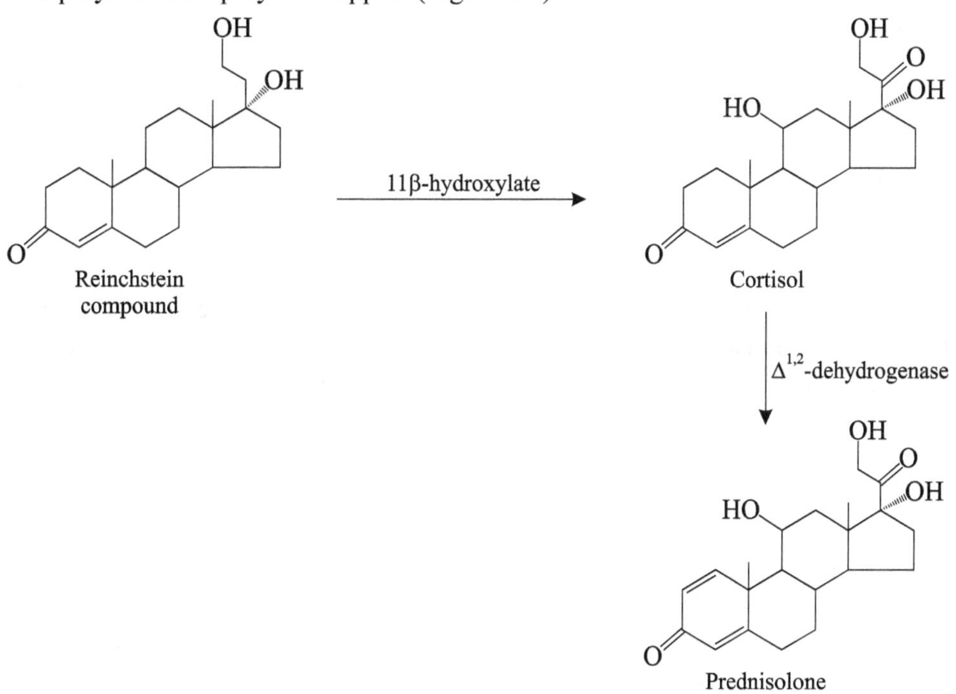

Fig. 13.50: Commercial Synthesis of Prednisolone

(*h*) Methane can be converted directly in methanol and its derivatives. These are the starting point for a number of industrial processes and is also incorporated into paints and plastics. The oxidation of methane to methanol involves C—H activation by soluble methane monooxygenase (sMMO). The reaction is represented as given below (C.E. Tinberg and S.J. Lippard, J. Biotech, 2010, **49**, 7902)

$$CH_4 + O_2 + NAD(P)H + H^+ \longrightarrow CH_3OH + NAD(P)^+ + H_2O$$

Other alkanes likes butane, pentane, hexane, heptane and 2-methylpropane or adamantane can also be oxidised. However, the rate of oxidation decreases from methane to adamantane (G.B. Shulpin, Org. Biomol. Chem., 2010, **8**, 4217.

(*i*) Omega hydroxyfatty acids have been synthesised using eukaryotic CYP from Candia tropicals (W. Lu, J.E. Ness, W. Xie *et al*, J. Am. Chem. Soc., 2010, 132, 15451). These fatty acids are best monomers for synthesising polyethylene-like biobased plastics. These can be converted into the corresponding alkyl ester monomer units gave biofuels similar is structures to biodiesels.

(*j*) Indirubin, a drug for chronic myelogenous leukemia and a novel anticancer agent has been prepared from indole as given below (Q.S.Li, U. Schwaneberg, P. Fisher and R.D. Schmid, Chem. Eur. J., 2000, **6**, 1531; S. Hua, J. Huangb, L. Meia et al., J. Mol. Catal. B: Enzymes, 2010, **67**, (29).

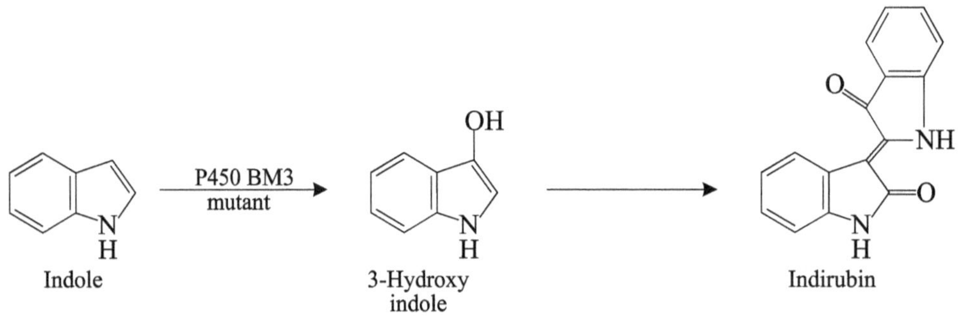

Conclusion

Biocatalytic transformations are the best green transformations. It has become one of the greenest technologies. These transformations are one step process and are carried out mostly in aqueous medium at ambient temperatures and pressures. A large number of processes like oxidations, reductions, hydrolysis, esterifications and synthesis of a number of bioactive compounds can be conveniently synthesised using biocatalysts. These are of special interest in pharmaceutical industry.

REFERENCES

1. R.L. Hinman, Chemtech, 1994, **24(6)**, 45.
2. M.A. Penalva, R.T. Rowlands, G. Turner, Trends Biotechnol., 1998, **16**, 483.
3. M.J. van der Werf, WJJ-Van der Tweel, J. Kamphuis, S. Hartman and J.A.M. de Bont, Trends Biotechnol., 1994, **12(3)**, 95.

4. D.Rozzell, Biocatalytic production of Amino Acids and derivatives, Oxford University Press, Oxford, UK, 1992.
5. A. Gabelman Ed., Bioprocess Production of Flavour, Fragrance and Color Ingredients, Willy, New York, 1994.
6. PH Nielsen, H. Malmos, T. Damhus, B. Doderichen, HK Nielson, M. Simonsen, HE Schiff, A. Oestergaad, HS Olsen, P. Eigtved, TK Nielsen. Kirk-Othmer Encyclopedia of Chemical Technology, 4th ed., 1994, **9**, 567-621.
7. S.M. Brown in Organic Synthesis: Theory and Practice; T. Hudlicky, Ed., JAI Press, Greenwich, CT, 1993, Vol. 2, P. 113; H.A.J. Carless, Tetrahedron Asymmetry, 1992, **3**, 795.
8. D.G.H. Ballard, A. Courtis, I.M. Shirley, and S.T. Taylor, J. Chem. Soc. Chem. Commun., 1983, 954.
9. T. Hudlicky, H. Luna, G. Barbieri and L.D. Kwart, J. Am. Chem. Soc., 1988, **110**, 4735.
10. A. Kiener, Chemtech, 1995, **25(9)** 31; M. Torimura, H. Yoshide, K. Kano, T. Ikeda, T. Nagasawa and T. Ueda, Chem. Lett., 1998, 295.
11. M. Held, W. Suske, A. Schmid, K.-H. Engesser, H.PE Kohler, B.Witholt and M.G. Wobbolts, J. Mol. catal B enzymol, 1998, **5**, 87; A. Smid, H. Pekohler, K.-H. Engesser, J. Mol cat B Enzymol, 1998, **5**, 311.
12. J.B. Jones and R.C. Raymind, Appl. Microbiol., 1961, **9**, 383.
13. J.D. Douros and J.W. Frankenfeld, Appl. Microbiol., 1968, **16**, 320.
14. A. Potthast, T. Rosenau, C.-L. Chen and JS Gratzl, J. Org. Chem., 1995, **60**, 4320; J. Mol cat A Chem., 1996, **108**, 5; T. Rosonai, A.Potthast, C.-L. Chen and J.S. Gratzl, Synth Commun., 1996, **26**, 315.
15. E. Fritz-Landhals and B. Kunath, Tetrahedron Lett., 1998, **39**, 5955.
16. M.H. Holicky, Oxidations in Organic Chemistry, ACS Monograph, 186, American Chemical Society, 1990, P. 147.
17. G. Grunwald, B. Wirz, M.P. Scollar and A.M. Klibanov, J. Am. Chem. Soc., 1986, **108**, 6732.
18. R. Raichstein, Helv. chim. Acta, 1934, **17**, 996.
19. A.B. Baeyer and V. Villiger, Ber., 1899, **32**, 3625.
20. J.M. Schwab, N.-b. Li and L.P. Thomas, J. Am. Chem. Soc., 1983, 105, 4800.
21. M.J. Taschner and D.J. Black, J. Am. Chem. Soc., 1988, **110**, 6892.
22. A. Willets, Trends Biotechnol, 1997, **15**, 55; R. Gagnon, G. Grogan, M.S. Levitt., S.M. Roberts, PWH Van and A.J. Wolletts, J. Chem. Soc., Perkin Trans I, 1994, 2537.
23. T.J. MC Curdy and R.D. Garrett, S.Org. Chem., 1968**, 33**, 660.
24. R.L. Prairie and Talalay, Biochemistry, 1963, **2**, 203.
25. J. Fried, R.W. Thoma and A. Klinbero, J. Am. Chem. Soc., 1953, **75**, 5764.
26. J. Friecl R.W. Thoma, J.R. Genke, J.E. Merz, M.N. Donin, D. Perlman, J. Am. Chem. Soc., 1952, **74**, 3962.
27. K.M. Mann, F.R. Hanson, P.W. O' Connell, H.V. Andersons, M.P. Brunner and J.N. Karnemaat, Appl. Microbiol, 1953, **3**, 14.
28. P.W. O'Connell, K.M. Mann, E.D. Nielson and F.R. Hanson, Appl.Microbiol, 1955, **3**, 17.
29. D.H. Peterson and H.C. Murray, J. Am. Chem. Soc., 1952, **74**, 1877.
30. D.H. Peterson, H.C. Murry, S.H. Eppstein, L.M. Reincke, A. Weintraub, P.D. Meister and H.M. Leigh, J. Am. Chem. Soc., 1952, 74, 593.
31. D.H. Peterson, A.H. Nathan, P.P. Meister, S.M. Eppstein, H.C. Murray, A. Weintraub, L.M. Reineke and H.M. Leigh, J. Am. Chem. Soc., 1953, **75**, 41.

32. S.H. Eppstein, D.H. Peterson, H.M. Leigh, H.C. Murray, A. Weintraub, L.M. Reinke and P.D. Meister, J. Am. Chem. Soc., 1953, **75**, 421.
33. D.H. Peterson, P.D. Meister, A. Weintraub, L.M. Reinke, S.H. Eppstein, H.C. Murroy and H.M.L. Osborn, J. Am., Chem. Soc., 1955, **77**, 4428.
34. Y.Y. Lin, M. Shipahara and L.L. Smith, J. Org. Chem., 1969, **34**, 3530.
35. J. Urech, E. Vischer and A. Wettstein, Helv. Chim. Actor, 1960, **43**, 1077.
36. A. Schubert, G. Langbein and R. Siebert, Chem. Ber., 1957, **90**, 2576.
37. C. Meyser, E. Vischer and A. Wettstein, Hel. Chem. Acta, 1955, **38**, 381; J. Fried, R.W. Thoma and A. Klingsberg J. Am. Chem. Soc., 1953, **75**, 5764.
38. C. Tamm, A. Gubler, G. Juhasz, E. Weiss-Berg and V.W. Zürcher, Helv. Chim. Acta, 1963, **46**, 889.
39. H.L. Herzog, M.J. Gentles, W. Charney, D. Sutter, E. Townley, M. Yudis, P. Kabasakalim and E.B. Hershberg, J. Org. Chin., 1959, **24**, 691.
40. W. Charney, A. Nobile, C. Federbush, D. Sutter, P.L. Perlamn, H.L. Herzog, C.C. Payne, M.E. Tully, M.J. Gentles and E.B. Hersberg, Tetrahedran,1962, 18, 591.
41. L. Manolic and A. Vercellone, Chem. Ber., 1938, **71**, 1686.
42. A.R. Hanze, O.K. Sebek and H.C. Murray, J. Org. Chem., 1960, **25**, 1968.
43. S. Kawai, T.Oshima and F. Egami, Boochem. Biophys. Acta, 1965, **97**, 391.
44. D.J. Siehr, J. Am. Chem. Soc., 1961, **83**, 2401.
45. R.A. Johnson, M.E. Herr, H.C. Murray and G.S. Fonken J. Org. Chem., 1968, **33**, 3187.
46. M.E. Herr, R.A. Johnson, H.C. Murray, L.M. Reineke and G.S. Fonken, J. Org. Chem., 1968, **33**, 3201.
47. A. Gupta, R. Nagarajan and A. Kilara, Ind. Eng. Chem., Res., 1995, **34**, 2910.
48. L.B. Davin, H.-B. Wang, A.L. Crowell, D.L. Bedgan, D.M. Martin, S. Sarkanen and N.G. Lewis, Science, 1997, **275**, 362: J. Kaiser, Science, 1997, **275**, 306.
49. MR gen Klaas, S. Warwel, J. Am. Oil Chem. Soc., 1996, **73**, 1453; J. Mol. Catal. A. Chen., 1997, **117**, 311; Synth Commun. 1998, **28**, 251.
50. G. Greenspan, H.W. Ruelius and H.E. Alburn, U.S. Patent, 3, 453, 179 (July 1, 1969)
51. L.L. Wallen, R.G. Benedict and R.W. Jackson, Arc. Biochem. Biophys., 1962, **99**, 249.
52. C.C. Levy and P. Frost, J. Biol. Chem., 1966, **241**, 997.
53. K. Decker and H. Blug, Bcochem Biophys. Acta, 1965, **105**, 312.
54. K. Kleinzeller and Z. Fencl, Chem. Listy, 1952, **46**, 300; Chem. Abstr., 1953, **47**, 4290.
55. V. Prelog, Pure Appl. Chem., 1964, **9**, 119.
56. B. Zhou, A.S. Gopalan, F. Van Middlesworth, W.-R. Shih and C.J.J. Sih, Am. Chem., Soc., 1983, **105**, 5925.
57. J. Bolte, J.G. Gourey and H. Veschambre, Tetrahedron Lett., 1986, **27**, 565.
58. H. Akita, A. Furuichi, H. Koshiji, K. Horokoshi and T. Oishi, Tetrahedron Lett., 1982, **23**, 4051.
59. A. Furuichi, H. Akita, H. Koshiji, T. Horikoshi and T. Oishi, Chem. Pharm., Bull., 1984, **32**, 1619; H. Akita, A. Furuichi, H. Koshiji, K. Horikosdi and T. Oishi, Chem. Pharm Bull, 1984, **32**,1 342; 1983, **31**, 4376.
60. J.K. Lieser, Syn. Commun., 1982, **13,** 765.

61. W.S. Zhou, D.Z. Hunag, Q.C. Dong, Z.P. Zhuang and Z.Q. Wang, Chem. Nat. Prod. Sino-Am. Sym., 1980, 299; Chem. Abstr, 1983, **98**, 9854w.
62. M. Desrut, A. Kergomard, M.F. Renard, and H. Veschambre, Tetrahedron, 1981, **37**, 385.
63. M. Bostmembrun-Desrut, G. Dauphin, A. Kergomard, M.F. Renard and H. Veschambre, Tetrahedron, 1985, **41**, 3679.
64. E. Kiean, E.K. Hateli, K.K. Sehl and R. Lamed, J. Am. Chem. Soc., 1986, **108**, 162.
65. G. Fronza, C. Fuganti, P. Grasselli, S.Servi, G. Zucchi, M. Barbeni and M. Villa, J. Chem. Soc. Chem. Commun, 1995, 939.
66. W. Baik, D.I. Kim, S. Koo, J.U. Rhee, S.H. Shin and B.H. Kim, Tetrahedron Lett., 1997, **38**, 845.
67. A. Svendsen, Int. News Fats Relat. Matter, 1994, **5**, 619; X.Y. Wu, S. Jaaskelainen, Y.-Y. Linko, Enzyme Microb Technat, 1966, **19**, 226.
68. K.-E. Jaeger, M.T. Reetz, Trends Biotechnol, 1998, **16**, 396; A.M.R. Yahya, W.A. Anderson and M. Mooyoung, Enzyme Microb Technol,, 1998, **23**, 438.
69. L.A. Nelson, T.A. Foglia and W.N. Marmer, J. Am. Oil Chem. Soc., 1996, **73**, 1191.
70. E. Ozaki, T. Uragaki, K. Sakashita and A. Sakimae, Chem. Lett., 1995, 539.
71. D.J. Ager and I. Prakash, Synth. Commun., 1995, **25**, 739.
72. O. Houille, T. Schmitterger and D. Uguen, Tetrahedron Lett., 1996, **37**, 625.
73. M. Holtzapple, The Presidential Green Chemistry Challange Awards Program EPA 744-K-96-001, Washington, Dec, 1996; July 7; Preprints ACS Div. Environ. Chem., 1997, 37(1), 344.
74. J.A. Ramsay, M.-C. Hassan and B.A. Ramsay, Enzyme Microb. Technol., 1998, **22**, 292.
75. K. Fouhy, Chem. Eng., 1997, **104(2)**, 25; R&D May., 1996, Dec. 64.
76. J.E. Gavagan, S.K. Fager, J.E. Seip, M.S. Payne, D.L. Anton and R. Dicosimo, J. Org. Chem., 1995, **60**, 3957.
77. D. Ura, T. Haskimukai and N. Fukuhara, US Patent 5, 382, 517, 1995.
78. R.S. Lees, M. Karel, Omega Minus Three Fatty Acids in Health and Disease, Dekker, New York, 1990; I.S. Newton, Chem. Ind. (Lond), 1997, 302.
79. O.P. Wand, Int. News Fats Oils Relat. Mater, 1995, **6**, 583; I. Gill, R. Valivety, Trends Biotechnol, 1997, **15**, 401, 470;
80. N. Totani and K. Oba, Appl. Microbiol. Biotechnol., 1988, **28(2)**, 135.
81. H.-C. Chen. and C-C. Chang, Bootechnol. Prog., 1996, **12**, 338.
82. P. Bajpai and P.K. Bajpai, J. Bootechnol, 1993, **30**, 161; N. Shirasaka and S. Shimizu, J. Am. Oil Chem. soc., 1995, **72**, 1545.
83. C.T. Hou, J. Am. Oil Chem. Soc., 1994, **71**, 995.
84. H.A. Wittcoff and B.G. Reuben, Industrial Org. Chemicals., Wiley, New York, 1996, 373.
85. E.-C. Chan and J. Kuo, Enzyme Microb Technol., 1997, **20**, 585.
86. E.S. Miller and S.W. Peretti, Abstracts of 2nd Green Chemistry and Chemical Engineering Conference, Washington, DC, 6/30/98-7/2/93, 3.
87. M. Wieser, Y. Yoshida and T. Nagaisawa, Tetrahedron Lett., 1998, **39**, 4309.
88. M.C. de Zoete, A.C. Kock- Van Dalen, F. Van Rantwijk and R.A. Sheldon, J. Mol. Catal B Enzymol, 1996, **1**, 109; 1996, **2**, 19, 141
89. S. Puertas, F. Rebolledo and V.Gotar, Tetrahedron, 1995, **51**, 1495.

90. E. Haslam, Shikimic acid metabolism and metabolies, Wiley, New York, 1993; N.J. Grinter, Chem. Tech., 1998, **28(7)** 333; K.M. Draths, D.R.Knop and J.W. Frost, J. Am. Chem. Soc., 1999, **121**, 1603; see also reference 92.

91. J. Frost, EPA J., 1994, **20(3-4)**, 22; J.W. Frosts and K.M. Daths, Chem. Ber., 1995, **31**, 201; J.W. Frost, and J. Lievense, New J. Chem., 1994, **18**, 341; K.W. Draths and J.W. Frorst, J. Am. Chem. Soc., 1995, **117**, 2395.

92. P.T. Anastas, T.C. Williams Eds., Green Chemistry Frontiers in Benign Chemical Synthesis and Processes, Oxford University Press, New York, 1998.

93. K.M. Draths and J.W. Frost, J. Am. Chem. Soc., 1994, **116**, 399.

94. Anon. Chem. Ind. (Lond) 1999, 689.

95. M.T. Reetz, W. Wiesenhoefer, G. Francio and W. Leitner, Chem. Commun, 2002, 992.

96. S. Saul, S. Corr and J. Micklefiein, Angew. Chem. Int. Ed., 2002, **43**, 5519.

97. B. Hungerhoff, H. Sonnenschein and F. Theil, Angew. Chem. Int. Ed., 2001, **40**, 2492.

98. B. Hungerhoff, H. Sonnenschein and F. Theil, J. Org. Chem., 2002, **67**, 1781.

99. P. Beier and D.O'H agan, Chem. Commun., 2002, 1680.

100. P. Beier and D.O' Hagan in Handbook of Flurous Chemistry, A. Gladyse, D.P. Curran and I.T. Hovath, Eds., Wiley VCH Verlag Gamb H and Co. KGag: Weinhein, Germany, 2004, PP. 333-340.

101. R.A. Sheldon, Chemtech., 1994, **24(3)**, 38; H. Waldman and A. Reidel, Anglow Chem. Int. Ed. Engl, 1997, **36**, 647.

102. L. Grawford, A.M. Stepan, P.C. McAda, J.A. Rambosek M.J. Conder, V.A. Vinci, C.D. Reeves, Biotechnology, 1995, **13(1)**, 58; Chem. Eng. News, 1996, June 3, 33.

103. H. Waldmann and A. Reidel, Angew Chem. Int. Ed. Engl., 1997, **36**, 647.

104. S. Forster, J.Rocs, F. Effenberger, H. Wajant and A. Sprauer, Angew. Chem. Int. Ed. Eng., 1996, **35**, 4370; M. Schmidt, S. Herye, N. Klempier and H. Gringl, Tetrahedron, 1996, **52**, 7833.

105. P. Boyavl, C. Corre and M.N. Made, Enzyme Microb. Technol., 1994, **16**, 883.

106. Chemtech, 1994, **24(7),** 52; P.J. Sliniger, E. Varcauwenberge, R.J. Bothhast, U.S. Patent, 4, 962, 027 (1990).

107. D.C. Cameron A.M. Held, M.-y. Zhu, F.A. Skarly, Anahein ACS meeting, April 1995, Biochem. Technol. Secretariat 131.

108. G. Parkinson, Chem. Eng. 1998, **105(10)**, 19.

109. K. Mossback and P.O. Larson, Biotech. Bioeng., 1970, **13**, 19.

14. ORGANIC TRANSFORMATIONS USING SUPPORTED CATALYSTS

14

Organic Transformations using Supported Catalysts

14.1 INTRODUCTION

Catalyst are known to be not used in organic transformations. These can be recovered after the reaction and reused. The catalysts can be bound to different types of carriers. These include Polymer-bound catalysts, polymeric super acid catalyst, polymeric esterolytic catalyst, polymer-supported phase-transfer catalysts, polymeric triphase catalysts and polymeric photosensitizers.

14.2 POLYMER-BOUND CATALYSTS

In polymer-bound catalysts, the catalyst is bound to a polymeric carrier. Examples include[1] ion-exchangers, esterolytic catalyst, phase-transfer catalysts and photosensitizers. Using these catalysts, the reaction can be stopped at any stage by removing the catalyst (by filtration). These catalysts can be reused a number of times.

14.2.1 Ion-exchange Resins

These are polymeric catalysts and are used in organic synthesis as acid-base catalysts. Some of these catalysts find industrial applications. As an example, isopropyl alcohol is manufactured by hydration of propene using a sulfonate resin[2] as a catalyst. The cation-exchangers normally contains sulfonic acid group, which are introduced by the sulfonation of polystyrene (3-5% cross-linked with divinylbenzene)[3]. On the other hand, anion-exchangers normally contain quaternary ammonium groups, which are introduced by chloromethylation of polystyrene followed by amination[3]. Both types of resins are commercially available. Following are given some of the reactions catalysed by polymeric and acid-base catalysts (Table 14.1).

Table 14.1: Reactions catalysed by polymeric acid-base catalysts

Type of Reaction	Reactants	Resin used	Reference
1. Acid-catalysed Reactions			
(a) Saponification of esters	Esters like esters of aceto acetic acid	Amberlite IR-100 [H^+] Amberlite IR-120 [H^+]	4
(b) Esterification	Fatty acids and alcohols α-amino acids and methanol	Ⓟ–SO_3H	5

(c)	Acetal formation synthesis of carbohydrates and alcohols	Glucose and methanol benzaldehyde and glycerol	Dowex-SO_3H^+	6
(d)	Hydrolysis of enamines	—	Amberlite IR 120 [H^+]	7
(e)	Epoxidation of fats followed by hydrolysis to give trans glycols	Fats and oils	Ⓟ–SO_3H	8
(f)	Amide formation	3-(p-nitrophenyl) propanoic acid and 2-(1-cyclohexenyl) ethylamine	Amberlite IR-120 (H^+)	9
(g)	Rearrangements	2-Ethynyl carbinol rearrangement to α-β-unsaturated ketones	DOWEX-50 [H^+]	10
(h)	Alkylation of phenols with alkenes	Phenol and isobutene	Ⓟ–SO_3H	11
(i)	Pechman reaction (Synthesis of Coumarins)	Resorcinol and ethylacetoacetate	Amberlite IR-120 [H^+]	12
(j)	Acetylation of Phenols	Pyrogallol and acetic anhydride	Amberlite IR-120 [H^+]	13
(k)	Hydrolysis of amides	Amides	Amberlite IR-100 [H^+]	14
(l)	Hydrolysis of terminal thiohydantion of terminal COOH group	Sequential degradation of peptides from COOH-terminus	Amberlite IR-100[H^+]	15
(m)	Hydrolysis of hydroxy amides	Peptides	Rexyn-100 [H^+]	16
(n)	Complete hydrolysis of peptides and proteins	Peptides and proteins	Ⓟ–SO_3H	8
(o)	Hydration of isoolefins	Propene and other olefins	Ⓟ–SO_3H	17
(p)	Hydration of acetylenes	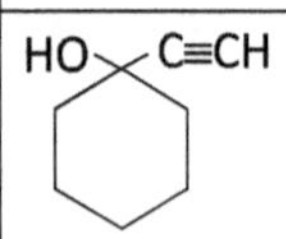	Dowex-50 (H^+) impregnated with 10% $HgSO_4$	10, 18
(q)	Hydrolysis of glucosylamine	N-(2-O-methyl-O-glycosyl)-piperidine	Amberlite IR-120 [H^+]	19
(r)	Inversion of sucrose	Sucrose	Ⓟ–SO_3H	20
(s)	Hydrolysis of carbohydrates and their derivatives	Maltose, glycogen methyl galacturonate and polymethyl galacturonate	Ⓟ–SO_3H	21
(t)	Dehydration	Glycols, various sugars etc.		8

2. Base-catalysed Reactions			
(a) Condensation of aldehydes and ketones with active methylene compounds	Acetone, cyclopentadiene Furfural and aliphatic aldehydes. C_2H_5CHO, CH_3NO_2	DOWEX I-X10 (P) – NH_3^+ – OH^- IRA – 400 (OH^-)	22 23 24
(b) Knoevenagel reaction	Aldehyde, cyanoacetic acid	IR-4B	25
(c) Hydrolysis of esters	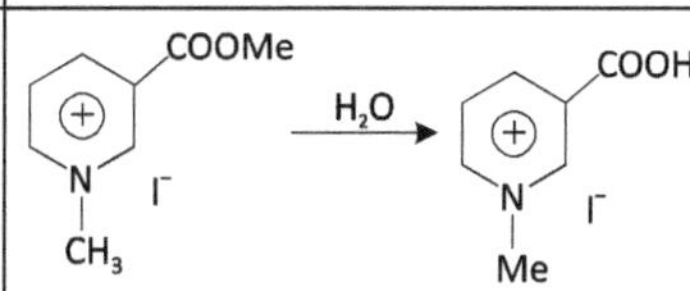	Dowex-I $(OH)^-$	26
(d) Hydration of nitriles to amides	Nicotinonitrile	IRA-400 $(OH)^-$	27
(e) Dehydrohalogenation	3-chloro-1,1-dimethyl indolinium chloride 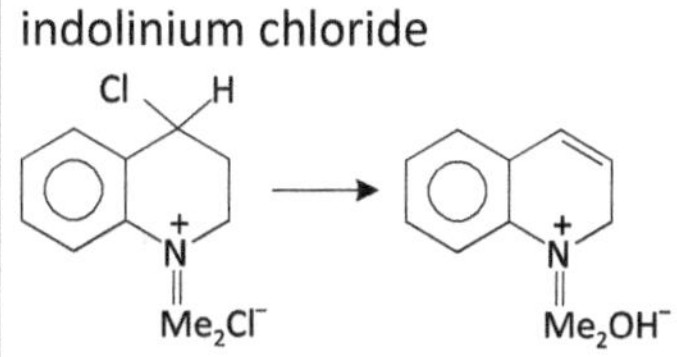	IR A-400 $(OH)^-$	28
(f) Cyclization of α- or β-haloamide to lactams	α or β-haloamides	IR A-400 $(OH)^-$	29
(g) Removal of peptide blocking group	Removal of fluorene-9-methoxy carbonyl (FMOC) group	Macroreticular piperazine contg. resign.	30
(h) Esterification of carboxylic acids	R–COOH + R′–X→RCOOR′	IRA-904	31

The advantage of using resin in acid-base catalyst in ease of their removal after the reaction by filtration, these can be used again, side reactions do not occur, give pure products and exhibit increased selectivity.

14.2.2 Polymer Supported Catalysts

In these catalysts, the conventional catalyst is linked to a polymer backbone and are useful in catalysing reaction. As in the case of ion-exchangers, the catalyst can be removed after the reaction by filtration and reused. In these catalysts polystyrene is most widely used polymer. Following are given some of the important polymer supported catalysts.

14.2.2.1 Polystyrene-Aluminium Chloride Catalyst

It is an effective Lewis acid catalyst and is prepared[32] by the treatment of co (polystyrene-divinyl benzene) (1.8% cross-linked, 50-100 mesh) with aluminium chloride in carbon disulfide, followed by addition of cold water. The catalyst thus obtained is filtered and washed successively with water, ether, acetone and hot isopropanol and dried in vacuum oven.

The polymer supported aluminium chloride in presence of some solvents like benzene, gets swelled and releases aluminium chloride, which can be used to catalyse the reactions. Thus, dicyclopropyl carbinol on treatment with polystyrene-$AlCl_3$ gives di(dicyclopropylcarbinyl) ether (Fig. 14.1)[32].

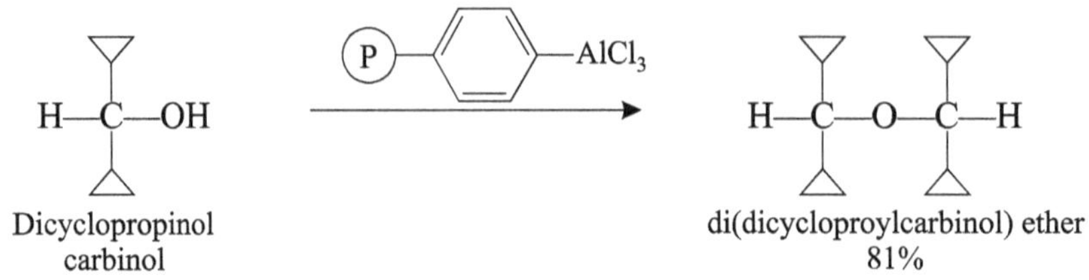

Fig. 14.1: Synthesis of ethers from Alcohols.

Polystyrene-$AlCl_3$ is also a useful catalyst for reactions requiring both a dehydrating agent and a lewis acid. Thus, treatment of an aldehyde with an alcohol in presence of polymeric-$AlCl_3$ in anhydrous benzene give the corresponding acetals in good yields.[33] In a similar way esterification of carboxylic acids with alcohols can be effected[34].

Polymeric-$AlCl_3$ is also an effective catalyst for the hydrolysis of acetals. Thus, the diethyl acetal of *o*-chlorobenzaldehyde with polymeric-$AlCl_3$ in benzene-methanol-water (2:6:1) for 17-18 hr gives *o*-chlorobenzaldehyde in 61% yield. Without catalyst, the yield is about 4%.

14.2.2.2 Polymeric Super Acid Catalyst

Super acids are a combination of Lewis acid and proton acids and are obtained by binding $AlCl_3$ to sulfonated, macroporous co(polystyrene-DVB)[35,36]. The catalyst brings about cracking and isomerisation of *n*-hexane at 357°C at atmospheric pressure and is much better than the usual Lewis acid catalysts.

The polymeric super acid catalysts are referred to as 'Organometallic super acid catalysts'. On the basis of electron microprobe analysis, it has been shown that in these catalysts the aluminium is uniformly distributed in polymer beads. Such catalyst are useful as Friedel-Craft catalyst. The cracking and isomerisations of alkanes can however, be done conveniently with zeolites (molecular sieves)[37].

14.2.2.3 Polymeric Esterolytic Catalyst

Ester hydrolysis are known to be performed with nucleophiles like imidazole, pyridine[38] and hydroxylamine. The polymeric analogues of these nucleophiles have been shown to be effective esterolytic catalysts[39-43].

An example of esterilytic catalyst is polymeric imidazole catalyst. These are synthetic prototype of esterolytic enzymes and have binding centers for the substrate and for the catalytic groups. In homopolymer polyvinylimidazole, the partially protonated groups act as binding sites, provided the substrate is negatively charged. At pH 7.5, 5% of the basic groups in polyvinylimidazole get protonated and at this pH, it acts as an effective catalyst for solvolysis of substrates carrying a negative charge, *e.g.*, for *p*-nitrophenyl hydrogen phthalate (1) or for polymeric substrates like a copolymer of acylic acid with *p*-nitrophenyl-*p*-vinylbenzoate (2*a*) or 2,4-dinitro-*p*-vinyl benzoate (2*b*)[44]

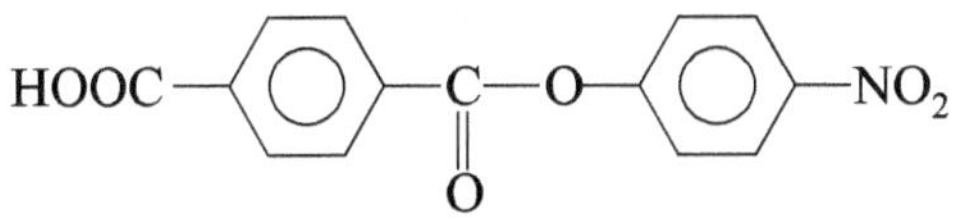

p-Nitrophenyl hydrogen phthalate
(1)

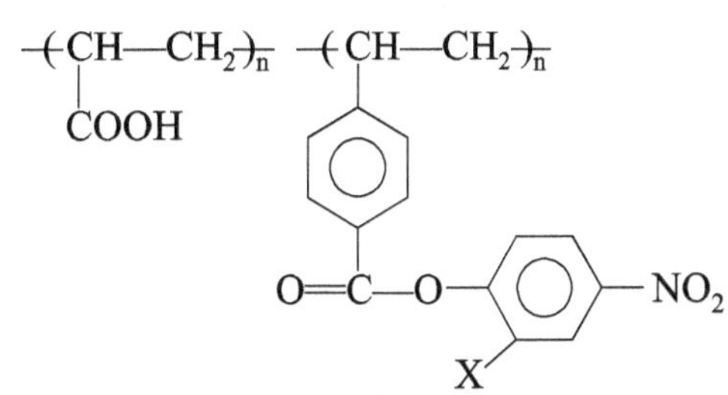

Copolymer of acrylic acid with
p-nitrophenyl-*p*-vinylbenzoate (2a, R = H)
OR copolymer of acrylic acid with
2,4-dinitro-*p*-vinylbenzoate (2b, R = NO_2)

It has been found[44, 45] that the copolymer of N-vinylimidazole and acrylic acid is an effective catalyst for the hydrolysis of *p*-nitrophenyl acetate. The COOH group in this polymer can bind a positively charged substrate, *e.g.* (3). In fact, the catalytic electrolytic activity of such a polymer for the positively charged 3, the neutral 4 and the negatively charged 5 is in the order.

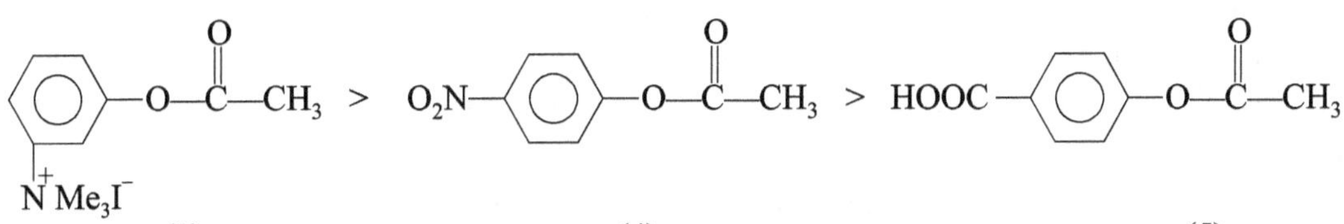

(3) (4) (5)

The catalytic activity of the polymeric catalyst substrate has been shown to increase due to its, non-ionic association. As an example, the rate of hydrolysis of 3-nitro-4-acetoxy benzoic acid (6) by polyvinyl imidazole increases with the increase in the bulk of the acetoxy group[46].

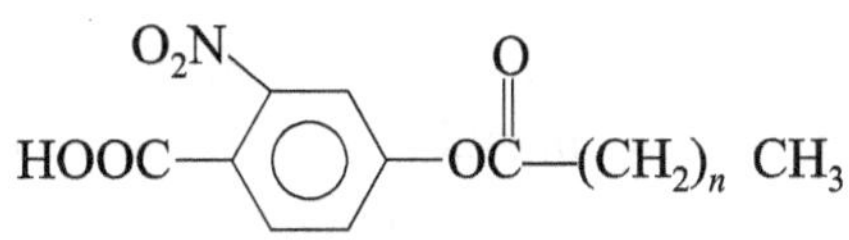

n-1, 6, 11, 17
3-Nitro-4-acetoxy benzoic acid
(6)

It has been shown[46] that the polymeric imidazole is about 10^2 times as reactive as the monomeric imidazole in the hydrolysis of *p*-nitrophenylacetate and some 3-nitro-4-acetoxybenzoic acids. The increased catalytic activity of the polymeric imidazoles is attributed to the interaction of the imidazole moieties arranged on the polymer chain at regular intervals (Fig. 14.2).

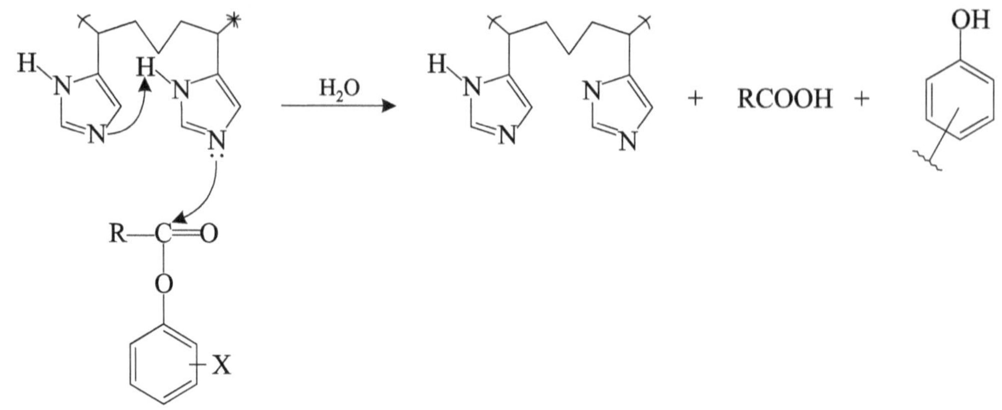

Fig. 14.2: Hydrolysis of *p*-nitrophenylacetate and 3-nitro-4-acetoxybenzoic acids with polymeric imidazole.

14.2.2.4 Polymer-supported Phase-transfer Catalysts

Phase-transfer catalysts are known to increase the yields considerably in an organic transformation (for details see Chapter 12). However, the catalytic activity of polymer-supported phase-transfer catalysts was found to be lower than that of the corresponding nonimmobilized PTCs[47-50]. However, it was shown that the catalytic activity of the parent group was retained on binding to the polymer. Insertion of the spacer-arms is found to increase the catalytic activity[51,52]. In case of PTCs based on tetraalkyl ammonium ions, the organophilicity of the polymer-bound group was dependent on the size of the alkyl group. The organo-bound catalyst could be recovered from the reaction mixture by filtration and could be reused.

The catalytic functional groups bound to the polymer were quaternary ammonium salts [Fig. 14.3(1*a*, *b*, *c*)], phosphonium salts [Fig. 14.3(1-d)], crown ethers [Fig. 14.3(1-e)] and cryptands [Fig. 14.3(1-f)]. The support polymer used is chloromethylated, 2-4% cross-linked polystyrene and silica gel and the catalyst group anchored by either reaction with the corresponding amine or phosphine or by absorption. Space-arms are used for linking the crown ethers and cryptand[50-54]. Polymer-supported asymmetric PTCs have also been prepared (For details see section 14.5).

Polymer supported crown ethers have also been synthesised[55,56] by the polymerisation of the corresponding vinyl monomer (4′-vinylmonobenzo-15-crown-5) (Fig. 14.4). These supported crown ethers, like the monomeric crown ethers can bind alkali metal, particularly potassium ions. These form complex with $KMnO_4$, which is used as an oxidising agent in non-aqueous media[57].

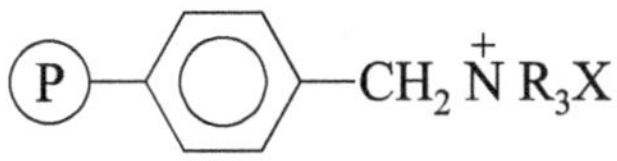

(1)

a, R = Me

b, R = *n*-Bu

c, R = *n* oct., X = halogen

Polymeric quaternary ammonium salt

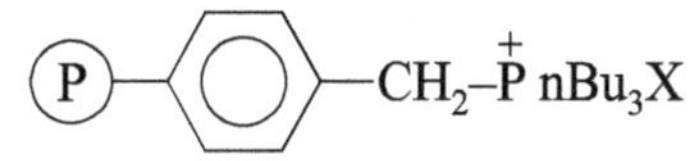

(1*d*)

Polymeric phosphonium salt

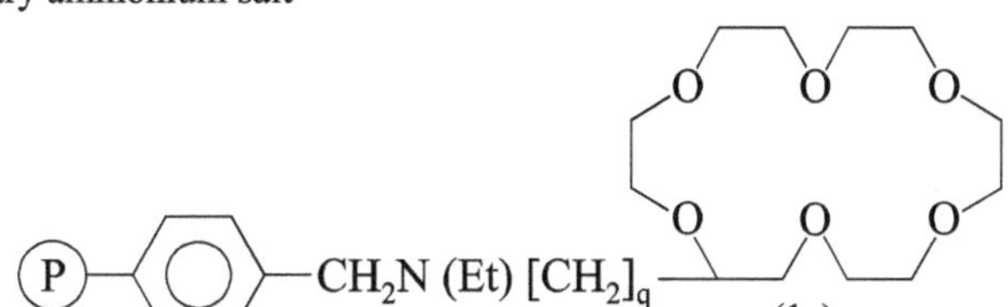

(1*e*)

Polymeric crown ethers

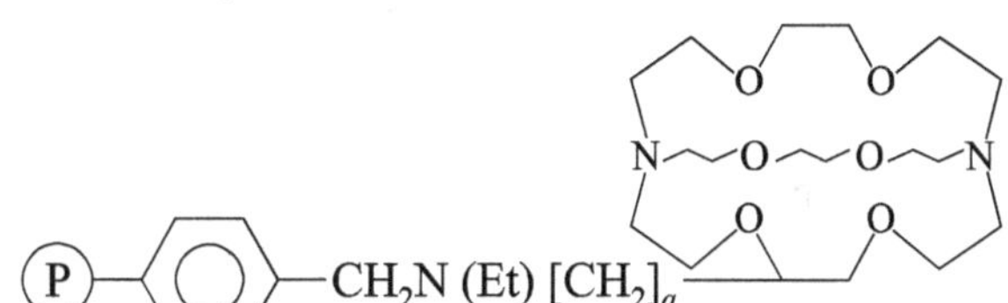

(1*f*)

Polymeric cryptands

Fig. 14.3: Examples of polymer supported PTC (ammonium and phosphonium salts), crown ethers and cryptand

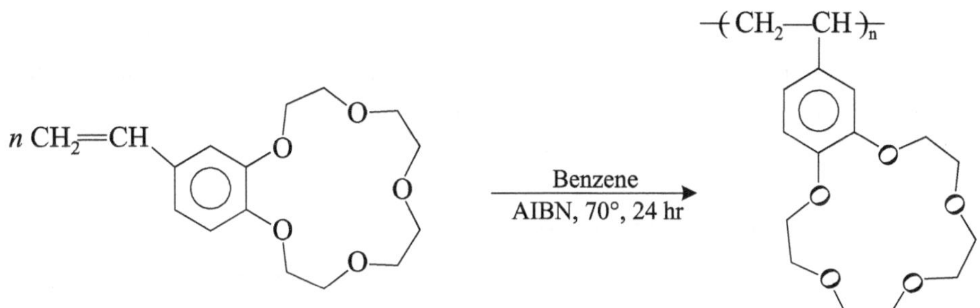

4'-Vinyl monobenzo –15 crown-5

Polymer supported crown ether

Fig. 14.4: Synthesis of polymer supported crown ether.

Polymeric quaternary ammonium salts, polyvinylpyridine dimethyl polyethylene glycols, have also been used as phase-transfer agents in a number of reactions like alkylations, displacement reactions, alkylations and carbenation reactions[58-61].

14.2.2.5 Polymeric Triphase Catalysts

The polymeric triphase catalysis procedure has been used for carrying out nucleophilic substitution reactions[62,63]. In these reactions, the resin acts as a polymer-bound PTC. The quaternary ammonium groups introduced into the resin contained large alkyl groups in order to act as the linked phase-transit catalyst. The procedure consist in keeping the resin is contact with an aqueous phase, such as a solution of NaCN and an organic phase, such as alkyl bromide. The product is alkyl cyanide obtained by nucleophilic substitution (Fig. 14.5).

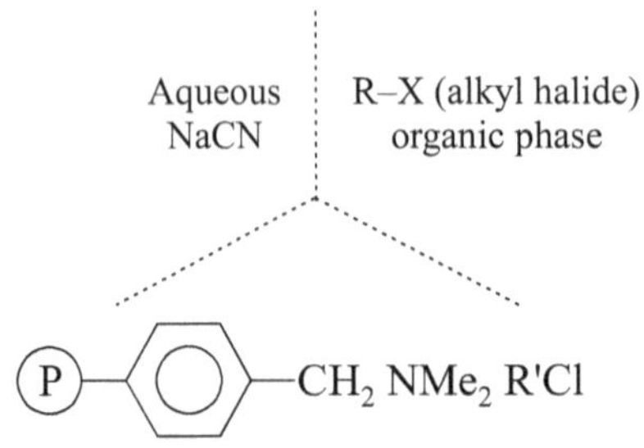

Fig. 14.5: Polymeric triphase catalysis procedure.

The basic difference between the above reaction and the common heterogeneous catalytic reaction is that in the former case, the catalyst and each of the reactants are located in separate phases.

In the above reaction, the catalyst acts like a PTC and brings together the reactants (Cyanide ion and alkyl halide) from the two phases.

The triphase catalysts have also been used for C-alkylations[64] and have been found to promote asymmetric addition in carbene addition reactions[59].

Alternate phase-transfer agents have been obtained by replacing ammonium group by phosphoric triamides[65], phosphonium groups[52] and polyethylene glycol[66] (for more details see S.L. Regen, Angew. Chem. Int. Ed. Engl, 1979, **18**, 421).

14.2.2.6 Polymeric Photosensitizers

Photosensitizers are commonly used in organic synthesis for the generation of molecular oxygen ($^1O_2$). A number of dyestuffs, *viz.*, Rose-Bengal, eosin-*y*, fluorescein have been used as photosensitizers. However, the use of these dyes as photosensitizers have some limitations as given below:

- Only limited solvents can be used for a reaction.
- The dye bleaches in case the reaction is carried out for longer time.
- The dye can sometimes react with the reactants or with the products.
- The separation of the product from the reaction mixture is difficult.

The above limitations can be overcome by using photosensitizers bound to Merrifield type resin via an ester bond[67,68]. (Fig. 14.6)

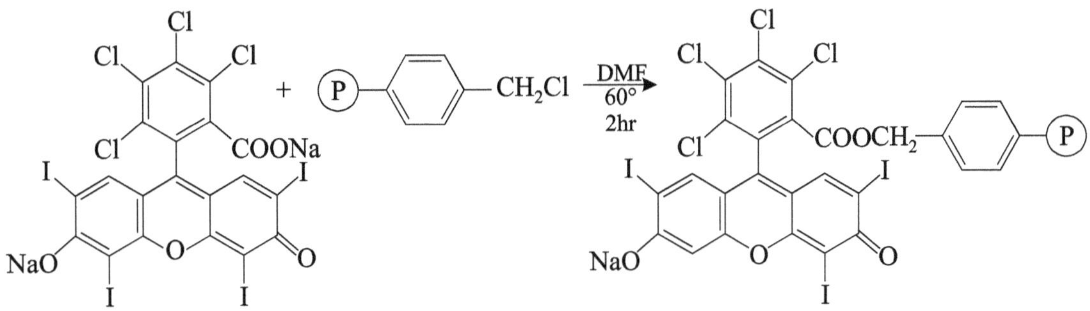

Rose Bengal Polystyrene Rose Bengal

Fig. 14.6: Synthesis of Polystyrene Rose Bengal.

Different types of reactions of singlet molecular oxygen have been carried out using polymer based Rose-Bengal (Table 14.2). These include 1,4-cyclo-addition with conjugated dienes to give cyclic peroxides, an 'ene'-type reaction to give allylic hydroperoxides and 1,2-cycloaddition to give 1,2-dioxetanes, which cleave thermally to give carbonyl-containing products.

Table 14.2: Photooxidation of organic compounds with (P)-Rose-Bengal

Organic compound	Product obtained	Yield (% isolated)
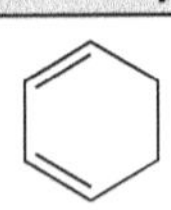	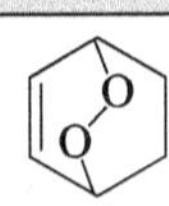	69
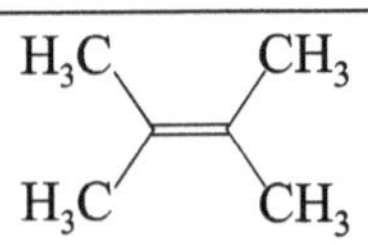	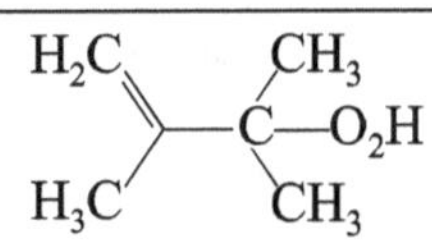	82
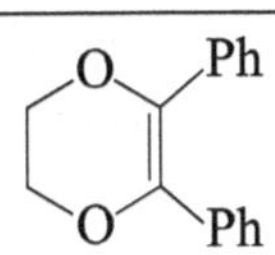	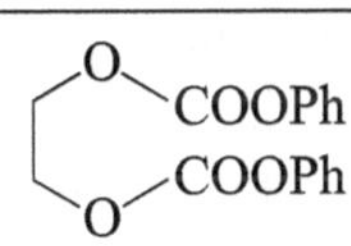	95

A polymer-bound Rose-Bengal under the name 'SENSITOX' is now commercially available. The advantages of using sensitox are:

- Being insoluble in most of the solvents, it is readily separated from the reaction mixture and can be used again.
- The polymer can be used in a number of solvents in which it can swell.
- A continuous process to carry out reactions can be developed.
- The efficiency of sensitox is about 65 percent than that of Rose-Bengal, but the high yields of the product and ease of isolation are more important.

In addition to Rose-Bengal, other phthalein dyes like eosin *y* and fluorescein and also porphyrins like chlorophyllin and hematoporphyrin can also be linked to the polymer.

Anthracene[69] and 9,10-diphenylanthracene[70] are also known to produce singlet oxygen. 9,10-di-*p*-styrylanthracene has been incorporated[71] with a polymer (Fig. 14.7).

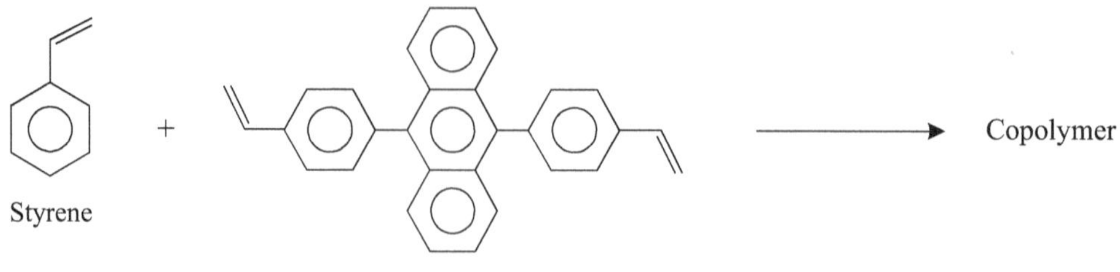

Fig. 14.7: Copolymer of Styrene and 9,10-di-*p*-styryl anthracene.

The copolymer of styrene and 9,10-di-*p*-styryl anthracene (Fig. 14.7) has been used for photooxidation of several compounds. Another polymer containing benzophenone group is prepared by the reaction of *p*-benzoylbenzoic acid with chloromethylated Co (polystyrene-DVB) (Fig. 14.8)[72].

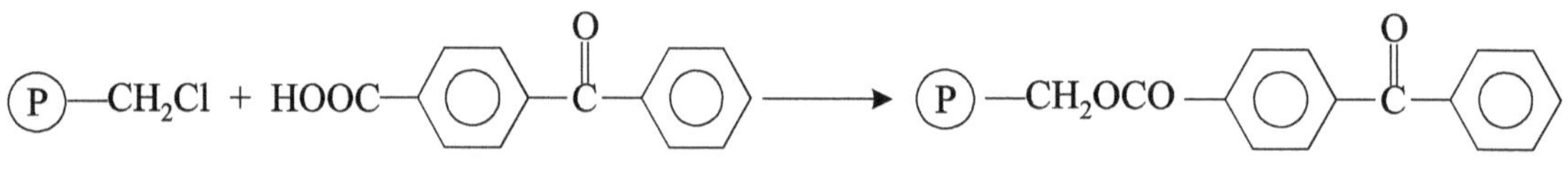

Fig. 14.8: Preparation of polymer containing benzophenone group.

The above reagent has been used to study the cycloaddition reactions of tetrachloroethylene to cyclopentadiene and the dimerisation of indene and coumarin.

14.3 SUPPORTED METAL COMPLEXES

The use of heterogeneous catalysts, a powerful tool for performing controlled reactions is well known. Typical reactions involving catalytic hydrogenations are the earliest known reactions. Such reactions are now considered as green reactions. The solid catalysts can be recovered easily by filtration and reused. The method is adopted for work on a large scale including industrial scale. The heterogeneous catalysts (like PtO, Pd-C, Ni suffer from a number of disadvantages like double bond migration, hydrogenolysis etc. Some of these difficulties have been overcome by using soluble catalysts, which permit reactions in homogeneous solution. Such catalysts are referred to as homogeneous catalysts and are difficult to remove after the reaction is over. This problem has been overcome by anchoring such metal catalyst on inert surface producing solid catalysts, which combine the benefits of homogeneous catalysts and solid catalytic materials. Thus, palladium complexes, immobilized on inorganic of organic support are found to be very effective.

14.3.1 Polymer-supported Palladium Complexes

Polymer-supported palladium complexes with C-N ligands have been prepared and used in Heck reaction of aryl halide with n-butyl acrylate (Fig. 14.9).

The catalyst (Fig. 14.9) can be recovered by filtration and used again and again with 100 percent conversion[73].

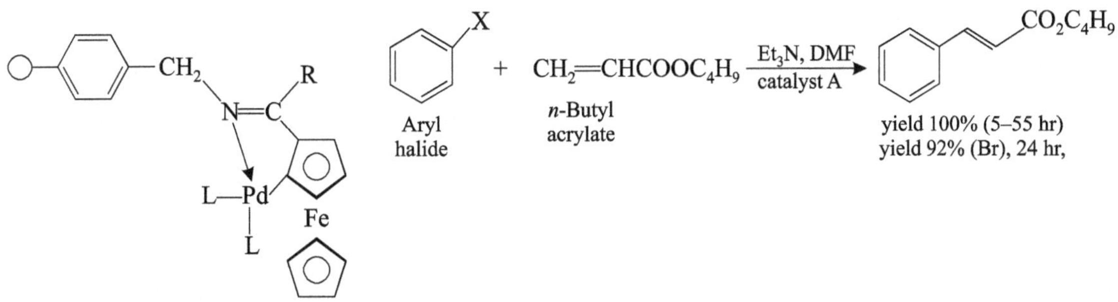

Catalyst A (R = H, CH_3, Ph)
Polymer-Supported Pd complex
with C—N ligands

Fig. 14.9: Heck reaction using polymer-supported Pd complex

Another polymer supported Palladium complex, *viz.* carbapalladacycle ion-exchanged MMT-clay-supported catalyst (catalyst B) has been used[74] for Mizoroki-Heck and Sonogashira couplings (Fig. 14.10).

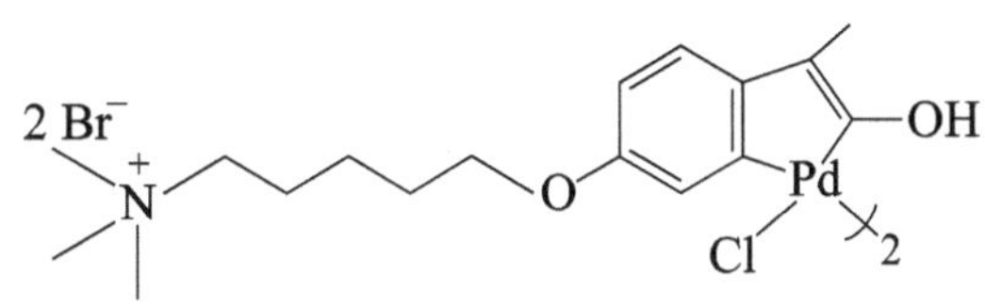

catalyst B
carbapalladacyle ion-exchanged
MMT-clay supported catalyst

Mizoroki-Heck coupling

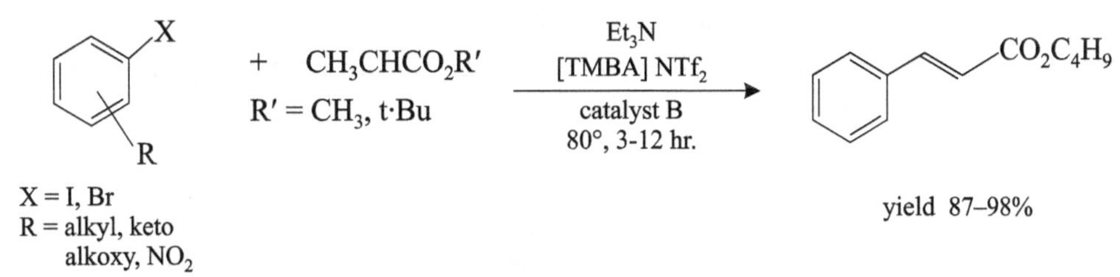

Sonogashira Coupling

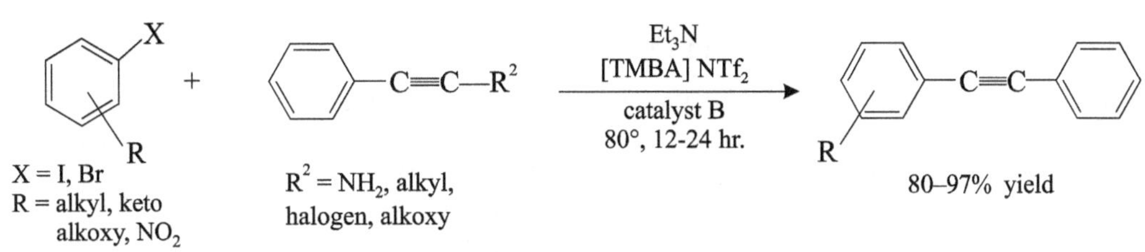

Fig. 14.10: Mizoroki-Heck coupling and Sonogastera coupling using carbapalladacycle ion-exchanged MMT-clay supported catalysts.

A number of other polymer-supported palladium complexes have been synthesised[75-80] and used for coupling reactions.

See also Section 14.6.1.

14.3.2 Supported Metal Catalysts

A number of supported metal catalysts (particularly palladium supported catalysts) have been prepared. Only a few are discussed.

A catalytic system based on Pd supported on polyether-functionalized mesoporous silica is prepared by the hydrolysis and polycondensation of tetraethyl orthosilicate and a bis-silylated triblock copolymer followed by decomposition of $Pd(OAc)_2$. The catalyst exhibited excellent properties in the Heck coupling of iodobenzene and *t*-butyl acrylate to yield t-butyl cinnamate. The catalyst could be recovered by filtration and reused[81].

Another supported Pd metal catalysts is mesoporous silica-carbon nanocomposite supported palladium catalyst. It was successfully used in the Heck coupling of chlorobenzene and styrene. The reaction produced E product in exclusive selectivity and the catalyst is valuable in green applications. In addition, the reactions are carried out in aqueous medium[82].

Carbon nanofiber (CNF) supported palladium nanoparticle catalysts find use in Heck coupling of different types of aryl halides and styrene derivatives (Fig. 14.11)[83].

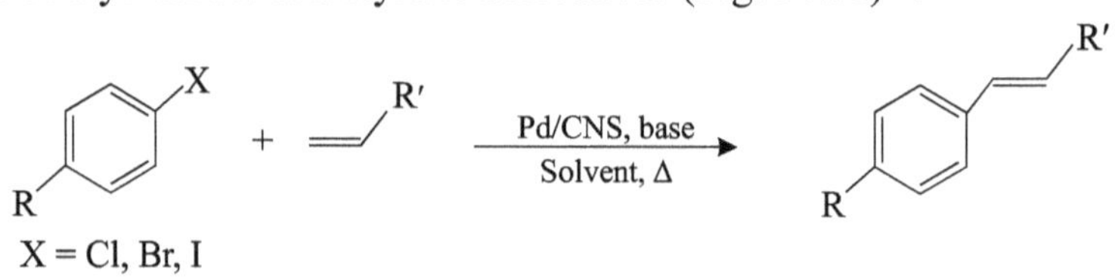

Fig. 14.11: Heck coupling using carbon nanofiber supported Pd nanoparticle catalyst.

14.3.3 Organocatalysts

Methods that do not need transition-metal catalysts are more-economical. This is especially important in the synthesis of drugs. The catalysts used are organocatalysts[83].

An organocatalyst, 1,10-phenathroline has been used for cross-coupling of aryl bromides or iodides and arenes in presence of potassium *tert*-butoxide as base yield biphenyls (Fig. 14.12)[84].

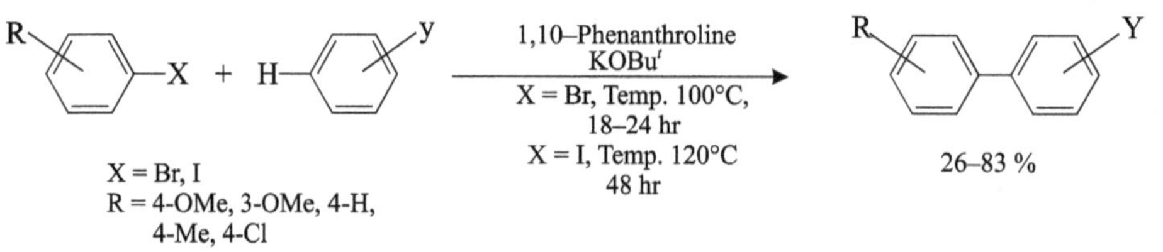

Fig. 14.12: Synthesis of biphenyls.

Biaryl derivatives have also been synthesized using N, N′-dimethylenediamine (DMEDA) as catalyst for direct C–H arylation of unactivated benzene with aryl halides under the assistance of an iodo group in presence of potassium *tert*-butoxide (Fig. 14.13)[85].

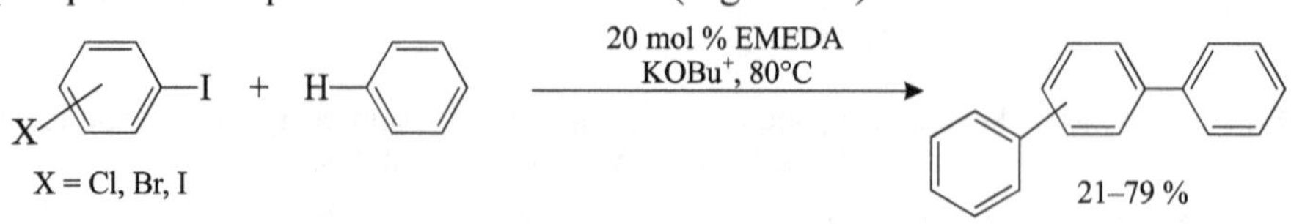

Fig. 14.13: Synthesis of biaryl derivatives from aryhalides and benzene in presence of EMEDA.

An highly active catalytic system containing boron trifluoride etherate as Lewis and ethylene glycol as organocatalyst has been developed. This system has been used[86] for the cyclisation of ethene derivative as shown in Fig. 14.14.

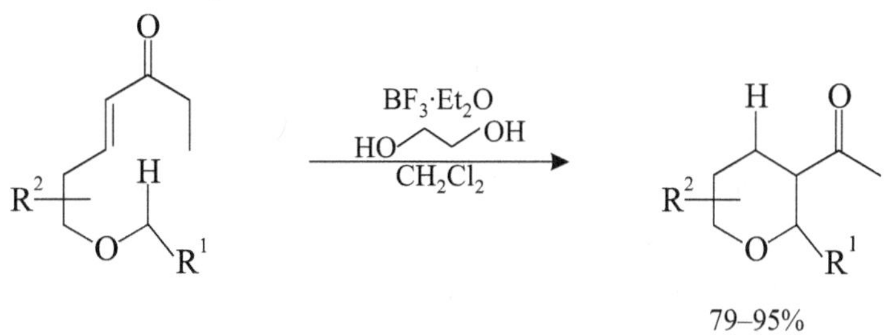

Fig. 14.14: Synthesis of pyran derivative.

A Mac-Millan-type organocatalyst (catalyst C) in presence of DDQ has been used for stereo selective alkylation of benzylic C–H together with an oxidative C–H activation (Fig. 14.15)[87].

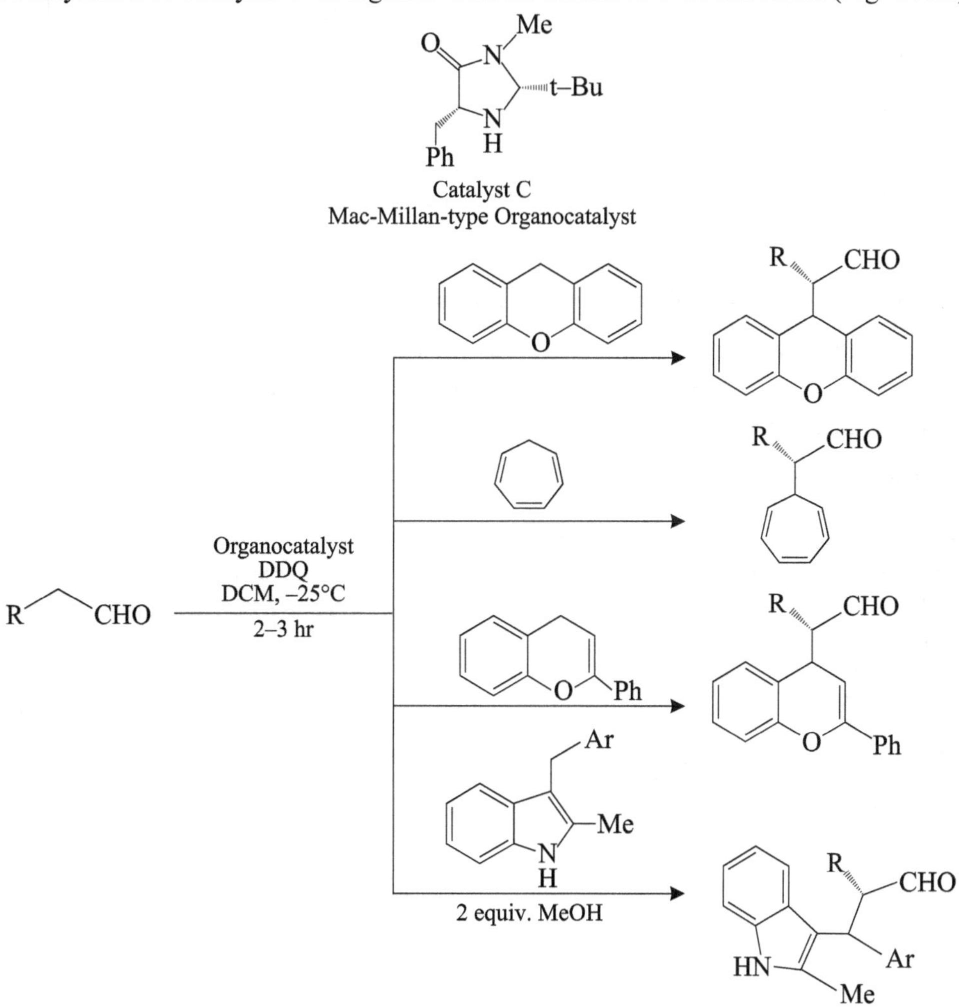

Fig. 14.15: Stereo selective alkylation of benzylic C–H using Mac-Millan type Organocatalyst.

The reaction of an aldehyde with terminal alkene in presence of an organocatalyst, 2-amino-3-picoline and Wilkinsos catalyst, $(Ph_3P)_3RhCl$ gave the coupled product (Fig. 14.16)[88]

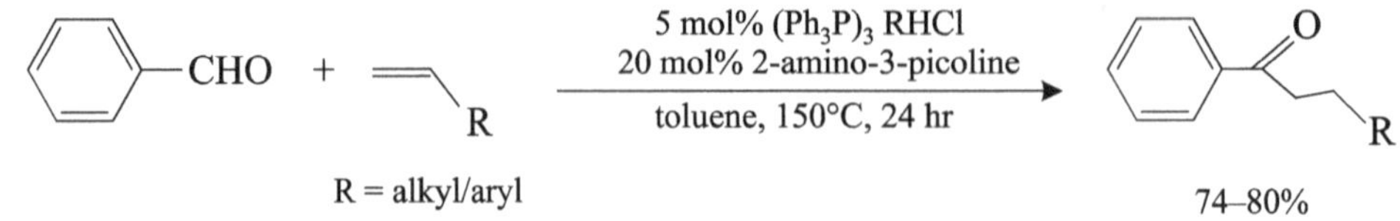

Fig. 14.16: The reaction of aldehydes with terminal alkenes.

An organocatalyst, dihydroquindine-2,5-dimethyl-4,6-pyrimidinediyl diether $[(DHQD)_2PYR]$ has been found to an effective catalyst in acetone at –40°C for the reaction of alkylidene malononitriles with trans-β-nitrostyrene gave the coupled product (Fig. 14.17)[89].

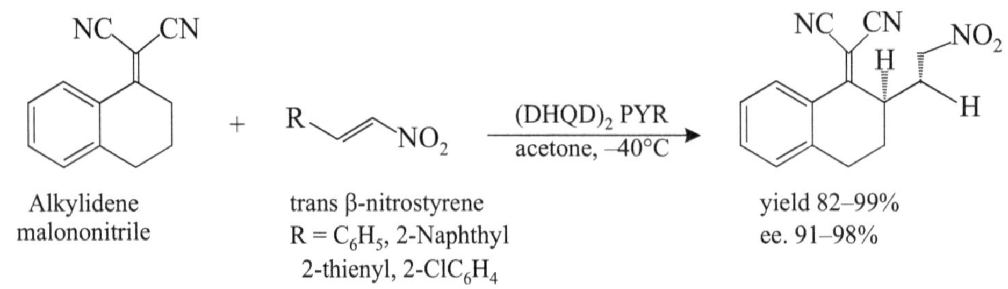

Fig. 14.17: The reaction of alkylidine malononitrile with trans β-nitrostyrenes using the organocatalyst $(DHQD)_2$ PYR.

14.4 SUPPORTED ASYMMETRIC ORGANOCATALYSTS

The supported organocatalysts (asymmetric) are considered to be environmentally benign synthetic methodology. The immobilized organocatalysts can be easily recovered and reused. Following are given some of the supported asymmetric organocatalysts.

14.4.1 Polymer-supported Chiral Amines

Polymers have been used in organic synthesis as supports for catalysts (see section 14.2). Proline and its derivatives have been commonly used. The earliest polymer-supported proline[90] [Merrifield resin-supported proline (catalyst D)] for **Hajos-Parrish-Elder-Sauer-Wilchert reaction** (Fig. 14.18).

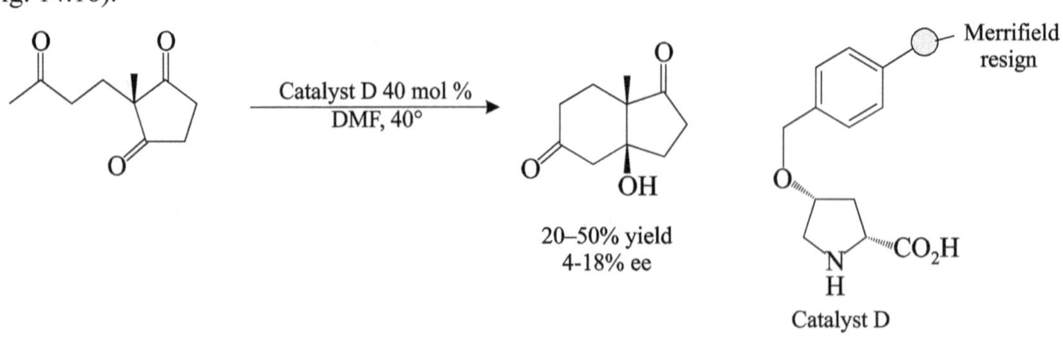

Fig. 14.18: Use of polymer-supported proline.

Immobilization of proline using PEG supports has also been effected[91-93]. These catalysts, unlike other polymer supported-Prolines are soluble in polar organic solvents such as DMF, DMSO and so the reaction can be performed in homogeneous phase. After the reaction, the supported catalyst can be precipitated by addition of a less polar solvent like ether. The PEG-supported proline (catalyst E) has be used in **aldol reaction**[91], **Henry reaction**[92] and **Michael addition reaction**[93] (Fig. 14.19).

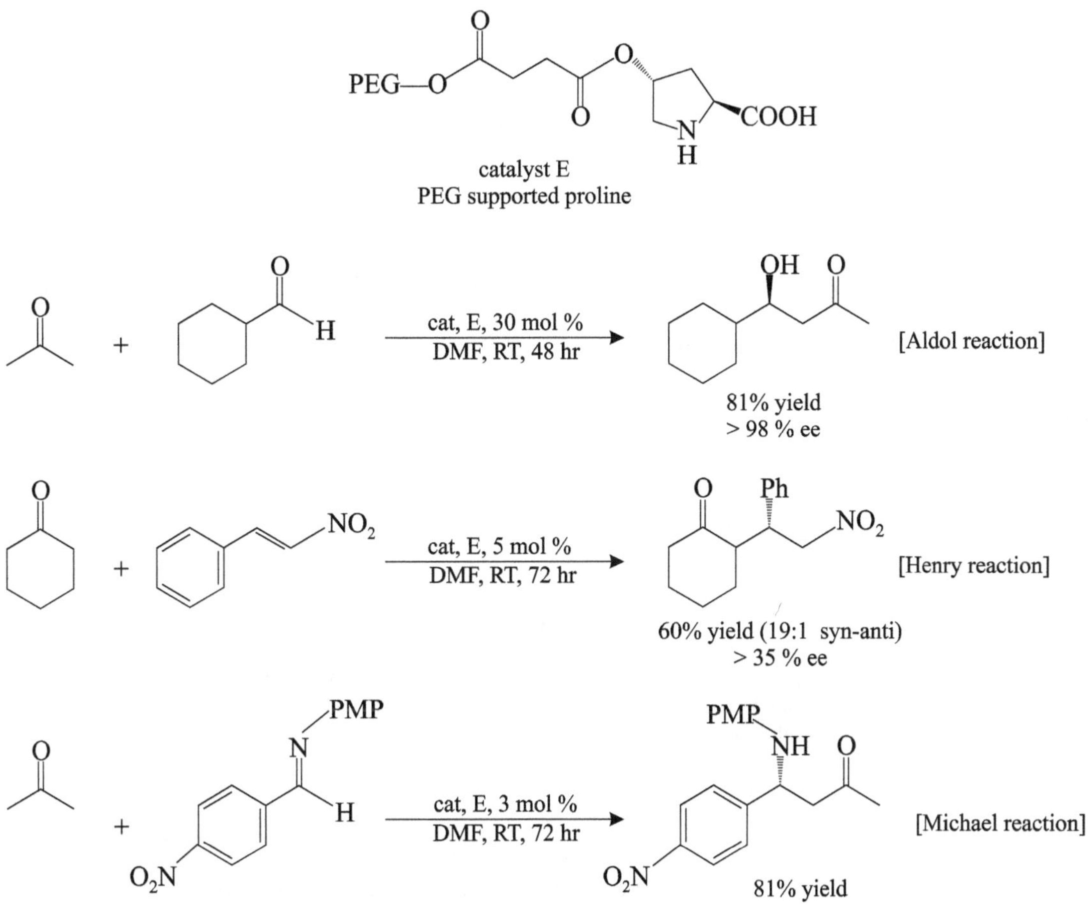

Fig. 14.19: Aldol reaction, Henry reaction and Michael addition reaction using PEG-supported proline.

A better PEG-supported proline catalyst bearing a sulfonamide linker has been developed and used for Michael addition reaction (Fig. 14.20)[94].

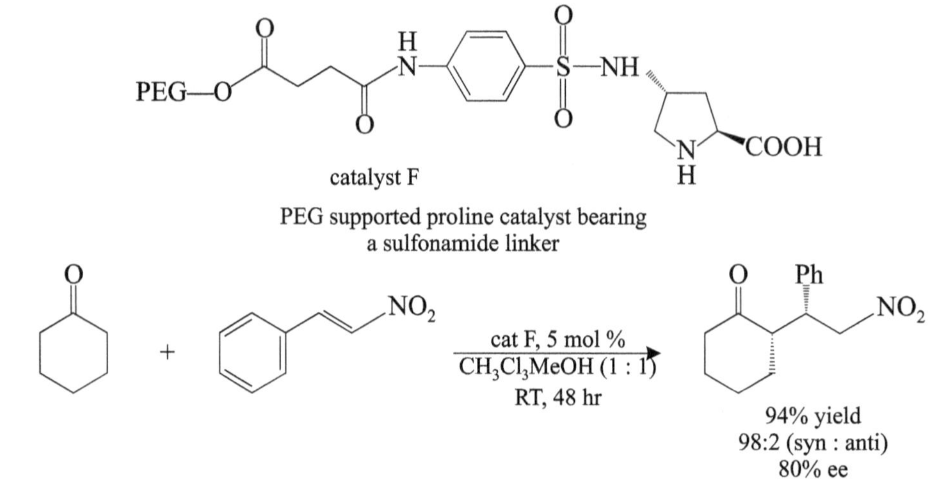

Fig. 14.20: Michael addition using a noval PEG-supported proline catalyst bearing a sulfonamide linker.

Polystyrene (PS) type resins and polyacrylic (PA) resins have also been used for supported proline. These exhibit good catalytic activities and enantioselectivities in direct aldol reactions (Fig. 14.21)[95].

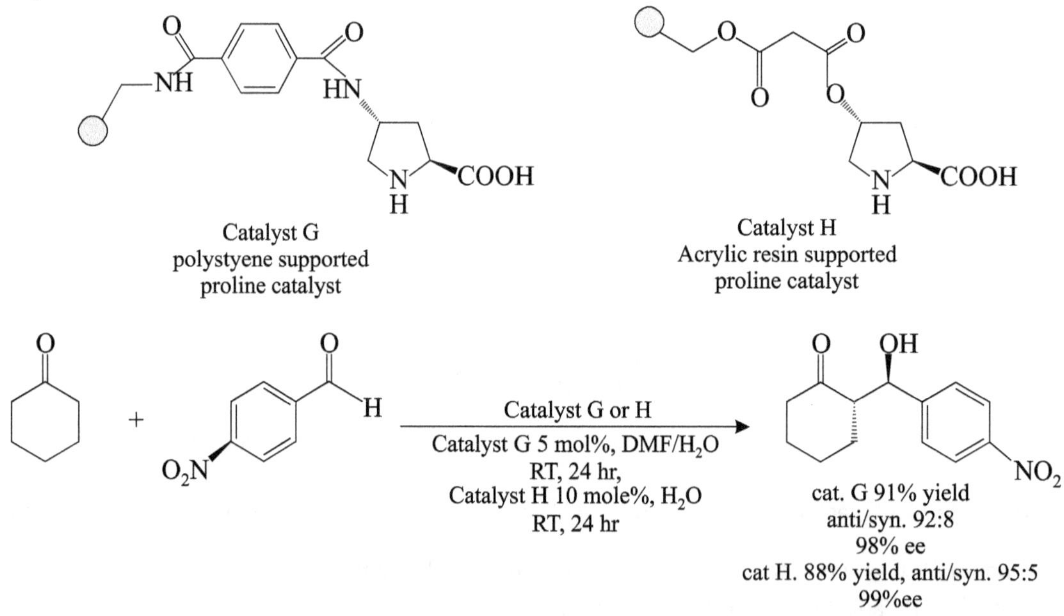

Fig. 14.21: Polystyrene and acrylic resin supported proline for use in Aldol reaction.

These catalysts (catalyst G and H) have also been used for catalysing other enamine based transformations such as Baylis-Hillmann reactions[96], Mannich reactions[97] and α-aminoxylations[98].

Besides the examples given above, a polymer supported proline with dendritic linker was developed[99] and used in direct aldol reaction (Fig. 14.22) .

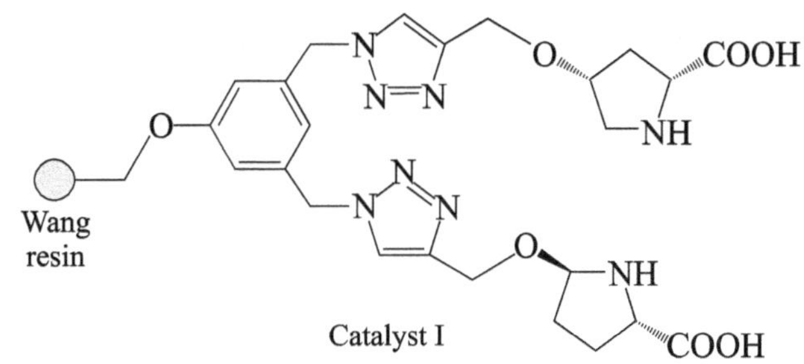

Polymer-supported proline with dendritic linker

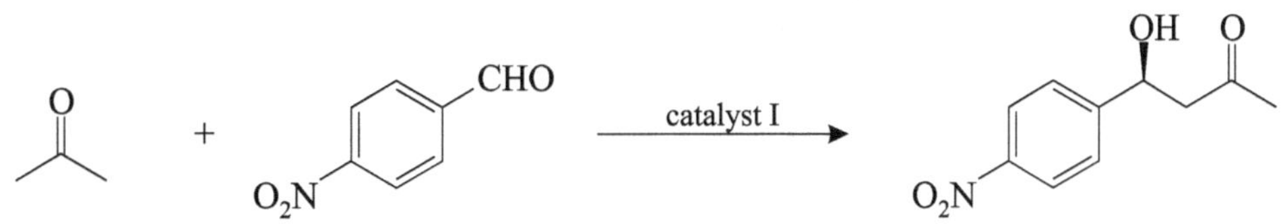

Fig. 14.22: Direct Aldol reaction using Polymer-supported proline with dendritic linker.

A number of proline derivatives such as proline-containing dipeptides and prolinamides have been prepared[99-103] (Fig. 14.23) and used as catalysts for aldol reactions.

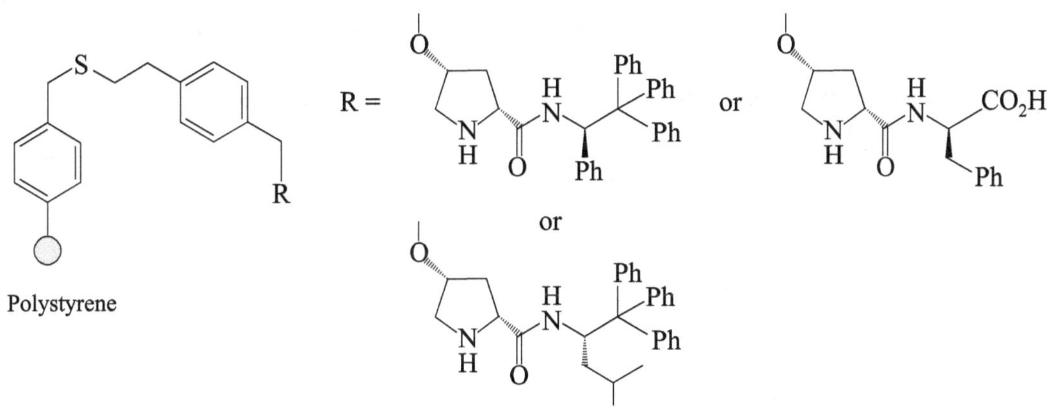

Polymer-supported prolenamides

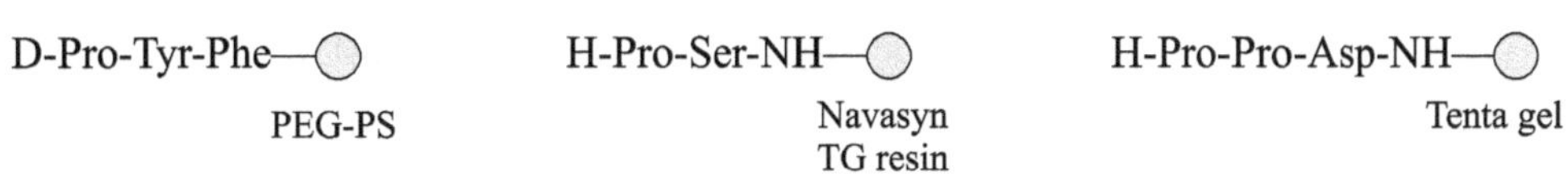

Polymor supported peptides

Fig. 14.23: Polymer-supported prolinamides and peptides.

Pyrrolidine-imidazolium salts[104] and pyrrolidine triazoles[105] are effective catalysts for the asymmetric Michael additions of ketones to nitroolefins. Their supported versions *viz.*, polymer supported pyrrolidine triazoles[106], polymer-supported pyrrolidine triazoles[107] and polymer supported pyrrolidine-imidazolium salts[108] have been prepared and used in asymmetric Michael addition of ketones to nitrostyrenes (Fig. 14.24).

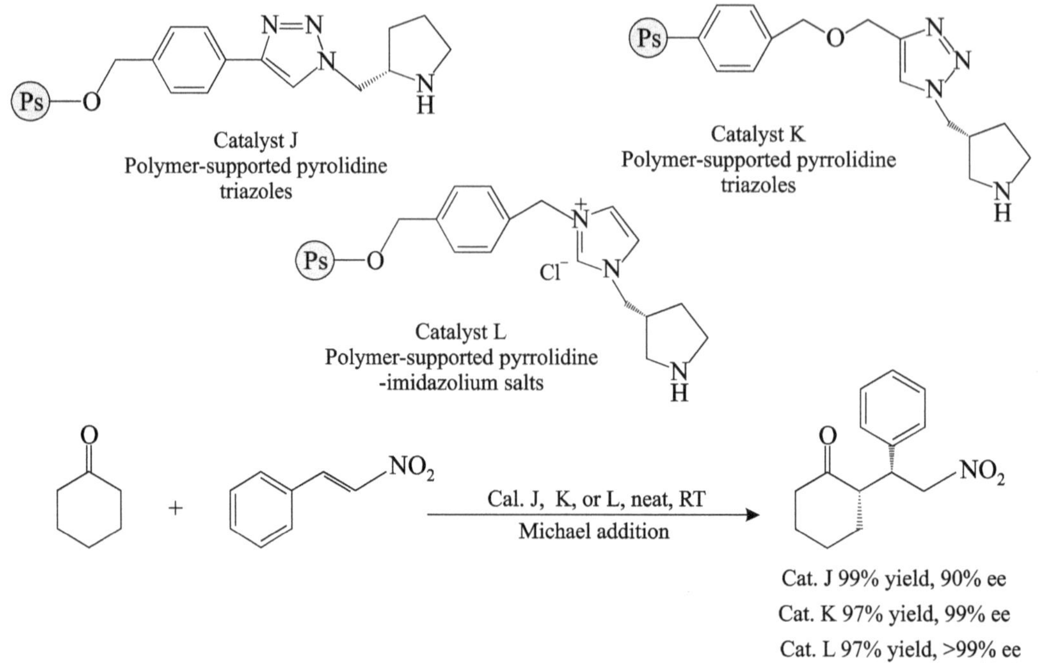

Fig. 14.24: Asymmetric Michael addition of Ketones to nitrostyrene.

14.5 POLYMER-SUPPORTEDASYMMETRICPHASETRANSFERCATALYSTS

Polymer-supported asymmetric Phase Transfer Catalyst, particularly based on cinchona alkaloids have been prepared[109-111] and used for the alkylation of glycine isopropyl ester derived Schiffs base (Fig. 14.25).

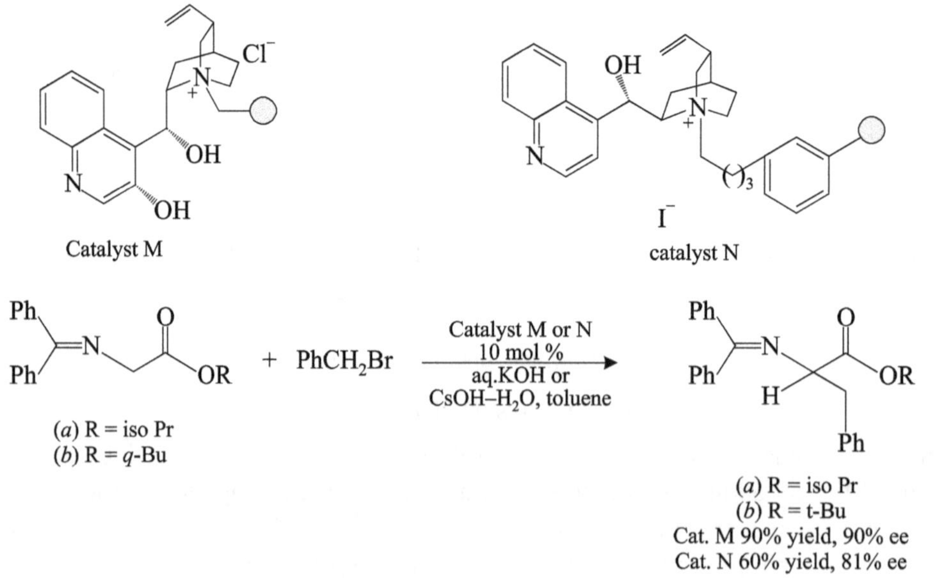

Fig. 14.25: Alkylation of glycine isopropyl ester derived Schiffs base using polymer supported asymmetric PTS.

14.6 POLYMER-SUPPORTED PHOSPHORIC ACID CATALYST

Only one example of polymer-supported chiral phosphoric acid has been reported[112]. It has been successfully used in multiple transfer hydrogenation of benzoxazine (Fig. 14.26).

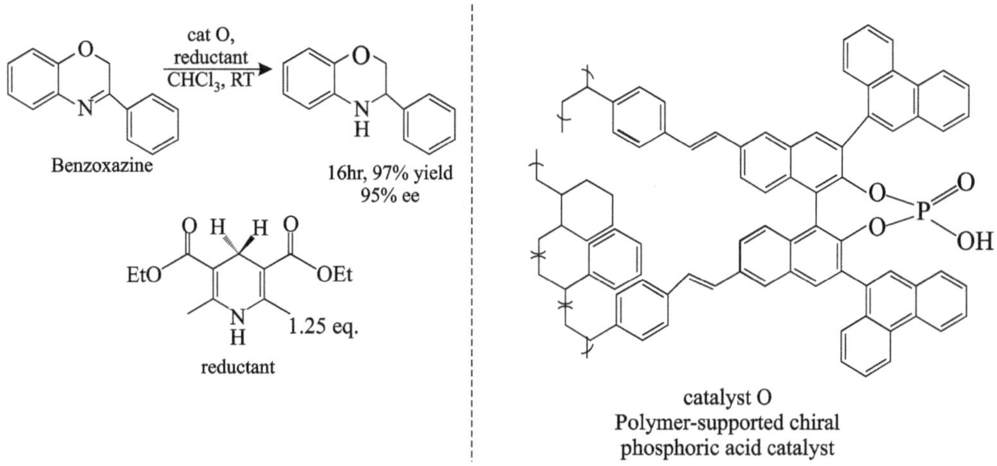

Fig. 14.26: Reduction of benzoxazine using catalyst O.

The polymer-supported phosphoric acid catalyst was prepared in the form of a polymer stick. After completion of the reaction, the stick is pulled out of the reaction mixture and used again.

Amongst the supported asymmetric organocatalysts, besides polymer-supported organocatalysts other supported catalysts include solid acid supported organocatalysis[113-119], ionic liquid supported organocatalysts[120-122] and silica supported asymmetric organocatalysts[123,124].

14.6 SOLID SUPPORTED CATALYSTS

It has already been stated that solid supported catalysts, which can be recycled are the best in the context of green chemistry. Additionally carrying out transformation in aqueous medium further adds to environmentally benign condition. A number of solid supported catalysts have been used. These include palladium, rhodium, ruthenium, cobalt, copper and irridium catalysts.

14.6.1 Immobilized Palladium Catalyst

***(a)* Suzuki Reaction:** The suzuki reaction, a powerful carbon-carbon bond forming reaction was initially conducted with homogeneous palladium catalyst in organic solvent. Subsequently polymer-supported palladium complexes were used (see section 14.3.1). It is now found that palladium catalyst on supported polyaniline (PANI) are a good choice for conducting cross-complying suzuki reaction in aqueous media[125,126] (Fig. 14.27).

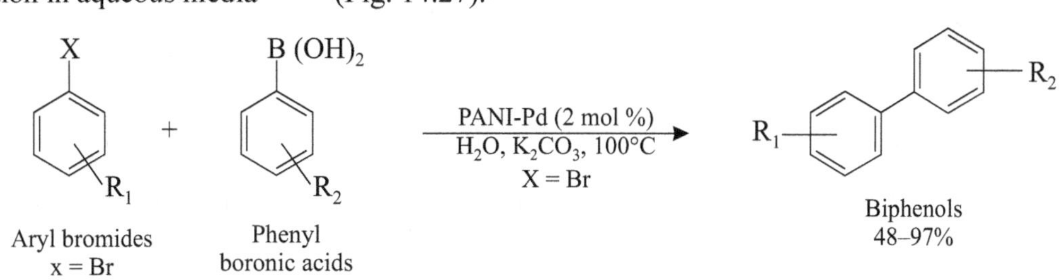

Fig. 14.27: Suzuki Reaction in water using PANI-Pd catalyst.

Aryl chlorides could also be used in the above Suzuki reaction (Fig. 14.27) but required the addition of 50 mol% of Tetrabutyl ammonium bromide (TBAD). The catalyst, PANI-Pd is recyclable and can be filtered and reused.

In an alternative method, $PdCl_2$ is used via complexation with immobilized phosphine ligands. Thus, complexed $PdCl_2$ with an immobilized diphenylphosphine ligand are used (in a modified suzuki coupling) to couple sodium tetraphenylborate with aryl bromides (Fig. 14.28)[127]. The method is atom economical since each molecule of sodium tetra phenylborate transfers four phenyl groups compared to one per molecule of phenyl boronic acid. Also, the reaction is promoted using microwave irradiation and the catalyst could be recycled.

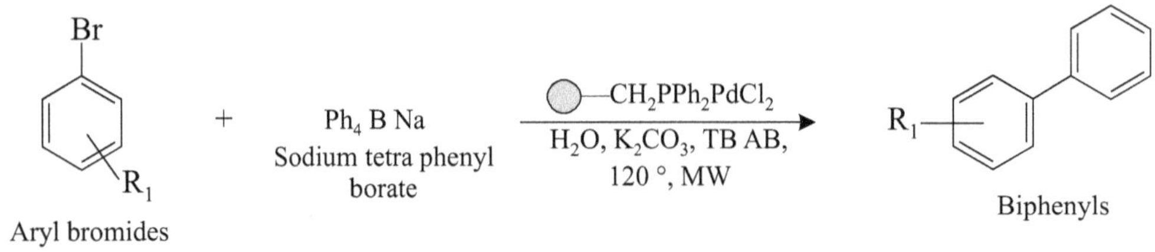

Fig. 14.28: Suzuki reaction employing immobilized $PdCl_2$.

A polystyrene-polyethylene glycol (PS-PEG) graft copolymer (Catalyst P) has also been used for Suzuki reaction at room temperature[128]. Also, a palladium supported catalyst (catalyst Q) was also used in Suzuki reaction (Fig. 14.29)

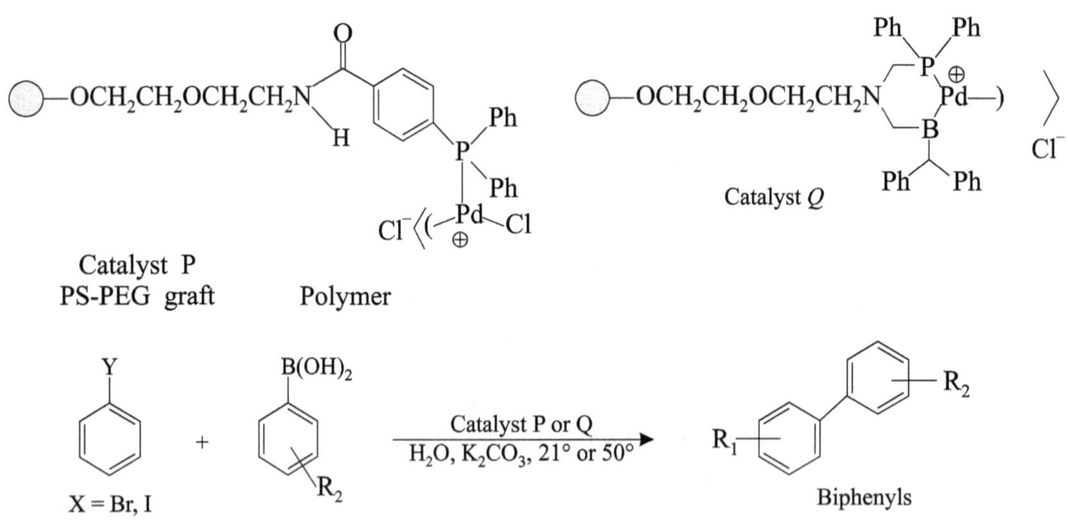

Fig. 14.29: Suzuki reaction by employing PS-PEG graft polymer or catalyse Q.

A number of other immobilized palladium catalysts were prepared and used for Suzuki reactions in water. Various catalysts are given below:[129-136] (Fig. 14.30).

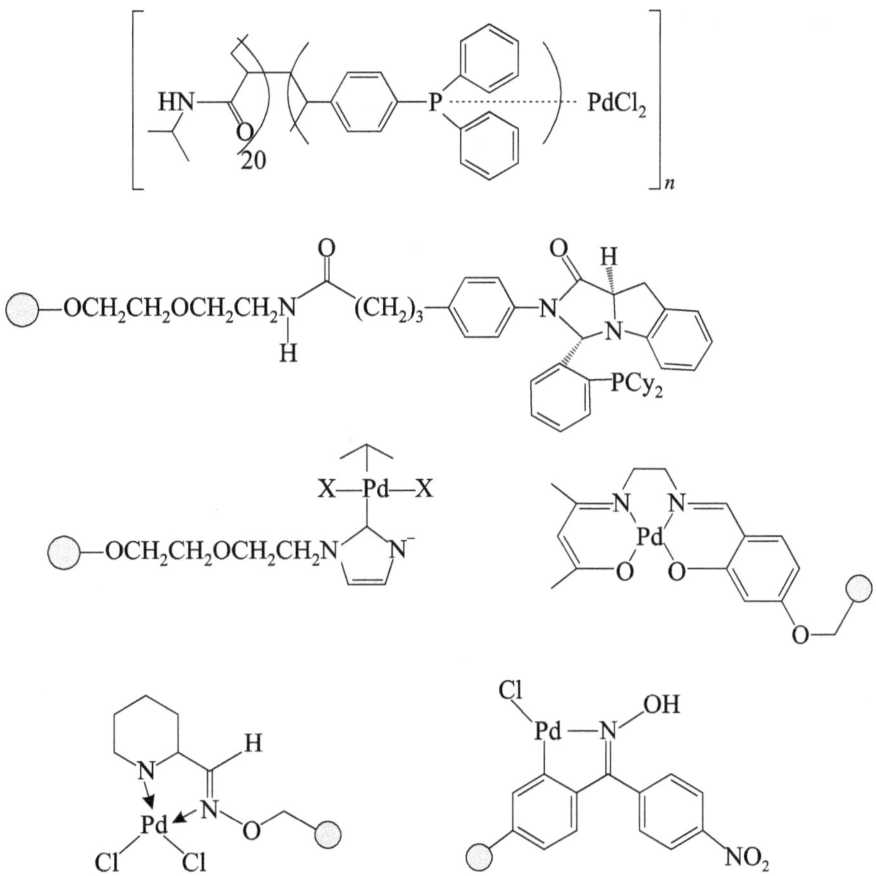

Fig. 14.30: Some other immobilized Pd catalysts.

(*b*) Heck-Mizoroki reactions: It involves Pd catalysed reaction between an alkene and an aryl or alkenyl halide and the reaction is conducted in aqueous media. The reaction is carried out in presence of catalyst P (see Fig. 14.29), which was found to be most effective (Fig. 14.31)[137,138].

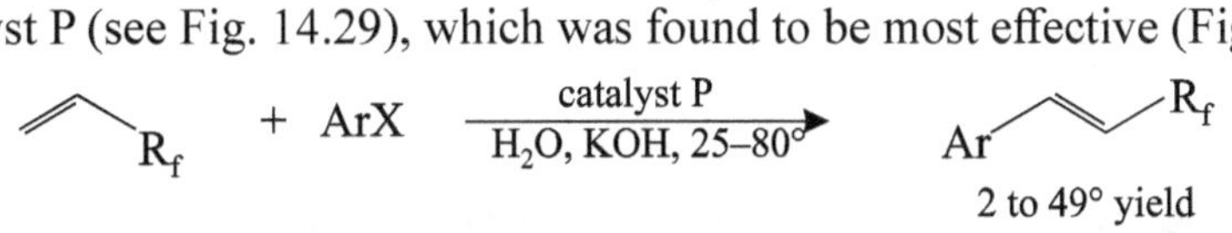

Fig. 14.31: Heck reaction in aqueous media in presence of catalyst P.

Oxime-based Pd catalyst that promote Suzuki reactions have also been used for the Heck reaction. Thus, the reaction of aryl halides (iodides or bromides) with alkenes (water soluble or water insoluble) in presence of palladium catalyst (catalyst Q, Fig. 14.32) gave good yields of the products (Fig. 14.32)[139,140].

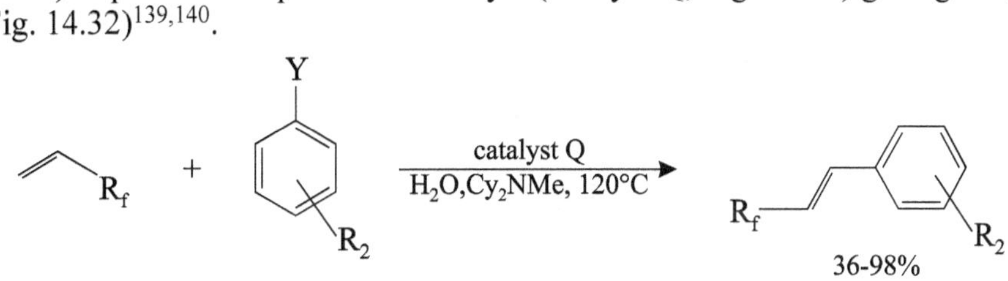

Fig. 14.32: Heck reaction in water in presence of immobilized oxime based catalyst Q.

Pyridine-based Pd catalysts have also been used for promoting Heck reaction in aqueous media. A typical example is given below (Fig. 14.33)[136].

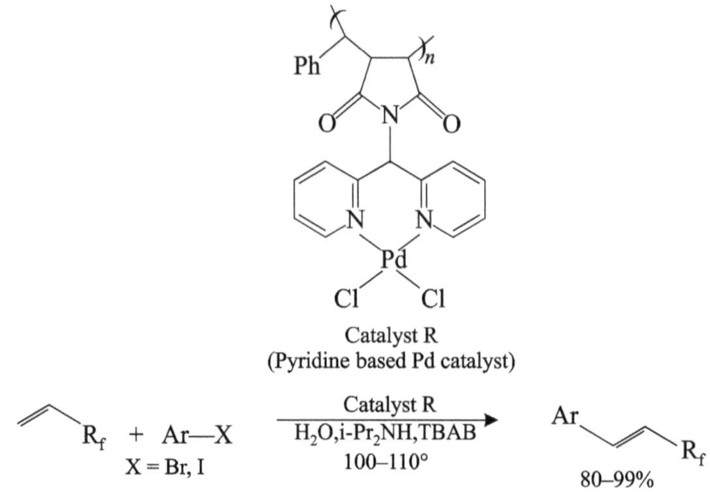

Fig. 14.33: Heck reaction promoted by immobilised pyridine based Pd catalyst (catalyst R).

(*c*) Sonogashira reaction: It involves Pd-catalysed cross-coupling reaction of an alkyne with aryl (or vinyl) halide.[141] The reaction is conducted in water in presence of catalyst P (Fig. 14.29). Alternatively, in case *o*-iodophenols are reacted with phenylacetylenes, the products are benzofurans (Fig. 14.34)[142].

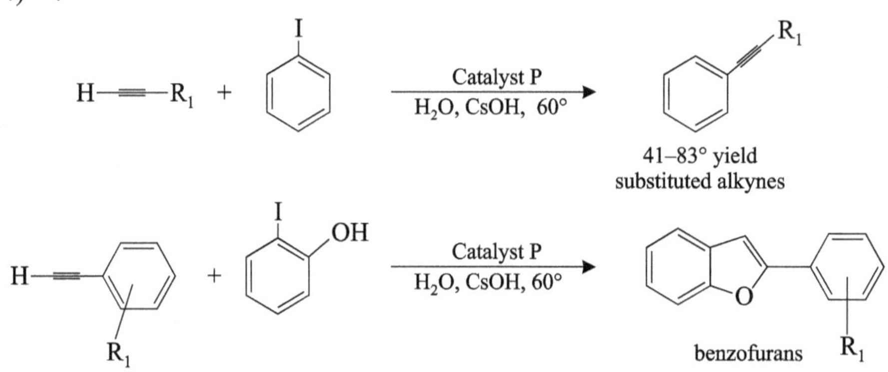

Fig. 14.34: Sonogashira Reaction in aqueous media using catalyst P.

Sonogashira reaction could also be conducted in water using NaOH or i-Pr_2 NH and TBAB (Fig. 14.35)[143] in presence of oxime-based catalyst.

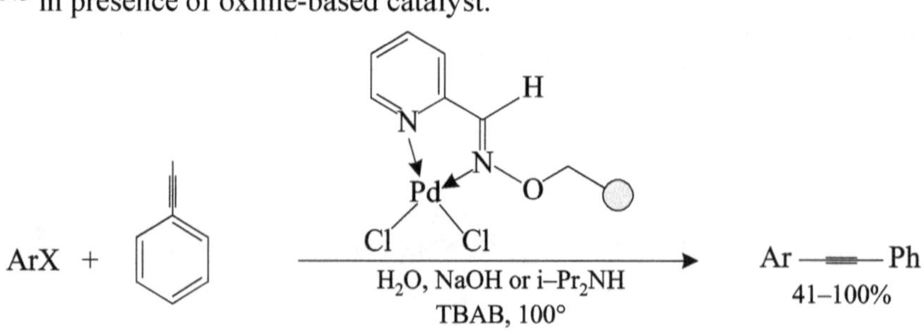

Fig. 14.35: Sonogashira reaction in water in presence of Oxime-based catalyst.

The above reaction could be conducted on microwave irridation at 160° (5-20 min, 35-100% yield) or on thermal heating at 100° (15 hr, 41-100% yield).

The reaction can also be conducted[136] in presence of immobilized dipyridyl-Pd complex (catalyst R). These reactions required TBAB and a base, pyrrolidine to proceed in H_2O at 100°.

(D) Tsuji-Trost reaction: It is a Pd-catalysed allylation[144,145] of allylic halides carbonates or acetates with nucleophiles like sodium azides, nitromethane, aryl boronic acids, malonic esters, β-ketoesters etc. (Fig. 14.36)[128, 146-150]. The reactions were prompted by Pd-phosphine catalyst supported on PS-PEG copolymer and conducted in water at room temperature.

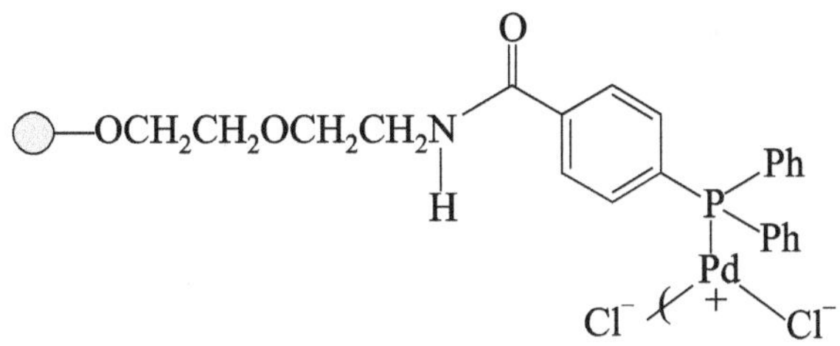

Pd-Phosphine catalyst supported on PS–PEG Copolymer

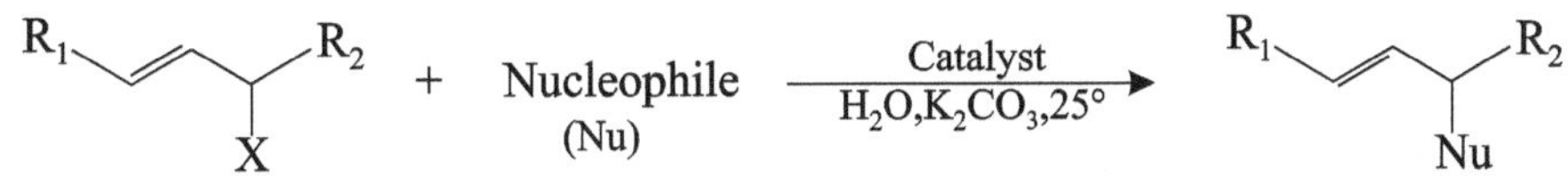

X = OAc, Cl
OCO_2 Me

45–100% yield

Fig. 14.36: Tsuji Trost Reaction is Water using Pd-Phosphine catalyst supported on PS-PEG copolymer.

Immobilised chiral phosphine catalysts promoted reaction between nucleophiles and cyclic and allylic species (Fig. 14.37)[151].

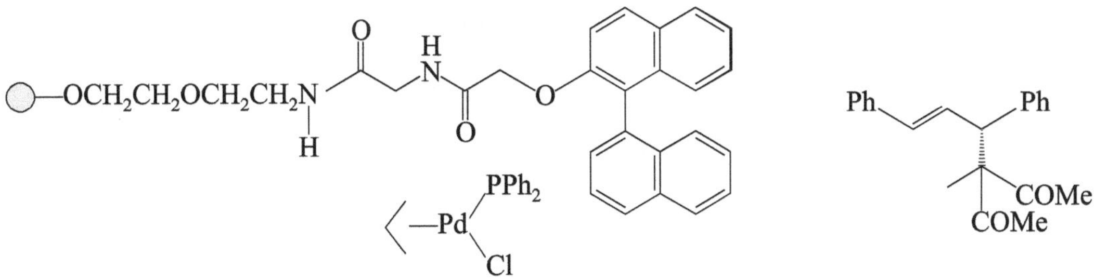

Chiral phosphine catalyst

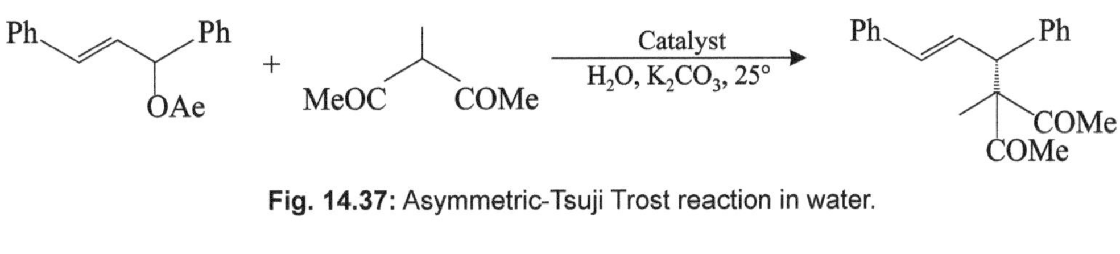

Fig. 14.37: Asymmetric-Tsuji Trost reaction in water.

14.6.2 Immobilized Rhodium Catalysts

Rhodium catalysts have been extensively used in synthetic organic chemistry. These have been immobilized on polymer supports and have been used[152,153] in greener synthetic processes, some of which are discussed below.

14.6.2.1 Cyclopropanation

A immobilized rhodium catalyst, dirhodium tetraprolinate catalyst Rh_2 (S-TBSP)$_4$ was immobilized via a pyridine linker on resins such as Argapore and used for the cyclopropanation of styrene with phenyldiazoacetate yielding cyclopropane derivative in good yield (Fig. 14.38)[154].

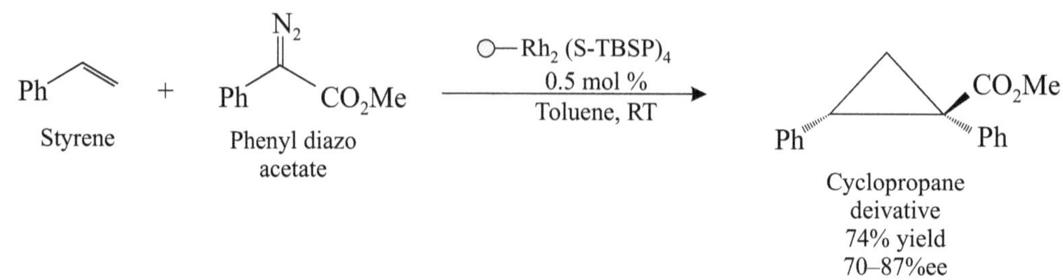

Fig. 14.38: Cyclopropanation using immobilised dirhodium tetraprolinate catalyst.

14.6.2.2 Conjugate Addition Reactions

A polymer supported rhodium complex has been used[155] for the addition of boronic acids to enonones and conjugate addition products obtained in good yields (Fig. 14.39).

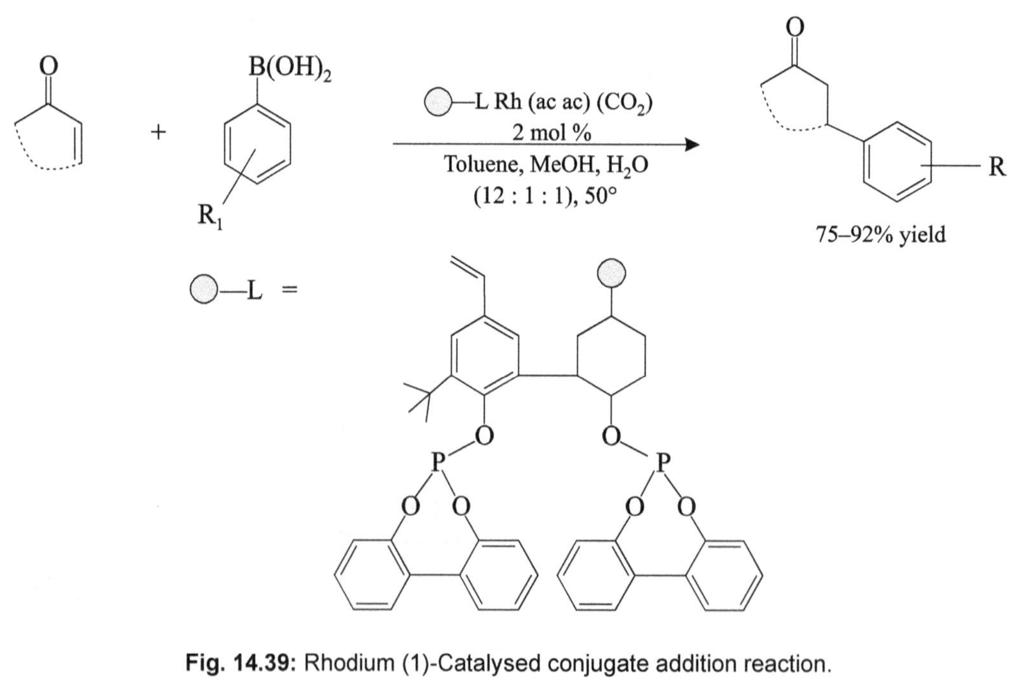

Fig. 14.39: Rhodium (1)-Catalysed conjugate addition reaction.

14.6.2.3 Hydrogenation Reactions

Asymmetric hydrogenation has been carried out using PS-supported aminophosphane-phosphite (and-phosphinite) ligands, premixed with $[Ph(Cod)_2]BF_4$. Thus, asymmetric hydrogenation of unsaturated methyl esters including α-acetamidoacrylate gave 2-acetamidopropanoate (Fig. 14.40)[156].

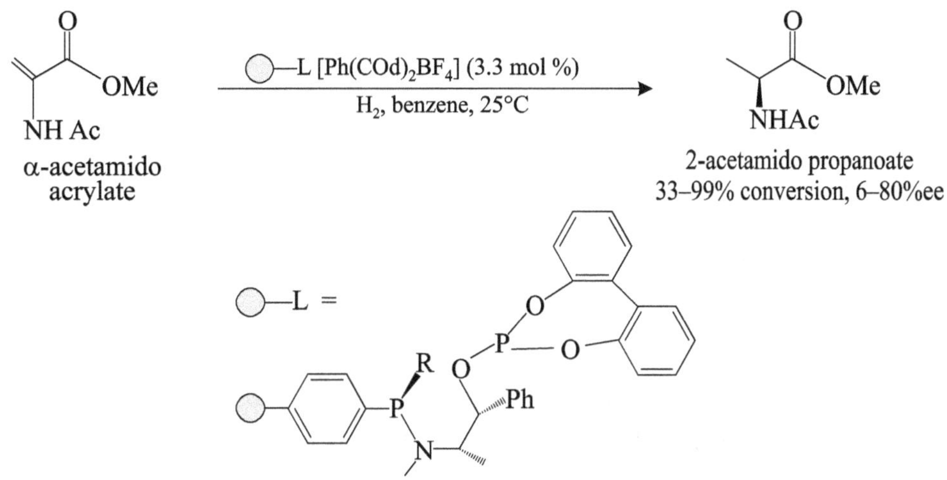

Fig. 14.40: Asymmetric hydrogenation reaction using immobilized rhodium catalyst.

14.6.2.4 Carbonylations

A rhodium complex supported on 2-vinylpyridine/ethylene glycol copolymer could affect carbonylation of methyl alcohol in presence of methyl iodide and CO at 3.2 MPa giving acetic acid along with other byproducts (Fig. 14.41)[157].

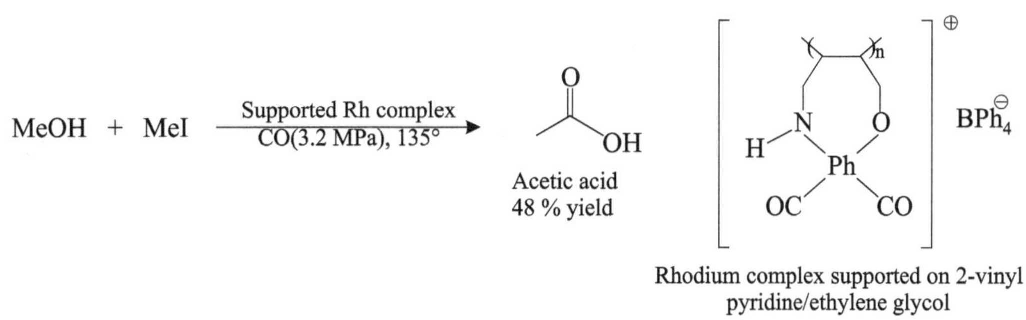

Fig. 14.41: Rhodium catalysed carboxylation reaction.

14.7 IMMOBILIZED RUTHENIUM CATALYSTS

The ruthenium catalysts like the rhodium complexes are very useful in modern organic synthesis. Some of the applications of ruthenium catalysis are described below.

14.7.1 Ruthenium Catalysed Metathesis Reactions

A number of procedures[158-162] have been described for Olefin metathesis reactions. These include ring closing metathesin (RCM), ring closing enyne metathesis (RCEM), cross-metathesis (CM) and polymerisation reactions like ring-opening metathesis polymerisation (ROMP). These relations are promoted by ruthenium catalysts of the type [$L_2X_2Ru = CHR$] (X = Cl; L = PPh_3, PCy_3, NHC, etc) (Fig. 14.42).

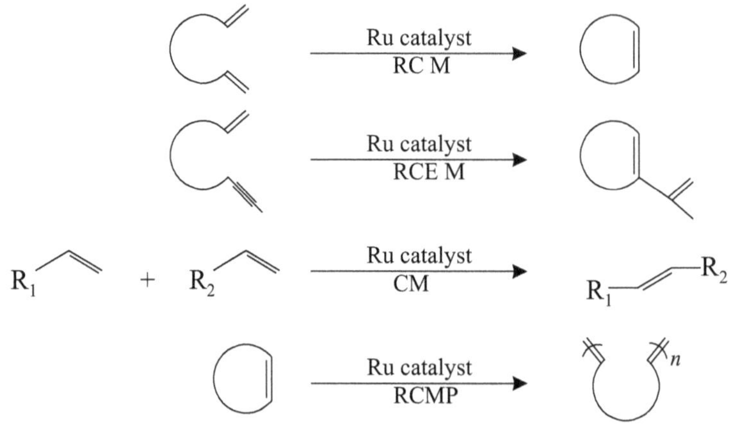

Fig. 14.42: Olefin and eyne metathesis reactions.

The early ruthenium catalysts being homogeneous exhibited hight activity and selectivity in Olefin metathesis reaction but in this case purification and recycling of the catalysts was very difficult. In view of this the catalysts were immobilized to render them greener heterogeneous catalysts[163,164]. From the green chemistry point of view the immobilized catalyst must be recyclable. One such catalyst is Hoveyda-type catalyst[165,166]. It was successfully used and recycled in carrying out reactions of RCM and CM type (Fig. 14.43)[167].

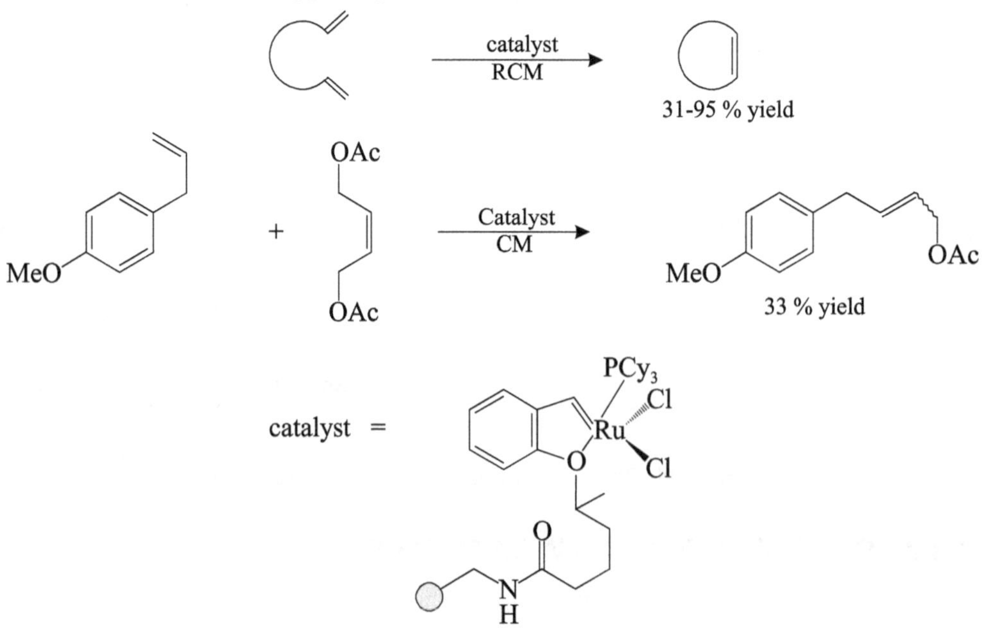

Fig. 14.43: RCM and CM reactions promoted by immobilized ruthenium catalyst.

Another immobilized ruthenium catalyst[168], has been utilized for CM, RCM and eyne metathesis reactions (Fig. 14.44).

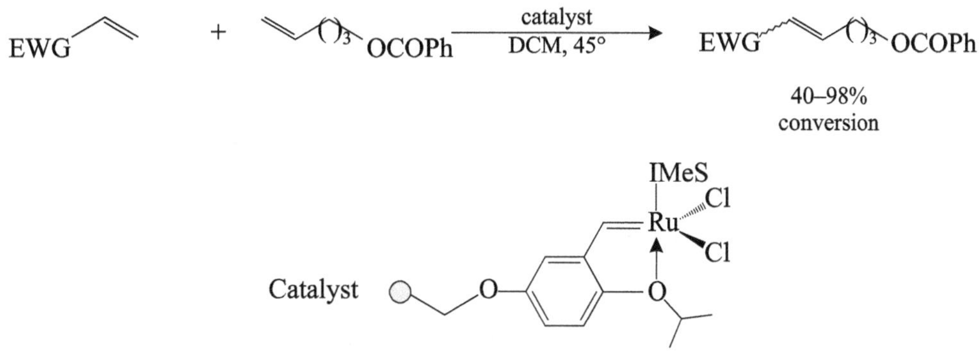

Fig. 14.44: CM reaction promoted by immobilised Ru catalyst.

Improved recyclability could be achieved by using another immobilised Hoveyda-type Rh catalyst (Fig. 14.45)[169].

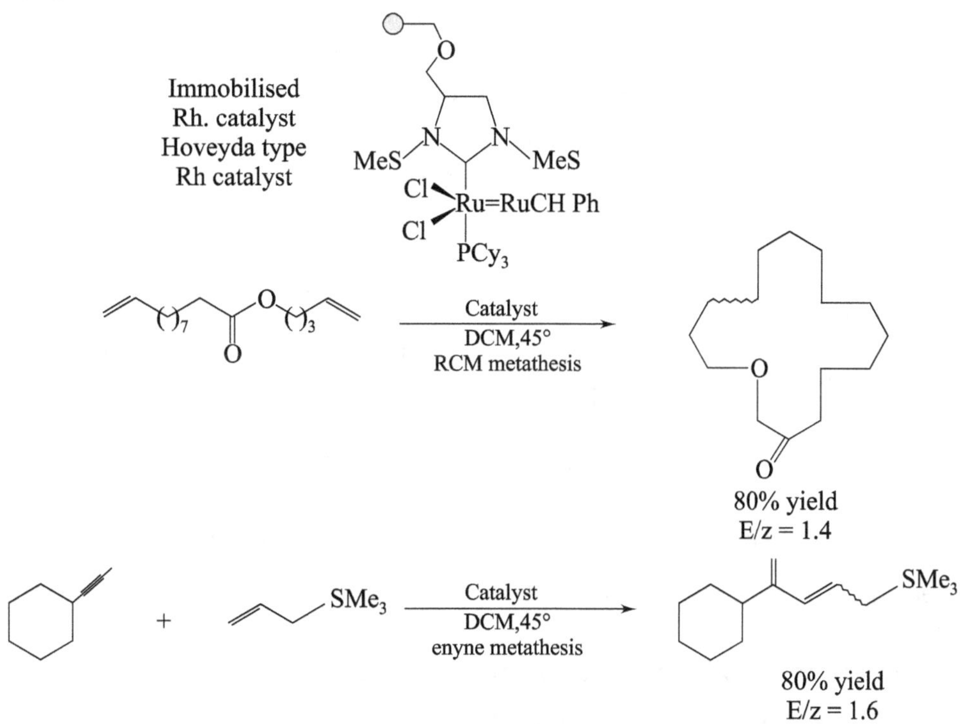

Fig. 14.45: RCM and enzyme metathesis promoted by Hoveyda type Rh catalyst.

14.7.2 Ruthenium Catalysed Transfer Hydrogenation

PS-supported N-toluenesulfonyl-1,2-diphenylethylenediamine ligands are functionalized with pendant quaternary ammonium sulphate groups. These ligands are used to immobilize a chiral rhodium catalyst. The resulting complex has been used in asymmetric transfer hydrogenation reactions.

Thus, the transfer hydrogenation of acetophenone in water at 40° gave (R)-phenylethyl alcohol (Fig. 14.46)[170,171]

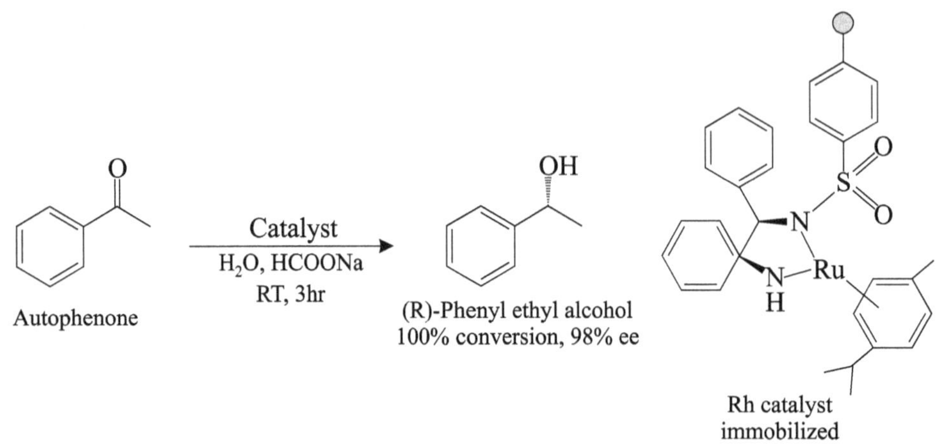

Fig. 14.46: Asymmetric transfer hydrogenation reaction promoted by immediated Ru catalyst.

14.7.3 Ruthenium-catalysed Ring Opening of Epoxides

Ruthenium complex immobilized via a *bis* (2-picolylamine) ligand has been used[172] for ring opening of epoxides. The catalyst could be recycled (Fig. 14.47).

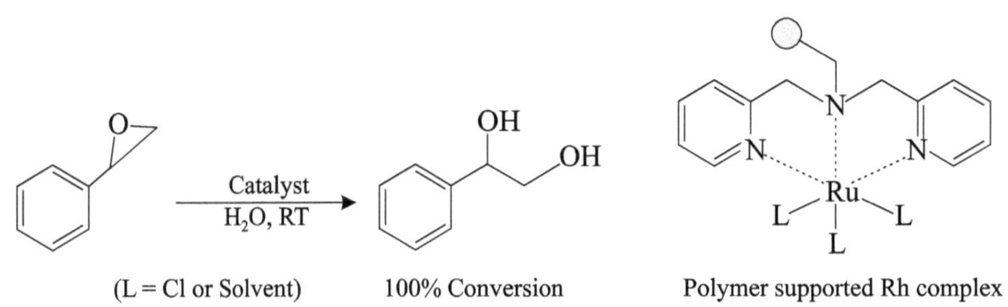

Fig. 14.47: Ruthenium catalysed ring opening of epoxide.

The epoxide ring opening can also be effected by immobilized cobalt and iron catalysts.

14.7.4 Ruthenium-catalysed Cyclopropanation Reaction

A immobilized ruthenium-pybox complex has been used[173] for cyclopropanation of alkenes with diazoacetate (Fig. 14.48). The catalyst, which is air-sensitive is packed inside monolithic microreactor so that cyclopropanation reaction could be carried out in a continuous flow approach.

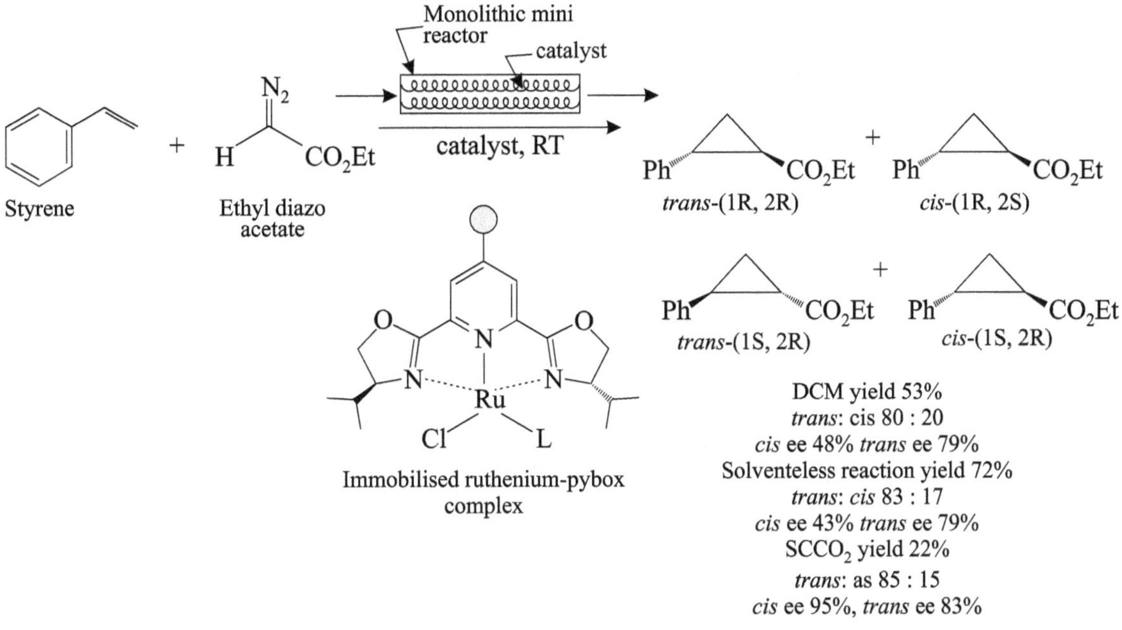

Fig. 14.48: Cyclopropanation reaction using immobilized ruthenium-pybox complex.

14.7.5 Ruthenium-catalysed Halogenation Reactions

Using a supported ruthenium catalyst, the reaction of styrene with CCl_4 in water at 60-70° gave the halogenated product (Fig. 14.49)[174]. The reaction is known as the **Kharasch reaction**, which was conducted[173] in presence of a radical initiar (AIBN).

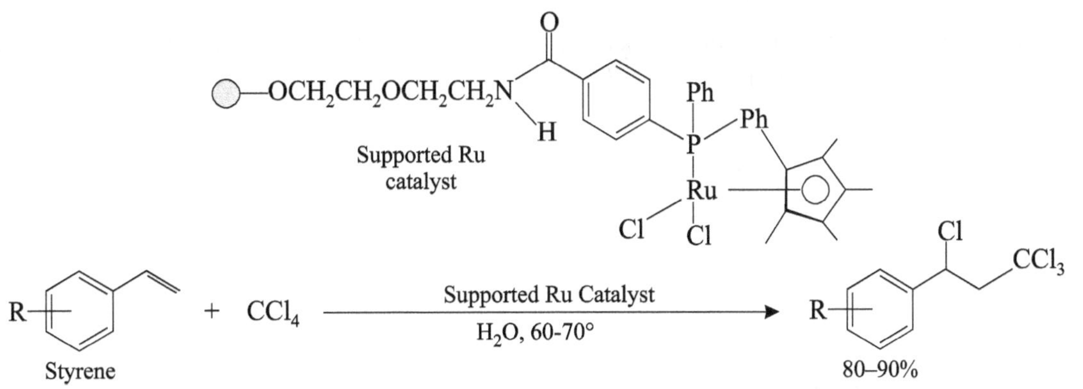

Fig. 14.49: Karasch reaction using Supported Ru catalyst in water.

14.8 IMMOBILIZED COBALT CATALYST

An immobilized Salen-cobalt complex has been used for kinetic resolution of epoxides via hydrolysis. Before use, the catalyst was activated using air and acetic acid, and adding water (Fig. 14.50)[176].

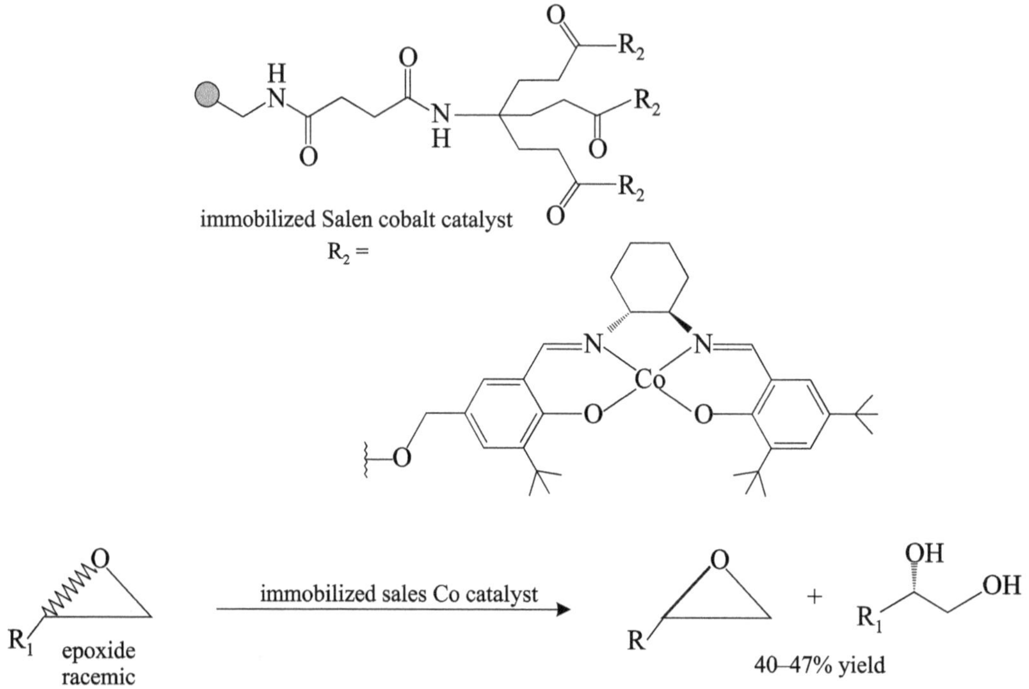

Fig. 14.50: Kinetic resolution of epoxides using immobilized Salen Co catalyst.

14.9 IMMOBILIZED COPPER CATALYST

A popular reaction involves the synthesis of 1, 2, 3-triazoles by 1, 3-dipolar cycloaddition between an azide and an internal or terminal alkyne in presence of a catalyst[177].

Subsequently a copper-mediated variant gave additional advantage of regiospecificity favouring formation of 1,4-substituted thiazoles[178] (Fig. 14.51).

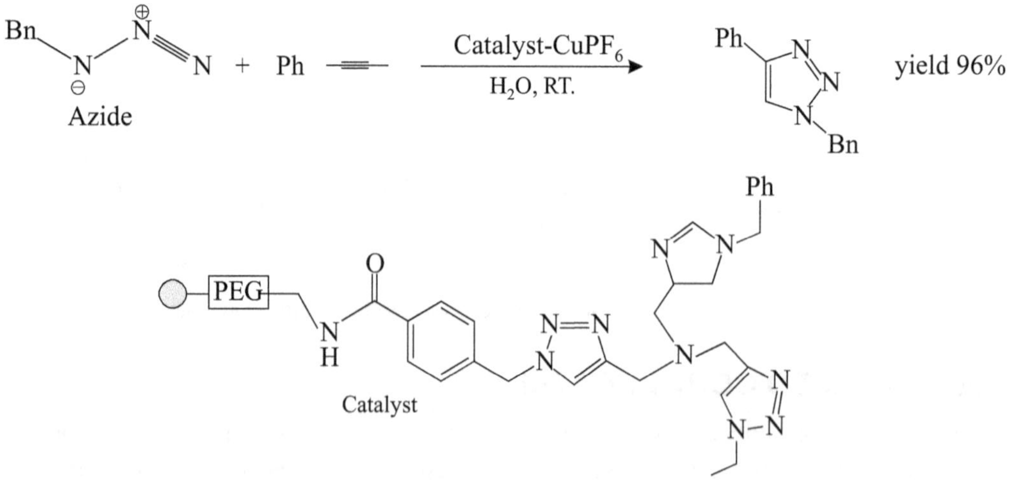

Fig. 14.51: Synthesis of 1,4-substituted, 1, 2, 3-triazoles using supported copper catalysts.

Prof. K. Barry Sharplus; 2001 Nobel laureate introduced **Click chemistry** for the above set of reactions. It is a chemistry philosophy created by Nature's ability to create complex molecular entities by 'clicking' together small molecular units[179-183].

14.10 IMMOBILIZED IRIDIUM CATALYSTS

An immobilized iridium catalyst (a supported pincer complex) has been used[184] for dehydrogenation of alkanes. The catalyst was attached to Merrifield resin and tert butylethylene was used as the hydrogen acceptor (Fig. 14.52).

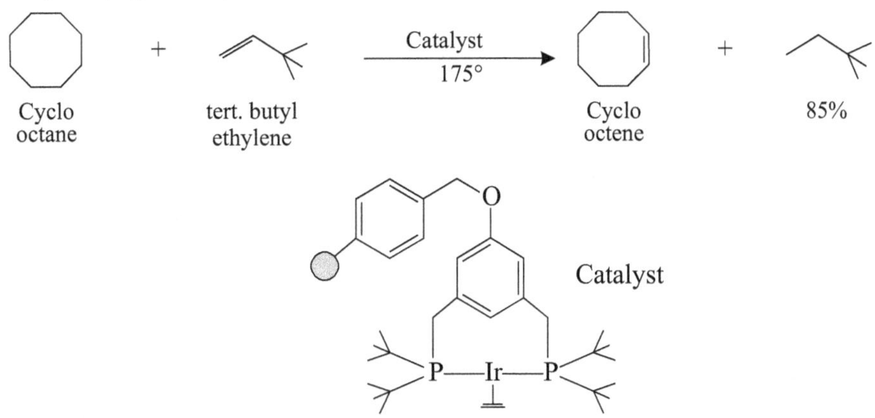

Fig. 14.52: Dehydrogenation of alkanes using immobilized iridium complex.

It was found[184] that the above catalyst adsorbed on to Al_2O_3 was more robust than the polymer-supported analog.

Conclusion

The use of catalyst immobilized on insoluble polymer support are a good choice in the context of green chemistry. Such catalysts can be recovered by filtration and reused. The procedure is useful for a number of organic synthesis on an industrial scale.

REFERENCES

1. G. Manecle and W. Storck, Angew. Chem. Int. Ed. Engl., 1978, **17**, 657.
2. W. Neier and J. Woellner, Chem. Technol., 1973, 95.
3. F. Helfferich, 'Ion Exchange, McGraw-Hill, New York, 1961.
4. C.W. Davies and G.G. Thomas, J.Chem. Soc., 1952, 1607; M.J. Astle and J.A. Oscar, J. Org. Chem., 1961, **26**, 1713.
5. F.V. George and I.S. Virgil, J. Org. Chem., 1971, **36**, 2548; P.J. Mill and R. Crimmin, Biochem: Biophys. Acta, 1957, **23**, 432; J.C. Jain, I.K. Shani, M.K. Gupta and N.K. Mathur, Indian J. Chem., 1977, **15B**, 766.
6. J.G. Moffatt, J.Am. Chem. Soc., 1963, **85,** 1118; D.F. Mowery, Jr., J. Org. Chem., 1961, **26**, 3484.
7. R.H. Hasek, P.G. Gott, R.H. Meen and J.C. Martin, J. Org. Chem., 1963, **28**, 2496.
8. R. Kunin, 'Ion Exchange Resins', Wiley, New York, 1958.

9. M. Walter, H. Besendorf and O. Schmider, Helv. Chim. Aeta., 1961, **44**, 1546.
10. M.S. Newman, J. Am. Chem. Soc., 1953, **75**, 4740.
11. B. Loev and J.T. Massengale, J. Org. Chem., 1957, **22**, 988.
12. E.V.O. John and S.S. Israelstam, J. Org. Chem., 1961, **26**, 240.
13. P. Price and S.S. Israelstam, J. Org. Chem., 1964, **29**, 2800.
14. R.F. Collins, Chem. Ind., 1957, 736.
15. S. Yamashita, Biochem. Biophys. Acta, 1971, **229**, 301.
16. J.M. Bora, A.K. Saund, I.K. Sharma and N.K. Mathur, Indian J. Chem., 1976, **14B**, 722.
17. W. Neier and J. Woellner, Chem. Tech., 1973, 95.
18. J.D. Billimoria and N.F. Maclagan, J. Chem. Soc., 1954, 3257.
19. J.E. Hodge and C.E. Rist, J. Am. Chem. Soc., 1952, **74**, 1498.
20. G. Bodamer and R. Kunin, Ind. Eng. Chem., 1951, **43**, 1082.
21. H. Deuel, Mitt. Geb. Lehensmittelunters Hyg., 1955, **46**, 12.
22. N.B. Lorette, J. Org. Chem., 1957, **22**, 346; G.H. McCain, J. Org. Chem., 1958, **23**, 632.
23. A. Mastagli, A. Floch and G. Durr, Comp. Rencl., 1952, 235, 1402.
24. M.J. Astle and F. Abbot, J. Org. Chem., 1956, **21**, 1228.
25. M.J. Astle and W. Gergel, J. Org. Chem., 1956, **21**, 493.
26. K.M. Kosower and J.W. Phatton, J. Org. Chem., 1961, **26**, 1318.
27. A. Galat, J. Am. Chem. Soc., 1948, **70**, 3945; J.M. Bobbitt and D.A. Scola, J. Org. Chem., 1960, **25**, 560.
28. R.L. Hinman and J. Lang, J. Org. Chem., 1964, **29**, 1449.
29. B.G. Chatterjee, V.V. Rao and B.N.G. Mazumdar, J. Org. Chem., 1965, **30**, 4101.
30. L. Carpino, J.R. Williams and A. Lopusinki, J. Chem. Soc. Chem. Cummun., 1978, 450.
31. G. Cainelli and F. Manescalchi, Synthesis, 1975, 723.
32. D.C. Neckers, D.A. Kooistra and G.M. Green, J.Am. Chem. Soc., 1972, 9284.
33. E.C. Blossey, L.M. Turner and D.C. Neckers J. Org. Chem., 1975, **40**, 959.
34. E.C. Blossey, E.C. Turner and D.C. Neckers, Tetrahedron Lett., 1973, 1823.
35. V.L. Magotta, B.C. Gates and G.C.A. Schutt, J. Chem. Soc. Chem. Commun, 1976, 342.
36. V.L. Magotta and B.C. Gates, J. Polym. Sci. Polym. Chem. Ed., 1977, **15**, 1341; J. Catal, 1977, **46**, 266.
37. W.M. Meier and J.B. Uytterhoeven, 'Molecular Sieves', Amer. Chem. Soc. Symposium No. 121, Washington, D.C., 1973.
38. R.L. Letsinger and T.J. Savereide, J.Am. Chem. Soc., 1962, **84**, 3112.
39. C.G. Overberger, R. Sitaramaiah, T. St. Pierre and S. Yaroslavsky, J. Am. Chem. Soc., 1965, **87**, 3270.
40. C.G. Overberger and T.W. Smith., Macromolecules, 1975, **8**, 401, 407, 416.
41. C.G. Overberg and J.C. Salamone, Macromolecules, 1969, **2**, 553.
42. C.G. Overberger, A.C. Guterl, Jr., Y. Kawakami, L.J. Mathias, A. Meenakshi and T. Tomono, Pure Appl. Chem., 1978, **50**, 309.
43. H. Morawitz, Ad. Catal, 1969, **24**, 341.
44. R.L. Letsinger and I.S. Klaus, J. Am. Chem. Soc., 1964, **86**, 3884.

45. T. Shimidzu, A. Furuta and Y. Nakamoto, Macromolecules,1974, **7**, 160.
46. C.G. Overberger and K.N. Sannes, Angew. Chem. Ed. Eng., 1974, **13**, 99.
47. J. Dockx, Science, 1973, 441.
48. D. Landini, F. Montanari and F.M. Pirisi, J. Chem. Soc. Chem. Commun., 1974, 879.
49. D. Landini, F. Montanari, F.M. Pirissi and H.M. Maia, Gazetta, 1975, **105**, 863.
50. M. Cinquini, F. Montanarri and P. Tundo, J. Chem. Soc. Chem. Commun. 1975, 393.
51. H. Molinari, F. Fontanari and P. Tundo, J. Chem. Soc. Chem. Commun., 1977, 639.
52. P. Tundo, Synthesis, 1978, 315.
53. M. Cinouini, S. Colonna, H. Molianari, F.Montanari and P. Tundo, J. Chem. Soc. Chem. Commun., 1976, 394.
54. P. Tundo, Synthesis, 1977, J. Chem. Soc. Chem. Commun., 1977, 641.
55. S. Kopolow, T.E. Hogen and J. Smid, Macromolecules,1971, **4**, 359.
56. S. Kopolow, T.E.H. Esch and J. Sinid, Macromolecules, 1973, **6**, 133.
57. D.J. Sam and H.E. Simmons, Chem. Tech., 1972, **2**, 450.
58. E. Chiellini and R. Solaro, J. Chem. Soc. Chem. Commun., 1977, 231.
59. E. Chiellini, R. Solaro and D'Antone, Makromol. Chem., 1977, 2545, 3165.
60. D.G. Lee and V.S. Chang, J. Org. Chem., 1978, **43**, 1532.
61. M.C. Vander Zwan, and F.W. Hartner, J. Org. Chem., 1978, **43**, 2655.
62. J.M. Brown and J.A. Jenkins, J. Chem. Soc. Chem. Commun., 1976, 458.
63. S.L. Regen, J. Am. Chem. Soc., 1975, **97**, 5956; 1976, **98**, 6270; J. Org. Chem., 1977, **42**, 875.
64. H. Komeili-Zadeh, H.J.-M. Dou and J. Metzger, J. Org. Chem., 1978, **43**, 156.
65. M. Tomoi, M. Ikeda and H. Kakiuchi, Tetrahedron Lett., 1978, 3757.
66. S.L. Regen and L. Dulak, J. Am. Chem. Soc., 1977, **99**, 623.
67. E.C. Blossey, D.C. Neckers, A.L. Thayes and A.B. Schapp, J.Am. Chem. Soc., 1973, **95**, 5820.
68. A.P. Schapp, A.L. Thayer, E.C. Blossey and D.C. Neckers, J. Am. Chem. Soc., 1975, **97**, 3741.
69. C.S. Foote and S. Wexler, J. Am. Chem. Soc., 1964, **80**, 3880.
70. E.J. Corey and W.C. Tayer, J. Am. Chem. Soc., 1965, **80**, 3881.
71. I. Rosenthal and A.J. Archer, Israel J. Chem., 1974, **12**, 897.
72. E.C. Blossey and D.C. Neckers, Tetrahedron Lett., 1974, 323.
73. Y.-X. Liu, Z.-W. Maa, J. Tia *et al.*, Appl. Organomet. Chem., 2010, **24**, 646.
74. V. Singh, R. Ratti and S. Kaur, J. Mol. Catal A: Chem., 2011, **334**, 13.
75. B.N. Lin, S.-H. Huang, W.-y. Wu et al, Molecules, 2010, **15**, 9157.
76. N. Iranpoor, H. Firouzabadi, A. Tarassoli and M. Fereidoonnezhad, Bull. Chem. Soc. Jpn., 2010, **83**, 1367.
77. D.-H. Lee, J.-y. Jung and M.-J. Jin, Green Chem., 2010, **12**, 2024.
78. M. Islam, P. Mondal, K. Tuhina *et al.*, J. Organomet. Chem., 2010, **695**, 2284.
79. M.-J. Jin and D.-H. Lee, Angew, Chem. Int. Ed., 2010, **49**, 1119.
80. W. Hao, J. Sha, S. Sheng *et. al.*, J. Mol. Catal. A: Chem., 2009, **298**, 94.
81. A.F. Grandsire, C. Laborde, F. Lamaty and A. Mehdi, Appl. Organomet. Chem., 2010, **24**, 179.

82. Y. Wan, H. Wang, Q. Zhao *et al*, J. Am. Chem. Soc., 2009, **131**, 4541.
83. X. Companyo, M. Viciano and R. Rios, Org. Chem., 2010, **7**, 1.
84. C.L. Sun, H. Li, D.G. Yu *et al.*, Nature Chem., 2010, **2**, 1044.
85. W. Liu, H. Cao, H. Zhang *et al.* J. Am. Chem. Soc., 2010, **132**, 16737.
86. K.M. McQuaid and D. Sames, J.Am. Chem. Soc., 2009, **131**, 402.
87. F. Benfatti, M.G. Capdevila, L. Zoli *et al.*, Chem. Commun., 2009, 5919.
88. Y.J. Park, J.W. Park and C.H. Jun, Acc. Chem. Res., 2008, **41**, 222.
89. T.B. Poulsen, M. Bell and K.A. Jorgensen, Org. Biomol. Chem., 2006, **4**, 63.
90. K. Kondo, T. Yamano and K. Takemoto, Macromol. Chem., 1985, **186,** 1781.
91. M. Banaglia, G. Celentano, F. Cozzi *et al.*, Adv. Synth. Catal. 2002, **344**, 171.
92. M. Banaglia, M. Cinquini, F. Cozzi *et al.*, Adv. Synth. Catal. 2002, **344**, 533.
93. M. Banaglia, M. Cinquini, F. Cozzi et al, J. Mol. Catal. A., 2003, 204.
94. L. Gu, Y. Wu, Y. Zhang and G. Zhau, J. Mol. Catal. A, 2007, 263, 186.
95. J.C. Tao, Catal. Lett., 2008, **120**, 281; F. Marculescu *et al.*, Eur. J. Org. Chem., 2007, 4688; Y.X. Liu, Y.N. Sun, H.H. Tan et al., Tetrahedron Asymmetry, 2007, **18**, 2656.
96. F. Giacalone, M. Gruttadauria, A.M. Marculescu *et. al.*, Catal. Commun., 2008, **9**, 1477.
97. E. Alza, C. Rodriguez-Escrich, S. Sayalero *et al.*, Chem. Eur. J., 2009, **15**, 10167.
98. D. Font, A. Bastero, S. Sayalero *et al.*, Org. Lett., 2007, **9**, 1943.
99. M. Gruttadauria, F. Giacolone and R. Noto, Chem. Soc. Rev., 2008, **37**, 1866; F. Giacalone, M. Gruttadauria, A.M. Marculescu and K. Noto, Tetrahedron Lett., 2007, **48**, 255.
100. A.B. Caballero, G. Guillena and C. Najera, Green Chem., 2010, **12**, 1599.
101. T.E. Kristensen, K. Vestli, M.G. Jakobsen, *et al.*, J. Org. Chem., 2010, **75**, 1620.
102. K. Akagawa, S. Sakamoto and K. Kudo, Tetrahedron Lett., 2005, **46**, 8185.
103. M.R.M. Andreae and A.P. Davis, Tetrahedron Asymmetry, 2005, **16**, 2487.
104. S. Luo, X. Mi, L. Zhang *et al.*, Angew. Chem. Int. Ed., 2006, **45**, 3093; S. Luo, X. Mi, S. Liu *et al.*, Chem. Commun., 2006, 3687.
105. S. Luo, H. Xu, X. Mi *et al.*, J. Org. Chem., 2007, **71**, 9244.
106. E. Alza, X.C. Cambeiro, C. Jimeno and M.A. Pericas, Org. Lett., 2007, **9**, 3717.
107. T. Miao and L. Wang, Tetrahedron Lett., 2008, **49**, 2173.
108. P. Li, L. Wang and Y. Zhang, Eur. J. Org. Chem., 2008, 1157.
109. R. Chinchilla, P. Mazón and C. Najera, Tetrahedron Asymmetry, 2000, **11**, 3277; Adv. Synth. Catal., 2004, **346**, 1186.
110. B. Thierry, J.C. Plaquevent and D. Cahard, Tetrahedron Asymmetry, 2001, **12**, 983.
111. B. Theirry, J.C. Plaquevent and D. Cahard, Tetrahedron Asymmetry, 2003, **14**, 1671.
112. M. Rueping, E. Sugiono, A. Steck and T. Theissmann, Adv. Synth. Catal., 2010, 352, 281.
113. S. Luo, J. Li, L. Zhang and J.-P. Cheng, Org. Lett., 2007, 3676.
114. J. Li, S. Hu, S. Luo and J.-P. Cheng, Eur. J. Org. Chem., 2009, 132.
115. J. Li, S.Luo and J.-P. Cheng, J. Org. Chem., 2009, **74**, 1747.
116. J.Y. Li, X. Li, P.X. Zhou et al., J. Org. Chem., 2009, 4486.
117. S.Z. Luo, J.Y. Li, L. Zhang et al. Chem. Eur. J., 2008, **14**, 1273.

118. N. Haraguchi, Y. Takemura and S. Itsuno, Tetrahedron Lett., 2010, **51**, 1205.

119. J. Gao, J. Liv, S. Bai et al, J. Mater Chem., 2009, **19**, 8580.

120. R. Sebesta, L. Kmentova and S. Toma, Green Chem., 2008, **10**, 484; B. Ni and A.D. Headley, Chem. Eur. J., 2010, **14,** 4426.

121. A. Cole, J. Jensen, L. Ntai *et al.*, J. Am. Chem. Soc., 2002, **124**, 5962; Y. Gu, F. Shi and Y. Deng, Catal. Commun., 2003, **4**, 597.

122. S. Luo, X. Mi, Li Zhang *et al.*, Angew. Chem. Int. Ed., 2006, **45**, 3093.

123. E.G. Doyagüez, F. Calderon, F. Sanchez and A. Fernandez-Mayoralas, J. Org. Chem., 2007, **72**, 9353.

124. E.A. Prastyanto, S.-C. Lee, S.-M. Jeong and S.-E. Park, Chem. Commun., 2008, 1995.

125. M.L. Kantam, M. Roy, S. Roy *et al.*, Tetrahedron, 2007, **63**, 8002.

126. B.J. Gallon, R.W. Kojima, R.B. Kaner and P.L. Diaconescu, Angew. Chem. Int. Ed., 2007, **46**, 7251.

127. L. Bai and J.-X. Wang, Adv. Synth. Catal., 2008, **350**, 315.

128. Y. Uozumi, H. Danjo and T. Hayashi, J. Org. Chem., 1999, **64**, 3384.

129. Y.M.A. Yamada, S. Takeda, H. Takahashi and S. Ikegami, Org. Lett., 2002, **4**, 3371.

130. Y.M.A. Yamada, K. Takeda, H. Takahasi and S. Ikegami, J. Org. Chem., 2003, **68**, 7733.

131. Y. Uozumi, Y. Matsurra, T. Arakawa and Y.M.A. Yamada, Angew. Chem. Int. Ed., 2009, **48**, 2708.

132. J.-W. Kim, J.-H. Kim, D.-H. Lee and Y.-S. Lee, Tetrahedron Lett., 2006, **47**, 4745.

133. N.T.S. Phan and P. Styring, Green Chem., 2008, **10**, 1055.

134. W. Solódenko, U. Schön, J. Messinger *et al.*, Synlett., 2004, 1699.

135. E. Alacid and C. Najera, J. Orgnomet. Chem., 2009, **694,** 1658.

136. J. Gil-Molto', S. Karlström and C. Nājera, Tetrahedron, 2005, **61**, 12168.

137. Y. Uozomi and T. Kimura, Synlett., 2002, 2045.

138. M. Shibaski, E.M. Vogl and T. Ohshima, Adv. Synth. Catal., 2004, **346**, 1532.

139. E. Alacid and C. Na'jera, Synlett, 2006, 2959.

140. E. Alacid and C. Na'jera, ARKIVOR, 2008, 50.

141. R. Chirchilla and C. Na'jera, Chem. Rev., 2007, **107**, 872.

142. Y. Uozomi and Y. Kobayashi, Heterocycles, 2003, **59**, 71.

143. K.M. Dawood, W. Solodenko and A. Kirschning, Arkivok, 2007, 104.

144. B.M. Trost and V.L. Van Vranken, Chem. Rev., 1996, **96**, 395.

145. B.M. Trost and M.L. Crawley, Chem. Rev., 2003, **103**, 2921.

146. Y. Uozomi, H. Danjo and T. Hayashi, Tetrahedron Lett., 1997, **38**, 3537.

147. H. Danjo, D. Tanaka, T. Hayashi and Y. Uozomi, Tetrahedron, 1999, **55**, 14341.

148. Y. Uozomi and T. Suzuka, Synthesis, 2008, **12**, 1960.

149. Y. Uozomi and T. Suzuka, J. Org. Chem., 2006, **71**, 8644.

150. Y. Uozomi, T. Suzuka, R. Kawade and H. Tekenaka, Synlett, 2006, 2109.

151. Y. Uozomi, H. Danjo and T. Hayashi. Tetrahedron Lett., 1998, **39**, 8303.

152. N. Kann, Molecules, 2010, **15**, 6306.

153. N. End and K.-U. Schöning, Top. Curr. Chem., 2004, **242**, 241.

154. H.M.L. Davies, A.M.Walji and T. Nagashima, J. Am. Chem. Soc., 2004, **126**, 4271.

155. R. Jana and J.A. Tunge, Org. Lett., 2009, **11**, 971.
156. S.R. den Heeten, B.H.G. Swennenhuis, D.W.N.M. Vanleuwen *et al.*, Angew. Chem. Int. Ed., 2008, **47**, 6602.
157. S. Zhang, C. Guo, Q. Qian and G. Yuan, Catal. Commun., 2008, **9**, 853.
158. S. Blechert and S.J. Connon, Angew. Chem. Int. Ed., 2003, **42**, 1900.
159. R.H. Grubbs, Tetrahedron, 2004, **60**, 7111.
160. A. Fürstner, Angew. Chem. Int. Ed., 2000, **39**, 3012.
161. R.H. Grubbs, R.R. Schrock and A. Fürstner (eds.), Adv. Synth. Catal (Olefin metathesis special issue), 2007, **349**, 1.
162. T.M. Trnka and R.H. Grubbs, Acc. Chem. Res., 2001, **34**, 18.
163. M.R. Bucmeiser, New J. Chem., 2004, **28**, 549.
164. H. Clavier, K. Grela, A. Kirschning *et al.*, Angew. Chem. Int. Ed., 2007, **46**, 6786.
165. S.B. Garber, J.S. Kingsbury, B.L. Gray and A.H. Hoyeda, J. Am. Chem. Soc., 2000, **122**, 8168.
166. J.S. Kingsbury, J.P.A. Herrity, P.J. Bonitatebus Jr., and A.H. Hoveyda, J. Am. Chem. Soc., 1999, **121**, 791.
167. J. Dowden and J. Savović, Chem. Commun., 2001, 37.
168. S.C. Schurer, S. Gessler, N. Buschmann and S. Blechert, Angew. Chem. Int. Edn. 2000, **39**, 3898.
169. S. Randl, N. Buschmann, S.J. Connon and S. Blechert, Synlett, 2001, 1541.
170. N. Haraguchi, K. Tsuru, Y. Arakawa and S. Itsuno, Org. Biomol. Chem., 2009, **7**, 69.
171. Y. Arakawa, Y. Chiba, N. Haraguchi and S. Itsuno, Adv. Synth Catal, 2006, **350**, 2295.
172. S.H. Lee., E.Y. Lee, D.W. Yoo *et al.*, New J. Chem., 2007, **31**, 1579.
173. M. Buguete, A. Cornejo, E. Garcia-Verdugo et al, J. Org. Chem., 2007, **72**, 4344.
174. Y. Oe and Y. Uozomi, Adv. Synth. Catal., 2008, **350**, 1771.
175. K. Severin, Adv. Synth. Catal, 2008, **350**, 1771.
176. P. Goyal, X.L. Zheng and M. Weck, Adv. Synth. Catal., 2008, **350**, 1816.
177. R. Huisgen, Proc. Chem. Soc., 1961, 357.
178. J.E. Hein and V.V. Fokin, Chem. Soc. Rev., 2010, **39**, 1302.
179. H.C. Kolb, M.G. Finn and K.B. Sharpless, Angew. Chem. Int. Ed., 2001, **40**, 2004.
181. H.C. Kolb and K.B. Sharpless, Drug Discov. Today, 2003, **8**, 1128.
182. R.A. Evans, Aust. J. Chem., 2007, **60**, 384.
183. V.D. Bock and J.H. van Maarseveen, Eur. J. Org. Chem., 2006, 51.
184. Z. Huang, M. Brookhart, A.S. Goldman *et al.*, Adv. Synth. Catal., 2009, **351**, 188.

15. ORGANIC TRANSFORMATIONS USING GREEN REAGENTS

15

Organic Transformations using Green Reagents

An organic transformation is carried out by the reaction of a substrate with a reagent. The reagent should be a green reagent. A reagent is considered to be green if it is not harmful to the environment. Some of such reagents are given below:

15.1 OXYGEN

A colourless and odourless gas, it makes up about one-fifth by volume of the atmosphere. Industrially, it is obtained from liquefied air by fractional distillation. It is routinely prepared by electrolysis of water. The ordinary oxygen is known as ground-state oxygen (triplet oxygen) and has two electrons with parallel spins. It behaves like a diradical and is paramagnetic.

$$[\uparrow\!.\text{O}:\text{O}.\!\uparrow]$$

Oxygen is used for various oxidations in presence of catalysts. This process called catalytic oxidation is useful for the industrial preparation of a number of important products, some of which are given below:

(*i*) Manufacture of formaldehyde by controlled oxidation of methane

$$\underset{\text{Methane}}{CH_4} + O_2 \xrightarrow[\text{MnO}_2,\,400°,\ 200\ \text{atm.}]{\text{Controlled oxidation}} \underset{\text{Formaldehyde}}{HCHO} + H_2O$$

(*ii*) Manufacture of methyl alcohol by oxidation of methane

$$\underset{\text{Methane}}{CH_4} + \frac{1}{2}O_2 \xrightarrow[260°,\ 100\ \text{atm.}]{\text{Cu catalyst}} \underset{\text{Methyl alcohol}}{CH_3OH}$$

(*iii*) Manufacture of acetaldehyde by oxidation of ethylene

$$\underset{\text{Ethylene}}{H_2C=CH_2} + \frac{1}{2}O_2 \xrightarrow[\text{CuCl}_2]{\text{PdCl}_2/\text{H}_2\text{O}} \underset{\text{Acetaldehyde}}{CH_3CHO}$$

(*iv*) Oxidation of toluene, naphthalene, *o*-xylene and furfural to give benzaldehyde, phthalic anhydride, phthalic anhydride and furoic acid respectively.

$$\underset{\text{Tolune}}{C_6H_5CH_3} + O_2 \xrightarrow[500°C]{MnO_2/Mo_2O_3/ZrO_2} \underset{\text{Benzaldehyde}}{C_6H_5CHO}$$

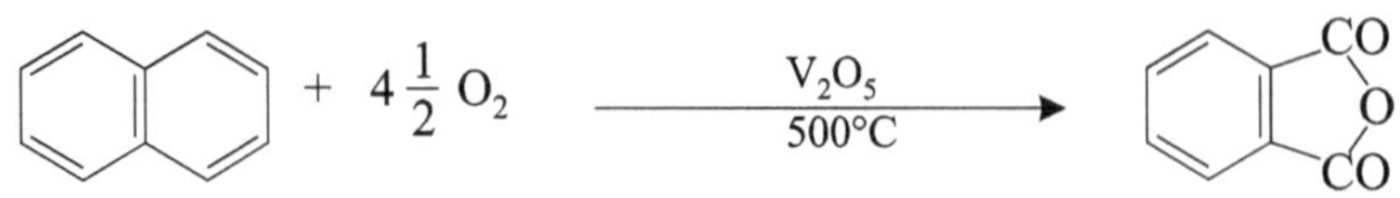

Naphthalene Phthalic anhydride

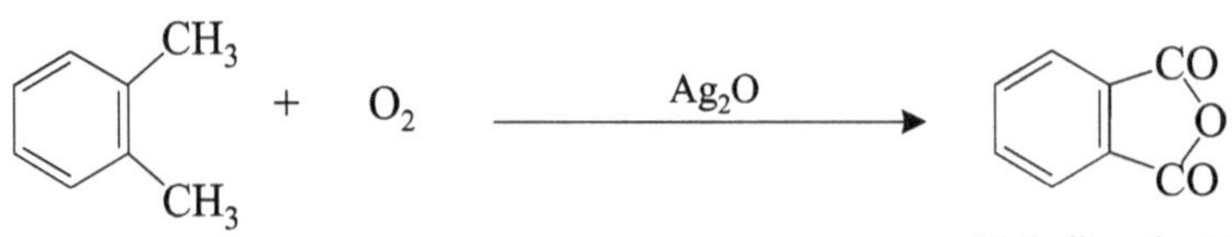

o-xylene Phthalic anhydride

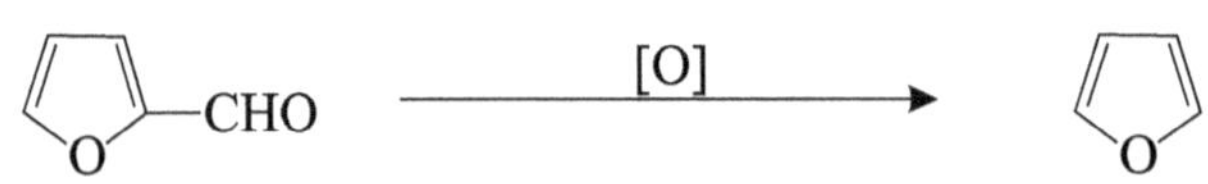

Furfural Furoic acid

(*v*) Manufacture of maleic acid by oxidation of benzene followed by hydrolysis

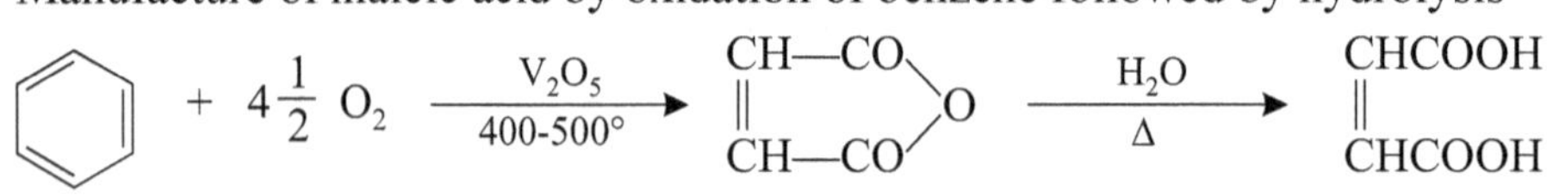

Benzene Maleic anhydride Maleic acid

(*vi*) Manufacture of chlorine (Decacon Process)

$$4HCl + O_2 \underset{440°}{\overset{CuCl_2}{\rightleftarrows}} 2Cl_2\uparrow + 2H_2O + \text{heat}$$

(*vii*) Dehydrogenation of primary amines to nitriles. The oxidation of *o*-phenylenediamine with oxygen in presence of cuprous oxide in pyridine at 25° gives 1,4-dicyano-1,3-butadiene[1].

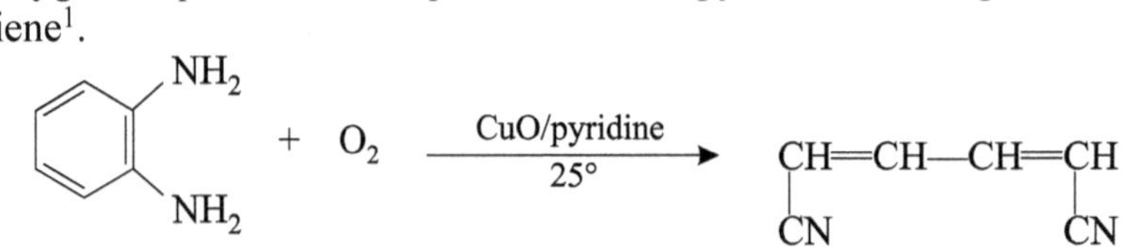

o-Phenylene diamine

77%
1,4-dicyanobutadine
(mucunitrile)

(*viii*) **Dehydrogenative coupling:** Treatment of diarylmethanes with oxygen in alcoholic KOH gives the coupled products[2]. Coupling takes place at benzylic position

$$\underset{\text{Diphenyl methane}}{(C_6H_5)_2CH_2} + O_2 \xrightarrow{\text{alcoholic KOH}} \underset{\text{70\% Coupled product}}{(C_6H_5)_2CHCH(C_6H_5)_2}$$

(*ix*) **Epoxidation of alkene:** Ethylene on reaction with oxygen in presence of silver oxide on alumina at 270° gives ethylene oxide. This is the industrial production of ethylene oxide[3]

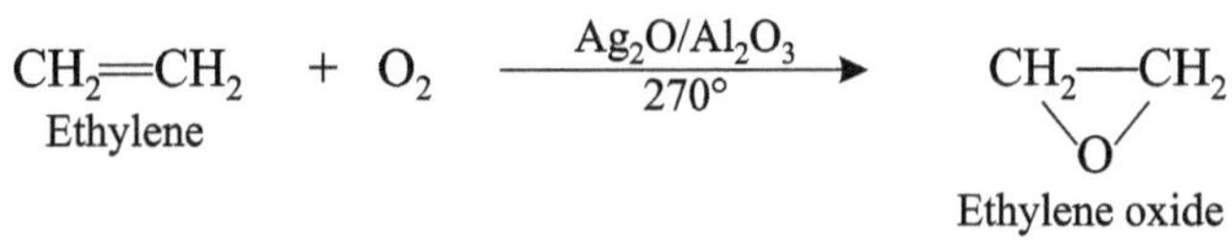

(*x*) **Oxidation of primary alcohols to aldehydes:** Vapour phase oxidation of primary alcohols (both saturated and unsaturated) to corresponding aldehydes takes place by heating with oxygen at 350°C in presence of copper catalyst[4].

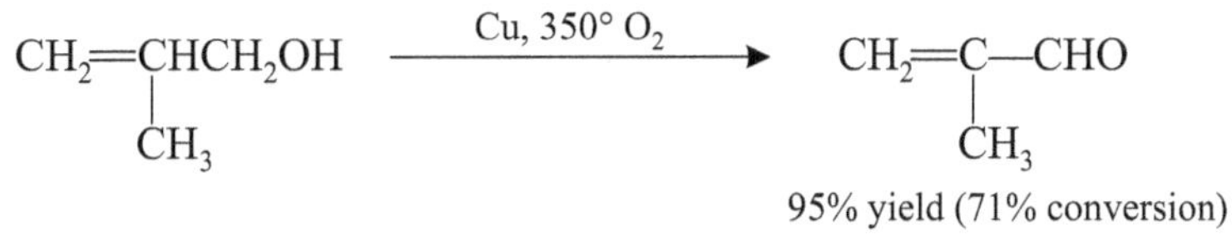

By this procedures only aldehydes that can tolerate high temperatures can be prepared. However, a catalytic oxidation of alcohols can be carried out by passing a current of air or oxygen in heptane[5] or ethyl acetate[6] in presence of platinum[5,6] or better still platinum oxide[5] or active cobalt oxide[7]. This method is applicable even to alcohols that cannot be vapourised and contain double bonds[5].

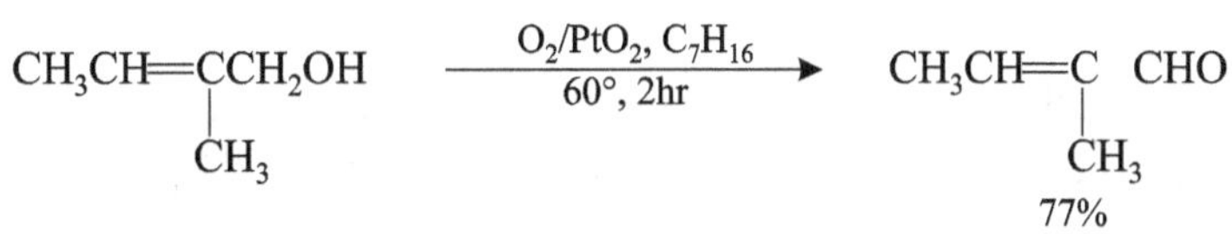

Allyl alcohol (CH_2=$CHCH_2OH$) can be oxidised to give the corresponding aldehyde (CH_2=CHCHO) by heating with air at 230-300° is presence of silver[8].

(*xi*) **Oxidation of aldehydes to carboxylic acids:** Furfural can be oxidised to furoic acid[9] in 86-90% yield by treatment with O_2 in presence of aq. NaOH and Cu_2O and Ag_2O. Similarly, methacrolein is oxidised to methacrylic acid in 99% yield with O_2 in presence of $Cu(OAc)_2$, $Ni(OAc)_2$ at RT[10].

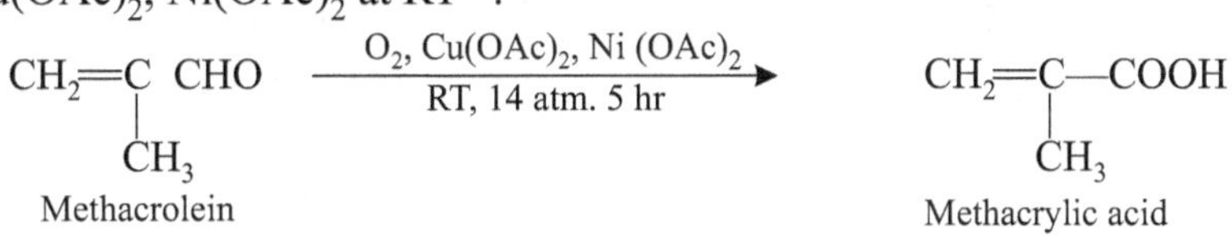

(*xii*) **Oxidation of aldoses to aldonic acids:** Aldoses like arabinose, xylose, glucose, galactose and mannose are converted into the corresponding aldonic acids by air or oxygen passed through their solution in dil. NaOH in presence of 5% Pd/C as catalyst[11,12]. The acids are isolated as Ca salts after addition of $CaCl_2$. This conversion was earlier affected by bromine water[13].

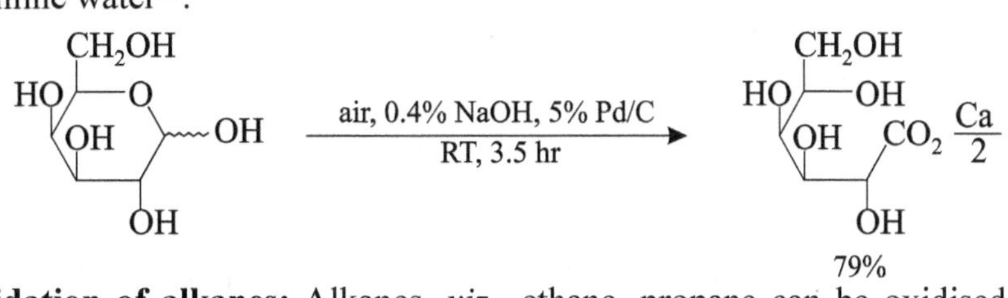

(*xiii*) **Oxidation of alkanes:** Alkanes, *viz*., ethane, propane can be oxidised by oxygen in presence of HBr as initator give acetic acid (64% at 220°) and acetone (72% at 189°) respectively[14]. However isobutane gives 69.5% yield of tert-butyl hydroperoxide, a 10% yield of tert-butyl alcohol and a 6% yield of di-tert-butyl peroxide at 163°.

(*xiv*) **Oxidation of alkyldichloroboranes:** Alkyldichloroboranes are oxidised[15] by oxygen to alkyl hydroperoxides

$$C_6H_{13}BCl_2 \xrightarrow[H_2O,\ RT]{O_2, Et_2O,\ -180°,\ 20\ min} \underset{94\%}{C_6H_{13}OOH}$$

(*xv*) **Oxidation of disulfides:** Disulfides on oxidation with oxygen in KOH solution in DMF at 23.5° to gives[16] the corresponding sulfonic acid in 90% yield

$$C_4H_9SSC_4H_9 \xrightarrow[23.5°]{O_2, KOH,\ DMF} \underset{90\%}{C_4H_9SO_3K}$$

(*xvi*) **Oxidation of mercaptans:** Mercaptans on oxidation with air in presence of Al_2O_3 in C_6H_6 gives[17] disulfides

$$C_6H_5SH \xrightarrow[RT,\ 6\ hr.]{air,\ Al_2O_3,\ C_6H_6} \underset{96\%}{C_6H_5SSC_6H_5}$$

(*xvii*) **Oxidation of thiols:** Thiols on oxidation with oxygen in KOH/DMF gives the corresponding sulfonic acid[16] in 91% yield.

(*xviii*) Oxidation of methylene groups adjacent to aromatic ring with O_2 in presence of chromium sesquioxide as catalyst gives[18] keto compounds ($CH_2 \rightarrow CO$)

$$C_6H_5CH_2CH_3 \xrightarrow{O_2/\text{Chromium sesquioxide}} C_6H_5COCH_3$$

(*xix*) **Oxidation of primary alcohols to carboxylic acids:** Primary alcohols are oxidised to carboxylic acids by O_2 in presence of PtO_2 in heptane at 60° for 2 hr. However if the reaction mixture is heated for 15 min, the corresponding aldehyde is obtained[5].

$$C_{11}H_{23}CH_2OH \xrightarrow[60°,\ 2\ hr]{O_2/PtO_2,\ C_7H_{16}} \underset{96\%}{C_{11}H_{23}CO_2H}$$

$$C_{11}H_{23}CH_2OH \xrightarrow[60°,\ 15\ min]{O_2/PtO_2,\ C_7H_{16}} \underset{72\%}{C_{11}H_{23}CHO}$$

(*xx*) **Oxidation of secondary alcohols to ketones:** Secondary alcohols on oxidation with O_2 in presence of PtO_2 in heptatane at 20° for 96 hr. give[5] ketones.

(*xxi*) **Oxidation of hydrocarbons containing tertiary H:** Tetralin on oxidation with O_2 at 110° gives [16,14] hydroperoxides

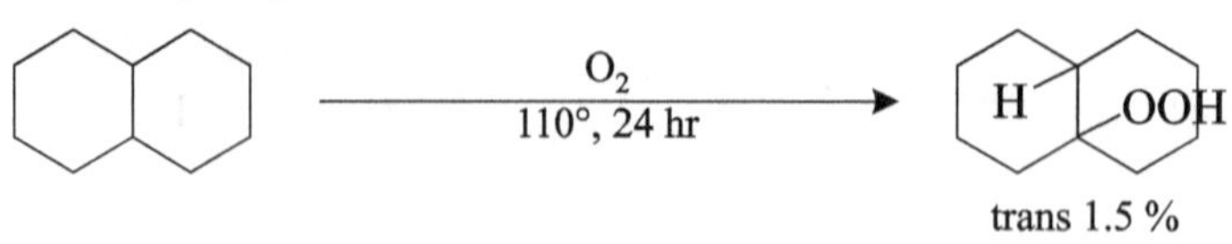

trans 1.5 %

$$(CH_3)_3CH \xrightarrow[163°]{O_2,\ HBr/\ catalyst} \underset{70\%}{(CH_3)_3COOH}$$

(*xxii*) **Oxidation of cyclic ketones:** Cyclic ketones, *e.g.* cyclopentanone on oxidation with O_2 in presence of NaOMe gives[19] glutaric acid.

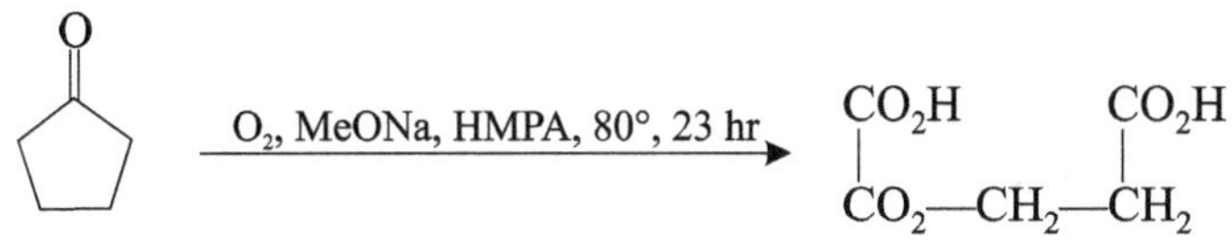

glutari acid 100%

(*xxiii*) **Oxidative coupling of acetylenic alcohols:** Acetylenic alcohols undergo oxidative coupling[20] with O_2 in presence of Cu $(OAc)_2$

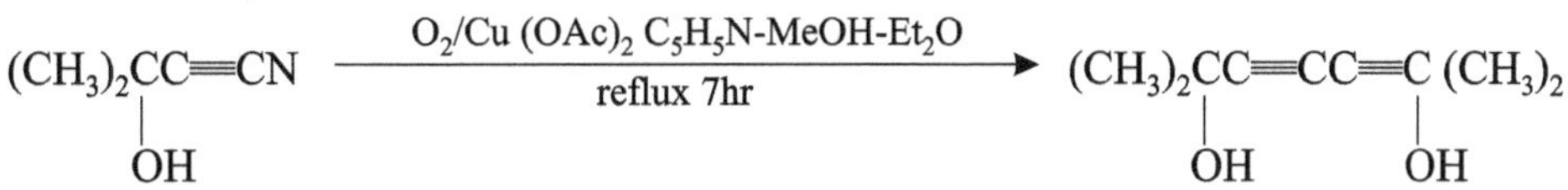

80%

15.2 SINGLET OXYGEN

It is the excited state of oxygen having two odd electrons possessing anti parallel spins.

$$[\uparrow .O:O. \downarrow]$$

It is formed[21] in situ by irridiation of gaseous oxygen with UV light in presence of a sensitizer like benzophenone, methylene blue, rose bengal, chlorophyll etc. Singlet oxygen is unstable with halt life of 10^{-6} seconds and is diamagnetic.

A number of chemical methods are also available for the generation of singlet oxygen. These include reaction of H_2O_2 in alkaline medium with sodium hypochlorite (equation 1)[22,23,24], calcium hypochlorites[23], bromine[25] or organic peroxyacids[25] (equation 2). Singlet oxygen can also be obtained by the decomposition aromatic endoperoxides such as 9,10-diphenyl anthracene endoperoxide[26] or rubrene peroxide[27] (equation 3) and by the decomposition of triphenyl phosphite ozonide[28,29] (equation 4)

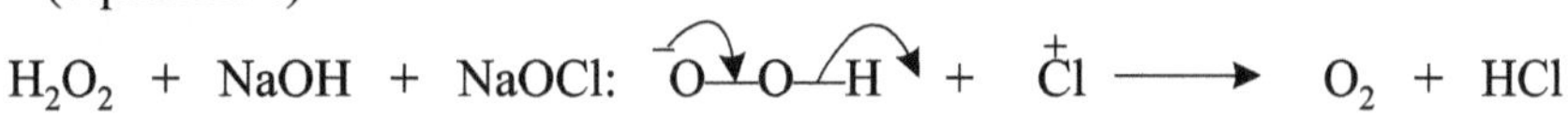

Singlet oxygen (eq. 1)

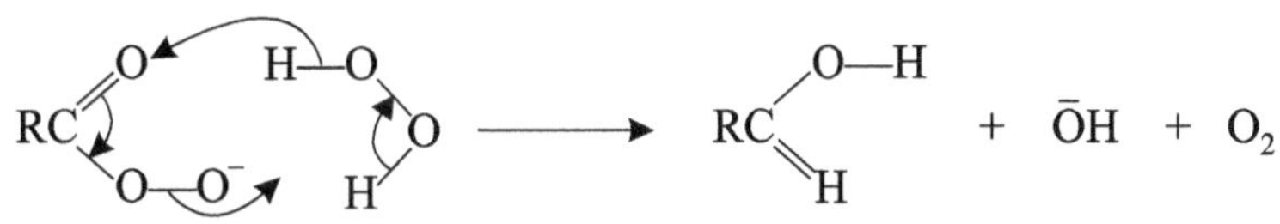

Peroxy acids (eq. 2)

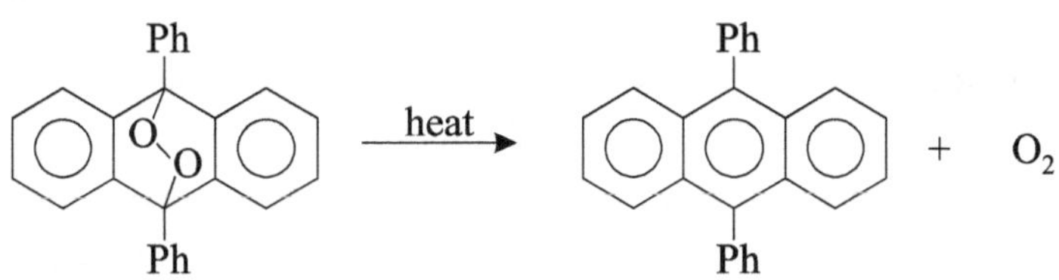

9, 10-diphenyl anthracene endo peroxide (eq. 3)

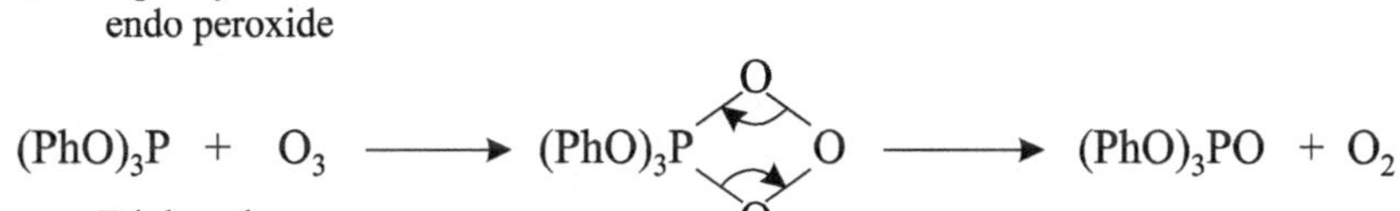

Triphenyl phosphine

Triphenyl phosphine ozonide (eq. 4)

However, the method of choice for the generation of singlet oxygen is the UV irradiation of gaseous oxygen.

Following are given some of the reactions in which singlet oxygen is used:

(*i*) **Formation of Cyclic peroxides:** Conjugated cyclic drienes like cyclohexadiene, α-terpinene, cyclopentadiene form cyclic per oxides (equations 5, 6, 7)[22,30,31]

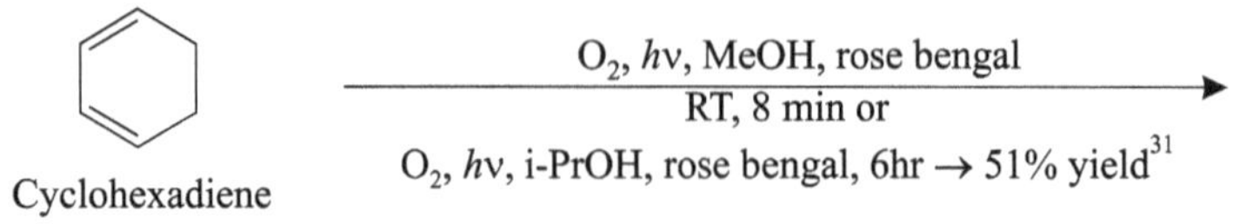

(eq. 5)

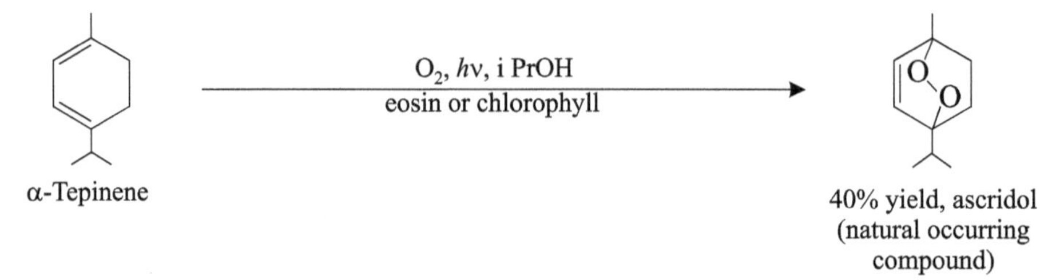

(eq. 6)

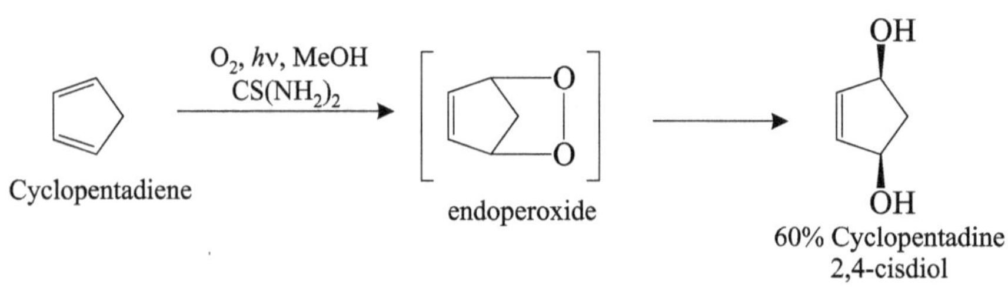

(eq. 7)

Open chain dienes like isoprene on reaction with singlet oxygen gives 4-methyl-1,2-dioxa-4-cyclohexene (eq. 8)[32]

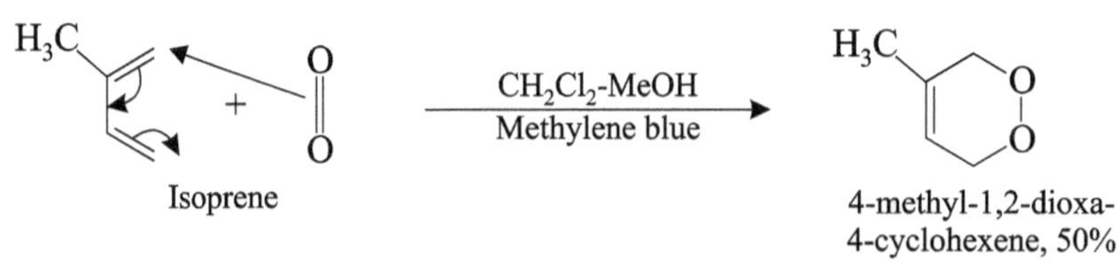

(eq. 8)

(*ii*) **Formation of Dioxetanes:** Dioxetanes are compounds containing two atoms of oxygen in a four membered ring. Thus, the addition of one molecule of oxygen across a double bond gives dioxetanes. The addition is believed to be stereospecific occurring in the syn mode[33] and takes place when singlet oxygen is used. The dioxetanes are unstable and decompose to give carbonyl compounds. However, some dioxetanes can be isolated or even distilled at low temperatures. As an example 1,2-diphenyl cyclobutane on reaction with singlet oxygen give the corresponding dioxetane, which being unstable decomposes to give dicarbonyl compound. It is found that different products may be formed depending on the solvent (eq. 9)[34].

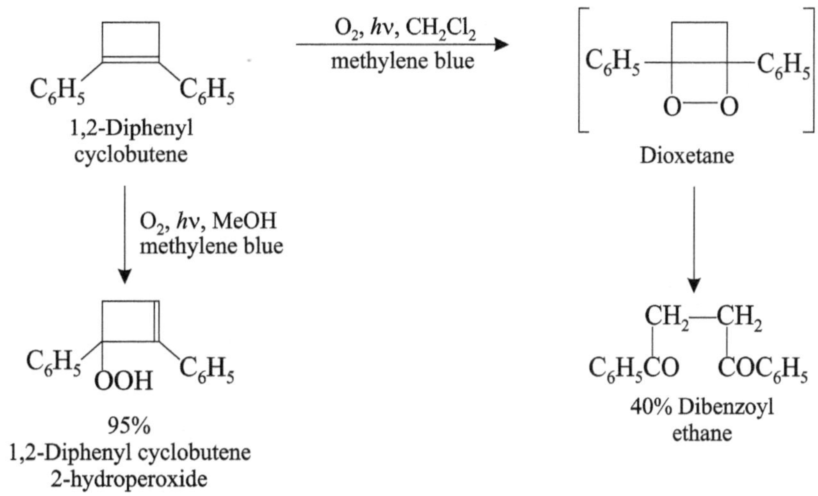

(eq. 9)

It is found that the addition of singlet oxygen to *cis*- and *trans*-1,2-diphenyl-1,2-dimethoxy ethylene gives cis- and *trans*- oxetanes, the addition is stereospecific (eq. 10)[35,36]

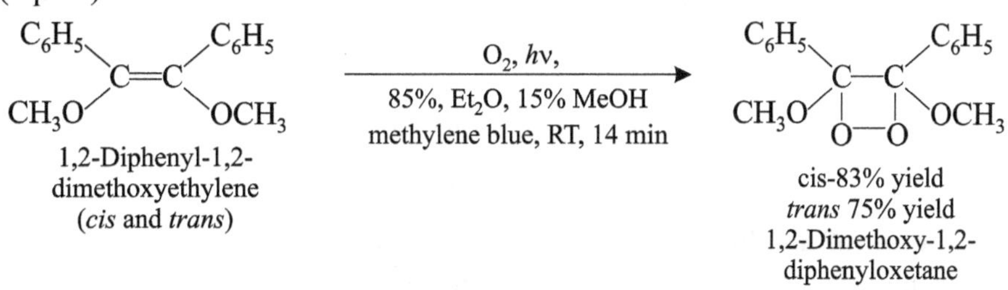

(eq. 10)

(*iii*) **Formation of epoxides:** Halogenated alkanes on reaction with singlet oxygen give the corresponding epoxides (eq. 11)[37]

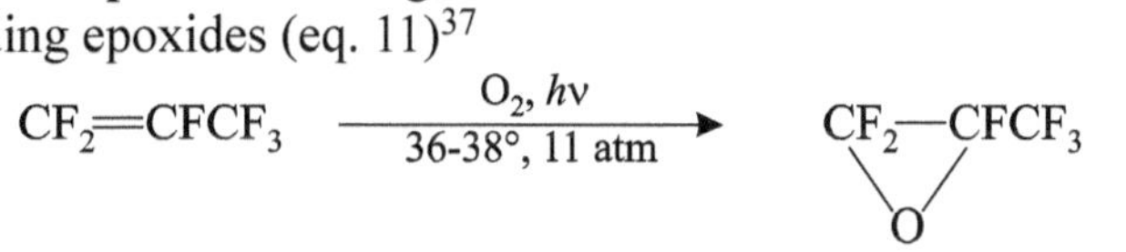

(eq. 11)

74% yield
24% conversion

(*iv*) **Formation of hydroperoxides:** Ethers having α-hydrogens form hydroperoxides, which are explosive in dry state. It is found that singlet oxygen (generated by irridation by UV light) on passing through THF in presence of a sensitizer gives 46% yield of tetrahydrofuran α-hydroperoxide. In a similar way, dioxane, methyl butyl ether, diisopropyl ether, dibutyl ether and diisoamyl ether give[38] the corresponding α-hydroperoxides in yields of 10, 15, 12, 6 and 8% respectively.

Hydroperoxides are also obtained from secondary alcohol (eq. 12)[38]

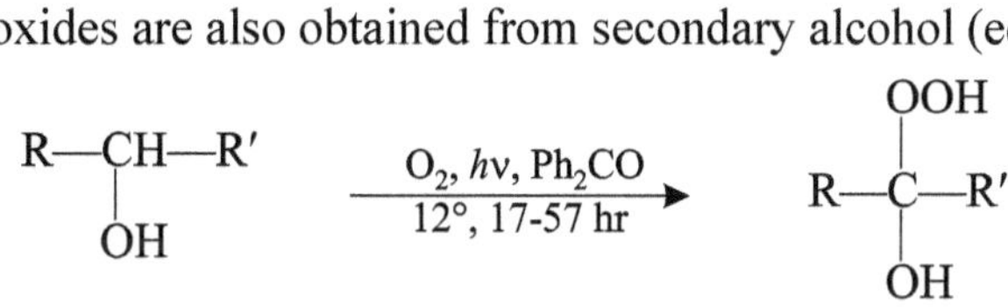

(eq. 12)

$R = CH_3$
$R_1 = CH_3, C_2H_5, C_3H_7$

Compounds containing primary, secondary or tertiary benzylic hydrogens on reaction with singlet oxygen results in a free radical chain to give hydroperoxides as the final product (eq. 13).

$$\begin{aligned}&ArCH_3 \xrightarrow{h\nu} ArCH_2^{\cdot} \xrightarrow{O_2} RCH_2-O-\dot{O}\\&RCH_2-O-\dot{O}+ArCH_3 \longrightarrow RCH_2-O-O-H+ArCH_2^{\cdot}\end{aligned} \qquad \text{(eq.13)}$$

Such reaction take place with *p*-xylene[39], ethylbenzene[39] and cumene (isopropyl benzene)[40]. In fact, oxidation of cumene to cumyl hydroperoxide (eq. 14) followed by treatment with acid is a basis for the manufacture of phenol[40].

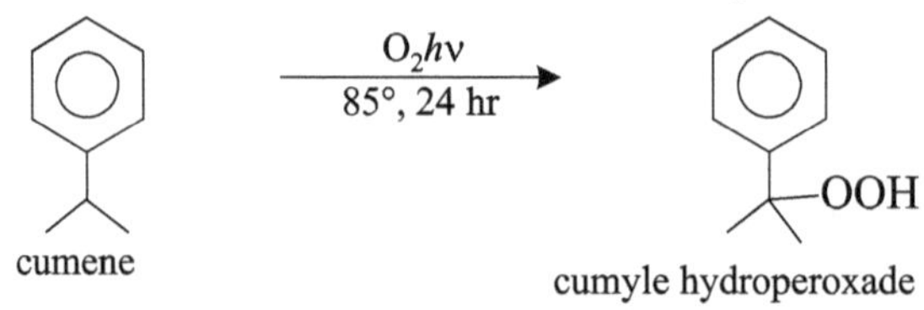

(eq.14)

Compounds containing saturated rings attached to aromatic nucleus, for example indane[41], tetralin[42,43] and octahydroanthracene[44], on oxidation with singlet oxygen takes place at carbons adjacent to the aromatic ring (eq. 15)

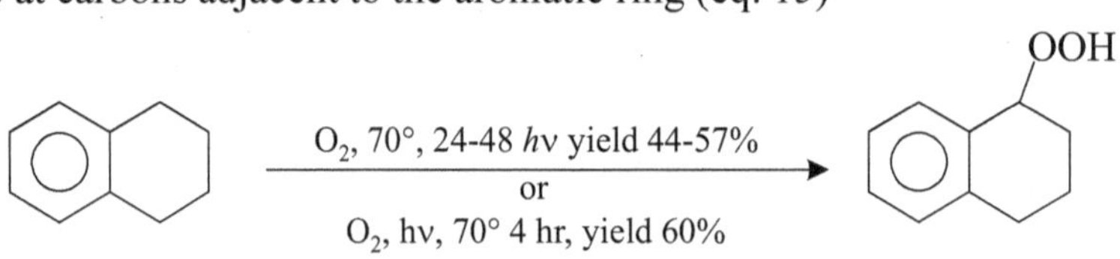

(eq. 15)

(*v*) **Oxidation of aldehydes to peroxy acids:** Aldehydes, *e.g.*, benzaldehyde on reaction with singlet oxygen gives peroxy acids (eq. 16)[45]

$$C_6H_5-CHO \xrightarrow{O_2, h\nu, Me_2CO} \underset{63\%}{C_6H_5CO_3H} \qquad \text{(eq.16)}$$

15.3 OZONE

It is a blue gas [O = O – O:] or dark blue liquid (b.p. –106°). It is used in a mixture with oxygen and is prepared by passing oxygen through silent electric discharge. Ozone is a powerful oxidising agent. Some of its reactions are given below:

(*i*) **Reaction with alkenes:** The most common reaction of ozone is with alkenes to give ozonide. The formation of the ozonide is as shown below (equation 1)[46,47]

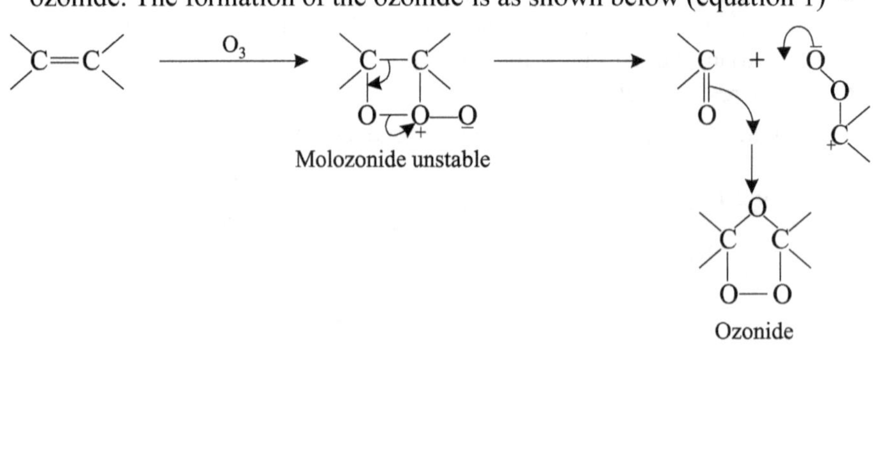

(eq. 1)

The ozonides are important intermediate for organic synthesis. In fact, conversion of ozonides to products is carried out in the same reaction vessel. Such reactions are discussed subsequently.

(*ii*) **Oxidative cleavage of double bonds:** Double bonds are oxidatively cleaved to alcohols, aldehydes, ketones or carboxylic acids. The method involves the conversion of the unsaturated compound with ozone followed by treatment of the ozonide formed. Thus, reduction of the ozonide with strong reducing agent like $LiAlH_4$ gives alcohols (eq. 2)[48]. Cycloalkenes yield diols (eq. 3)[49]

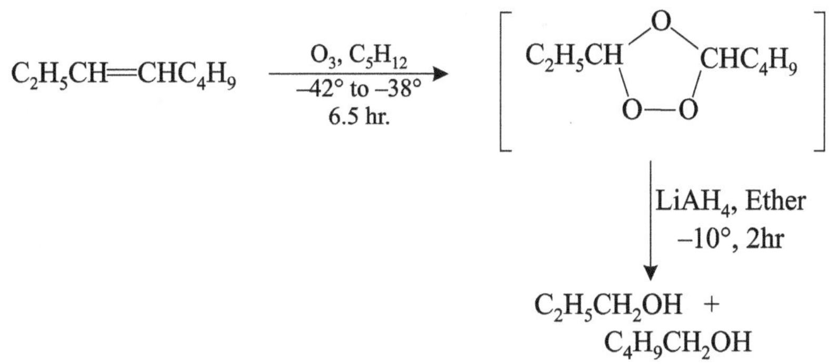

(eq. 2)

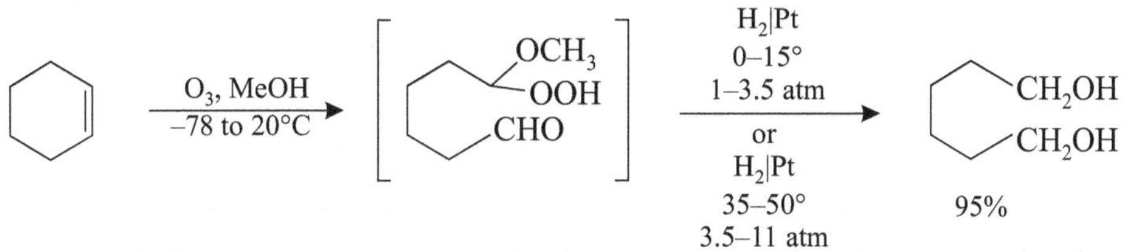

(eq. 3)

(*iii*) **Oxidative cleavage of alkenes/cycloalkenes to aldehydes:** The alkenes/cycloalkenes are converted into ozonide followed by reduction (eq. 4)[50] and (eq. 5)[51]

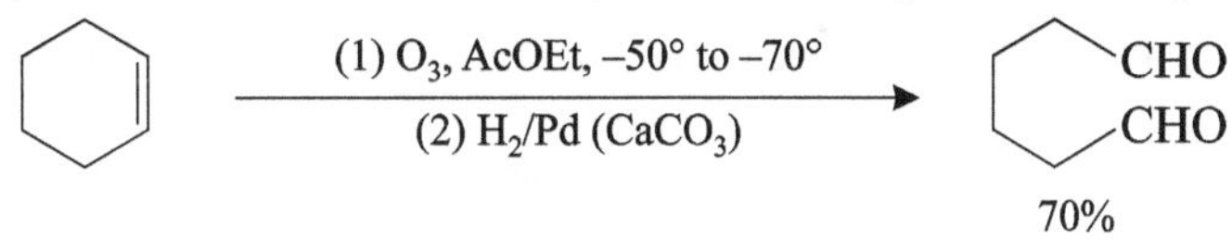

(eq. 4)

$$CH_3(CH_2)_7CH=CH(CH_2)_7CO_2CH_3 \xrightarrow[\substack{-20°, 3.25hr \\ (2)\ Zn-HOAC, \\ MeOH, 30-35°}]{(1)\ O_3,\ MeOH} \underset{81\%}{CH_3(CH_2)_7CHO} + \underset{94\%}{OHC(CH_2)_7CO_2CH_3}$$

(eq.5)

(*iv*) **Oxidative cleavage of alkenes/cycloalkenes to carboxylic acids:** The alkenes/cycloalkenes are converted into ozonide followed by treatment with H_2O_2/HCO_2H (eq. 6)[52] or (eq. 7)[53]

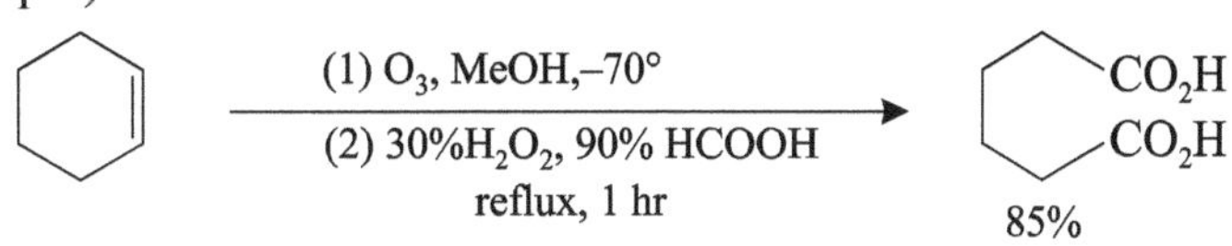

(eq. 6)

$$C_{10}H_{21}CH{=}CH_2 \xrightarrow[\substack{(2)\ Ag_2O,\ NaOH,\ 95°,\ 6hr \\ (3)\ HNO_3,\ 60°,\ 1\ hr}]{(1)\ O_3,\ C_5H_{12},\ -10°} \underset{98\%}{C_{10}H_{21}COOH} \qquad \text{(eq. 7)}$$

(*v*) **Oxidation of aldehydes to carboxylic esters:** The aldehydes are first converted into their acetals and then reacted with ozone in CH_2Cl_2 (eq. 8)[54,55]

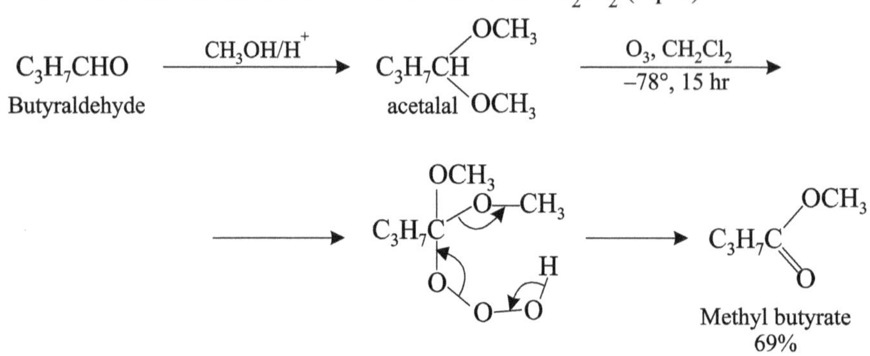

(eq. 8)

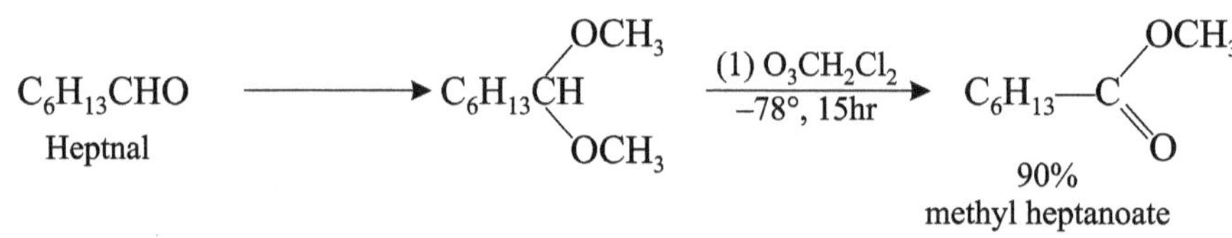

(*vi*) **Oxidation of aldehydes to peroxy acids:** Passing a current of O_3-containing O_2 into a solution of benzaldehyde in ethyl acetate gives perbenzoic acid (eq. 9)[56].

$$C_6H_5CHO \xrightarrow{O_3/O_2,\ AcOEt,\ 20-22°,\ 1.7\ hr} \underset{90\%}{C_6H_5CO_3H} \qquad \text{(eq.9)}$$

(*vii*) **Oxidation of aromatic side chain to carboxylic acids:** Aromatic compounds like 2-methoxytoluene on treatment with O_3 in presence of $Co(OAc)_2$ give *o*-methoxybenzoic acid (eq. 10)[57]

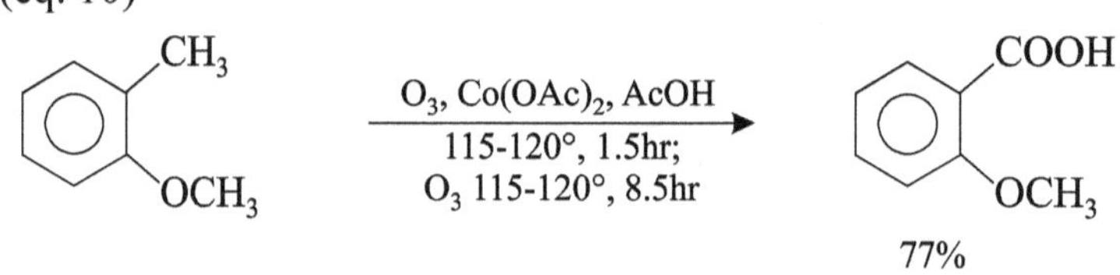

(eq. 10)

(*viii*) Oxidation of azides to aldehydes and nitro compounds. Alkyl azides are converted into aldehydes and nitro compounds[58] by a sequence of reaction as shown in eq. 11.

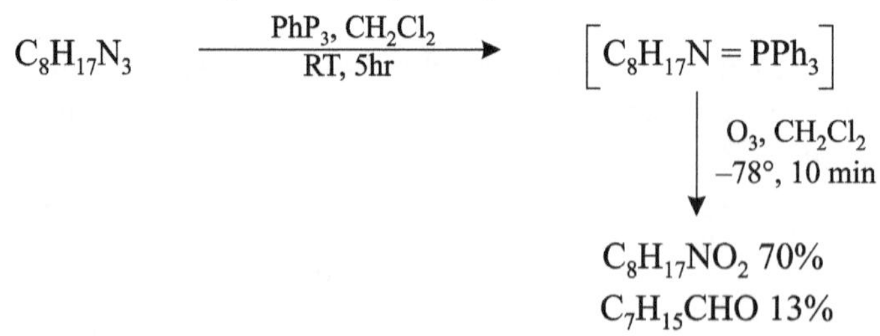

(eq. 11)

(*ix*) **Conversion of ethers into esters:** Ethers are converted into esters to passing ozone at 0° (eq. 12)[59]

$$C_3H_7CH_2OC_4H_9 \xrightarrow{O_3,\ 0°,\ 7\ hr.\ 25\ min} \underset{30\%}{C_3H_7CO_2C_4H_9} \qquad \text{(eq. 12)}$$

(*x*) **Oxidation of glycosides to esters:** The oxidation of glycosides to esters with ozone in presence of $Ac_2O/NaOH$ gives esters. The reaction is stereo selective, only the β-anomers are affected (eq. 13)[60]

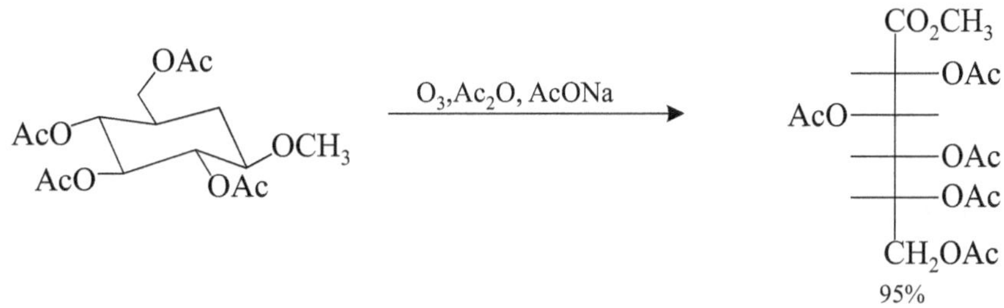

(eq. 13)

(*xi*) **Oxidation of organomercury compounds to ketones:** Organomercury chlorides are oxidized to ketones by ozone (eq. 14)[61]

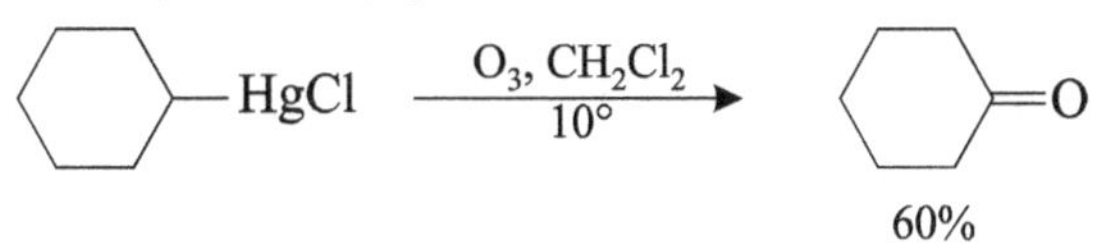

(eq.14)

(*xii*) **Oxidation of tertiary phosphines to phosphine oxides:** Tertiary phosphine like triphenyl phosphene are oxidised to the corresponding oxides by ozone adsorbed on silica gel at –70° (eq. 15)[62]

$$(C_6H_5)_3P \xrightarrow{O_3,\ SiO_2,\ CH_2Cl_2,\ -70^\circ} (C_6H_5)_3P=O \qquad \text{(eq.15)}$$

(*xiii*) **Oxidation of phosphites to phosphates:** Phosphites are oxidised to phosphates by ozone in CH_2Cl_2 at 5–10°C (eq. 16)[62]

$$(C_6H_5O)_3P \xrightarrow[5-10^\circ]{O_3,\ CH_2Cl_2} \underset{95\%}{(C_6H_5O)_3P=O} \qquad \text{(eq.16)}$$

This conversion is important in the chemistry of nucleoside phosphites.

15.4 HYDROGEN PEROXIDE

It is commercially available in aqueous solution of 30 percent concentration, in which form it a colourless liquid. It is stabilised against decomposition which occurs in presence of traces of iron, copper, aluminium, platinum and other transition metals.

Hydrogen peroxide is used in a number of oxidations. A molecular complex of H_2O_2 with 1, 4-diazabicyclooctane (Dabco) can be prepared (m.p. 112° (*d*)) and can be used as a source of H_2O_2 in oxidations[63]. Following are given some of the reactions in which H_2O_2 is used.

(*i*) **Oxidation of aldehydes to carboxylic acids:** Aldehydes are converted into carboxylic acid with H_2O_2. In presence of SeO_2 as catalyst, acrolein is oxidised to acrylic acid is 90% yield with a 12% solution of H_2O_2 in tert. amyl alcohol at 40° for 3 hr[64]. Methacrolein is converted into methacrylic acid[64] by 90% H_2O_2 in presence of SeO_2 at 60° (eq. 17)

$$\underset{\text{Acrolein}}{CH_2=CHCHO} \xrightarrow[SeO_2,\ 40°.\ 3\ hr]{12\%\ H_2O_2,\ \text{tert. amyl alcohol}} \underset{\text{Acrylic acid}}{CH_2=CHCOOH}$$

$$\underset{\text{Methacrolein}}{CH_2=C(CH_3)\ CHO} \xrightarrow[60°,\ 1.75\ hr.]{90\%\ H_2O_2/SeO_2} \underset{\text{Methacrylic acid}}{CH_2=C(CH_3)-COOH} \quad \text{(eq. 17)}$$

(*ii*) **Oxidation of alkenes into ketones:** Alkenes are oxidised to ketones by H_2O_2 in presence of Pd $(OAc)_2$ (eq. 18)[65]

$$CH_2=CHR \xrightarrow[AcOH,\ 80°,\ 6\ hr]{30\%\ H_2O_2,\ Pd(OAc)_2} CH_3COR \quad \text{(eq. 18)}$$

$$R = C_{10}H_{21}\ \text{yield } 92\%$$
$$R = C_6H_{13}\ \text{yield } 96\%$$

(*iii*) **Dakins Oxidation:** Aromatic aldehydes containing OH groups are oxidised into the corresponding phenols (–CHO→OH) with H_2O_2 and sodium hydroxide (eq. 19)[66]

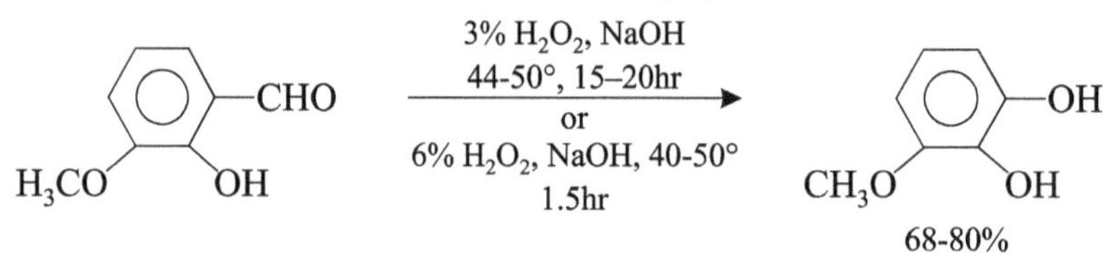

(eq. 19)

(*iv*) **Oxidation of alkenes to alcohols:** Alkenes on reaction with borane (BH_3, THF) to give a tertiary borane, which on oxidation with H_2O_2/alkali gives alcohols (eq. 20)[67]

$$\underset{R = C_4H_9 \text{ or } C_6H_{13}}{3\ RCH=CH_3} \xrightarrow[0°,\ 30\ min,\ RT]{BH_3.THF,\ \text{diglyme}} \underset{\text{tri alkyl borane}}{(R\ CH_2CH_2)_3B} \xrightarrow[NaOH]{THF,\ 30\%\ H_2O_2} \underset{95\%}{3R\ CH_2Cl_2OH} \quad \text{(eq. 20)}$$

(*v*) *Oxidation of carboxylic acids to peroxy acids* Benzoic acid is converted into peroxybenzoic acid[68]. ($C_6H_5CO_3H$) in 90% yield by treatment with 70% H_2O_2 in presence of methansulfonic acid at 25-30° for 2.5 hr.

(*vi*) **Oxidation of dicarbonyl compounds to carboxylic acids:** The reaction of benzil with 95% H_2O_2 in presence of 70% perchloric acid (refluxing 15 min) gives benzoic acid. Similarly, anisil gives anisic acid[69].

(*vii*) **Oxidation of α-diketones to acid anhydrides:** α-Diketones like diphenylcyclobutene-1,2-dione is oxidised with H_2O_2 in CCl_4 to yield 79% yield of diphenyl maleic anhydride (eq. 21)[70]

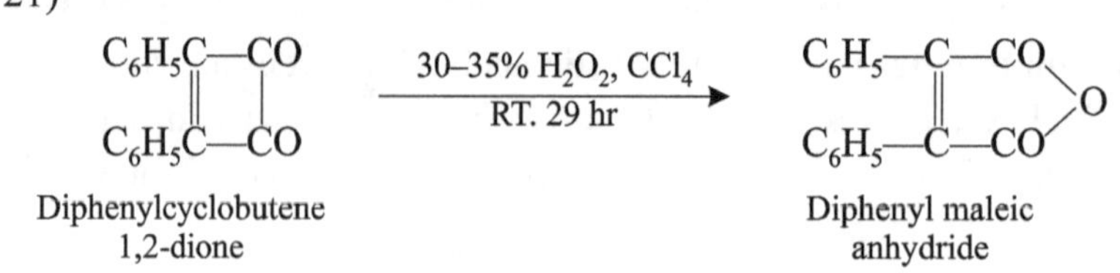

Diphenylcyclobutene 1,2-dione — Diphenyl maleic anhydride (eq. 21)

(*viii*) **Oxidation of epoxides (Oxiranes):** Epoxides on reaction with H_2O_2 gives hydroxy hydroperoxides (eq. 22)[71]

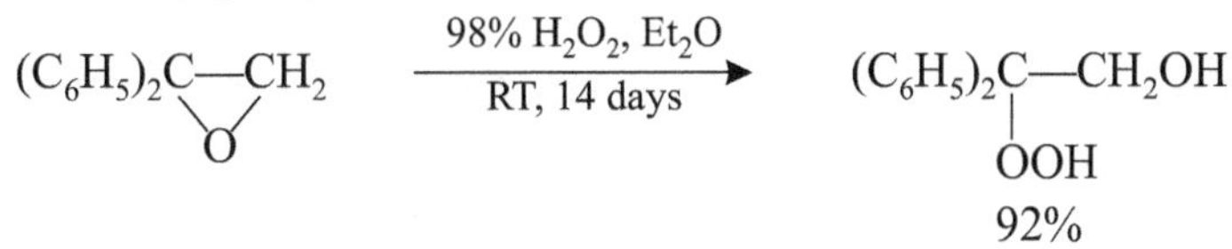

(eq. 22)

(*ix*) **Balyer-Villiger Reaction:** It involves the oxidation of ketones to esters by H_2O_2. However, cyclic ketones give lactones[72,73].

(*x*) **Oxidation of Thiols (mercaptans):** Oxidation of thiols with H_2O_2 gives coupled products, the disulphides[74] (eq. 23)

$$R - SH \xrightarrow[60°,\ 2hr]{H_2O_2(30\%)} R - S - S - R \qquad \text{(eq. 23)}$$

Another examples of oxidation of thiols is given in (eq. 24)

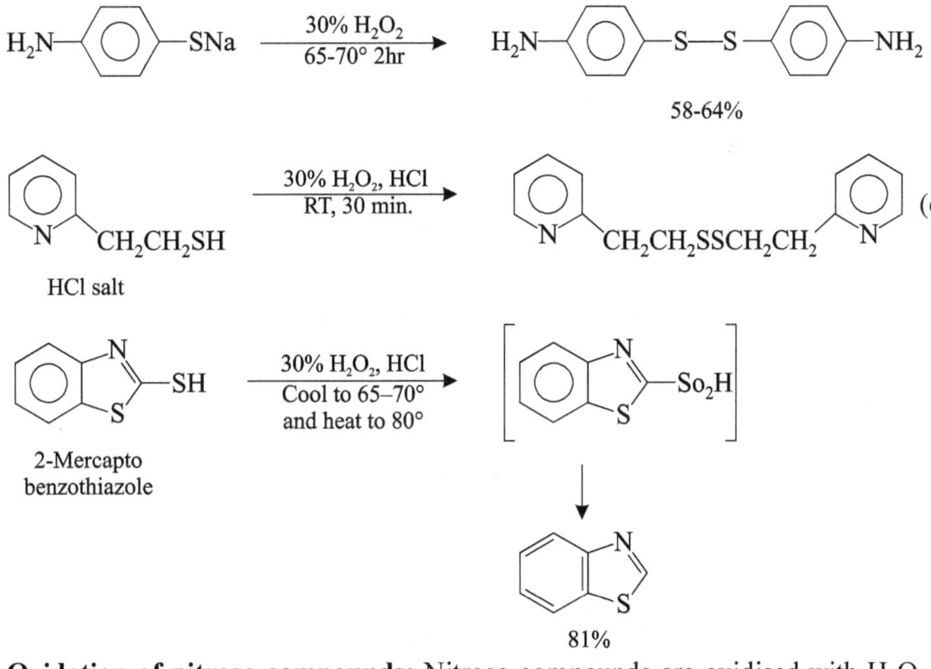

(eq. 24)

(*xi*) **Oxidation of nitroso compounds:** Nitroso compounds are oxidised with H_2O_2 to the corresponding nitrocompounds (eq. 25)[78]

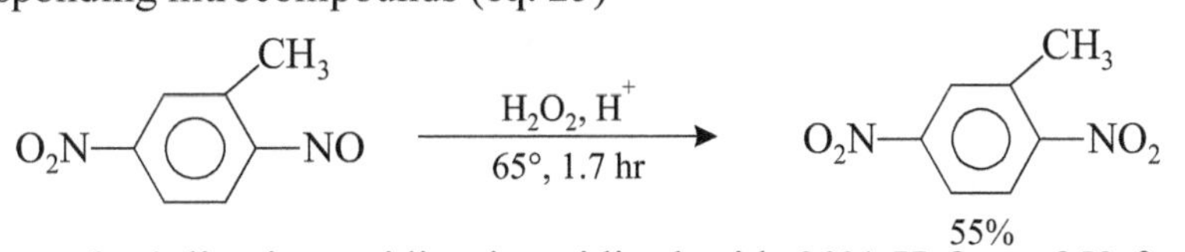

(eq. 25)

3-Nitroso-2, 6-diaminopyridine is oxidised with 30% H_2O_2 at 35° for 1.5 hr. to give 3-nitro-2,6-diaminopyridine in 96% yield[79]. Under these conditions primary amino groups remain uneffected.

Nitrosoamines are oxidised in H_2O_2 in presence of trifluoroacetic anhydride to give the corresponding nitro compound (eq. 26)[80].

$$(C_2H_5)_2NNO \xrightarrow[\text{RT, 30 mim; reflux 1 hr.}]{90\%\ H_2O_2,(CF_3CO)_2O,\ CH_2Cl_2} \underset{70\%}{(C_2H_5)_2NNO_2} \quad \text{(eq. 26)}$$

(*xii*) **Oxidation of amines:** Both aliphatic and aromatic primary amines are oxidized with H_2O_2 to the corresponding nitrocompounds[81,82] (eq. 27)

$$\underset{\text{2-Aminobutane}}{C_2H_5-CH(CH_3)-NH_2} \xrightarrow[\text{reflux 2 hr}]{90\%\ H_2O_2,\ Ac_2O,\ H_2SO_4,\ CH_2Cl_3} \underset{\text{2-Nitrobutane 65\%}}{C_2H_5-CH(CH_3)NO_2} \quad \text{(eq. 27)}$$

$$C_2H_5NH_2 \xrightarrow{\substack{90\%\ H_2O_2,\ Ac_2O,\ H_2SO_4,CHCl_3 \\ \text{reflux 80 min}}} \underset{83\%}{C_2H_5NO_2}$$

Secondary amines are oxidised with 6% H_2O_2 at room temp. to give dialkyl hydroxylamines[83].

Tertiary amines are oxidised with H_2O_2 to N-oxides[84].

(*xiii*) **Oxidation of Sulfides:** Sulfides are oxidised to yield sulfoxides in acetic anhydride at room temperature (eq. 28)[85]

$$C_6H_5-SCH{=}CH_2 \xrightarrow[\text{RT, 5-6 hr}]{30\%\ H_2O_2,\ Ac_2O} C_6H_5-\overset{\overset{\displaystyle O}{\|}}{S}CH{=}CH_2 \quad \text{(eq. 28)}$$

Under the above conditions, alkene double bonds are not epoxidised.

Some other examples of oxidation of sulfides with H_2O_2 are given below (eq. 28 to eq. 30)[86,87]

$$C_6H_5S\ CH_3 \xrightarrow[\text{H}_2\text{O, MeOH, RT 5 min}]{30\%\ H_2O_2,\ H_2O/TiCl_3} \underset{100\%}{C_6H_5\overset{\overset{\displaystyle O}{\|}}{S}CH_3} \quad \text{(eq. 28a)}$$

$$C_6H_5\ S\ C_6H_5 \xrightarrow[\text{RT. 5 min.}]{30\%\ H_2O_2/TiCl_3,\ MeOH,\ H_2O} \underset{100\%}{C_6H_5\ \overset{\overset{\displaystyle O}{\|}}{S}\ C_6H_5} \quad \text{(eq. 29)}$$

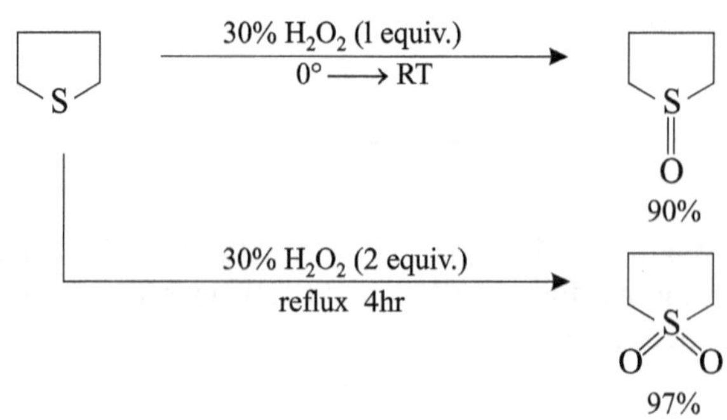

(eq. 30)

(*ix*) **Oxidation of tertiary alcohols:** Only a few tertiary alcohols are converted into the corresponding hydroperoxides on treatment with H_2O_2 in H_2SO_4[88,89]. Dimethylphenyl carbinol on adding to a mixture of 87% H_2O_2 and H_2SO_4 at room temp. below 0° gives 94% yield of cumyl hydroperoxide[88]. In a similar way, tertiary OH group attached to a triple bond is converted into the corresponding hydroperoxide (eq. 31)[89]

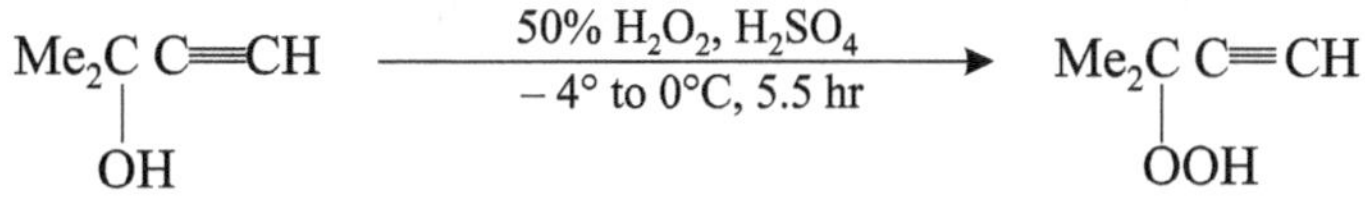

98% yield (eq. 31)

(*x*) **Oxidation of Unsat'd aldehydes:** Oxidation of unsat'd aldehydes has been shown to give the corresponding acid (eq. 17). However, under alkaline conditions (pH 8-8.5) the corresponding epoxy aldehydes are obtained (eq. 32)[90]

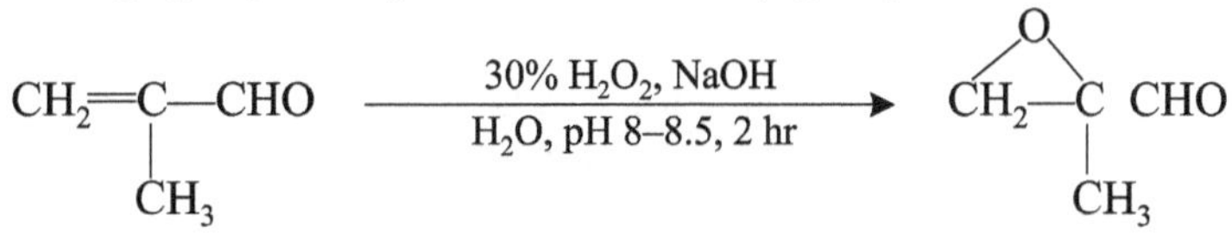

88% yield (eq. 32)

(*xi*) **Epoxidation:** Epoxidation of alkenes with H_2O_2 in alkaline medium give the corresponding epoxides (eq. 33)[90-92]

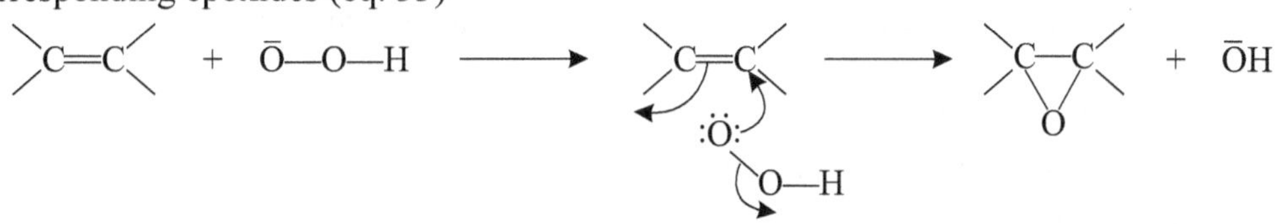

(eq. 33)

Nitriles on reaction with alkaline H_2O_2 at pH 8 gives the corresponding epoxides and a carboxamide (eq. 34)[93]

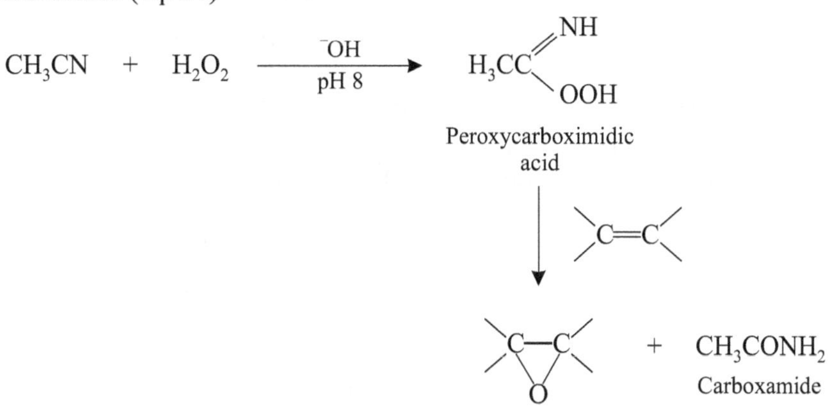

(eq. 34)

Halogenated alkenes are converted into halogenated epoxides (eq. 35)[94]

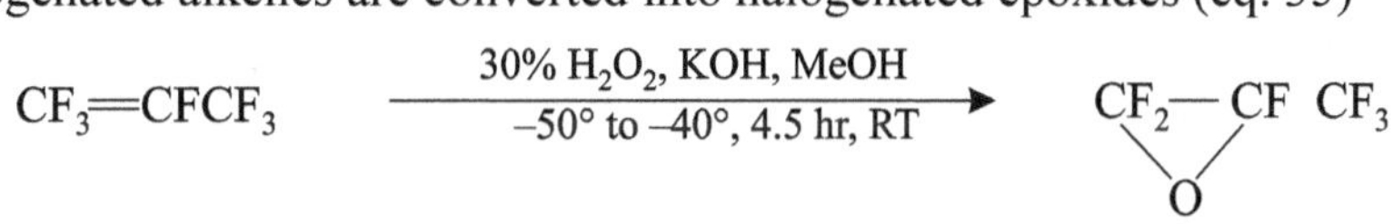

23.5–51.5% (eq. 35)

In a similar way cyclohexene gives the corresponding epoxide (eq. 36)[95]

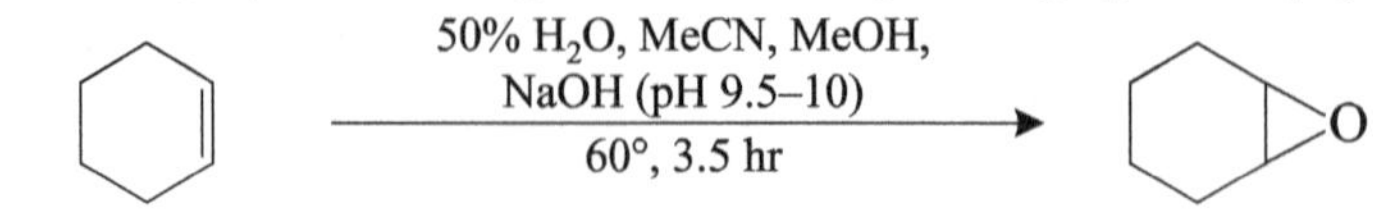

(eq. 36)

Unsaturated carboxylic acid give the corresponding epoxides (eq. 37)[96]

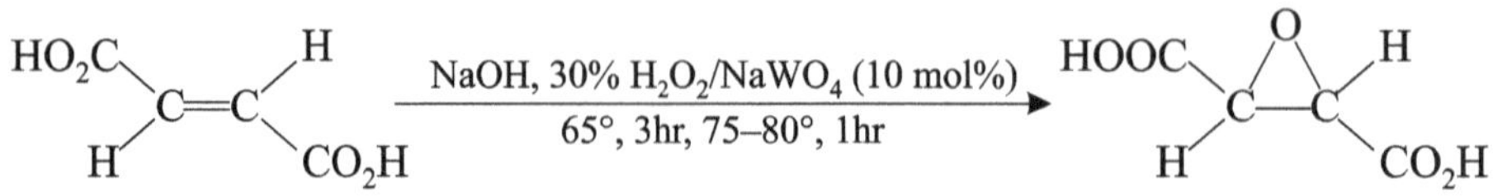

(eq. 37)

60%

2-Alkyl cyclohexanone also gives the corresponding epoxide (eq. 38)[97]

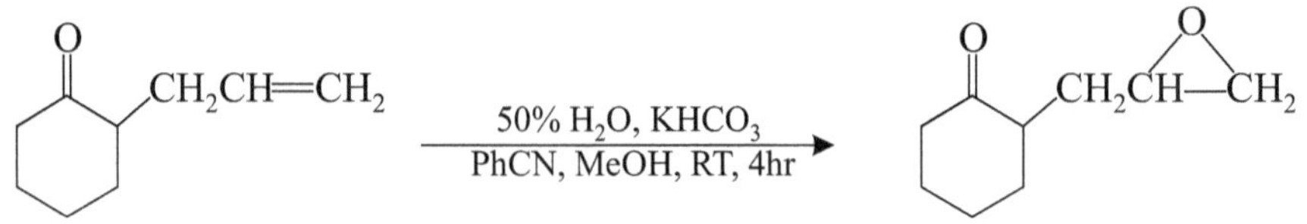

54% (eq. 38)

Unsaturated nitriles are epoxidised with H_2O_2. Thus tetracyanoethylene on treatment with 30% H_2O_2 in acetonitrile at 10-12° gives tetracyanoepoxyethane in 68% yield[98]. Similarly, acrylonitrile gives[99] expoxypropionamide (glycidimide) is 62% yield.

15.5 DIOXIRANES

These are obtained from acetone and other aliphatic ketones by treatment with oxone ($2KHSO_5$, $KHSO_4.K_2SO_4$) (eq. 39)

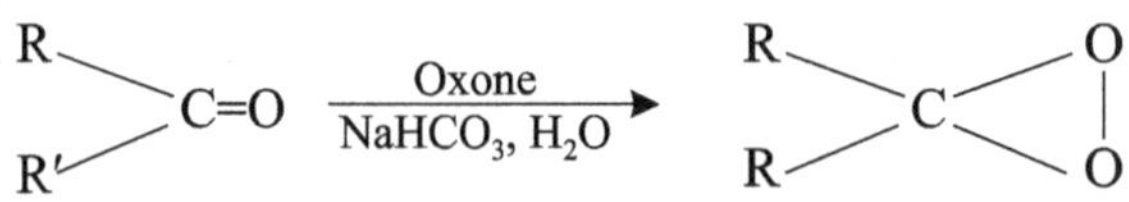

(eq. 39)

Dioxiranes

It can be used for oxidations that are not usually accomplished by oxone itself[100,101] following are given some of the oxidations which can be accomplished by dioxiranes.

(*i*) **Epoxidation of alkanes:** Alkenes on reaction with dioxiranes in acetone at 22° for 1 hr. give the corresponding epoxides (eq. 40)[100] and (eq. 41)[100].

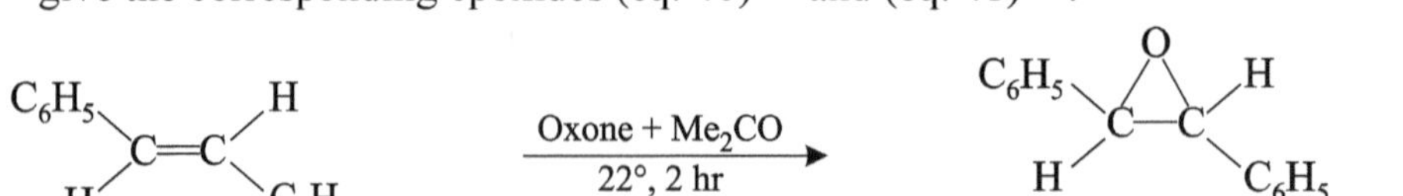

70-75% (eq. 40)

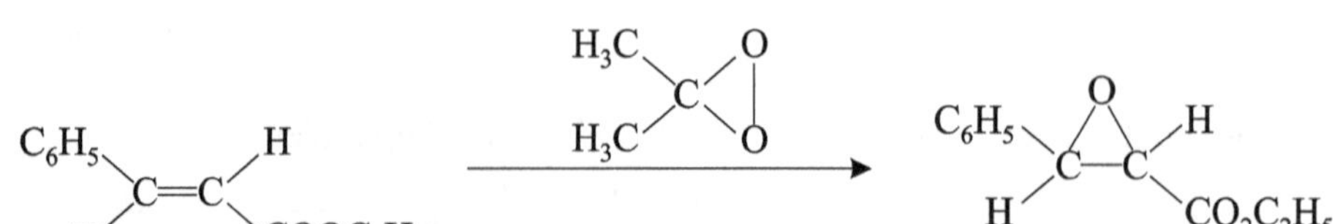

(eq. 41)

ethyl trans-cinnamate

63%
Ethyl 2,3 epoxy cinnamate

(*ii*) **Oxidation of primary amines to nitro compounds:** Primary amines are oxidised to the corresponding nitro compounds by dimethyldioxirane[101] (eq. 42)

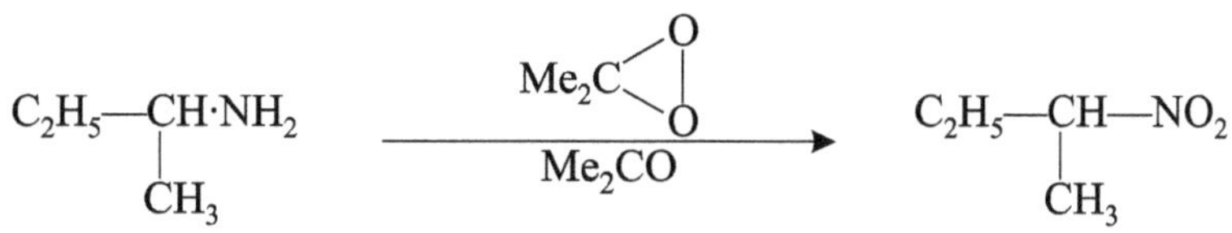

2-Aminobutane 87% 2-Nitrobutane (eq. 42)

(*iii*) **Oxidation of sulfides to sulfoxides:** Dimethyldioxirane oxidises sulfides to sulfoxides at room temperature (eq. 43)[100]

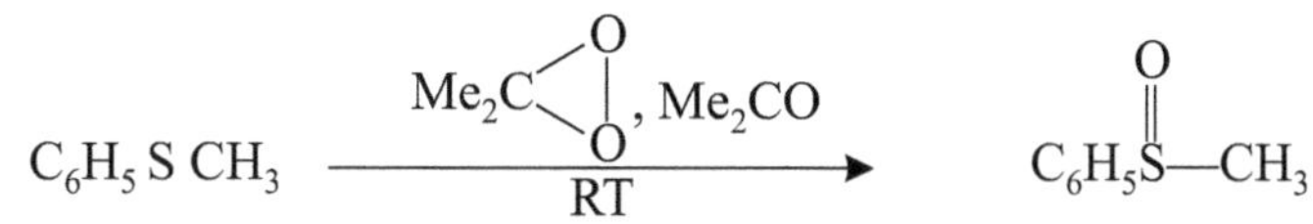

65% (eq. 43)

(*iv*) **Oxidation of tertiary amines to amine oxides:** Tertiary amines are oxidised to the corresponding amine oxides[100] (eq. 44)

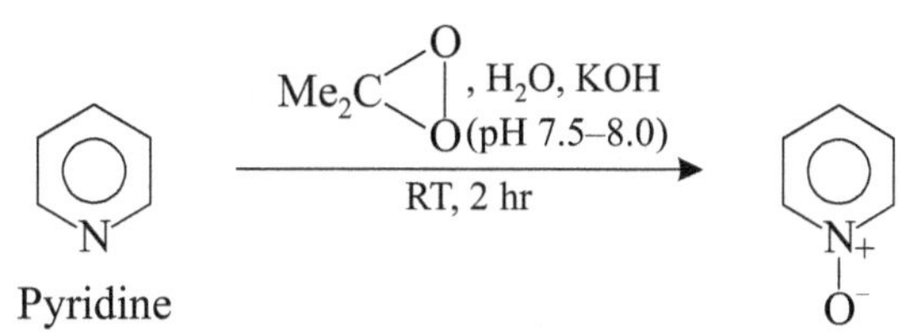

Pyridine 93% (eq. 44)

In the above reactions, dimethyl dioxione to be used can be generated in situ by the reaction of oxone with acetone (see eq. 39)

15.6 OXONE

It is a registered trade mark of E.I. du Pont de Numours and company and has the composition $2KHSO_5 \cdot KHSO_4 \cdot K_2SO_4$ and is a versatile oxidising agent[102]. Some of its reactions are given below:

(*i*) **Epoxidation of alkenes:** Alkenes are epoxidised by oxone in presence of acetone. The dimethyldioxirane generated in situ epoxidises the alkenes (see eq. 39).

(*ii*) **Oxidation of aromatic side chain to carboxylic acids:** Toluene on refluxing with oxone gives benzoic acid (eq. 45)[102]

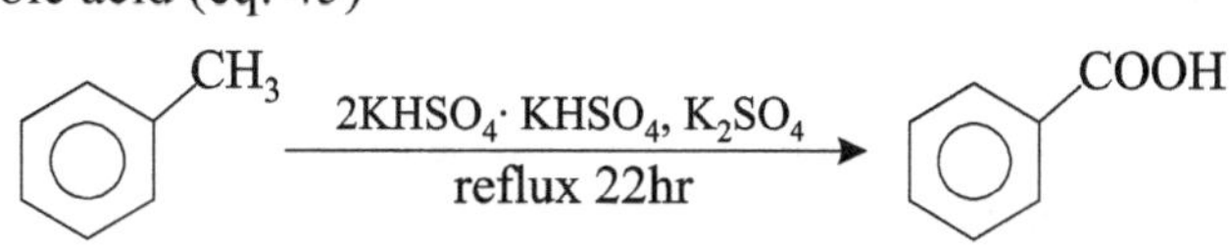

50% (eq. 45)

(*iii*) **Oxidation of iodo compounds to iodoxy compounds:** Iodobenzene on reaction with Oxone in presence of H_2SO_4 at 0° gives iodosxy benzenc (cq. 46)[102]

$$\underset{\text{Iodo benzene}}{C_6H_5I} \xrightarrow[H_2SO_4,\ \text{ice, 4 hr.}]{2KHSO_5.KHSO_4.K_2SO_4} \underset{\substack{\text{Iodoxy benzene}\\ 72\%}}{C_6H_5IO_2} \qquad \text{(eq. 46)}$$

(*iv*) **Oxidation of primary alcohols to ester:** Primary alcohols on treatment with oxone in presence of Conc. H_2SO_4 give quantitative yield of ethyl acetate[102].

(*v*) **Oxidation of primary amines to nitroso compounds:** Primary amines on reaction with oxone give the corresponding nitroso compounds[103]. *p*-Phenylene diamine (1,4-diamino benzene) is oxidised by oxone in aqueous suspension at room temperature gives *p*-dinitrobenzene in quantitative yield[102].

(*vi*) **Oxidation of secondary alcohols to ketones:** Secondary alcohols on treatment with oxone give ketones[102].

(*vii*) **Oxidation of Sulfides to Sulfones:** Phenyl methyl sulphide on reaction with oxone (3 equivalents) in MeOH, H_2O at 0° for 4 hr. at room temp. gives the corresponding sulfones (eq. 47)[104]

$$C_6H_5S\,CH_3 \xrightarrow[\text{MeOH, } H_2O,\ 0°,\ 4\text{ hr RT}]{\text{oxone (3 equiv)}} \underset{95\%}{C_6H_5S(=O)_2{-}CH_3} \qquad \text{(eq. 47)}$$

Diphenylsulphide on oxidation with oxone gives 97% yield of corresponding sulfone (eq. 48)[102]

$$C_6H_5S\,C_6H_5 \xrightarrow[\text{AcOH, RT, 5 hr}]{\text{oxone, EtOH, } H_2SO_4} \underset{97\%}{C_6H_5S(=O)_2{-}C_6H_5} \qquad \text{(eq. 48)}$$

15.7 PEROXY ACIDS

Peroxy acids can be obtained by passing oxygen under irridation through a solution of benzaldehyde in acetone (see section 15.2 equation 13) or by passing a current of ozone containing oxygen through a solution of benzaldehyde in ethyl acetate of 20-22° for 1 hr. (see section 15.3, equation 9).

Peroxyacids can also be obtained by the oxidation of carboxylic acids with H_2O_2 (see section 15.4, sub section V).

m-Chloroperoxy benzoic acid is obtained[105] by the reaction of m-chloro benzoyl chloride with 30% H_2O_2 in a solution of NaOH in aq. dioxane at 15-25° is 80-85% yield.

Following are given some of the reactions of peroxy acids.

(*i*) **Epoxidation of alkenes and cycloalkenes:** Alkenes are epoxidised with peroxy acids (eq. 49)[106,107]

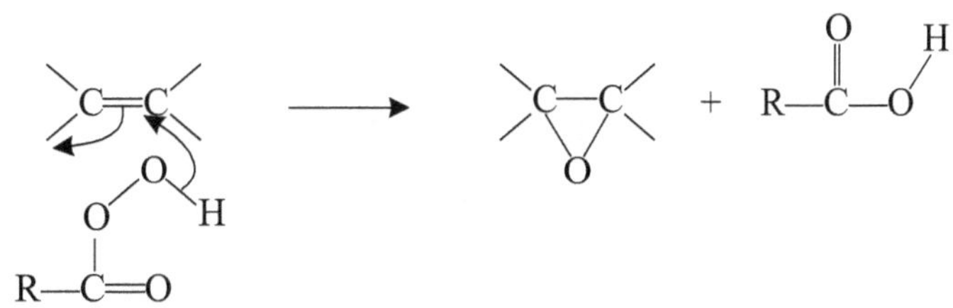

(eq. 49)

It is found that epoxidation takes place preferentially at electron rich (*i.e.*, tetraalkylated) double bond[108]. The reaction is stereospecific. Thus *cis*- and *trans*-alkenes give *cis* and *trans* epoxides respectively[108]. 1,2-Dimethyl cyclopentene is oxidised with perbenzoic acid to give 1,2-dimethyl cyclopentene oxide in 85% yield.[109]

1,2-dimethylcyclohexa-1, 4-diene on reaction with *m*-chloroperoxybenzoic acid gives the 1,2-oxide (eq. 50)[110]. In this case only the electron-rich tetrasubstituted double bond is epoxidised.

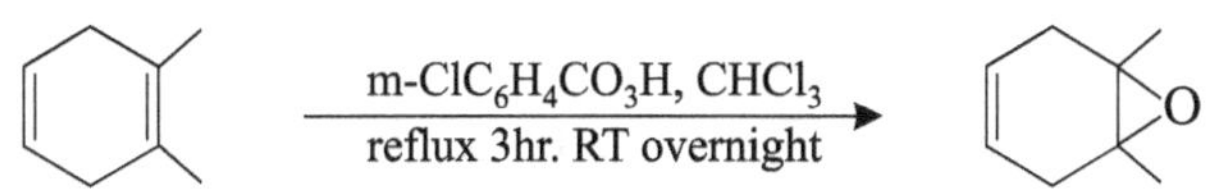

1,2-Dimethylcyclohexa- -1,4 diene   1,2-oxide   (eq. 50)

Cyclohexene is epoxidised with m-chloroperbenzoic acid in KF/CH_2Cl_2 at RT, overnight to the corresponding epoxide in > 95% yield.[111]

(*ii*) **Epoxidation of unsaturated ethers:** Unsaturated ethers are epoxysed at the double bond. Thus, vinyl ethers and peroxy acids give the epoxide, which can be isolated[112] or can react further with the reaction medium[113] (equations 51 and 52)

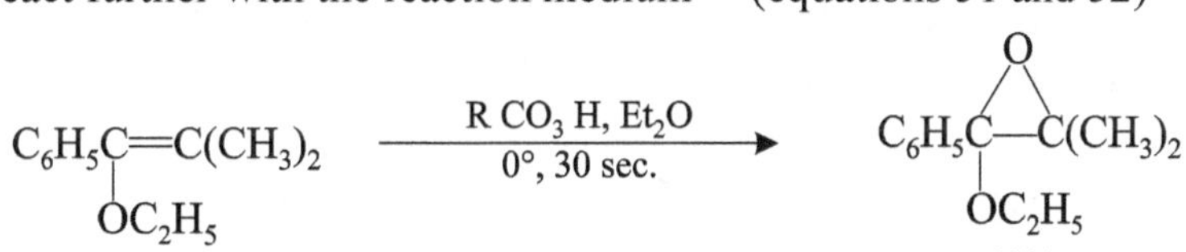

90%   (eq. 51)

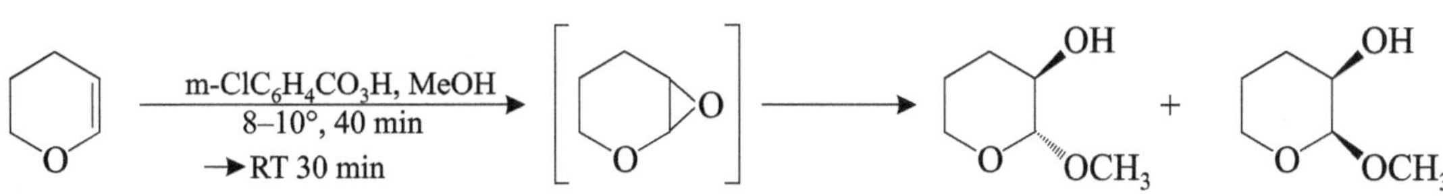

67%   8%

(eq. 52)

(*iii*) **Epoxidation of unsaturated ketones:** Unsaturated ketones can be epoxidised with peroxy acids (eq. 53)[114,115] though a Baeyer-villiger reaction is possible.

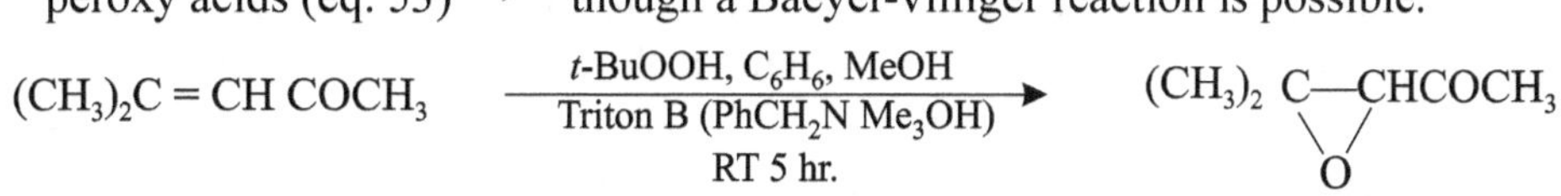

56%   (eq. 53)

2-cyclohexen-1-one on reaction with *tert*-butyl hydroperoxide gives 2,3-cyclohexene-1-one oxide[116] (eq. 54)

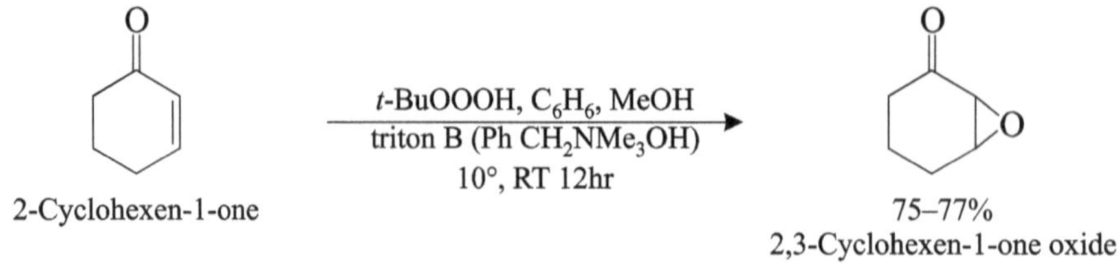

(eq. 54)

α-Ionone can be epoxidised to give 3,4-epoxy-α-ionone with peroxybenzoic acid or peroxyphthalic acid in 96.5 and 66% yields respectively[114] (eq. 55)

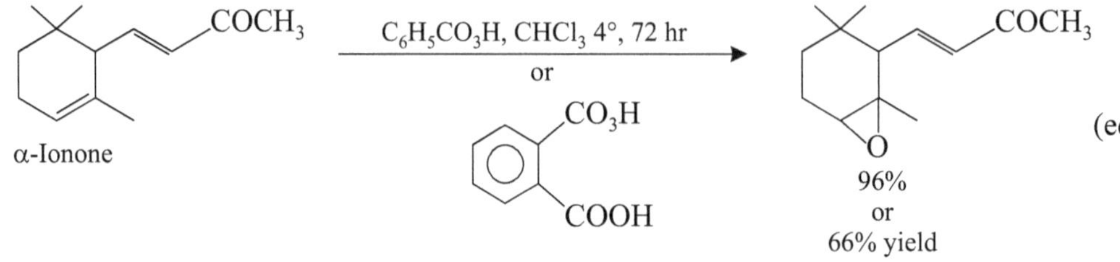

(eq. 55)

However, epoxidation with alkaline H_2O_2 gives α,β-epoxide (eq. 56)[115]

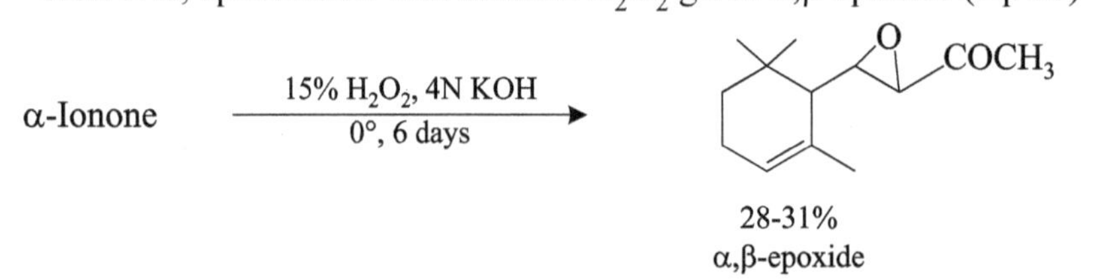

(eq. 56)

(*iv*) **Hydroxylation of alkenes:** Alkenes on reaction with peroxyacids give the corresponding epoxides, which are hydrolytically cleaved by back-side nucleophilic attack (inversion) to give anti hydroxylation products (eq. 57)

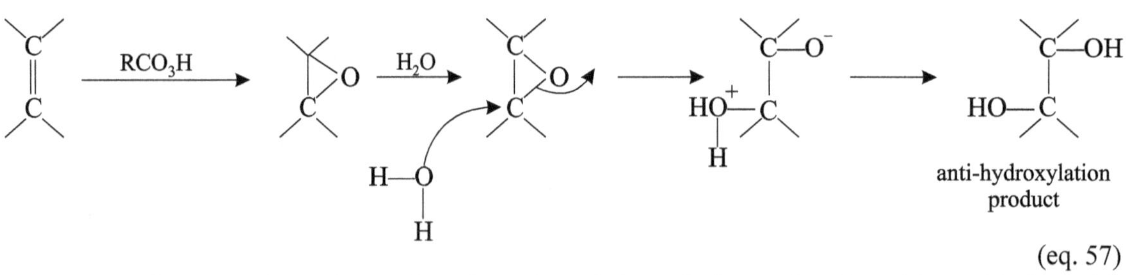

(eq. 57)

It is appropriate to state that cis hydroxylation of alkenes is effected by oxidation with osmium tetroxide or potassium permanganate.

(v) **Oxidation of acetals to esters:** Acetals of aldehydes, *e.g.*, butyraldehyde diethyl acetal on treatment with peroxy acids and H_2SO_4 as catalyst give esters[117], *e.g.* ethyl butyrate in 69% yield after 11 hr at 40°. Unsaturated aldehyde acetals are transformed into either unsaturated esters or epoxy esters.

(*vi*) **Oxidation of aromatic aldehydes:** Aromatic aldehydes on oxidation with peroxyacetic such as *m*-chloroperoxybenzoic acid followed by hydrolysis of the formed formyl ester give phenols (eq. 58)[111,118]

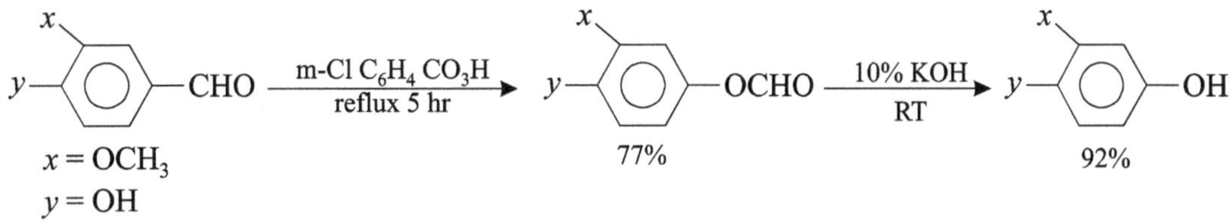

(eq. 58)

(*vii*) **Oxidation of ketones to esters or lactones:** As in the case of oxidation of ketones with H_2O_2 (see section 5.4. subsection IX), peroxy acids also oxidise ketones to esters or lactones (eq. 59)[119,120]

$$C_6H_5\,CO\,C_6H_5 \xrightarrow[\text{RT}\cdot 0.5\text{ hr}]{\text{R }CO_3H,\ HOAc,\ H_2SO_4} C_6H_5\,COOC_6H_5\ (92\%)$$

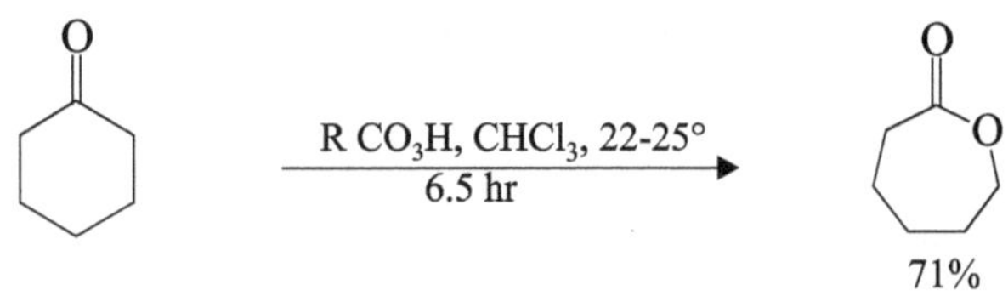

(eq. 59)

(*viii*) **Oxidation of primary amines to azoxy compounds:** Primary amines are oxidised to azoxy compounds with peroxy acids (eq. 60)[121]

$$C_6H_5NH_2 \xrightarrow[40°,\ \text{RT, overnight}]{45\%\ RCO_3H,\ H_2O} C_6H_5N(\rightarrow O)=NC_6H_5$$

azoxy compound
60%
+ 11% $C_6H_5NO_2$

(eq. 60)

15.8 DIMETHYL CARBONATE

Dimethyl carbonate, a selective methylating agent is useful for the methylation of active methylene compounds[122]. As an example, aryl acetonitrile on reaction with dimethyl carbonate in presence of K_2CO_3 gives the corresponding methylated product (Fig. 15.1)

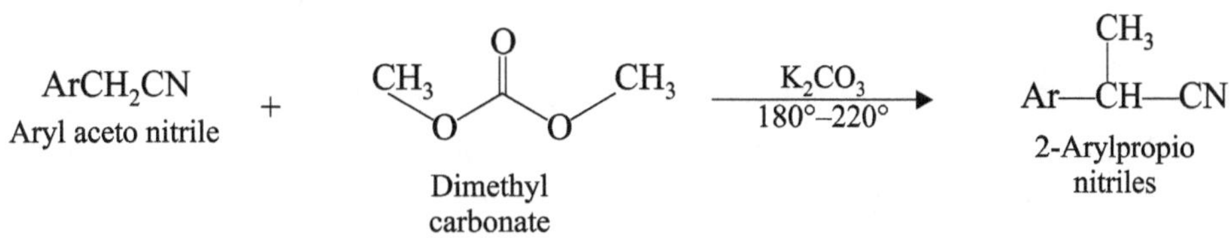

Fig. 15.1: Methylation of active methylene nitrile compounds.

It is very convenient to use dimethyl carbonate instead of methyl halides or dimethyl sulfate, which are toxic.

Dimethyl carbonate on reaction with aniline (aliphatic and aromatic amines) form urethane which on pyrolysis gives isocyanates (Fig. 15.2)[123]

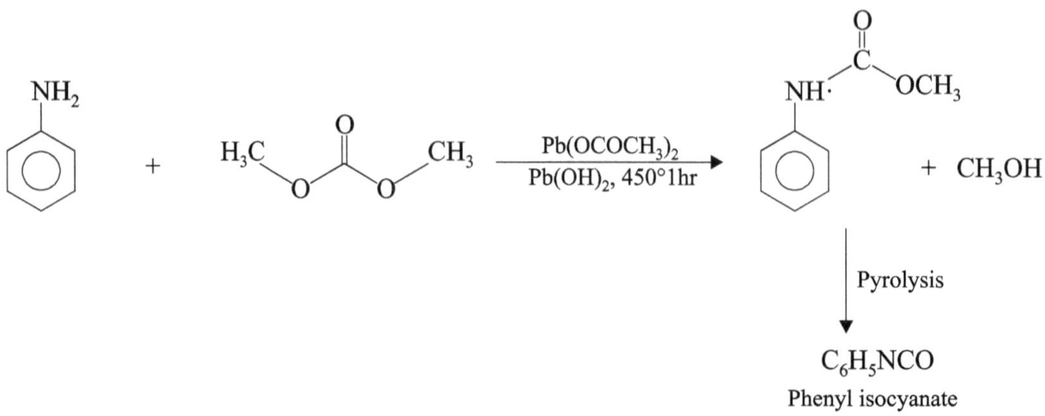

Fig. 15.2: Synthesis of isocyanates.

The above method can also be used for the preparation of methyl isocyanate. In fact, due to tradegy at Bhopal, India, it is not necessary to store huge quantities of MeNCO; it can be prepared whenever needed.

A method for the preparation of methyl isocyanate has been developed[124] by du Pont CO (Fig. 15.3)

$$CH_3NH_2 + CO \longrightarrow CH_3NHCHO \xrightarrow[>240\%]{O_2} \underset{84\text{–}89 \text{ yield}}{CH_3NCO}$$

Fig. 15.3: Du Pont CO method for the preparation of methyl isocyanate.

Dimethyl carbonate has also been prepared from urea and methanol in triglyme using the technique of reactive distillation[125] (Fig. 15.4)

$$CH_3OH + NH_2CONH_2 \xrightarrow[\text{heat}]{\text{triglyme}} [NH_3COOCH_3] + NH_3$$

$$\downarrow CH_3OH$$

$$\underset{\substack{\text{dimethyl carbonate} \\ 98.3\ \%\ \text{conversion}}}{CH_3OCOOCH_3} + NH_3$$

Fig. 15.4: Preparation of dimethyl carbonate from urea and methanol.

Diphenyl carbonate is preferred over dimethyl carbonate for the production of polycarbonates. It is made[126] from dimethyl carbonate by reaction with phenol (Fig. 15.5)

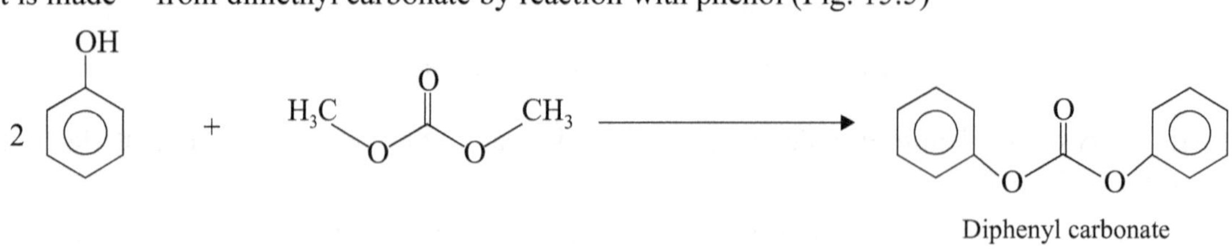

Fig. 15.5: Synthesis of diphenyl carbonate.

See also Chapter 16, section 16.3 sub section (*i*).

15.9 POLYMER SUPPORTED REAGENTS

Normally in an organic reaction, a substrate is reacted with a reagent to give the product, which is isolated from the reaction mixture by procedures like extraction, distillation or chromatography steps. In this way the isolation of the product involves one or more steps. However, using polymer supported reagents the isolation of the product involves only filtration and the supported organic reagent is obtained as a solid product which can be reused.

The procedure involves the linking of the polymeric material to the reagent to form polymer reagent [(P)–R]. Next steps involves addition of the substrate (A). On completion of the reaction, the spent polymer (byproduct polymer, (P-B along with the unused polymer reagent [(P)–R] is removed from the mixture by filtration and the formed product isolated from the filtrate (Fig. 15.6).

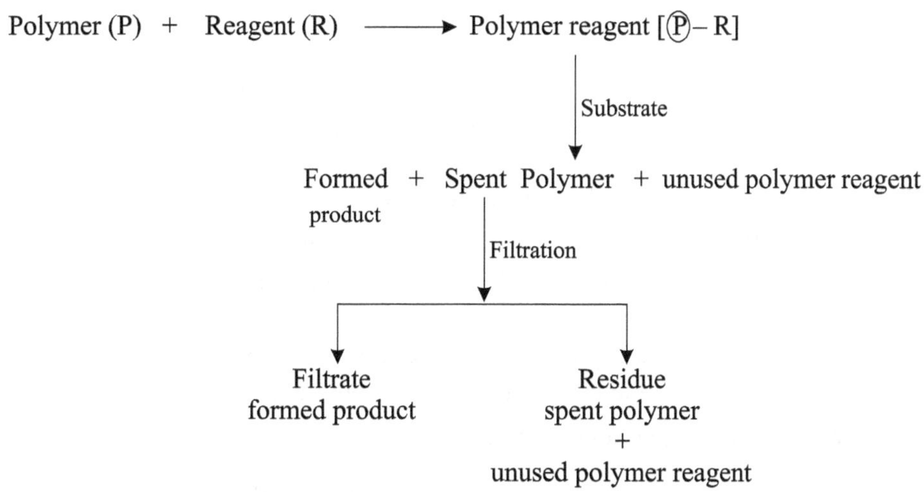

Fig. 15.6: Steps involved in organic synthesis using polymer supported reagent.

Following are given some of the common polymer supported reagents and their uses.

15.9.1 Polymeric-N-Bromosuccinimide (P-NBS)

It is prepared by the addition of bromine to a suspension of polymaleimide polymer in aqueous NaOH solution (Fig. 15.7)[127]. The polymaleimide is obtained by free radical polymerisation of maleimide in presence of 2.5-5% divinyl benzene.

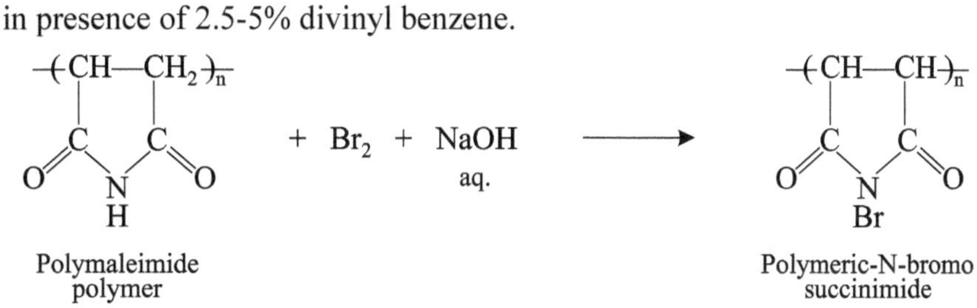

Fig. 15.7: Preparation of P-NBS.

P-NBS has been used for the allylic bromination of cyclohexene. Also cumene on bromination with P-NBS in CCl_4 gave α, β, β-tribromo cumene as the main product. However, bromination with NBS gives α, β-dibromocumene and α-bromocumene (Fig. 15.8)[127,128]

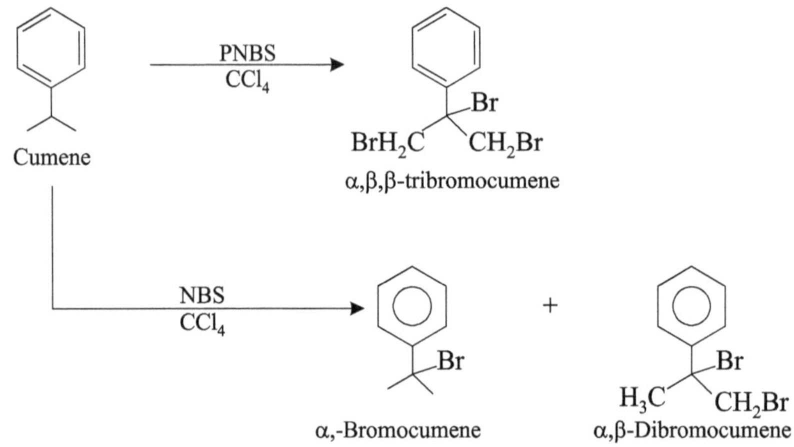

Fig. 15.8: Bromination of cumene with PNBS and NBS

15.9.2 Polymeric Organotin Dihydride Reagent

Polymeric organotin dihydride reagent a well known reducing agent has been used[129] for the conversion of aldehydes and ketones to alcohols (80-90% yield) and for the reduction of halides to hydrocarbons in presence of other functional groups. The reduction by polymeric organotin dihydride reagent is superior to that by Li AlH_4. The reagent can also be used for selective reduction of one functional group of a symmetrical difunctional aldehyde (tere-phthaldehyde)

The reagent, polymeric organotin dihydride reagent in prepared as follow (Fig. 15.9)[129]

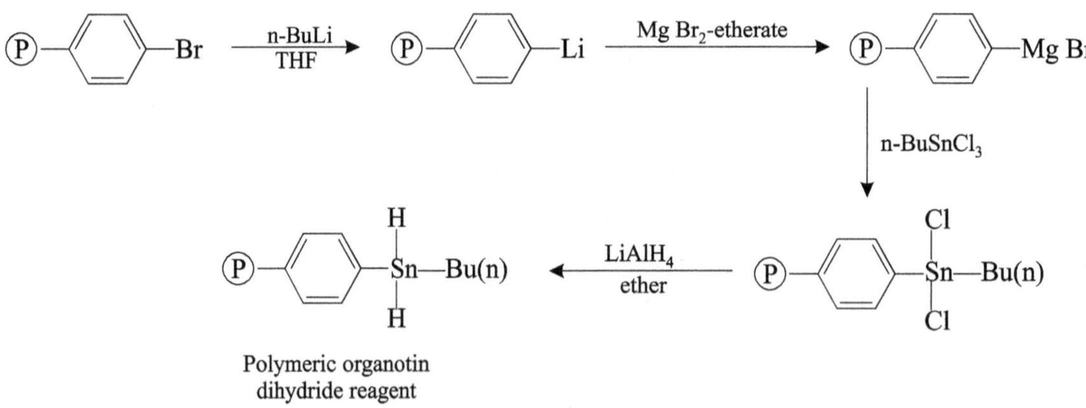

Fig. 15.9: Synthesis of Polymeric organotin dihydride reagent.

15.9.3 Poly (4-vinylpyridine) borane

The polymeric 4-vinylpyridine borane, also a polymeric hydride is used for the reduction of carbonyl compounds. It is synthesised by the reaction of poly (4-vinylpyridine) with $NaBH_4$ (Fig. 15.10)[130]

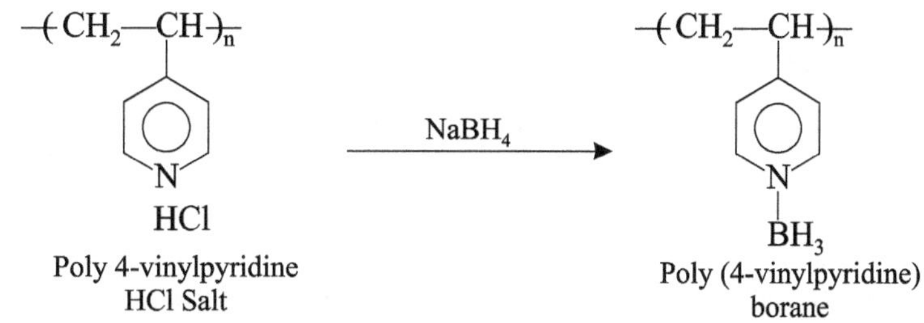

Fig. 15.10: Synthesis of poly (4-vinylpyridine) borane.

15.9.4 Polymeric Carbodiimide

A coupling agent, polymeric carbodiimide is used in the synthesis[131] of anhydrides from carboxylic acids. Thus, anhydrides from acetic acid, stearic acid and glutaric acid can be prepared using polymeric carbodiimides in benzene-ether (2 : 1). The formed anhydrides are obtained in quantitative yield by simple filtration and evaporation of the filtrate.

The **Moffatt oxidation** of a number of alcohols to aldehydes and ketone can be carried out using polymeric carbodiimide[131]. Using this procedure even labile prostaglandin intermediate (A) is easily converted to the desired aldehyde (Fig. 15.11).

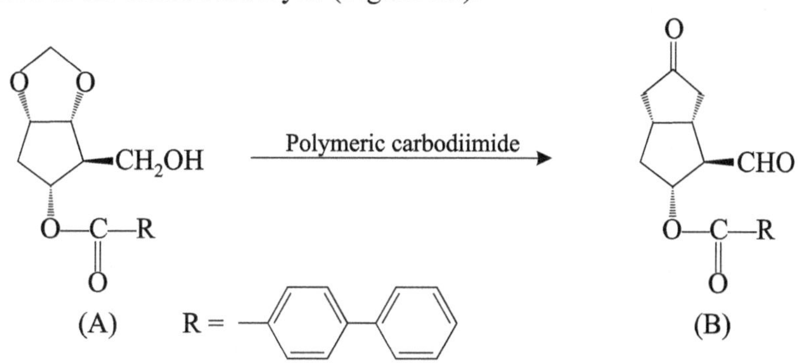

Fig. 15.11: Moffatt oxidation.

Using this method, the product obtained are free from contamination of ureas, a major purification problem in the usual synthetic method.

The required polymetric carbodiimide is obtained[131] as given below (Fig. 15.12).

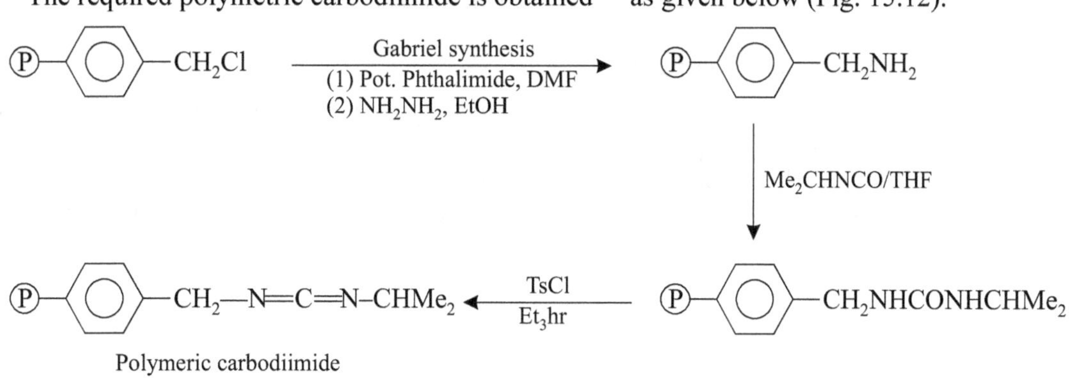

Fig. 15.12: Synthesis of polymeric carbodiimide.

15.9.5 Polymeric Benzene Sulfonyl Chloride

The polymeric benzene sulfonyl chloride has been used as a coupling reagent[132] for the synthesis of peptides via the polymeric mixed carboxylic sulfonic anhydrides. A hindered polymeric arylsulfonyl chloride(A) was prepared[133] by copolymerisation of 3,5-diethylvinylbenzene, styrene and DVB, followed by chlorination. This reagent was used as a coupling reagent in the synthesis of oligonucleotides[133]. By using this reagent, the usual side reaction of sulfonation of alcoholic group and difficulty in complete removal of sulfonic acid byproduct is avoided.

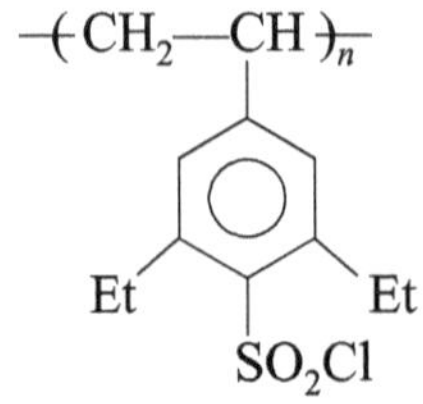

(A)

15.9.6 Polymer Supported Peptide Coupling Reagent EEDQ

A polymeric peptide coupling reagent based on 2-ethoxy-1-ethoxycarbonyl-1,2-dihydroquinoline (EEDQ) was prepared as follows (Fig. 15.13)[134,135]

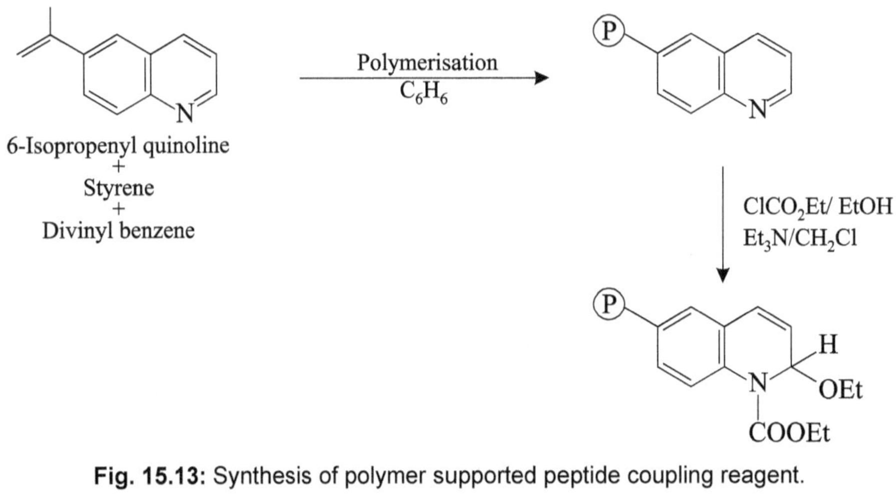

Fig. 15.13: Synthesis of polymer supported peptide coupling reagent.

15.9.7 Polymer Supported Peracid

Polymer supported peracid (like the ordinary peracid) is useful for the epoxidation of olefinic compounds (Fig. 15.14). The epoxides are obtained in good yields[136,137] (80-95%). Even monosubstituted olefins, which are known to be comparatively unreactive towards organic peracids, can be epoxidised. The oxidation property of polymeric peracid is quite comparable to that of commercial 85% *m*-chloroperbenzoic acid.

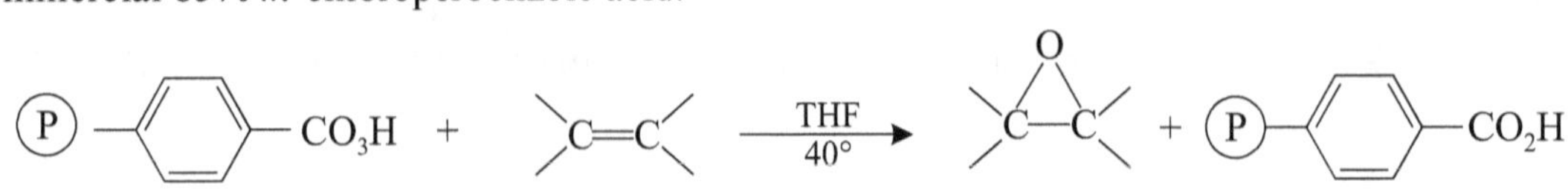

Fig. 15.14: Epoxidation of alkenes using polymeric peracids.

Polymer supported peracid is prepared from polymer supported carboxylic acid[138], polymer supported acid chloride[138], polymeric carboxylic acid chloride or from polystyrene aldehyde[139] (Fig. 15.15)

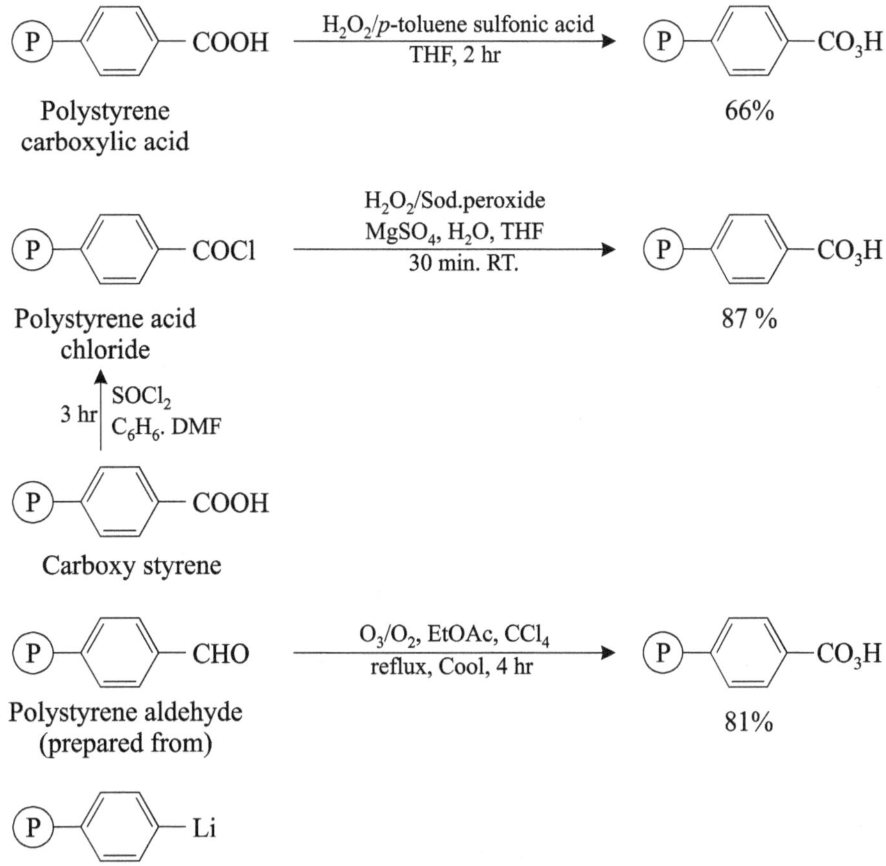

Fig. 15.15: Synthesis of polymeric peracid.

15.9.8 Polymer Supported Chromic Acid

The polymer-supported chromic acid (Amberlyst A-26, H CrO_4^- form) is commercially available. It has been used for the oxidation of primary and secondary alcohols to carbonyl compounds[140] and for the oxidation or allylic and benzylic halides to aldehydes and ketones[141].

Besides polymer-supported chromic acid, other chromium-supported reagents are also available. These include (*i*) Chemisorbed chromyl chloride (CrO_2Cl_2 on silica-alumina). It has been used for the oxidation of primary and secondary alcohols to carbonyl compounds in 75-100 percent yield[142]; (*ii*) An insersion complex of chromium trioxide and graphite. It oxidises[143] primary alcohols to aldehydes. However, secondary alcohols and uneffected. It has been suggested that the oxidizing agent consist of a surface deposit of Cr_3O_8[144]. The complex is commercially available; (*iii*) Polyvinyl pyridinium chlorochromate has oxidizing properties similar to polymer-supported chromic acid[140]. It is prepared by the reaction of cross-linked polyvinylpyridine with chromic acid is acid solution[145].

15.9.9 S-chlorosulfonium Chloride Resin

It is a polymer thioanisoyl resin and is prepared by the reaction of polystyrene methyl sulfide with chlorine in presence of triethylamine (Fig. 15.16)[146,139]

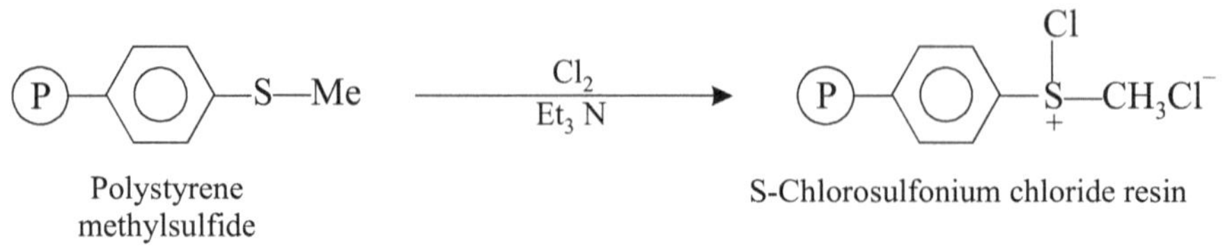

Fig. 15.16: Preparation of S-chlorosulfonium chloride resin.

The chlorinated thioanisole polymer acts as a selective oxidant for alcohols[147]. It is used for the oxidation of α, ω-diols to give monoaldehydes in higher yields (50.2%) compared to the formation of dialdehydes (2.2%) (Fig. 15.17)

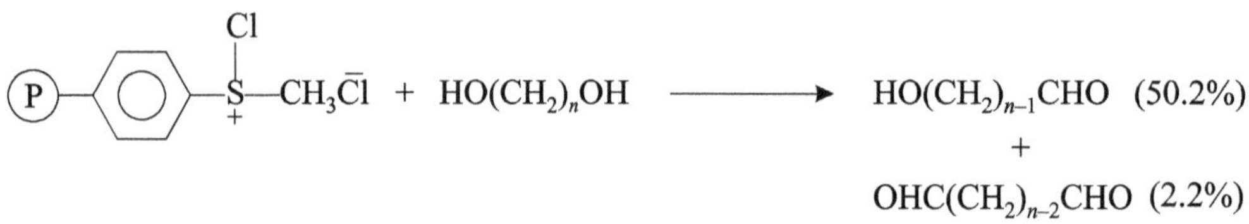

Fig. 15.17: Oxidation of α, ω-diols with s-chlorosulfonium chloride resin

A polymeric resin, polymeric thioanisoyl resin lithium salt has been used for an homologation reaction. Thus, the Li salt of polymeric thioanisoyl resin on reaction with iodooctane at –70° and cleavage of the higher homologue carried out via quaternization with CH_3I followed by cleavage with NaI (Fig. 15.18)[148] gave the higher homologue of iodooctane.

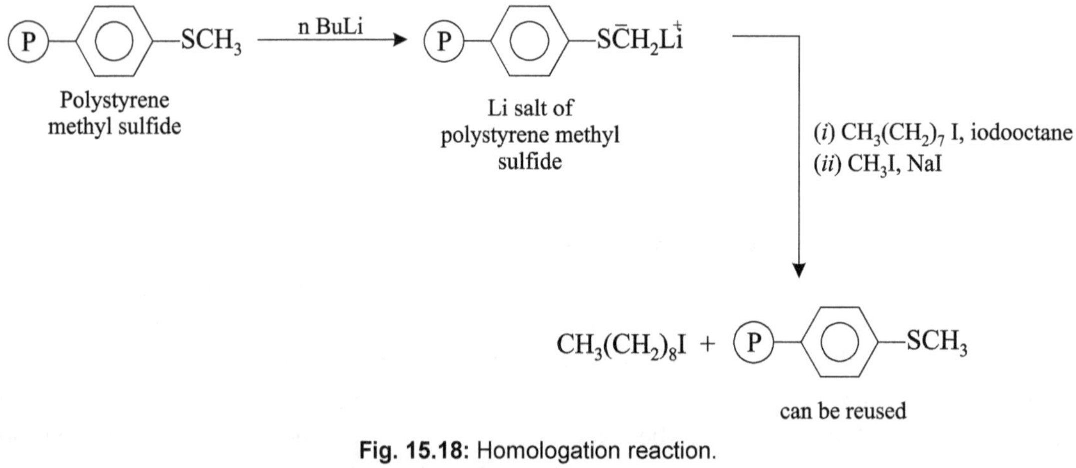

Fig. 15.18: Homologation reaction.

The above homologation reaction (Fig. 15.18) was an extension of another reaction[149] originally developed by using low molecular-weight phenyl thiomethyl lithium.

15.9.10 Polymeric Wittig Reagent

Polymeric Wittig reagent for the well known Wittig Reaction is synthesised as given below (Scheme 15.19)[150].

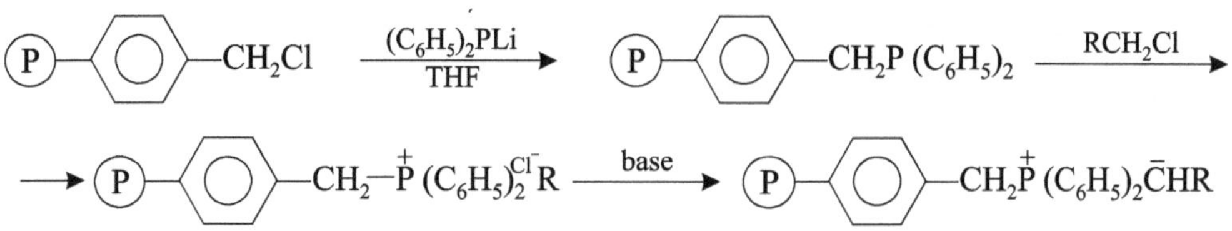

Fig. 15.19: Synthesis of Polymeric Wittig Reagent.

15.9.11 Polymeric Tosyl Azide

Polymeric tosyl azide has been prepared[151] (Fig. 15.20) and used to convert β-diketones into diazoketones. The reagent, unlike its low molecular weight analog is not explosive.

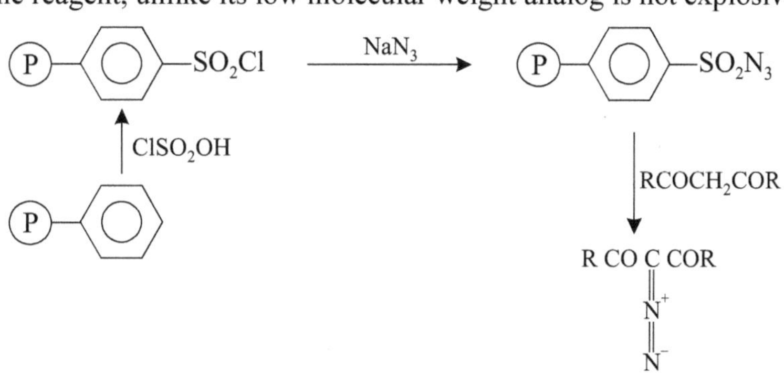

Fig. 15.20: Synthesis of Polymeric tosyl azide and its use as diazo transfer reaction.

15.9.12 Polymeric Nucleophiles

These are anion-exchange resins, IRA-400 cyanide form or resin bound phenoxide. These resins are used as nucleophiles. For example the reaction of benzyl bromide with an ion-exchange resin, IRA-400 (cyanide form) gives benzyl cyanide (Fig. 15.21)[152]. The reaction avoids the isolation of product from impurities in case inorganic cyanides are used.

$$C_6H_5CH_2Br + \text{Ⓟ}\!-\!CH_2\overset{+}{N}Me_3\,CN^- \longrightarrow C_6H_5CH_2CN + \text{Ⓟ}\!-\!CH_2\overset{+}{N}Me_3\,Br^-$$

1 RA 400
(Cyanide form)

Fig. 15.21: Conversion of benzyl bromide into benzyl cyanide.

The procedure is very simple and involves passing of benzyl bromide through a column packed with IRA-400 (cyanide form).

In a similar way, benzyl ethers are prepared by the reaction of benzyl bromide with resin bound phenoxide as the nucleophile (Fig. 15.22)[153].

$$C_6H_5CH_2Br + \text{Ⓟ}\!-\!CH_2\overset{+}{N}Me_3\overline{O}C_6H_5 \longrightarrow C_6H_5CH_2OC_6H_5 + \text{Ⓟ}\!-\!CH_2\overset{+}{N}Me_3\,Br^-$$

Benzylphenyl
ether

Fig. 15.22: Conversion of benzyl bromide into benzylphenyl ether.

15.9.13 Polymer Supported Trisubstituted Phosphine Dichlorides

The reagent polymer-supported trisubstituted Phosphine dichloride is useful for the following conversions[154]

$$C_6H_5CH_2COOH \longrightarrow C_6H_5CH_2COCl$$

$$CH_3C_6H_4COOH(p) \longrightarrow CH_3C_6H_4COCl$$

$$HOOC-C_6H_4-COOH(p) \longrightarrow ClOCC_6H_5COCl$$

$$C_6H_5\,CONHC_6H_5 \longrightarrow C_6H_5\,\underset{\displaystyle Cl}{\underset{|}{C}} = N-C_6H_5$$

$$C_6H_5CO\,NH_2 \longrightarrow C_6H_5\;C \equiv N$$

$$C_6H_5CH_2OH \longrightarrow C_6H_5CH_2Cl$$

$$C_6H_5COCH_3 \longrightarrow C_6H_5\,\underset{\displaystyle Cl}{\underset{|}{C}} = CH_2$$

The reagent is prepared as follows (Fig. 15.23)[154].

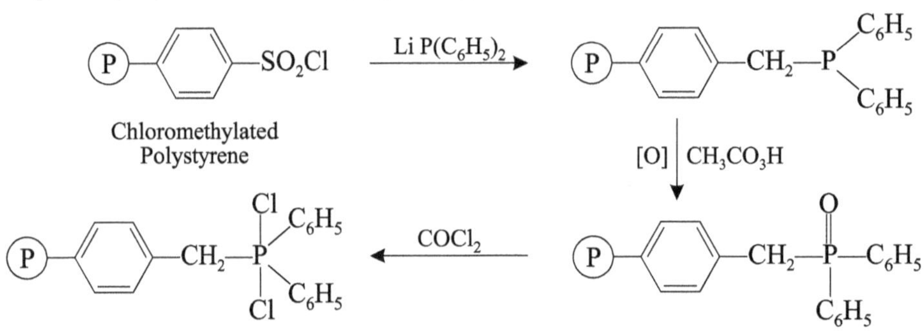

Fig. 15.23: Synthesis of Polymer supported tri substituted phosphine dichloride.

15.10 N-HALO REAGENTS

A number of N-halo reagents have been commonly used in organic transformations[155]. These reagents include N-haloimides, N-haloimides, N-halo sulforamides, N-haloamides and N-halosulfonimides and find use in promoting important reaction like halogenation, oxidation and protection and also for the formation of C–X, C–O, and C=O bonds. However, all reagents are not green reagents. Only one reagent has be used as green reagent, which is polymeric-N-bromosuccinimide (P-NBS) (see section 15.9.1). In case other N-halo reagents could be developed in supported reagents, these could be used as green reagents.

15.11 UREA-HYDROGENPEROXIDE COMPLEX

Urea-hydrogen peroxide complex (UHP) is commercially available and can be easily prepared from urea and hydrogen peroxide[156].

$$NH_2CO\,NH_2 + H_2O_2 \longrightarrow \begin{array}{c} NH_2CONH \\ | \\ HOOH \\ (UHP) \end{array}$$

This complex has been used for a number of environmentally friendly synthesis[156], some of the transformations include:

Oxidation of hydroxy aldehydes to hydroxy phenols (CHO→OH) is solid state.

Oxidation of sulfides to sulfoxides and sulfones.

Oxidation of nitriles to amides.

Oxidation of nitrogen heterocycles to the corresponding N-oxides.

Conclusion

Using green reagents are the best choice in the context of green chemistry. Specially polymer supported reagents are particularly useful. Using these reagents, the product can be obtained by simple filtration of the reaction mixture and the reagent could be reused.

REFERENCES

1. J. Thsuji and H. Takayangi, Org. Syn. Collective Volume, 1988, **6**, 662.
2. H.O. House, Org. Synth. Collective Volume, 1963, **4**, 367.
3. R.L. Heider, U.S. Patent 2,554,459,(1951), CA, 1951, **45**, 9076e.
4. M.Y. Sheikh and G. Eadon, Tetrahedron Lett., 1972, 257.
5. K. Heyns and L. Blazejewicz, Tetrahedron, 1969, **9**, 67.
6. R.P.A. Sneeden and R.B. Turner, J. Am. Chem. Soc., 1955, **77**, 190.
7. J.S. Belew, C. Garza and J.W. Mathieson, Chem. Commun., 1970, 634.
8. C. Moureu and G. Mignonac, C.R. Hebd. Seances Acud. Sci., 1920, 170, 258.
9. R.J. Harrisson and M. Moyle, Org. Synth. Collective Volume, 1963, **4**, 493.
10. K. Heyns and O. Stöckel, Justus Liebigs Ann. Chem., 1947, **558**, 192.
11. R. Heyns and O. Stöckel, Justus Liebigs Ann. Chem., 1947, **558**, 187.
12. K. Heyns and R. Heinemann, Justus Liebigs Ann. Chem., 1947, **558**, 187.
13. M. Hönig and W. Ruziczk, Chem. Ber., 1930, **63**, 1648.
14. F.F. Rust and W.E. Vaugian, Ind. Eng. Chem., 1949, **41**, 2594.
15. M.M. Midland and H.C. Broun, J. Am. Chem. Soc., 1973, **95**, 4069.
16. T.J. Wallace and A. Schrieshein, Tetrahedron Lett., 1963, 1131.
17. K.-T. Liu and Y.-C. Tong, Synthesis, 1978, 669.
18. W.S. Emerson and G.F. Deebel, Org. Synth. Collective Volume, 1963, **4**, 579.
19. T.J. Wallace, H. Pobiner and A. Schriesheim, J. Org. Chem., 1965, **30**, 3768.
20. K. Heyns and L. Blazejewicz, Tetrahedron, 1960, **9**, 67.

21. C.S. Foote, Ace. Chem. Res., 1968, **1**, 104.
22. C.S. Foote, S. Wexler, W. Ando and R. Higgis, J. Am. Chem. Soc., 1968, **90**, 975.
23. C.S. Foote, S. Wexler and W. Ando, Tetrahedron Lett., 1965, 4111.
24. C.S. Foote and S. Wexler, J. Am. Chem. Soc., 1964, **86**, 3879.
25. E. Mckeown and W.A. Waters, J. Chem. Soc. B, 1966, 1047.
26. H.H. Wasserman and J.R. Scheffer, J. Am. Chem. Soc., 1967, **89**, 3073.
27. C. Moureu, C. Dufraisie and P.M. Dean, C.R. Hebd. Seances Acid. Sci., 1926, **182**, 1584.
28. R.W. Murray and M.L. Kaplan, J. Am. Chem. Soc., 1969, **91**, 5358.
29. N.J. Turro, Y. Ito, M.-F. Adam, O. Rodriquez and F. Yany, J. Am. Chem. Soc., 1977, **99**, 5836.
30. F. Kaulen and H.J. Schäfer, Synthesis, 1977, 516.
31. C. Kaneko, A. Sugimoto and S. Tanaka, Synthesis, 1974, 876.
32. K. Kondo and M. Matsumoto, Chem. Commun., 1972, 1332.
33. P.D. Barlett and A.P. Schaap, J. Am. Chem. Soc., 1970, **92**, 3223.
34. A.G. Schultz and R.H. Schlessinger, Tetrahedron Lett., 1970, 2731.
35. P.D. Barlett and A.P. Schaap, J. Am. Chem. Soc.,1970, **92**, 3223.
36. G. Rio and J. Berthelot, Bull. Soc. Chim. Fr., 1971, 3555.
37. A. Posetti and D. Sainsesi, Gazz. Chim. Ital., 1968, **98**, 265.
38. G.O. Schenck, H.-D. Becker, K.-H. Schulte-Elte and C.H. Kraunch, Chem. Ber., 1963, **96**, 569.
39. H. Hock and S. Lang, Chem. Ber., 1943, **76**, 169.
40. H. Hock and S. Lang, Chem. Ber., 1944, **77**, 257
41. H. Hock and S. Lang, Chem. Ber., 1942, **75**, 1051.
42. M.S. Kharasch and J.G. Burt, J. Org. Chem., 1951, **16**, 150.
43. H.B. Knight and D. Swern, Org. Synth. Collective Volume, 1963, **4**, 895.
44. H. Hock and S. Lang, Chem. Ber., 1943, **76**, 1130.
45. P.W. Jorissen and P.A.A. Van Der Beek, Rec. Trav. Chim. Pays-Bas, 1926, **45**, 245.
46. R. Criege, G. Blust and H. Zinke, Chem. Ber., 1954, **87**, 766.
47. R.W. Murray, R.D. Youssefyeh and P.R. Storey, J. Am. Chem. Soc., 1967, **89**, 2429.
48. F.L. Greenwood, J. Org. Chem., 1955, **20**, 803.
49. R.W. White, S.W. King and J.L. O'Brian, Tetrahedron Lett., 1971, 3587.
50. W.S. Knowles and Q.E. Thompson, J. Org. Chem., 1960, **25**, 1031.
51. W.J. Gensler and H.N. Schlein, J. Am. Chem. Soc., 1955, **77**, 4846.
52. P.S. Bailey, Ind. Eng. Chem., 1958, **50**, 993.
53. F. Asinger, Chem. Ber., 1942, **75**, 656.
54. D.L. Heywood and B. Phillips, J. Org. Chem., 1960, **25**, 1991.
55. P. Deslongchamps and C. Moreau, Can. J. Chem., 1971, **49**, 2465.
56. C.R. Dick and R.F. Hanna, J. Org. Chem., 1964, **29**, 1218.
57. A.S. Hay, J.W. Eustane and H.S. Blanchard, J. Org. Chem., 1960, **25**, 616.
58. E.J. Corey, B. Samuelsson and F.A. Luzzio, J. Am. Chem. Soc., 1984, **106**, 3682.
59. F.G. Fischer, Justus Liebigs Ann. Chem., 1929, **476**, 233.

60. P. Deslongchamps and C. Moreau, Can. J. Chem., 1971, **49**, 2463.
61. P.E. Pike, P.G. Marsh, R.E. Erickson and W.L. Waters, Tetrahedron Lett., 1970, 2679.
62. O.E. Thompson, J. Am. Chem. Soc., 1961, **83**, 845.
63. P.G. Cookson, A.G. Davils and N. Fazal, J. Organomet. Chem., 1975, **99**, C31.
64. C.W. Smith and R.T. Holm, J. Org. Chem., 1957, **22**, 746.
65. M. Roussel and H. Mimoun, J. Org. Chem., 1980, **45**, 5387.
66. H.D. Dakin, Org. Synth. Collective Volume, 1932, **1**, 149.
67. G.W. Kobalka and H.C. Hedgecock, Jr., J. Org. Chem.,1975, **40**, 1776, H.C. Broun and S.K. Gupta, J. Am. Chem. Soc., 1975, **97**, 5249.
68. L.S. Silbert, E. Siegel and D. Swern, Org. Synth. Collective Volume, 1973, **5**, 904.
69. J.F. Leffler, J. Org. Chem., 1951, **16**, 1785.
70. J.A. King and F.H. McMillan, J. Am. Chem. Soc.,1946, **68**, 525.
71. M.V. Bhatt, G.V. Rao and K.S.R. Rao, Chem. Commun, 1971, 822.
72. P.S. Starcher and B. Phillips J. Am. Chem. Soc., 1958, **80**, 4079.
73. S.L. Friess, J. Am. Chem. Soc., 1949, **71**, 2571.
74. C.C. Price and G.W. Stacy, Org. Synth. Collective Volume, 1955, **3**, 86.
75. F. Korte and K.H. Löbmer, Chem. Ber., 1957, **90**, 1290.
76. L. Baver and L.A. Gardella, Jr., J. Org. Chem., 1961, **26**, 82.
77. J. Cech, Collect. Czech. Chem. Commun., 1949, **14**, 558.
78. C.D. Hurd, J.W. Garrett and E.N. Osborne, J.Am. Chem. Soc., 1933, 55, 1082.
79. E.C. Taylor and A. McKillop, J. Org. Chem. 1965, **30**, 3153.
80. W.D. Emmons, J.Am. Chem. Soc., 1954, **76**, 3468.
81. M. Carmack and M.A. Spielman, Org. React (N.y.) 1946, **3**, 83.
82. W.D. Emmons, J. Am. Chem. Soc., 1957, **79**, 5528.
83. W.R. Dunstan and E. Goulding, J. Chem. Soc., 1899, **75**, 1004.
84. S.M. Roberts and H. Suchitzky, J. Chem. Soc. C., 1969, 1485.
85. A.V. Sviridova, V.I. Laba and E.N. Prilezhaeva, Zh. Org. Khim., 1971, **7**, 2487 (Engl. Trans 2577].
86. Y. Watanbe, T. Numata and S. Oac, Synthesis, 1981, 204.
87. D.S. Tarbell and C. Weaver, J. Am. Chem. Soc., 1941, **63**, 2939.
88. H. Ross and R. Hüttel, Chem. Ber., 1956, **89**, 2641.
89. N.A. Milas and O.L. Magelis J. Am. Chem. Soc., 1952, **74**, 1471.
90. G.B. Payne, J. Am. Chem. Soc., 1958, **80**, 6461.
91. G.B. Payme and P.H. Williams, J. Org. Chem., 1961, **26**, 651.
92. G.B. Payne, Tetrahedron, 1962, **18**, 763.
93. P.S. Bailey and R.E. Erickson, Org. Synth. Collective Volume 1973, **5**, 489, 493.
94. A. Pasetti and D. Sianesi, Gazz. Chim. Ital., 1968, **98**, 265.
95. G.B. Payne, P.H. Deming and P.H. Williams, J. Org. Chem. 1961, **26**, 69, 225.
96. G.B. Payne and P.H. Williams, J. Org. Chem.,1959, **24**, 54.
97. G.B. Payne, Tetrahedron, 1962, **18**, 763.

98. W.J. Linn., Org. Synth. Collective Volume, 1973, **5**, 1007.
99. G.B. Payne and P.H. Williams, J. Org. Chem., 1961, **26**, 651.
100. R.W. Murray and R. Jeyaraman, J. Org. Chem., 1985, **50**, 2847.
101. R.W. Murray, R. Jeyaraman and L. Mohan, Tetrahedron Lett., 1986, **27**, 2335.
102. R. Kennedy and A.M. Stock, J. Org. Chem., 1960, **25**, 1901.
103. D. Swern, T.W. Findley and J.T. Secanlan, J. Am. Chem. Soc., 1944, **66**, 1925.
104. B.M. Trost and D.P. Curran, Tetrahedron Lett., 1981, **22**, 1287.
105. R.N. McDonald, R.N. Steppel and J.E. Dorsey, Org. Synth. Collective Volume, 1988, **6**, 276.
106. W. Dittmann and F. Stürzenhofecter, Justus Liebigs Ann. Chem., 1965, **688**, 57.
107. D.J. Recf and H.O. House, Org. Synth. Collective Volume, 1963, **4**, 860.
108. N. Indictor and W.F. Brill, J. Org. Chem., 1965, **30**, 2074.
109. P.D. Barlett and A. Bavley, J. Am. Chem. Soc., 1938, **60**, 2416.
110. L.A. Paquette and J.H. Barrett, Org. Synth. Collective Volume, 1973, **5**, 467.
111. F. Camps, J. Coll, A. Messeguer and M.A. Pericās, Tetrahedron Lett., 1981, **22**, 3895.
112. C.L. Stevens and J. Tazuma, J. Am. Chem. Soc., 1954, **76**, 713.
113. A.A. Frimer, Synthesis, 1977, 578.
114. Y.-R. Naves, O. Schwarzkof and A.D. Lewis, Helv. Chem. Act 1947, **30**, 880.
115. P. Karrer and H. Stürzinger, Helv. Chim. Acta, 1946, **29**, 1829.
116. N.C. Yang and R.A. Finnegan, J. Am. Chem. Soc., 1958, **80**, 5845.
117. D.L. Heywood and B. Phillips, J. Org. Chem., 1960, **25**, 1699.
118. I.M. Godfrey, M.V. Sargent and J.A. Elix, J. Chem. Soc., Perkin Trans, I 1974, 1353.
119. V.W.E. Doering and L. Speers, J. Am. Chem. Soc., 1950, **72**, 5515.
120. S.L. Friess, J. Am. Chem. Soc., 1949, **71**, 2571.
121. F.P. Green Span, Ind. Eng. Chem.,1947, **39**, 847.
122. P. Tudo, F. Trotta and G. Morgilo, J. Org. Chem., 1987, **52**, 1300; P. Tudo, F. Trotta, G. Morgilo and F. Ligontti, Ind. Eng. Chem. Res., 1998, **27**, 1565.
123. Z.-H. Fu and Y. Ono, J. Mol. Catal, 1994, **91**, 399.
124. J.R. Thomen, Chem. Eng. News, Feb. 6, 1995, 2; V.N.M. Rao and G.E. Heinsohn, U.S. Patent 4, 537, 726 (1985).
125. J.Y. Ryer, U.S. 5,902,894 (1999); 6,010,976 (2000).
126. G.E. Harrison, A.J. Dennis and M. Starif, U.S. patient, 5,426,207 (1995).
127. Y. Yanagisawa, M. Akiyama and M. Okawara, J. Polym. Sci. Part A-1 (Chem., 1969, **7**, 1905.
128. Y. Yanagisawa, M. Akiyame and M. Okawara, Kogyo Kagaku Zasshi, 1969, **72**, 1399.
129. N.M. Weinshenker, G.A. Crossby and J.Y. Wong, J. Org. Chem., 1975, **40(13),** 1966.
130. M.L. Hallenleben, J. Polymer Sci., Symposium No. 47, P1.
131. N.M. Weinshenker and C.M. Shen, Tetrahedron Lett., 1972, 3281.
132. M. Fridkin, A. Patchornik and E. Katchalski, Pept. Proc. Eur. Pept. Symp., 10th, 1969, p. 166, North-Holland, Amsterdam (1971).
133. M. Rubenstein and A. Patchornik, Tetrahedron Lett., 1972, 2881; Tetrahedron, 1975, **31**, 1517.

134. J. Broun and R.E. Williams, Can. J. Chem., 1971, **49**, 3765.
135. R.E. Willams, J. Broun and D.R. Lauren, Polym. Prepr., Am. Chem. Soc. Div., Polym. Chem., 1972, **13(2)**, 823.
136. J.M.J. Frechet and K.E. Haque, Macromolecules, 1975, **8**, 130.
137. C.R. Harrison and P. Hodge, J. Chem. Soc. Chem. Commun.,1974, 1009.
138. J.M.J. Frecher and K.E. Haque, Macromolecules, 1975, 8, 130.
139. M.J. Farrall and J.M.J. Frechet, J. Org. Chem., 1976, **41**, 3877.
140. G. Cainelli, G. Cardillo, M. Orena and S. Sandri, J. Am. Chem. Soc., 1976, **98**, 6737.
141. G. Cardillo, M. Orena and S. Sandri, Tetrahedron Lett.,1976, 3985.
142. J. San Filippo Jr., and C.I. Chern, J. Org. Chem., 1977, **42**, 2182.
143. J.M. Lalancette, G. Robbin and P. Dumas, Can. J. Chem., 1972, **50**, 3058.
144. L.B. Ebert, R.A. Huggins and J.I. Brauman, Carbon, 1974, **12**, 199.
145. J.M.I. Frechet, J. Warnock and M. Farrall, J. Org. Chem.,1978, **43**, 2618.
146. S. Tanimoto, J. Horikawa and R. Oda, Kogyo Kagaku Zasshi 1967, **10**, 1269.
147. G.A. Crosby, N.M. Weinshenker and H.S. Uh, J. Am. Chem. Soc., 1975, **97**, 2232.
148. G.A. Crosby and M. Kato, J. Am. Chem. Soc., 1977, **99**, 278.
149. E.J. Corey and M. Jautelat, Tetrahedron Lett., 1968, 5787.
150. R.H. Grubbs. and L.C. Kroll, J. Am. Chem. Soc., 1971, **93**, 3062.
151. W.R. Roush, D. Feilter and J. Rebek, Tetrahedron Lett., 1974, 1391.
152. M. Gordon, M.L. De Pambhilis and C.E. Grifin, J. Org. Chem., 1963, **28**, 698.
153. E.J. Rowe, K.L.Kaufman and C. Piantadosi, J. Org. Chem., 1958, **23**, 1622.
154. H.M. Relles and J. Schluenz, J. Am. Chem. Soc., 1974, **96**, 6469.
155. E. Kolvari, A. Ghorbani-choghamarani, P. Salehi, F. Shirini and M.A. Zolfigol, J. Iran. Chem. Soc., 2007, **4(2),** 126-174 and the references cited therein.
156. R.S. Varma and K.P. Naicker, Organic Letters, 1999, **1(2),** 189.

16. ORGANIC TRANSFORMATIONS USING RENEWABLE STARTING MATERIALS

16

Organic Transformations using Renewable Starting Materials

16.1 INTRODUCTION

Most of the starting materials for Organic Synthesis are derived from non-renewable sources like petroleum, natural gas and coal. All these starting materials are mostly organic chemicals, which end up as carbon dioxide, the main green house gas well known for global warming. It is necessary that for sustainable future these must be derived from renewable sources.

The renewable sources mostly include Biomass and gases like carbon dioxide and methane.

16.2 BIOMASS AS RENEWABLE STARTING MATERIAL

The term Biomass, in general refers to waste materials from fields and forests. More specifically, biomass includes organic matter that is available on renewable basis from crops and trees, agricultural food and food crops, residues, aquatic plants, wood, animal waste and other waste materials[1]. The biomass is converted into a variety of valuable chemicals[2]. Biomass comprise of five categories, *viz*., starch, cellulose, hemicellulose, lignin and oils (Fig. 16.1)[3]. The biomass present in wood, grasses and straw comprise of cellulose, hemicelluloses and lignin. Starch and cellulose are polysaccharides consisting of hexose units. Hemicellulose is a hetropolysaccharide and contains a mixture of pentose and hexose monomers. Lignin is a complex three-dimensional polymer and oils are triglycerides. Some other products present in biomass include alkaloids, carotenoids, flavonoids, phenols, resins, sterols, tannins, terpenes etc.[2]

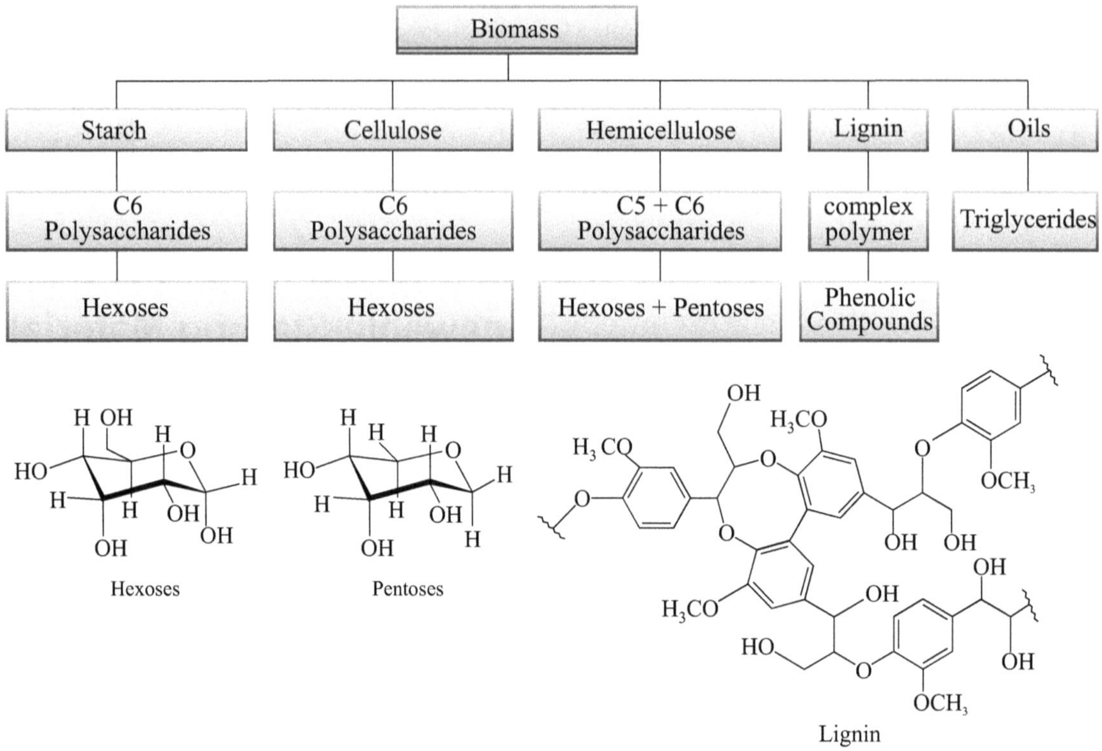

Fig. 16.1: Main components of Biomass (Source: Reference 4).

A number of methods for transforming Biomass into useful products have been reviewed[5,6-9]. Following is a brief account of the various products obtained from biomass.

16.2.1 Olefins from Biomass

Ethanol obtained by fermentation process[10] on dehydration gives ethylene which is most widely used for the manufacture of polymers. For dehydration of ethanol (20% aqueous ethanol is used) over HZSM-5 Zeolite gives ethylene (76-83%), ethane (2%), propylene (6.6%), propane (2%), butenes (4%) and *n*-butane (3%)[11].

Ethylene can be dimerised using a nickel catalyst to give 1-butene[12] and can be trimerised to 1-hexene using a chromium catalyst[13]. 1-Hexene on copolymerisation gives low-density polyethylene[14]. Some other olefins like 4-vinyl cyclohexene and styrene are obtained from butadiene which is obtained from 1-butene by catalytic dehydrogenation[15] (Fig. 16.2).

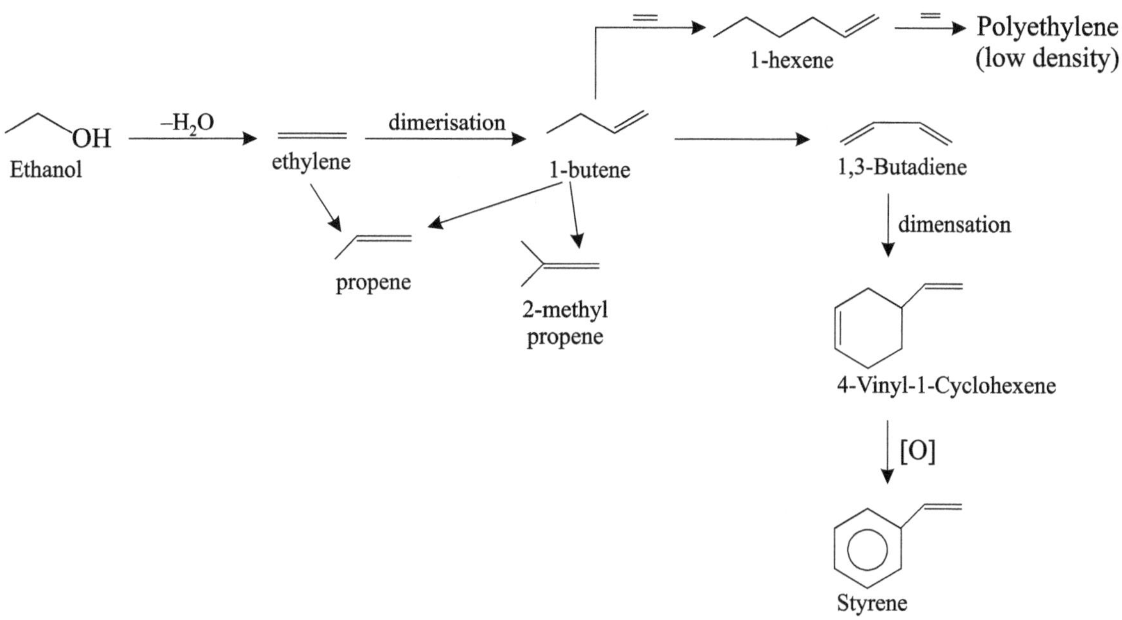

Fig. 16.2: Some olefins obtained from biomass.

Fermentation of sugars has also been used for making a mixture of acetone and 1-butanol. Catalytic hydrogenation of acetone followed by dehydration of the intermediate isopropyl alcohol gives propylene. 1-Butanol could be dehydrated to 1-butene, which is tern can be isomerised to 2-butene. 2-Butene can be reacted with ethylene to produce more of propylene. Thus, both ethylene and propylene could be obtained from sugars obtained from the hydrolysis of carbohydrates. The two olefins, ethylene and propylene are used to manufacture plastics (Fig. 16.3).

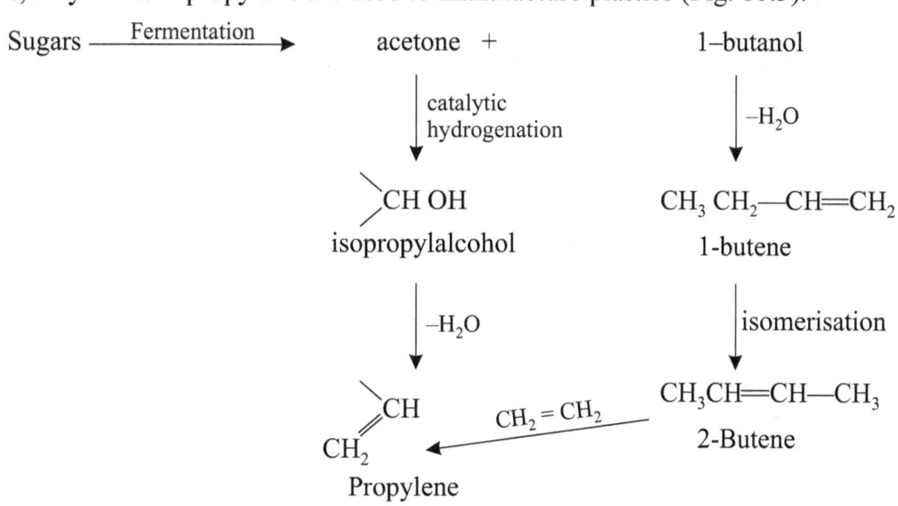

Fig. 16.3: Manufacture of propylene by the fermentation of Sugars.

Co-polymerisation of ethylene with carbon monoxide gives thermoplastic engineering resins (Fig. 16.4)[16]. The process also works with higher olefins.

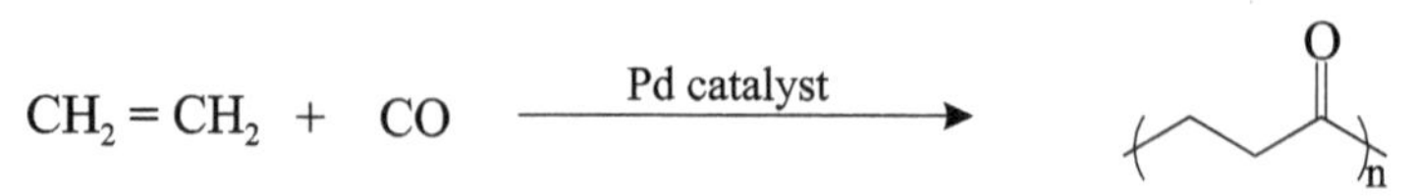

Fig. 16.3a: Co-polymerisation of ethylene with carbon monoxide.

A number of compounds on large scale *viz*., ethylene oxide, propylene oxide, vinyl chloride, vinyl acetate and acetylene can be prepared from ethylene, propylene and ethyl alcohol (Fig. 16.4)

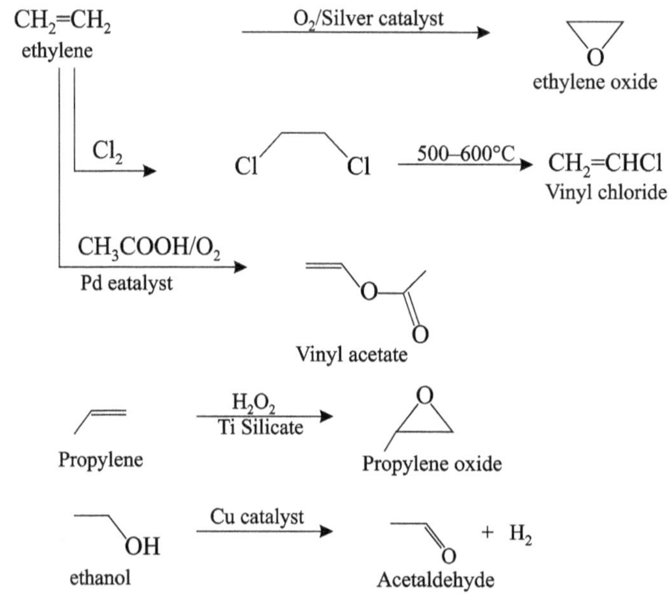

Fig. 16.4: Large scale preparation of some compounds from ethylene, propylene and propyl alcohol

Acrylic acid is prepared by air oxidation of propylene to acrolein, which is converted into acrylic acid[17]. Similarly methacrylic acid is obtained in the same way from isobutylene[18]. Methacrylic acid can also be made by fermentation of glucose[19] to give citric acid, followed by conversion into methacrylic acid (Fig. 16.5).

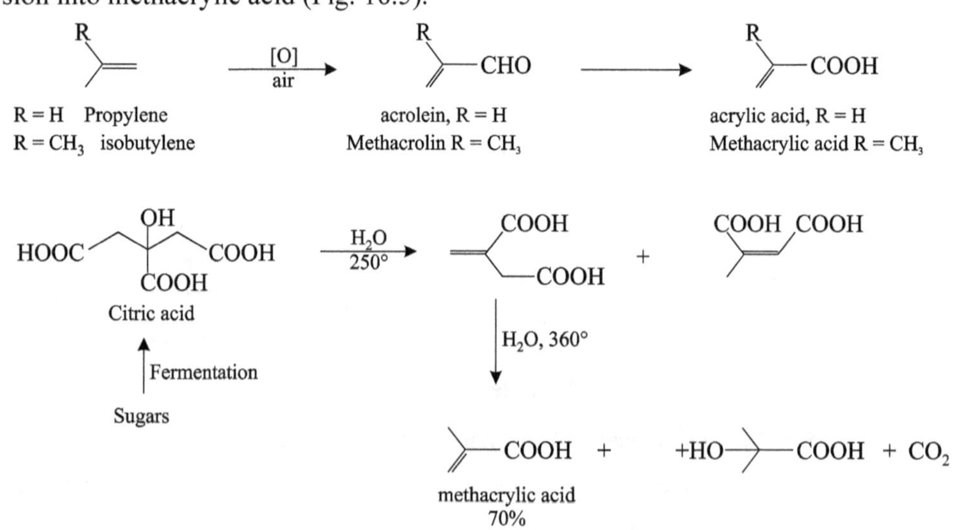

Fig. 16.5: Synthesis of acrylic acid and methacrylic acid.

Propylene on reaction with ammonia and oxygen gives acrylonitrile[20], a carcinogenic monomer. 1,3-Butadiene (obtained as given in Fig. 16.2) is converted into various commodity chemicals[21] (Fig. 16.6).

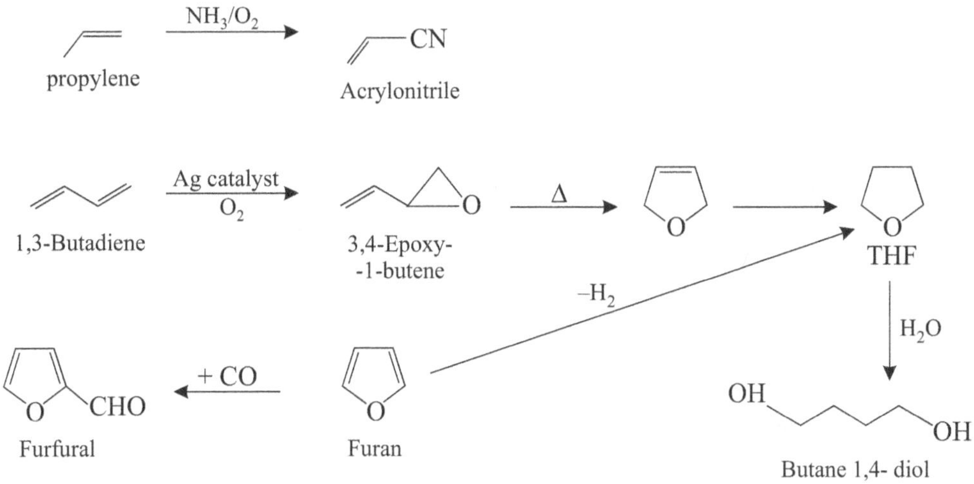

Fig. 16.6: Synthesis of some commodity chemicals.

16.2.2 Alcohols and Carboxylic Acids from Biomass

It is known that glucose on treatment with enzymes is converted into 1,3-propanediol. Some other diols like ethylene glycol and 1,2-propanediol can be obtained by hydrolysis of the corresponding epoxides. Hydrogenolysis of sorbitol (obtained by reduction of glucose) with Ru/Sn as catalyst in presence of NaOH[22] gives a mixture of ethylene, propylene and butylene glycols alongwith a small amount of glycol.

Anaerobic fermentation of carbohydrates gives[23] succinic acid, which on heating gives succinic anhydride, which on subsequent oxidation with O_2 in presence of iron phosphate catalyst gives maleic anhydride; it is used for conversion into maleic acid and also for the synthesis of phthalic anhydride (Fig. 16.7).

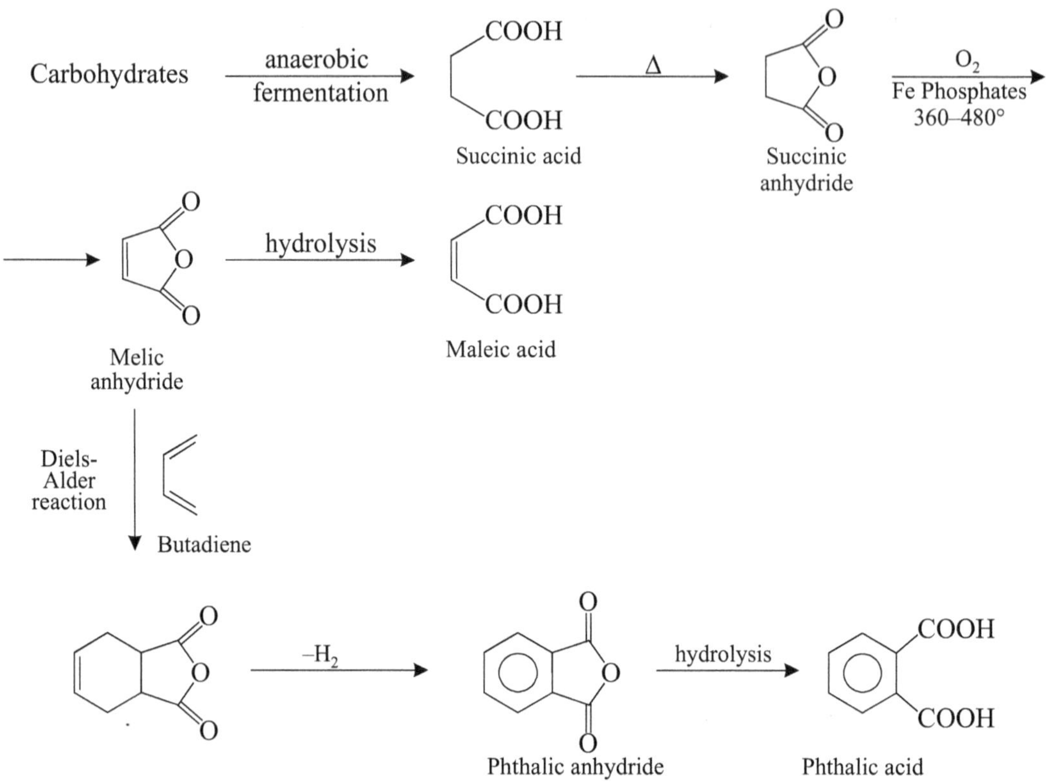

Fig. 16.7: Synthesis of maleic acid and phthalic acid.

Fermentation of starch wastes or cheese whay gives[24] lactic acid, which can be converted into an analogue of maleic anhydride[25] (Fig. 16.8)

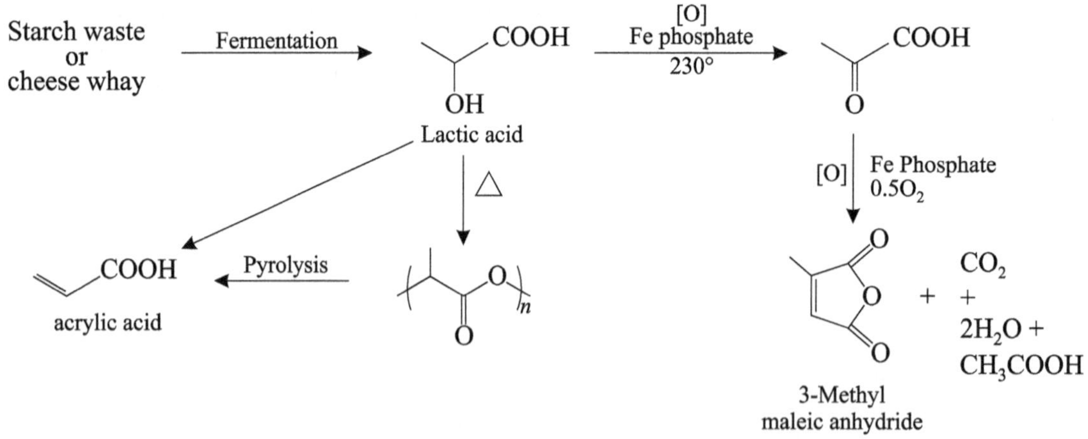

Fig. 16.8: Synthesis of Lactic acid and 3-methyl maleic anhydride.

Adipic acids, can be prepared from glucose (For details see chapter 13, section 13.2.6)

Adipic acid can also be prepared from glucose by a chemical route[26,27] (Fig. 16.9).

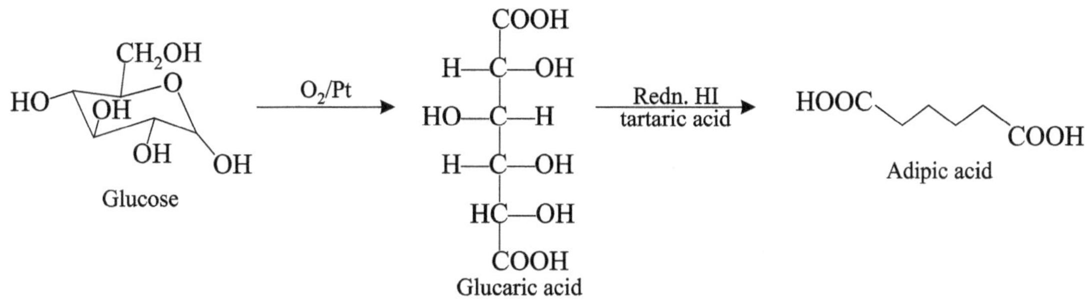

Fig. 16.9: Synthesis of adipic acid from glucose by a chemical route.

It is well known that cyclohexene (obtained by Diel-Alder reaction of butadiene with ethylene) on oxidation with H_2O_2 or ozone gives adipic acid[28], which in turn can be converted into 1,6-hexanediol[29], adiponitrile, hexamethylene diamine and adipaldehyde[30] (Fig. 16.10). Dimethyl adipate can also be made by tail-to-tail dimensation of methyl acrylate using Pd catalyst[31] (Fig. 16.10).

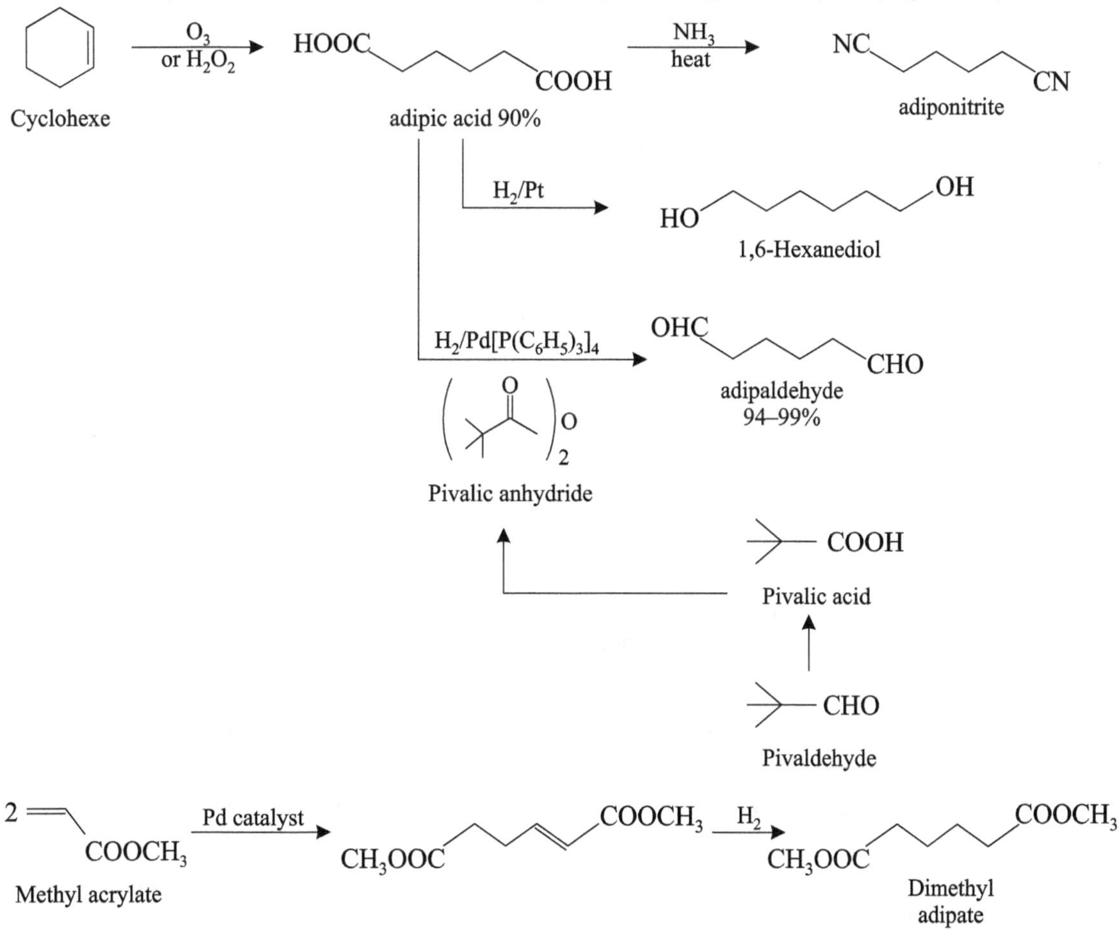

Fig. 16.10: Conversion of Cyclohexene into adipic acid, adiponitrile, 1,6-hexanediol, and adipaldehyde.

A keto acid, Levulinic acid can be obtained[32] from cellulose (or paper mill sludge) (Fig. 16.11), which has been used for the manufacture of tetrahydrofuran, 2-methylfuran.

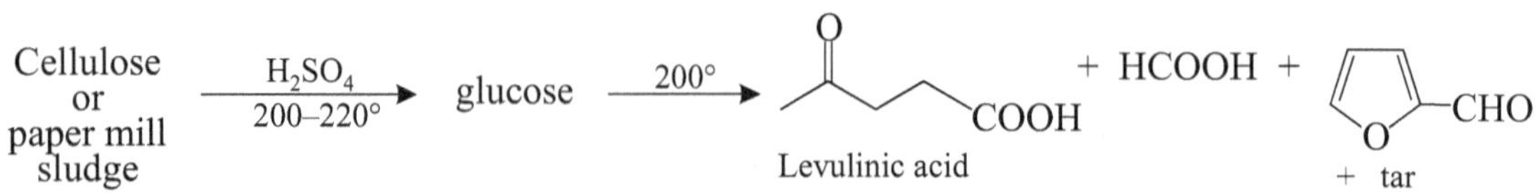

Fig. 16.11: Conversion of cellulose into Levulinic acid.

A number of carboxylic acids, *viz.* acetic acid, formic acid, lactic acid, propionic acid, butyric acid and some long chain carboxylic acids can be obtained by biocatalytic transformations (see Chapter 13, Section 13.2.5).

16.2.3 Aromatic Compounds from Biomass

Phenol can be obtained from styrene which is obtained from renewable source (see Fig. 16.2)] by reduction followed by oxidation to give hydroperoxide and subsequent treatment with acid. Phenol thus obtained on reaction with acetone gives bisphenol A (Fig. 16.12)[33]

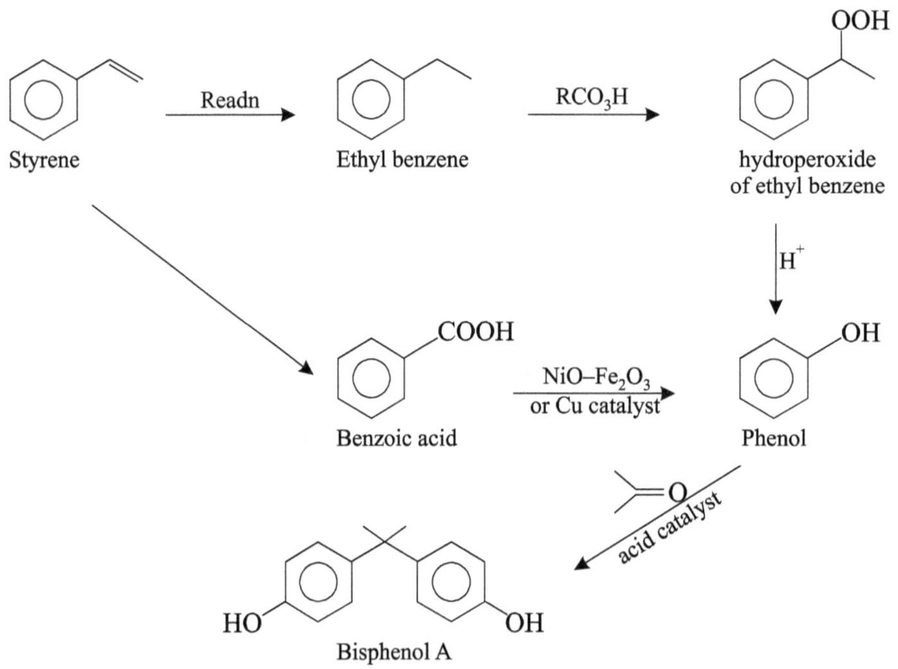

Fig. 16.12: Synthesis of Phenol and bisphenol-A.

Benzoic acid can also be obtained by Diels-Alder reaction of 1,3-butadiene with acrylic acid followed to catalytic dehydrogenation. Phenol can be converted into aniline by treatment with ammonia at high temperature. Ethylbenzene can be rearranged to xylenes with zeolite resins. The xylenes (*o*-, *m*- and *p*) in turn are a good source for convertion into phthalic, isophthalic and terephthalic acids by oxidation. The xylenes can also be made by dehydrocyclisation of ethylene, propylene and butenes[34]. Methane can be used to make benzene[35].

Catechol and hydroquinone can be obtained from glucose to using enzymes (see Chapter 13, Section 13.2.6).

Tannins which are polyphenols obtained from plants[36] can be hydrolysed in some cases to give gallic acid, which can be converted into pyrogallol (Fig. 16.13).

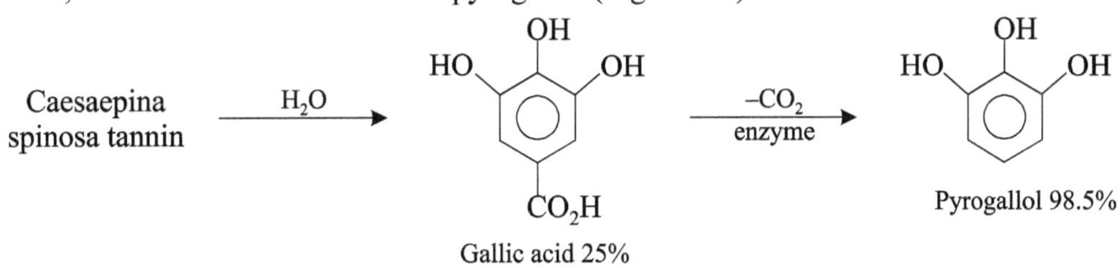

Fig. 16.13: Preparation of Pyrogallol.

16.2.4 Fats and Oils as a Source of Chemicals

Fats and oils are simple lipids and are esters of fatty acids and alcohols. These are obtained from plants as well as from animals. Hydrolysis of fats and oils gives a mixture of glycerol and fatty acids (depending on the nature of fat or oil). Following are given some of the important fatty acids obtained by the hydrolysis of fats and oils (Table 16.1). Major source of the fatty acids are also given.

Table 16.1: Fatty acids obtained by hydrolysis of fats and oils

Acid	Structure	Major Source
Saturated Fatty acids	$CH_3(CH_2)_n$ COOH	
Butyric	n=2	Tributyrin
Caproic	n=4	
Caprylic	n=6	Butter oil
Capric	n=8	Coconut oil
Lauric	n=10	
Myristic	n=12	
Palmitic	n=14	Butter oil
Stearic	n=16	Tristearin
Archidic	n=18	
Lignoceric	n=22	Soyabean oil
Unsaturated fatty acids		
Palmitoleic acid (cis-Hexadecenoic acid)		
$CH_3(CH_2)_5CH = CH(CH_2)_7COOH$		
Oleic acid (cis-9-octadecenoic acid)		
$CH_3(CH_2)_7CH = CH(CH_2)_7COOH$		
Linoleic acid (cis, cis-9, 12-octadecadienoic acid)		
$CH_3(CH_2)_4CH = CHCH_2CH = CH(CH_2)_7COOH$ (Peanut oil, corn oil)		
α-Linolenic acid (cis, cis, cis, 9, 12, 15-octadecatrienoic acid)		
$CH_3CH_2CH = CHCH_2CH = CHCH_2CH = CH(CH_2)_7$ COOH		
Archidonic acid (cis-cis, cis, cis-5, 8, 11, 14-(cosatetraenoic acid)		
$CH_3(CH_2)_4CH = CHCH_2CH = CHCH_2CH = CHCH_2CH = CH(CH_2)_3COOH$		

Hydroxy Fatty Acids

Ricinoleic acid

$CH_3(CH_2)_5CHOH.CH_2.CH = CH(CH_2)_7COOH$ (Castor oil, 90%)

Cerebroic acid

$C_{22}H_{45}$ CH(OH) COOH

Dihydroxy-Stearic acid

$C_{17}H_{33}$ $(OH)_2COOH$ (Castor oil)

Glycerol, which is obtained by hydrolysis of fats and oils can be converted into acrolein and acrylic acid

$$\underset{\text{Glycerol}}{CH_2OH\,CHOH - CH_2OH} \xrightarrow[-H_2O]{\Delta} \underset{\text{Acrolein}}{CH_2 = CHCHO} \xrightarrow{[O]} \underset{\text{Acrylic acid}}{CH_2 = CHCOOH}$$

Besides fatty acids, a number of oil contains other chemicals particularly terpenoids. These are given below along with their sources.

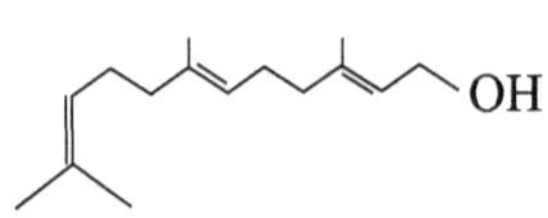

Farnesol
(oil of ambrette seeds)

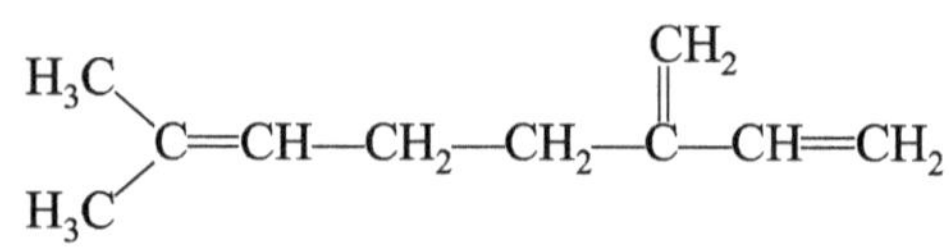

Myrcene
(Verbena and bay oils)

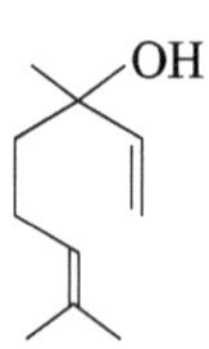

Linalool
[Rose oil (–) Linalool
orangeoil (+) Linalool]

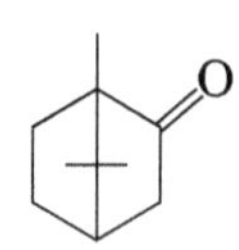

Camphor
(camphor tree bark)

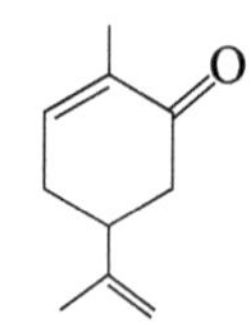

Carvone
(Spearmint and caraway oils)

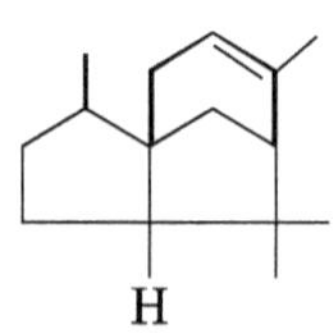

and

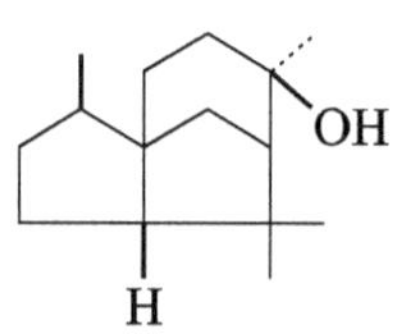

cedrene

cedrol

(Both cedrene and cedrol occur in cedar wood oil)

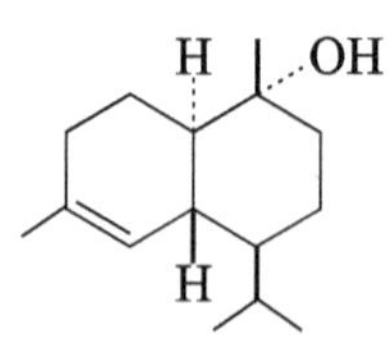

and

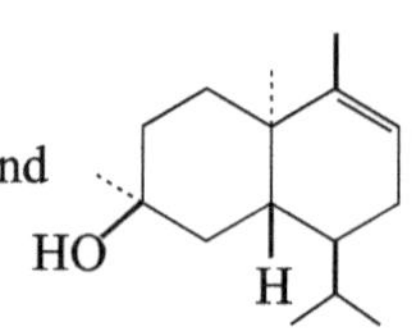

α-cadinol

δ-cadinol

(Both α-cadinol and δ-cadinol occur in celery oil)

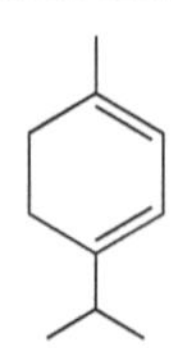

Ascaridole
(Chenopodium oil)

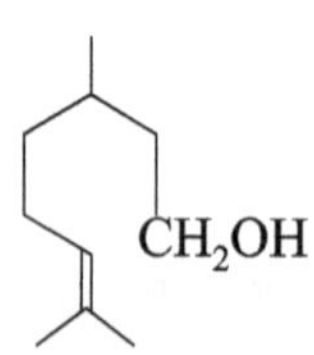

citronellol
(citronella oil)

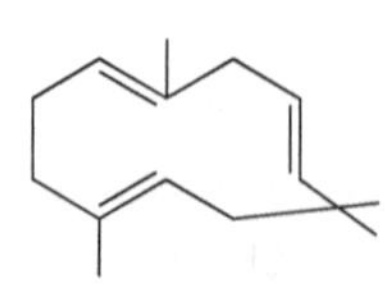

Humulene or
α-caryophyllene
(oil of hops and
oil of cloves)

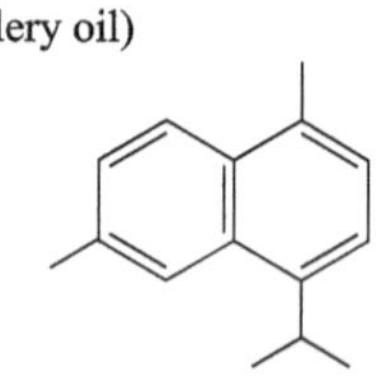

α-cadelene
(oil of cubebs)

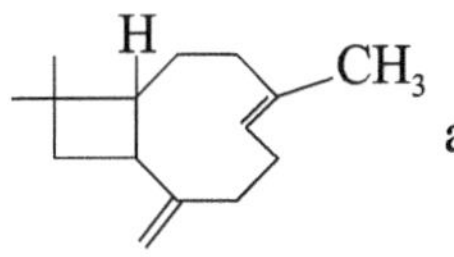

and

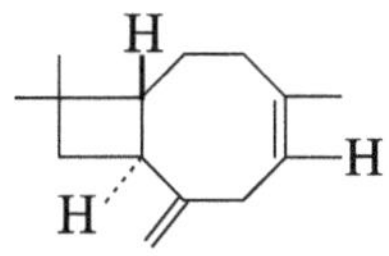

caryophyllene isocaryohyllene

(Both caryophyllene and isocaryophyllene occur in oil of cloves)

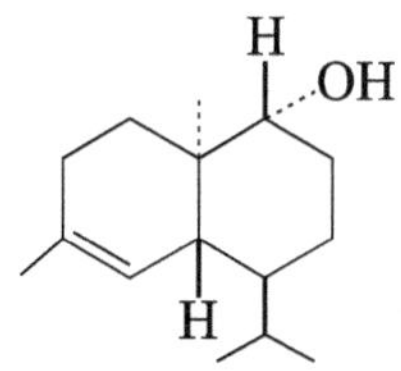

and

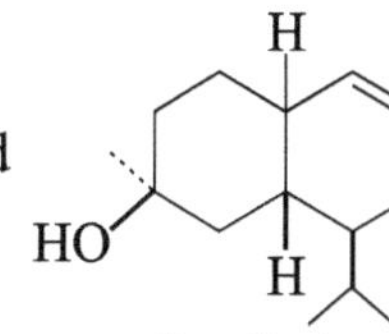

α-cadinol δ-cadinol

(Both α-and δ-cadinol occur in various essential oils)

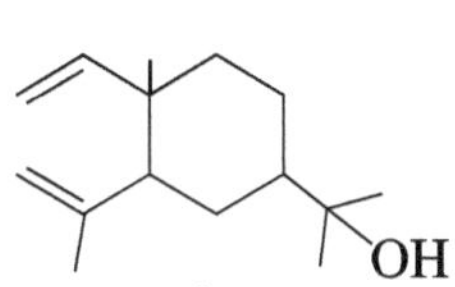

Etemol
(oil of elemi)

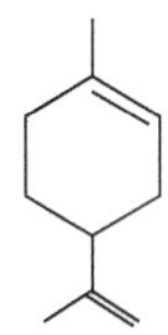

Li monene
(Lemon and orange oils (+ form)
Peppermint oil (– form)
Turpentine oil (± – form)

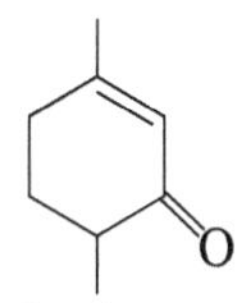

Piperitone
[Eucalypus oils (–) form]

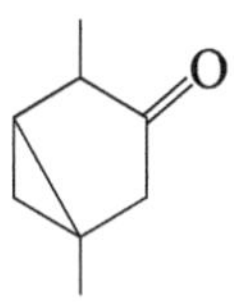

and

(–)-Thujone

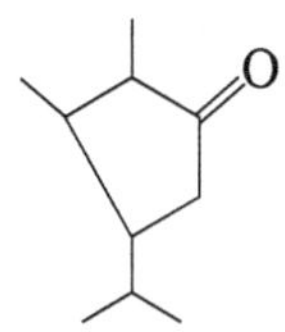

(+)-isothujone

Both (–)-thujone and (+)-ISO thujone occur in oils of sagle sage etc.]

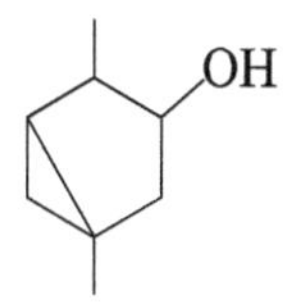

Thujyl alcohol
(worm seed oil)

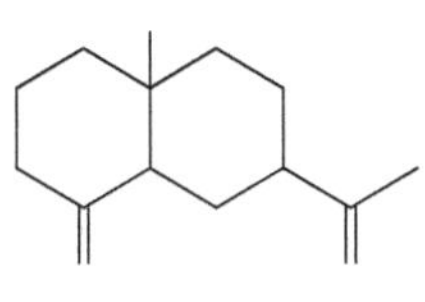

β-selinene
(celary oil)

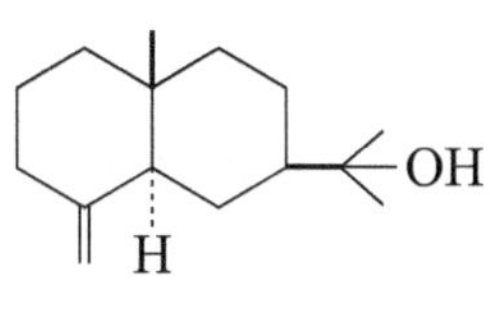

β-Eudesmol
(Eucalyptus oil)

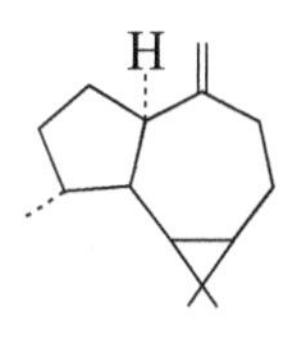

Aromadendrene
(Eucalyptus oil)

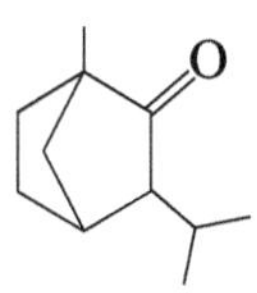

Fenchone
(oil of fennel)

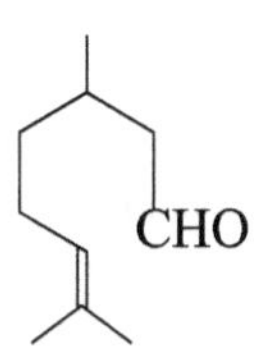

citronellal
(citronella oil)

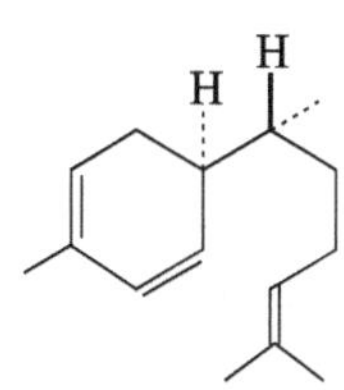

Zingiberene
(ginger oil)

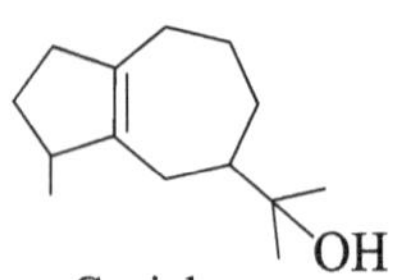

Guaiol
(wood oil)

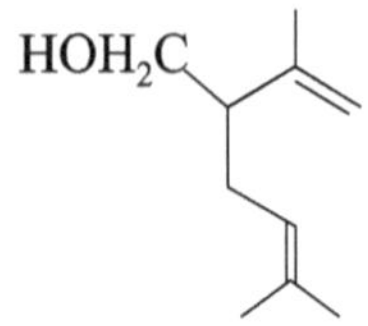

Lavandulol
(French lavenderoil)

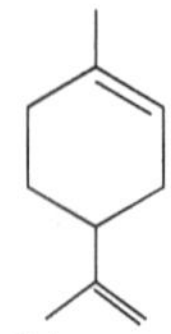

Limonene
Lemon and orange oils (–) form
Pipperment oil (–) form
tarpentine oil (±) –form

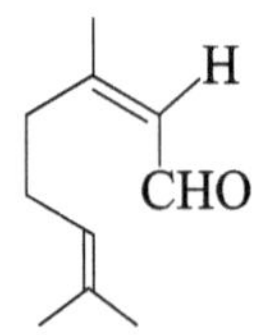

citral
(lemon grass oil)

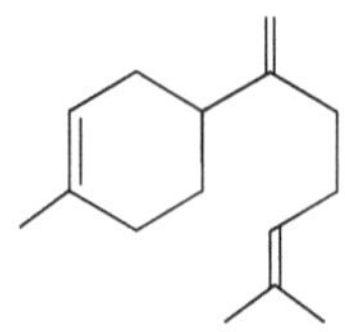

β-Bisabolene-
(Myrrh and other essential oils)

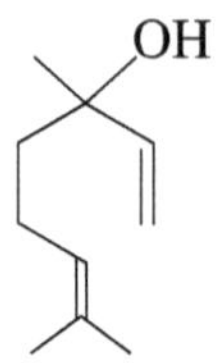

Linalool
[Rose oil (–) form
orange oil (+) – form]

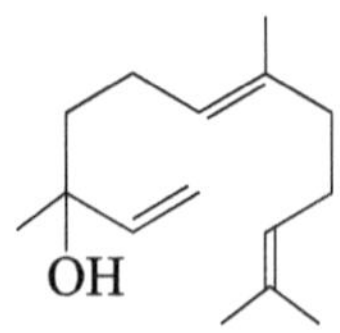

Nerolidol
[oil of nerole (+)-form]

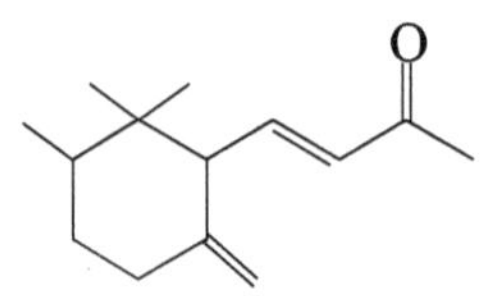

γ-Irone
(orris roots)

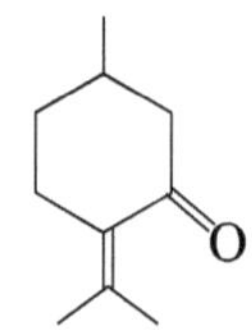

Pulegone
(Pennyroyal oils)

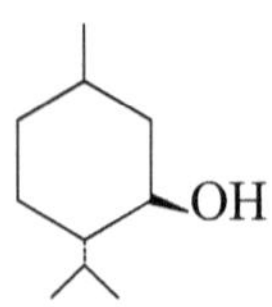

Menthol
[Pippermint oils
(–)-form

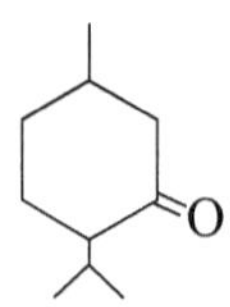

Menthone
(Pippermint oil)

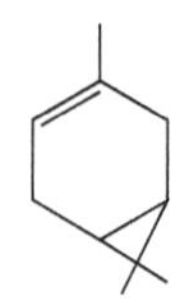

Car-3-ene
(Swedish pine oil)

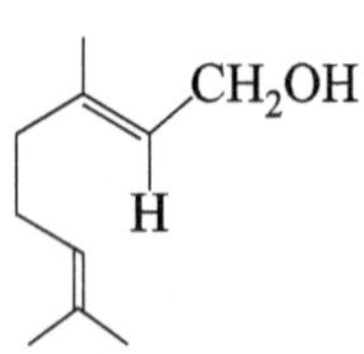

Geraniol
[Rose oil (trans or E)]

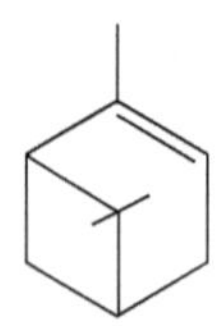

α-Pinene
(Turpine oils)

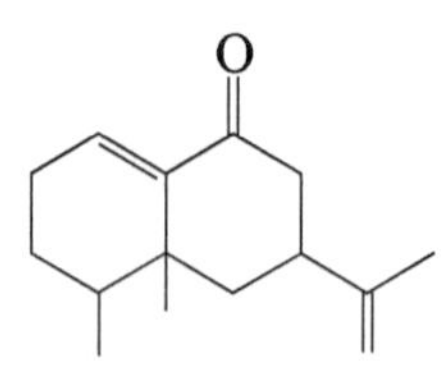

Eremophilone
(wood oil of Eremophila mitchelli)

[**Source:** I.L. Finer Organic Chemistry Vol. 2, 5th edn., Longman, Page 354-462.

16.2.5 Synthesis Gas from Biomass

Though synthesis gas is made from natural gas or petroleum oil, but can also be made from biomass[37] and from municipal waste[38]. The synthesis gas is a mixture of carbon monoxide and hydrogen. It can be converted to a number of useful chemicals (Fig. 16.14)[39].

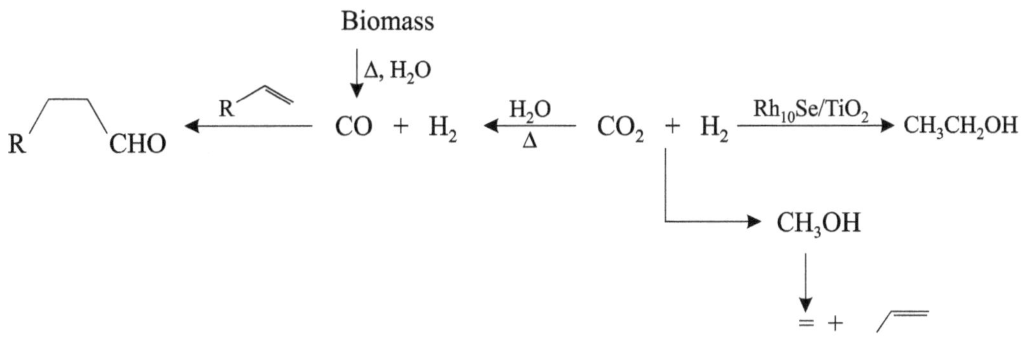

Fig. 16.14: Conversion of Synthesis gas into useful products.

16.2.6 Conversion of Cellulose into useful Chemicals

Cellulose represents largest part of wood biomass (55-80%)[40]. Hydrolysis of cellulose gives monomeric sugar units, which can be converted into numerous value-added chemicals[41]. Reactions for the production of various products involves hydrolysis, dehydration, isomerisation, aldol condensation, reforming, hydrogenation and oxidation etc.[42] Fig. 16.15 lists some of the fine chemicals via cellulose chemocatalytic conversion of biomass.

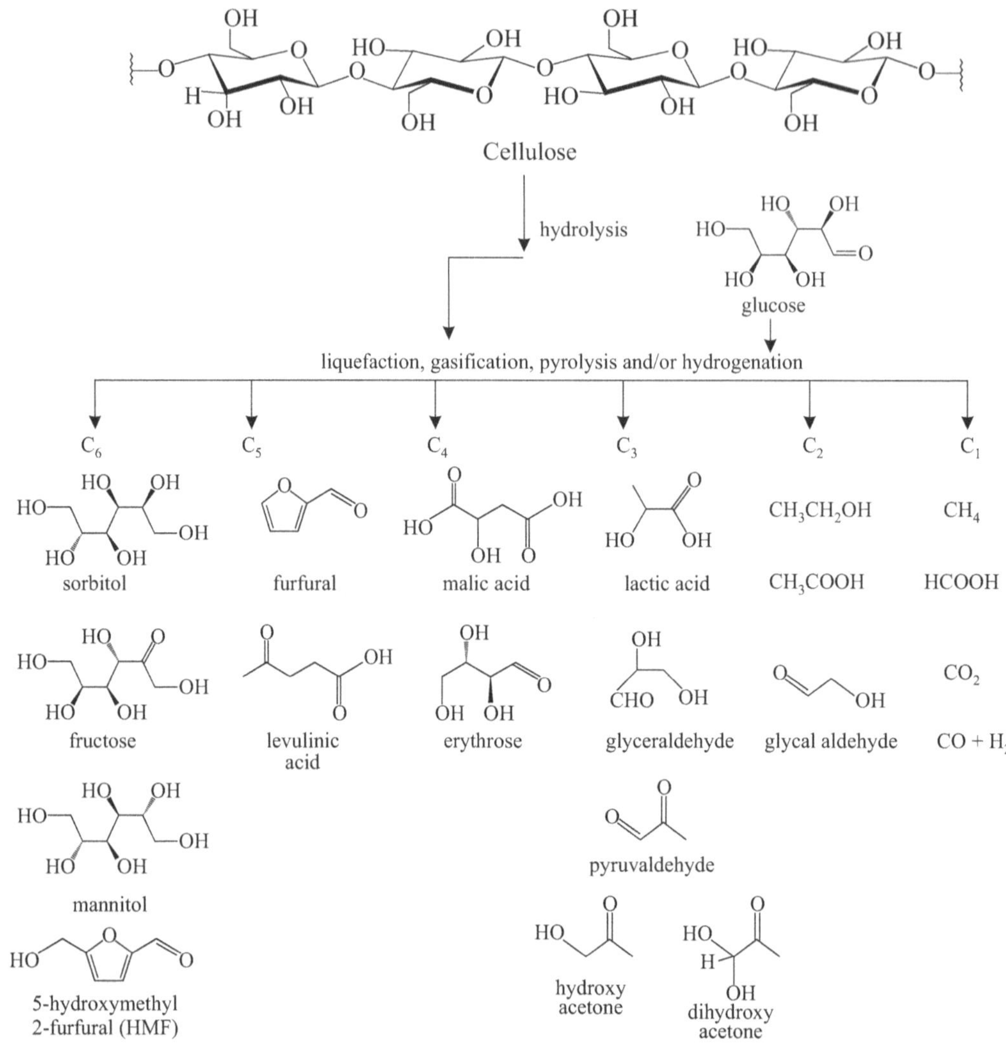

Fig. 16.15: Some fine chemical derived from cellulose.

[Data taken from reference 4]

The 5-hydromethyl-2-furfural (HMF) (Fig. 16.15) on aldol condensation with acetone give the corresponding aldol products[43] (Fig. 16.16)

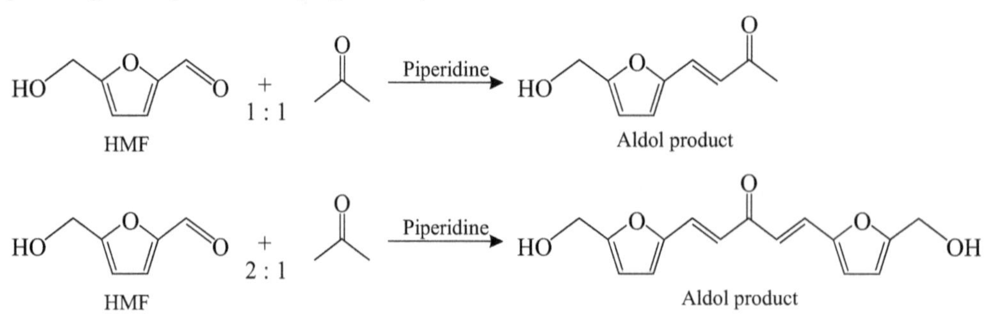

Fig. 16.16: Aldol products obtained from HMF.

An interesting example for the conversion of biomass to end-products via one-pot catalytic process[44,45]. Figure. 16.17 gives a one-pot conversion of wheat bran syrup to a mixture of alkyl-pentoside surfactants[46] by the reaction of the syrup with *n*-decanol in presence of dil. H_2SO_4 at 90°. The mixture of alkyl-pentoside is used for laundry purposes and dishwashing[47].

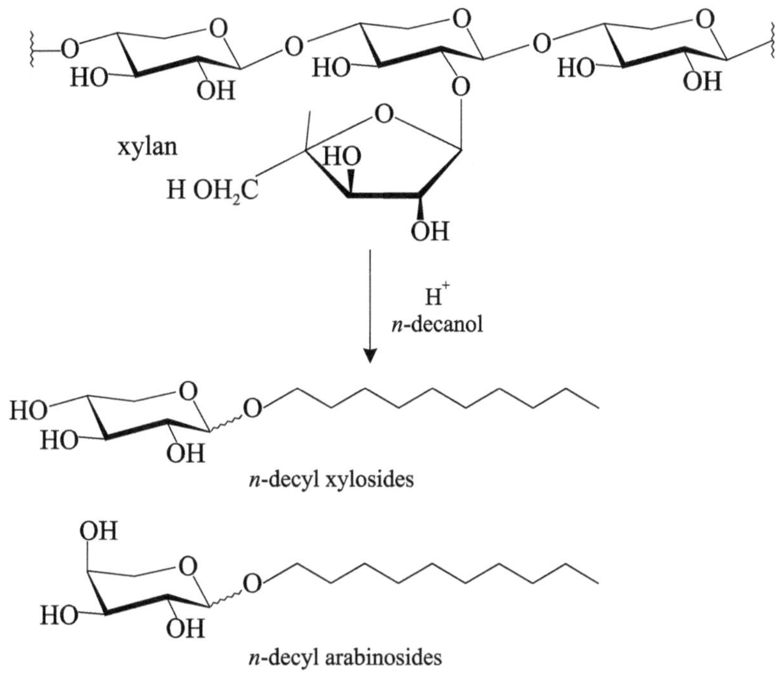

Fig. 16.17: Preparation of alkyl-pentoside from wheat bran syrup

16.3 SYNTHESIS OF USEFUL PRODUCTS FROM CARBON DIOXIDE

Carbon dioxide, a green-house renewable gas is non-toxic, abundant and inexpensive. It is obtained in large quantities by combustion of organic compounds, burning agricultural wastes and metallurgic processes. It has been used for a number of transformations, some of which are given below:

(*i*) **Synthesis of Organic Carbonates:** Four industrially important carbonates are ethylene carbonate (EC), propylene carbonate (PC), Dimethyl carbonate (DMC) and Diphenyl carbonate (DPC). These carbonates (except PC) are useful intermediates for the manufacture of polycarbonates via a non-phosgene process[44,45]. Additionally, EC, PC and DMC find use as electrolytes in lithium ion batteries and are commonly used as aprotic solvents. Also, DMC is used as a fuel additive[45]. The linear carbonates are useful alkylating agents.

Dimethyl carbonate (DMC) is synthesised from CO_2 and methanol (Fig. 16.18)

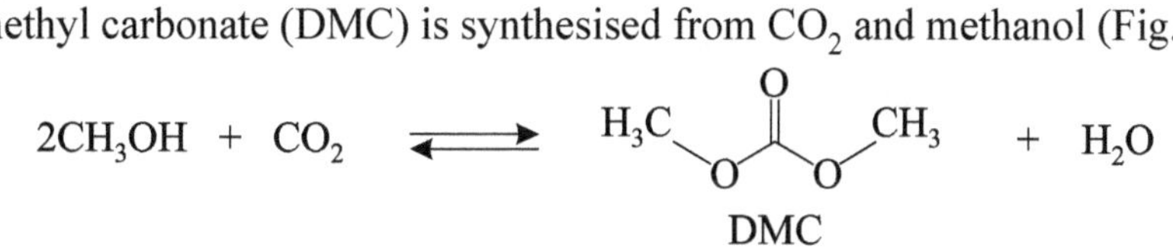

Fig. 16.18: Synthesis of DMC.

Best yields are obtained either by removing H_2O soon after it is formed of by increasing CO_2 concentration by pressurising CO_2. This method (Fig. 16.18) can be used for the preparation other linear carbonates. The procedure avoids the use of highly toxic phosgene for the synthesis of alkyl carbonates.

See also chapter 15, Section 15.8.

On a commercial scale, it is not practicable to remove H_2O or increasing CO_2 pressure (Fig. 16.18). The problem has been solved by using dimethyl ketal (obtained by the reaction of acetone with methanol) and reacting with CO_2 and methanol. Alternatively, dimethyl ketal can be prepared from methanol and CO_2 (Fig. 16.19)[46]. The reaction is carried out in presence of a catalysis.

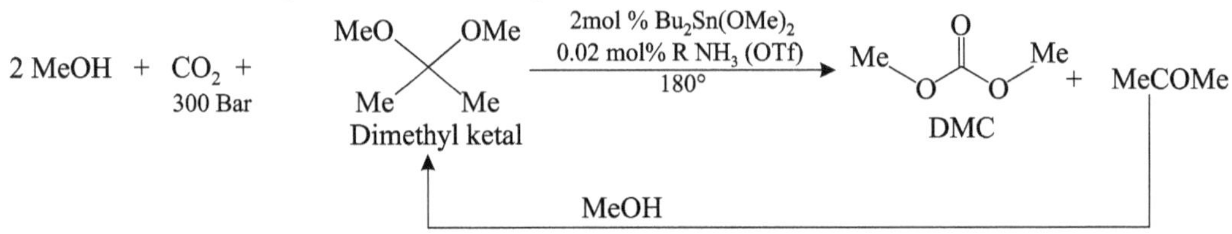

Fig. 16.19: Synthesis of DMC from dimethyl ketal.

The aliphatic carbonates have been prepared on industrial scale in China[46a].

(*ii*) **Synthesis of Cyclic Carbonates:** The cyclic carbonates (as in linear dialkyl carbonate) are useful intermediate for polycarbonates, electrolytes in lithium ion batteries and as green solvents. These can be prepared[7,47,48] from oxiranes and oxetanes by reaction with CO_2 (Fig. 16.20).

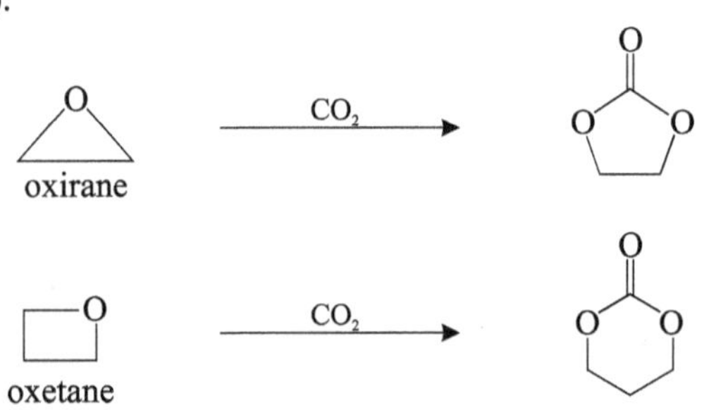

Fig. 16.20: Preparation of cyclic carbonate from oxirane and oxetane.

The production of five membered cyclic carbonate from CO_2 has been industrialised[7]. In industrial process, KI has been found to be one of the best catalyst. Five membered cyclic carbonates can also be synthesised by oxidative carboxylation of olefins (Fig. 16.21)[49] and by dehydrative reaction of 1,2-diols (Fig. 16.21)[50-52]

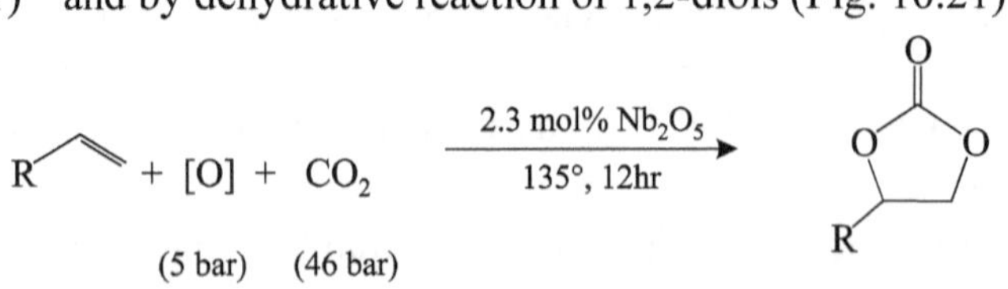

Fig. 16.21: Synthesis of 5-membered carbonates by oxidative carboxylation of olefins.

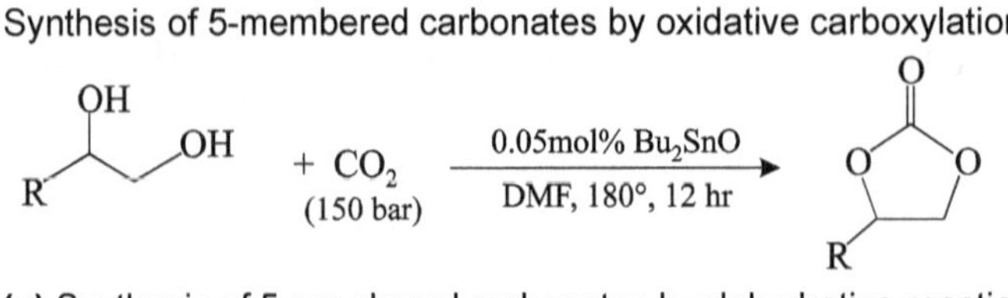

Fig. 16.21(a) Synthesis of 5 membered carbonates by dehydrative reaction of diols.

(*iii*) Synthesis of Urea and Substituted Ureas: Urea is prepared industrially from CO_2 by reaction with ammonia (Fig. 16.22)[53-55]

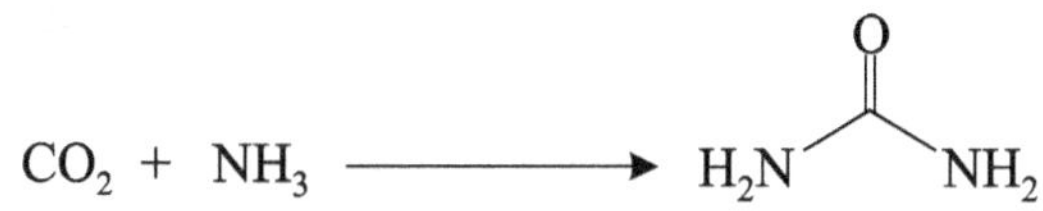

Fig. 16.22: Synthesis of urea.

Substituted ureas (N, N′-dialkyl urea) has been synthesis from CO_2 and secondary amines ($R'R_2NH$) by removing H_2O using ionic liquids as reaction media (Fig. 16.23)[53]

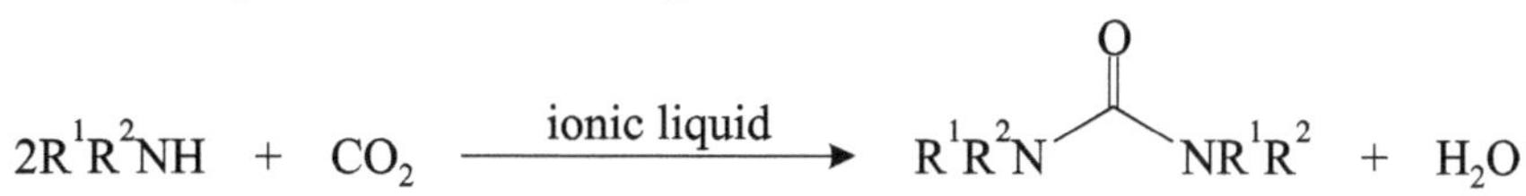

Fig. 16.23: Synthesis of Substituted Ureas.

(*iv*) Synthesis of Urethane: Urethane is prepared from primary amines with CO_2 and alkyl halide in presence of a base. The reaction proceeds via the formation of carbamate (Fig. 16.24).

$$RNH_2 + R^1X + CO_2 \xrightarrow{\text{Base}} RNHCO_2R^1 + HX$$

Fig. 16.24: Synthesis of Urethane.

A number of homogeneous and heterogeneous catalysts are used to synthesise urethane[56-58]. In the above scheme, if organic halide is replaced by alcohol than the byproduct is H_2O not HX and the process becomes environmentally friendly (Fig. 16.25)[59,60]

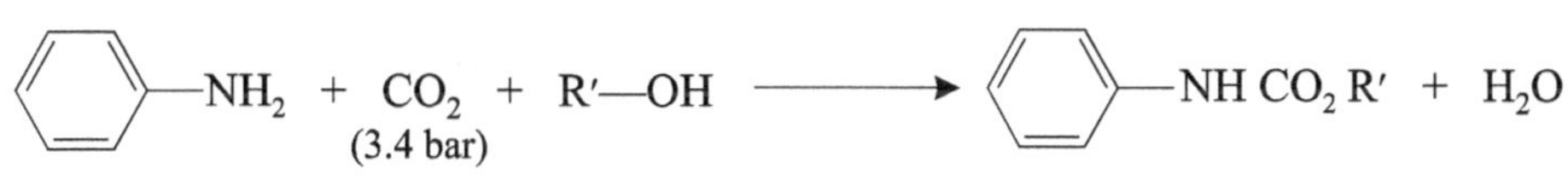

Fig. 16.25: Synthesis of Urethane with an alcohal.

It is found that the yield of urethane can be considerably increased if in place of alcohol, an ketal is used (Fig. 16.26)[61]

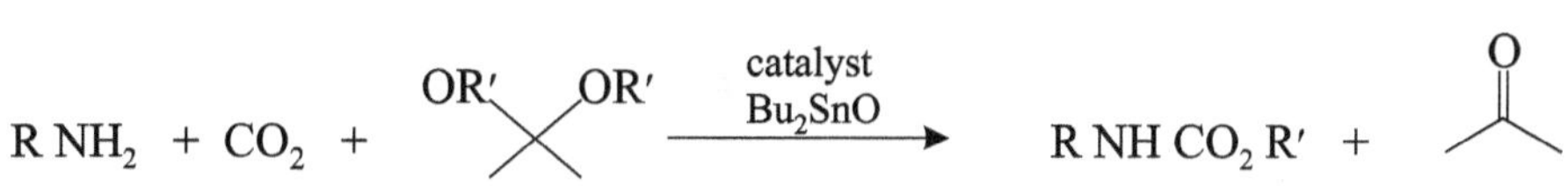

Fig. 16.26: Synthesis of Urethane using Ketal.

(*v*) Synthesis of isocyanates: Isocyanates are prepared by thermolysis of carbamates (Fig. 16.27)[62]

$$\underset{\text{Carbamates}}{RNHCO_2R'} \xrightarrow{\Delta} \underset{\text{Isocyanates}}{R-N=C=O} + R'OH$$

Fig. 16.27: Synthesis of isocyanates.

The required carbamates are obtained by the process already described (see Fig. 16.24, 16.25, 16.26)

(*vi*) **Synthesis of quinazolines:** The reaction of o-amino nitriles with carbonic acid gives quinazolines (Fig. 16.27)[63-65].

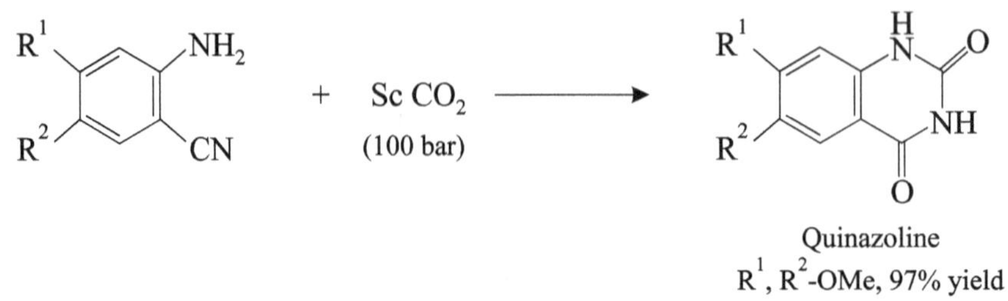

Fig. 16.28: Synthesis of Quinazolines.

(*vii*) **Synthesis of Oxazolidinones:** Aziridines (a nitrogen containing three membered ring) on reaction with CO_2 gives oxazolidinones (Fig. 16.29)[66-68]

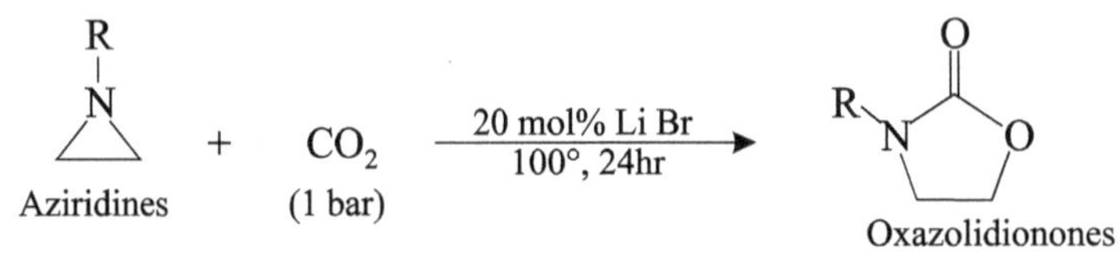

Fig. 16.29: Synthesis of Oxazolidinones.

(*viii*) **Synthesis of Carboxylic acids:** CO_2 reacts with organometallics to give carboxylic acids (Fig. 16.30)[69-71]

$$\underset{\text{m = Mg, Zn, Cu, Li}}{R-M-X+CO_2} \xrightarrow{H^+} RCO_2H$$

Fig. 16.30: Synthesis of Carboxylic acids.

Another carboxylic acid, salicylic acid (used for the preparation of Aspirin) is obtained by the reaction of CO_2 with phenol in presence of NaOH followed by addification (Fig. 16.31).

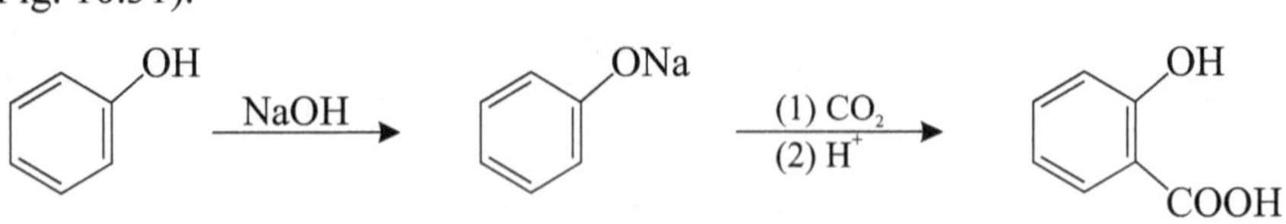

Fig. 16.31: Synthesis of Salicyclic acid.

This is a commercial method for the preparation of salicyclic acid.

Carboxylic acids can also be obtained from hydrocarbons like naphthalene[72], methane[73] and *p*-xylene[74] by C–H bond activation (Fig. 16.32).

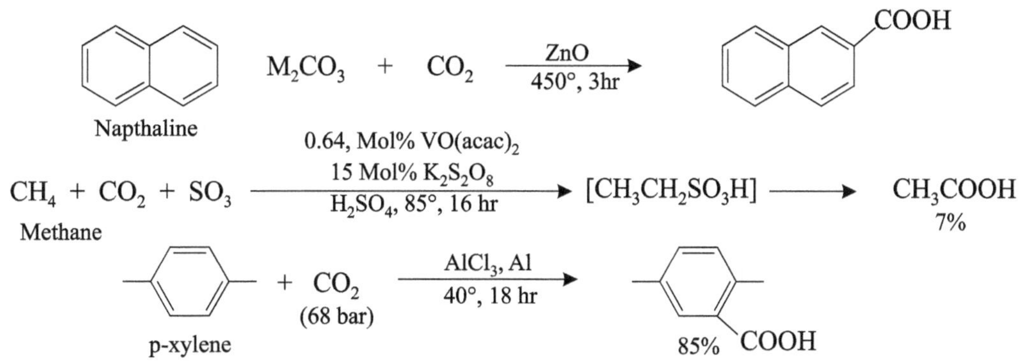

Fig. 16.32: Synthesis of carboxylic acid by C–H bond activation.

Carbon dioxide has also been used for carboxylation of methyliodide[75-77] and ethylene[78] (Fig. 16.33).

$$\underset{}{CH_3I} + \underset{(20\ bar)}{CO_2} + \underset{(20\ bar)}{H_2} \xrightarrow[150°,\ 24\ hr]{[Ru_3(CO)_{12}/10CO_2(CO)_8]} CH_3CO_2H$$

$$CH=CH_2 + \underset{(140\ bar)}{CO_2} \xrightarrow[4\ mol\ \%\ HBr,\ 180°,\ 12\ hr]{0.04\ mole\ \%\ Rh\ (PPh_3)Cl} \underset{38\%}{CH_3CH_2CO_2H} + \underset{15\%}{ECOH}$$

Fig. 16.33: Carboxylation of methane and ethylene.

Acrylic acid which is used for commercial production of water-absorbing polymers is produced by oxilation of propylene via acrylein. It is also made by the reaction of CO_2 with ethylene (Fig. 16.34)[79].

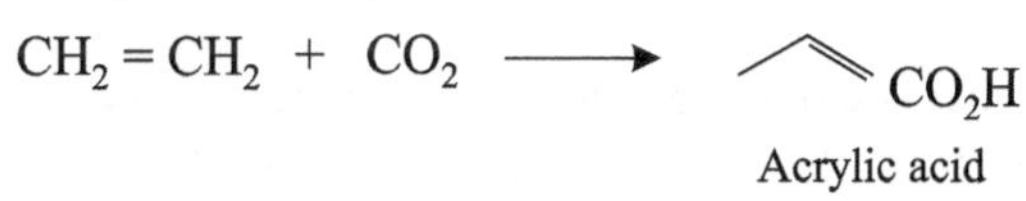

Fig. 16.34: Preparation of acrylic acid from ethylene.

(*ix*) **Synthesis of isocyanates:** The reaction of primary amines with CO_2 give carbonic acids, which on dehydration gives isocyanates (Fig. 16.35)[80,81] (see also section 16.3 sub section V)

$$R\,NH_2 + CO_2 \rightleftharpoons R\,NH\,CO_2H \xrightleftharpoons{Et_3N} [R\,NH\,CO_2]^- \ [Et_3NH]^+ \xrightarrow{POCl_3\ or\ P_4O_{10}} R{-}N{=}C{=}O$$

Fig. 16.35: Synthesis of isocyanates

(*x*) **Synthesis of methylformate and N,N-dimethylformamide:** Methyl formate[82] has been synthesized by the reaction of ScCO_2 with methanol and H_2. However, the reaction of Sc-CO_2 with dimethylamine and H_2 gives N,N-dimethylformamide[83] (Fig. 16.36).

$$\underset{(85\ \text{bar})}{CO_2 + H_2 + MeOH} \xrightarrow[\substack{ScCO_2\,(210\ \text{bar})\\ Et_3N,\ 80°}]{\text{Cat. } RuCl_2[P(CH_3)_3]_4} \underset{\text{Methyl formate}}{HCO_2Me + H_2O}$$

$$\underset{(85\ \text{bar})}{CO_2 + H_2 + Me_2NH} \xrightarrow[\substack{ScCO_2\,(210\ \text{bar})\\ 100°}]{\text{Cat. } RuCl_2[P(CH_3)_3]_4} \underset{\text{N,N-Dimethyl formamide}}{HCONMe_2 + H_2O}$$

Fig. 16.36: Synthesis of methyl formate and N,N-dimethylformamide.

(*xi*) **Carboxylation of Organic Halides:** Electrochemical reaction of allylic halides in presence of CO_2 give the carresponding carboxylic acids[84]. In a similar way carboxylation of benzylic chloride is possible. The later method has been used for the synthesis of **Ibuprofen**[85] (Fig. 16.37)

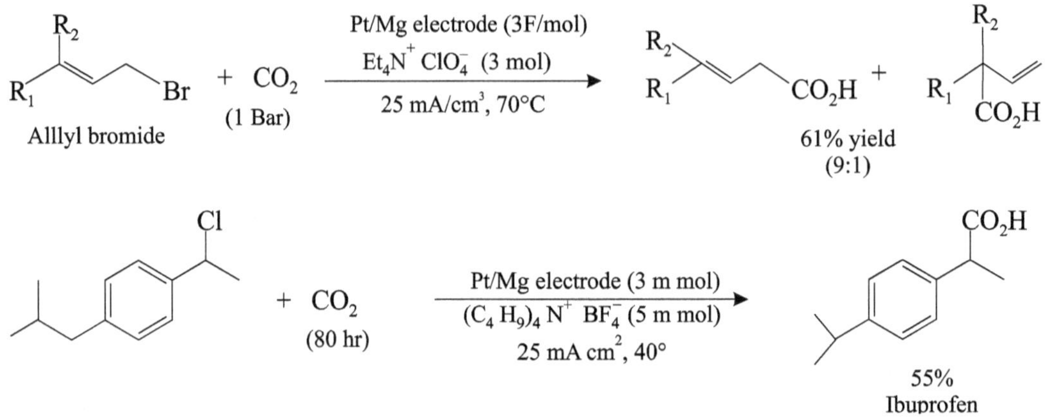

Fig. 16.37: Carboxylation of Organic Halides.

(*xii*) **Carbonic acid-catalysed reactions:** It has already been stated that CO_2 dissolves in water to give mildly acidic carbonic acid (Fig. 16.38)

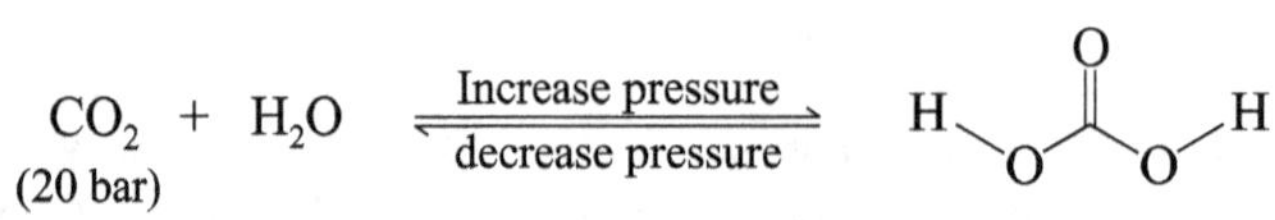

Fig. 16.38: Carbonic acid formation.

The formed carbonic acid can catalyse a number of reactions. The release of pressure after the reaction, raise the pH back to levels requiring minimum neutralisation[86]. Some typical examples of simple transformations are given below (Fig. 16.39)[87].

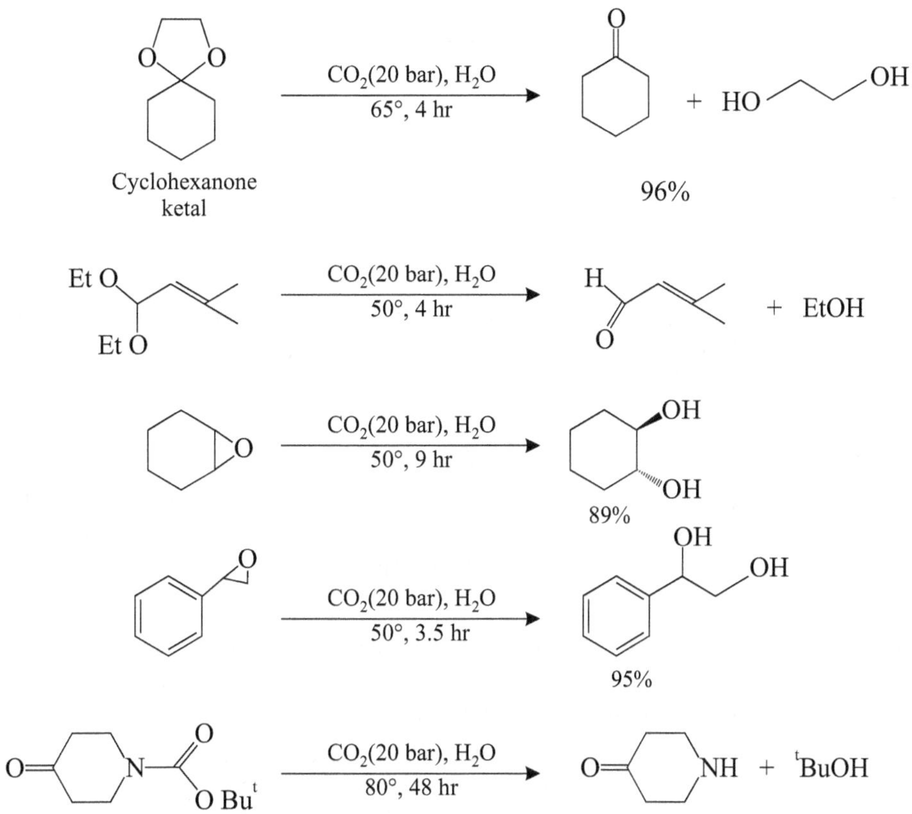

Fig. 16.39: Typical examples of simple transformations using Carbonic Acid.

Besides the above reactions of CO_2, supercritical carbon dioxide has been used as solvent in a number of transformation (For details see Chapter 4).

16.4 USEFUL PRODUCTS FROM METHANE

Methane, also known as marsh gas is obtained from petroleum. The natural gas or the gobar gas contains 70-80% methane. It is obtained as a result of bacterial decomposition of dead vegetable matter.

Methane is used as raw material for the manufacture of methyl alcohol, formaldehyde, acetylene and hydrogen cyanide.

$$CH_4 + \frac{1}{2}O_2 \xrightarrow[260°,\ 100\ \text{atm.}]{\text{Cu Catalyst}} \underset{\text{Methyl alcohol}}{CH_3OH}$$

$$CH_4 + O_2 \xrightarrow[MnO_2,\ 400°,\ 200\ \text{atm.}]{\text{Cantrolled oxidn.}} \underset{\text{Formaldehyde}}{HCHO} + H_2O$$

$$CH_4 \xrightarrow[\text{high temp.}]{\text{Craking}} HC{\equiv}CH$$

$$2CH_4 + N_2 \xrightarrow{\text{electric arc.}} 2\,HCN + 3H_2$$

$$2CH_4 + N_2 \xrightarrow[Al_2O_3]{NH_3,\ 100°C} HCN + 3H_2$$

It is also used for industrial production of H_2

$$CH_4 + H_2O \xrightarrow[Ni]{1000°} CO + 3H_2$$

Benzene can also be made from methane (y. xu and L.Lin, Appl. Catal. A, 1999, **185**, 53. Methane reacts with CO in water in presence of catalyst to generate acetic acid (For details see Chapter 7, Section 2.1.1.4 sub Section [C]).

16.5 BIOSOLVENTS FROM BIOMASS

Biosolvents as replacements of other VOC solvents can be obtained from biomass. However, the biosolvents do retain some of the hazards of many petroleum solvents. Some of the bio solvents include Ethanol, Ethylacetate, 2-methyl THF, THF, Diterpene, *p*-cymene, ethyl lactate, fatty acid methyl esters etc.,

The use of 2-methyl THF to replace THF and DCM is an important example of biosolvent[88,89] 2-Me-THF is obtained[90] as decomposition product of sugar via the formation of levulinic acid or from 2-methyl furturaldehyde obtained from agricultural waste[91].

The fatty acid methyl esters (FAMS) (known to be used as fuel) can also be used as solvents[92]. Also glycerol which is obtained as a byproduct during the formation of biodiesel can be used as a biosolvent. The preparation of the biosolvent is given in (Fig. 16.39)[93, 94-96]

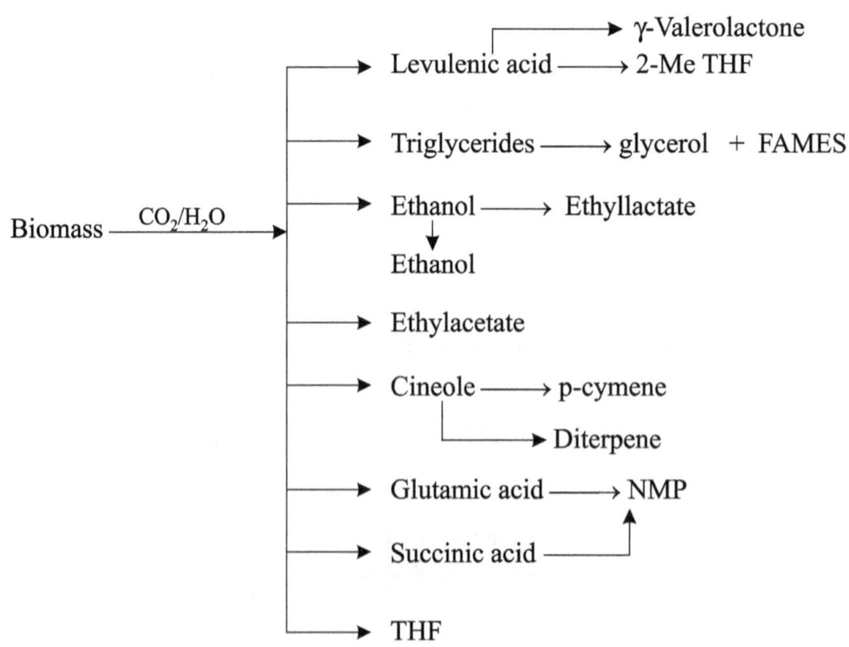

Fig. 16.39: Biosolvents from Biomass.

Conclusion

The biomass, a renewable natural resource has been used to obtain a large number of valuable chemicals as replacements of products from petroleum. Besides biomass other renewable sources include methane and carbon dioxide, which can also be used for the synthesis of organic chemicals. Some of the biosolvents have also been obtained from biomass.

REFERENCES

1. Biorefineries, Industrial processes and products in: Ullmanns' encyclopedia of industrial chemistry, 2005, volume 4, Wiley-VCH, Weinheim, pp. 101-1333.
2. J. Clark, F. Desware (eds). Introduction to chemicals from biomass, 2008, Wiley, Chichester.
3. E.M. Rubin, Genomics of cellulose biofuels. Nature, 2008, **454**, 841.
4. Paulo M. Donate, Green Synthesis from biomass. Donate chemical and biological Technologies in Agriculture, 2004, 1:4 http/www.chernbioagro.com/content/1/1/4.
5. P. Mäki-Anvela, B. Hombom, T. Salmi and Dy. Muzin, Catal Rev., 2007, **49**, 197.
6. A. Corma, S. Iborra and A. Veity, Chem. Rev., 2007, **107**, 2411.
7. G. Budroni and A. Corma, J. Catal., 2008, **257**, 403.
8. J.N. Chheda and A. Dumesic, Catal. Today, 2007, **123, 59.**
9. IIMoiseev, Russ. Chem. Rev., 2013, **82**, 616.
10. C.B. Phillips and R. Datta, Ind. Eng. Chem. Res. 1997, **36**, 4466.
11. A.K. Talukdar, K.G. Bhattacharya and S. Sivasanker, Appl. Catal. A., 1997, **148**, 357.
12. P. Andrews, J.M. Corker, J. Evans and M. Webster, J. Chem. Soc. Dalton Trans., 1994, 1337; H.K. Hall Jr., Chemtech, 1997, **27(6)** 20; G.R. Lappin, L.H. Nemec, J.D. Sauer and J.D. Wanyer, Kirk-othmer Encyclopedia of Chemical Technology 4th ed., 1996, **17**, 839.
13. G. Parkinson Chem. Eng., 1997, **104(4)**, 17; M.E. Lashear, J.W. Freeman and R.D. Knudsen, U.S. Patent 5, 689,027 (1947); F.-J. U/S, U.S. Patent 5,811, 61 (1998)
14. L.K. Johnson, S. Mecking and M. Brookhart, J. Am. Chem. Soc., 1996, **118**, 267; C.M. Killian, D.J. Tempel, L.K. Johson and M. Brookhart. J. Am. Chem. Soc., 1996, 118, 11664.
15. M.L. Moran, Chem. Ind. (Lond.), 1998, 90.
16. Chem. Eng. News, Nov. 4, 1996, 17; Y. Koide and A.R. Barron, Macromolecules, 1996, **29**, 1100; E. Drene and P.H.M. Budzelaar, Chem. Rev., 1996, **96**, 633.
17. N. Nojiri, Y. Sakai and Y. Watanabe, Catal. Rev. Sci. Eng. Chem. Br., 1998, **34(3),** 44.
18. G. Parkinson, Chem. Eng., 1997, **104(6),** 21; T. Jinbo, T. Pendo, M. Murakami, T. Matsushira, K. Kawahara and N. Horicuchi, Chem. Abstr, 1994, **121**, 281422.
19. M. Carlsson, C. Habenicht, L.C. Kan, M.J. Antal Jr., N. Bian, R.J. Cunningham and M. Jones Jr. Ind. Eng. Chem. Res., 1994, **33**, 1989.
20. S.J. Ainsworth, Chem. Eng. News, May 22, 1995, 13.
21. S. Stinson, Chem. Eng. News, Aug. 21, 1995, 7; C. Ondrey, Chem. Eng., 1995, **102(9)** 17; M.L. Morgan, Chem. Ind. (Lond), 1997, 166.
22. G. Gubitosa and B. Casal, U.S. Patient 5,354,914 (1994).
23. S.K. Yedur, J. Dulebohn, T. Werpy and K.A. Berglud, Ind. Eng. Chem. Res., 1996, **35**, 663.

24. I.-K. Yoo, H.N. Chang, E.G. Lee, Y.K. Chang and S.-H. Moon, Biotechnol. Lett., 1997, **19**, 237; D. Porro, L. Brambilla B.M. Ranzi, E. Martegani and L. Alberglhina, Biotechnol Prog., 1995, II, 294.
25. M. Ai and K. Ohdan, Chem. Lett., 1996, 247.
26. C.L. Mehlretter, C.E. Rist and B.H. Alexander, U.S. Patent 2,472,168 (1949).
27. S. Coffey ed., Rodd's Chemistry of Carbon compounds. 2nd ed., Elsevier Amsterdam, 1976, 1G, 21.
28. F. Fringulli and A. Taticchi, Dienes in Diels-Alder Reaction, Wiley, New York, 1990, 65.
29. G. Parkinson, Chem. Eng., 1998, **105(11)** 19.
30. K. Nagayama, I. Shimizu and A. Yamamoto, Chem. Lett., 1998, 1143.
31. G.L. Tembe, A.R. Bandyopadhyay, P.A. Ganehpure and S. Satish, Cat. Rev. Sci. Eng., 1996, **38**, 299; G.M. di Renzo, P.S. White and M. Brookhart, J. Am. Chem. Soc., 1996, **118**, 6225.
32. R. Dagani, Chem. Eng. News, July 5, 1999, 30; G. Parkinson, Chem. Eng., 1998, **105(10)**, 25; P. Anastas, M. Kirchhoff and T. Williams, Green Chem., 1999, **1**, G124.
33. J. Mike, M. Asanuma, Y. Tachibana and T. Shikada, J. Chem. Soc. Chem. Commun., 1994, 1685.
34. P. Meriaudeau and C. Naccache, Catal. Rev. Sci. Eng., 1997, **39**, 5; V.R. Choudhary and P. Devadas, J. Catal., 1997, **172**, 475.
35. Y. Xu and L. Lin, Appl. Catal. A., 1999, **188**, 53.
36. R.W. Hemingway and P.E. Laks eds., Plant polyphenols. Synthesis, properties, significance, plenum, New York, 1992.
37. D. Wang, D. Czernik, M. Montane, M. Mann and E. Chornet, Ind. Eng. Chem. Res., 1997, **36**, 1507; X. Xu, Y. Matsumara, J. Stenberg and M.J. Antal, Jr., Ind. Eng. Chem. Res., 1996, **35**, 2522.
38. G. Parkinson, Chem. Eng., 1998, **105(10)**, 23.
39. G. Braca, Oxygenates by Homologation or carbon monoxide hydrogenation with metal complexes Kluwer Academic, Dordrecht, 1994.
40. P. Mäkl-Arvela, B. Holmoon, T. Salmi, and D.Y. Murzin, Ratal. Rev., 2007, **49**, 197.
41. F.W. Lichtenthaler and S. Peters, C.R. Chimie, 2004, **7**, 65.
42. J.M. Chheda, G.W. Huber and J.A. Dumesi, Anglew Chem. Int. Ed., 2007, **46**, 7164.
43. A.D. Sutton, F.D. Waldie, R. Wu, M. Schlaf, L.A. Silks and J.C. Gordon, Nature Chem., 2013, **5**, 428.
44. S. Furkuoka, M. Kawamura, K. Komiya, M. Togo, H. Hachiya, K. Hasewaga, M. Aminaka, H. Okamato, I. Fukawa and S. Konno, Green Chem., 2003, **5**, 497.
45. M.A. Pacheco and C.L. Marshall, Energy Fuels, 1997, **11**, 2.
46. K. Kohno, J.-C. Choi, H. Yasuda and T. Sakakura, 53th Symposium on Organometallic chemistry, Osaka, Japan, Sept. 8-9, 2006, Kinki Chemical Society, Abstract PB 261.
46a. S. Inoue, Gerdai Kagaku, 2005, **408**, 42.
47. A. Baba, H. Kashiwagi and H. Matsuda, Organometallics, 1987, **6**, 137.
48. A. Baba, H. Kashiwagi and H. Matsuda, Tetrahedron Lett., 1985, **26**, 1323.
49. M. Aresta and A. Dibenedetto, J. Mol. Catal A: Chem. 2002, **182**, 399.
50. T. Tomishige, H. Yasuda, Y. Yoshida, M. Nurunnabi, B.T. Li and K. Kuniomori, Green Chem., 2004, **6**, 206.
51. K. Tomishige, H. Yasuda, Y. Yoshida, M. Nurunnabi, B.T. Li and K. Kuniomori, Catal. Lett., 2004, **95**, 45.

52. Y. Du, D.L. Kong, H.Y. Wang, F. Cai, H.S. Tian, J.Q. Wang and L.N. He, J. Mol. Catal. A., Chem. 2005, **241**, 233.
53. F. Shi, Y.Q. Deng, T.L. SiMa J.J. Peng, Y.L. Gu and B.T. Quiao, Angew. Chem. Int. Ed., 2003, **42**, 3257.
54. C.C. Tai, M.J. Huck, E.P. Mckoont, T. Woo and P.G. Jessop, J. Org. Chem., 2002, **67**, 9070.
55. R. Nomura, Y. Hasegawa, M. Ishimoto, T. Toyosak and H. Matsuda, J. Org. Chem., 1992, **57**, 7339.
56. R.N. Salvatore, S.L. Shin, A.S. Nagle and K.W. Jung, J. Org. Chem., 2001, **66**, 1035.
57. M. Yoshida, N. Hara and S. Okuyama, Chem. Commun., 2000, **151**.
58. W. Mcghee, D. Riley, K. Chirst, Y. Pan and B. Parmus, J. Org., Chem., 1995, **60**, 2820.
59. M. Alba, J.-C. Choi and T. Sakakura, Chem. Commun., 2001, 2238.
60. M. Alba, J.-C. Choi and T. Sakura, Green Chem., 2004, **6**, 524.
61. T. Sakakura, J.C. Choi, P. Saito, T. Masuda, T. Saka and T. Oriyama, J. Org. Chem., 1999, **64**, 4506.
62. V.L.K. Valli and H. Alper, J. Org. Chem., 1995, **60**, 257.
63. T.J. Connolly, P. McGarry and S. Sukhtankar, Green Chem., 2005, **7**, 586.
64. T. Mizuno and Y. Ishino, Tetrahedron, 2002, **58**, 3153.
65. T. Mizuno, T. Iwai and Y. Isino, Tetrahedron Lett., 2004, **45**, 7073.
66. H. Kawanami, H. Matsumoto and Y. Ikushima, Chem. Lett., 2005, **34**, 60.
67. A.W. Miller and S.T. Nguyen, Org. Lett., 2004, **6**, 2301.
68. A. Sudo, Y. Morioka, F. Sanda and T. Endo, Tetrahedron Lett., 2004, **45**, 7889.
69. N.P. Mankad, T.G. Gray, D.S. Laitar and J.P. Sadighi, Organometallics, 2004, **23**, 1191.
70. K. Chiba, H. Tagaya, S. Miura and M. Karasu, Chem. Lett., 1992, 923.
71. K. Chiba, H. Tagaya, M. Marasu, M. Ishizuka and T. Sugo, Bull. Chem. Soc. Jpn., 1994, **67**, 452.
72. A. Solafani, L. Palmisano and G. Farneti; Chem. Commun., 1997, 529.
73. C.G. Jia, T. Kitamura and Y. Fujiwara, Acc. Chem. Res., 2001, **34**, 633.
74. G.A. Olah, A. Torok, J.P. Joschek, I. Buchi, P.M. Esteves, G. Rasul and G.K.S. Prakash, J. Am. Chem. Soc., 2002, **124**, 11379.
75. A. Fukuoka, N. Gotoh, N. Kobayashi, M. Hirano and S. Komiya, Chem. Lett., 1995, 567.
76. C. Amatore, A. Jutand, F. Khalil and M.F. Nielsen, J. Am. Chem. Soc., 1992, **114**, 7076.
77. K. Osakada, R. Sato and T. Yamamoto, Organometallics 1994, **13**, 4645.
78. A.L. Lapidus, S.D. Pirozhokov and N.D. Zelinsky, Proc. 9th Int. Congr. Catal, 1988, **3**, 1028.
79. M. Aresta and A. Dibenedetto, Catal. Today, 2004, **98**, 455.
80. T.E. Waldman and W.D. McGhee, J. Chem. Soc. Chem. Commun., 1994, 957.
81. D. Saylik, M.J. Horvath, P.S. Elmes, W.R. Jackson, C. Lovel and K. Moody, J. Org. Chem., 1999, **64**, 3940.
82. P.G. Jessop, Y. Hsiao, T. Ikariya and R. Noyori, J.Am. Chem. Soc., 1996, **118**, 344.
83. P.G. Jessop, Y. Hsiao, T. Ikariya and R. Nigori, J. Am. Chem. Soc., 1994, **116**, 8851.
84. M. Tokuda, T. Kabuki and H. Suginome, Tetrahedron Lett., 1995, **36**, 3343.
85. A. Sasaki, H. Kudoh, H. Senboku, M. Tokuda. In Novel Trends in Electroorganic Synthesis; S. Torii Ed.; Springer-Verlag, Tokyo, 1998.
86. R.R. Weikel, J.P. Hallett, C.L. Liotta and C.A. Eckert, Top. Catal., 2006, **37**, 75.

87. D.A. Clarke, J. Ammens, C.M. Raynes, Univ. of Leed, Unpublished results. Cited in C.M. Rayner, Org. Process Res. and Development 2007, **11**, 121.
88. D.H. Brown-Ripin and M. Vitelino, Synlett, 2003, 15, 2353.
89. D.R. Spring, S. Krishnan and S.L. Schreiber, J. Am. Chem. Soc., 2000, **122**, 5656.
90. T. Werpy and G. Petersen. The value added. Chemicals from Biomass. Volume I-Results of screening for potential candidates from sugars and synthetic gas, US Department of Energy.
91. R. Kottke, Furan derivative, in Kirk-Othmer Encyclopedia of Chemical Technology (ed. M. Howe Grant), John Wiley and Sons, Ltd., New York, Page 259-286.
92. M. Knise and M.A.R. Meier, Green Chem., 2010, **12**, 169.
93. D.F. Aycock, Org. Process Res. Dev., 2007, **11**, 156.
94. Y. Gu and F. Jerome, Green Chem., 2010, **12**, 1127.
94a. D.G. Blackmond, A Armstrong, V. Combe and A. Wells, Ang. Chem. Int. Ed., 2007, **46**, 3798.
95. T.R. Cartsen, G.A. Tompsett, W.C. Conner and G.W. Huber, Top. Catal, 2009, 52, 241.
96. J.N. Chheda and M.W.M. Tuck and J.A. Dumesic, Angew. Chem. Int. Ed., 2007, **46**, 7164.

17. ELECTROCHEMICAL SYNTHESIS

17

Electrochemical Synthesis

17.1 INTRODUCTION

The well known **Kolbe reaction**[1] is one of the earliest electrochemical reaction for organic synthesis. The reaction involves electrochemical oxidation of a carboxylic acids in forming decarboxylated coupling product (Fig. 17.1).

$$2RCOO^- \xrightarrow{2e} 2R^\circ + 2CO_2$$
$$2R^\circ \longrightarrow R - R$$

Fig. 17.1: Kolbe reaction.

The reagent in the above reaction is an electron. The reaction proceeds via free radical mechanism (Fig. 17.2)

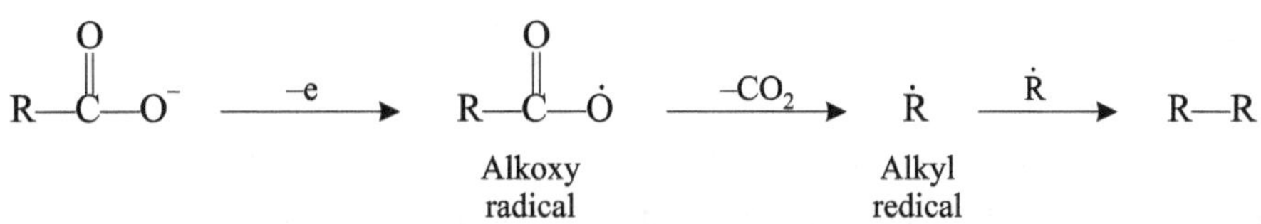

Fig. 17.2: Mechanism of Kolbe reaction.

At present, electrochemical synthesis has become an independent discipline. For performing electrochemical synthesis, the conductivity of the medium is an essential requirement. Use of water as reaction medium has distinct advantage over organic solvents.

The electrochemical synthesis, in general are of two types, *viz*., anodic oxidation and cathodic reduction. A number of compounds can be electrochemically oxidized. The product formed in oxidation reactions depends on the solvent, the pH of the medium and the oxidative potential. However, for the reduction of organic compounds an electron is provided by the cathode. The rate of reduction increases with the acidity of the medium[2].

The electrochemical reactions are as diverse as non-electrochemical reactions. These reactions can be conducted in presence of catalysts (electrochemical catalytic process)[3], photochemical irradiation (photoelectro chemical process)[4] and enzymes (electro enzymatic reactions).[5]

In electrochemical reactions. There is selective introduction and removal of electrons from organic molecules, and so there is reversal of the polarity of functional groups. So the electro chemical reactions are considered to be **umpolung reactions**[6]. In these reactions electrons can be added to electron-poor functional groups in order to convert them from electrophiles to nucleophiles or removal of electrons from electron-rich functional groups in order to convert them from nucleophiles into electrophites. Some example are given below (Fig. 17.3)[6].

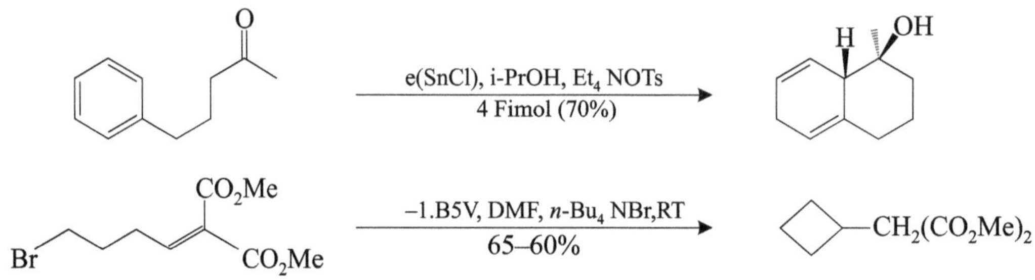

Fig. 17.3: Some examples of Electrochemical Synthesis.

Electrochemical alternatives have been developed for the well known Heck and Still reactions as well as Sonogashira and Suzuki reactions[7]. The Nozaki-Hiyama-Kishi coupling between allyl, Vinyl or aryl halides and aldehydes, for example is normally achieved using excess $CrCl_2$ in presence of catalytic amount of nickel chloride. It has been now found possible to generate catalytic amounts of chromium and nickel from a pre-electrolysis of a stainless steel rod. That material, used in the presence of 3% 2,2′-bipyridine and a sacrificial iron anode has made it possible to achieve coupling (Fig. 17.4)[8].

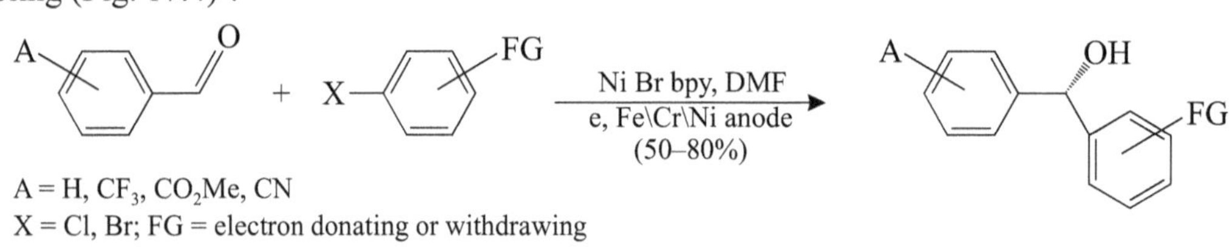

Fig. 17.4: Stainless-steel used in presence of 3% 2, 2′-bypyridine and a sacrificial iron anode used for the coupling of aryl halides and aldehydes.

17.2 APPLICATIONS OF ELECTROCHEMICAL SYNTHESIS

Following are given some Industrial applications of electrochemical synthesis.

17.2.1 Synthesis of Adiponitrile

Adiponitrile is produced (3000 tons/year) by an important industrial process involving electrohydro dimerization of acrylonitrile. It is an important precursor for hexamethylenediamine and adipic acid, which are used for the synthesis of Nylon-66 polymer. Adiponitrile is prepared by the electroreductive coupling of acrylonitrile. By this process about 90% of adiponitrile is obtained[9]. In process, a concentrated solution of certain quaternary ammonium salts (QASs), such as tetraethylammonium-*p*-toluene sulfonate is used along with lead or mercury cathode (Fig. 17.5).

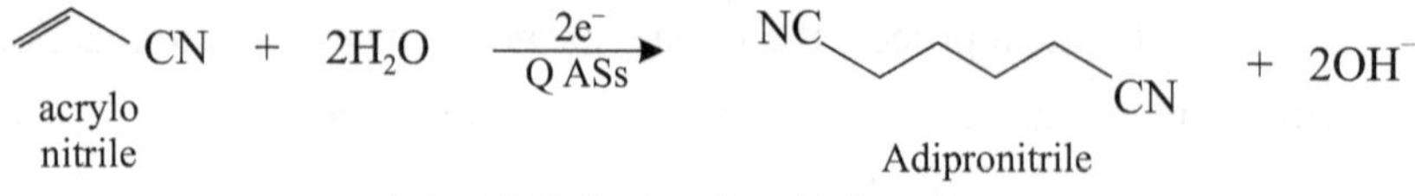

Fig. 17.5: Synthesis of Adiponitrile.

Much better yields of adiponitrile are obtained[10] by using an undivided cell (instead of divided-cell) using a dilute solution of acrylonitrile in a mixed sodium phosphate borate electrolyte using cadmium cathode and a carbon steel anode in presence of quaternary ammonium salt.

17.2.2 Synthesis of Sebacic Acid

Sebacic acid is an important intermediate for the manufacture of polyamide resins. The conventional method of alkaline hydrolysis of caster oil has been replaced by electrochemical process[11]. The method involves three steps in which the key step is based on the anodic coupling of the monomethyl ester of adipic acid. Much better results are obtained by using a very high electrolyte velocity through the cell (3-4 m/s^{-1}). The electrolyte used is a 20% aqueous solution of monomethyl ether of adipate neutralised with sodium hydroxide. A platinum-plated titanium anode and a steel cathode is used. The cells are usually operated at a current density of $20 A/dm^2$ and 14V. The electrolysis is carried out at 55°C (Fig. 17.6).

$$\underset{\text{Adipic acid}}{HO_2C(CH_2)_4CO_2H} \xrightarrow{CH_3OH} \underset{\text{Adipic acid monomethyl ester}}{CH_3O_2C(CH_2)_4CO_2H} \xrightarrow{\text{electrolysis}}$$

$$\underset{\text{Sebacic acid dimethyl ester}}{CH_3O_2C(CH_2)_8CO_2CH_3} \xrightarrow{\text{hydrolysis}} \underset{\text{Sebacic acid}}{HOOC(CH_2)_8COOH}$$

Fig. 17.6: Electrochemical route to the synthetic of Sebacic acid.

17.2.3 Miscellaneous Applications

(*i*) Electrolysis of aliphatic alcohols on platinum or carbon electrodes is aqueous KI at room temperature yields the corresponding aldehydes in 80-83% yields[12].

(*ii*) Electrolysis of secondary alcohols in a solution of KI in aqueous *tert*-butyl alcohol results in 57-98% yield of ketones at room temperature when 4-15F/mol electricity in cells with platinum or carbon electrodes in used[12]. Some examples are given below[12].

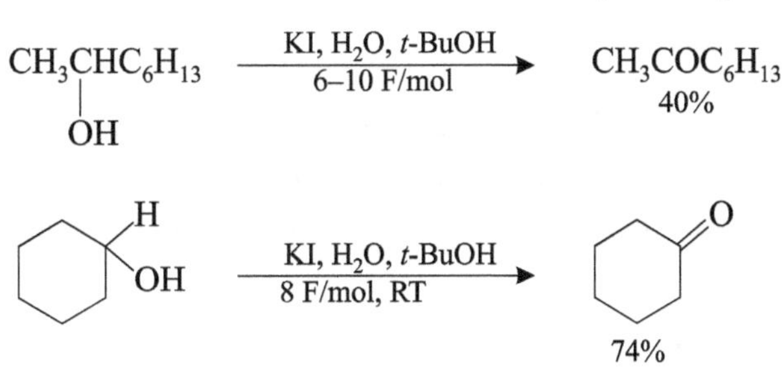

(*iii*) **Oxidation of alkenes to epoxides or ketones:** The electrolytic oxidation takes place at the anode of an electrolytic cell, which may or may not contain a diaphragm to separate cathode and anodes spaces. The anode must be made of a metal that resists oxidation, such as lead[13], nickel[114], or, more frequently platinum. The anode is usually in the shape of a cylinder made of wire gauze[15,16]. The possible electrolytes are dilute sulphuric acid[15] or sodium methoxide prepared in situ from methanol and sodium[16]. The direct-current voltage is 3-100 V, the current density is 10-20 A/dm^2 and the electrolysis temperature is 20-80°C.

Thus, electro-oxidation of 2-octene in DMF gives 73% yield of 1-methyl-2-pentyloxirane[17].

1-Methyl, 2-pentyl oxirane (73%)

Another example of epoxidation of alkene is given below[18]

$$\underset{R=C_8H_{17}}{CH_2 = CHR} \xrightarrow[\substack{Et_4NBF_4\text{,pt anode,} \\ 0.83\ A/dm^2\text{, RT}}]{\text{electrolysis, DDQ, Pd(OAc)}_2} \underset{63\%}{CH_3COR}$$

(*iv*) **Oxidation of homologues of pyridine:** Homologues of pyridine are converted into pyridine carboxylic acids by electro-oxidation[15]

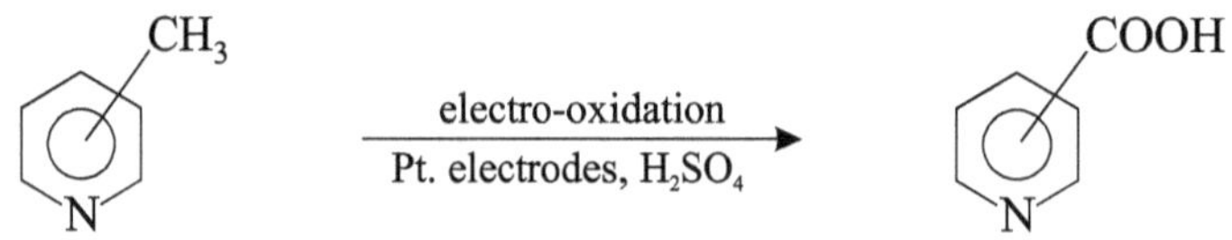

60%

(*v*) **Oxidation of Quinoline:** In heterocyclic aromatic compounds, the ring containing electron withdrawing substituents is more resistant to oxidative degradation. Thus both quinoline and 8-hydroxy quinoline on electrolysis give quinolinic acid (pyridine-2,3-dicarboxylic acid)[15]

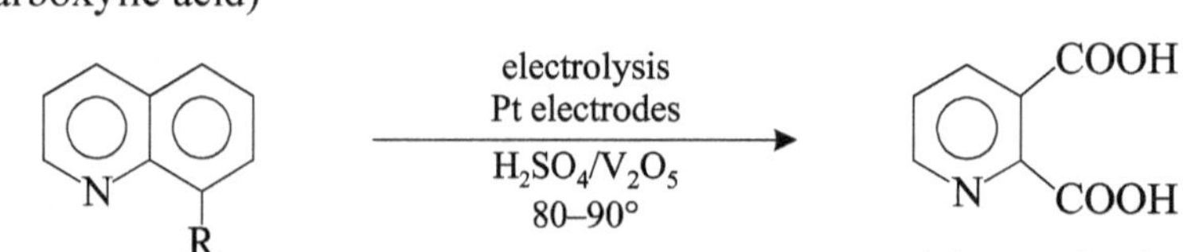

Quinolune, R = H.
8-hydroxyquinoline
R = OH

Pyridine 2,3-dicarboxylie acid (77–89%)

(*vi*) **Oxidation of primary alcohols to Carboxylic acids:** On electrolytic oxidation saturated, unsaturated, acetylenic and aromatic alcohols are converted into the corresponding carboxylic acids using nickel anodes and stainless steel cathodes and NaOH or KOH as electrolytes[14].

$$RCH_2OH \xrightarrow[\substack{\text{Stainless steel cathode} \\ 4A,\ 25°,\ 5\ hr}]{\text{Ni anode, 1M NaOH}} RCO_2H$$

$R = C_6H_{11}$ $\qquad$ $R = C_6H_{13}$ yield 84%

$R = CH \equiv C$ $\qquad$ $R = CH \equiv C$ yield 51%

(*vii*) **Oxidation of Vicinal diols to Carboxylic acids:** Cyclic vicinal diols having 6-8 carbons in the ring are oxidised[19] to the corresponding dicarboxylic acids in 79% yield by carrying out electrolysis at 20-30° in aqueous K_2CO_3 with nickel electrodes in an undivided cell and a current of 0.3 A.

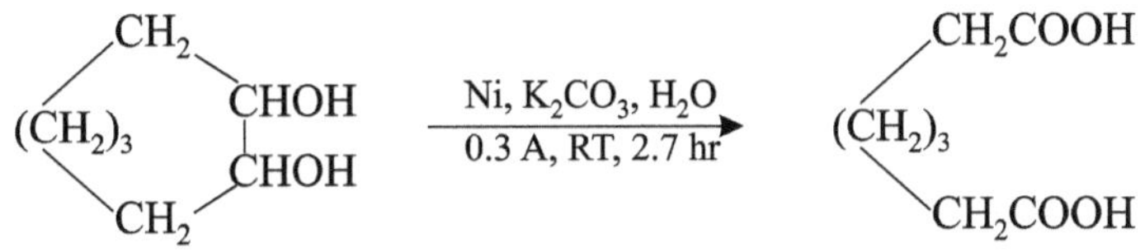

79%

(*viii*) **Electroreduction:** Reduction as we know is defined as acceptance of electrons. In case electrons are supplied by an electrode-cathode, the reduction is called electroreduction. The most important factor in electroreduction is the nature of the metal used as cathode. Metals of low over voltage, *e.g.*, platinum (0.005 to 0.09 V), palladium, nickel and iron give results similar to catalytic hydrogenation[20]. Cathodes made of metals of high overvoltage, e.g. copper (0.23 V), cadmium (0.48 V), lead (0.64 V), Zinc (0.70 V) or mercury (0.78 V) give results similar to those of dissolving metal reductions.

The electrolyte used in electro reduction also affects the nature of the product obtained. Generally, electroreductions are carried out in dilute H_2SO_4. However, some are carried out in alkaline electrolytes such as alkali hydroxides, alkoxides of solutions of salts like tetramethyl ammonium chloride in methanol[21] or lithium chloride in alkyl amines[22,23].

The results of electro reduction also depends on whether the electrolytic cell is divided (by a diaphragm separating the cathode and anode) or undivided[22].

The current density in electroreduction is determined by the size of the electrodes and the current passed through the electrolytic cell. In most reductions the current density is in the range of 0.05-0.2A/cm^2. The current is determined by the imposed voltage and conductivity of the electrolyte. The amount of electricity required for electroreduction of one gram equivalent of a compound is 96,500 coulombs (26.8 ampere-hours). However, in practice about twice the amount is used.

(*ix*) **Reduction of carboxylic acids:** Saturated aliphatic acids are reduced to the corresponding aldehydes by electrolysis using lead electrodes[24].

However, aromatic carboxylic acids *e.g.*, anthranilic acid is electro reduced in H_2SO_4 at 20-30° to give[25] the corresponding alcohol in 67-78% yield.

Unsaturated aromatic carboxylic acid (*e.g.*, Cinnamic acid, $PhCH=CHCO_2H$) on electro reduction[26] produces dihydro acids (*e.g.*, hydrocinnamic acid, $PhCH_2CH_2CO_2H$).

(*x*) **Reduction of aldehydes:** Unsaturated aliphatic aldehydes on electro reduction give the corresponding saturated alcohols[27].

$$\underset{\text{Crotonaldehyde}}{CH_3CH=CHO} \xrightarrow[\text{Pb cathode, EtOH, } H_2SO_4, H_2O]{\text{electro reduction}} CH_3CH_2CH_2OH$$

Reduction of both the double bond and the carbonyl group in unsaturated aromatic aldehydes is accomplished by electrolysis[28].

$$\underset{\text{Cinnamaldehyde}}{PhCH=CHCHO} \xrightarrow[\text{Pb cathode 20-25°, } H_2SO_4, \text{ EtOH, AcOEt, } H_2O]{\text{electro reduction}} \underset{55\%}{PhCH_2CH_2CH_2OH}$$

It is found that pyridine α-carboxyaldehyde on electrolytic reduction using lead cathode in 20% H_2SO_4 gives[29] a mixture of 41% α-picoline 25% α-picolinic acid and 11% 2-methyl-1,2,3,6-tetrahydropyridine.

(*xi*) **Reduction of alkenes:** Alkenes containing isolated double bond can be reduced electrolytically[30].

(*xii*) **Reduction of alkynes:** Alkynes can be reduced electrolytically. Internal alkynes on reduction is 10% H_2SO_4 in ethanol at spongy nickel cathode[20] give cis alkenes. In case electrolysis is carried out in methylamine solution of lithium chloride, trans-alkenes are obtained. The yields of the alkenes and the ratio of trans to cis alkenes depends on whether the electrolysis is carried out in divided or undivided cells (yields 24-80%, composition of product 89-99% of trans alkene)[31].

(*xiii*) **Reduction of amides:** Electrolytic reduction of aliphatic tertiary or secondary amides in undivided cells fitted with platinum electrodes and filled with solutions of lithium chloride in methylamine give the corresponding aldehydes. However some secondary and especially primary amides gave 51-97% yield of alcohols under the same conditions[23].

Electro reducton of phthalimide on a lead cathode in H_2SO_4 gave 72% yield of isoindoline[32].

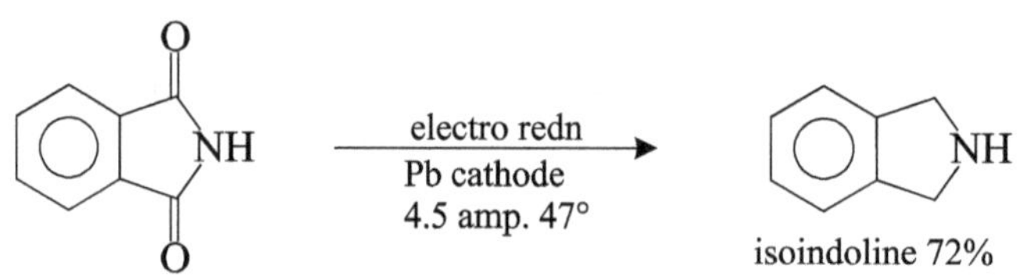

Phthalimide

(*xiv*) **Reduction of aromatic hydrocarbons:** Electrolysis of benzene and its homologues using platinum electrodes and lithium chloride in anhydrous methylamine yields partially hydrogenated products. When the electrolytic cell is undivided, the products are 1,4-cyclohexadiene and its homologues.

However, when the cell is divided by an asbestos diaphragm, cyclohexene and its homologues are obtained in high purities in 44-85% yields[22,33].

Double bonds conjugated with benzene rings are reduced electrolytically[30].

Triple bonds not conjugated with benzene rings can be partially reduced electrochemically in a solution of lithium chloride in methyl amine[31]. Electrochemical reduction of diphenyl acetylene in LiCl and $MeNH_2$ give 2% trans alkene and 78% completely reduced product[31].

$$\text{PhC} \equiv \text{CPh} \xrightarrow[\text{LiCl, MeNH}_2]{\text{electro redn.}} \underset{2\%\ \text{trans}}{\text{PhCH} = \text{CHPh}} + \underset{78\%}{\text{PhCH}_2\text{CH}_2\text{Ph}}$$

(*xv*) **Electrolytic reduction of Cephalosporin C:** A step in the conversion of Cephalosporin C to third generation cephalosporin antibiotics involves electrochemical reduction[34] with an aqueous phosphate buffer, pH 7-8 and a temperature of about 280K. The preferred cathode material is tin mesh.

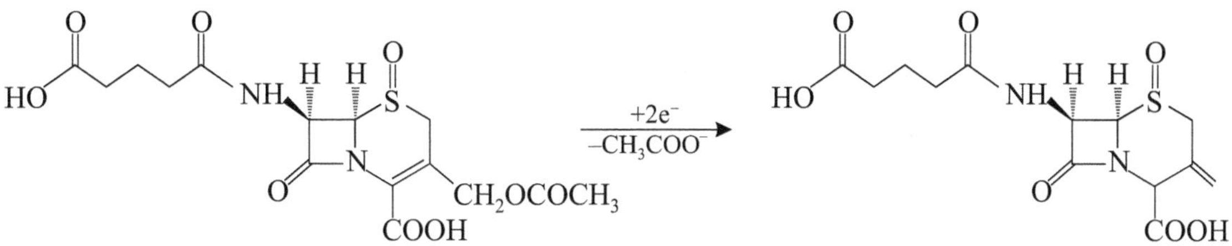

Conclusion

Electrochemical synthesis is now an independent discipline. The synthesis in general is of two types, *viz.*, anodic oxidation and cathodic reduction. The technology has been used for industrial preparation of a number of organic nitrites like adiponitrile, sebacic acid etc. A large number of oxidations and reductions have been performed using the electro oxidation or electroreduction technique. In fact electrochemical synthesis is considered to be the best choice in green chemistry.

REFERENCES

1. H. Kolbe, Ann., 1849, 257; B.C.L. Weedon, Quart. Rev., 1952, **6**, 380.
2. A.N. Frumkin in Advances in Electrochemistry and Electrochemical Engineering, Vol. 3, P. Delahayand, C.W. Tobias, Eds., Interscience, New York, 1963.
3. Topics in current chemistry, Vol. 142, E. Stekhan ed., springer-verlag, Berlin, 1987.
4. M.A. Fox, in Topics in current chemistry, Vol. 142, E. Stekhan, ed., Springer-Verlag, Berlin, 1987.
5. E. Steckhan in topics in Current Chemistry, Vol. 170, E. Steckhan, ed., Springer-Verlag, Berlin 1994.
6. R.D. Little and K.D. Moeller, The Electrochemical Society Interface, Winter 2002.
7. K.W.R. de Franca, M. Navarro, E. Leonel, M. Durandetti and J.-y. Nedelec, J. Org. Chem., 2002, **67**, 1838; C. Cannes, S. Condon, M. Durandetti. J. Perichon and J.-y. Nedeles, J. Org. Chem., 2000, **65**, 4575; C. Gosmini, Y. Rollin, J.y. Nedeles and J. Perichon, J. Org. Chem., 2000, **65**, 6024.
8. M. Durandetti, J.-y. Nedeles and J. Perichan, Org. Lett., 2001, **3**, 2073.
9. D.E. Danly and C.J.H. Ring, in electrochemistry, 3rd ed., H. Lund and M.M. Baizer eds, Marcel Dekker, New York, 1991; M.M. Baizer and D.E. Danly, Chemtech, 1980, **10(3)**, 161.
10. D.E. Danly, A/ChE Symp. series 77(204), **39** (1981).
11. U.S. Patent, 3, 652, 430 (1972), F. Beck, W. Himmele, J. Haufe and A. Brunold (to BASF AG).
12. T. Shono, Y. Matsumura, J. Hayashi and M. Mizoguchi, Tetrahedron Lett., 1979, 165.
13. F. Fichter and F. Ackermann, Helv. Chim. Acta, 1919, **2**, 563.
14. J. Kaulena and H.J. Schäfer, Synthesis, 1979, 513.
15. M. Kulka, J. Am. Chem. Soc., 1946, **68**, 2472.
16. F.L.M. Pattison, J.B. Stothers and R.G. Woolford, J. Am. Chem. Soc. 1956, **78**, 2255.
17. J. Yoshida, J. Hashimoto and N. Kawabata, J. Org. Chem., 1982, **47**, 3575.
18. J. Tsuji and M. Minato, Tetrahedron Lett., 1987, **28**, 3683.
19. H. Ruholl and H.J. Schäfer, Synthesis, 1988, 54.
20. K.N. Campbell and E.E. Young, J. Am. Chem. Soc., 1943, **65**, 965.
21. L. Horner and H. Röder, Chem. Ber., 1968, **101**, 4179.

22. R.A. Benkeser, E.M. Kaiser and R.F. Lambert, J. Am. Chem. Soc., 1964, **86**, 5272.
23. R.A. Benkeser, H. Watanabe, S.J. Mels and M.A. Sabol, J. Org. Chem., 1970, **35,** 1210.
24. J. Tafel and G. Fridrichs, Chem. Ber., 1904, **37**, 3187.
25. G.H. Coleman and H.L. Johnson, Org. Syn. Coll. Vol., 1955, **3**, 60.
26. A.W. Ingersoll, Org. Syn. Coll. Vol., 1932, **1**, 311.
27. H.C. Brown and G.G. Pai, J. Org. Chem., 1982, **47**, 1606.
28. H.D. Law, J. Chem. Soc., 1912, **101**, 1016.
29. M. Ferles and A. Tesarova, Collect. Czech. Chem Commun, 1967, **32**, 1631.
30. K. Tungnans, Chem. Ber., 1973, **106**, 3465; 1974, **107**, 319.
31. R.A. Benkeser and C.A. Tinchner, J. Org. Chem., 1968, **33**, 2727.
32. A. Dunet, R. Rollet and A. Willemart, Bull. Soc. Chem. Fr, 1950, 877.
33. R.A. Benkeser and E.M. Kaiser, J. Am. Chem. Soc., 1963, **85**, 2858.
34. G.D. Zappi, 12th International forum on Electrolysis in the Chemical Industry–clean and Efficient Processing, Sheraton Sand Key, Cleanwater Beack Florida, October, 1998.

18. PHARMACEUTICAL INDUSTRY: NOW AND THEN

18

Pharmaceutical Industry: Now and Then

18.1 INTRODUCTION

Pharmaceuticals have not only given us relief from suffering but also increased our average life span. The average life span of human beings was ~50 years in 1950–51 and has increased to ~ 68 years in 2015. Pharmaceutical industry is one of the most important industries for us but unfortunately is one of the most polluting one today. We not only need medicines but also need a clean and healthy environment to live as many diseases are caused by unhealthy air, water and soil. The importance of pharmaceutical industry is because

- There is a constant need of medicines that are used today and sometimes an urgent need of new pharmaceutical. The challenges are numerous
  (*i*) The infecting microbes evolve and develop resistance to the available drugs.
  (*ii*) New diseases are noticed.
  (*iii*) Toxic effects of some drugs come to light hence their use is discontinued and these have to be replaced by new drugs.
  (*iv*) Chemists are always exploring for more potent and less toxic coumpounds.
  (*v*) The cost of the life saving drugs should be affordable.
- The R&D plays a crucial role in all the above mentioned factors but the major issues are:
  - Most pharmaceuticals have complex structure, are produced by multistep synthesis and conventionally require purification of the intermediate compound at every stage.
  - The consumption of chemicals, especially the solvents are exorbitant because most reactions are carried out in a single or a mixture of organic solvents. Solvents are used for extracting, washing and purification of the desired product (like chromatography etc.).
  - The E factor (defined as the kilogram of waste per kilogram of the required compound) therefore is normally 25–100 in pharmaceutical industry and it is not uncommon to find E factor in excess of 100 in some pharmaceutical synthesis.
  - Many synthetic chemists are even today adopting the principle of hit and trial method for discovery of new drugs and thus synthesise a large number of structurally related compounds. In the process use enormous amounts of chemicals and generate huge amounts of chemical waste.

- ♦ Scaling up the procedure for commercial production is a challenge.
- ♦ Out of a large number of synthetic compounds, only few are found to posses some biological activity.
- ♦ Only a small number of active compounds meet the conditions laid down for human trials.
- ♦ And finally a very small percentage of these is found to be suitable for human consumption as medicines.
- ♦ Out of ~ 5000 molecules that are synthesised and used for preclinical testing, only ~5 reach the stage of human trial and may be only one makes it to the market.

- There are three phases of human trials. In phase I trial, which takes about one year, the drug is tested on a small sample of healthy volunteers (>80). In phase II trials, the drug is tested on a larger sample of volunteers (100–300) which takes about two years. In final phase III trial, the drug is tested on patients (1000–3000) and it takes about three years. After this the NDA (New Drug Application) with reports of all trials is submitted to the authorities like US FDA (Food and Drug Administration in USA) and CDSCO (Central Drug Standard Control Organisation in India) or similar regulatory bodies in other countries.
- After the use of drug is granted, post marketing studies (phase IV studies) are mandatory and have also to be submitted.
- On an average, development of a new drug costs 1.3–1.8 million dollars and it could be as high as 4 million to 11 million dollars in a few cases. And it takes 12–15 years to bring a new drug to the market.

All the above mentioned factors add to the cost and because of the very high E factor, harm the ecosystem.

The development of a new drug is cost and time intensive.

18.2 Green Chemistry in Pharmaceutical Industry

The demand for new and existing drugs will always be there but a few steps in the right direction (greening the processes in pharmaceutical production) for the discovery and synthesis of drugs can actually help save the environment and mother earth. In the past companies had reservations in changing the existing technologies as there were concerns of adverse economic impact. Challenge was first accepted by a very few pharmaceutical companies, as a result of which a large number of alternative green synthesis have been achieved and implemented on a commercial scale. All companies should make efforts to go green. It is almost a well established fact now that green options for synthesis have proved to be low cost options.

Greening of the pharmaceutical processes is already catching up fast but much more needs to be done. Here is a brief account of what can be and has already been acheived.

18.2.1 Computer Aided Drug Designing

There is a slow down in the new medical therapies because drug discovery, its human trials and scaling up the production are all time requiring processes.

Logical, well discussed and thought of approach should be adopted for synthesis of drugs. Time should be spent in finding useful literature and planning a synthesis on paper. Computational simulations are a handy tool. Any number of compounds can be synthesised on the computer screen!

Designing simple molecules computationally helps to eliminate the time, resources that are spend in the synthesis of enormous number of compounds that turn out to be either inactive or fail during human trials. A number of programs and software are available for drug development. Computational methods of designing a lead compound requires a very detailed knowledge of drug-target interactions and is a challange because of the inherent complexicity of biological systems.

Drug Designing includes

- Identification of biological targets.
- Designing a series of molecules that fit into the active sites and identifying the active compounds.
- Study of pharmacokinetics, pharmacodynamics and potential side effects of the lead/ designed molecules.

18.2.2 Redesigning The Drug Synthesis

Existing methods of some life saving drugs should be modified partially or completely keeping in mind the atom economy, E factor, eliminating the use of hazardous chemicals and solvents, in short cutting down the production of waste.

18.2.3 Reducing the Consumption of Solvents and Choosing Benign Solvents

As already stated solvents are used for synthesis, extraction and purification etc., in the pharmaceutical industry. These constitute ~ 60–85% of the raw material used in the pharmaceutical production.

Eliminating the use of a solvent altogether in the pharmaceutical synthesis may be a dream today, tomorrow it may be a reality!! Reducing the use of solvents and hence the E factor and replacement of hazardous solvents (red solvents) by benign solvents (green solvents) is **Yes** definitely possible and has been achieved by a number of pharmaceutical companies. Here are a few points:

(*i*) If possible, a synthesis should be performed without the use of a solvent (solvent less synthesis).

(*ii*) Water is one of the most benign and low cost solvent. The problem in using water as a solvent in organic synthesis is that most of the organic compounds are insoluble in it. Efforts are being made to carry out reactions on the surface of water.

Another problem in using water for the synthesis of a water soluble compound is that removal of water to obtain the product is a high energy process.

(*iii*) Ionic liquids are being used as alternative solvents.

(*iv*) Supercritical liquids are another category of solvents being used in pharmaceutical industry. Supercritical carbon dioxide is one of the solvent of choice as it is available in plenty, can be recycled and leaves no solvent residue.

(*v*) Data is available in literature, for example GSK solvent profile, to make a suitable choice of the solvent which are green or desirable and can replace the red or undesirable solvents used in the past for pharmaceutical synthesis. A number of other pharmaceutical industries have also developed their own solvent selection guides.

18.2.4 Biocatalysts

Biocatalysts can play a tremendous role in greening the pharmaceutical industry. Though some of the potential biocatalysts have been explored by synthetic chemists, a lot are still unutilized.

Biocatalyst can be used in pure form or as complete cells containing these. The main advantages of using them is that the procedures are energy efficient because the reactions which otherwise require drastic conditions can generally be carried out at much lower temperature and pressure etc. Under mild reaction conditions, the yields may be high because fewer side products are formed. Another major advantage of using biocatalyst in pharmaceutical industry is that the required active stereo isomer can be obtained in higher yield because of the specificity of the enzymes. An inactive isomer can be isomerised to the active isomer by use of biocatalyst there by increasing the yield. Advancements made in bioengineering can be used to genetically enhance the catalytic activity of some enzymes or tailor new enzymes for specific reactions. All these green modifications will not only reduce the waste (reduce the E factor) but save energy, reduce the use of non-renewable chemicals and above all reduce the cost of life saving drugs and thus make them affordable to common man.

Oxidation and reduction reactions which are very common in the production of pharmaceuticals are generally carried out using hazardous chemicals or catalyst. Replacing these with benign process can help in making pharmaceutical companies eco friendly. One of the options is to use oxygenases and reductases as biocatalysts for these reactions and oxygen or hydrogen peroxide as oxidising agent. Reductases can replace chemical catalyst for reduction reactions.

Development of biopharmaceuticals can also help in greening the pharma industries.

18.2.5 Continuous Flow Process

Traditional drug designing and synthesis was a time consuming process and sometimes the supply rate of drug did not match with the sudden increased demand due to sudden epidemic. The continuous flow process systems developed in the last twenty years which includes a range of tools, equipment and technologies are now available to a synthetic organic chemist for a given transformation or the entire sequence of reactions. It is now possible in certain cases to synthesise a drug in a few hours. The continuous flow process has really proved to be a boon for the pharmaceutical industries. Traditionally the drugs were synthesised in batches and the steps of isolation, purification and resolution etc. of the intermediate and the final product required time. In the continuous process, the reaction/s, the work up, purification, inline analysis of the intermediates and the final product (using techniques like FTIR) is performed automatically. The reactants flow through tubes or equipment and in a multistep synthesis, the new reactants, other ingredients or solvent etc. are added at specific point during the flow. There are innumerable advantages of this technique.

(*i*) Human handling of chemicals and therefore exposure is drastically reduced.

(*ii*) Human error is minimised as there are multiple check points.

(*iii*) Procedure once standardised is exactly reproducible.

(*iv*) It is a time efficient technique, a lot of time and labour is saved.

(*v*) Rapid diffusion mixing of reactants avoids the issues which are associated with batch reactors.

(*vi*) A better control of reaction conditions. For example, rapid cooling and heating can be achieved or there is a better control of reaction temperature.

(*vii*) Loss of valuable material is reduced since the intermediates are not isolated.

(*viii*) Labour cost is reduced.

(*ix*) Last but not the least, the solvent consumption is reduced phenomenally.

18.2.6 Semi Synthetic Approach

It is an interesting approach that has been developed for certain pharmaceuticals. It involves a combination of isolation of the intermediate from natural sources and its synthetic conversion to the active pharmaceutical ingredient (API). This technique has proved beneficial in cases where

- The API can be extracted from its natural source but the yield is extremely low (sometimes < 0.0005%).
- Damaging the trees for isolation adversely affects the ecosystem.
- The total synthesis is very tedious, time consuming and the cost of the drug thus escalates.
- In some cases the intermediate is found in much higher amounts in nature or can be obtained by fermentation process.
- The intermediate is isolated and chemically converted to the drug.

18.2.7 Use of Renewable Resources

Resources that can be replenished at a rate equal to or faster than consumed should be used wherever possible. These include some agricultural and marine feedstock. Resources become nonsustainable if the rate of consumption is faster than production.

Currently, some solvents from renewable sources that are used are ethanol, methanol, glycerol, ethyl acetate, ethyl lactate, acetic acid and acetone etc. Use of non-renewable solvents from natural resources such as oil, natural gas and coal etc. should be minimised or elimnated altogether.

18.2.8 Reduction in the Use of Non-renewable Energy

Attempts should be made to reduce use of energy from non-renewable sources. Energy from sun and wind etc. should be harnest and used in pharmaceutical manufacture. There should be an earnest effort to reduce energy use in energy intensive processes.

Some of the measures that can be used are:

- It is a well known fact that catalytic processes are energy efficient.
- Use of microwaves for heating can help save energy.
- Removal of solvents to isolate the products are energy intensive steps. Eliminatingg or reducing the use of solvent during pharmaceutical manufacture will have multifold advantages.

18.3. Success Stories of Green Pharmaceutical Synthesis

18.3.1 Synthesis of Amoxicillin[1]

Amoxicillin (I) is a β-lactam antibiotic (a class of antibiotics most widely used worldwide) and is used to treat a wide variety of bacterial infections. Amoxicillin is an outcome of search for more effective and less toxic β-lactam antibiotics. Amoxicillin is an analogue of penicillin G (II) and is obtained from it by a semi synthetic route. It is preferred to penicillin because of it has broad spectrum of activity, high absorption rate and stability in acidic medium.

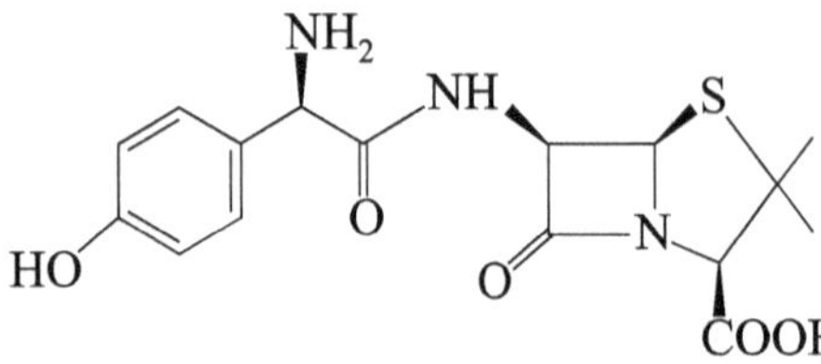

Amoxycillin (I)

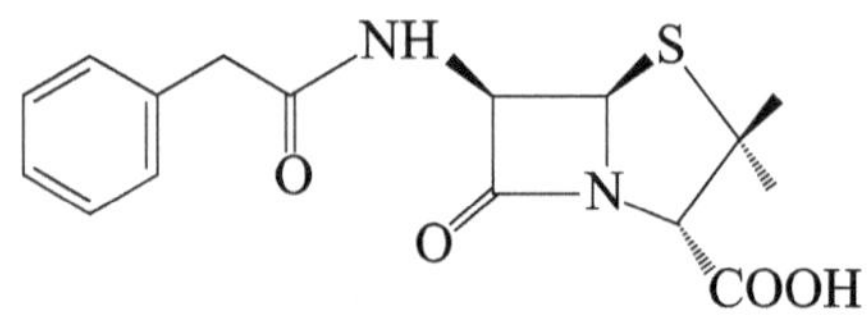

Penicillin G (II)

Penicillin itself is commercially produced biosynthetically from L-α-amino adipic acid, L-cystein and L-valine by the Scheme (18.1) outlined below.

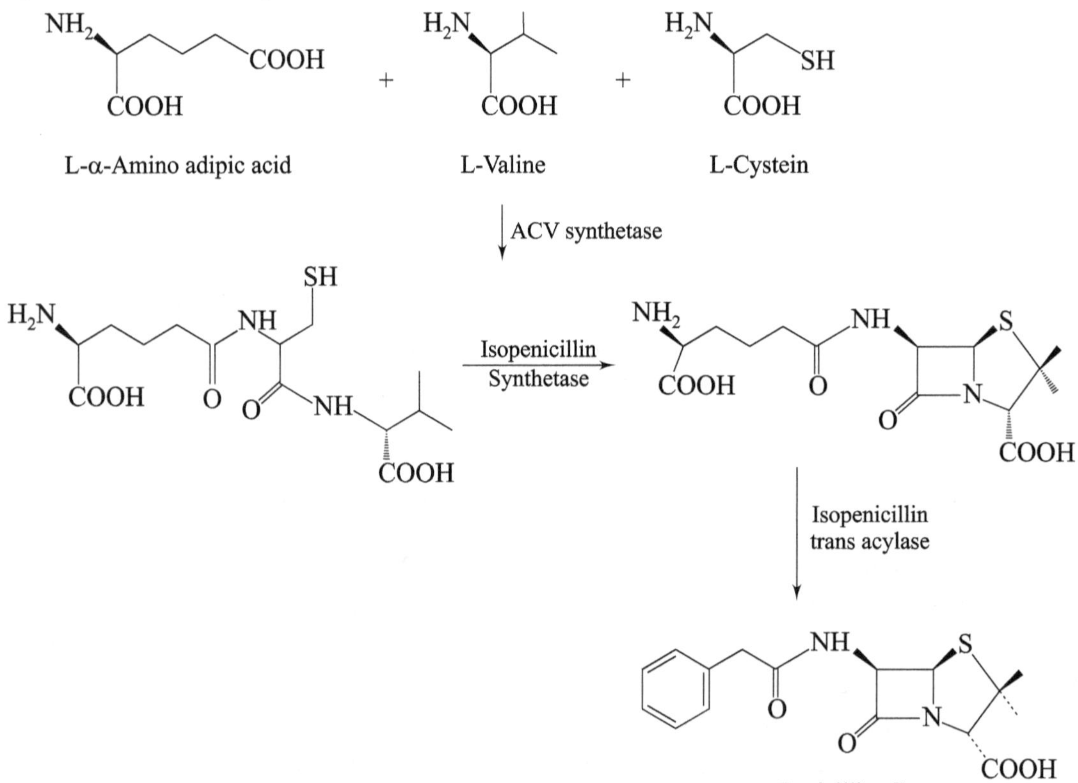

Penicillin G

Scheme 18.1: Biosynthetic Route to Penicillin G

Amoxycillin is conventionally prepared from penicillin G by the following sequence of reactions (*i*) the –COOH group of penicillin is protected by silylation with dimethyl silyl dichloride in CH_2Cl_2 (*ii*) the protected Penicillin G is converted to imino silylchloride by reaction with PCl_5 in CH_2Cl_2

at –40 °C (*iii*) the C_6H_5CO– group is then removed (deacylated) and the –COOH is deprotected and the resultant intermediate (6–APA) (111) (Scheme 18.2) is then substituted with the required side chain at the $–NH_2$ group.

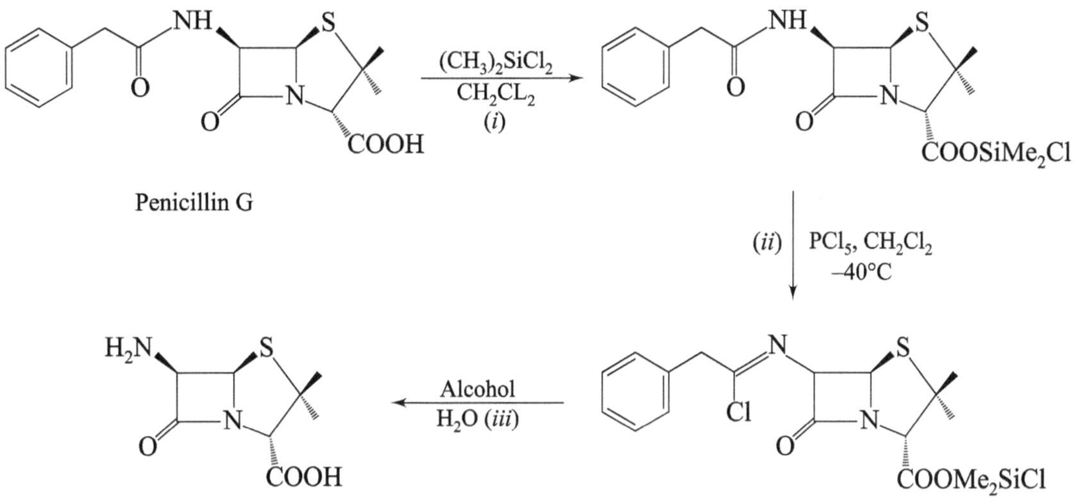

Penicillin G

6-APA(III)

Scheme 18.2: Conventional Synthesis of 6-Aminopenicillin Acid from Penicillin G

The three steps (*i*)–(*iii*) outlined above have a lot of environmental issues. The protection and deprotection of –COOH is unnecessary and involves use of stotiometric amounts of silylating agent, CH_2Cl_2, PCl_5. One of the reactions carried out at –40 °C is an energy intensive step.

Synthesis of amoxicillin was first modified by M. Cole in 1969 in which the steps of protection and deprotection were eliminated and deacylation was smoothly carried out by using Penicillin G acylase (PGA) at room temperature in water. The steps are outlined in Scheme (18.3). PGA is obtained from *Escherichia coli*.

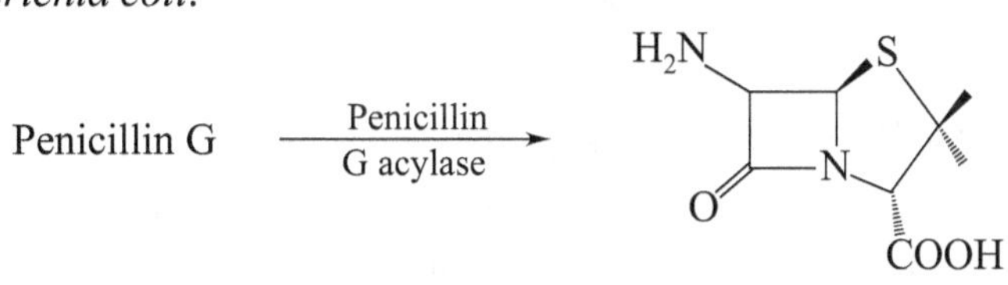

6-Aminopenicillin acid (6-APA)

Scheme 18.3: Modified Synthesis of 6-APA

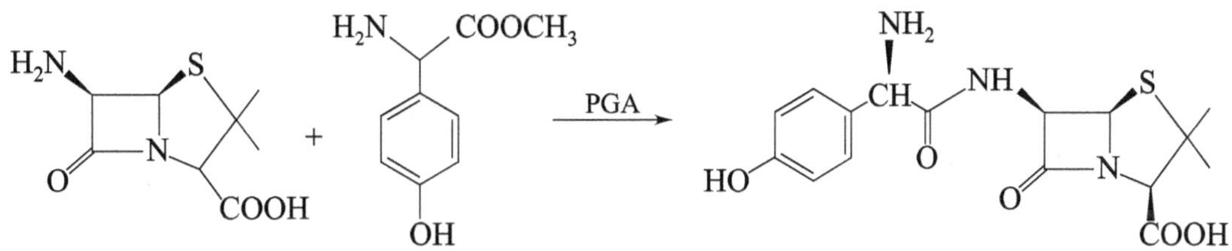

6-APA PHPGME Amoxycillin

Scheme 18.4: Synthesis of Amoxycillin

Though the above synthesis is more sustainable than the first synthesis, it suffered from the undesired hydrolysis reaction that occured during the synthesis.

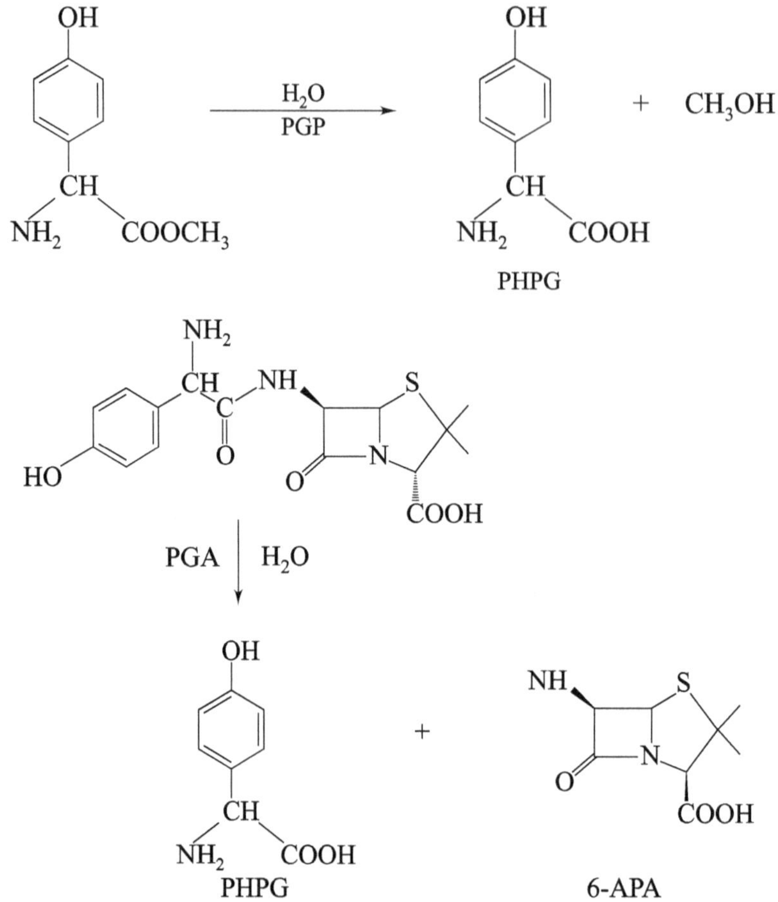

Scheme 18.5: Undesired Hydrolysis Reaction during Amoxycillin Synthesis

As a result of these undesired hydrolysis reactions, the yield of amoxycillin obtained is low. The yield of amoxycillin by chemical method is higher.

Efforts are being continuously made to improve the yield of Amoxycillin by biosynthetic routes.

Water used as a solvent during the synthesis is believed to be responsible for undesirable hydrolysis reactions (Scheme 18.5)

Studies have been done to replace water by ionic liquids. Best results were obtained in BMI-PF_6 as a cosolvent at 75% (V_{IL}/V_{water}) as compared to a purely aqueous system. (BM1-1-Butyl-3-methylimidazolium cation).

Combination of PGA with other enzymes like AEH (α-Amino Ester Hydrolase) have been used in the final coupling step (Scheme 18.4). The two enzyme system yields better result than the single enzyme system.

18.3.2 Synthesis of Aprepitant[2]

Aprepitant (IV) is an active component of Emend, a drug developed by Merck & Co. to combat nausea and vomitting often resulting from chemotherapy.

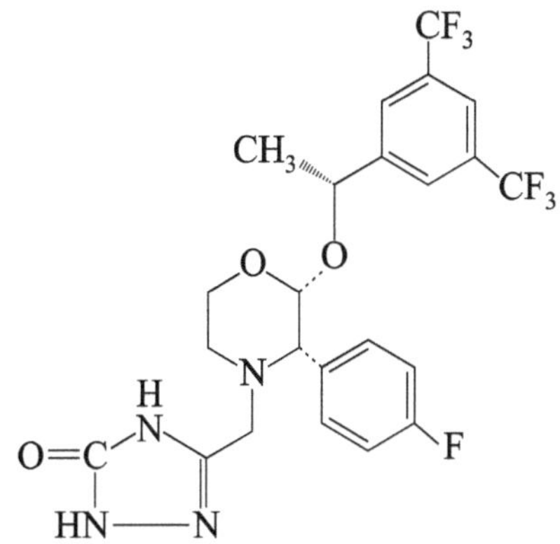

Aprepitant (IV)

First synthesis of Aprepitant was carried out by Merck. This synthesis used hazardous chemicals like sodium cyanide, dimethyl titanocene, hydrogen and ammonia, required cryogenic temperature for some steps and produced certain undesirable by products like magnesium chloride and methane. Merck company decided to look for an environmentally sustainable solution and developed a new synthetic route for the commercial production of Aprepitant for which they were awarded Presidential Green Chemistry Challenge Award for Greener Synthetic Pathway in 2005.

Apart from the adverse environmental impact of the first synthesis, it was also cost intensive requiring stoichiometric amount of an expensive chiral acid to set the stereochemistry of aprepitant. In the second generation synthesis, a chiral alcohol is used for crystallisation–induced asymmetric transformation to set the stereogenic centers in the subsequent steps. All these modification resulted in reduction in (*i*) quantities of raw materials, (*ii*) adverse environmental impact and (*iii*) cost of the drug.

Steps involved in the green synthesis of aprepitant

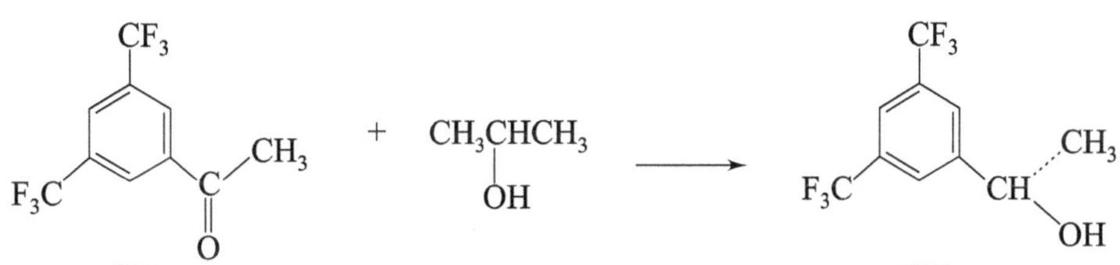

(V) (VI)

Ketone (V) is reduced to (VI) with isopropanol instead of hydrogen used in the first generation synthesis. Assymetric hydrogenation is carried out using Noyori catalyst under mild conditions.

The other fragment (VII) is produced as outlined below:

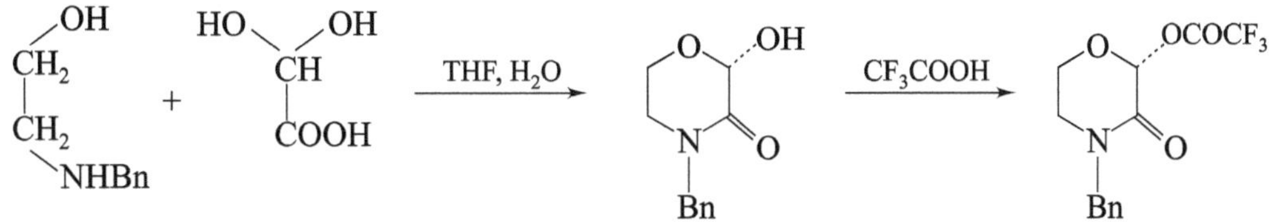

(VII)

The two fragments (VI) and (VII) are then converted to aprepitant by the scheme outlined below:

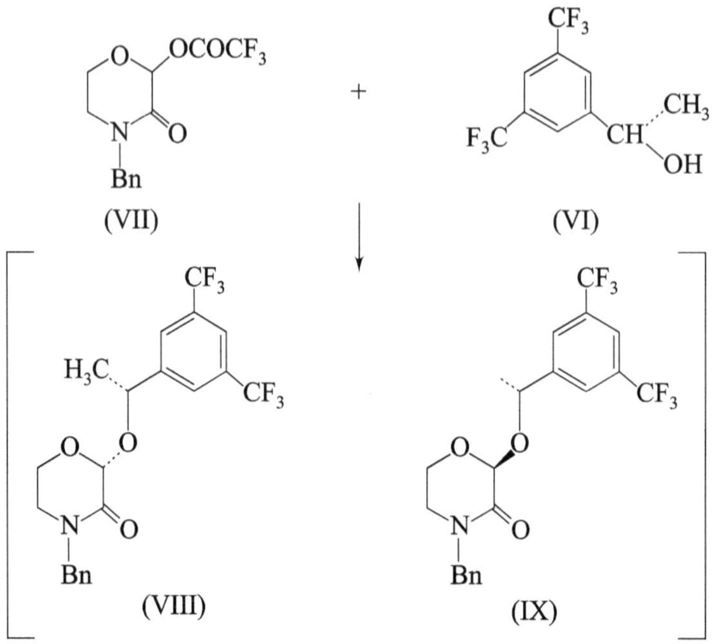

The yield of the required stereoisomer (VIII) increased by base-catalysed equilibriation of the mixture and crystallisation.

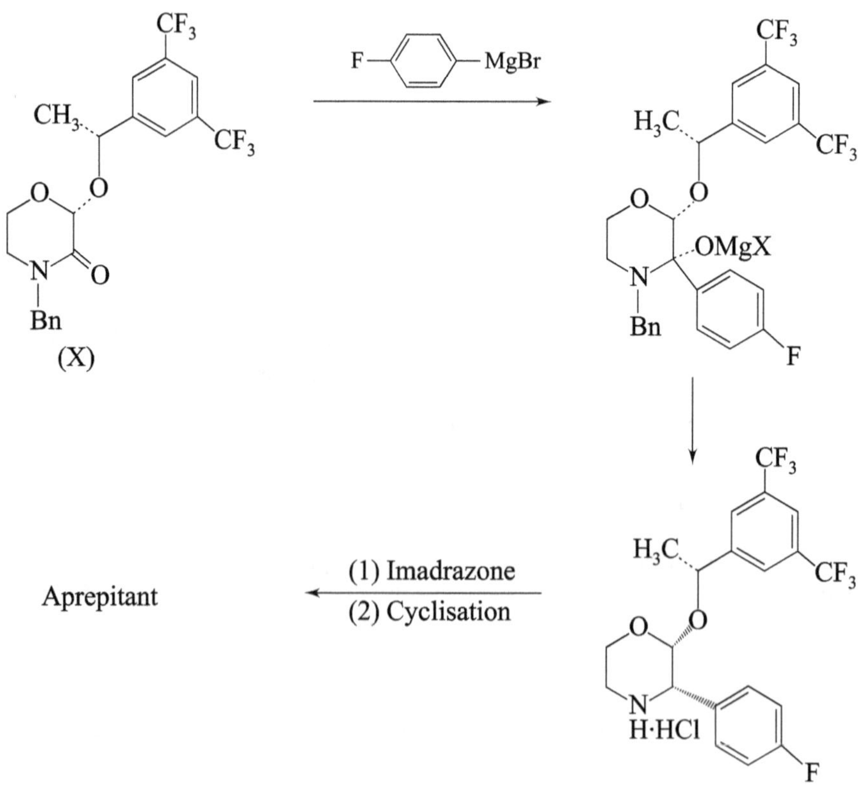

18.3.3 Synthesis of Artemisinin[3,4,5]

Artemisinin (X) a sesquiterpene lactone endoperoxide is an effective antimalarial (active against drug resistant strains of *Plasmodium falciparum*). WHO recommended Artemisinin based combination therapies for the treatment of malaria. Artemisinin is isolated from *artemisia annua,* commonly known as sweet wormwood. The yield of artemisinin is only 0.01–0.6% of dry weight of the natural source. The highest yield of 4.85–4.90% has been reported from the same plant found in Egyptian desert. The synthesis of the drug is complex. Low yield from natural sources and complicated laboratory synthesis are the reason for high cost of artemisinin.

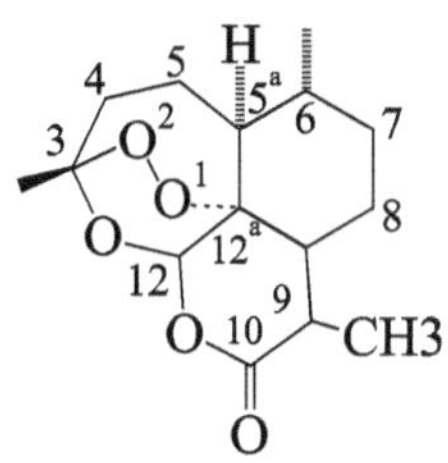

Artemisinic (X)

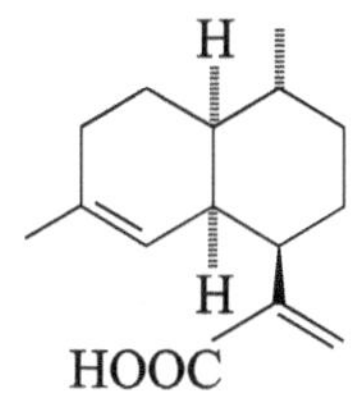

Artemisinic Acid (XI)

Artemisinic Acid, a precursor of artemisinin is obtained in good amount as a by product during the isolation of artemisinin from *Artemisia annua.*

Several attempts have been made to synthesise artemisinin from artemisinic acid. In one of the earlier attempt, dihydroartemisinic acid (XII) (obtained by reduction of artemisinic acid) was photooxidised at –78 °C in methylene chloride or at 0 °C in acetone and the product thus obtained is air oxidised in petroleum ether to give 17–32% isolated yield of artemisinin (Scheme 18.6)

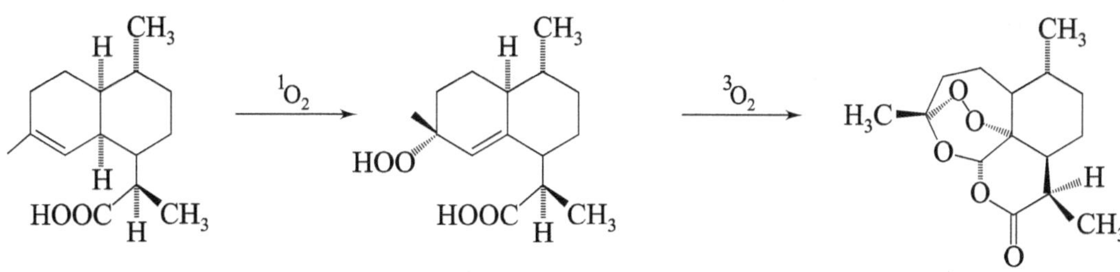

(XII) Artemisinin

Scheme 18.6: Artimisinin from Artemisinic acid

Green developments in the synthesis of artemisinin include the following

- Artemisinic acid is obtained by fermentation in high yield.
- Artemisinic acid from the fermentation broth is converted to artemisinin (scheme 18.6) using a chemical source of oxygen rather than the photochemical use of oxygen thereby eliminating the need for specialised photochemical equipment.

Strains of *Saccharomyces cerevisiae* (baker's yeast) have been developed which give a very good yield of artemisinic acid. This is then converted to artemisinin by the following scheme.

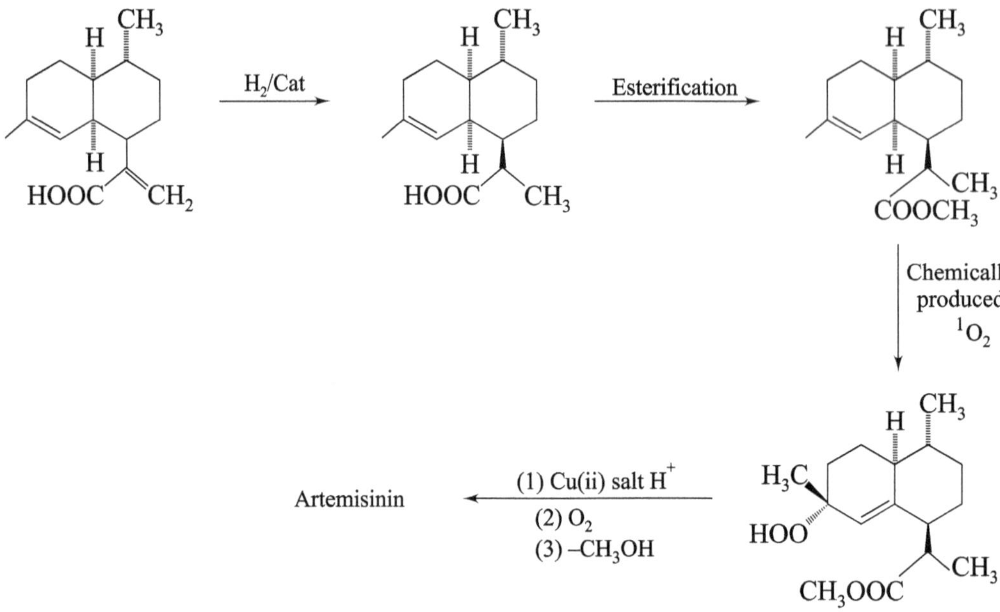

18.3.4 Synthesis of Ibuprofen[6,7]

One of the earliest examples of a green pharamceutical manufacture is that of Ibuprofen (XIII) [2-(4-isobutylphenyl) propionic acid].

Stewart Adams *et al.* working for Boot's company in UK prepared ~600 synthetic compounds in 1960 for anti-inflamatory properties, one of which later called Ibuprofen was found to be suitable and was patented in 1961. It was first launched in UK in 1969 and is now available over the counter without prescription.

The boot's synthesis of ibuprofen is outlined in Scheme 18.7.

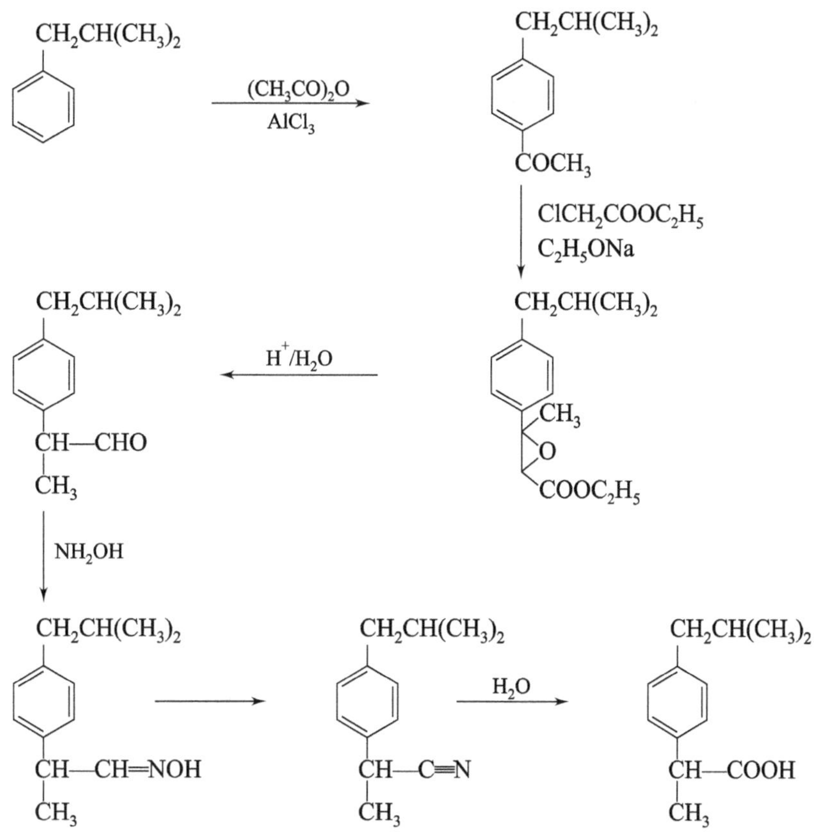

Scheme 18.7: Boots' Synthesis of Ibuprofen (XIII)

There were a lot of environmental issues linked with the Boot's synthesis. The synthesis was modified by BHC in 1992 for which the company was bestowed with Kirkpatrick Achievement Award for outstanding advances in chemical engineering technology in 1993 and Presidential Green Synthetic Pathway Award in 1997.

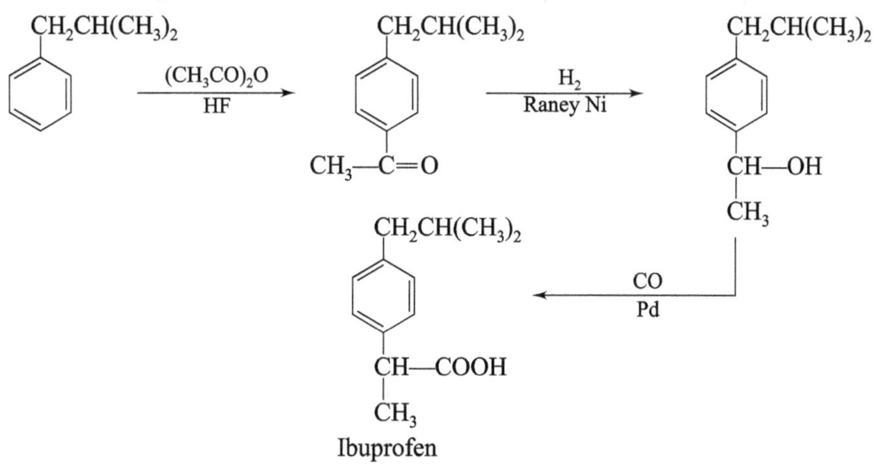

Scheme 18.8: BHC Synthesis of Ibuprofen

The modified synthesis is a three step synthesis compared to a six step synthesis. It uses HF both as solvent and catalyst. Virtually all starting materials are used because the byproducts are recovered and reused. Another modified synthesis[7] reported in 2014 is outlined below

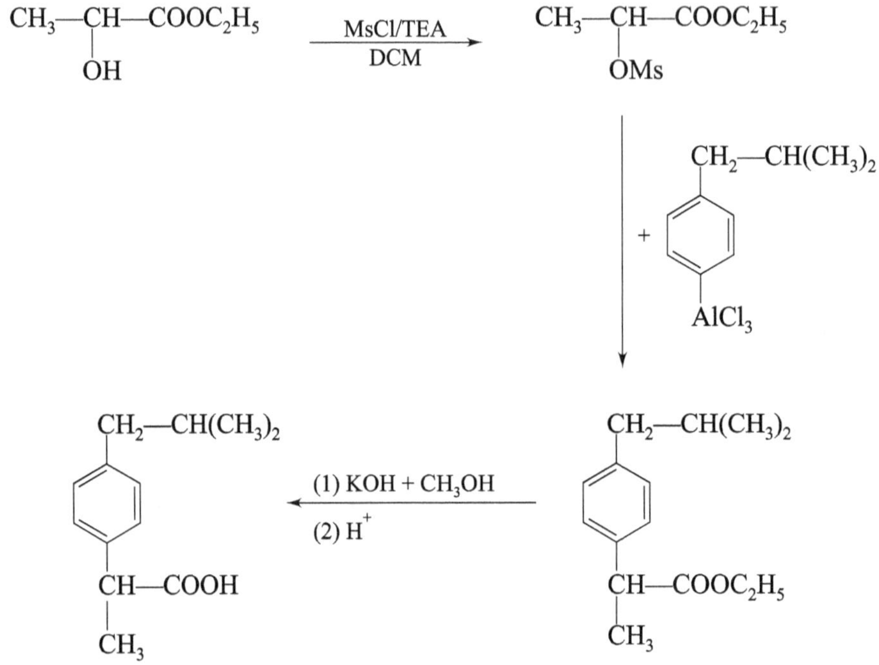

Scheme 18.9: Modified Synthesis of Ibuprofen

18.3.5 Synthesis of Imatinib[8,9,10]

Imatinib (XIV) is a product of rational drug design and was invented in late 1990s. The first clinical trial of the drug was carried out in 1998 and was approved by FDA in 2001. However, there are claims that it was first synthesised in 1993.

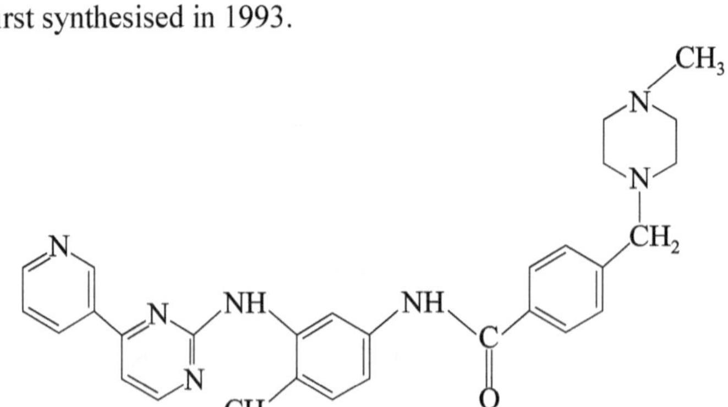

Imatinib (XIV)

Imatinib is used to treat a number of malignancies but is specifically used for chronic Myelogenous Leukemia (CML) and stomach tumors.

Imatinib was first synthesised by Zimmermann in 1993 by the Scheme (18.10) given below

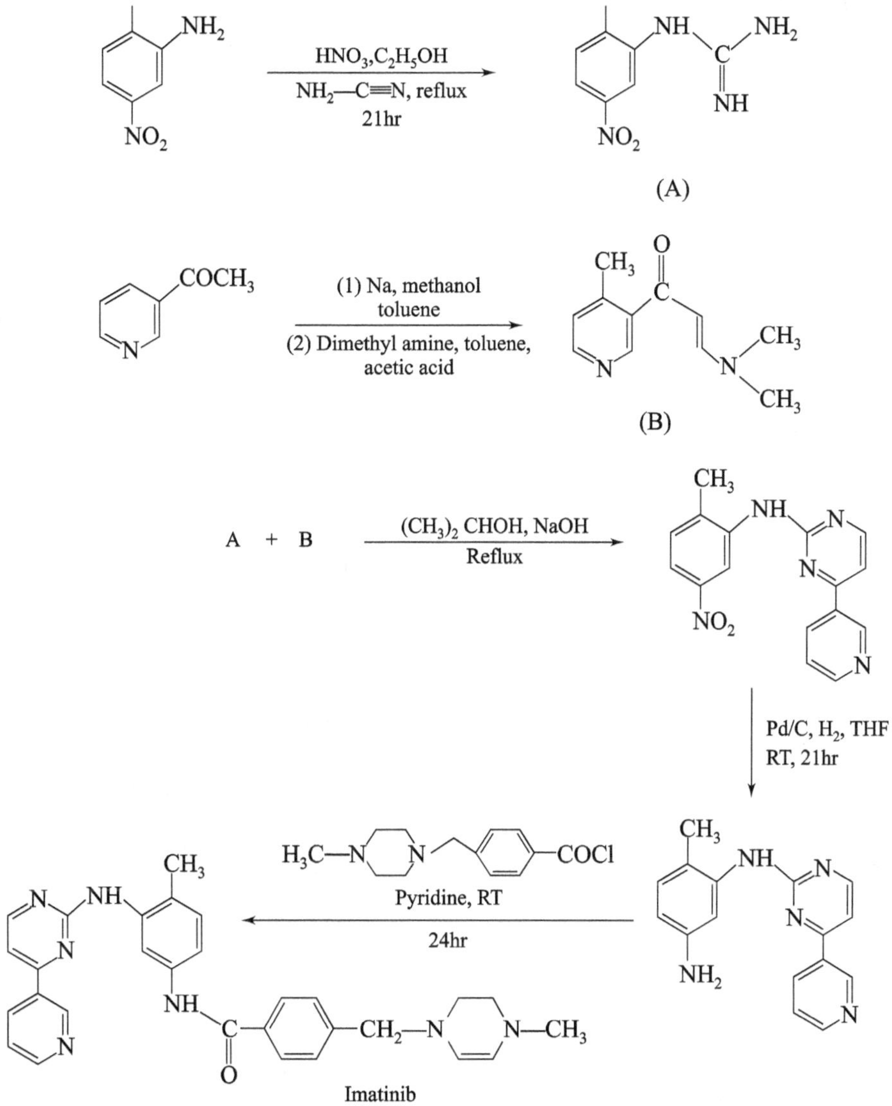

Scheme 18.10:Synthesis of Imatinib (XIV)

In 2003, Loiseleur et. al developed another route for the synthesis of imatinib.

Another synthesis was developed by Cui-Ling Wang[9] in 2008 outlined in scheme 18.11.

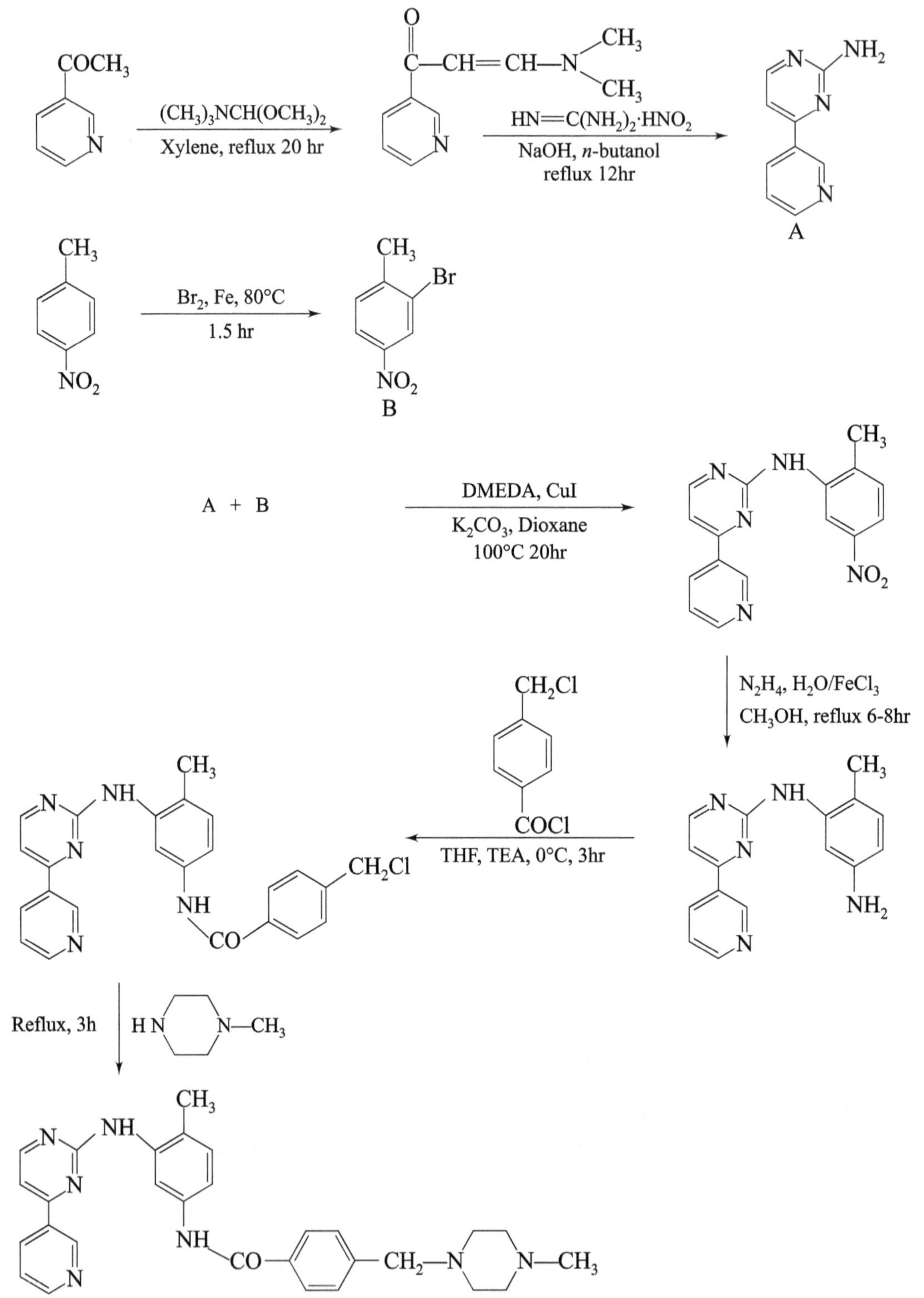

Scheme 18.11: Modified Synthesis of Imatinib

A flow based synthesis of imatinib was developed by Hopkin *et al*[10] in 2010 which is outlined below:

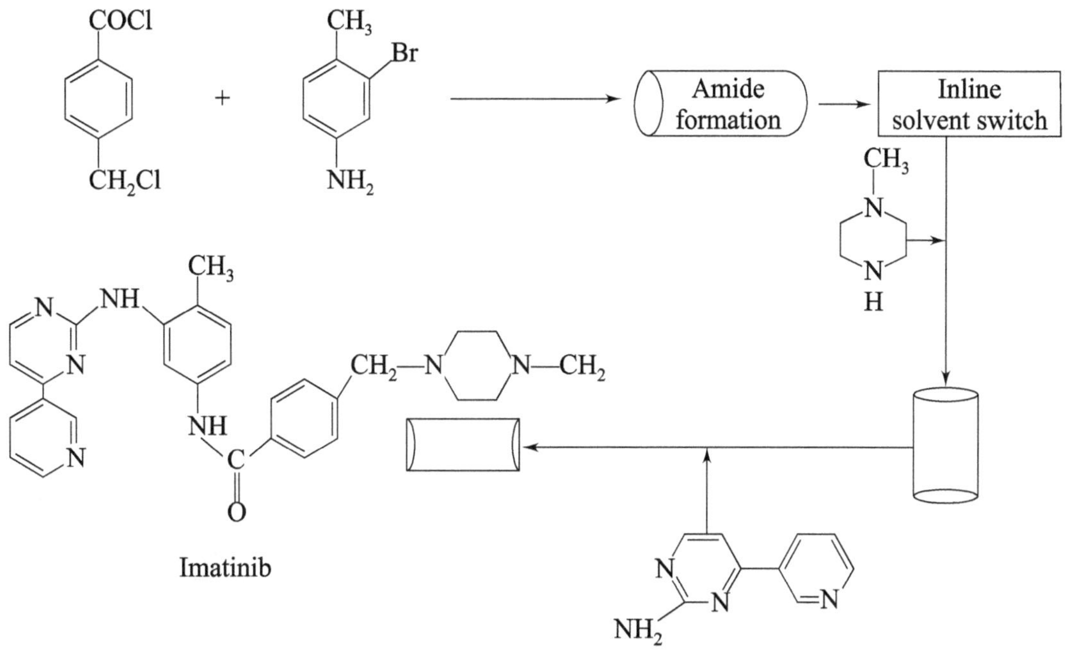

Scheme 18.12: Continuous flow synthesis of imatinib

The total yield of imatinib by the flow process is ~32%. Much higher than the previous synthesis. Thus the E factor has been improved.

18.3.6 Synthesis of Paracetamol (Acetaminophen)[11]

Paracetamol (XV) is a safe antipyretic and analgesic drug available over the counter without prescription and is even used in paediatric formulations. Unlike aspirin it is "gentle to the stomach" and does not exhibit toxicity like acetanilide etc. It was first sold in 1950 (though was synthesised much earlier) in USA in combination with aspirin and caffeine, this formulation was discontinued from the market because three of its users developed a blood disease. Later it was confirmed that the diseases was not due to usage of paracetamol. Paracetamol was finally launched in 1955. It was first synthesised from *p*-nitro phenol with tin in glacial acetic acid followed by acetylation and has been commercially produced since then by a number of methods.

A few other methods were developed during the course of time

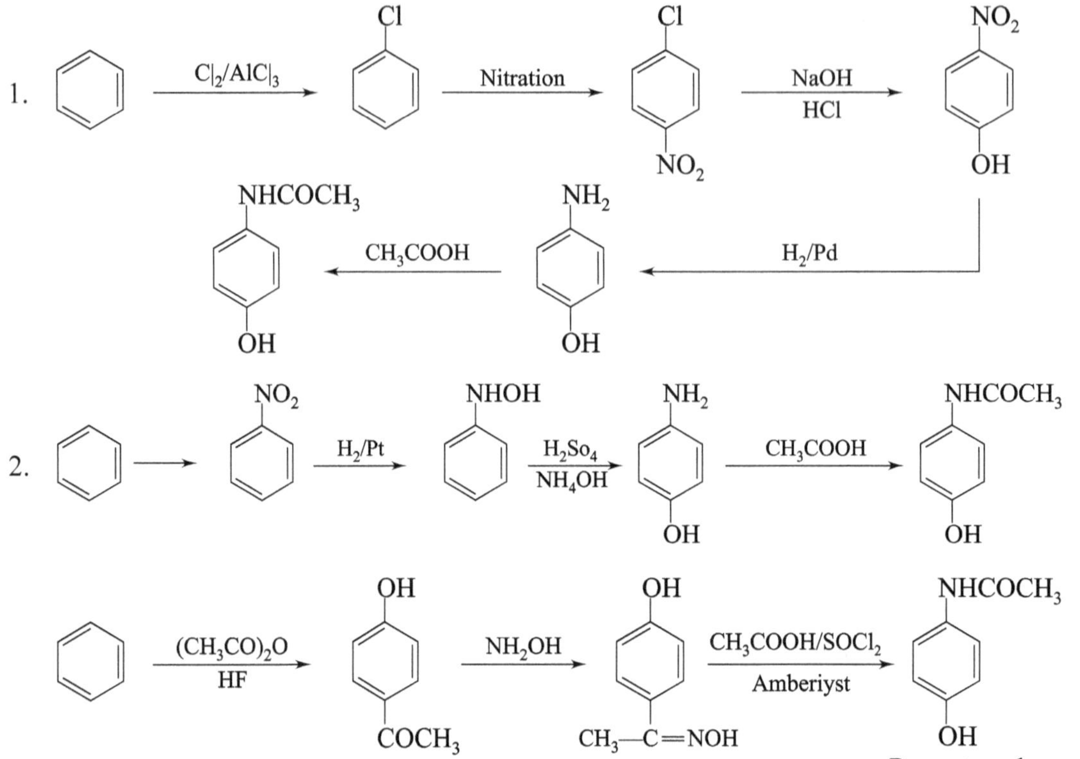

Paracetamol
(XV)

A green alternative synthesis" has been recently developed by Marc Lemaire *et al.* in 2014 in which paracetamol is directly obtained by amidation of hydroquinone with ammonium acetate at elevated temperature without any metallic catalyst or solvent. A yield selectivity of >90% is obtained. The method has been scaled up to 44g of hydroquinone. Best yield is reported at 180°C.

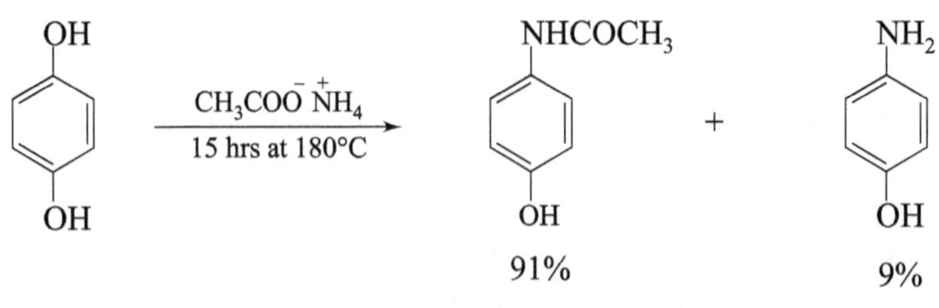

Scheme 18.13: Green Synthesis of Paracetamol

18.3.7 Synthesis of Pregabalin[12]

Pregabalin (XVI) is used as an antiepileptic and anticonvulsant drug. It was first launched in 2004 and was prepared by an all chemical method. The yield of the product was low (~20%). A modified biocatalysed synthesis has now been developed in which the yield has been doubled which can further be improved by recycling the inactive isomer to the active one.

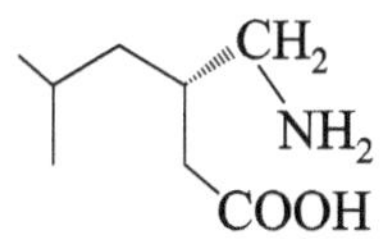

Pregabalin (XVI)

In the conventional commercial synthesis of pregabalin, isovaleraldehyde was condensed with diethyl malonate (knoevenagel condensation) to produce an unsaturated diester (*i*), addition of potassium cyanide to it produced a racemic cyanoester (abbreviated as CNDE). The CNDE was then hydrolysed and decarboxylated with potassium hydroxide, the –CN group reduced to $-CH_2NH_2$ with hydrogen in presence of nickel to produce a racemic γ-amino acid. The γ-amino acid was resolved with (S) + mandelic acid. An overall 20% yield of pregabalin (S-enantiomer) was obtained. The inactive (R) enantiomer of γ-amino acid could not be used, which added to the high E factor as it was a waste byproduct at that time. THF was used for recrystallisation during the resolution process.

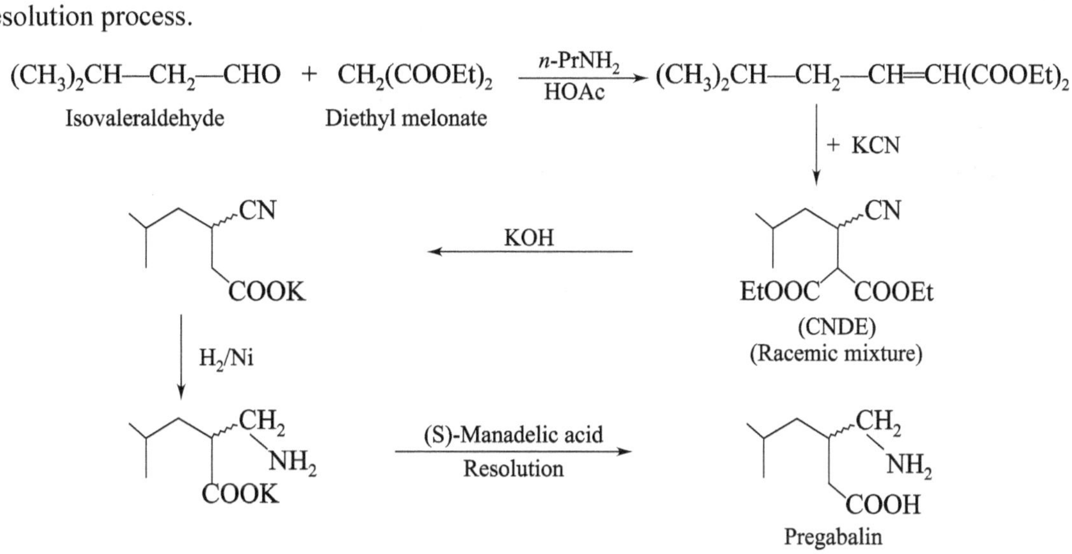

Scheme 18.14: Synthesis of Pregabalin

In the modified method, all the chemical steps involved in the synthesis were same as in the conventional synthesis. However, an enzyme catalysed (lipolase obtained from *Thermomyces lanigunosus*) conversion of the (R) CNDE to (S) enantiomer was developed in 2008 which resulted in doubling the yield of pregabalin (~45%) after the first cycle, above it the use of THF was eliminated and all the subsequent steps to form pregabalin could be carried out in water!

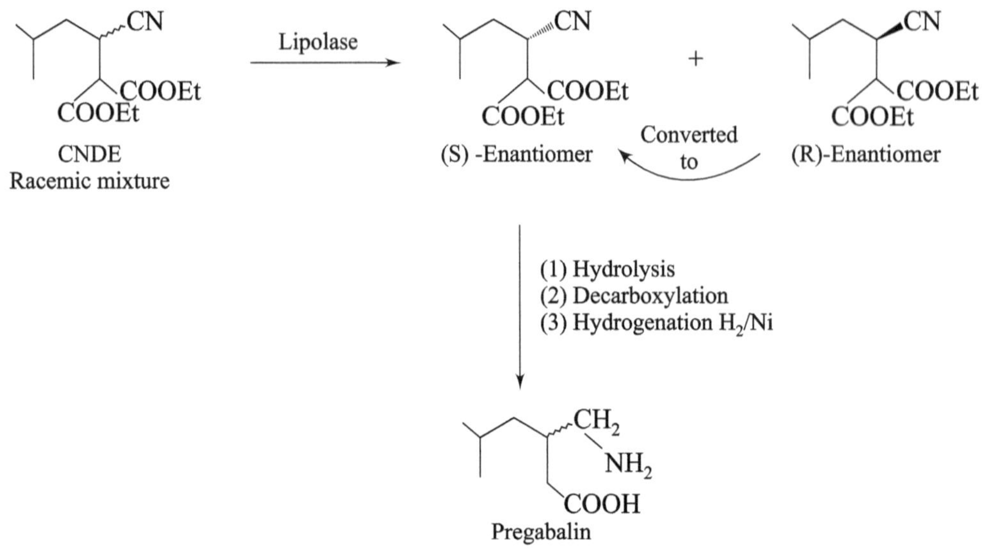

Scheme 18.15: Modified Synthesis of Pregabalin

18.3.8 Synthesis of 13-*cis*-Retinoic Acid[13]

13-*cis*-Retinoic Acid (XVII) is used for the treatment of acne.

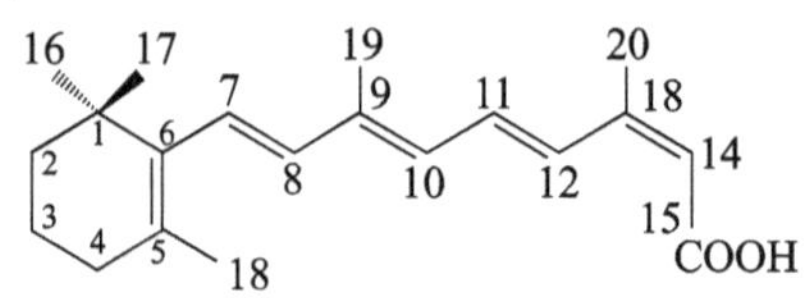

(XVII)

It is produced by the following sequence of reactions.

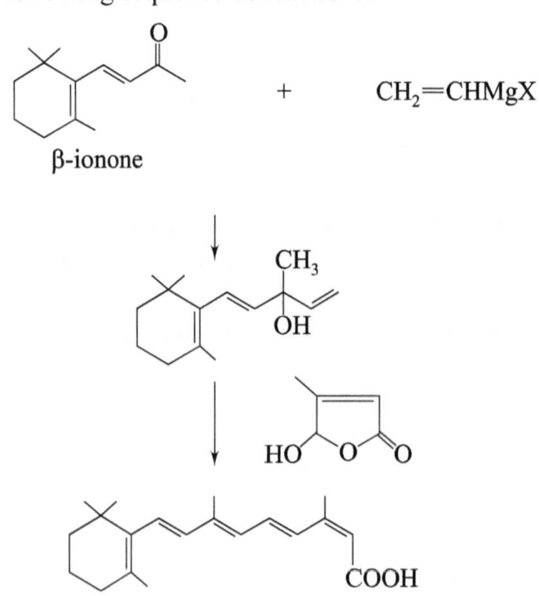

Scheme 18.16: Synthesis of 13-*cis*-Retinoic Acid

Along with the desired compound, two other compounds *i.e.* 11, 13-*di-cis*-retinoic acid (XVIII) and 11, 13-*di-cis*-trans retinoic acid (XIX) are produced.

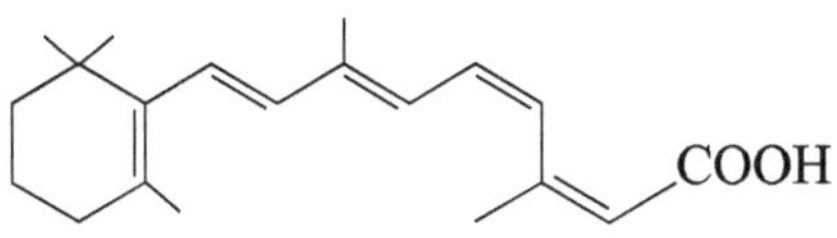

(XVIII)

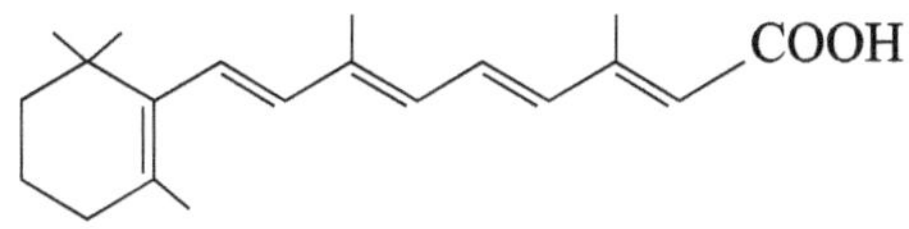

(XIX)

The required compound is obtained in low yield and separated by chromatography or fractional crystallisation.

There have been a number of attempts since then to improve the yield of ApI.* In one of such report, the required isomer was obtained in 10–30% yield and 70-–90% of (XIX) was obtained. (XIX) was subsequently isomerised to 13-*cis*-retenoic acid in presence of rhodium or palladium catalyst. Rh and Pd are not only very expensive but are difficult to separate.

A process for the isomerisation of the undesired compounds XVIII and XIX to 13-*cis*-retinoic acid has been developed[13]. The isomerisation is carried out photo-chemically in an aqueous solution of the alkali metal salts of the acids thereby

- The 13-*cis*-retenoic acid is obtained in good yield and does not require the purification from the inactive compounds.
- No heavy metals are used thus the process of removal of these is not required.

18.3.9 Synthesis of Sertraline[14]

Sertraline (XX), a drug of choice for depression, panic disorder and obsessive compulsive disorder was developed by Pfizer and was approved by FDA in 1991. Its first commercial synthesis is outlined in scheme 18.17. Many other routes have been proposed since then by synthetic organic chemists.

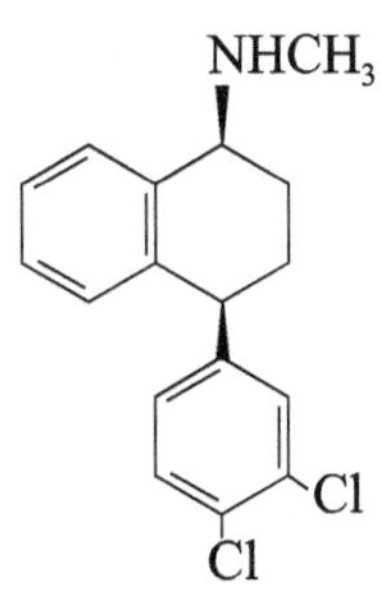

Sertraline (XX)

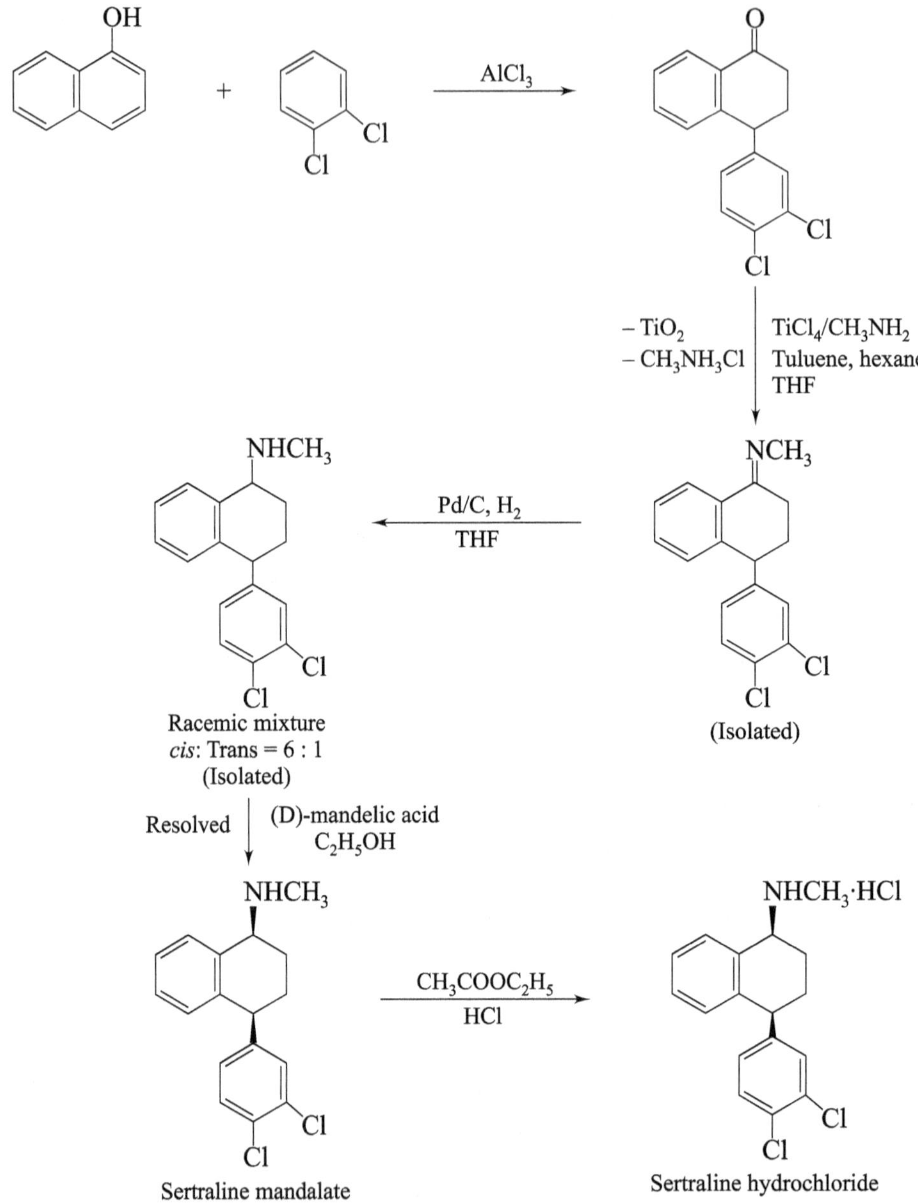

Scheme 18.17: First Commercial Synthesis of Sertraline

The major issue has been the use of $TiCl_4$, enormous amounts of undesirable solvents like CH_2Cl_2, THF, Toluene and Hexane and production of titanium Dioxide as a solid waste product.

A few steps of the synthesis were modified in Pfizer by Taber *et. al.* in which the use of all the red solvents and titanium tetrachloride is eliminated, instead ethanol is used as a solvent and titanium dioxide is obviously not produced. Also the amount of acid and base used during the synthesis are reduced in the green route. Pfizer was awareded US president Green Chemistry Challenge Award for green synthetic pathways in 2002.

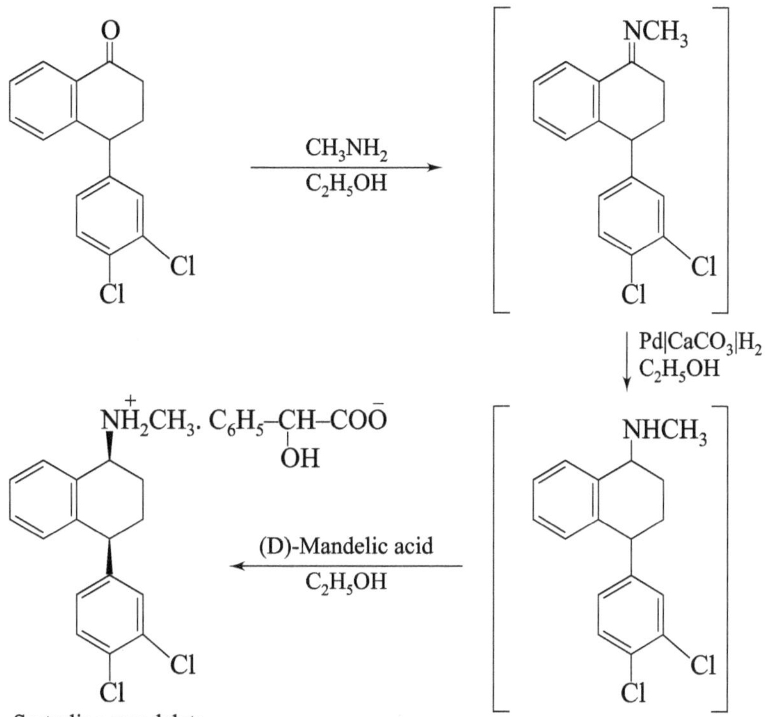

Scheme 18.18: Modified Synthesis of Sertraline

A comparison of solvent used in modified synthesis Vs the first synthesis is given in Table 18.1.

Table 18.1: Comparison of solvents used per 1000 kg of Sertraline

Solvent	1st Synth	Modified
C_2H_5OH	34,000 L	15,000 L
$CH_3COOC_2H_5$	28,400 L	9,000 L
THF	19,000 L	-
Toluene	8,000 L	-
Hexane	12,000 L	-
CH_2Cl_2	101,400	24,000/-

18.3.10 Synthesis of Sildenafil[15,16]

Sildenafil (XXI) an active ingrediant of Viagra was first prepared by a group of scientists at Pfizer to be tried for angina pectoris and hypertension. During the trials, it was realised that the drug had little effect on angina but induced marked penile erections. Since then it is marketed for erectile dysfunction. The drug was approved for sale in 1998.

Sildenafil was first synthesised[15] by the route outlined in scheme 18.19.

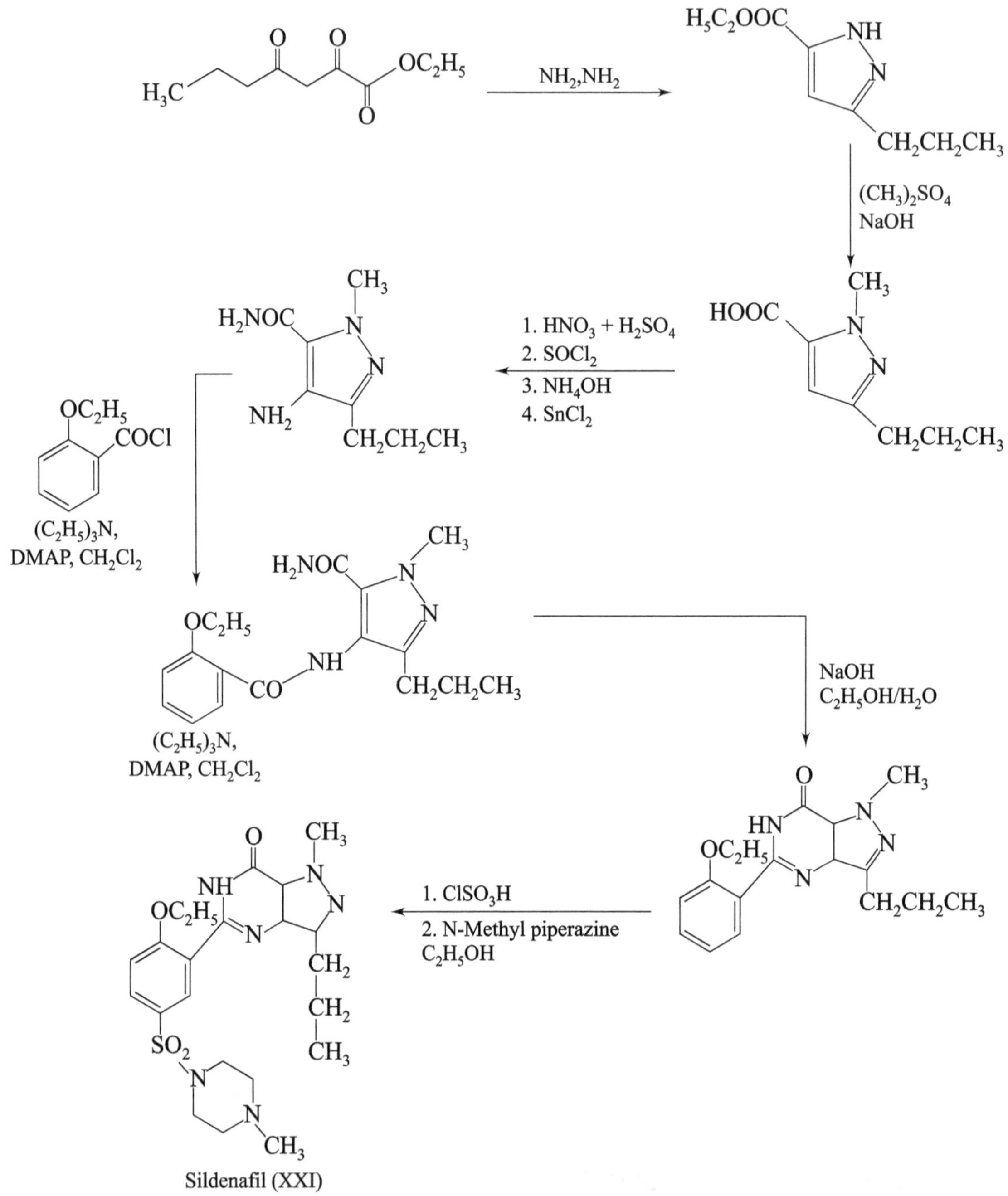

Scheme 18.19: Synthesis of sildenafil

Peter J Dunn *et. al.* working for Pfizer global research and development improved the method used for synthesis by reducing the waste production from 100 to 0.5%, yield enhancement from 9.8 to 75% and enormous reduction in the consumption of solvents as no isolation of the intermediates was required in the last seven steps of sildenafil citrate synthesis. Pfizer received 2003 UK Award for green chemical technologies. The two intermediate compounds are prepared first and reaction completed as indicated in scheme.

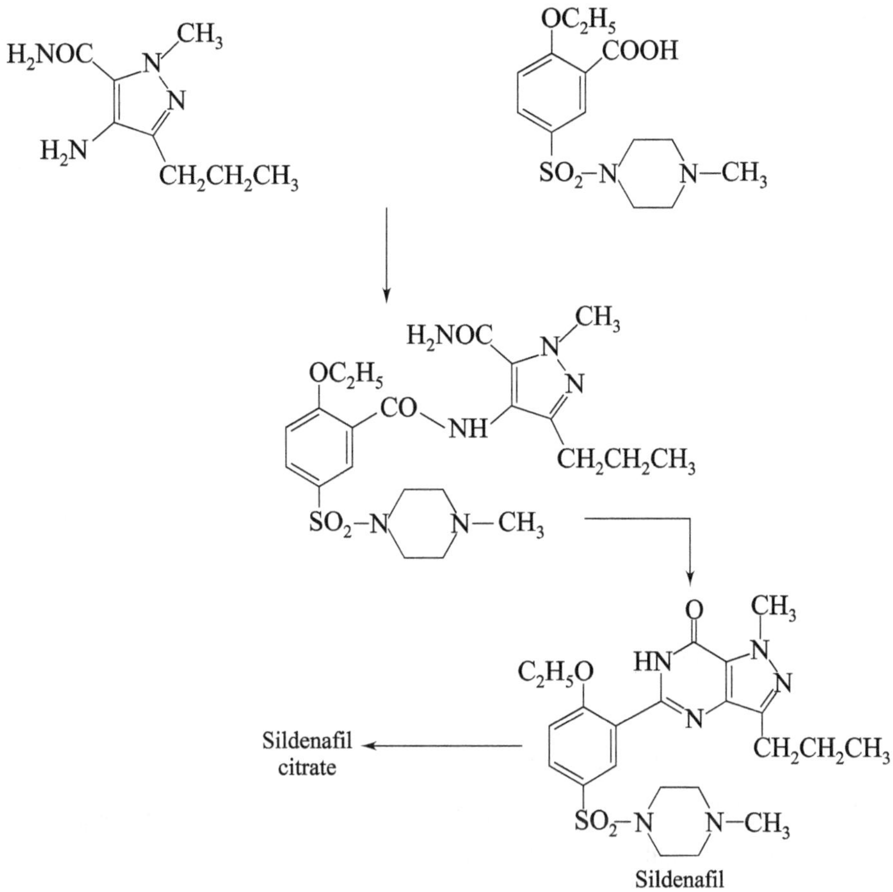

Scheme 18.20: Green Synthesis of Sildenafil

1300 litre of the solvents (Dichloromethane, Acetone, Ethanol, Methanol, Ether, Ethyl Acetate, Toluene, Pyridine, 2-butanone, I-butanol etc.) were used for the production of one kg of sildenafil in 1990. The consumption was reduced to 22 litres/kg in 1997 and only 7 litre of solvent per kg of the drug is used now.

18.3.11 Synthesis of Sevoflurane[17]

Seveflurane (1, 1, 1, 3, 3, 3-hexafluoro-2-(fluoromethoxy)-propane) (XXII) is a nonflamable general anesthetic used worldwide.

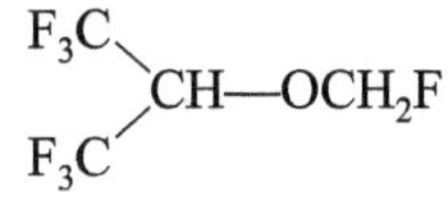

Sevoflurane (XXII)

It has been synthesised by a number of methods with an effort to reduce environmental impact, cost and hazard. A few of these are discussed in this section.

1. Hexafluoisopropanol is converted to sevoflurane in one pot synthesis by reaction with stoichiometric excess of para-formaldehyde, HF in presence of fuming sulphuric acid.

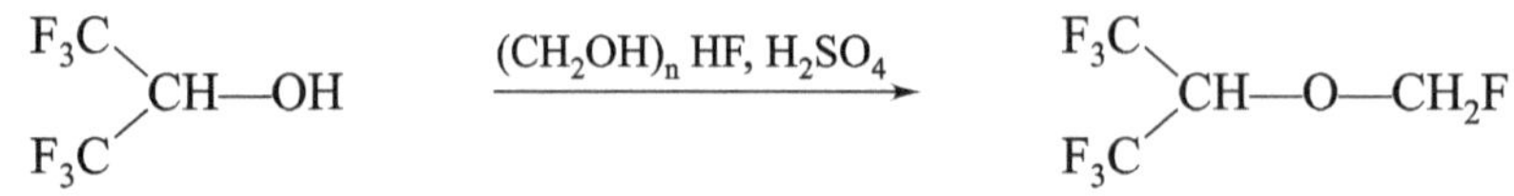

Formaldehyde is a beleived to be human-carcinogen, HF and H_2SO_4 are both hazardous and corrode equipment.

2. In this method hexafluoro-isopropanol is converted to sevoflurane by the following sequence of reactions

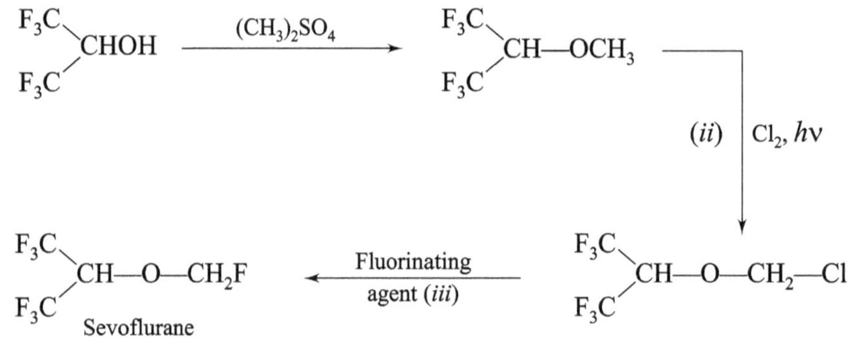

Scheme 18.21: Synthesis of Sevoflurane

Use of dimethylsulphate and chlorine and low yield in the chlorination step (*ii*) in the above scheme were the main issues in the above synthesis of sevoflurane.

3. Sevoflurane is obtained as given below:

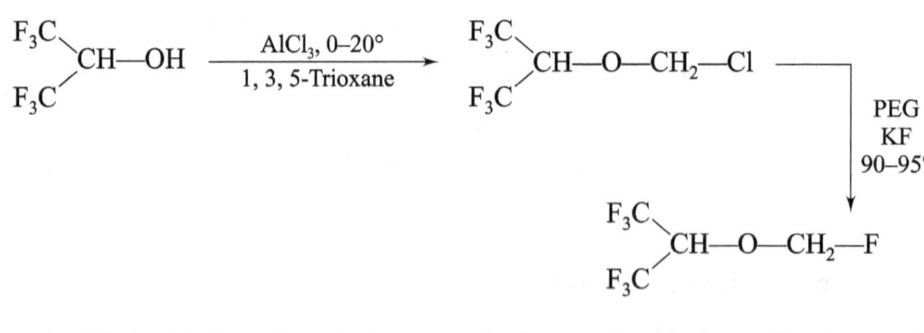

This method is better than the other two methods mentioned above. Further modifications[17] have been made to improve the yield, reduce time, chemicals and save energy compared to original two step synthesis of Sevoflurane, the reaction time in the first step is reduced from ~24 hrs to 4 hrs. This is achieved by marginal increase in $AlCl_3$ ratio. The aqueous phase used for quenching and washing the organic layer is reused.

In the second step, the amount of KF was reduced by 1.2 mole and use of PEG was reduced by 3.7 times which resulted in a major advantage. This is accompanied by a marginal reduction in the yield of the products. The benefits overweigh the loss and hence these modifications can be used for commercial production of Sevoflurane.

18.3.12 Examples of Computer Aided Drug Discovery[18,19]

Introduction

As has already been stated, out of ~5000 compounds that are synthesised with a hope to be used as medicines, only 1–2 may pass the stringent trial procedures and find a place in the market to be used as drugs. One can only imagine the human time and effort, the cost, chemicals and solvents used, chemical waste produced and the energy required for the synthesis of the rest of ~4998 compounds.

A synthetic chemist thus must think first as to "what to synthesise" and only then "How to synthesise".

The reasonable, suitable solution as to what to synthesise can be obtained computationally. Though designing of the drug molecules on computer is a complex and time consuming process and requires a detailed knowledge of the drug target interation. It saves all the inputs that go into the synthesis of a very large number of compounds that are either not active or toxic and not suitable as drugs.

Advancements in engineering, more sophisticated and computational methods, bioinformatics and pharmacogenomics in future will be more helpful to a synthetic chemist to decide as to "what should be synthesised" to be used as drug molecules.

This section deals with computer-aided finding of a few structures which can/have been found to be potential drug molecules.

(*a*) Triosephosphate isomerase inhibitor

Structure based virtual double screening has been used to indentify an inhibitor of triosephosphate isomerase (TIM) which might have the potential to reverse or inhibit aerobic glycolysis in cancer.

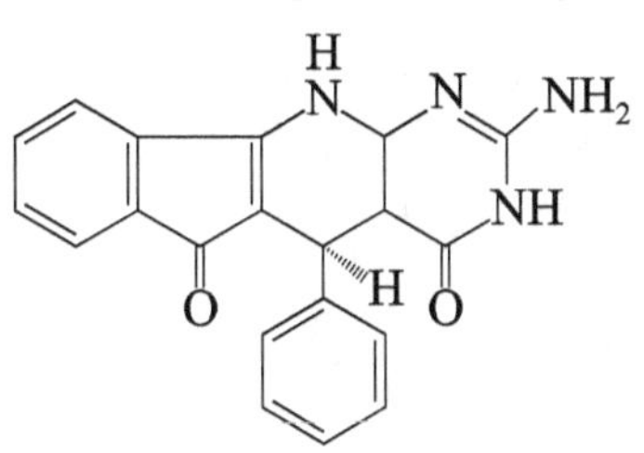

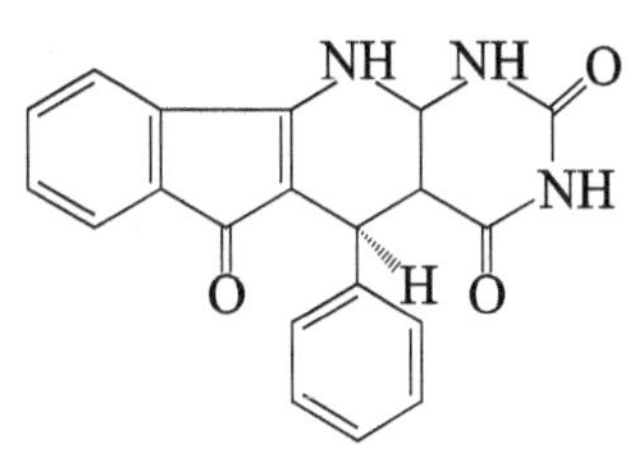

Six novel inhibitors of Protein Tyrosine Phosphate Receptors type Q (PTPRQ) have been recently identified (on the basis of structure-based virtual screening) which can exhibit antiobestic activities.

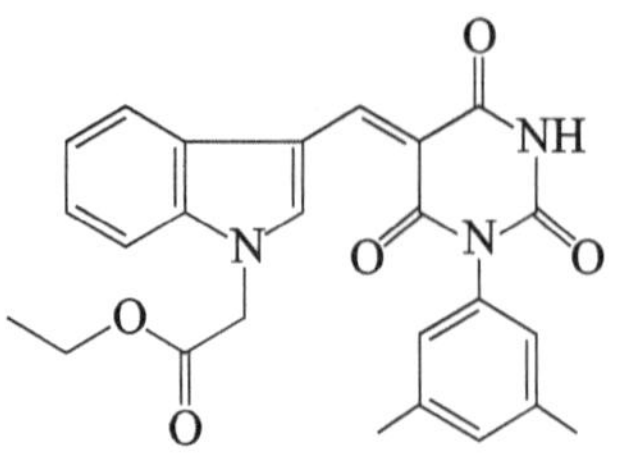

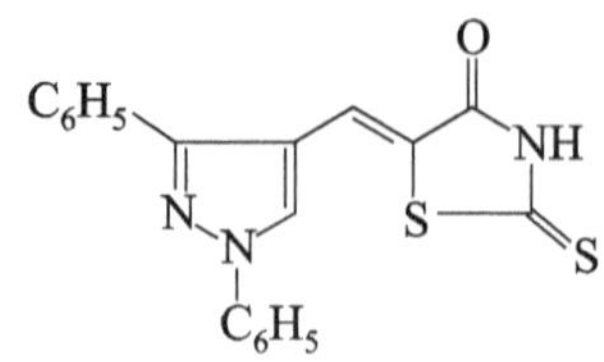

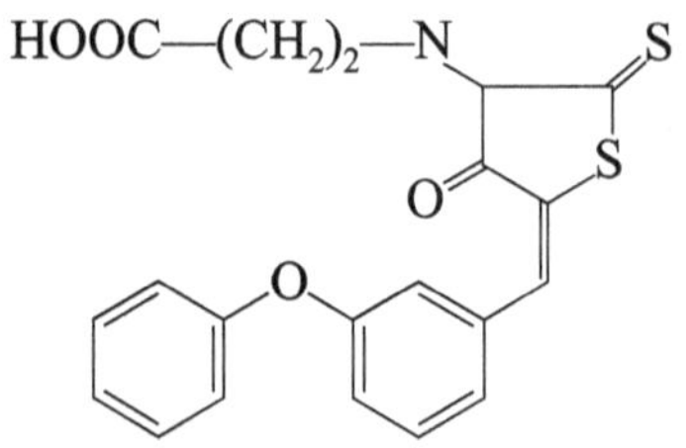

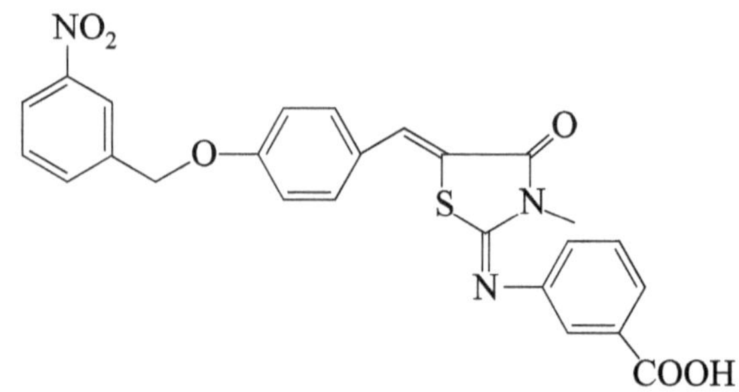

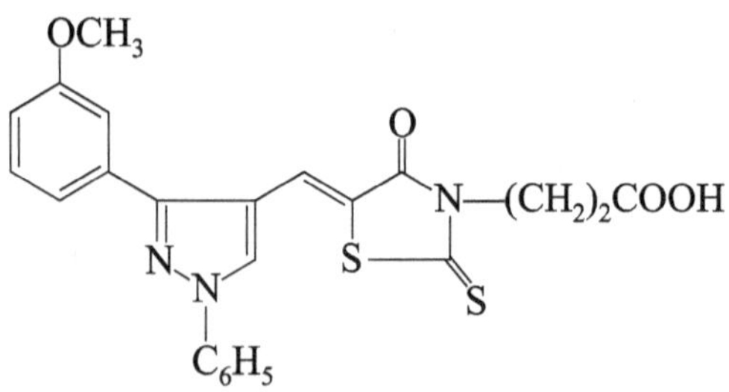

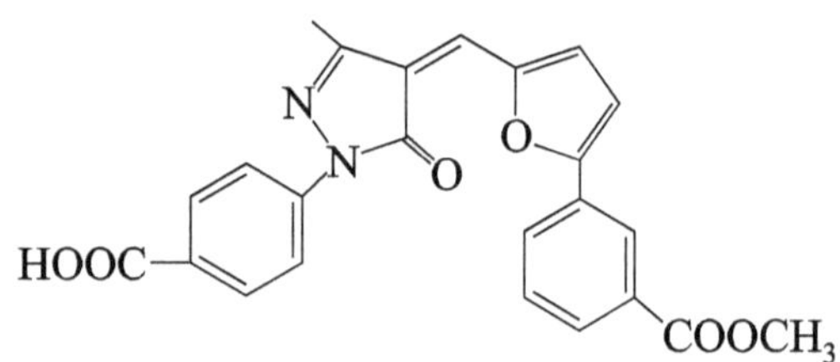

18.3.13 Paclitaxel (Taxol)[20]

Taxol (XXIII) is a drug of choice for the treatment of ovarian, breast, lung and pancreatic cancer. It was first isolated from *Taxus brevifolia* (yew tree) in 1960s by wall and wani under the US government plant screening programme for anticancer activitiy. Its complete structure was elucidated in 1971. The trials for the anti tumor activity were carried out in 1980s.

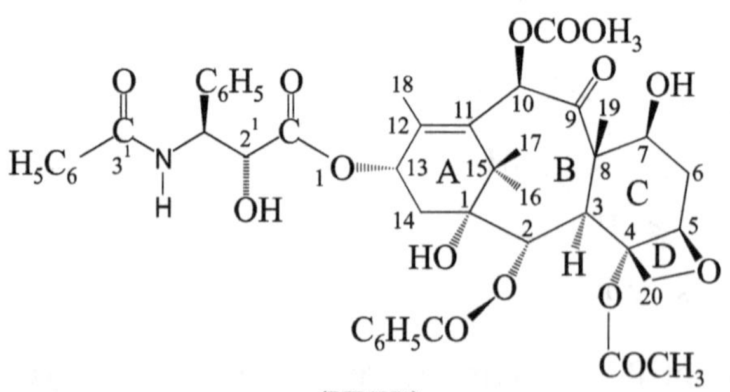

(XXIII)

Taxol is present only to the extent of 0.0004% in the bark of the yew tree. Another major problem in extracting taxol from it is that the yew tree is a very slow growing tree and takes ~200 years to mature and once the bark is removed, the tree dies causing major environmental concerns. Hence this method of obtaining taxol was not commercially viable and could not meet the demand of the drug.

Since the isolation of taxol from the natural sources was not a viable option for the commercial production, its synthesis was taken up as a challange by a number of research groups. Holton *et. al.* and Nicolaou *et. al.* reported the total synthesis of taxol in 1994. The synthesis involved ~40 steps as the structure of taxol is very complex and contains eleven chiral centres. A number of synthetic approaches have been adopted since then but none is used for commercial production.

A semisynthetic route was then adopted for the production of taxol. 10-Deacetyl baccatin III (XXIV), a precursor of taxol was isolated from leaves and twigs of European Yew Tree (Taxus baccata), which was then synthetically converted to taxol. The advantages of using Baccatin III were

- Unlike taxol, Baccatin III was isolated without harm to the trees.
- It is present to the extent of 0.1% in comparison to only 0.0004% of taxol in the bark.
- Since Baccatin contains most of the complexities of taxol structure, its conversion to taxol is now used for commercial production.

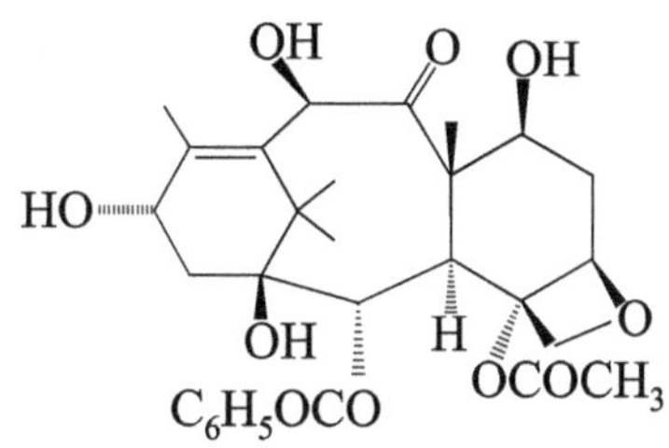

10-Deacetyl Baccatin III (XXIV)

18.3.14 Sterilisation: Greener Alternatives

Ethylene oxide and gamma radiation are presently used for sterilisation of soft tissues for transplant of organs etc. because contaminated donar tissues can result in serious illnesses and infections. Both ethylene oxide and gamma radiations have safety and health issues because ethylene oxide is carcinogenic, inflammable gas and gamma radiations adversely effect healthy cells.

Novasterilis, a US based private company has developed new sterilising technique for sterilisation of soft tisues. It uses super critical carbon dioxide, small amount of peracetic acid and water for sterilisation.

Conclusion

Pharmaceutical industry is one of the most pollution causing industry. E factor is generally between 25-100 and in some cases above 100. It is therefore most essential to green this industry. A few steps have been taken in this direction but more needs to be done.

REFERENCES

1. Sandra Cerqueira Pereira, Green Chemistry, 2012, 14, 3146.
2. US Patent 8133994B_2.
3. Nancy Acton et al, J. Org. Chem. 1992, 57(13), 3610–3614.

4. Dae-kyun, Nature Latters, 2006, Volume 440.
5. C.J. Paddon, Nature, 2013, Volume 496.
6. N.J. Stuart *et. al.*, US Patent 3, 385886 (1961)
7. Abbas Ahmed, *et. al.*, Journal of Applied Chemical Research, 2014, 83, 91–95.
8. Zimmermann J. *et. al.*, Bio-organic Medical Chemistry Lett. 1996, **11**, 1221.
9. Yi-Feng Liu, Org. Proc. Res. Devel. 2008, **12**, 490–495
10. Mark D. Hopkin *et. al.*, Chemical Communication, 2010, **46**, 2450–2452.
11. Marc Lemaire *et. al.*, Green Chemistry 2014, **16**, 2997.
12. Martinez, C.A. Org. Proc. Res. Dev. 2008, **12**, 392-398.
13. US Patent 6177579B.
14. Geraldine P. Taber et al. Org. Proc. Res. Dev. 2004, **8**, 385–388.
15. Bio-organic and Medicinal Chemistry Letters 1996, Vol. 6, 1819–1824
16. Peter J. Dunn Green Chemistry 2004, **6**, 43–48.
17. Moghimi A. *et. al.*, Iranian Journal of Pharmaceutical Research, 2015, **14(3)**, 733–738.
18. Lorraine Marsh et al. International Journal of Medicinal Chemistry, 2014, Article 1D 469125.
19. Park *et. al.*, Theoretical Biology and Molecular Modelling, 2013, 10: 40.
20. Wani *et. al.*, J. Am. Chem. Soc. 1971, 93, 2325.
21. US Patent 7108832.

19. POLYMERS

19

Polymers

19.1 INTRODUCTION

Polymers are macromolecules made by chemical combination of a large number of small molecules called monomers/building units. Polymers have become inevitable companions of human beings. Today the life of elite, common or poor population of the world cannot be imagined without polymers. Natural polymers meet only 1% of the total demand of polymers required, the rest is met by synthetic or manmade polymers. This explosion has occurred because

- The polymer structure can be tailored to meet specific requirements.
- Additives added to the polymers further enhance their utility.
- The cost of production of the polymers is within reasonable limits.
- Polymer made products are durable and have a much longer life.

Polymers are used for a wide range of applications in our day-to-day life ranging from construction material, packaging, textiles, footwear, parts of vehicles, machines, tires and tubes, laminated wood, paper, foams, table and kitchenware, toys, adhesives, sports goods to bullet proof windows and jackets, liquid crystal display, conducting polymers, nano-particles for drug delivery etc.

A number of Journals are published to report the latest developments in the field of polymer science and a number of scientists have been awarded various honours including Nobel Prize for new discoveries in this field.

There are major environmental issues concerning production and end used of the polymers

- Most of the monomers used in the production of polymers are derived from natural non-renewable resources which are depleting fast.
- Most of the systhetic polymers are non-biodegradable and hence the used polymer waste is accumulating.
- Leaching of hazardous chemicals in the soil, air and water occurs continuously from the polymer waste.

19.2 GREENING THE POLYMER INDUSTRY: REMEDIAL MEASURES

There is an urgent need to green the polymer industry and the guidelines or the answer can be found in the green chemistry principles.

- Non-renewable resources should be replaced by renewable resources as far as possible for production of raw materials.
- Hazardous chemicals should be replaced by safer chemicals.

- The entire process of polymer production should be so designed that it is benign, time, resource and energy efficient.
- The use of metal catalyst should be eliminated or at least reduced as metallic catalyst contaminate the polymer and are harmful because these leach out during the use of polymer product.
- The use of solvents and other undesirable chemicals should be minimised. Hazardous and volatile organic solvents should be replaced by safer solvents.
- Recycling of the non-biodegradable polymers should be encouraged.
- Biodegradable polymers should be developed.

19.3 GREEN MONOMER SYNTHESIS

Almost all the monomers required for polymer synthesis are derived from fossil resources. Ever increasing demand of polymers and use of millions of tonnes of polymers is fast depleting the non-renewable natural resources. There is an urgent need to shift from use of these to renewable resources for a sustainable future. Use of indigenous raw materials and the waste produced by other manufacturing units should be encouraged.

Some of the monomers are now produced from bio-mass[1]. Ethanol, for example, produced from hexoses and pentoses is the source of compounds like ethylene and ethylene glycol [used for the production of polyethylene (LDPE and HDPE) and polyethylene terephthalate (PET)] respectively.

One of the natural source of 3 carbon compounds is glycerol. It is used for production of 1, 2 and 1, 3-propanediol (versatile building blocks of polyesters). Lactic Acid (used for production of polylactic acid).

Dicarboxylic acids like succinic, fumaric, adipic, terephthalic and furan-2, 5-dicarboxylic acid, diamines, alcohols like butanol and butanediols, caprolactam and many others can be derived from bio-mass.

Synthesis of few monomers derived from renewable resources are given in this section.

19.3.1 Synthesis of Adipic Acid[2]

Adipic Acid (I), a six carbon dicarboxylic acid is copolymerised with hexamethylene diamine (II) to produce Nylon 66 (III).

$$\underset{\text{Adipic acid (I)}}{HOOC{-}(CH_2)_4{-}COOH} \quad + \quad \underset{\text{Hexamethylene diamine (II)}}{H_2N{-}(CH_2)_6{-}NH_2}$$

$$\downarrow$$

$$\underset{\text{Nylon 66 (III)}}{-NH{-}CO{-}(CH_2)_4{-}CO{-}NH{-}(CH_2)_6 - NH{-}CO{-}(CH_2)_4{-}CO-}$$

Adipic acid has been produced from benzene in the past Benzene is obtained from petroleum and coal tar (both being non-renewable resources) and is a human carcinogen. Nitric acid was being used for oxidation resulting in the production of oxides of nitrogen, which contribute to green house effect and also depletion of ozone layer.

A number of modifications have been exploed for a greener synthesis of adipic acid.

In one of the modification, cyclohexene is used as a starting material. It is oxidised using hydrogen peroxide to adipic acid. Aliquat, 336, a phase transfer catalyst is used along with sodium tungstate and potassium hydrogen sulphate as catalyst.

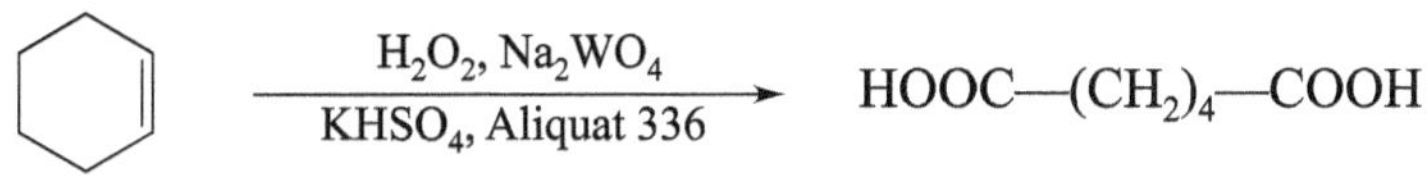

A further modification of the catalyst improves the yield of adipic acid.

Glucose (a renewable starting compound) has also been converted to adipic acid by using genetically modified strain of Escherichia coli known as Klebsiella pneumoniae.

The sequence of biochemical reactions by which glucose in the biomass is converted to adipic acid is given in scheme 19.1.

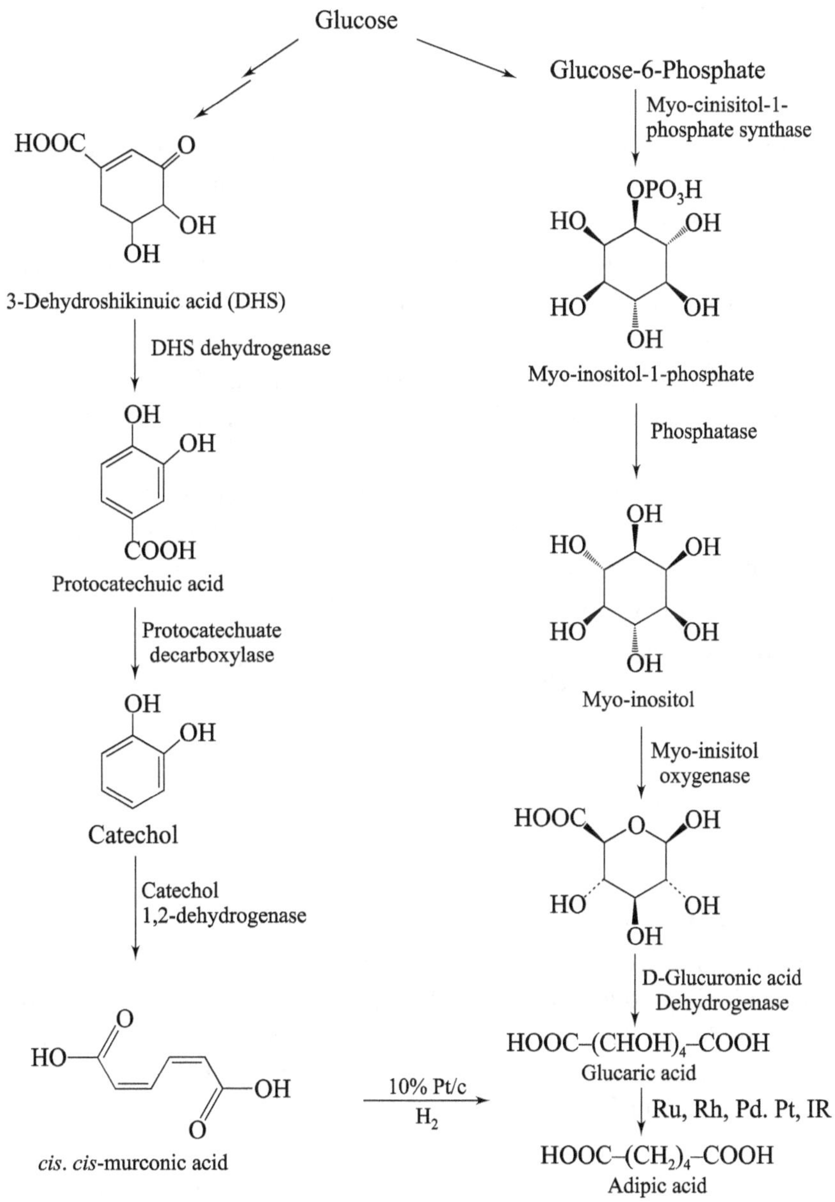

Scheme 19.1: Glucose to adipic acid by Klebsiella pneumoniae

Both Adipic acid and hexamethylene diamine can be prepared from adiponitrile (see sec. 17.2.1).

19.3.2 Synthesis of Caprolactam[3]

Caprolactam has been mainly used for the production of Nylon-6 and is produced from Cyclohexanone as given below.

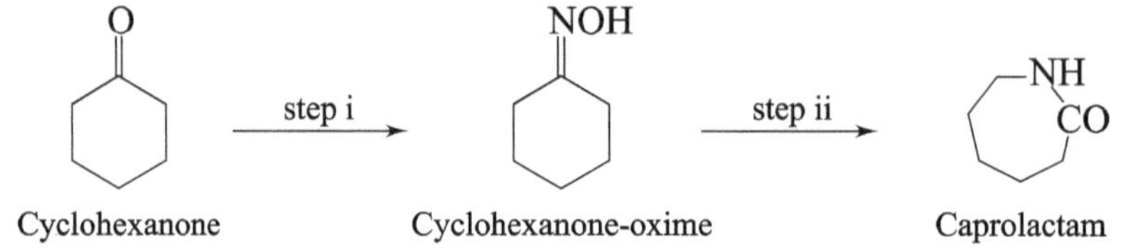

Conversion of cyclohexanone oxime to caprolactam (step *ii*) is carried out using concentrated sulphuric acid, which is neutrilised after the reacion by ammonia producing huge amounts of ammonium sulphate as a byproduct, the main drawback of the process.

A few modifications to the above method have been developed to eliminate (*i*) the use of sulphuric acid for Beckmann Rearrangement (*ii*) production of huge amounts of ammonium sulphate.

Beckmann rearrangement has been reported using (*a*) solid acid as a catalyst (*b*) super critical water (*c*) Aluminium containing catalyst for both liquid and vapour phase reaction (*d*) ionic liquids etc.

In a method developed by Chong Yan *et al.* ε-Caprolactam has been prepared from 6-Aminocapronitrile in a continuous process using High Temperature Water (HTW). No catalyst or solvent is used for the conversion. A temperature of 400 °C , pressure 400 bar and reaction time (96 seconds) has been found to give best results

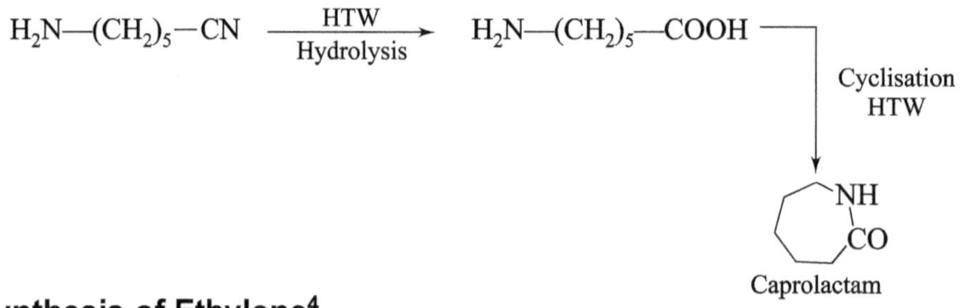

19.3.3 Synthesis of Ethylene[4]

Ethylene, a building block of polyethylene and a starting material for the synthesis of many other industrial compounds, is produced via ethanol from bio-mass. Bio-ethanol can be produced by fermentation process. A number of microbes like Zymomonas mobilis, Saccharomyces cerevisiae, etc. are used for fermentation to produce ethanol from crops that do not threaten food security, grow fast and do not require too much of water.

In one of the process corn stover is used as the bio-resource. Corn stover is milled and subjected to acid hydrolysis to break down complex carbohydrates. Excess acid is neutrilised and the resultant mass fermented with *Zymomonas mobilis*. Ethanol thus produced is purified and dehydrated using a fixed bed catalytic reactor train consisting of aluminium oxide on gamma alumina. Ethylene thus obtained is separated from water using triethylene glycol as the entrainer.

19.3.4 Synthesis of Ethylene Glycol[5]

Ethylene glycol is conventionally produced from ethylene (obtained from fossil fuel) in two steps as shown below

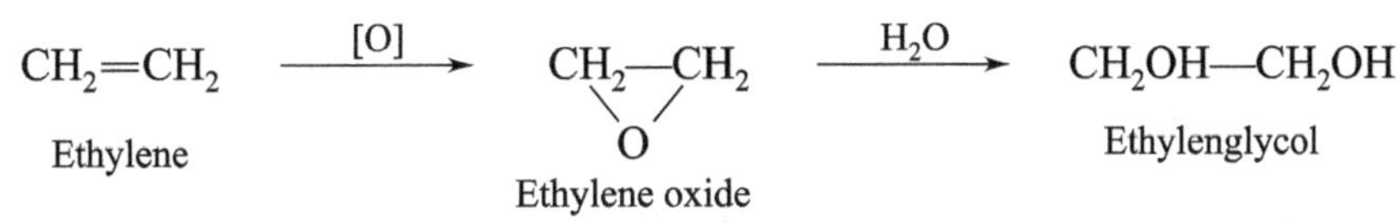

Bio-mass is now being used for production of ethylene and ethylene glycol. A number of renewable starting materials are being used for the purpose and in a few countries these are used for commercial production. The main problem is the cost factor. In one of the method used in many countries in the world, ethylene is obtained by dehydration of ethanol but the prices depend on the cost of indigenous natural resource. For example ethylene obtained from sugar cane is more expensive than ethylene derived from fossil fuel and ethylene from sorghum is more costly than that obtained from sugar cane. Various methods that use bio-mass for ethylene glycol production are outlined below:

- **From Ethanol:**
  Bio ethanol obtained from a number of resources is converted to ethylene glycol by dehydration, oxidation to ethylene oxide followed by its hydrolysis.

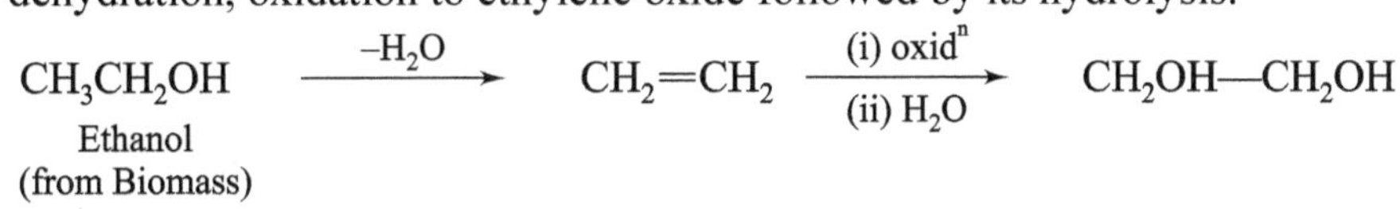

- **From Glycerol:**
  Glycerol is a by product of biodiesel and soap industry. Efforts are on to convert it to commercially useful products. Hydrogenolysis of glycerol has been studied under different conditions specially by use of different metal catalyst. The product obtained is a mixture of 1,2- and 1,3-propylene glycol (abbreviated as 1,2-PG and 1,3-PG), ethylene glycol (EG), dihydroxyacetone, acrolein etc. Yield of EG has been improved by use of Ru and Ni as catalyst.

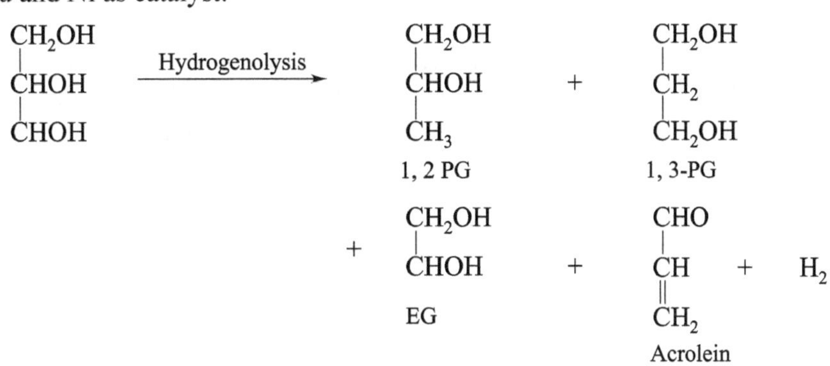

- **From Sorbitol and Xylitol**

  Sorbitol is obtained by catalytic hydrogenation of glucose. Sorbitol can be converted to 1, 2-PG and EG in different yields by use of different metal catalyst.

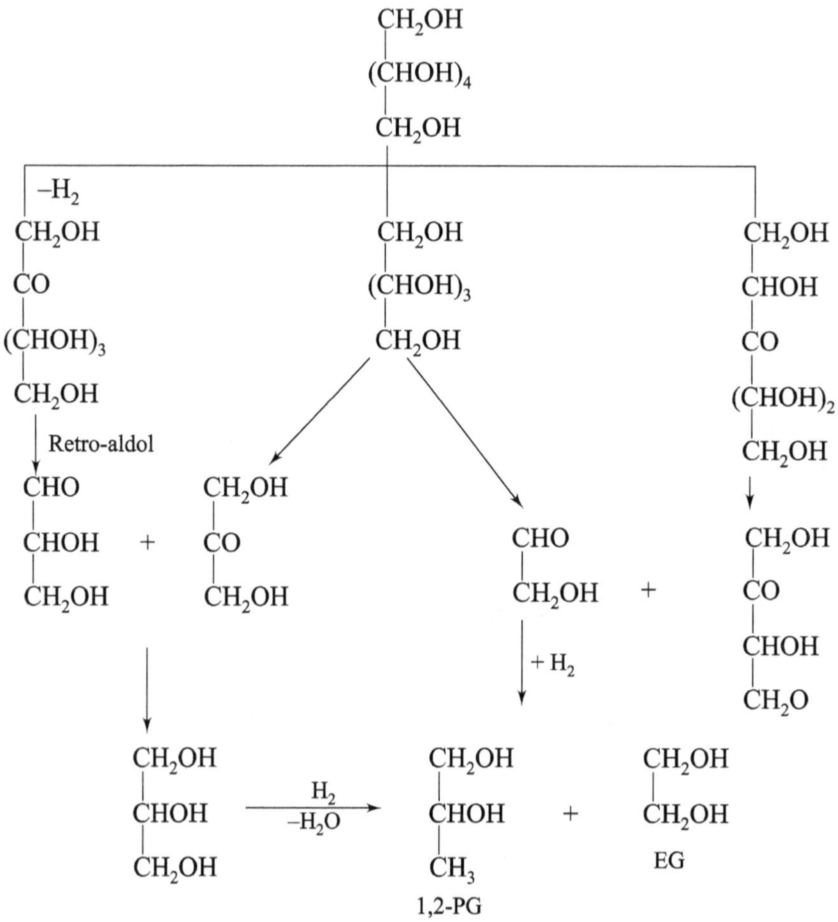

Scheme 19.2: Chemicals obtained from sorbitol and xylitol

19.3.5 Synthesis of Terephthalic Acid[6]

Terephthalic acid is another important monomer used in the production of Polyethylene Terephthalate (PET).

It has been obtained by oxidation of *p*-xylene. *p*-Xylene itself has been obtained from petroleum or coal tar, both of which are non-renewable resources. The other problem was the use of nitric acid for oxidation of *p*-xylene to terephthalic acid. Efforts are now being made

(*i*) To obtain p-Xylene from renewable resources.

(*ii*) Carry out oxidation of *p*-xylene with a benign oxidising agent.

(*iii*) Improve the yield, purity and cost of the monomer.

Some of the greener alternatives for the production of *p*-xylene/terephthalic acid are outlined below.

- **From Ethylene :**

Ethylene can be converted to *p*-xylene by following sequence of reactions

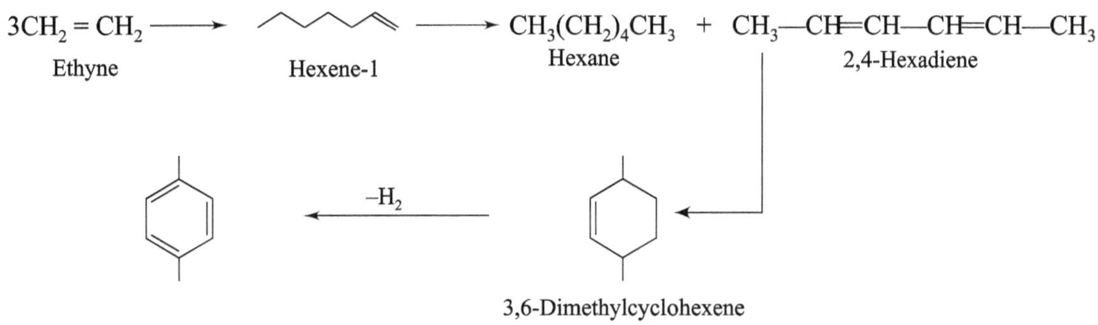

- **From Hydroxyl Methyl Furfural (HMF)**

HMF which is obtained by dehydration of hexoses can also be converted to p-Xylene

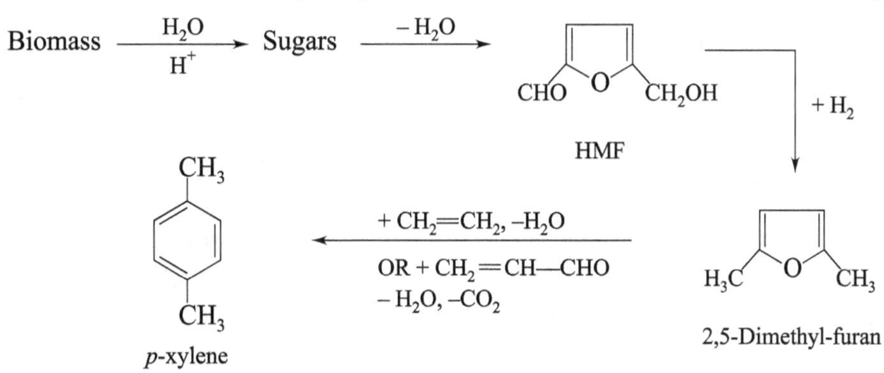

- **From Isobutanol**

Isobutanol which is obtained by fermentation of bio-mass has been converted to p-xylene by the following scheme

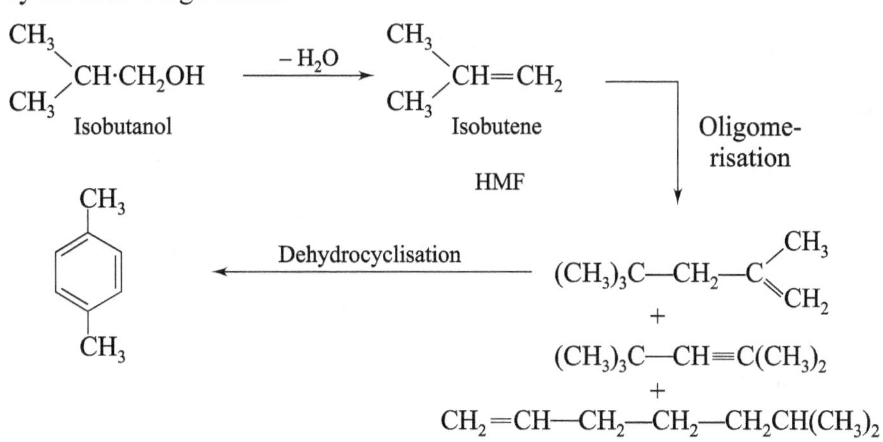

- **From Isoprene**

Cycloaddition reaction of isoprene and acrylic acid, dehydrogenation of the product yields 4-methyl benzoic acid which on oxidation gives terephthalic acid.

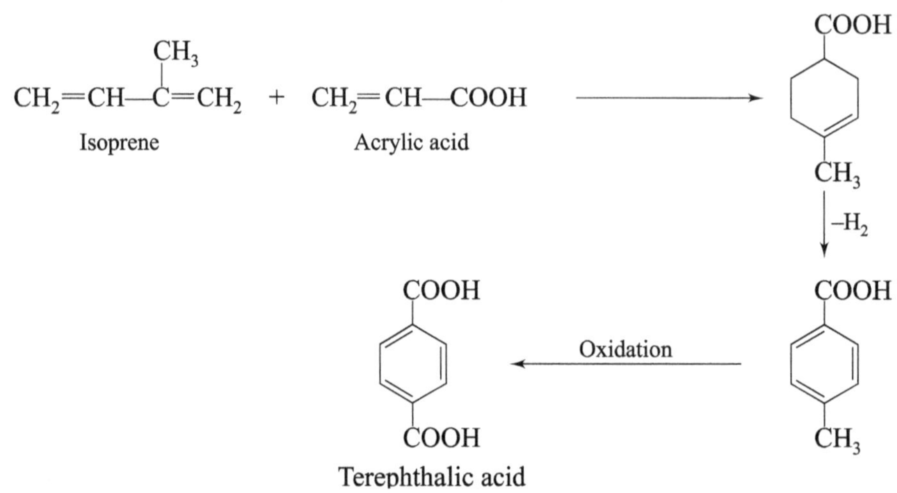

- **From Limonene**

Limonene, which is present in orange peel is first dehydrogenated to p-cymene, which is then oxidised to terephthalic acid. This method is of theoretical significance because limonene has limited availability and is used as food additive.

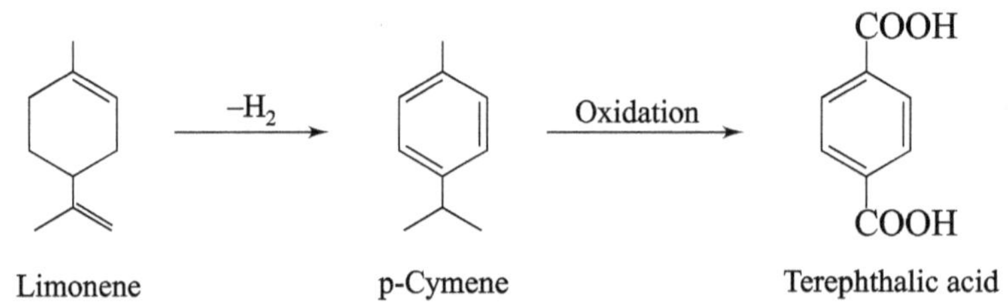

- **Pyrolysis of Bio-mass**

Bio-mass from various sources has been pyrolysed thermally or catalytically to a variety of useful compounds. Efforts have been made to produce aromatics by pyrolysis. Pyrolysis of bio-mass to p-xylene is being studied for commercial production.

HMF method produces the highest yield of *p*-xylene (~85%). Bio-ethylene method gives good yield of *p*-xylene but the reaction conditions are harsh and energy intensive, use of isobutanol for production of p-xylene gives only ~20% yield but the cost of feedstock is low as compared to limonene, isoprene etc. Pyrolysis of bio-mass may prove to be a better technology for production of terephthalic acid.

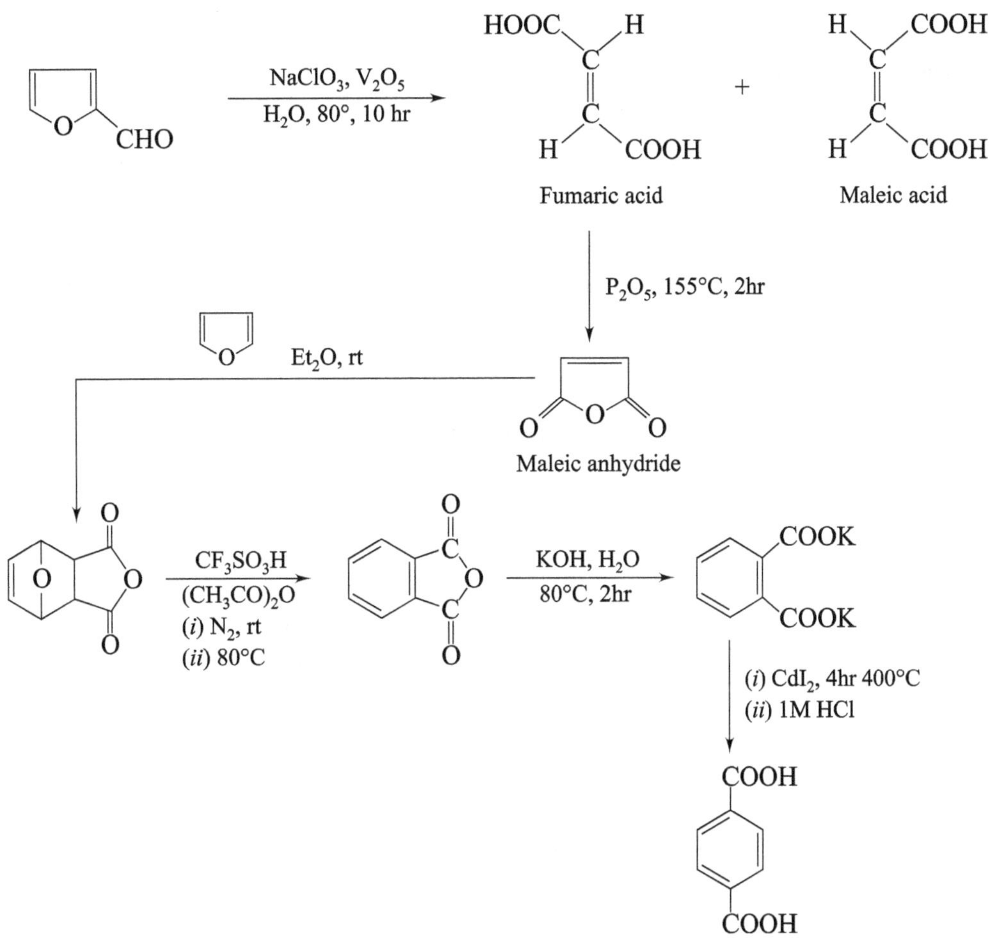

Scheme 19.3: Terephthalic acid from biomass

Oxidation of p-xylene to terephthalic acid : Terephthalic acid has been obtained by oxidation of *p*-xylene. A number of oxidising agents have been developed so as to reduce the adverse effects of the oxidising agents and the byproducts of the reaction. Terephthalic acid was first prepared by oxidation of *p*-xylene with dilute nitric acid and then with air. Terephthalic acid thus obtained had to be purified by esterification to dimethyl terephthalate. Attempts have been constantly made to improve the yield, purity and cost of terephthalic acid by oxidation with air. Both heterogeneous and homogeneous catalysts have been studied. In one of the modern process, terephthalic acid is obtained by liquid phase oxidation of p-xylene by air in acetic acid and manganese or cobalt acetate as a catalyst. Acetic acid and catalyst are recycled.

In another process (Amoco process), the oxidation of *p*-xylene is carried out in acetic acid with air in presence of cobalt, manganese and bromide ions HBr (corrosive in nature) or NaBr are used as source of bromide ions. HBr has been replaced by cobalt/manganese bromide. Super critical water has also been explored as a green solvent.

A number of alternatives are still being explored like solvent free oxidation, use of CO_2, less expensive catalyst and milder reaction conditions etc.

19.3.6 Styrene and Acrylates from Bio-mass[7]

Styrene and esters of acrylic acid have been made from bio-mass left after recovery of bio diesel. Other sources are DDGS (distillers dried grains with solubles), vinasses (residue left after distillation of alcoholic beverages) and leaves, depending on the source, the residual biomass contains 20-40% protein. Efforts are now being made to convert it to important industrial chemicals. In one of such efforts phenyl alanine in this waste is first converted to cinnamic acid ester which is then subjected to ethenolysis (metathesis reaction using ethylene) to yield styrene and acrylate. Cross metathesis reaction is carried out by using 2nd generation Hoveyda-Grubbs catalyst (1,3-Bis-(2,4,6-trimethylphenyl)-2-imadazolidinylidene)dicloro—(*o*-isopropoxyphenylmethylene)ruthenium and low pressure of ethylene.

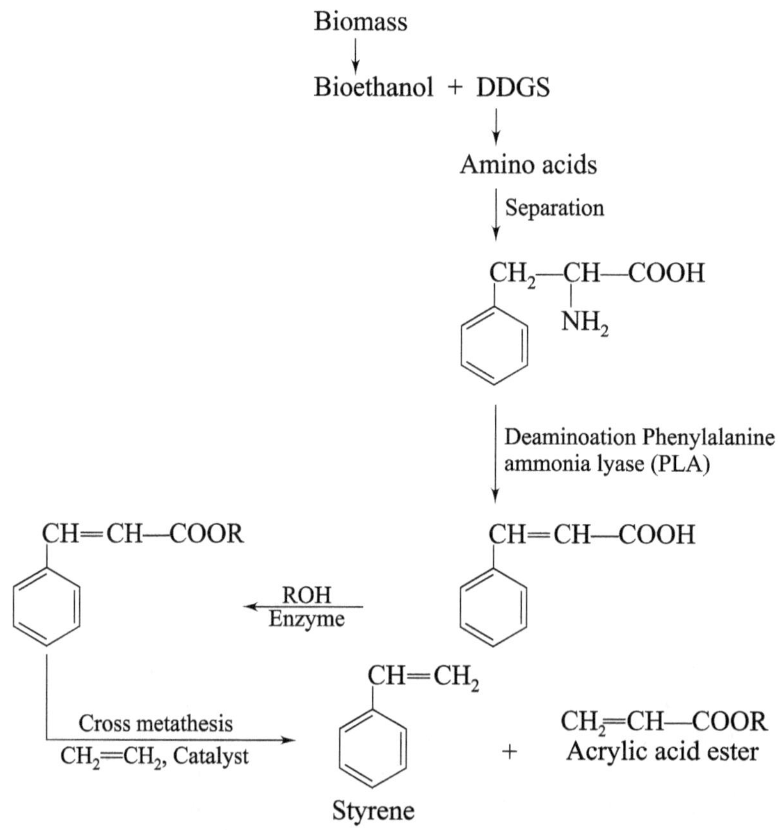

Other amino acids like Tyrosine can be converted to useful compounds by a similar method.

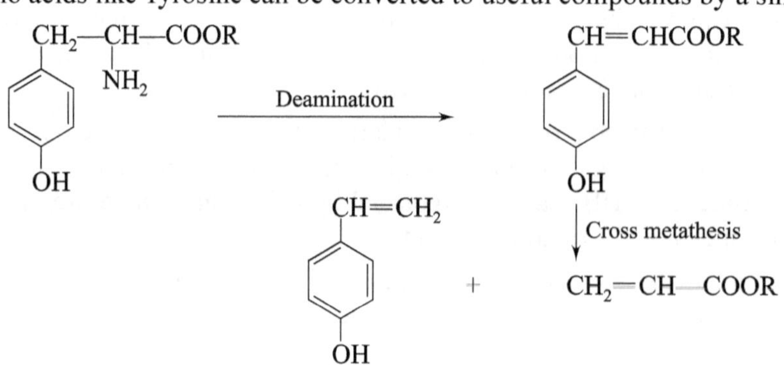

OR

The amino acid mixture is first de-aminated, cinnamic acid is purified and then converted to styrene and acrylic acid by cross-metathesis as given above.

Acrylates from Sugars

The sugars present in biomass are converted to itaconic acid by fermentation, which is then converted to acrylates by the following scheme.

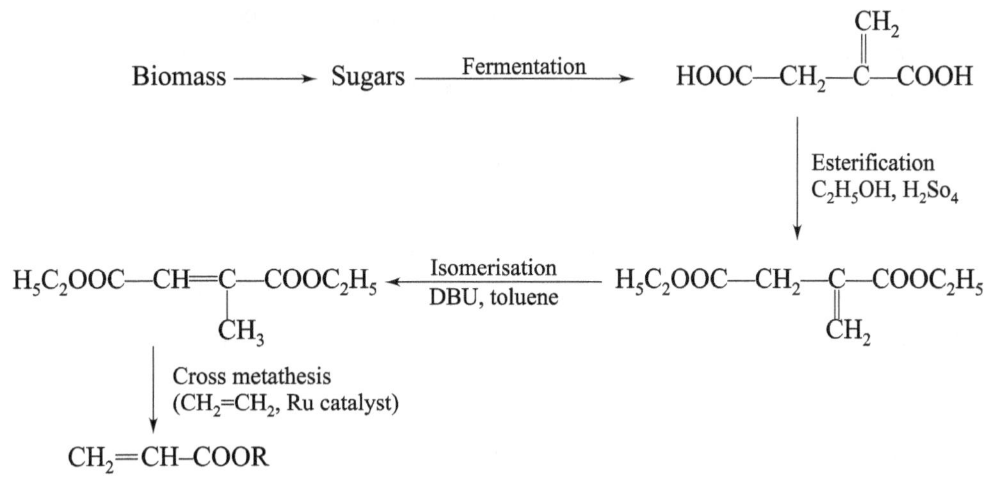

19.4 GREEN POLYMERISATION

A number of techniques are used today for polymerisation reaction. Many of these are carried out in presence of metals/complexes as catalyst and organic solvents as medium for the reaction. Constant efforts are being made to replace metallic catalyst by organic catalyst or enzymes and solvents by bulk polymerisation or by using benign solvents like water, ionic liquids, $ScCO_2$ etc. some polymers have been made by trapping methane, carbon dioxide, carbon monoxide etc. from air. Efforts are also being made to make these reactions energy efficient. All these will lead to lower cost of end product and minimum adverse effect on environment, producer and the user. It is not possible to give detailed account of all the alternative processes in this chapter. We have tried to give a glimpse of some of greener aspects of these developments.

19.4.1 Enzyme Catalysed Polymerisations[8(a-e)]

Use of enzymes for polymer synthesis is relatively new. The first generation catalyst used for polymerisation reactions were normally acids, basis or free radicals (used from around 1920s to 1950s), From 1950s, transition metals and rare earths were used as catalyst for polymerisation. Purified enzymes or whole cells are now being used and explored for polymerisation reactions. These enzymes work under normal temperature and are therefore energy efficient and also produce polymers, which have controlled stereochemistry. The enzymatic polymerisation reactions are chemo and regioselective also.

- Enzyme assisted polymerisations have been used to prepare polyesters, polyamides, polyphosphates, polycarbonates, polysaccharides, polyphenols and polyamines etc.
- Conducting polymers like polyaniline, and poly(*p*-aminobenzoic acid) have been produced by using horseradish peroxidase and hydrogen peroxide. A water soluble copolymer of aniline and sulphonated styrene has also been made by use of same catalyst.
- A number of polyesters are being made by enzyme catalysed reactions. Lipases/ estrases from various sources like *Candida antarctica, Candida rugosa, Pseudomonas cepacia, Psedomonas fluorescence* have been explored for ring opening polymerisation of lactones of different ring size. Many of these enzymes have been immobilised for recovery and reused. Novozym 435 (abbreviated as CALB, lipase B from *candida Antarctica immobilised on Lewatit*) for example is the most used and studied of all types. ε–caprolactone and many other monomers have been polymerised using CALB.
- Polycarbonate from trimethylene carbonate and its co-polymers with other carbonates have been synthesised using CALB and porcine pancreatic lipase
- CALB has been used in ionic liquids for synthesis of poly (L-lactide) and poly(L-lactide-co-glycolide).

19.4.2 Green Solvents fo Polymerisation[9(a-f)]

Polymerisation is basically carried out in three different ways : (*i*) solution polymerisation (*ii*) suspension polymerisation (*iii*) bulk polymerisation each of these has certain advantages and disadvantages. Solution polymerisation is one of the best techniques for polymerisation as there is a good control on the reaction and average molecular weight of the polymer but hte major disadvantage of this technique has been the use of volatile organic solvents, many of which are hazardous, have adverse effect on environment and living beings. Bulk polymerisation is a benign technique but the quality of the polymer is compromised.

Constant efforts are being made to replace the VOS by greener solvents, a few of these are given in this section :

(a) Supercritical Solvents

Supercritical Carbon Dioxide

It is one of the most benign solvent and can be recycled number of times. It is easily available, can be removed easily after the completion of reaction and does not contaminate the final product. There are a few complications associated with its use like high pressure equipment, developing new technologies rather than extending existing technologies and complex phase behaviour of the solvent etc. However development of new equipment and technologies in near future can overcome the difficulties and make use of $ScCO_2$ for polymerisation cost competitive.

Poly (L-lactide) has been synthesised in $ScCO_2$, it is difficult to obtain this polymer in organic solvents as reaction medium.

Polyacrylic acid has been produced by precipitation polymerisation in $ScCO_2$ using ethyl mercaptan as chain transfer agent.

Flouropolymers have also been produced using $ScCO_2$.

Supercritical carbondioxide has been used for synthesis of cross linked partially neutralised Poly(Acrylic Acid) by suspension polymerisation.

Similarly Poly(Vinylidene fluoride) and tetrafluoro ethylene have also been polymerised by Free Radical Polymerisation.

(b) Biodiesel

Biodiesel, a green additive to fuels which is obtained by transesterification of natural oils (a renewable resource) with methanol is commonly known as fatty acid methyl ester (FAME) and has been tried as an alternative to volatile organic solvents for solution polymerisation. Styrene, Methyl methacrylate, Butyl acrylate and Vinyl acetate have been polymerised at 60°C to produce the corresponding homopolymers by free radical polymerisation. The poly(butyl acrylate) is precipitated out by addition of methanol and the other three polymers by addition of hexane after completion of polymerisation.

(c) Ioinc Liquids

Organic salts that are liquids at ambient temperature are known as ionic liquids. These have replaced volatile organic solvents in a number of reactions because of their negligible volatility, non-inflammable, highly polar nature, capacity to dissolve a number of organic, inorganic and metal containing compounds etc. Most enzymes remain active in these solvents and hence ionic liquids can be used in enzyme catalysed reactions. Another advantage is that many of these can be recycled.

The common cationic components in an ionic liquid are

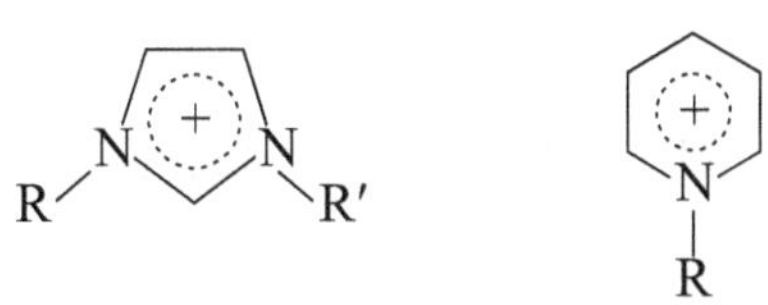

$R_4\overset{+}{N}$ $R_4\overset{+}{P}$

Some of the common anionic components are

$$AlCl_4^-, BF_4^-, CF_3COO^-, NO_3^-, ClO_4^-, PF_6^- \text{ etc.}$$

Polymerisation in ionic Liquids

A number of free radical polymerisation have been carried out in chemically inert ionic liquids and those which do not undergo hydrolysis during the reaction.

Methyl methacrylate has been traditionally polymerised in benzene (a human carcinogen), it can now be polymerised in ionic liquid ($[b_{min}][PF_6]$) where the rate of propagation (K_p) is faster than rate of termination (K_t) producing a high molecular weight polymer.

Methyl methacrylate → Poly(methyl meth acrylate)

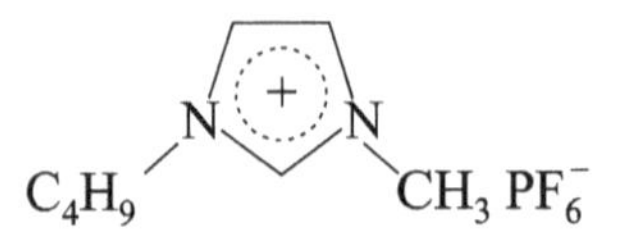

($[b_{min}]\ [PF_6^-]$)

Similarly butyl, hexyl and dodecylacrylates have been polymerised in ionic liquids.

Block copolymers of styrene and methyl methacrylate have also been prepared in the ionic liquid $[b_{min}][PF_6]$.

Atom Transfer Radical Polymerisation (ATRP) and Radical Addition Fragmentation Transfer Polymerisation (RAFTP) have also been carried out in ILs.

Coordination polymerisation of ethene and phenyl acetylene has also been reported in $[b_{min}]$ $[AlCl_4]$ and $[b_{min}][BF_4]$ respectively.

Ionic polymerisation of styrene, vinyl ethers, propylene oxide etc. have also been carried out in ILs.

19.4.3 Atom Tranfer Radical Polymerisation[10]

Atom Transfer Radical Polymerisation (ATRP) is a new technique developed in 1995 for the polymerisation reaction. Started as an extension of Atom Transfer Radical Addition (ATRA), it has begun a new era in the history of polymerisation technique. This technique has evolved exponentially in the last few years. In the past, it was difficult to keep a control on the molecular dispersity and structure of the polymers, especially during co-polymerisation by free radical polymerisation. It is now possible to produce macromolecules having well-defined composition, functionalities and architecture using ATRP.

Types of polymers which have been prepared by ATRP are given below and classified into three basic types based on their composition, functionality and topology.

(*i*) Composition of Polymers

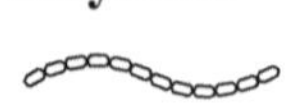

Homopolymer

Block copolymer

Random copolymer

Periodic (alternate) copolymer

Gradient/tappered copolymer

***(ii)* Functionality of Polymers**

Macro-molecules with different functionalities organised in a number of ways have also been made, two examples are shown below:

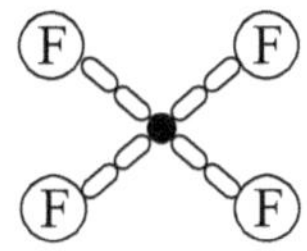

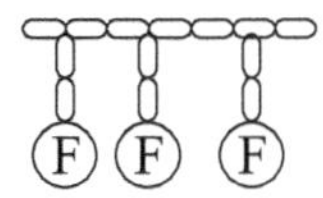

F = Functional group

***(iii)* Topology of Polymers**

Linear

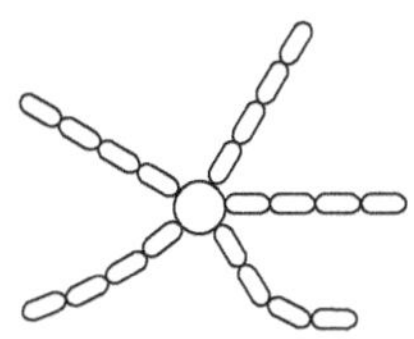

Star

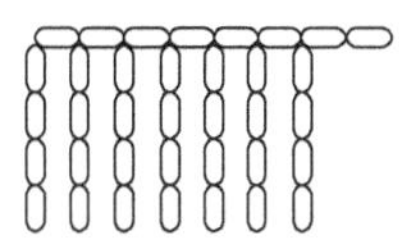

Comb|Gragf

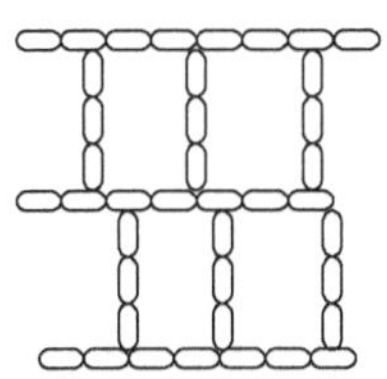

Network

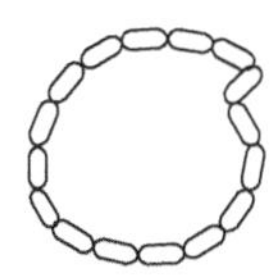

Cyclic

Scheme 19.4: Types of polymers obtained by use of ATRP technique

Polymerisation by ATRP is a multi-component system consisting of the monomer, a copper[1] complex and an initiator.

Monomers which have been polymerised are methyl acrylate, methyl methacrylate, methyl acrylamide, acrylonitrile, substituted acrylic acids and styrene etc. Both homo and hetero polymers have been successfully synthesised.

$CH{=}CH_2$

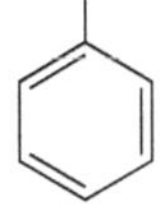

Styrene

$CH_2{=}CH{-}COOH$

Acrylic acid

$CH_2{=}CH{-}CONH_2$

Acrylamide

$CH_2{=}CH{-}CN$

Acrylonitrile

$CH_2{=}C(CH_3){-}COOCH_3$

Methyl methacrylate

Halogenated alkanes are the most commonly used initiators. Benzylic halides, α-haloester, α-haloketones, α-halonitriles and sulphonyl halides etc. have also been used as initiators.

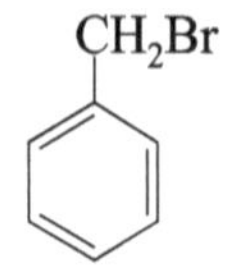

Benzyl bromide

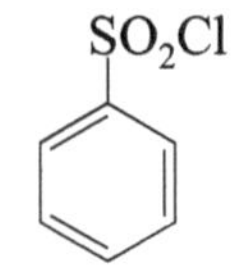

Sulphonyl chloride

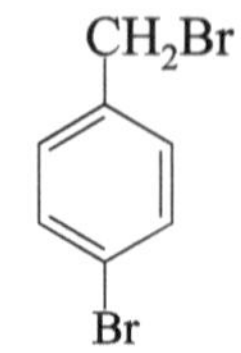

p-Bromo-benzyl-bromide

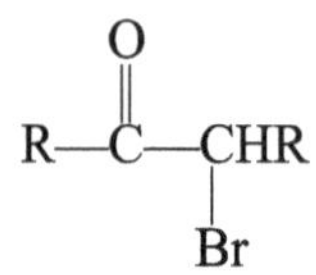

α-Bromoketone

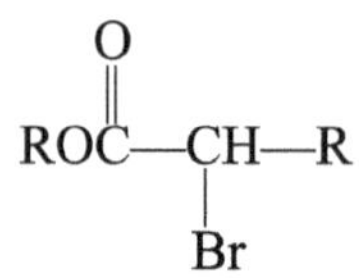

α-Bromoester

Copper complexes with nitrogen containing ligands (some examples are given below) are the most studied catalyst used in ATRP. It is one of the most important component of ATRP as it regulates the dynamic equilibrium between the dormant and propagating species during polymerisation.

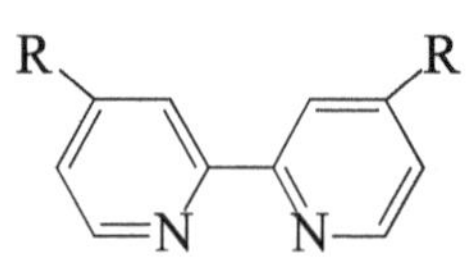

Bidendate ligand

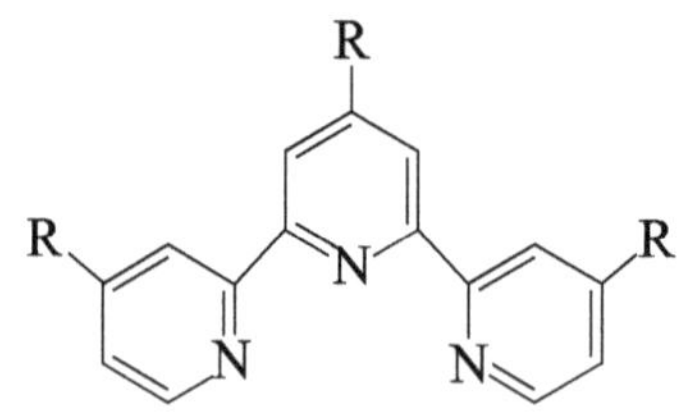

Tridendate ligand

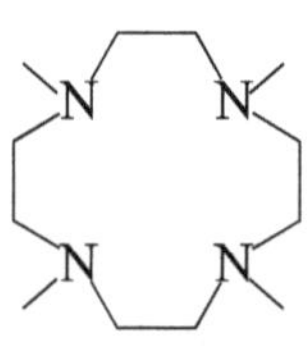

Tetra dentate ligand

Other organometalic compounds like complexes of iron have also been used as catalyst in ATRP.

In the earlier studies, the amount of copper complex used was high (0.1-1 mole%), now only ppm quantities of it are used. This has been made possible by incorporating a reducing agents in the polymerisation which converts copper^2 complex formed during the reaction back to copper^1 complex. Because of this, besides the reduction in the amount of catalyst, the removal of copper complex from the polymer is also simplified.

Various reducing agents which have been used are : Glucose, phenols, hydrazine,ascorbic acid etc. Reducation has been carried out electrochemically.

Polymerisation Reactions

The copper[1] complex abstracts the halogen from the initiator to form a free radical $R^{\bullet}$, which initiatives the polymerisation

$$Cu^{1}L + RX \rightarrow R^{\bullet} + XCu^{2}L$$

$$R^{\bullet} + CH_2{=}CH\text{-}Y \rightarrow RCH_2\text{–}CH^{\bullet}Y$$

Propagation and termination follow as in free radical polymerisation.

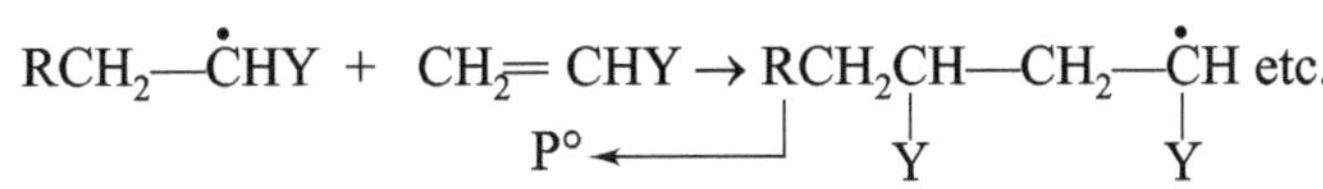

Termination occurs by dimerisation or disproportionation.

A growing polymer $P^{\bullet}$ can abstract a halogen from the copper[2] complex and form PX. PX can again generate $P^{\bullet}$ and begun propagation.

$P^{\bullet} + XCu^{2}L \rightarrow PX + Cu^{1}L$

$XCu^{2}L$ is converted back to active catalyst ($Cu^{1}L$) by the reducing agent.

19.4.4 Synthesis of Polystyrene[11(a,b)]

Polystyrene has been made from styrene by polymerisation using cationic, anionic or free radical catatyst. In most cases solvents like THF, toluene, cyclohexane, benzene etc. are used during polymerisation and separation of polymer from solvent and catalyst is an expensive process.

In one of the greener approaches by Kannan C. *et al*, styrene has been polymerised in vapour phase over alumina (basic and acidic), no solvent is used during polymerisation

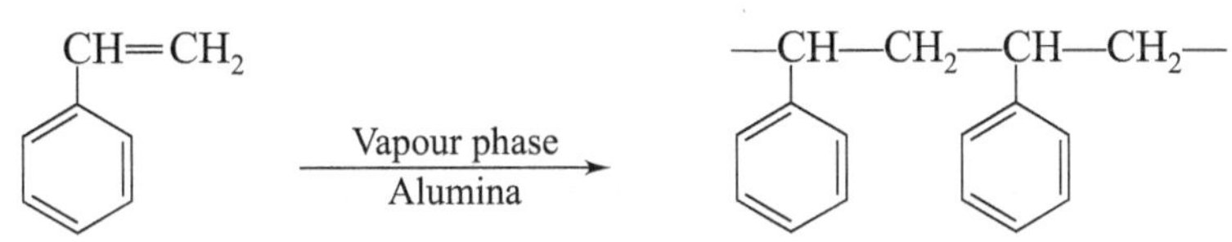

Styrene Polystyrene

Styrene has also been polymerised in 1,8-Diazabicyclo [5.4.0] undec-7-ene (DBU) or *n*-propanol in presence of an initiator.

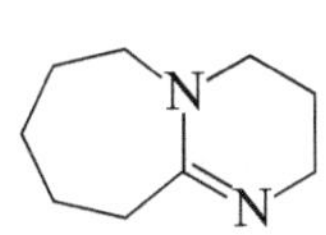

DBU

Styrene and an initiator are taken in DBU/n-propanol at 50°C. On completion of polymerisation, carbon dioxide is bubbled into the reaction mixture. Polystyrene is precipitated, which is removed by filteration and the solvent is recovered and reused by bubbling nitrogen or mild heating. The process is depicted in Fig. 19.1.

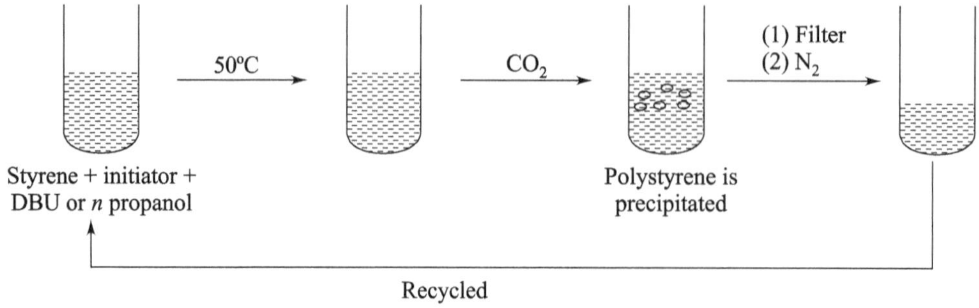

Fig 19.1 : Synthesis of Polystyrene

Expanded Polystyrene

Expanded polystyrene is extensively used as a packaging material. It is conventionally obtained by blowing styrene using hard CFC–12 (Dichloro-diflouromethane) and HCFC–22 (Chlorodiflouromethane).

DOW has developed a process for producing expanded polystyrene using 100% carbon dioxide thereby reducing the cost of production, ozone layer depletion, ground level smog and global warming.

Expanded polystyrene contains only 5% polymer and the rest is air.

19.4.5 Green Synthesis of Polyurethane

Polyurethanes are conventionally prepared from a diisocyanate and a diol.

$$\underset{\text{Diisocyanate}}{\text{OCN—(CH}_2)_n\text{—NCO}} \quad + \quad \underset{\text{Diol}}{\text{HO—(CH}_2)_n\text{—OH}}$$

$$\downarrow$$

$$\underset{\text{Polyurethane}}{\text{—}\overset{\text{O}}{\overset{\|}{\text{C}}}\text{—NH—(CH}_2)_n\text{—NH—}\overset{\text{O}}{\overset{\|}{\text{C}}}\text{—O—(CH}_2)_n\text{—O—}}$$

Exposure to isocyanate is known to cause skin and respiratory problems in humans, these are also toxic to wild life. Hybrid Coating Technologies (HCT) have produced green polyurethane by reaction between a carbonate (mono or polycyclic), epoxy oligomers and polyamines (aliphatic or cyclopolyaliphatic) containing a primary amino group. HCT has been awarded Presidential Green Chemistry Challenge Award 2015 for designing greener chemicals.

19.4.6 Nylon 6,12

Nylon 6,12 is a synthetic polyamide obtained by condensation polymerisation between hexamethylene diamine and dodecanedioic acid (DDDA).

$$\underset{\text{Nylon 6 12}}{\text{–NH–(CH}_2)_6 \text{ – NH–CO–(CH}_2)_{10}\text{–CO–NH–(CH}_2)_6\text{–NH–}}$$

Nylon 6,12 possesses high abrasion, moisture and chemical resistance and is used for making brushes, adhesives and in automotive and aviation oils.

A new process has been developed by Verdezne (a US based company) for the production of DDDA for which a genetically modified yeast of genus candide is used to convert lauric acid (a 12 carbon fatty acid) to DDDA by aerobic fermentation process

$$\underset{\text{lauric acid}}{CH_3(CH_2)_{10}COOH} \xrightarrow[\text{fermentation}]{\text{aerobic}} \underset{\text{DDDA}}{HOOC-(CH_2)_{10}-COOH}$$

Verdezyne has been awarded the Presidential Green Chemistry Challenge Award 2016.

19.4.7 Air Carbon

Scientist at Newlight Technologies have developed a high performance thermoplastic called "AirCarbon". A biocatalyst is used to combine methane (emitted from natural resources), a green house gas with oxygen to produce Air Carbon. The polymer is carbon-negative and reduces the green house gases.

This thermoplastic is used to make bags, cell phone cases, furniture and containers etc. Air Carbon can be used as a substitute for polyethylene, polypropylene, polystyrene and ABS etc. another major advantage of using it is its biodegrability.

Newlight Technologies was awarded Presidential Green Chemistry Challenge Award in 2016 for designing greener chemicals.

19.5 SYNTHESIS OF POLYESTERS

A number of new polyesters have been produced in the recent past. Polyesters like polylactic acid and polyethylene furanoates have a saturated backbone, their monomers are produced from renewable resources and are recycled like PET.

Some unsaturated polyester that have been produced have the advantage that these can be cross linked also. These are also recyclable.

Polyhydroxy alkonates are produced by fermentation of renewable resources and another equally important fact is that these are biodegradable.

19.5.1 Polylactic Acid/Polylactide (PLA)[12]

Polylactic acid is similar in properties to polystyrene (a non biodegradable polymer) but is a biodegradable polymer and is widely used in the production of medical devices and electronic equipment.

The monomer lactic acid or lactide is obtained from renewable starting material, which are conventionally polymerised in organic solvents in presence of metallic catalyst containing metals like tin, aluminium, zinc and rare earths etc. The major drawback of using these catalyst are (*i*) the metals are extracted from minerals (non-renewable resources) (*ii*) traces of metals are invariably found in the product which raise health and environment issues. The use of organic solvents is also undesirable and cost intensive. Though PLA has been produced without the use of solvents (bulk polymerisation), there is little control over mass dispersity in this method.

A number of efforts have been made in the past to substitute metal catalyst by organic catalyst, a recently developed green alternative method for the production of PLA by ring opening polymerisation (ROP) gives a good quality product and has the following advantages:

- Metal catalysts have been replaced by organic catalyst.
- $ScCO_2$ is used in place of organic solvents.
- Benzyl alcohol is used as an initiator, water has also been used as an initiator.
- The polymerisation is carried out at 80°C compared to bulk polymerisation which is carried out at 130°C.

The rate of reaction is however slower as compared to polymerisation in organic solvents.

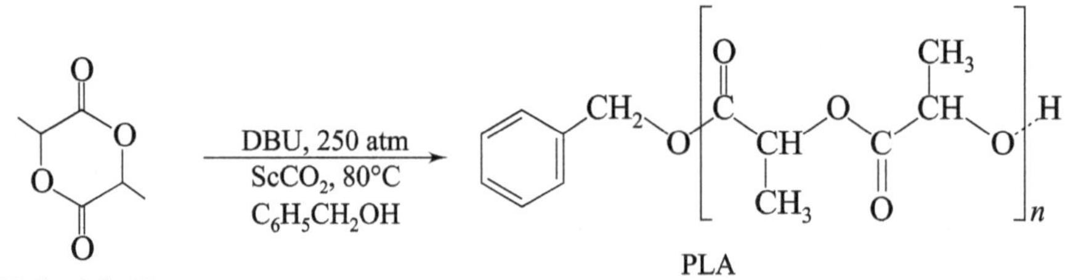

DL Lacitde 13

19.5.2 Polypropylenefumarate (PPF)[13]

It is a linear polyester of fumaric acid and propene 1, 2-diol and is obtained in two steps.

The first step involves the reaction between diethylfumarate and propylene glycol, the *bis* (hydroxypropyl) fumarate thus obtained undergoes transesterification in the second step to produce poly (propylenefumarate).

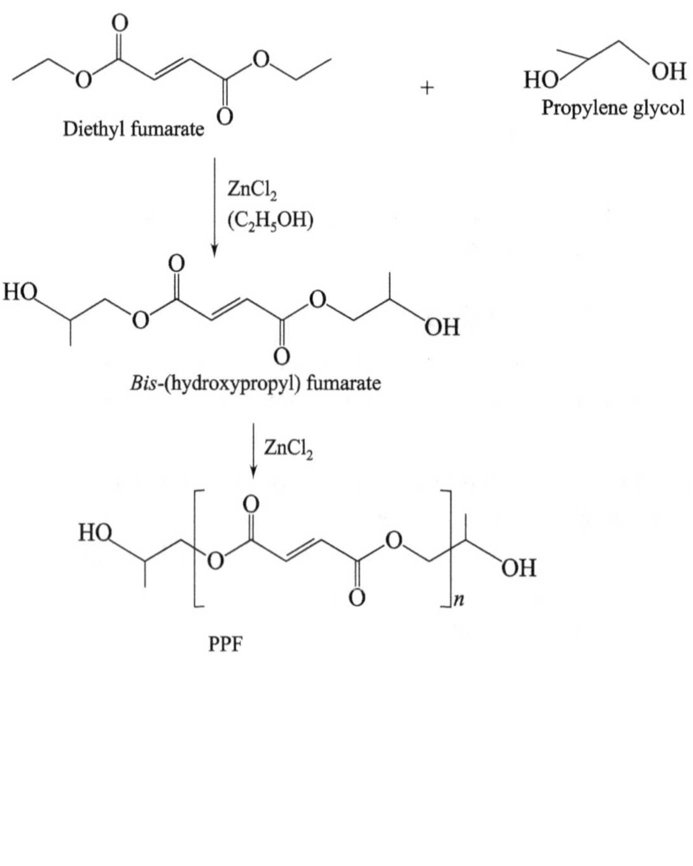

The linear PPF contains double bonds in the polymer chain. It has been cross linked using a number of reagents. For example, in one of the methods (US patent : US 20020022676A1). PPF has been cross linked by reaction with poly(ethylene glycol)-dimethylacrylate (PEG-DMA). The cross linked polymer can be used for replacing or reconstructing a semi-rigid living tissue.

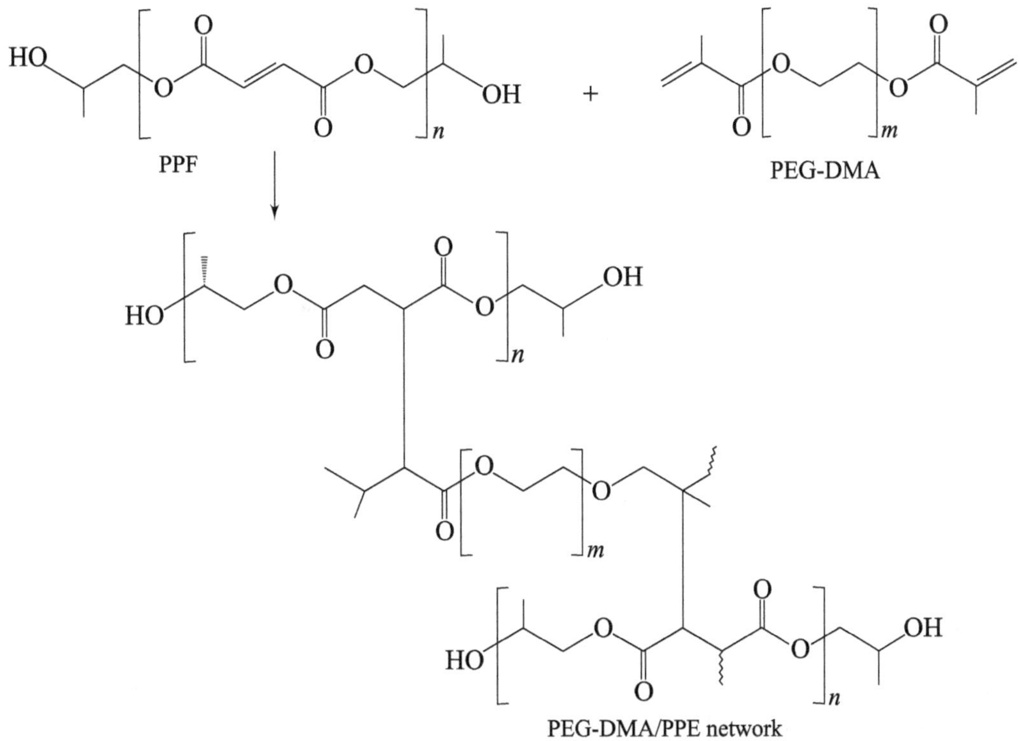

19.5.3

A copolymer of Diethyl succinate, dimethyl itaconate and 1,4-butanediol[14(a,b)].

Succinic acid, itaconic acid and 1,4-butanediol, all obtained from renewable biomaterials are copolymerised using *Candida antarctica* Lipase B (CALB in imbobilised form as Novozyme 435) to obtain this polyester.

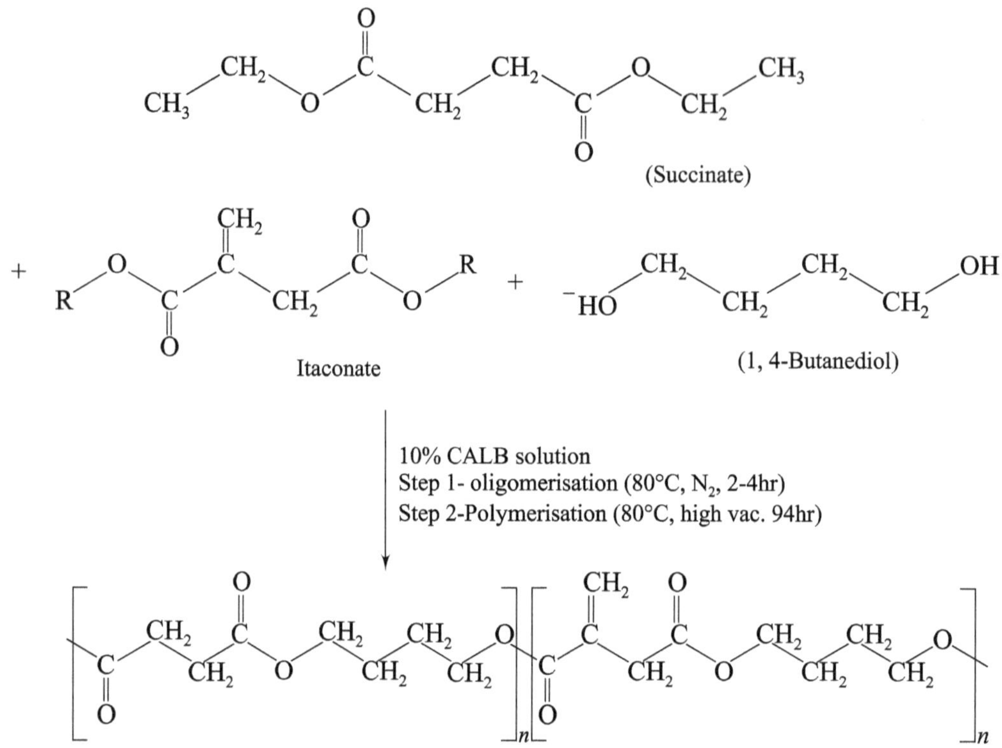

19.5.4 Polyethylene Furanoate (PEF)

Polyethylene Furanoate (IV), a 100% biobased polymer has been produced by copolymerisation of ethylene glycol and furan dicarboxylic acid or its diester.

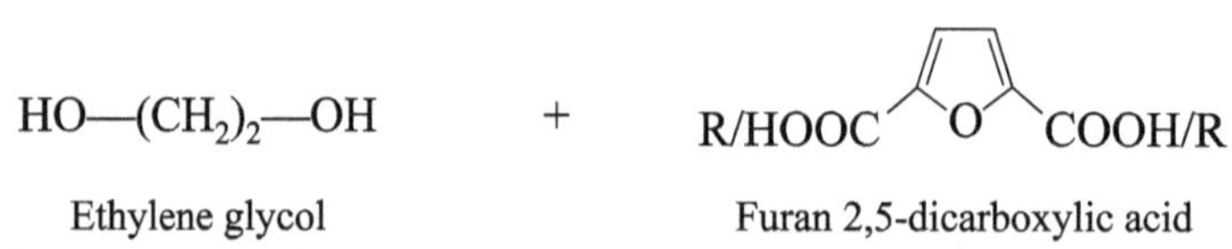

Ethylene glycol is produced from bioethanol (page 561) and furan dicarboxylic acid is obtained from hexoses.

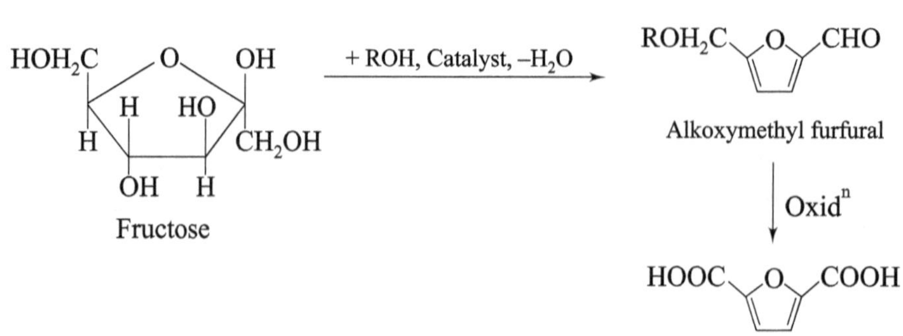

The co-polymer (PEF) thus produced has oxygen, CO_2 and H_2O barrier 6, 3 and 2 times more than that of PET. PEF also has better mechanical properties than PET.

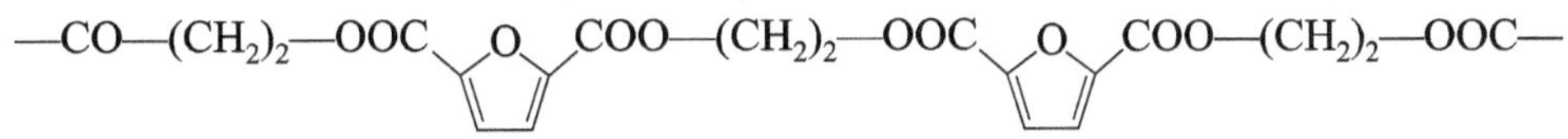

PEF (IV)

PEF can be recycled and can be blended with rPET.

PEF is used for production of bottles, textiles, carpets, electronic material, automotive parts. It is also used for packaging.

Furan dicarboxylic acid can also be used for production of polyamides and polyurethanes.

19.5.5 Biodegradable Polymers

Long utility time of synthetic polymer products is their main advantage but at the same time a major disadvantage. Because of their non-biodegradable nature, these persist after use and add to the volume of ever increasing after use polymer waste. Efforts are therefore being constantly made by the scientific community to look for better alternatives. One such effort is to obtain all types of building units for polymerisation from renewable resources which has been greatly successful. Another major step has been the developmnet of biodegradable polymers. These polymers as the name suggests are biodegraded under suitable conditions in the environment. These have different biodegradation time, are useful for one time applications and are extensively used in medical field. Their utility has further been widened by copolymerisation or blending with other polymers. Some of the polymers of this type are:

Polyhydroxy Alkonates (PHAs)[15(a,b,c)]

These are biodegradable thermoplastic polyesters having a general formula —O—CH(R)—(CH_2) xCO– where R is an alkyl group. Some bacteria (Prokaryotic organisms) accumulate these types of polymers from 30–80% of their dry body weight as stored nutrients to obtain energy during crises. These polymers have been produced biogenetically as well as by chemical synthesis. Because of their promising properties like a wide spectrum of structures, processing versatility, compatibility with other natural polymers and high biodegradability etc., a large number of natural bacteria, fungi, moulds, genetically modified bacteria, culture media and other parameters are being investigated to produce PHAs at a competitive price. The possibilities of using transgenic plants (rapeseed and cress) are also being explored for production of PHAs. PHAs can replace PE, PP, PET and PVC etc. in near future. It has now become possible to precisely design and obtain PHAs *i.e.* design based on demand polymers.

PHAs have been known for almost 100 years. Polyhydroxyl butyrate (PHB) was first discovered by Mauris Lemoigne (a French microbiologist) in 1920s in a bacterium (*Bacillus megaterium*), which stored PHB to the extent of ~30% of its body weight. PHB is highly crystalline, brittle and difficult to process. Its copolymer with Poly 3-hydroxyvalerate has been produced, the copolymer is a thermoplastic, flexible and easy to process. 3-PHB-co-3PHV is commercially produced by *Alcaligenes eutrophus* and other bacteria, which accumulates the polymer to the extent of ~80% of their dry body weight.

3-PHB-co-3PHV is extensively used in medical field for making sutures, wound dressing, bone replacement, for controlled drug delivery. 3-PHB-co-3PHV is also used in the manufacture of drink cups, disposable utensils, containers for food and shampoo bottles etc. *Cupriavidus eutrophus* B10646, a recombinant bacterial strain has been used to produce P(3HB-co-3HHx) with varying compositions of the monomers. The substrate consisted of glucose, nitrogen, sodium hexanoate and sodium acrylate.

Depending on the type of microorganisms, cultivation medium etc., it has been possible to produce a large variety of different types of PHAs ranging from simple polymers (homopolymer), random and block copolymers. Functional and graft PHAs. *Pseudomonas putida, Pseudomonas oleovorans, Pseudomonas entomophila, Alcaligenes eutophus, Alcaligenes latus, Ralstonia eutropha, Aeromonas hydrophilia*, modified E. coli etc. are used for different types of PHAs. A number of rich sources of carbon are being explored to reduce the cost of PHAs. Carbon rich by products of industries like glycerine and vegetable oil. Glucose, fructose, sugar (sucrose) and other biomass etc. A number of countries like USA, UK, China, Japan, Italy and Germany are producing PHAs commercially.

P(3HB-4HB) is produced in a medium containing sodium butyrate and γ-butyrolactone, P(3HB-3HHx) is produced in a medium containing sodium butyrate and sodium hexanoate, P(3HHx-3HD3HDD) in hexanoate and dodecanoic acid and PHB-P(3HV3HHp) in a medium containing sodium butyrate and sodium heptanoate. All these can be produced by *Pseudomonas putida*. E. coli has been used to produce P(3HP-4HB).

As already stated a large number of PHAs have been produced, some of these are shown in table 19.1 below.

Table 19.1: Some examples of PHAs (Homopolymers)

Chemical name and abbreviation	Structure
Poly(3-hydroxypropionate) P(3HP)	$-[-CO-CH_2-CH_2-O-]_n-$
Poly(3-hydroxybutyrate) P(3HB)	$-[-CO-CH_2-CH(CH_3)-O-]_n-$
Poly (3-hydroxyvalerate) P(3HV)	$-[-CO-CH_2-CH(C_2H_5)-O-]_n-$
Poly (3-hydroxyhexanoate) P(3HH)	$-[-CO-CH_2-CH(C_3H_7)-O-]_n-$
Poly (3-hydroxyoctanoate) P(3HO)	$-[-CO-CH_2-CH(C_5H_{11})-O-]_n-$
Poly (3-hydroxynonanoate) P(3HN)	$-[-CO-CH_2-CH(C_6H_{13})-O-]_n-$
Poly (3-hydroxydecanoate) P(3HD)	$-[-CO-CH_2-CH(C_7H_{15})-O-]_n-$
Poly(4-hydroxybutyrate P(4HB)	$-[-CO-CH_2-CH_2-CH_2-O-]_n-$
Poly (5-hydroxyvalarate) P(5HV)	$-[-CO-CH_2-(CH_2)_3\text{-}O-]_n-$

Some examples of copolymers of PHAs

Name	Abbreviation and microoganism
Poly(hydroxybutyrate-co-hydroxyvalerate)	P(3HB-co-3HV) *C.necator*
Poly(hydroxybutyrate-co-hydroxyhexanoate)	P(3HB-co-3HH_x) *P.putida*
Poly(hydroxybutyrate-co-hydroxyoctonoate)	P(3HB-co-3HO)
Poly(hydroxybutyrate-co-hydroxydecanoate)	P(3HB-co-3HD)
Poly(hydroxybutyrate-co-hydroxoctadecanoate)	P(3HB-co-3HOd)

Uses of PHAs

PHAs are used in packaging, agriculture, leisure, medical field and fast food.

Medical Applications

Surgical Sutures

Absorbable synthetic sutures were developed in 1960s. Many of these are compatible with body tissues and are therefore used in surgery. Sutures made from PGA, PLA and their copolymers are commercially available. Polydioxanones and polyglyconates are flexible and more suitable. Sutures made from some other common ones are:

Polycaprolactone are also under study.

Dexon is made from poly glycolic acid

Polygalactin 910 is a copolymer of glycolide and lactide (9:1).

Mexon is a copolymer of glycolic acid and trimethyl carbonate.

Nylon66 and Nylon6 mixed with other polyamides are used for making sutures. These are longer lasting as compared to others threads used in surgery but have mild reaction with tissue.

Bone Fixations

Metals have been used to fix bones in the past these fixing devices are removed by a second operation in most cases after healing. The use of biodegradable polymers in bone fixation can help to eliminate the second operation. PGA, PLA, PHD and polydioxanone can potentially be used.

19.6 GREEN POLYMER ADHESIVES

Wood composite industry has been using synthetic polymers as adhesives for production of plywood, fiberboard and particle board etc. Urea-formaldehyde and phenol formaldehyde resins have been used in the past as adhesives. Both are made from formaldehyde, which is believed to be a human carcinogen.

Professor Kaichang Li of Oregon State University developed a protein based wood adhesive for which Greener Synthetic Pathway Award (US) was given in 2007. A soy-based adhesive was developed during 2006. The work was inspired by the superior adhesive properties of protein used by mussels to adhere to rocks. The soy proteins were modified to get proteins similar to those used by mussels.

Mussels achesive proteins (MAPs) contain 3,4-Dihydroxyphenyl-alanine (DOPA) which is believed to be the amino acid responsible for the adhesive properties of these proteins.

19.7 POLYMER ADDITIVES

Additives are compounds that are added to the polymer to perform various functions. These enhance the properties of polymers, protect the products made from polymers from degradation, fire etc. For example fillers like carbon black increase the strength of polymer product and saves it from degradation by UV light. Antioxidants delay the oxidative degradation of polymers. Plasticizers convert hard and rigid polymers to soft and flexible once and so on.

Some of these were not benign and some greener additives are now used in polymer industry.

19.7.1 Flame Retardents

This is a group of chemicals added to the plastics for protection against fire. Compounds like polybrominated diphenyl ethers (PBDEs) and polychlorinated biphenyls (PCBs) have been used as flame retards. PCBs were banned for their toxic nature and PBDEs also have health and environmental issues. Polybrominated polymers have been replaced by low molecular PBDEs as fire retardants.

Now a number of non-halogenated compounds have been produced which might replace all bromine and chlorine containing compounds. One group of such compounds are oligomeric/polymeric phosphonates. Oxygen containing oligomers/polymers such as carbonates, esters, epoxides etc. are most effective.

Fyrol HF-5 is one such flame retardant which contains high degree of phosphorus.

Infra red reflective coating of thermoplastics is also being used for flame protection. This type of coating on the plastic prevents it from getting heated and hence catching fire.

19.7.2 Stabilizers

Inorganic or organic salts of metals are added as stabilisers to PVC. Cadmium and lead salts have been replaced by calcium and zinc salts. Metal free organic compounds are also used as stabilizers.

19.7.3 Plasticizers

PVC is hard and rigid, it is converted to soft and flexible form by addition of plasticizers. Low molecular weight esters of phthalic acid were used to produce soft PVC. These have now been replaced by high molecular weight esters of phthalic acid.

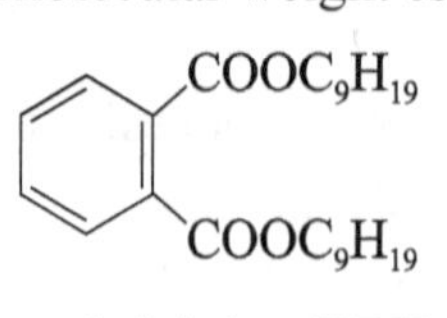

Diisononyl phtihalate (DINP)

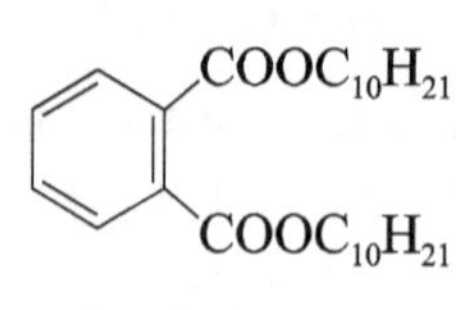

Diisodecyl phtihalate (DIDP)

Adipates and terephthalates are also used in place of low molecular weigh phthalates. Hexamoll DINCH is another non-aromatic compound used as plasticizer in PVC.

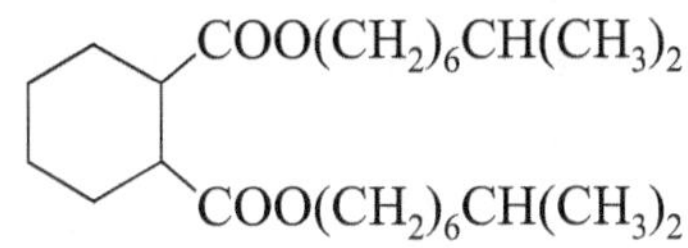

1,2-Cyclohexane dicarboxylic acid disononyl ester

19.7.4 Fillers

Talc, carbon black, calcium carbonate etc. have been used as fillers in polymers. Hydrophobic lignocellulose and xylite are now being preffered as fillers for polymers.

19.8 RECYCLING OF POLYMERS

19.8.1 Introduction

Products made from plastics are accumulating because of their ever increasing use and non-biodegradable nature. Many of these products are used once and discarded (single use application) like packaging material and account for ~50% of the total plastic use in the world. 20-25% plastic is used for long term application like production of pipes, cable cover, electronic goods, car parts etc. but after use waster produced due to these is also increasing. Recycle, reduce and reuse the polymer products is one of the solutions to reduce the burden of non-biodegradable polymer waste on the earth. It is a method of reducing resource depletion and environmental impact. It is a better option than producing more of virgin polymers that are known and produced today. Recent trends in recycling indicate a substantive increaser but an ideal situation will be to reuse and recycle 100% of the used polymer waste. There are a few inherent problems in recycling.

- The cost of virgin polymers is not so high in most cases and many business houses thus do not find recovery and recycling of polymers a high profit proposition.
- Sorting out, cleaning, transport to the site for recycling is another set of problems, though a number of innovative methods have evolved in the recent past.
- Paucity of technologies for recycling.

19.8.2 Polymer Recycling

Recycling of polymers has been known and done since around 1970s but was limited to melting and remoulding the thermoplastic into poor quality products in the past.

Now with increasing awareness for need of clean environment, many efforts have been made to develope techniques for polymer recycling but much more needs to be done. Today less than ~15% of the polymers are recycled.

Polymers are mechanically recycled or degradation or depolymerisation chemically by one of the following method.

- Thermal Degradation
- Chemical Degradation

- Biological Degradation
- Degradation by using radiations

Addition polymers like polyethylene, polystyrene and polypropylene are degraded by thermal methods whereas condensation polymers like polyesters, polyamides are depolymerised by chemical methods. The cost involved in chemical recycling is more than that used for mechanical recycling.

The terms used for the polymer recycle are

Primary Recycle	(mechanical method, close loop recovery re-extrusion)
Secondary Recycle	(mechanical method, down grading)
Tertiary Recycle	(chemical method, feed stock recovery)
Quartenary Recycle	(incineration to recover energy)

Increasing Polymer Recycle

The main steps which can help increase the polymer recycle are

- **Increasing Public Awareness**

  Educating people about the adverse effects of plastics can encourage plastic reuse or minimise its use. Participation by public can also help in initial cleaning, drying and sorting out of plastic waste.

- **Contribution by the scientific community for development of new technology**

  The new technologies that are developed for collection, sorting and recycling should be energy and cost effective. The polymer waste can be segregated on the basis of their plastic identification number, but many scientific identification techniques like FTNIR, optical colour recognition, X-ray detection etc. are used today. The separation of mixture of polymers is carried out by a number of methods like (*a*) difference in densities (*b*) froth flotation (*c*) laser sorting (*d*) by selective dissolution from the mixed polymer waste etc.

- **Removal of paint from plastics is another major issue. Some methods that can be used are**
- Solvent stripping of paint.
- Abrasion of the surface to remove paint.
- High temperature aqueous-based removal-some paints get hydrolysed by water and hence get removed.
- Grinding of the polymer sometimes helps remove paint etc.
- **Participation of the industry to set up plants for recycling.**
- **Govt. Subsidies to encourage recycling.**
- **Recognition of exemplary work in the field.**

• using recycled polymers

If big names in industries can initiative and start using good quality recycled polymers for their products, it will widen the scope of recycling.

Dell, for example is using 100% of virgin plastic waste of their own company.

Hewlett-Packard Co., a producer of various types of printers, desktops, notebooks has taken an initiative to recycle plastic used in its own products and use them.

19.8.3 Recycling of PET

Polyethyleneterephathalate is one of the most widely used plastic world-wide. Fortunately it is not discarded in the land fills because techniques have been developed to recycle and reuse it. A small amount which is not recycled, can be incinirated to obtained energy.

A very small amount of PET that reaches land fills is not an environmental issue because it neither leaches nor degrades.

Two types of methods are used to recycle PET

(*i*) Mechanical recycling of polymer.

(*ii*) Chemical methods of depolymerisation and subsequent repolymerisation of the monomers.

(*i*) Mechanical Recycling: The PET waste (mainly bottles which were used for storage of water, soft drinks, juices, beverages, oils, shampoos and other household items) are sorted out.These bottles are identified by their plastic identification number (which is ♳ or a dot or a circular gate).The caps are not made from PET and hence are removed. The bottles are then segreggated into transparent and coloured (green). A few others which are light brown etc. are not mixed.

PET is a porous material and absorbs the products stored in it. These absorbed materials are difficult to remove. Heating destroys the polymer and hence sterlization can not be achieved by heating.

However, the transparent and green bottles are separately washed, dried and shredded. Special washing, drying and shredding machines have been developed for this purpose.

PET chips are then used for the production of bottles for shampoos, some cosmetics etc. but not for storage of food. It is also used for production of carpets, some car parts and filling for pillows etc.

- In past, the containers made from recycled PET were not used for storage of food items like milk and juices etc. but technical advancements developed for decontamination of PET allow these containers from r PET to be used for food items.

In the EREM super clean technology, dry bottles obtained from used PET bottles (after collecting, sorting , hot caustic washing and drying) are heated in a continuous reactor at high temperature and then heated in another reactor at high temperature under vacuum;

(*ii*) Chemical Methods: A number of chemical methods have been developed for depolymerisation of PET. Two of these are outlined below:

(a) Glycolysis–PET is depolymerised by treatment with glycol and hence the name glycolysis.

In one of the methods, the clean, dry and shredded PET is heated with excess glycol at its boiling point in presence of a catalyst. Bis(hydroxyethyl) terephthalate (BHET) is obtained in good yield.

In another method, PET is heated with ethylene glycol in presence of hydrotalcite as catalyst at 198°C for two hours. Monohydroxy ethylterephthalate (99% pure) is obtained. Both solvent and catalyst are recycled.

(*b*) Methanolysis: In this method, developed by Mitsubishi Heavy Industries (MHI), PET is treated with super critical methanol. Dimethyl terephthalate and glycol are produced, which are repolymerised.

Recycled PET is being used for production of polyester filament yarn in frankfurt.

19.8.4 HDPE Recycle

High density polyethylene finds a number of applications, one of the major being production of containers for milk.

In the past, the containers made from rHDPE were not used for storage of mild etc. as HDPE is even more porous than PET and diffusion and sorption of food occurs in HDPE containers which were difficult to remove.

Now techniques have been developed for decontamination of HDPE which allows use of rHDPE containers for food items. EREM process that is used for PET is also now being used for HDPE.

Supercritical carbon dioxide is also being tried as an alternative solvent for decontamination of HDPE.

Crates and bins are made from HDPE which cannot be recycled to make bottles.

19.8.5 PVC Recycle

- PVC is separated from the mixed waste by selective dissolution in a suitable solvent, filteration and drying by bubbling steam into the solution.
- The unmixed waste is used directly back in production.
- Pure HCl is obtained as feedstock which goes back into Vinyl Chloride production on thermal treatment of PVC. The hydrocarbon part is then incinirated to generate electricity or heat.

19.8.6 Nylon-6 Recovery from Carpets

The carpets are made from polypropylene, polyester, wool, nylon6 and nylon66. Nylons which are expensive as compared to other polymers are segregated and collected from used carpets by one of the following method.

In one of the method the carpets are subjected to dry process consisting of reduction and separation steps. Water is then added to wash the shredded fibres which are then separated from the fillers, nylon and polypropylene are then separated on the basis of their density.

Solvent extraction is also used to separate nylon from used carpets.

Supercritical CO_2 is use to extract various polymers in the carpets sequentially by varying pressure and temperature.

Nylon 6 collected from carpet industry is depolymerised with superheated steam in presence of a catalyst. Crude caprolactam thus obtained is purified by distillation and repolymerised.

19.8.7 Recycling of Polycarbonate

CDs and DVDs which are made from polycarbonate (plastic identificatin number 7) are converted back to bis-phenol by heating with glycerine, sorbitol and water (6:1:3) at 180°C for 45 minutes in presence of sodium hydroxide (1%). bis phenol thus obtained is purified by crystallisation from water.

19.8.8 Recycling of Polyethylene (PE) and Polypropylene (PP)

- PE and PP are recycled by dissolution followed by precipitation.
- Thermal pyrolysis for feed stock recovery.
- Heat and electricity can be generated by heating.

Conclusion

Products made from polymers play a very important role in our day to day life. Most of the earlier polymers are non-biodegradable and hence adding to an unmanagable waste which is becoming difficult to handle. There is need to substitute non-biodegradable by biodegradable polymers, reduce the use and recycle the non-biodegradable polymers.

REFERENCES

1. Yu-sin Jang *et al.*, Biotechnology an Bioengineering, 2012, **109(10)**, 2437–2459
2. US Patent 5487987A.
3. Shu Giuo *et al*, Green Chemistry, 2006, **8**, 296-300
   Yan Chang *et al.*, Green Chemistry, 2008, **10**, 98–103.
4. Processdesign. mccormick. northwestern, edu/index. php/Biomass
5. Banu M, Top. Catal., 2012, **55**, 897–907.
   Li N, J. Catal., 2010, **270**, 48–59
6. Pang J., Green Chemistry, 2016, 18, 342.
   Lyons T.W. J. Am. Chem. Soc. 2012, 134, 15708–15711
   Wang F., RSC Adv, 2014, 4, 6314-6317.
7. Spekreijse J. *et al.*, Green Chemistry 2012, 14, 2747.
8. (a) Yang Yan, Process Biochemistry 2011, **46**, 1900–1908
   (b) Anagnostopoulos A. Macromolecules, 1998, 31, 4376

(c) Zemel H, Chemical Abstracts, 1995, 123, 199, 816

(d) Bisht KS., Macromolecules, 1997, 30, 3122–3124

(e) Chanfreau S, Biopro. Biosyst. Eng. 2010, 33, 629-638.

9. (a) Romack TJ, Macromolecules, 1995, 28, 912–916

(b) Parkinson G, Chem. Eng. 1999, 106, May 17

(c) Hussain Yazon A, Ind. and Eng. Res., 2012, 51, 11401–11406

(d) Anhouxu, European Polymer Journal, 2012, 48(8), 1431–1438

(e) Somaieh Salehpour, Green Chem. 2008, 10, 321–326

(f) Przemyslaw K. Prog. Polym. Sci. 2004 (29), 3–12

10. Krzysztof M. Chemical Soc. Rev., 2008, 17, 1087–1097

11. (a) Kanan C *et al.*, Res. J. Chem. Sci. 2012, 2(7), 27–35.

(b) Phan L Ind. Eng. Chem. Res., 2008, 47, 539

12. Blakey I *et al.,* Green Chem. 2011, 13, 2032

13. Kurtis Kasper, Nat. Protocol, 2009, 4(4), 518–525

14. (a) Bozell JJ., Green Chem., 2010, 12, 539–554

(b) Jiang Y., Biomolecules, 2013, 3, 461-480.

15. (a) Volova TG, Wiley Online Library, 2014, 14 Nov.

(b) Guo-Qiang Chem. Trands in Biotechnology, October 2015, 3, 10

(c) Alvarez V, Express Polym. Lett., 2014, 8, 791–808.

20. AGROCHEMICALS

20

Agrochemicals

20.1 INTRODUCTION

World population, which is about 6.5 billion in the year 2015 is increasing at the rate of ~80 million per annum leading to an increased requirement of food by 26 million ton every year. To meet the food requirement of ever increasing population, there is a need to increase productivity, reduce the loss of crops due to pests, infections and weeds etc.

Scientific community has played a major role in increasing the land productivity by understanding the soil conditions needed for various coprs, developing feritilizers, persticides and herbicides etc.

Increase in productivity was achieved but the consequences were serious.

Some of these agrochemicals did not kill only the targets but had adverse effect on humans, others polluted water, air, land and still others had direct as well as indirect effect on useful animals and insects. Honeybees for example, which help in pollination were damaged by some insecticides.

20.2 SYNTHETIC PESTICIDES

A number of synthetic pesticides were developed to reduce loss of crops due to pests. The use of synthetic pesticides began in 1930s and became widespread by 1950s. Synthetic agrochemicals specially insecticides are used even today to control pests.

20.2.1 Organochlorines

Organochlorines were the first major group of synthetic pesticides, a number of these were developed around World War II. Most of these were banned by the end of the century mainly because these did not undergo biodegradation and hence persisted and accumulated in the environment. Some of these were found to be carcinogenic and bad for the central nervous system etc.

Prior to organochlorines compounds of arsenic, nicotine and related compounds, amides and thiocyanates have been used as insecticides.

Some of the chlorine containing pesticides were

1. **Aldrin:** An organochlorine insecticide which was widely used until 1970s but was banned in most countries in the world as it was found to be persistent and a potential occupational carcinogen.

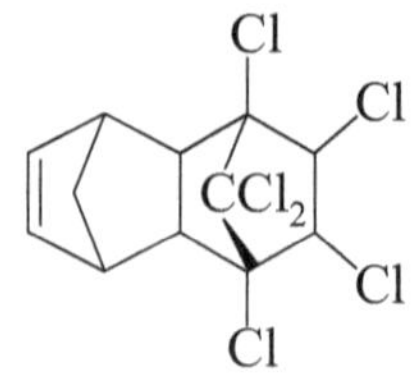

2. **Chlordane:** Another chlorine containing pesticide which does not break down fast after use and therefore persists in the agricultural land, is known to be a carcinogen and hence banned in most parts of the world.

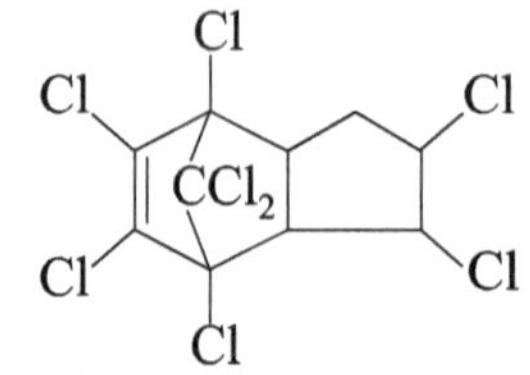

3. **DDT:** It was the first chemical used worldwide as a pesticide. It has been banned in many countries because of its adverse effect on human health like breast cancer, damage to nervous system and liver and developmental delay in children.

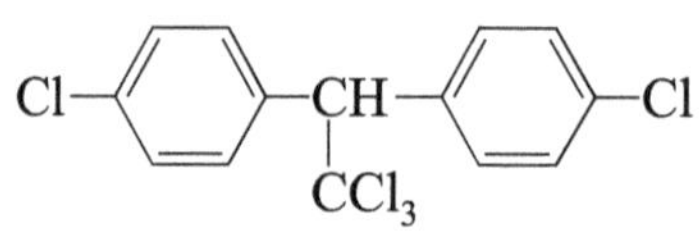

DDT

4. **Dieldrin:** An effective chlorine containing pesticide had to banned because of its persistence and accumulation.

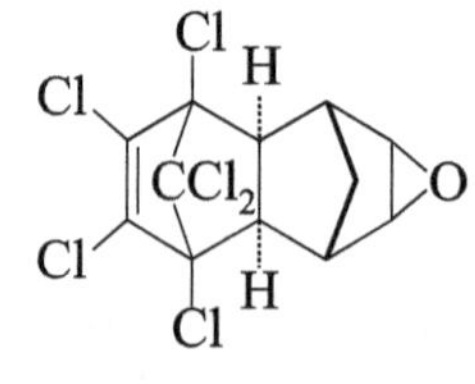

5. **Endrin:** An organochlorine compound was produced in 1950s and used as insecticide and rodenticide but its use was discontinued in many parts of the world around 1986. It is persistent, is present in water and a health hazard for human beings.

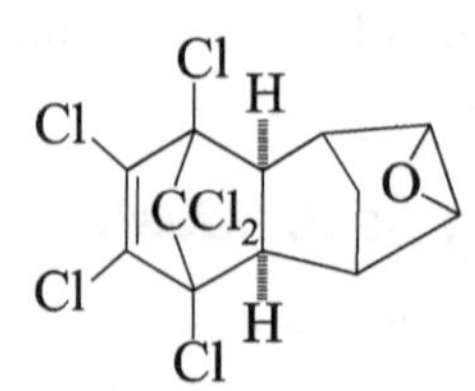

6. **Heptachlor:** was synthesised in 1946 but banned after ~40 years of use as insecticide.

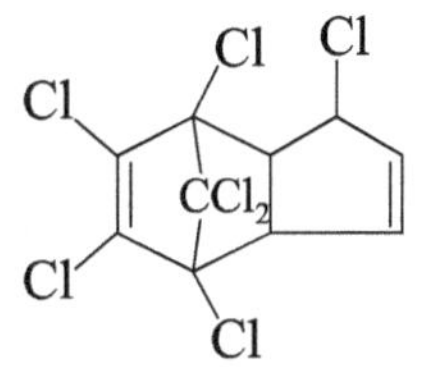

7. **BHL:** Which is now banned in many countries was used as an insecticide. It adversly effects human nervous system.

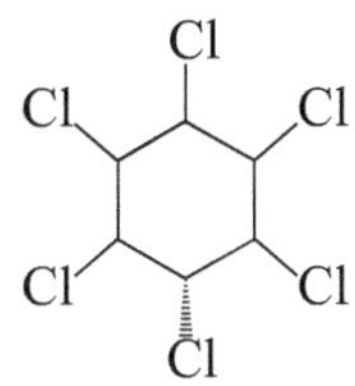

8. **Mirex:** A pesticide used against fire ants was banned because of its persistent and bioaccumlating nature.

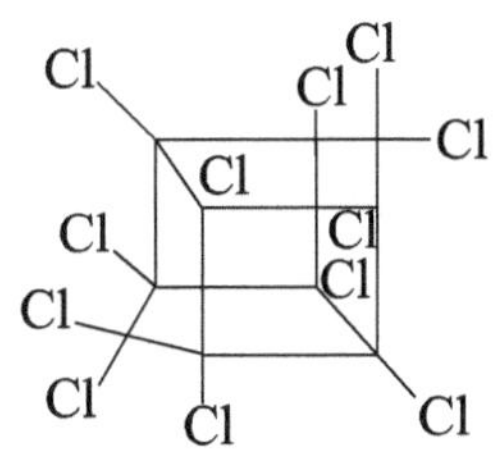

9. **Toxaphene:** A mixture of compounds obtained by reaction of chlorine with camphene was used as an insecticide from 1960s, was banned in 1990 in USA and worldwide in 2001 because of its non-degrading nature. It was found to be very toxic to the fish and some birds.

20.2.2 Organophosphate Pesticides

Organophosphate pesticides were the second generation chemicals, these were less persistent but some of them were found to be more toxic to human beings. Some are toxic to aquatic organisms.

Some examples are

1. **Melathion:** It is an organophosphate used as insecticides to control a variety of insects that attack fruits and vegetables, can be absorbed through skin and adversely effects nervous system, stomach and gastro-intestinal track and skin.

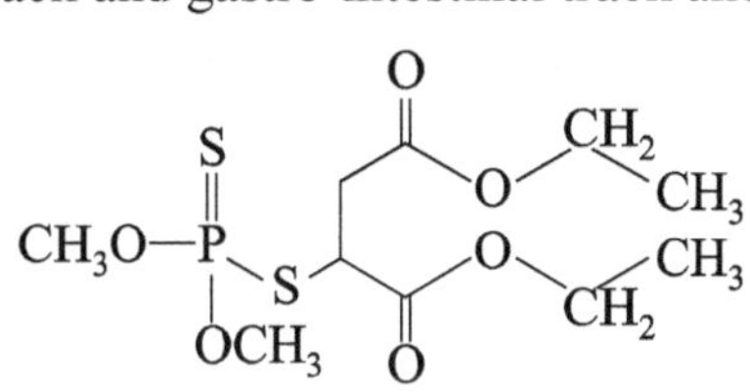

Parathion–Developed in 1940s and was mainly used as pesticide for cotton, rice and fruit trees. It has high human toxicity and is banned in many parts of the world.

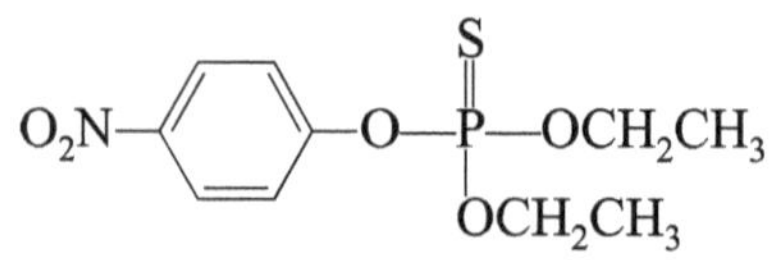

Tetraethyl Pyrophosphate: It was used as rhodenticide and pesticide (for control of aphids and red spider) but is extremely poisonous to human beings.

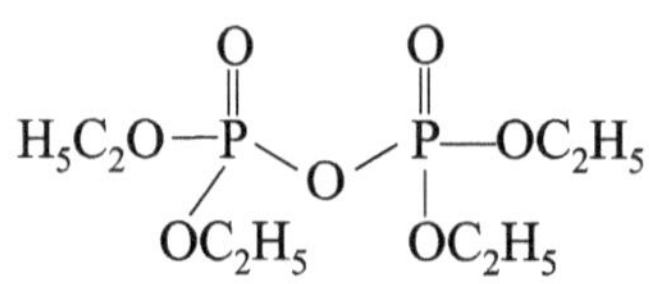

Other organcophosphate insecticides are.

Methyl Schradan 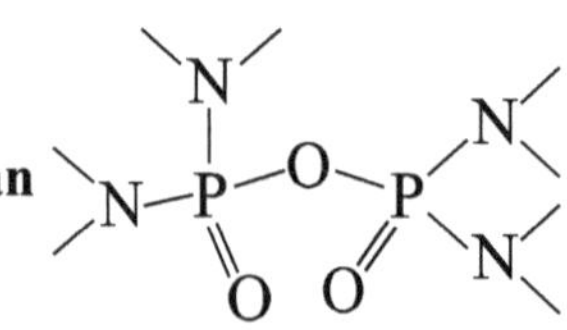

Phorate: 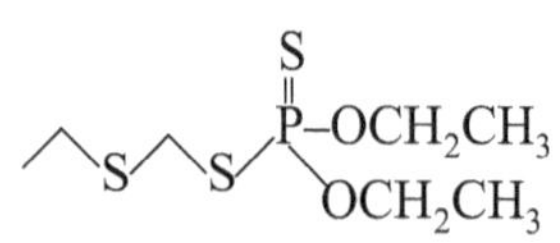 –Very toxic both to target organisms and mammals.

Diazinon: 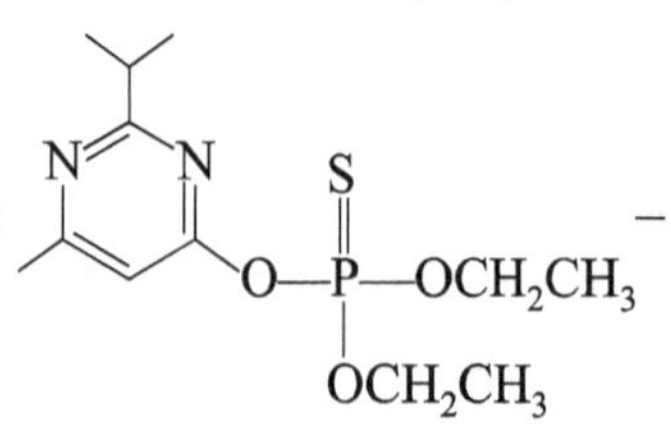 – High toxicity to vertebrates

Dimethoate: 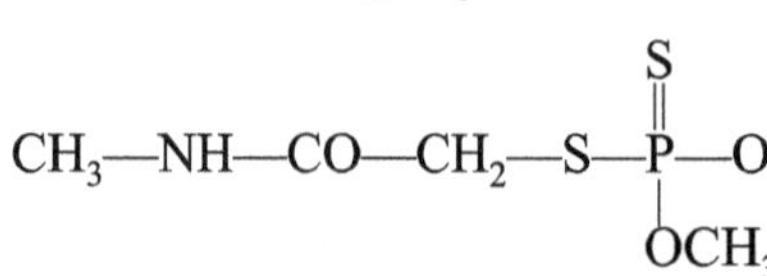 –Degrades relatively rapidly.

Mevinphos: 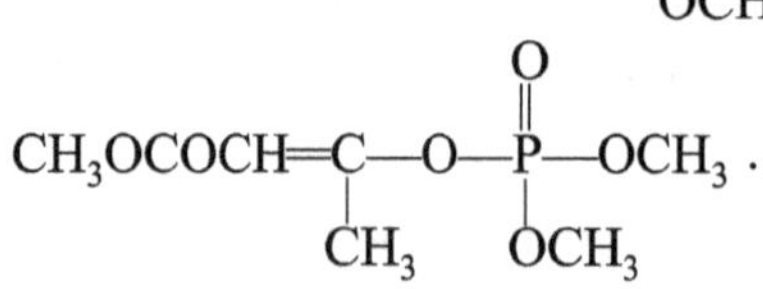

20.2.3 Pyrethroids

Pyrethroids like Fenpropanthrin, deltamethrin and cypermethrin belong to this class of pesticides. Their structure is similar to pyrethrin.

Some of the Pyrethroids have adverse effect on male reproductive health but do not persist and do not contaminate ground water

Fenpropanthrin.

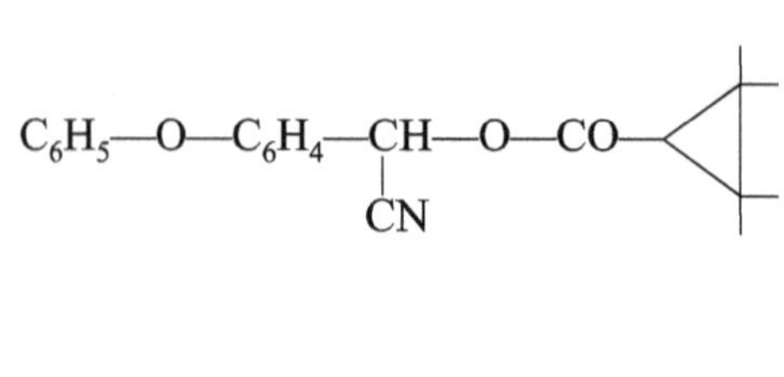

Deltamethrin

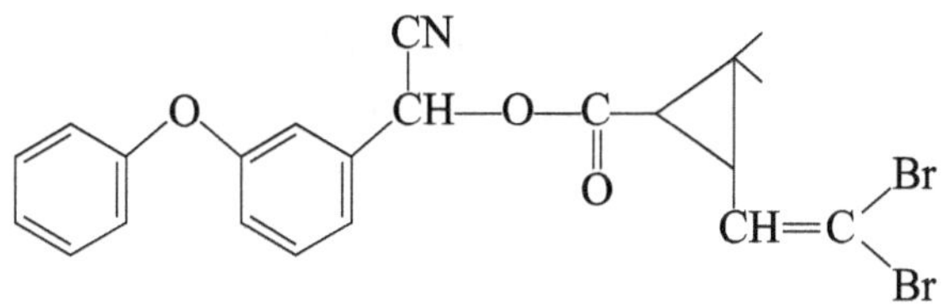

Cypermethrin

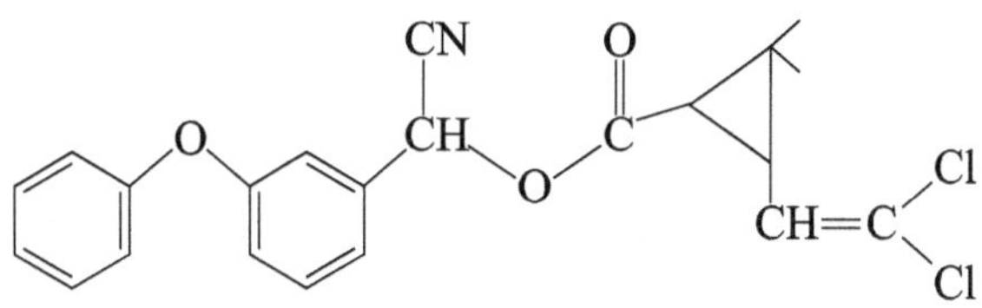

20.2.4 Carbamates

Carbamates like carbofuran, aldicarb and carbaryl were the next series of pesticide.

Carbofuran: It is one of the most toxic carbamate.

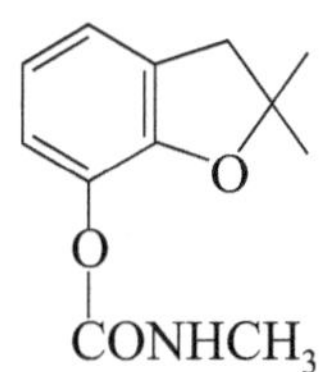

It is particularly toxic to birds and has high mammalian toxicity. Its use in restricted only to potatoe, maize, spinach grown for seed production, pumpkin and sunflower etc.

Aldicarb: It is mainly used to control aphids, spider mites, white flies and nematodes. It is used to protect crops like citrus fruits, strawberry, banana, soy, potatoes, onions, coffee etc.

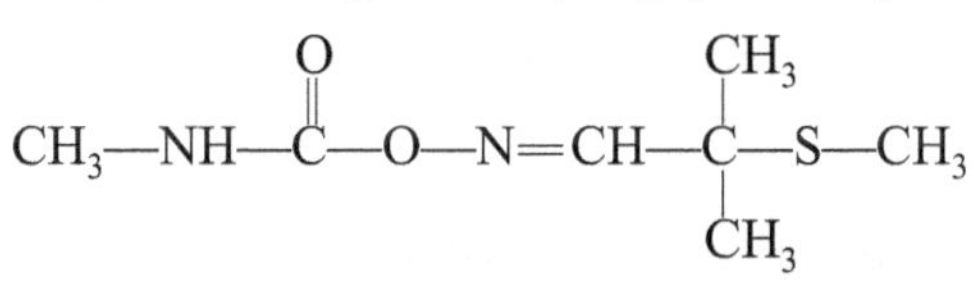

It also kills beneficial insects like earthworm and honeybees.

It is highly water soluble and thus contaminates water. It has negative effect on soil, environment in general. It causes a number of health problems in human.

Carbaryl: It is a widely used insecticide. Methyl isocyanate was first used for its

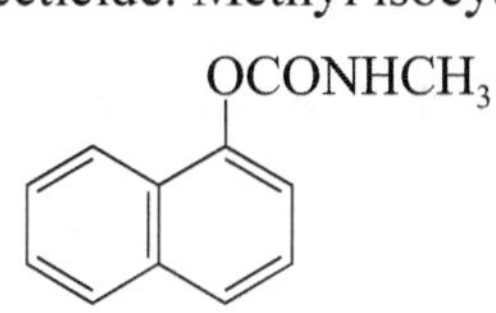

production. A major accident happened in Bhopal, an alternative synthetic route is used to produce carbaryl. It can be used as an insecticide on all crops. It adversely effects beneficial insects, it has tendency to bioaccumulate in some snails and crawfish and may cause mutagenic effects in human.

20.2.5 Neonicotinoids

These are insecticides which are chemically similar to nicotine.

Imidacloprid: As of 1999, it was the most widely used insecticide in the world. Recent research has indicated that it may be responsible for decline of honeybee colonies.

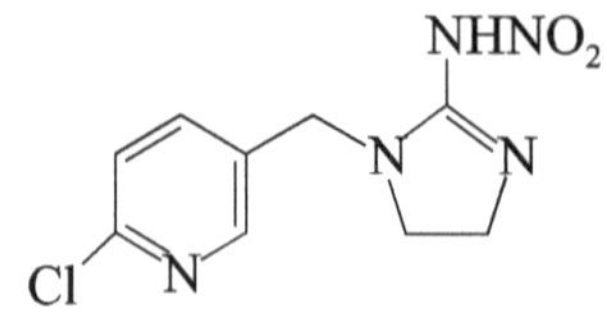

Clothianidin: It has adverse effect on honeybee colonies.

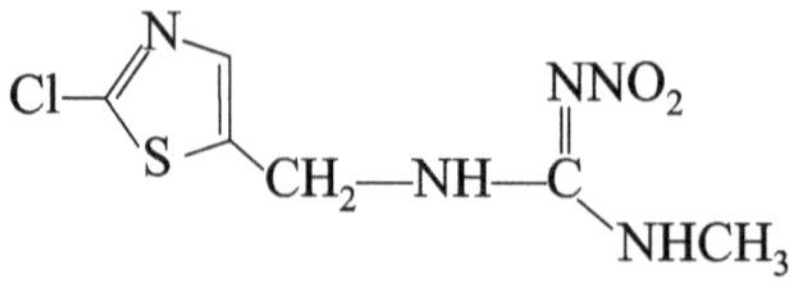

Thiamethoxam: It is a broad spectrum systemic insecticide moderately hazardous to humans.

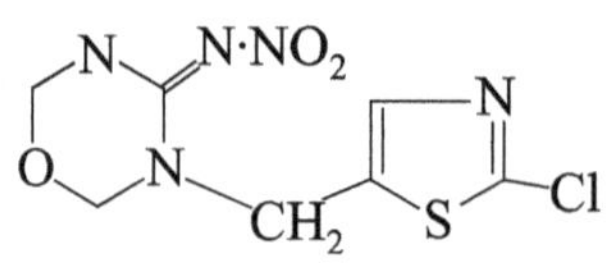

20.3 HERBICIDES

The first herbicide which was synthesised to kill weeds with broad leaves was 2, 4-Dichlorophenoxy acetic acid (2,4-D)

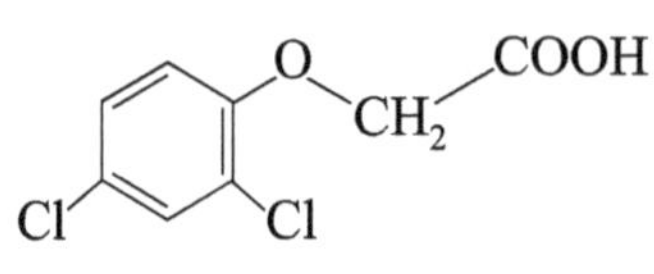

2, 4-D

2, 4-Dichlorophenoxy Propionic Acid (2, 4-DP)

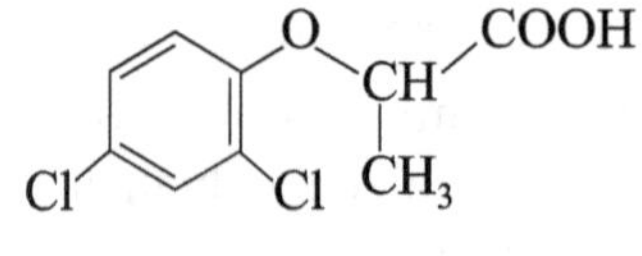

2, 4-DP

Atrazine: The triazine group of herbicides were introduced in 1950s. Atrazine was the first herbicide in this class

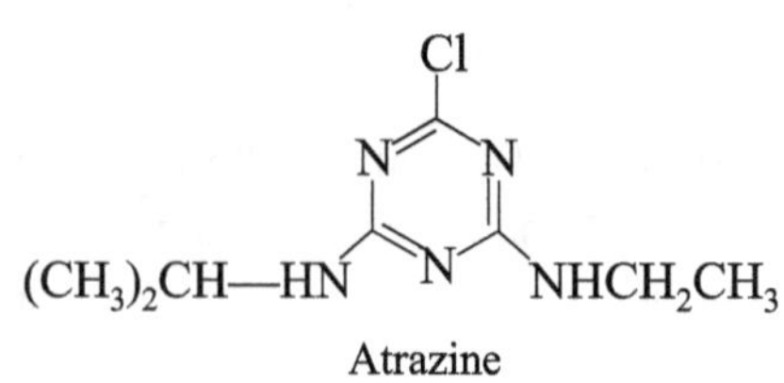

Atrazine

Atrazine does not break down easily and hence contaminates water.

Metsulfuron:

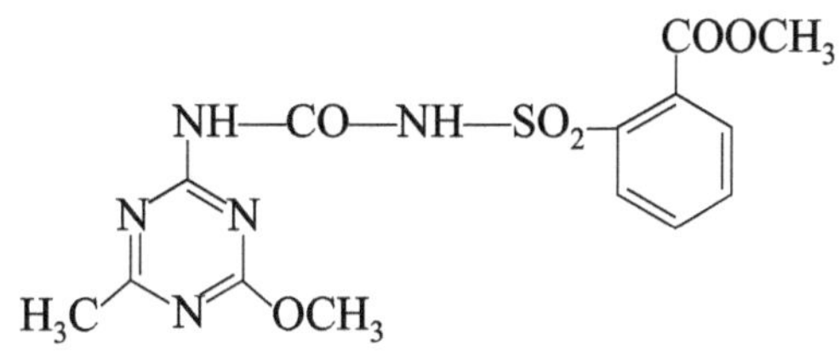

It is a herbicide used to control weeds with broad leaves. It has low toxicity to mammals.

Hexazinone:

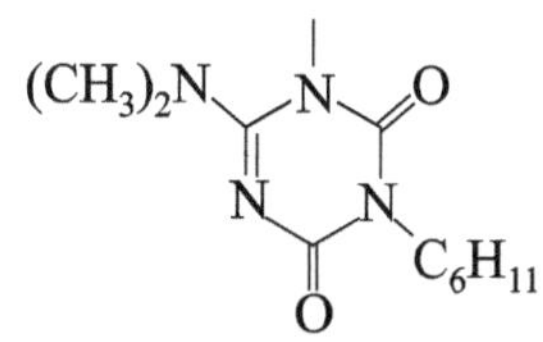

Synthesised around 1970s is a broad spectrum herbicide used to control grasses, broad leaf weeds and woody plants. Ground water contaminant.

Glyphosphate–Glyphosphate was discovered in 1970

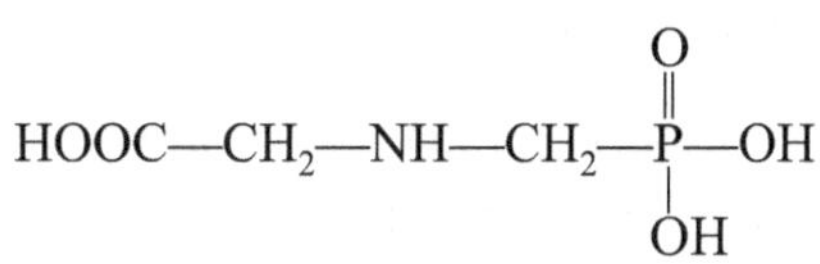

It is used as selective herbicide because many weeds have developed resistance to it. Contradicting reports have been published about its carcinogenic side effects.

20.4 SYNTHETIC FUNGICIDES

A large number of synthetic compounds have been developed and used as fungicides. Inorganic compounds were the earliest fungicides, some of these were Arsenic, Copper Sulphate, Cuprous oxide, Sulphur dust, Lime Sulphur etc. This was followed by development of organic fungicides like Dithio-carbamates, Phalimide, Triazines, Nitroanilines, Benzimadazole, Carboxanilide and Morpholine etc. Pyrimidines, Imadazoles and Azolone derivatives were subsequently produced and used as effective fungicides. A few herbicides with their structure and uses are outlined below.

Tricycloazole : It is used for control of blast disease of rice.

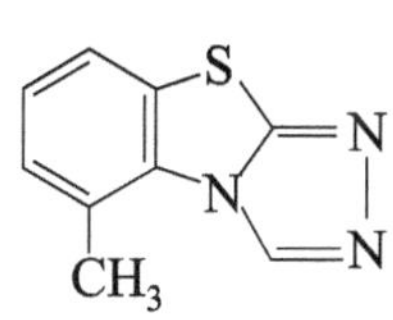

Cyprodinilin :

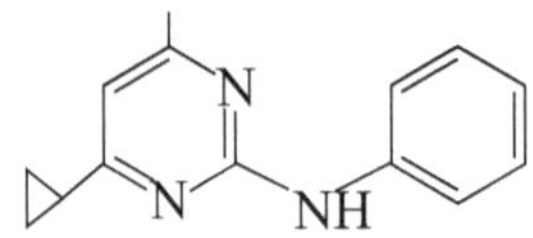

Quinoxyfen : It is used against powdery mildew in cereals, mangoes, tomato and grapes etc.

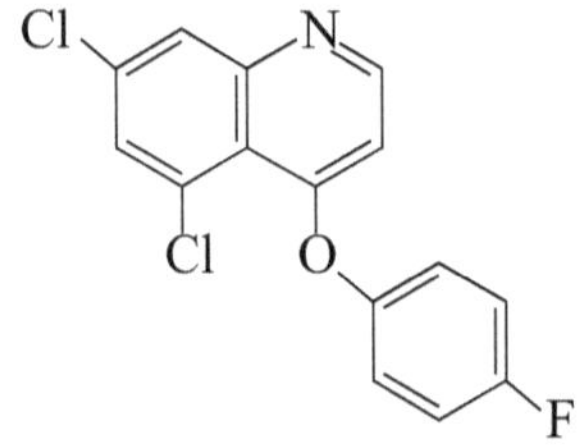

Captan : It used to control disease on fruits and vegetables etc.

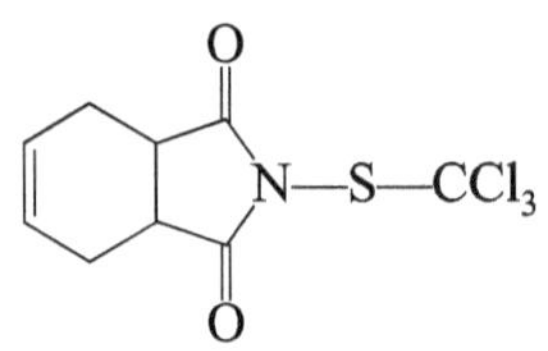

20.5 GREEN AGROCHEMICALS

There is a tendency by agriculturalist to move towards natural fertilizers, pesticides and weedicidesete. There are a few green developments in this direction but it may not be possible to replace all synthetic chemicals used in agriculture in near future to meet the food requirements of world population.

It is therefore very important to judicially use least hazardous chemicals as pesitcides etc., stop use of banned agrochemicals and simultaneously use and produce more and more benign agrochemicals.

20.6 GREEN CHEMISTRY AND SUSTAINABLE AGRICULTURE

Green Chemistry can help in sustainable agriculture:

- By providing environmentally friendly chemicals for agricultural use. These agrochemicals should be produced by green chemical synthesis.
- Green chemists use agricultural waste and other agricultural products as renewable starting materials for the synthesis of chemicals useful to mankind.
- Green chemists use plants and other microbes to clean water and soil from contaminants/ pollutants (Phytoramediation).
- Green chemists use biocatalyst (either as whole cell or pure enzymes isolated from plants) instead of chemical catalyst to carry out chemical transformation.

- Instead of using chemicals as pesticides, fungicides, weedicides etc. naturally occuring agrochemicals can be identified and used.
- Green chemists can provide formulations containing active agrochemicals such that exessive use of these is reduced. Following section gives an account of some of these green developments.

20.7 CONTROLLED RELEASE AGROCHEMICAL FORMULATIONS

Agrochemicals are conventionally applied at frequent intervals during the crop growth to maintain their optimum required concentration. This has to be done because a considerable amount of these is lost during application and afterwards. Frequent application of the agrochemicals besides being cost intensive is also bad for environment, plants and animals. Some of these also destroy or adversively effect useful insects etc. required for pollination.

Scientists are developing new ways to reduce the cost and adverse effects of the agrochemicals on workers and environment in general.

In one of such development, the active agrochemical is released in a controlled manner.

The pesticide is (*i*) covalently bound to a polymer (*ii*) linked to the monomer and the monomer is then polymerised (*iii*) dispersed uniformly in a ploymeric matrix (*iv*) encapsuled or laminated for slow release.

If the pesticide cannot be chemically linked to the polymer or monomer it is fixed to them via a chemical bridge.

Efforts are also being made to reduce the cost by using relatively less expensive polymers or polymer waste like saw dust, rice husk, lignin, pine kraft lignin or starch etc. Another aspect that is being explored is that the polymer not only provides an inert support but can perform a dual function *i.e.* as support and a fertilizer.

Pentachlorophenol pesticide was linked to vinyl monomers via ester linkage which was then polymerised to produce a slow releasing pesiticide system. The pesticide is released by slow hydrolysis of ester link. Vinyl Ester of 2, 4-Dichlorophenoxy acetic acid (2, 4-D) was polymerised to produce slow releasing herbicide.

Systems have been prepared from rice husk lignin + 2, 4-D for slow release, chitin is used as a carrier for metribuzin (a herbicide).

Neem seed oil (a natural pesticide) has been encapsuled using sodium alginate as a polymer.

Polymeric beads made from sodium algenate, gelatin and glutraldehyde have been used for slow release of chlorpyrifos

Cypermethrin can be similarly loaded on beads for controlled release.

Clay minerals which have large surface area and ion exchange capacity can absorb and desorb pesticides for slow release.

Starch-alignate beads have also been prepared and used for controlled release of agrochemicals.

20.8 NITRIFICATION INHIBITORS FOR NITROGEN FERTILIZERS

Nitrogen fertilizers are used to increase agricultural productivity to meet the food requirements of ever increasing population. The soil bacteria convert the fertilizers (Urea and Ammonium Salts)

into Nitrate which leaches out of the plant root zone and thus the applied fertilizer is not available for plant growth.

Scientist at DOW Chemical Company. have developed a powerful **Nitrification Inhibitor** that inhibits soil bacteria from converting the ammonical fertilizers into nitrate. These Nitrogen stabilizers retain the applied fertilizers for plant growth and also prevent pollution of water and soil (by Nitrate) and pollution of air by oxides of nitrogen.

The company has won presidential Green Chemistry Challenge Award in 2016.

20.9 BIO-AGROCHEMICALS

In the past, the need for increasing food demand was met by use of synthetic fertilizers and the losses of the crops by pests were reduced by use of pesticides, the weeds were controlled by chemical compounds and so on. All this was achieved with a cost. Many of these chemicals had a harmful effect on living beings and environment. Before discussing the alternatives to synthetic agrochemicals, it is important to mention here that the advantages of these chemicals can outweigh the disadvantages by their judicial and scientifically evaluated use.

Bio-pesticides and bio-fertilizers (formulations made with naturally occurring organisms) are now emerging as greener alternatives to synthetic agrochemicals. These bio-based products are safe, eco friendly, target specific, where as the synthetic chemicals used in agriculture generally accumulate, are harmful to the environment and many a times non-specific *i.e.* these kill useful insects also. The lack of profile of bio-pesticides and technology required for production are the two major obstacles in the wide spread use of these alternatives. Another practical disadvantage is that the bio-pesticides have to be applied more frequently than the chemical compounds.

20.9.1 Bacillus thuringiensis (Bt)

It was discovered by a Japanese biologist. It is being used as a pesticide since ~1920. A number of strains of Bt have been discovered since then. Some of these have insecticidal activity against lepidoteran (moth) larvae, some against beetles and others against flies. These are active only against Larvae Table given below lists some of the common strains of Bt.

Bt Sandiego	Certain bettle species and boll weevil
Bt Kurstaki (BtK)	Lepidopterous insects, moth, cabbage looper.
Bt Aizawai (BtA)	Wax moth Larvae
Bt Israelenesis (BtI)	Mosquitoes, blackflies and midges.

Bt strains are permitted for formulation used in organic farming. Bt are safe for mammals and do not persist in atmosphere.

20.9.2 Spinosad

Developed by Dow Agro-Science LLC in 1999 is made by a soil micro-organism.

It is a mixture of basically two compounds Spinosyn A and D. It is obtained by fermentation using *Saccharopolyspora spinosa*. It was first registered as a pesticide in 1997, has low mammalian toxicity and a broad insect pest activity.

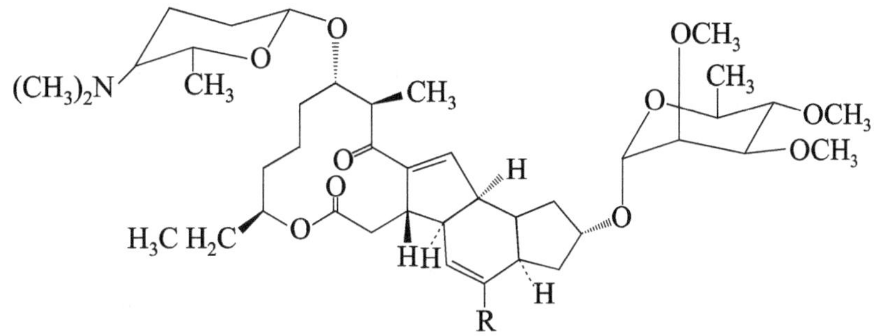

Spinosyn A R = (H)
Spinosyn D R = CH_3

It is used as an insecticide against lepidoptera (Moths and Butterflies)
Diptera (True Flies)
Thysanoptera (Insects with slender body and small wings.)
Coleoptera (Beetles)
Hymenoptera (Bees, ants, wasps, saw flies)
It is used to control insect pests in cotton, fruits, vegetables and ornamental plant etc.

20.9.3 Spinetoram

It is a broad spectrum insecticide produced by fermentation of sugars in cotton seed, corn, soya, sorgham etc. by *Saccharopolyspora spinosa*. The product is a mixture of two chemical compounds namely 3'-o-Ethyl-5, 6-Dihydro Spinosyn J and 3'-O-Ethyl Spinosyn L. Its production process was developed by Dow Agro Sciences for which the company was bestowed with Presidential Green Chemistry Challenge Award. Spinetoram is more effective than Spinosad.

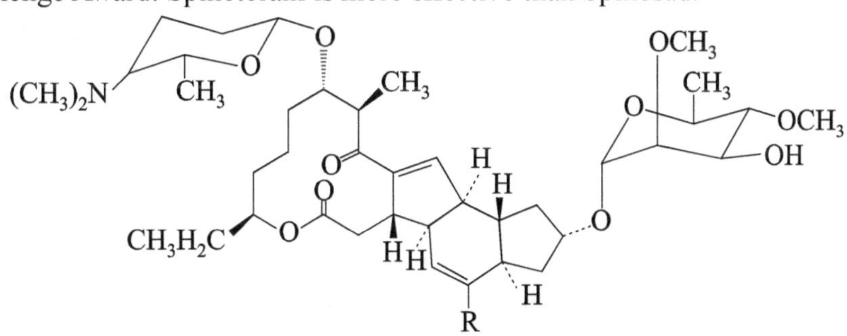

Spinosyn J (R = H)
Spinosyn L (R = CH_3)

It is used as an insecticide for apple, citrus, grapes, melons, caneberries, corn, soyabeans, leafy vegetables etc. and is effective against leafminers, apple magot, pearpsylla, grape berry moth, orange worm, tomato fruit worm, cut worm, loopers etc.

20.9.4 Serenade

A biofungicide, contains a unique strain (QST713) of *Bacillus subtilis*. It is used as a fungicide on vegetables, fruits, roses and flowering plants. It controls diseases like fire blight, rust, powdery mildew, white mold etc.

20.9.5 Rhamnolipid

Rhamnolipid biosurfactant is a natural glycolipid found in soil and plants. It has been produced by fermentation using Pseudomonas Aeruginosa. It is used as a fungicide in addition to cleaning contact lenses, crude oil tanks etc.

20.9.6 Harpins

Harpins are a class of naturally occuring proteins which are non-toxic and activate a plant's defence and growth mechanism. It is produced in water based fermentation system and is easily biodegraded.

20.10 BIO-FERTILISERS

These are preparations containing living cells or latent cells of microorganisms which are used to provide macronutrients like nitrogen and phosphorus to the plants. Production of synthetic frtilizers is an energy intensive process, for example ammonia required to produce nitrogen fertilizers is made by using natural gas (a non-renewable source), at high temperature using metal catalyst as per the following reactions.

$$CH_4 + H_2O \longrightarrow CO + 3H_2$$
$$CO + H_2O \longrightarrow CO_2 + H_2$$
$$N_2 + 3H_2 \longrightarrow 2NH_3$$

Ammonia gas thus obtained has to be cooled to –10 to –20 °C (another energy intensive step).

Some examples of nitrogen fixing microbes are *Rhizobium, Frankia, Annabeena Azoliae, Azobacter, Clostridium* etc.

Baccillus Magterium, B. Subtilis, B Circulans, Pennicilum and Aspergillus Awamori are some phosphorus solubilising bio-fertilisers and *Rhizoctonia Solani, Species of Glomus*, Gigaspora, *Laccaria, Pisolithus, Boletus* etc. are phosphorus mobilising microbes.

Azospirillum cultures synthesise considerable amount of compounds like vitamins, geberllins, indole subtituted acetic acids etc. which help in plant growth.

Conclusion

Demand of food is increasing with ever increasing population which can only be met with use of agrochemicals. It is difficult, at present, to bridge the gap between demand and supply by natural agrochemicals. Therefore it will be wise to judicially use synthetic agrochemicals and weigh advantages to their harmful effects and simultaneously develope bening chemicals for agricultural use.

APPENDIX

1

Green Chemistry Academic Programs

Alabama:

Research center focused on ionic liquids, biorenewables, novel separation strategies and more.

Arkansas:

Toad Suck Institute for Green Organic Chemistry.

Green chemistry incorporated into bachelor's curriculum.

California

Innovative courses for chemistry and engineering majors and nonmajors, green labs, and applied research.

Courses focusing on global environmental issues and solutions and biocatalysis research.

Colorado

Green chemistry incorporated into bachelor's curriculum.

Connecticut

Courses in green engineering and chemistry; research in greener materials, energy, water, systems, and metrics.

District of Columbia

Emphasizes both environmental chemistry and green chemistry, the design of new chemicals and chemical processes with minimal environmental impact.

Florida

Green chemistry in organic labs and research opportunities including biopolymers and catalysts.

Georgia

Interdisciplinary approach, courses at undergraduate and graduate level, and research opportunities.

Illinois

Offers a BS in Environmental Engineering and a Certificate in Energy and Sustainability.

Offers Environmental Chemistry-focused undergraduate curriculum.

Indiana

Graduate courses in green chemistry, toxicology, risk assessment; Master's in Environmental Chemistry.

Iowa

Green chemistry research including biorefinery.

Kentucky

Research areas in green chemistry and biocatalysis.

Maryland

Green chemistry courses, curriculum, and senior research for undergraduates.

Massachusetts

Offers green chemistry courses, labs, research, and a BS with Environmental Chemistry concentration.

Green chemistry incorporated into bachelor's curriculum and lecture series.

Offers a PhD in Green Chemistry; Research in chemical fate, renewable energy, benign synthesis, and more.

Offers a PhD in Green Chemistry.

Offers professional training in green chemistry for scientists and engineers (non-academic).

Michigan

Undergraduate courses in green chemistry; Offers a Certificate in Green Chemistry.

BS in Environmental Chemistry; Engineering Courses on biofuels and fuel cells.

Undergraduate courses in green chemistry and research in greener organic synthesis and green catalytic pathways.

Research areas in Environmental and Sustainable Chemistry including green catalysis.

Minnesota

Green chemistry incorporated into bachelor's curriculum.

Environmental focus and research in synthetic polymers, renewable materials, and bio-inspired oxidation catalysts.

Missouri

Offers an undergraduate Certificate in Renewable Energy and the Environment.

New York

Research areas in bio-inspired catalysis, green nanostructure synthesis, and sustainable materials.

North Carolina

Undergraduate research in green chemistry; Offers a Master's in Biomanufactuhng.

Ohio

Graduate courses in environmental chemistry; Research in biobased polymers.

Courses in green chemistry, engineering and biofuels; MS and BS with minor in green chemistry and engineering.

Oregon

BS in Environmental Chemistry; Research in thin films, water-based chemistry, solar, and scaling.

Green chemistry lab, multidisciplinary courses; Research in green nanomaterials, photoactive materials, and more.

Pennsylvania

Classes and research center focused on sustainable chemistry.

MS in Green Chemistry; Chemists learn entrepreneurial skills.

Green chemistry and engineering courses; Offers Sustainable Engineering Certificate.

Green chemistry incorporated into curriculum; Offers a BS with concentration in green chemistry.

Puerto Rico

Research includes biocatalysts and environmental analytical chemistry.

South Dakota

Research in supercritical fluids, analytical separations, bioprocessing, and green chemistry education.

Texas

Undergraduate lab in green chemistry.

Virginia

Research includes LCA, ecotoxicology, nanobiosensors; Offers a PhD in Sustainable Nanotechnology.

Washington

One year, online professional certificate program in Green Chemistry and Chemical Stewardship.

Australia

Green chemistry is part of the bachelor's and master's degrees; Offers a PhD with focus in green chemistry.

Research in inorganic green chemistry, green nano, biopolymers, clean tech, process intensification.

Research in biorenewable chemicals, earth abundant materials; Offers PhD in Green Chemistry.

Canada

Courses in green chemistry; Research in green materials, manufacturing, solvents; Offers Environmental Chemistry focus.

Green chemistry courses; Research in biocatalysts, alternative solvents, photocatalytic reactions.

Interdisciplinary communication of sustainability and green chemistry in the business program.

Research in chemicals from renewable freedstocks, non-precious metal catalysis, biodegradable polymer synthesis.

China

Research in green catalysis, biological catalysts, ionic liquids.

Denmark

Offers a MSc in Chemistry with a Green and-Sustainable Chemistry focus.

European Union

Multi-institution partnership offering doctoral programs in sustainable industrial chemistry.

France

Offers a MSc in Chemistry with a specialization in Green Chemistry.

Greece

Offers a MSc in Green Chemistry and Clean Technologies.

Hong Kong

Research in green oxidation and materials; PhD with research focus in catalysis.

Italy

A network of academic research chemists working in green chemistry and environmental protection.

Offers a MSc double degree in Environmental and Sustainable Chemical Technologies.

Netherlands

Research in catalytic alternatives and sustainable energy technology.

Portugal

Offers a PhD in Sustainable Chemistry hosted at Aveiro, NOVA Lisbon and Porto.

Singapore

Offers a Green Chemistry and Sustainable Energy research track.

Spain

Interuniversity postgraduate green chemistry programs

Research institution affiliated with the Green Chemistry Institute of Spain and annual green chemistry conferences

Research in organic, medicinal, biocatalysis and biotransformations.

Courses in catalysis, toxicology, and solvents; Offers Master's in Green Chemistry.

Sweden

Research in fuel cells, green electricity, combustion of biomass, and CO_2 capture.

Course in green chemistry and biotechnology; Research in biocatalysts and biorenewable speciality chemicals.

Switzerland

Courses include chemical processes for renewable energy; Research in catalysis for energy and environment.

United Kingdom

Courses and seminars include green chemistry catalysts, sustainable materials; Wide variety of research areas.

Research in ionic liquids, catalysis, advanced materials; Offers MRes in Green Chemistry, Energy, and the Environment.

Research in ionic liquids and alternative solvents.

Diverse research areas; Offers MRes in Sustainable Chemical Technologies.

Research in sustainable synthesis and catalysis; MSc in Chemical Research - Green Chemistry.

Green chemistry research areas; Offers MSc in Green and Sustainable Chemistry.

Graduate courses and research in sustainable energy chemistry; Oxford Catalysis Network.

Research in renewable materials, clean synthesis, platform molecules; Offers masters,PhD and professional ed.

2

Green Chemistry Organizations

Advancing Green Chemistry

American Chemical Society Green Chemistry Institute (ACS-GCI)

ACS-GCI Chemical Manufacturer's Roundtable

ACS-GCI Formulator's Roundtable

ACS-GCI Pharmaceutical Roundtable

American Institute of Chemical Engineers (AIChE): Institute for Sustainability

Berkeley Center for Green Chemistry

Beyond Benign

Bioeconomy Institute (Michigan State University)

Blue Green Alliance

Business-NGQ Working Group for Safer Chemicals and Sustainable Materials

California Green Chemistry Initiative

Canadian Green Chemistry Network

Center for Green Chemistry at University of Massachusetts Boston

Center for Green Chemistry (Monash University) - Australia

Center for Green Chemistry and Green Engineering at Yale University

The Center for Occupational and Environmental Health (University of California): Program in Green Chemistry & Chemicals Policy Center for Sustainability at Aquinas College.

Clean Production Action

Delaware Sustainable Chemistry Alliance

Ecology Center

European Technology Platform for Sustainable Chemistry (SusChem)

Environmental Council of States

EQRNT Centre in Green Chemistry and Catalysis (CGCC)

Grand Valley State University-Green Chemistry

Green Product Design Network

Green & Sustainable Chemistry Network (GSCN)—Japan
Green Chemistry and Commerce Council (GC3)
Green Chemistry Centre of Excellence (University of York)
Green Chemistry Commitment
Green Chemistry Connection (New England Green Chemistry Initiative)
Green Centre Canada
Institute for Green Science (Carnegie Mellon University)
Interstate Chemicals Clearinghouse
Investor Environmental Health Network
Journal of Chemical Education (JCE): JCE chemical Education Exchange (Chem Ed X)
Lowell Center for Sustainable Production (University of Massachusetts Lowell)
Massachusetts Coalition for Occupational Safety & Health (Mass COSH)
Michiagan Green Chemistry Clearinghouse
Michigan Green Chemistry Program
Minnesota Green Chemistry Forum
Minnesota Pollution Control Agency (MPCA): Green Chemistry & Design
National Center for Manufacturing Sciences
National Conversation on Public Health and Chemical Exposures
National Pollution Prevention Roundtable
National Science Foundation (NSF): SusChem
New York State Department of Environmental Conservation: Green Chemistry
New York State Pollution Prevention Institute (NYSP2I)
NSF International
NSF Science and Technology Center: Environmentally Responsible
Solvents and Processes
Ontario BioAuto Council
Oregon Environmental Council
Oregon University - Green Chemistry
Organization for Economic Co-operation and Development (OECD): Sustainable Chemistry
SAE International: Green Technology Systems Group
South African Chemical Institute - Green Chemistry
Suppliers Partnership for the Environment
Sustainable Chemistry Alliance (Canada)
United States Environmental Protection Agency (US EPA) Green Chemistry Program
University of Scranton - Green Chemistry
University of Toledo: School of Green Chemistry and Engineering
Warner Babcock Institute for Green Chemistry

3

Green Chemistry Journals

1. Green Chemistry—published by Royal society of Chemistry 1999.
2. Sustainable Chemistry and Engineering: American Chemical Society. The Journal is available from 2013.
3. Current Green Chemistry.
4. Green Chemistry Letters and Reviews.

4

Green Chemistry Awards

Some Green Chemistry Awards

1. Presidential Green Chemistry Challenge Awards:

   Six awards are given every year for outstanding contribution in following categories.

   (a) Greener Synthetic Pathways

   (b) Greener Reaction Conditions

   (c) Designing Greener Chemicals

   (d) Small Business

   (e) For Academic

   (f) Specific Environmental Benefits: Climate Change.
2. Royal Society of Chemistry gives Green Chemistry Award for design, development or implimentation of novel chemical products or processes which have the potential to reduce or eliminate the use and generation of hazardous substances.
3. Green Chemistry Challenge Award by Royal Australian Chemical Institute.
4. Industrial Green Chemistry World (IGCW) Awards.
5. Candian Green Chemistry and Engineering Network Awards.

Index

www.ingramcontent.com/pod-product-compliance
Lightning Source LLC
LaVergne TN
LVHW080800011026
849160LV00015B/1492